Principles of
Animal Behavior

THIRD EDITION

Principles of Animal Behavior

THIRD EDITION

Lee Alan Dugatkin

UNIVERSITY OF LOUISVILLE

W. W. NORTON & COMPANY | NEW YORK | LONDON

W. W. Norton & Company has been independent since its founding in 1923, when William Warder Norton and Mary D. Herter Norton first published lectures delivered at the People's Institute, the adult education division of New York City's Cooper Union. The firm soon expanded its program beyond the Institute, publishing books by celebrated academics from America and abroad. By midcentury, the two major pillars of Norton's publishing program—trade books and college texts—were firmly established. In the 1950s, the Norton family transferred control of the company to its employees, and today—with a staff of four hundred and a comparable number of trade, college, and professional titles published each year—W. W. Norton & Company stands as the largest and oldest publishing house owned wholly by its employees.

Editor: Betsy Twitchell
Development Editor: Beth Ammerman
Project Editor: Amy Weintraub
Electronic Media Editor: Carson Russell
Editorial Assistant: Courtney Shaw
Marketing Manager, Biology: John Kresse
Production Manager: Eric Pier-Hocking
Photo Editor: Stephanie Romeo
Permissions Manager: Megan Jackson
Book Design: Leelo Märjamaa-Reintal / Rubina Yeh
Design Director: Rubina Yeh
Composition: TSI Graphics
Manufacturing: Courier Kendallville

The text of this book is composed in Fairfield LT with the display set in Meta Plus.

Library of Congress Cataloging-in-Publication Data

Dugatkin, Lee Alan, 1962-
  Principles of animal behavior / Lee Alan Dugatkin. -- Third edition.
     pages cm
  Includes bibliographical references and index.
  ISBN 978-0-393-92045-1 (pbk.)
1. Animal behavior. I. Title.
  QL751.D748 2013
  591.5--dc23
                                                        2013004071

W. W. Norton & Company, Inc., 500 Fifth Avenue, New York, NY 10110-0017
                 wwnorton.com
W. W. Norton & Company Ltd., Castle House, 75/76 Wells Street, London W1T 3QT

1 2 3 4 5 6 7 8 9 0

For Jerram L. Brown, my mentor and friend.

Contents in Brief

Contents

Preface

Now is an exciting time to be participating in the field of animal behavior—whether as a researcher, an instructor, or a student. In particular, students taking courses in animal behavior today are getting their first glimpses of the field at a dynamic point in its history. The third edition of *Principles of Animal Behavior* aims to show why—by building on the work in the first two editions of this book and adding the latest, best, cutting-edge research being done in animal behavior. Much has happened in the field of animal behavior since the last edition of this book was published in 2009. Recent research findings have given me ample opportunity not only to update and expand on the studies presented in the book but also to reinforce the previous editions' focus on ultimate and proximate causation, as well as the book's unique emphasis on natural selection, learning, and cultural transmission. But there is more to this new edition of *Principles of Animal Behavior* than that.

The third edition greatly expands the discussion of proximate causation, so much so that I have added a new second "primer" chapter on this subject. Chapter 3 is now devoted to hormones, neurobiology, and behavior, while Chapter 4 focuses on molecular genetics, development, and behavior. This discussion of proximate causation introduces a line of inquiry that is sustained throughout the book, alongside ultimate causation. My goal is to weave together the most current knowledge on proximate and ultimate factors and present an integrated approach to animal behavior.

The process of natural selection produces the vast diversity of behavior we see within and across animal species. As such, I delve deeply into the adaptationist approach to animal behavior. In this edition of *Principles of Animal Behavior,* I have also added a great deal of new material on another way to study behavior in an evolutionary context—the phylogenetic approach to the study of behavior. Again, the aim is to produce an integrative overview of animal behavior: The tapestry of animal behavior is created from weaving all of its components into a beautiful whole.

A completely new feature in this edition is the Conservation Connection boxes in Chapters 2–17. Many students taking a course in animal behavior are interested in the course, in part, because they care about the natural world and the creatures that inhabit it. They want to make a difference, and some may even pursue careers in conservation biology. But most animal behavior textbooks barely touch on the subject of conservation biology, or they discuss it only in passing. The Conservation Connection boxes that run throughout the third edition of *Principles of Animal Behavior* give the topic of conservation and animal behavior the space it deserves. Each box focuses on a specific conservation issue related to the chapter topic—such as migration or foraging—and shows how ethology and conservation biology can inform each other in addressing that issue.

From the first edition of this book, my aim has been to explain underlying concepts in a way that is scientifically rigorous but, at the same time, accessible to students. Each chapter in the book provides a sound theoretical and conceptual basis upon which the empirical studies rest. The presentation of theory, sometimes in the form of mathematical models, is not meant to intimidate students but rather to illuminate the wonderful examples of animal behavior in that chapter. My goal has been to produce a book that students will actually enjoy and will recommend to their friends as a "keeper." I also hope that instructors will find this book useful in their research programs, as well as in their courses.

MAJOR FEATURES

The book is written for both biology and psychology students. Its core strengths are:

▶ A BALANCED TREATMENT OF PROXIMATE AND ULTIMATE FACTORS. A comprehensive understanding of animal behavior requires a balanced and integrated approach to proximate and ultimate causation. Each of these perspectives informs the other, and both are necessary for an understanding of animal behavior. This book weaves together these two perspectives in ways that other books do not. In the third edition, coverage of proximate factors has been expanded from one chapter to two, allowing for greater depth of material in neurobiology, endocrinology, genetics, and development. Once these topics are thoroughly introduced, examples of proximate and ultimate factors are then integrated into every chapter that follows, reinforcing how modern ethologists study behavior.

▶ LEARNING AND CULTURAL TRANSMISSION PRESENTED ALONGSIDE NATURAL SELECTION AND PHYLOGENY. This book has always been distinctive in that it integrates learning, cultural transmission, natural selection, and phylogeny throughout the book, bringing together perspectives and research from various subdisciplines in biology, psychology, and anthropology. In recent years, these topics have only become more important to the study of animal behavior. The third edition's coverage of them has been expanded to reflect this.

- ▶ **AN EXTENSIVE DISCUSSION OF PHYLOGENY.** Darwin spoke of two "great laws": one centered on natural selection and the other on phylogeny. An emphasis on phylogeny has become more evident in animal behavior research in the last few years, so this edition delves more deeply into the role that phylogeny plays in understanding fundamental issues in animal behavior. Chapter 2 provides an overview of phylogenetic approaches to ethology, including a detailed description of how to build a phylogenetic tree, and later chapters include comprehensive discussions of the phylogeny of specific animal behaviors, including learning, parental care, cooperation, foraging, migratory behavior, and play.

- ▶ **A THOROUGHLY UPDATED ART PROGRAM.** The art program in this book has always included extensive data graphics, as well as photographs that convey the beauty of the natural world. But students often struggle to interpret the graphical representations of data that are so widely used for reporting results across the sciences. The third edition's art program therefore has been updated to include a new element—extensive bubble captions that help students identify and interpret information conveyed in the figure.

- ▶ **NEW CONSERVATION CONNECTION BOXES.** Increasingly, conservation biologists and environmental scientists are using animal behavior research to maintain and improve ecosystems around the world. Chapters 2–17 in this book now each include a Conservation Connection box that describes both a current research inquiry and an application of that inquiry in nature.

- ▶ **EXTENSIVE VIDEO CLIPS OF ANIMAL BEHAVIOR.** To illustrate animal behavior in its entirety and to show students the behaviors about which they are learning, the text includes hundreds of beautiful photos and line drawings. But students in the twenty-first century have the opportunity to *see* animal behavior in action through video, as well as print. That is why, in addition to the photos and line art in the text, we provide a collection of over 200 wonderful videos—from the BBC, the Cornell Lab of Ornithology, and researchers cited in the book—that capture the beauty of studying animal behavior.

These clips are offered through two resources, the Norton Animal Behavior DVD, which includes descriptions of each clip and references to the book, and 60 new video clips, which are on the Web at wwnorton.com/college/biology/animalbehavior. Each of these clips is accompanied by assignable quizzes that test students' grasp of core concepts, as well as their ability to analyze examples of animal behavior.

INSTRUCTOR RESOURCES

THE NORTON ANIMAL BEHAVIOR DVD

Available to instructors who adopt *Principles of Animal Behavior*, Third Edition. A resource of 200 video clips, accompanied by a booklet written by Jim Hare of the University of Manitoba, providing short descriptions of each clip. The footage is drawn from three sources:

1. RESEARCHERS CITED IN THE TEXT. Numerous adopters of the first and second editions of this book expressed a desire to show their students the studies described in the text. Many researchers generously provided their lab and field videos to make this desire a reality.

2. BBC. Most people who are familiar with the BBC's offerings rank their collection of animal behavior videos as among the best in the world. In reviewing the clips that are included on *The Norton Animal Behavior DVD*, I am inclined to agree. We are fortunate to be able to offer so many BBC video clips of animal behavior in this book.

3. CORNELL LIBRARY OF ORNITHOLOGY. The Cornell Library of Ornithology has an unparalleled collection of footage done by animal behavior researchers. The quality of both the production and the science in the CLO's collection is remarkable.

WEB-BASED VIDEO QUIZZES

Sixty new video clips, obtained from researchers around the world, serve as the basis to test students' ability to analyze examples of animal behavior and their mastery of core concepts. Students watch each clip and answer up to five questions on the specific behavior being illustrated or on the underlying theoretical concept being demonstrated. Quiz results report to an instructor grade book, making them easy to assign and grade.

All 60 clips and quiz questions have been converted to PowerPoint format for use in lecture as clicker questions.

INSTRUCTOR'S MANUAL

Ryan Earley of the University of Alabama has updated the *Instructor's Manual* to reflect changes in the third edition of the text. This resource includes in-depth answers to the end-of-chapter discussion questions in the text. It also includes a bank of multiple-choice questions, as well as review and challenge questions, from which instructors can draw when creating tests. The IM is available for download at wwnorton.com/books/Principles-of-Animal-Behavior.

NORTON MEDIA LIBRARY

Digital files of all drawn art and most photographs are available to adopters of the text at wwnorton.com/books/Principles-of-Animal-Behavior.

ACKNOWLEDGMENTS

I wish to thank my gifted editor, Betsy Twitchell, for shaping this third edition. Her editorial skills took the third edition to new heights. I would also like to thank Jack Repcheck, my editor on the first edition of this book, for all the time and effort that he invested in this project, and Michael Wright, who did a great job as editor for the second edition. Beth Ammerman's work as the developmental editor has been nothing short of fantastic. The same holds true for project editor Amy Weintraub's work. My thanks also go to Ryan Earley, who has been involved in all three editions of this book, producing a wonderful *Instructor's Manual* for each edition. I would also like to thank Jim Hare for his outstanding work on *The Norton Animal Behavior DVD* and the Web-based video quizzes. Jim not only selected every clip on the DVD but also wrote useful and succinct descriptions for each clip that will aid instructors in presenting the clips in their lectures. Jim's extensive field experience, and his deep understanding of the conceptual underpinnings of animal behavior, are evident in every description. I also extend my thanks to associate editor extraordinaire, Carson Russell, and production associate, Ashley Polikoff, for improving an already excellent DVD for this edition.

Each of the seventeen chapters in the book ends with an illuminating, in-depth interview with a leader in the field of animal behavior. I am deeply indebted to these seventeen brilliant (and busy) animal behaviorists who took time to allow me to interview them. So I extend a huge thank you on this front to E. O. Wilson, Alan Grafen, Geoffrey Hill, Gene Robinson, Sara Shettleworth, Cecilia Heyes, Anne Houde, Nick Davies, Francis Ratnieks, Kern Reeve, John Krebs, Anne Magurran, Rufus Johnstone, Judy Stamps, Karen Hollis, Marc Bekoff, and Sam Gosling.

The production of the text itself has benefited from the artistic skills of Dartmouth Publishing and the composition skills of TSI Graphics. The keen eyes of my photo editors, Stephanie Romeo and Julie Tesser, have taken the text and brought it to life through the beautiful new photos that they found. Production manager Eric Pier-Hocking and pinch-hitter Sean Mintus deserve thanks for managing the transformation of the manuscript files into a beautiful book and for coordinating the many aspects of the book's production. I am also grateful to Courtney Shaw for her assistance in helping us keep track of all the important details of the project. And all of this—the whole book—might have turned out differently had it not been for my remarkable agent, Susan Rabiner.

Literally dozens of my colleagues have read all or parts of *Principles of Animal Behavior*, and I extend my thanks to them all.

The manuscript of the third edition of the book was reviewed by:

Noah Anderson
University of Wisconsin Baraboo Sauk County

Andrea Aspbury
Texas State University-San Marcos

Marin Beaupré
University of California, Irvine

Brett Beston
McMaster University

H. Jane Brockmann
University of Florida

Sarah F. Brosnan
Georgia State University

Anne B. Clark
Binghamton University

Ann Cleveland
Maine Maritime Academy

Cathleen Cox
University of California, Los Angeles

Mary Dawson
San José State University

Robert Gerlai
University of Toronto Mississauga

Harold Gouzoules
Emory University

Peter Henzi
University of Lethbridge

Kurt Hoffman
Virginia Polytechnic Institute and State University

Clint Kelly
Iowa State University

Scott L. Kight
Montclair State University

Catherine Lohmann
University of North Carolina

Suzanne E. MacDonald
York University

Maria Maust
Hunter College

Kevin McGraw
Arizona State University

John Pastor
University of Minnesota Duluth

Terry F. Pettijohn
The Ohio State University at Marion

Jesse Purdy
Southwestern University

Gwynne Rife
University of Findlay

John Rosenkoetter
Missouri State University

Olav Rueppell
University of North Carolina at Greensboro

Roger D. Santer,
Aberystwyth University

Andrea Schnitz
Southwestern College

Carol Shearer
University of Illinois at Urbana-Champaign

Joseph Sisneros
University of Washington

Mark Spritzer
Middlebury College

John Swallow
University of South Dakota

Elizabeth Tibbetts
University of Michigan

Carolyn Walsh
Memorial University of Newfoundland

Scott R. Wersinger
University at Buffalo, State University of New York

David White
Wilfrid Laurier University

David J. White
University of Pennsylvania

Reviewers for the second edition of the book were:

Andrea Aspbury
Marc Bekoff
Thore Bergman
Richard Buchholz
Terry Christenson
Anne Clark
Reuven Dukas
George Gamboa
Harold Gouzoules
David Gray
Douglas Grimsley
Teresa Horton
Melissa Hughes
Eileen Lacey

John Maerz
Jill Mateo
Jennifer Mather
Kevin McGraw
Roger Mellgren
Peter Nonacs
Shawn Nordell
Dan Papaj
Aras Petrulis
Stephen Pruett-Jones
Rick Relyea
Christoph Richter
Bruce Schulte
Con Slobodchikoff

Jeanette Thomas Stim Wilcox
Kaci Thompson Sarah Wooley
Sean Veney

Reviewers for the first edition of the book were:

Marc Bekoff   Anne Magurran
Samuel Beshers   Michael Mesterton-Gibbons
Anne Clark   Manfred Milinski
Fred Dyer   Allen Moore
Susan Foster   Dan Papaj
Nick Fuzessery   Geoff Parker
Deborah Gordon   David Pfennig
Ann Hedrick   Naomi Pierce
Geoff Hill   Locke Rowe
Anne Houde   Michelle Scott
Rudolph Jandler   Max Terman
Curt Lively   Jerry Wilkinson

The manuscript in each edition benefited from these reviewers' close reading and sound advice. Please credit these folks with all that is good about this book, and assign any problems you have to my hand.

Last, special thanks go to my wife Dana, who helped with almost every aspect of this project, and to my son Aaron for being such a special young man, and for keeping me smiling. Also thanks to 2R, who knows who he is.

L.A.D.
January 2013

Principles of
Animal Behavior

THIRD EDITION

1

Principles of Animal Behavior

FIGURE 1.1. American cockroach.
Almost everyone is familiar with the American cockroach, often a pest in households around the world. *(Photo credit: Bates Littlehales/Animals Animals–Earth Scenes)*

I grew up in the heart of New York City. One animal that my family and I encountered on a fairly regular basis was the American cockroach (*Periplaneta americana*) (Figure 1.1). Much to my mother's chagrin, we seemed locked in a never-ending battle with these creatures—a battle that we usually lost. And we probably lost because cockroaches have been subject to this sort of problem—other organisms trying to kill them—for tens of millions of years. As a result, they have evolved an exquisite set of antipredator behaviors, which have had the side effect of making them a thorn in the side of modern apartment dwellers.

As a very young boy, I had, of course, never heard of the **scientific method**—which, according to the Oxford English Dictionary, involves "scientific observation, measurement, and experiment, and the formulation, testing, and modification of hypotheses." Nevertheless, I was able to draw some inferences and formulate some hypotheses about cockroach behavior by watching my mother put out the bug traps. First, it seemed to me that roaches liked to spend their time in dark places, and second, it appeared that most roaches agreed on what was a good place for roaches to be, as we kept putting the traps out in the same place. These two thoughts on cockroach behavior could easily be developed into the following hypotheses: (1) cockroaches will choose dark places over light places, and (2) roaches will return to the same places over and over, rather than moving randomly through their environment. Of course, as a child, I didn't formally sit down and generate these hypotheses, and I surely didn't run the controlled experiments that a scientist studying animal behavior would run to test these ideas, but I was nonetheless dabbling with scientific hypotheses about animal behavior—a field technically known as **ethology**.

Many people think like ethologists: from my mother, who understood roach behavior, to the farmer who has detailed knowledge about pigs, cows, chickens, and other domesticated farm animals. The girl who works to train her dog, and the outdoorsman who, on his camping vacation, searches for some animals and tries to avoid others also think like ethologists. Indeed, humans have always thought and acted like ethologists. If our hunter-gatherer ancestors had not thought like ethologists, and hadn't, for example, understood the prey they were trying to catch, as well as the behavior of the predators that were trying to catch them, we humans wouldn't be here today.

The study of animal behavior appears to have been so fundamental to human existence that the earliest cave paintings tended to depict animals. This choice of subject matter was certainly not inevitable—early cave drawings might have focused on any number of things, but apparently understanding something about the other life forms surrounding our ancestors was fundamental enough that they chose animals as the subjects for the earliest art. This focus on animals, and their behaviors, continued as humans began developing other types of art. For example, using artifacts from 4,000-year-old Minoan cultures, Marco Masseti argues that the Minoans had an advanced understanding of some aspects of animal behavior (Masseti, 2000). One fascinating example supporting this claim is a golden pendant from a Cretan cemetery that depicts two wasps transferring food to one another (Figure 1.2). Masseti hypothesizes that this kind of knowledge of insect food-sharing behavior could only have come from people who observed and studied the details of wasp life. A similar sort of argument is offered regarding a beautiful Minoan wall painting of "white antelopes."

FIGURE 1.2. Art captures animal behavior. This pendant from the Chrysolakkos funeral complex in Crete suggests that some members of the ancient culture had a detailed knowledge of wasp behavior. *(From Gianni Dagli Orti/ The Art Archive at Art Resource, NY)*

This painting probably depicts gazelles in the early stages of an aggressive interaction (Figure 1.3), and again it is the sort of art that is associated with an in-depth knowledge of the subject in question (Voultsiadou and Tatolas, 2005).

Spanning the millennia between ancient Cretan civilization and the present, literally thousands of amateur and professional naturalists have made some contribution to the study of animal behavior. These contributions have enabled ethologists to draw on a rich trove of information that has greatly expanded our understanding of animal behavior (Figure 1.4). Aristotle's work on animals, for example, though 2,500 years old, is a veritable treasure chest of ethological tidbits. Indeed, with Aristotle's books, *Physics* and *Natural History of Animals*, the field of natural history was born. In these and other works, Aristotle distinguished among 500 species of birds, mammals, and fish, and he wrote entire tracts on the behavior of animals.

In many ways, a course in animal behavior is where all the other biology and psychology classes that you have sat through up to this point in your academic career come together. Evolution, learning, genetics, molecular biology, development, neurobiology, and endocrinology congeal into one grand subject—animal behavior. The field of ethology is integrative in the true sense of the word, in that it combines the insights of biologists, psychologists, anthropologists, and even mathematicians and economists.

FIGURE 1.3. Minoan wall paintings of "white antelopes." The drawing may depict a "lateral intimidation" during an aggressive encounter between the animals. *(From Masseti. Courtesy Ministry of Culture, Hellenic Republic)*

Types of Questions and Levels of Analysis

As you will learn in this book, ethologists have asked questions about almost every conceivable aspect of animal behavior—feeding, mating, fighting, and so on. At a broad level, however, ethologists pose four distinct *types of questions*,

FIGURE 1.4. Fantastic images from a cave. A drawing of a herd of antelope found on the walls of a cave at Dunhuang, China. *(Photo credit: Pierre Colombel/Corbis)*

which Niko Tinbergen outlined in a classic paper entitled "On the Aims and Methods of Ethology" (N. Tinbergen, 1963). These questions center on:

▸ Mechanism—What stimuli elicit behavior? What sort of neurobiological and hormonal changes occur in response to, or in anticipation of, such stimuli?
▸ Development—How does behavior change as an animal matures? How does behavior change with the ontogeny, or development, of an organism? How does developmental variation affect behavior later in life?
▸ Survival value—How does behavior affect survival and reproduction?
▸ Evolutionary history—How does behavior vary as a function of the evolutionary history, or **phylogeny**, of the animal being studied? When did a behavior first appear in the evolutionary history of the species under study?

Thousands of studies have been undertaken on each of these four types of questions. Tinbergen's four questions can be captured in two different kinds of analyses—proximate analysis and ultimate analysis (Alcock and Sherman, 1994; Dewsbury, 1992, 1994; Hailman, 1982; Hogan, 1994; J. Huxley, 1942; Mayr, 1961; Orians, 1962; Reeve and Sherman, 1993). **Proximate analysis** focuses on *immediate causes*, whereas **ultimate analysis** is defined in terms of the *evolutionary forces* that have shaped a trait over time. As such, proximate analysis incorporates Tinbergen's first two types of questions, whereas ultimate analysis covers the latter two types (Figure 1.5). We could ask, for example, the following questions: Why do some bird chicks peck at red stimuli but not stimuli of other colors? Does red trigger a set of neuronal responses that are not triggered otherwise? If so, exactly which neurons and when? An analysis at the ultimate level, on the other hand, would ask: What selective forces in the birds' evolutionary past would have favored individuals that had responses to red stimuli? Was the color red associated with a particular food source? Do other closely related bird species show similar responses to red stimuli?

Every chapter of this book examines animal behavior from both proximate and ultimate perspectives.

What Is Behavior?

What do ethologists mean by the word *behavior*? It turns out that this is not a trivial question, and it is one that ethologists have grappled with for some time. Early on, ethologists such as Niko Tinbergen defined behavior as "the total movements made by the intact animal," but that definition seems far too general, incorporating almost everything an animal does. But if a definition proposed by Tinbergen—who shared a Nobel Prize as a founder of the study of animal behavior—doesn't work, how can a satisfactory definition be achieved?

FIGURE 1.5. Tinbergen's four types of questions. A diagrammatic representation of the four different types of questions asked by ethologists. Two of these types of questions are proximate and two are ultimate.

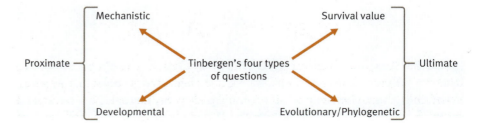

One solution is to survey ethologists to get a discipline-wide view of the way the term *behavior* is employed. In a review paper on definitions of behavior, Daniel Levitis and his colleagues surveyed 174 members of three professional societies that focus on behavior to try and determine what researchers meant when they used the term (Levitis et al., 2009). What they found was a great deal of variation among ethologists on how behavior was defined. Based on their survey results, Levitis and his colleagues argued that many of the definitions that ethologists use can be captured by a few published, but quite dated, definitions already in the literature. These include Tinbergen's 1952 definition of behavior, as well as the following:

▶ "Externally visible activity of an animal, in which a coordinated pattern of sensory, motor and associated neural activity responds to changing external or internal conditions" (Beck et al., 1981).
▶ "A response to external and internal stimuli, following integration of sensory, neural, endocrine, and effector components. Behavior has a genetic basis, hence is subject to natural selection, and it commonly can be modified through experience" (Starr and Taggart, 1992).
▶ "Observable activity of an organism; anything an organism does that involves action and/or response to stimulation" (R. Wallace et al., 1991).
▶ "Behavior can be defined as the way an organism responds to stimulation" (D. Davis, 1966).
▶ "What an animal does" (Raven and Johnson, 1989).
▶ "All observable or otherwise measurable muscular and secretory responses (or lack thereof in some cases) and related phenomena such as changes in blood flow and surface pigments in response to changes in an animal's internal and external environment" (Grier and Burk, 1992).

As with all definitions, each of these has its pluses and minuses. If "behavior has a genetic basis," as it certainly does in many instances, does that mean that we should exclude all actions that have not been studied from a genetic perspective when we speak of behavior? Surely not. For any of the definitions above we could pose equally strong challenges. That said, I needed to adopt a consistent definition of behavior in this book. I chose one that is a slight modification of a suggestion by Levitis and his colleagues—namely, that behavior is the coordinated responses of whole living organisms to internal and/or external stimuli. This definition is appropriate for a number of reasons (all of which are somewhat subjective): (1) it seems to capture what most modern ethologists and behavioral ecologists mean when they use the term behavior, (2) it works fairly well for the behaviors covered in detail in Chapters 6–17 of this book, and (3) it makes an important distinction between organism and organ. What this third point means is that, as Levitis and his colleagues note, sweating in response to increasing body temperature is not generally thought of as a behavior per se. But when an animal moves to the shade in response to heat and its own sweating, most ethologists would agree that this is a behavioral response.

Three Foundations

Incredible tales and fascinating natural history make a textbook on animal behavior different from a textbook on organic chemistry or molecular genetics. What links animal behavior to all scientific endeavors, however, is a structured system for developing and testing hypotheses and a bedrock set of foundations on

Fill out the Four Q's

which such hypotheses can be built. Throughout this book, the force of natural selection, the ability of animals to learn, and the power of transmitting learned information to others (cultural transmission) will serve as the foundations upon which we build our approach to ethology.

In his classic book, *On the Origin of Species*—a text widely regarded as the most important biology book ever written—Charles Darwin laid out general arguments for how evolutionary change has shaped the diversity of life and how the primary engine of that change is a process that he dubbed **natural selection** (Darwin, 1859). In a nutshell, Darwin argued that any trait that provided an animal with some sort of reproductive advantage over others in its population would be favored by natural selection. Natural selection is, then, the process whereby traits that confer the highest relative reproductive success on their bearers and that are heritable—that is, can be passed down across generations—increase in frequency over generations.

Whereas natural selection changes the frequency of different behaviors over the course of many generations, **individual learning** can alter the frequency of behaviors displayed within the lifetime of an organism. Animals learn about everything from food and shelter to predators and familial relationships. If we study how learning affects behavior *within the lifetime* of an organism, we are studying learning from a proximate perspective. If we study how natural selection affects the *ability* of animals to learn, we are approaching learning from an ultimate perspective. Later in this chapter, an example is used from a study on learning and foraging (feeding) behavior in grasshoppers. When we ask what sort of cues grasshoppers use to learn where to forage, we are addressing learning from a proximate perspective. When we examine how a grasshopper's learning about food sources affects its reproductive success, we are studying learning from an ultimate perspective. Both approaches can shed light on animal behavior, and this book employs both of these complementary approaches to learning throughout.

Cultural transmission also affects the type of behavior animals exhibit and the frequency with which behaviors occur. While definitions vary across disciplines, this book uses the term *cultural transmission* to refer to situations in which animals learn something by copying the behavior of others, through what is typically referred to as **social learning**. Cultural transmission can allow newly acquired traits to spread through populations at a very quick rate, as well as permit the rapid transmission of information across generations. As with individual learning, natural selection can also act on animals' ability to transmit, acquire, and act on culturally transmitted information.

NATURAL SELECTION

Darwin recognized that his theory of natural selection applied to behavioral traits as well as morphological, anatomical, and developmental traits. Indeed, morphological traits are often the physical underpinning for the production of behavior, so morphology and behavior are linked at many levels. More detail about this linkage is provided below and throughout the book, but for the moment, the key point is that Darwin's ideas on evolution, natural selection, and behavior were revolutionary, and ethology today would look very different were it not for the ideas that Darwin set forth in *On the Origin of Species*. A fascinating example involving mating and parasites in Hawaiian crickets illustrates how natural selection operates on animal behavior in the wild.

In the evening on the Hawaiian Islands, male crickets sing to attract their mates. This "singing" results when the male cricket rapidly moves the smooth scraper on the front of one wing against the serrated file on the other wing. Females cue in on male songs, and they typically will not mate with males that do not produce songs. But as with many behavioral traits associated with attracting mates, male singing is not cost free. Just as females are attracted to male song, so are potentially very dangerous parasites (Zuk and Kolluru, 1998).

Marlene Zuk and her colleagues have been studying this trade-off in male song production—between attracting females and attracting parasites—in the field cricket *Teleogryllus oceanicus* (Zuk et al., 2006). These crickets are parasitized by the fly *Ormia ochracea*, a species that is attracted to the singing male *T. oceanicus*. If a fly finds a singing cricket, it lays its eggs on the cricket, and then the fly larvae burrow their way into the cricket and grow. When the flies emerge from the larvae, they kill the cricket.

Parasitic flies are found on three of the Hawaiian Islands—Oahu, Hawaii, and Kauai—that are also home to *T. oceanicus*. The flies are most prevalent on the island of Kauai, where 30 percent of the crickets are parasitized. Zuk and her team have been studying the relationship between crickets and parasitic flies since 1991, and over time, they noted what appeared to be a significant decline in the cricket population on Kauai. Over the years, they heard fewer and fewer singing males on this island, and they assumed that the parasitic fly was slowly causing the extinction of *T. oceanicus* on Kauai. Indeed, in 2003 they heard only a single male singing. Nonetheless, when they got down on their hands and knees and searched for crickets, Zuk and her team found *T. oceanicus* in abundance. How can we explain these seemingly contradictory findings?

What Zuk and her team found was that most of the males on Kauai had modified wings that were not capable of producing song (Figure 1.6). The file section of the wings of these Kauai males (called "flatwing males") was significantly reduced compared to that of normal males, and its position on

A **B** **C**

FIGURE 1.6. Natural selection in crickets. Marlene Zuk and her colleagues have been studying the field cricket *Teleogryllus oceanicus*. Pictured here are **(A)** a field cricket with normal wings (the arrow points to the file on its outstretched wing); **(B)** a field cricket with flat wings, in which the file section on the outstretched wing has evolved to a much smaller size and is visible only under a high-powered microscope; and **(C)** fly larvae in a parasitized cricket. *(Photo credits: Robin Tinghitella)*

the wings changed, such that song production was no longer possible. These changes were likely the result of mutations of one, or possibly, a few genes associated with wing development and song production. Once such mutations arose, natural selection should strongly favor such flatwing males, that would virtually never be parasitized by very dangerous flies. Or should it?

Flatwing males should have a huge survival advantage, but they might also be at a severe disadvantage with respect to attracting females that hone in on singing males as potential mates. For flatwing males to be favored by natural selection, they must somehow still secure opportunities to mate. Zuk and her colleagues hypothesized that flatwing males do this by staying near the handful of singing males still on Kauai, and mating with females as they approach singers. This sort of "satellite" male mating behavior has been seen in many *T. oceanicus* populations (Tinghitella et al., 2009). To test their hypothesis, they collected 133 Kauai males—121 of which were flatwings, and 12 of which were singers. They then used "playback" experiments, in which male songs were broadcast over loudspeakers. What they found was that flatwing males were drawn to playbacks more strongly than normal males, suggesting that flatwing males stay near singer males in order to secure chances to mate with females drawn in by the singers. With both a huge survival advantage and the continued ability to obtain matings, flatwing males should be strongly favored by natural selection. And indeed, Zuk and her colleagues suggest that the mutation leading to the loss of song occurred only fifteen to twenty generations ago and has quickly increased in frequency, so that now most males on Kauai are flatwing males.

As a second example of natural selection acting on animal behavior, let's examine how individuals in social groups respond to strangers. For animals that live in stable groups, strangers—unknown individuals from outside one's group—represent a threat. Such individuals may compete for scarce resources (including food and mates), disrupt group dynamics that have long been in place, and so on. Because of such costs, ethologists have examined whether animals from group-living species display a fear of strangers, a phenomenon technically known as **xenophobia**. In particular, ethologists hypothesize that xenophobia may be especially strong when resources are scarce, since competition for such resources will be intense under such a scenario, and keeping strangers away may have a strong impact on the lifetime reproductive success of group members.

To examine the effect of resource scarcity on the evolution of xenophobia, Andrew Spinks and his colleagues examined xenophobia in the common mole rat (*Cryptomys hottentotus*) (Spinks et al., 1998; Figure 1.7). Common mole rats live in South Africa in underground colonies made up of two to fourteen individuals. They are an ideal species in which to examine xenophobia and its possible connection to resource availability for two reasons: First, all populations of common mole rats are "tightly knit" in the sense that each group typically has a single pair of breeders that produce most of the offspring in a colony, which means that most group members are genetic relatives (J.M. Bishop et al., 2004). Second, populations of common mole rats differ in terms of the amount of resources in their environments. Some common mole rat populations inhabit mesic (moderately moist) environments that present only mild resource limitations, while other populations live in arid (dry) environments and face intense limitations on their resources. Variation in resource availability between arid and mesic populations is largely due to the fact that mesic environments have about four times as much rainfall as arid environments.

FIGURE 1.7. Common mole rat. This xenophobic common mole rat *(Cryptomys hottentotus)* is showing an aggressive stance in response to a stranger. *(Photo credit: Graham Hickman)*

Spinks and his colleagues examined whether populations from arid areas were more xenophobic than those from mesic environments, as one might predict based on the discussion above about natural selection, resources, and xenophobia. To do so, they conducted 206 "aggression" trials in which two mole rats—one from the arid and one from the mesic environment—were placed together, and any aggression that occurred between them was recorded. Results were clear-cut: When the pair of individuals were both males or both females, aggression toward such strangers was much more pronounced in the common mole rats from the arid environment, where resources were limited, than it was in the common mole rats from the mesic environment. This result was not a function of individuals from arid populations just being more aggressive in general. Control experiments demonstrated that when two individuals that knew each other from the arid population were tested together, aggression disappeared—thus it was the identification of a stranger that initiated the aggression. Natural selection has favored stronger xenophobic responses in common mole rats whose resources are more limited.

The ecology of common mole rats is such that some individuals leave their home colony to find a mate. What this means is that some strangers that are encountered by members of a social group are potential mates, and perhaps worth tolerating. Natural selection then should not simply favor all xenophobia, but a xenophobia that is sensitive to the sex of the stranger. In trials in which the two individuals tested were a male and a female, Spinks and his colleagues found that while aggression was still observed in the low-resource, arid population, the level of aggression decreased dramatically when compared with aggression in same-sex interactions (Figure 1.8). Natural selection has favored common mole rats that temper their fear of strangers as a function of both where they live and the sex of the strangers.

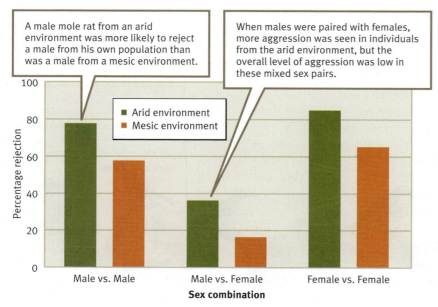

FIGURE 1.8. Xenophobia in common mole rats. Spinks and his colleagues found that mole rats from an arid environment (green bars) were more likely to reject a potential partner from their own population than were mole rats from a resource-rich mesic environment (orange bars). *(From Spinks et al., 1998, p. 357)*

INDIVIDUAL LEARNING

As Chapter 4 explores in detail, individual learning can take many forms. Let's begin by considering a hypothetical case of learning in the context of mate choice. Suppose that we are studying a species in which female birds mate with numerous males throughout the course of their lifetime and are able to keep track of how many chicks fledged their nest when they mated with male 1, male 2, male 3, and so forth. Further suppose that older females prefer to mate with the males that fathered the most successful fledglings. If we found that females changed their mating behavior as a result of direct personal experience, these results might lead us to conclude that learning had changed the behavior of an animal within the course of its lifetime (Figure 1.9).

The learning example above highlights an important relationship between learning and natural selection. In the hypothetical example, females changed their preference for mates as a result of prior experience, and so learning affected mating behavior within a generation. But just because the frequency of a behavior is changing within the course of an individual's lifetime does not mean that natural selection is not occuring. It is certainly possible for natural selection to operate on the *ability* to learn. That is, natural selection might favor the ability to learn which individuals make good mates over, say, the lack of such an ability. If this were the case in the example above, learning would change behaviors within a generation, and natural selection might change the frequency of different learning rules across generations.

You can also see how learning and natural selection can be intimately tied together in Reuven Dukas and Elizabeth Bernays's ingenious experiment examining the fitness consequences of learning in insects (Dukas and Bernays, 2000). While learning in insects is well documented, documenting the potential fitness-related benefits of learning has proved to be more difficult (Dukas, 2006). To address the question of learning-related benefits directly, Dukas and Bernays examined the potential fitness-related benefits of learning

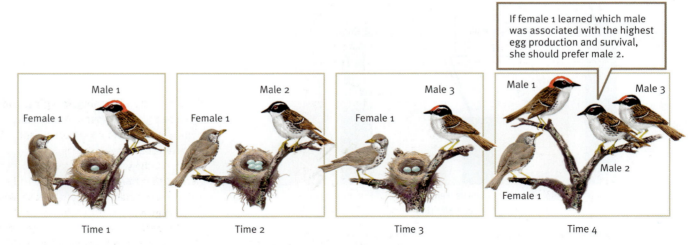

FIGURE 1.9. **A role for learning.** Imagine a female that mates with different males over the course of time. Such a female might learn which male is a good mate by keeping track of the number of eggs she laid after mating with each male.

FIGURE 1.10. Some aspects of foraging in grasshoppers are learned. *Schistocerca americana* grasshoppers learned to associate various cues with food sources. *(Photo credit: Stephen Dalton/Naturepl.com)*

in the context of feeding behavior in the grasshopper, *Schistocerca americana* (Figure 1.10).

In their experiment, Dukas and Bernays placed two food dishes in a grasshopper's cage. The food in one dish provided a "balanced diet (b)" that included proteins and carbohydrates—a diet that promotes maximal growth rates in S. *americana*. The food in a second dish was labeled a "deficient diet (d)." This diet contained flavoring and protein, but no carbohydrates. Specific odors and colors were associated with each of the two diets. Diets were supplemented with either citral (odor 1) or coumarin (odor 2), and food dishes were placed near either a brown-colored card (color 1) or a green-colored card (color 2). This created an opportunity for the grasshoppers to pair balanced and deficient diets with both odor cues and color cues.

Dukas and Bernays's experiment contained a "learning" treatment and a "random" treatment (Figure 1.11). In the learning treatment, the balanced

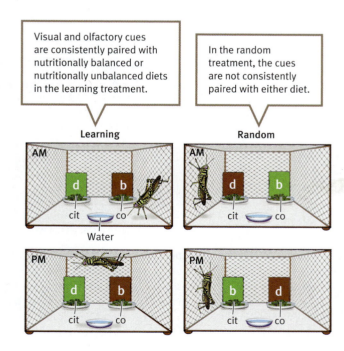

Visual and olfactory cues are consistently paired with nutritionally balanced or nutritionally unbalanced diets in the learning treatment.

In the random treatment, the cues are not consistently paired with either diet.

FIGURE 1.11. Learning, foraging, and fitness in grasshoppers. A schematic of the set-up showing the learning and random conditions. In the learning condition, the set-up consisted of a water dish in the center of the cage and a nutritionally balanced dish (b) on one side of the cage and a nutritionally deficient dish (d) on the other side of the cage. Each dish was paired with one odor (citral [cit] or coumarin [co]) and one colored card (brown or green). *(Based on Dukas and Bernays, 2000)*

diet dish was always paired with one specific odor and one specific colored card. Twice a day, a grasshopper was presented with the two food dishes and allowed to choose one from which to forage. For example, grasshopper A might be placed in a cage in which the balanced diet was always paired with the brown color and the odor of coumarin. In principle, grasshopper A could learn that together the cues coumarin and brown color meant that a food dish contained the balanced diet. In contrast, in the random treatment, the odor and color cues associated with the balanced diet would be randomly assigned. For example, in the morning, grasshopper B might have the balanced diet dish paired with the color green and the odor of coumarin, but in the afternoon, the balanced diet dish might be paired with the color green and the odor of citral, while the next morning the balanced diet dish might be paired with the color brown and the odor of coumarin. In this treatment, the grasshopper would be prevented from learning to pair the balanced diet with specific color and odor cues.

Significant differences between the grasshoppers in the learning and random treatments were uncovered. Grasshoppers in the learning treatment ate a much greater proportion of their food from the balanced diet dish than did the grasshoppers in the random treatment (Figure 1.12). That is, grasshoppers learned to pair diet type with color and odor cues when the situation allowed for such learning. Over the course of the experiment, individuals in both treatments increased the proportion of time they spent feeding on the balanced diet, but grasshoppers in the learning treatment did so more quickly than did those in the random treatment. This difference was most likely due to the fact that grasshoppers in the learning treatment went to the balanced diet dish almost

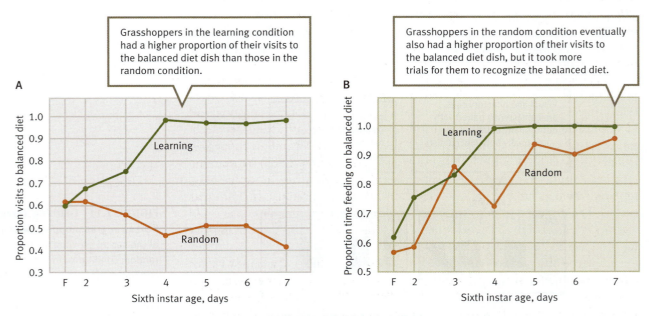

FIGURE 1.12. A balanced diet in grasshoppers. Grasshoppers in the sixth instar stage of insect development were given a choice between a balanced diet or a deficient diet, and researchers recorded the proportion of visits and feeding times of those in a learning treatment and those in a random treatment. In the learning condition, the food was presented in a way in which grasshoppers could learn to associate colored background cards and odors with balanced and unbalanced diets. In the random condition, grasshoppers could not make such associations. *(From Dukas and Bernays, 2000)*

immediately when feeding, while those in the random treatment ended up at the balanced diet dish, but only after much sampling of the deficient diet dish. Perhaps most important of all, the individuals in the learning treatment had a growth rate that was 20 percent higher than that of the grasshoppers in the random treatment (Figure 1.13).

The ability to learn about food in *S. americana* translated into an important gain: a significant increase in growth rate observed in individuals in the learning treatment. This difference in growth rate likely translates into greater reproductive success later in life, as growth rate is positively correlated with the number and size of eggs laid over the course of an individual's life (Atkinson and Begon, 1987; Slansky and Scriber, 1985).

CULTURAL TRANSMISSION

Cultural transmission has received much less attention in the ethological literature than natural selection or individual learning, but work in this area is growing very quickly (Danchin et al., 2004; Galef and Laland, 2005; Galef et al., 2005; Kendal et al., 2005; Mesoudi et al., 2006).

For an interesting case study illustrating the importance of cultural transmission and social learning in animals, consider Jeff Galef's work on foraging behavior in rats. Rats are scavengers and often encounter new foods (Figure 1.14A). This has probably been true for most of the rat's long evolutionary history, but it has been especially true over the last few thousand years, during which time humans and rats have had a close relationship. Scavenging presents a foraging dilemma. A new food source may be an unexpected bounty for rats, but it may be dangerous, either because it contains elements inherently bad for rats, or because rats have no experience with the odor of that food, so they may not be able to tell if some piece of this new food type is fresh or spoiled. One possible way to get information about new food types is through the cultural transmission of information.

Galef began his study of cultural transmission and food preferences in rats by testing what is known as the information-center hypothesis, which posits that foragers may learn critical pieces of information about the location and identity of food by interacting with others that have recently returned from foraging bouts (Figure 1.14B; Ward and Zahavi, 1973). Galef and his colleagues tested this hypothesis in the Norway rat (Galef and Wigmore, 1983). To examine whether cultural transmission via social learning played a role in rat foraging, rats were divided into two groups—observers and demonstrators (also known as tutors). The critical question that Galef examined was whether observers could learn about a new, distant food source by interacting with a demonstrator that had recently encountered such a food source.

After the observer and demonstrator had lived together in the same cage for a few days, a demonstrator rat was removed and taken to another experimental room, where it was given one of two new diets—either rat chow flavored with Hershey's cocoa (eight demonstrators) or rat chow mixed with ground cinnamon (eight demonstrators). The demonstrator was then brought back to its home cage and allowed to interact with the observer for fifteen minutes, at which time the demonstrator was removed from the cage. For the next two days, the observer rat—that had no personal experience with either of the novel foods, and had never *seen* the demonstrator eat anything—was given two food bowls, one with rat chow and cocoa, the other with rat chow and cinnamon. Galef found that

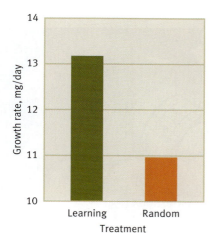

FIGURE 1.13. Fitness and foraging. Not only did grasshoppers in the learning condition approach the balanced diet dish more often, but this translated into quicker growth. Growth rate in grasshoppers is positively correlated with egg size and number. *(From Dukas and Bernays, 2000)*

FIGURE 1.14. Scavenging rats and cultural transmission. (A) When a rat scavenges in the trash, it may encounter new food items that are dangerous or spoiled and that can lead to illness or even death. **(B)** Smelling another rat provides olfactory cues about what it has eaten. This transfer of information from one rat to another about safe foods is a form of cultural transmission. *(Photo credits: Stephen Dalton/Mindon Pictures; Tom McHugh/Photo Researchers Inc.)*

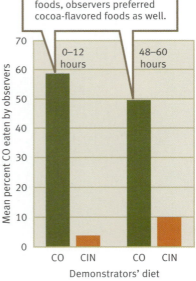

FIGURE 1.15. Social learning and foraging in the Norway rat. Observer rats had a "tutor" (demonstrator) that was trained to eat rat chow containing either cocoa (CO) or cinnamon (CIN) flavoring. After the observer rats had time to interact with a demonstrator rat, the observer rats were much more likely to add their tutor's food preferences to their own. *(From Galef and Wigmore, 1983)*

through the use of olfactory cues, observer rats were influenced by the food their tutors had eaten, and they were more likely to eat that food (Figure 1.15).

Cultural transmission is more complicated than individual learning. The actual information acquired via individual learning never makes it across generations. In contrast, with cultural transmission, if a single animal's behavior is copied, it can affect individuals many generations down the road (see Chapter 6).

Suppose adult rat A (in generation 1) adopts a new, formerly uneaten, type of food into its diet after it smelled this food on a nestmate. Now suppose young individuals (generation 2) in the same colony as rat A add this new food to their diet after they smell it on rat A. When individual A eventually dies, the cultural transmission chain it began may still be in force, as the young individuals that copied rat A will still be around. In other words, a culturally learned preference in generation 1 may make it to generation 2 (Figure 1.16). If generation 3 individuals learn from generation 2 individuals, then the culturally derived preference will have been transmitted across two generations, and potentially so on down the generations (Mesoudi et al., 2006). Cultural transmission itself, in other words, has both within- and between-generation effects (see Chapter 6). Understanding the dynamics of cultural transmission can be very complicated. In addition to the within- and between-generation effects just discussed, if there is variation in the tendency to copy the behavior of others, and that variation is due to certain types of genetic variation, then natural selection can act on copying behavior as well.

An understanding of natural selection, learning, and cultural transmission is critical to any broad-based overview of ethology. Yet, by necessity, each chapter of this book does not give equal weight to each of these three forces. In fact, some chapters do not even touch on all three. Historically, many more studies have been conducted on the role of natural selection on nonlearning-related behaviors (see Chapter 2), and hence this approach takes up a significant proportion of most chapters. Conversely, while social psychologists have been studying social learning in humans for more than a century (see Chapter 6), it is only over the last few decades that animal behaviorists have seriously investigated social learning, or more generally cultural transmission.

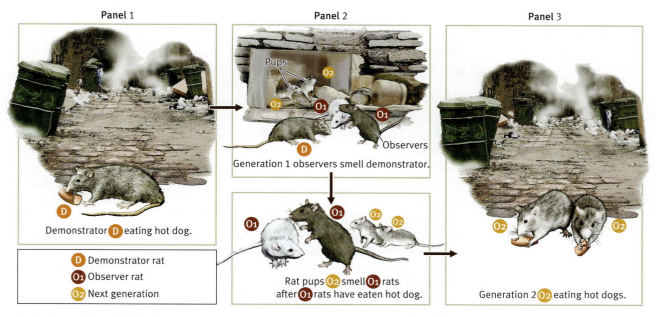

Panel 1

Panel 2

Pups

O2

O2

O1

O1

D

Observers

Generation 1 observers smell demonstrator.

Panel 3

D

Demonstrator D eating hot dog.

D Demonstrator rat
O1 Observer rat
O2 Next generation

O1

O2

O2

O1

Rat pups O2 smell O1 rats
after O1 rats have eaten hot dog.

O2

O2

Generation 2 O2 eating hot dogs.

FIGURE 1.16. A role for cultural transmission. In panel 1, a rat eats a new food type (hot dog). When this rat (D for demonstrator rat) returns to its nest (panel 2), observer rats (O1) smell the rat and then are more likely to add hot dogs to their diet when they encounter such an item. Multigenerational cultural transmission occurs when rats from the next generation (O2) smell generation O1 rats after they have eaten hot dogs and subsequently add hot dogs to their own diet (panel 3).

Conceptual, Theoretical, and Empirical Approaches

In ethology, as in all sciences, every question can be studied using conceptual, theoretical, and empirical approaches (Dugatkin, 2001a, 2001b; Figure 1.17). In fact, the best studies in animal behavior tend to use all three of these approaches to one degree or another. In addition to the focus on natural selection, learning, and cultural transmission, the empirical/theoretical/conceptual axis also plays an important role in almost every chapter of this book.

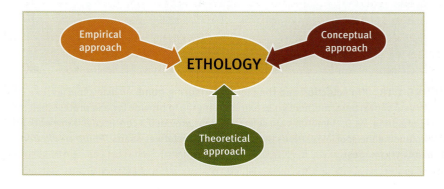

Empirical approach

Conceptual approach

ETHOLOGY

Theoretical approach

FIGURE 1.17. Different approaches to ethology. Ethology can be studied from a conceptual, theoretical, or empirical approach.

CONCEPTUAL APPROACHES

Conceptual approaches to ethology typically involve integrating formerly disparate and unconnected ideas and combining them in new, cohesive ways to make advances in the field. Generally speaking, natural history and experimentation play a role in concept generation, but a broad-based concept itself is not usually directly tied to any specific observation or experiment.

Major conceptual advances tend not only to generate new experimental work, but also to reshape the way that a discipline looks at itself. One conceptual breakthrough that has made animal behaviorists rethink the basic way they approach their science is the late W. D. Hamilton's ideas on **kin selection** (Chapter 9). Kin selection expanded the bounds of classic natural selection models by demonstrating that natural selection not only favors behaviors that increase the reproductive success of individuals expressing that behavior, but also favors behaviors that increase the reproductive success of those individuals' close genetic kin (Figure 1.18). Hamilton's work has a strong theoretical component to it as well, but here I will focus on the conceptual nature of this idea.

Hamilton hypothesized that individual 1's total fitness is not simply the number of viable offspring it produces (Hamilton, 1964; Figure 1.19). Instead, Hamilton proposed that fitness is composed of two parts: direct fitness and indirect fitness. **Direct fitness** is measured by the number of viable offspring produced, plus any effects that individual 1 might have on the direct descendants of its own offspring: for example, any effect individual 1 might have on the reproductive success of its grand-offspring. **Indirect fitness** effects are measured by the increased reproductive success of individual 1's genetic relatives—not including its offspring and any lineal descendants of offspring—that are *due to individual 1's behavior*. These actions *indirectly* get copies of

FIGURE 1.18. Kin selection and the mother-offspring bond. In many species, like the vervets shown here, mothers go to extreme lengths to provide for and protect their young offspring. W. D. Hamilton's kin selection ideas provided a conceptual framework for understanding the special relations that close genetic relatives share. *(Photo credit: Ben Cranke/Getty Images)*

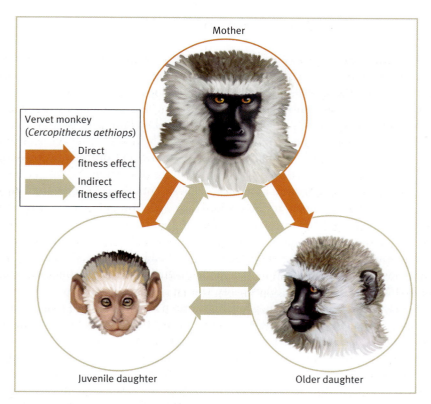

FIGURE 1.19. Two components to fitness. Three vervet monkeys—a mother, her juvenile offspring, and her older female offspring. Mother helping either daughter would be an example of a direct fitness effect. Siblings helping each other, or helping their mother, would represent indirect fitness effects. *(Based on J.L. Brown, 1987)*

individual 1's genes into the next generation. An individual's **inclusive fitness** is the sum of its direct and indirect fitness (J.L. Brown, 1980; Hamilton, 1964).

Chapter 9 explores the logic of inclusive fitness in detail, but the kernel of this powerful idea is that, evolutionarily speaking, close genetic relatives are important because of their shared genes—genes inherited from some common ancestor. Imagine for a moment a Mexican jay, a species of birds that has been the subject of much work on inclusive fitness (J.L. Brown, 1987). A jay's inclusive fitness is a composite of the number of offspring it has, plus some fraction of the number of offspring it helps a relative raise. Let's say that a jay helps its parents raise an additional brood of two birds, above and beyond what its mother and father could have raised on their own. Our helper is related to its siblings by a value of 0.5 (see Chapter 9 for more on this calculation). By helping mother (and perhaps father) raise two additional offspring, it has contributed 2 × 0.5 or the equivalent of 1 offspring to its inclusive fitness. If this is the only help that it gives, our jay's inclusive fitness is calculated by adding its indirect fitness (from helping its parents to raise its siblings) to its direct fitness (measured by its success at raising its own offspring).

Today, one of the first things that ethologists consider when studying social behavior is whether the individuals involved are close genetic kin. This is a direct result of Hamilton's conceptual breakthrough.

THEORETICAL APPROACHES

In the late 1960s and early 1970s, ethologists' understanding of how natural selection operates on animal behavior was greatly advanced with the appearance of sophisticated, usually mathematical, models of the evolution of social behavior in animals and humans. This work is most closely associated with George C. Williams, John Krebs, William D. Hamilton, John Maynard Smith, Robert Trivers, and Richard Alexander. The models that these animal behaviorists developed in the 1960s and 1970s revolutionized the way that ethologists look at almost every type of behavior they study.

A **theoretical approach** to animal behavior entails the generation of some sort of mathematical model of the world. During the formative years of modern ethology, much theoretical work focused on animal foraging behaviors (Kamil et al., 1987; Stephens and Krebs, 1986). One foraging-related question of particular interest was "which food items should an animal add to its diet, and under what conditions?" To tackle this question, a mathematical tool called optimality theory was used (see Chapter 11). Optimality theory searches for the best (optimal) solution to a particular problem, given that certain constraints exist in a system.

For example, you might be interested in building a model that examines how animals choose which prey to add to their diet to maximize the amount of energy they take in per unit time foraging. In that case, the amount of daylight could be a constraint (for some foragers), and your mathematical model could include the total amount of time an animal has to search for food (let's label that T_s), the energy (e) provided by a prey type, the time it takes to handle (h) the prey (e.g., to kill and then eat it), and the rate at which prey are encountered (λ). You would then examine how these variables affect foraging decisions made by animals (Figure 1.20). These variables

$$\text{Value of fawn} = \frac{\lambda_{\text{Fawn}} \times e_{\text{Fawn}}}{1 + \lambda_{\text{Fawn}} \times h_{\text{Fawn}}}$$

Cheetah

Gazelle fawn

FIGURE 1.20. Mathematical optimality theory and foraging. Cheetahs can feed on many different prey items, including a gazelle fawn. Ethologists have constructed mathematical models of foraging that determine which potential prey items should be taken. The value assigned to each prey is a composite of energy value (e), handling time (h), and encounter rate (λ).

would be placed in a fairly complex algebraic inequality, and solving this inequality would produce numerous testable, and often counterintuitive, predictions (see Chapter 11). For example, one such model predicts that the decision to add certain prey types into a forager's diet does not depend on how often a predator encounters that prey, but on how often it encounters more preferred prey types. Without a mathematical model of foraging behavior, it is unlikely that anyone could have used intuition to come up with that hypothesis.

It is important to realize that theoreticians, including those who work on ethological questions, are not interested in mimicking the natural world in their models, but rather in condensing a difficult, complex topic to its barest ingredients in an attempt to make specific predictions. In that sense, the criticism that a particular theory doesn't match the details of any given system will often be true, but irrelevant. A good theory will whittle away the details of specific systems, but just enough to allow for general predictions that can apply to many systems.

EMPIRICAL APPROACHES

Much of this book will be devoted to empirical studies and the **empirical approach**, which involves gathering data and drawing conclusions generating new, testable predictions, from that data. Empirical work in ethology can take many forms, but essentially it can be boiled down to one of two types—either observational or experimental studies. Both have been, and continue to be, important to the field of animal behavior.

While empirical studies in ethology preceded the work of Karl von Frisch, Niko Tinbergen, and Konrad Lorenz, modern ethological experimentation is often associated with these three Nobel Prize winners, each of whom was an extraordinary naturalist who had a fundamental understanding of the creatures with whom he worked and the world in which these creatures lived. They were able to ask fundamentally important questions about animal behavior— questions that could be addressed by a combination of observation and experimentation.

Observational work involves watching and recording what animals do, without attempting to manipulate or control any ethological or environmental variable. For example, I might go out into a marsh and record every action that I see red-winged blackbirds doing from 9 a.m. to 5 p.m. I might record foraging behavior, encounters with predators, the feeding of nestlings, and so forth, and be able to piece together how red-winged blackbirds in my study population spend their time. Next, from my observations, I might hypothesize, for example, that red-winged blackbirds seem to make few foraging bouts when predators are in the vicinity. To empirically examine the relationship between foraging and predation pressure, I might make detailed observations on how much food red-winged blackbirds eat and how many predators I can spot. I could then look for a relationship between these two variables and test the hypothesis that they are correlated.

Let's say that when I graph foraging behavior against predation pressure, I find that they are correlated. Redwings do increase and decrease their foraging behavior as a function of the number of predators in their environment. During periods when lots of predators are around, redwings forage infrequently, but when predators are few and far between, redwings forage significantly more often. What then can I conclude? Is it fair to say that increased predation pressure causes decreased foraging? No, the data we have so far do not demonstrate causation. I can say that predation and foraging are correlated, but from the existing data, I can't speak to the subject of what caused what—in other words, correlation does not equal causation. It might be that some other variable is causing both greater predation pressure and less redwing foraging behavior. For example, it might be that when the temperature rises, redwing predators become more active, but redwings themselves become less active, and hence forage less frequently. Increased predation pressure and foraging would still be correlated, but now the former wouldn't be seen as causing the latter; rather, they would both be associated with changes in weather.

In order to examine causality, I must experimentally manipulate our system. I might, for example, experimentally increase the number of redwing predators in area 1, but not in area 2, and see how redwing foraging is affected in these populations (Figure 1.21). I could do so by using trained predators or by simulating increased predation pressure by flying realistic

FIGURE 1.21. Observation and experimentation. Imagine your observations led you to believe that red-winged blackbirds decrease foraging when under predation pressure. To experimentally examine causality, you could allow a trained falcon to fly over a red-winged blackbird area and observe how its presence affects the amount of foraging.

predator models in area 1, but not in area 2. In either case, if redwing foraging behavior decreases in area 1 but not in area 2, I can feel much more confident that increased predation pressure causes decreased foraging in red-winged blackbirds.

Before completing this section on conceptual, theoretical, and empirical perspectives in ethology, we need to address one last question—whether there is any natural ordering when it comes to the theoretical and empirical approaches. Does theory come before or after empirical work? The answer is, "It depends." Good theory can precede or postdate data collecting. On some occasions, an observation or experiment will suggest to a researcher that the results obtained call for a mathematical model of behavior to be developed. Models of reciprocity and cooperation, for example, originally emerged from anecdotal evidence that many animals appeared to sacrifice something in order to help others. Given that natural selection should typically eliminate such unselfish actions, the anecdotes cried out for mathematical models to explain their existence. Mathematical models were then developed, and they provided some very useful insights on this question, as well as stimulating more empirical work.

In turn, theoretical models can inspire empirical studies. The foraging models discussed earlier in the chapter preceded the large number of empirical studies on foraging that ethologists and behavioral ecologists continue to undertake. While it is true that ethologists have long studied what and when animals eat, controlled experimental work designed to test specific predictions about foraging were initially spurred on by the theoretical work in this area. Regardless of whether theoretical work predates or postdates empirical work, however, a very powerful feedback loop typically emerges wherein advances in one realm (theoretical or empirical) lead to advances in the other realm.

An Overview of What Is to Follow

Following this chapter, are five "primers" that provide an overview of natural selection, phylogeny, and animal behavior (Chapter 2); hormones, neurobiology, and animal behavior (Chapter 3); molecular genetics, development, and animal behavior (Chapter 4); learning and animal behavior (Chapter 5); and cultural transmission from an ethological perspective (Chapter 6). The topics reviewed in the primer chapters are intertwined in the remaining eleven chapters, which cover sexual selection (Chapter 7), mating systems (Chapter 8), kinship (Chapter 9), cooperation (Chapter 10), foraging (Chapter 11), antipredator behavior (Chapter 12), communication (Chapter 13), habitat selection and territoriality (Chapter 14), aggression (Chapter 15), play (Chapter 16), and animal personalities (Chapter 17). In addition, studies of our own species, *Homo sapiens*, are woven into the fabric of each chapter. In this way, the reader receives a truly integrative view of animal (nonhuman and human) behavior.

Dr. E. O. Wilson

The 25th anniversary edition of your classic book *Sociobiology*, a landmark book in the field of animal behavior, was published in 2000. What prompted you to write *Sociobiology*?

In the 1960s, as a young researcher working in the new field of population biology, which covers the genetics and ecology of populations of organisms, I saw the logic of making that discipline the foundation of the study of social behavior in animals. At that time a great deal was known about societies of bees, ants, fish, chimpanzees, and so forth, but the subject largely comprised descriptions of each kind of society in turn, and with few connections. There had been little effort to tie all that information together. I had the idea of analyzing animal societies as special kinds of populations, with their characteristics determined by the heredity of behavior of the individual members, the birth rates of the members, together with their death rates, tendency to emigrate or cluster, and so forth—in other words, all the properties we study and put together in analyzing ordinary, nonsocial populations.

Sociobiology as a discipline grew from this idea and was born, not in my 1975 book with that name (*Sociobiology: The New Synthesis*), but in my 1971 *The Insect Societies*. In this earlier work I synthesized available knowledge of the social insects (ants, termites, the social bees, and the social wasps) on the base of population biology. I defined

the term "sociobiology" that way, and predicted that if made a full unified discipline it would organize knowledge of all animal societies, from termites to chimpanzees. In *Sociobiology: The New Synthesis* I added the vertebrates to the social insects (and other invertebrates) to

substantiate this view, then in the opening and closing chapters, the human species. In the latter chapters, I suggested that sociobiology could (and eventually would) serve as a true scientific foundation for the social sciences. This was a very controversial notion then, but it is mainstream today.

What do you see as *Sociobiology's* legacy to date?

The legacy of *Sociobiology,* which took hold and generated interest and discussion as *The Insect Societies* never could, is indeed the discipline of

sociobiology, with journals and many new lines of research devoted to it. This advance was greatly enhanced by the rapid growth of studies on animal communication, behavioral ecology, and, in population genetics, kin selection. Of ultimately equal and probably even greater importance, it showed how to create a link of cause-and-effect explanation between the natural sciences, including especially the study of animal social behavior on the one side and the social sciences on the other.

What sort of debt do ethologists owe Charles Robert Darwin?

Ethologists owe an enormous debt to Darwin, by encouraging the deep and now well-established concept that instincts are biological traits that evolved by natural selection. A word on terminology is worth introducing here. Ethology is the systematic (i.e., scientific) study of the behavior of animals (including, by extension, humans) under natural conditions. Sociobiology is the study of the biological basis of all forms of social behavior and social organization in all kinds of organisms, including humans, and organized on a base of ethology and population biology. Evolutionary psychology is a spin-off of both ethology and sociobiology, including both social and nonsocial behavior with special links to traditional studies of psychology.

After Darwin, whose work has had the most profound impact on the scientific study of animal behavior?

In 1989 the Fellows of the International Animal Behavior Society voted *Sociobiology: The New Synthesis* the most influential book on animal behavior of all time. The most important individual discoveries of all time would have to include sign stimuli, ritualization, the multiple modalities of nonhuman communication, the neurological and endocrinological basis of many forms of behavior, and the amazingly diverse and precise manifestations of kin selection.

Why should a talented undergraduate studying biology care about animal behavior?

Animal behavior is of course a fundamental and extraordinarily interesting subject in its own right. But it is also basic to other disciplines of biology, all the way from neuroscience and behavioral genetics to ecology and conservation biology.

Why should social scientists pay attention to what is happening in the field of animal behavior? What can they gain by doing so?

The social sciences desperately need biology as their foundational discipline, in the same way and to the same degree as chemistry needed physics and biology needed chemistry. Without biology, and in particular genetics, the neurosciences, and sociobiology, the social sciences can never penetrate the deep wells of human behavior; they can never acquire the same solidity and explanatory power as biology and the other natural sciences.

You and Bert Hölldobler won a Pulitzer prize for *The Ants*. Why have you devoted so much time and effort to studying this taxa?

There are two kinds of biologists: those who select a scientific problem and then search for the ideal organism to solve it (such as bacteria for the problems of molecular genetics), and those who select a group of organisms for personal aesthetic reasons and then search for those scientific problems which their organisms are ideally suited to solve. Bert Hölldobler and I independently acquired a lifelong interest in ants as children, and added science to that fascination later.

> **Sociobiology is the study of the biological basis of all forms of social behavior and social organization in all kinds of organisms, including humans, and organized on a base of ethology and population biology.**

You have written much on the subject of conservation biology. How does work in animal behavior affect conservation biology studies, and vice versa?

The understanding of animal behavior is crucial to conservation biology and its applications. Consider how important to ecosystems and

species survival are the behaviors of mating, territorial defense, dispersal, pollination, resource searching, and predation. To be successfully grasped, these phenomena have to be studied in an organized, scientific manner, not just added haphazardly to conservation strategies.

What do you believe will be the most important advance in animal behavior in the next twenty-five years?

My prediction: the complete linkage of a number of complex behavior patterns from genes to proteonones to sensors and neuron circuits to whole patterns of behavior. Biologists will learn how to scan the whole range of levels of organization to account for each animal behavior in turn.

Will animal behavior be a discipline fifty years from now, or will it be subsumed by other disciplines?

Today the study of animal behavior is the broad gateway to a wide array of different modes of study. But in fifty years—who knows? It may well be subsumed by other disciplines, some as yet undefined.

Dr. E. O. Wilson is an emeritus professor at Harvard University and a member of the National Academy of Sciences. He is the recipient of two Pulitzer prizes, and his book Sociobiology *(Harvard University Press, 1975) is regarded as one of the most important books on evolution and behavior ever written.*

SUMMARY

1. The scientific study of animal behavior, which dates back hundreds (if not thousands) of years, is called ethology.
2. The process of natural selection, the ability of animals to learn, and the process of cultural transmission are all important concepts for developing an integrated view of animal behavior.
3. Niko Tinbergen suggested that ethologists ask four types of questions: What are the immediate causes of behavior? How does behavior change as an animal develops and matures? How does behavior affect survival and reproduction? How does behavior vary as a function of evolutionary history?
4. Ethologists examine behavior from a proximate perspective by examining immediate causes of behavior, and from an ultimate perspective by examining evolutionary factors that lead to a behavior.
5. Work in ethology, like in all scientific fields, can be conceptual, theoretical, or empirical. Empirical work can be further subdivided into observational and experimental studies.

DISCUSSION QUESTIONS

1. Take a few hours one weekend day and focus on writing down all the behavioral observations you've made, as well as any, even indirect, behavioral hypotheses you have constructed. Think about your interaction with both humans and nonhumans. How has your very brief introduction into ethology reshaped the way you observe behavior?
2. Why do we need a science of ethology? What insights does this discipline provide both the scientist and the layperson?
3. Imagine that you are out in a forest, and you observe that squirrels there appear to cache their food only in the vicinity of certain species of plants. Construct a hypothesis for how this behavior may have been the product of (a) natural selection, (b) individual learning, and (c) social learning.
4. Why do you suppose that mathematical theories play such a large part in ethology? Couldn't hypotheses be derived in their absence? Why does mathematics force an investigator to be very explicit about his or her ethological hypotheses?
5. Discuss the pros and cons of each of the bulleted definitions of behavior on page 7.

SUGGESTED READING

Alexander, R. D. (1974). The evolution of social behavior. *Annual Review of Ecology and Systematics, 5,* 325–383. This paper, published just before E. O. Wilson's *Sociobiology*, provides the reader with a good overview of how one approaches behavior using "natural selection thinking."

Dewsbury, D. (Ed.) (1985). *Studying animal behavior: Autobiographies of the founders.* Chicago: University of Chicago Press. A fascinating introduction to the lives of early ethologists.

Galef, B. G. (2009). Strategies for social learning: Testing predictions from formal theory. *Advances in the study of behavior*, 39, 117–151. In this review, Galef formalizes predictions from social learning models, and puts those models to the test.

Heinrich, B. (1999). *Mind of the raven*. New York: HarperCollins. Heinrich is a top-notch scientist and science writer who gives the reader a sense of both the science of animal behavior and the beauty of nature.

Levitis, D. A., Lidicker, W. Z., & Freund, G. (2009). Behavioural biologists do not agree on what constitutes behaviour. *Animal Behaviour*, 78, 103–110. A review of different definitions of the word *behavior*.

Tinbergen, N. (1963). On aims and methods of ethology. *Zeitschrift fur Tierpsychologie*, 20, 410–440. A classic paper that outlines Niko Tinbergen's approach to animal behavior.

2

The Evolution of Behavior

Every biology and psychology student should read Charles Darwin's *On the Origin of Species*, considered one of the greatest, if not *the* greatest, science books ever written. What may surprise you most about Darwin's book is not only the ease with which it can be read, but also the subject material of the first chapter. The opening chapter of the most significant book ever written in biology talks at some length about pigeon breeding in Victorian England. Victorians and others, Darwin noted, had bred many varieties of pigeons that looked and behaved dramatically differently from one another. Indeed, Darwin himself was a pigeon breeder, informing the readers of *On the Origin of Species,* "I have kept every breed [of pigeon] I could purchase or obtain. . . ." (p. 23). Some of these various breeds have exquisite color and very elaborate tail feathers (Figure 2.1).

Darwin was fascinated by both the morphological and the *behavioral* varieties of pigeons. "Tumbler" pigeons, for example, seem to somersault over themselves as they fly. Homing pigeons can be released long distances from home yet somehow find their way back (Figure 2.2). All of these morphological and behavioral varieties of pigeons were the product of many generations of breeding, primarily by amateur pigeon breeders.

The reason for this seemingly odd subject matter for the opening chapter of *On the Origin of Species* was strategic—Darwin was preparing the reader for what was to come. He knew that his readers would feel at home with a discussion of pigeon breeding, a popular pastime in Victorian days. If Darwin could convince them that the process leading to the creation of extraordinary breeds of pigeons was similar to the process leading to new varieties and species in nature, his task would be a little simpler. The process leading to new pigeon breeds—tumbler pigeons, homing pigeons, and so on—is called **artificial selection**, which is defined as the process of humans deliberately choosing certain varieties of an organism over others by implementing breeding programs that favor one variety over another. Darwin's discussion of artificial selection led directly to his introducing readers to his ideas on natural selection.

As opposed to artificial selection, the evolutionary process leading to the extraordinary variation—including behavioral variation—that we see in nature

A **B**

FIGURE 2.1. Natural and artificial selection. Both natural and artificial selection have produced many morphological varieties of the pigeon, including **(A)** bright colors and **(B)** elaborate tail feathers. *(Photo credits: Brandon Borgelt/Agefotostock; Rashad Zainal–Bahrain)*

A  **B**

FIGURE 2.2. Artificial selection on pigeon behavior. Pigeon breeders have selected for behavioral varieties of pigeons, including **(A)** tumbler pigeons (here we see one bird tumble as it flies) and **(B)** homing pigeons. *(Photo credits: James Whitmore/Time Life Pictures/Getty Images; Derrick Francis Furlong/Alamy)*

is called natural selection, which, you will recall from Chapter 1, is the process whereby traits conferring the highest reproductive success to their bearers increase in frequency over time. This chapter serves as an introduction or "primer" to the manner in which ethologists think about evolution and animal behavior.

Once Darwin's ideas were widely disseminated and integrated into the heart of biology during what is called "the modern synthesis" (J. Huxley, 1942), animal behaviorists possessed a theory that helped explain not only *what* animals do, but *why* they do it. These sorts of "why" questions—that is, questions that deal with how evolutionary processes shape traits—are often also labeled "ultimate" questions (see Chapter 1). The term "ultimate" does not imply a greater importance attached to such questions than to any other questions in animal behavior. Instead, the word *ultimate* simply refers to a focus on evolutionary forces per se.

This chapter outlines, step-by-step, how ethologists tackle ultimate questions. Following Darwin's strategy, begin by discussing how selection operates when humans, rather than "nature," are the selective force shaping behavior; that is, we begin with artificial selection. From there, we will move directly to the discussion of natural selection and animal behavior, and then delve into the role of phylogenetic history in understanding behavioral evolution.

Artificial Selection

For at least 10,000 years, humans have been shaping the way that animals and plants look, as well as the way that animals behave, by the process of artificial selection (Denison et al., 2003). Ever since we selected some varieties of wheat, corn, and rice over others, and systematically planted their seeds, we have been involved in artificial selection. The question "Why do we see particular forms of grain today?" can only be understood in terms of the process of artificial selection. The same can be said of our systematic preference for breeding certain varieties of animals over others, as well as hundreds of other examples. For example, in an experiment that has been ongoing from the late 1950s until today, a team of Russian scientists has been systematically choosing the tamest, most docile foxes from a population of foxes in Siberia. Over the course of the experiment, they have tested more than 40,000 foxes. In each generation, they allow only the tamest to breed and become parents for the subsequent generation. The results have been remarkable: this artificial selection program has produced foxes that not only can be held and petted by humans but also seek out human contact (Trut 1999; Trut et al., 2009).

Let's examine the process of artificial selection in more detail. Suppose we begin an artificial selection program with a group of collies, and we wish to produce a variety of dog that is useful in herding sheep. What we want is a dog breed in which individuals will circle around our flock of sheep, keeping the sheep together and also keeping predators away from our economically valuable flock. Assuming that herding behavior is heritable, how would we go about this through artificial selection? First, we would begin our breeding program by choosing the individuals that are already closest to our ideal herding dog—that is, those that behaved the way we wanted our dogs to behave. We could determine which male and female collies were best at herding sheep and keeping flocks safe, and then preferentially breed those individuals. That is, in every generation we would sort the dogs we had, choose those that met our breeding criteria (herding ability) and allow them the chance to mate, while denying breeding opportunities to those dogs that failed to meet our criteria. With each new generation of dogs, we would repeat this process, producing individuals that were coming closer and closer to our ideal dog variety. Eventually we would recognize that we were as close as we were ever going to get to our ideal herding dog. We can answer the question, Why do we see the herding breeds of dog we see today? by referring to a selection process—in this case, artificial selection (Figure 2.3).

With a basic understanding of the nuts and bolts of artificial selection and behavior in hand, we, like Darwin's original readers, are ready to move on to natural selection and animal behavior.

Natural Selection

Despite the monumental impact that Darwin's two greatest works, *On the Origin of Species* and *The Descent of Man and Selection in Relation to Sex* have had, Darwin's theory of natural selection is straightforward to understand, particularly once artificial selection is understood (G. Bell, 1997; Darwin, 1859, 1871; Endler, 1986; Williams, 1966). Indeed, Darwin came up with his theory of natural selection before Mendel's work on genetics was disseminated. But Darwin didn't need to know about genes per se for his theory to work; all he needed to realize was that somehow behavioral traits that affected reproductive success were passed from parents to offspring. Any Victorian naturalist would have known that offspring resemble their parents, and Darwin was an excellent naturalist (Darwin, 1845).

SELECTIVE ADVANTAGE OF A TRAIT

Consider any trait—height, weight, fur color, foraging behavior, mating, and so on—and instead of imagining humans as the selective agent, allow the selective agent to be nature itself. When nature is the selective agent, traits,

In generation 1, the population contains very good herders, good herders, and poor herders.

After four generations of selecting for better herders, the population contains very good herders and good herders.

After many generations of selecting for better herders, the population contains primarily very good herders.

Generation 1

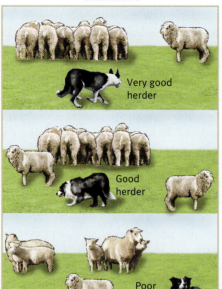

Generation 5

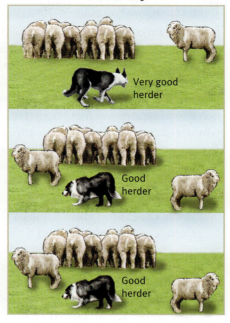

Generation N

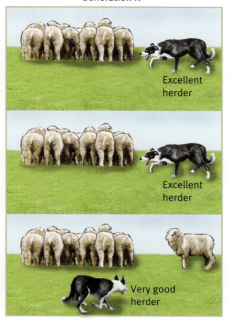

including behavioral traits, increase or decrease in frequency as a function of how well they suit organisms to their environments. If one variety of a trait helps individuals survive and reproduce better in their environment than another variety of the same trait, and if the trait can be passed down across generations, then natural selection will operate to increase its frequency over time. To see this, let's examine how natural selection might have favored "pack hunting" behavior in wild dogs (*Lycaon pictus*). Modern wild dogs tend to hunt in packs of ten individuals, but what would a hypothetical scenario for the evolution of the trait "hunt in packs" look like (Creel, 2001; Figure 2.4)?

FIGURE 2.3. Artificial selection on herding behavior. An example of how herding behavior might be selected in dogs. In each generation, the dogs that displayed the "herding" traits that a breeder was interested in would be allowed to breed, with preferential access to breeding given to the best herders. Over many generations, breeding can lead to dogs that are excellent herders—dogs that will circle around a flock of sheep, keeping the sheep together and also keeping predators away from the flock of sheep.

FIGURE 2.4. Group hunting in wild dogs. Wild dogs are ferocious predators and often hunt in groups. Here they are shown capturing a wildebeest. *(Photo credit: © Bruce Davidson/naturepl. com/ARKive)*

First, imagine a simplified scenario during the early stages of wild dog evolution, in which two types of foragers existed: one type that hunted in packs, and one type that hunted prey alone.

Hunting behavior—alone or in groups—represents one component of a wild dog's **phenotype**, typically defined as the observable properties of an organism. An individual's phenotype itself is the result of its **genotype**—that is, its genetic makeup—and the way that a particular genotype manifests itself in the environment. In our wild dogs hunting scenario, suppose that in the evolutionary past, individuals that hunted as part of a group got more meat, on average, than wild dogs that hunted alone, and that the more food a wild dog took in, the more offspring it could produce (Figure 2.5). If wild dogs that prefer to hunt in groups produce offspring that also like to hunt in groups, this behavioral variant will increase in frequency over time, helping us to understand why modern wild dogs display this behavior. This increase will occur even if group-hunting behavior produces only a very slight advantage in terms of the number of offspring an individual raises. To see why, we need to think about the implication of small fitness differences that are magnified over long periods of time; that is, we need to conceptualize natural selection in terms of an evolutionary time frame.

Even a fitness advantage of 1 percent per generation is sufficient for one behavior to replace another over evolutionary time. For example, for the sake of simplicity, let's assume that hunting preference—alone or in groups—is controlled by a single gene. In reality, of course, there are probably dozens of genes that work together to control this trait. This logic would work equally well for traits controlled by dozens of genes, but the math would be more difficult. For our purposes, let's imagine a trait largely controlled by variation associated with a single gene.

If an **allele**—that is, a gene variant, one of two or more alternative forms of a gene—that codes for group hunting provides its possessors with an average of just 1 percent more offspring per generation than the allele associated with hunting alone, then all else being equal, the group-hunting allele will eventually increase in frequency to the point where virtually all dogs in the

FIGURE 2.5. Natural selection for group hunting. A schematic of how natural selection could favor wild dogs that prefer to hunt in groups.

population have it. That is, natural selection will result in a population of individuals that almost always hunt in groups, because the fitness advantage conferred by group hunting makes those with our hypothetical group-hunting allele more likely to survive and produce offspring. These offspring, that in turn will have the allele coding for hunting in groups, are more likely to survive and reproduce, and so on, down through the generations. Over evolutionary time, small differences in fitness can accumulate into large changes in gene frequencies. (To see how evolutionary time allows small differences in fitness to translate into large-scale changes in the frequency of traits, see Table 2.1.)

In a breeding population of about 100 wild dogs, if group hunting provided a fitness benefit of just 1 percent, it would increase to a frequency of 100 percent in 1,060 generations. If we assume an average generation time of about five years for a wild dog, that amounts to just 5,300 years. If the selective advantage were 5 percent—that is, if those that hunted in groups had reproductive success that was 5 percent higher than that of other individuals—in just 211 generations, or about 1,055 years, our entire hypothetical population of wild dogs would be composed of animals that hunted in groups. Even the longest time period mentioned—5,300 years—is tiny in the context of evolutionary time.

HOW NATURAL SELECTION OPERATES

The example of hunting in groups gives you a flavor for how natural selection operates on behavior. But what exactly does it take for the process of natural selection to operate, and what is the end product of this process (G. Bell, 1997; Endler, 1986; Mousseau et al., 1999; Williams, 1966)?

To understand how natural selection operates, the first thing any ethologist must do is to be specific about which behavior is being studied. That is, we don't so much speak of "natural selection" as we do of "natural selection operating

TABLE 2.1. Fitness benefits and frequency of traits. When population size equals 100, wild dogs that hunt in groups will increase in frequency, and they will eventually make up 100 percent of the population. In this model, we are assuming no mutation and no migration in or out of our population. The number of generations for group hunting to go to 100 percent is calculated as follows:
(2/the selective advantage to group hunting) × the natural log of population size. *(Based on Carroll, 2007)*

FITNESS BENEFIT TO GROUP HUNTING	POPULATION SIZE	GENERATIONS BEFORE GROUP HUNTING REACHES 100% OF POPULATION
1%	100	1,060
5%	100	212
10%	100	106
20%	100	53

on foraging behavior," or "natural selection operating on fighting behavior," and so on. Once a trait is specified, the process of natural selection requires three prerequisites to operate:

▶ Variation in the trait—different varieties of the trait.
▶ Fitness consequences of the trait—different varieties of the trait must affect reproductive success differently.
▶ A mode of inheritance—a means by which the trait is passed on to the next generation.

Technically speaking, a fourth requirement also exists, and that is that resources must be limited with respect to the trait being studied. So, if one is studying natural selection and foraging, food in some sense must be limited. If one is studying natural selection and mating, there must be a finite set of individuals available to mate with, and so on. In practice, the limited resources requirement is almost always met.

To better understand natural selection, let's follow this process with respect to a specific behavioral trait: how quickly an animal will approach a novel object in its environment, which is a behavior that has been well studied in birds (Drent et al., 2003; van Oers, Drent, de Goede, et al., 2004; van Oers, Drent, de Jong, et al., 2004; van Oers et al., 2005). "Object" here is used in the broadest sense—for example, a novel object might be an individual from a species that has not been encountered before or a trap put out by humans or a new type of food in the environment. Approaching novel objects can be dangerous (if they turn out to be predators), but it might also yield benefits (if the novel object is a new type of prey). For simplicity, let's designate the time it takes for an individual to approach a novel object in its environment—the time from when the object is first spotted to the time the animal interacts with this object— as the "approach" score.

We now examine approach behavior by stepping through the three prerequisites mentioned—variation, fitness consequences, and mode of inheritance. We focus on a hypothetical population in a simplified example designed to illustrate the process of natural selection in the clearest possible manner. But you should note that the details of both novel object approach behavior in birds and group hunting in wild dogs (discussed earlier), have been well studied by ethologists in more complicated experiments.

VARIATION. For natural selection to act, there must be variation in the trait under investigation (Mousseau et al., 1999). If every animal in our hypothetical population of birds displayed the same approach score, there would be nothing for natural selection to select between (Dingemanse et al., 2002). For natural selection to act, individuals in our population must differ from one another with respect to their approach score. If, for example, average approach scores in our birds range from 30 seconds (approach quickly) to 60 seconds (approach moderately) to 120 seconds (approach slowly), we have behavioral variation in approach score (Figure 2.6).

Variation in a trait can be caused by either environmental or genetic factors. We shall return to these two different forms of variation a bit later, but here we focus on **genetic variation**—in our case, behavioral variation in approach score that correlates with genetic differences between birds in our population. Genetic variation in a population can be generated in a number of ways. For example, **mutation**—which is defined as any change in genetic structure— creates new variation in a population.

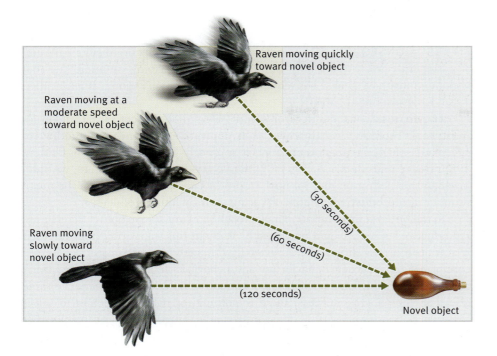

Raven moving quickly toward novel object

Raven moving at a moderate speed toward novel object

Raven moving slowly toward novel object

(30 seconds)

(60 seconds)

(120 seconds)

Novel object

FIGURE 2.6. Natural selection and variation. For natural selection to act on a behavior, behavioral variation must be present in the population under study. In the case of novel object approach behavior, different birds may approach an object they have not previously encountered (for example, a new food source or a brightly colored glass vial, as shown here) slowly (taking 120 seconds to reach the object), moderately slowly (taking 60 seconds to reach the object), or quickly (taking 30 seconds to reach the object).

Mutations can manifest themselves in many different ways. Addition and deletion mutations occur when a single nucleotide is either added or deleted from a stretch of DNA. Genes code for the production of enzymes and other kinds of proteins (which are made up of amino acids). So this type of mutation typically causes the production of an inactive enzyme, which may, in turn, affect an animal's behavior. Base mutations occur when one base in a nucleotide replaces another. Many base mutations can potentially affect protein function, and hence they may have an impact on an individual's reproductive success often, but certainly not always, by affecting behavior. Some base mutations do not cause changes in which amino acids are produced. These mutations are known as silent mutations.

In addition to mutation, another factor that produces variation in a population is **genetic recombination**. In sexually reproducing organisms— and the vast majority of animals reproduce sexually, as opposed to asexually— when pairs of chromosomes (one of which comes from the mother and one from the father) line up during cell division, sections of one chromosome may "cross over" and swap positions with sections of the other chromosome. This swapping creates new genetic variation. Crossing-over points are essentially random, and so virtually any crossover between a pair of chromosomes is possible in principle. As such, crossing over creates huge amounts of new genetic variation—including genetic variation in behavioral traits—in sexually reproducing organisms.

Although it may sound paradoxical, new genetic variants of a trait can enter a population via nongenetic pathways. The most common way for this to occur is through **migration**. Migration can increase genetic diversity in a given population because individuals coming from other populations can introduce new trait variants. In our bird population, the range of approach scores is from 30 to 120 seconds, but suppose that in a neighboring population some birds have approach scores of 150 seconds. If the differences in approach scores between the migrants and those in our population are a function of underlying genetic differences, then migration increases genetic variation in the time it takes to approach novel objects. When migrants mate with individuals in their new population, this generates even more genetic variation.

FITNESS CONSEQUENCES. Variation alone is insufficient to allow natural selection to operate. Not only must there be behavioral variation but that variation must have fitness consequences (Darwin, 1859; Dejong, 1994; Endler, 1986; Kingsolver et al., 2001).

The **fitness** consequences of a trait refer to the effect of a trait on an individual's reproductive success—for example, the difference in reproductive success associated with slow versus fast approach (R. A. Fisher, 1958; Grafen, 1988; Reeve and Sherman, 1993; Williams, 1966). In later chapters, we shall broaden this definition, but for now, "reproductive success" refers to the mean number of reproductively viable offspring an individual produces.

In the bird example, variation in approach score must map onto differences in reproductive success, even if only weakly. Without this translation from variation in a trait to fitness differences associated with such variation, natural selection cannot act on approach behavior (or any behavior). To understand why, think about it like this: If we have behavioral variation, but all variants have the same effect on reproductive success, there is nothing for natural selection to select between. Suppose we have 100 birds, and to make things simple, let's imagine that 50 of them show approach scores of 30 seconds, and 50 display approach scores of 120 seconds—in this population, we clearly have variation in approach score. Let's assume that individuals with approach scores of 30 seconds produce an average of 4 offspring, and birds with approach scores of 120 seconds also produce an average of 4 offspring. Since individuals with different approach scores produce the same average number of offspring, variation in approach score (30 versus 120 seconds) does not translate into variation in fitness, and natural selection cannot act on approach score. In contrast, if all individuals with lower approach scores of 30 had greater reproductive success than all individuals with higher approach scores, then behavioral variation does translate into fitness differences (Figure 2.7).

Conventional wisdom in ethology is that fitness differences are almost always present if the investigator searches hard enough, and experimental work has supported this assumption (Endler, 1999; S. J. Gould and Lewontin, 1979; Mayr, 1983). The odds that two behavioral variants would have the exact same effect on reproductive success are very low, and we have already seen that over evolutionary time even small differences in reproductive success can have important consequences on the evolution of behavior.

MODE OF INHERITANCE. For natural selection to act on a trait, that trait must be passed down from one generation to the next. That is, in addition to variation and fitness consequences, there must be some mode of inheritance in place for natural selection to act on a trait. Without a mode of inheritance, any fitness differences that exist within one generation are washed away, and natural selection cannot move our population of birds toward longer or shorter approach scores. To understand why, imagine that birds with approach scores of 30 seconds have (on average) five offspring and that birds with approach scores of 120 seconds have (on average) three offspring. In this case, both variation and fitness consequences are associated with behavioral variation. If there is no mode of inheritance in place, however, offspring will not resemble their parents with respect to approach score. That is, without a mode of inheritance, individuals that have low approach scores are no more likely to produce offspring with low approach scores than are individuals that have high approach scores, and vice versa. Any fitness associated with approach score would be lost in the next generation and natural selection would not be able to operate on this behavior.

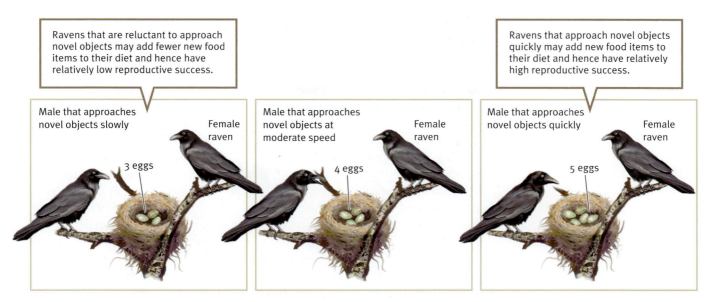

> Ravens that are reluctant to approach novel objects may add fewer new food items to their diet and hence have relatively low reproductive success.

> Ravens that approach novel objects quickly may add new food items to their diet and hence have relatively high reproductive success.

Male that approaches novel objects slowly — Female raven — 3 eggs

Male that approaches novel objects at moderate speed — Female raven — 4 eggs

Male that approaches novel objects quickly — Female raven — 5 eggs

FIGURE 2.7. Natural selection and fitness consequences. If approaching novel objects quickly enables a bird to be the first to reach a new food source, this may contribute to its survival, as it may get more of that food than birds with slower approach times. This variation in approach time will have fitness consequences if it leads to those approaching quickly having more eggs and hence more offspring.

Because genes are passed down from generation to generation, they are the most obvious candidate for a method of transmission. We shall focus on genetic transmission here, holding off a discussion of other modes of transmission until an in-depth discussion of cultural transmission in Chapter 6.

One way to study genes as a mode of transmission is by calculating narrow-sense **heritability**—a measure of the proportion of variance in a trait that is due to genetic variance (Hartl and Clark, 2006; Mousseau and Roff, 1987). Recall that for natural selection to operate on a trait, there must be variation in that trait—in our example, variation in approach scores. But these *differences* in approach scores—indeed, in any behavior—can come about in many ways. Individuals displaying different approach scores may have been raised on different diets, been exposed to different learning opportunities, and so on. Behavioral differences can also be the result of genetic differences. It is this genetic variance that is measured in heritability experiements.

TRUNCATION SELECTION. One means for measuring narrow-sense heritability is by designing a **truncation selection experiment**. In step 1 of a truncation selection experiment examining heritability in approach scores, we measure the approach score of every bird in our population when it reaches twelve months of age. Suppose this gives us a mean approach score of 60 seconds for generation 1. Let's label this mean value x_0. Step 2 of our experiment is to *truncate*, or cut off, the population variation in approach scores by allowing only those individuals with approach scores greater than some value—for example, 80 seconds—to breed. We then calculate the mean approach score of those individuals that we have allowed to breed. Let's label that mean as x_1 and say that x_1 equals an approach score of 90 seconds in our population (Figure 2.8).

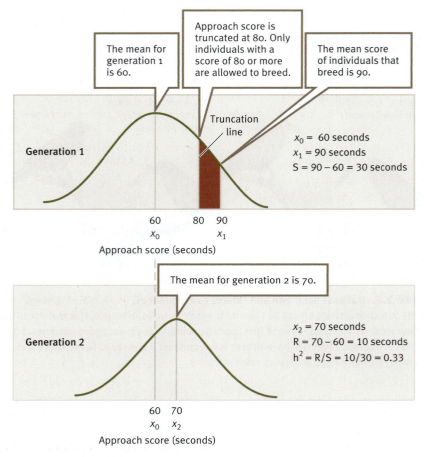

FIGURE 2.8. Heritability of novel object approach. Hypothetical results from a heritability experiment on novel object approach. The mean of *all* generation 1 individuals (60 seconds) is labeled x_0, and the mean of those generation 1 individuals that were allowed to breed (90 seconds) is labeled x_1. The mean of *all* generation 2 individuals (70 seconds) is labeled x_2. $S = x_1 - x_0 = 30$ seconds, $R = x_2 - x_0 = 10$ seconds, $h^2 = R/S = 0.33$.

The difference between x_1 and x_0 is referred to as the selection differential, or S. In our case, $S = 30$ seconds. One way to think about S is as the maximal amount we could expect natural selection to change approach scores—the amount of change that we might expect if all the variation in approach score was genetic variation upon which natural selection could act.

In step 3 of our truncation selection protocol, we raise the offspring produced by generation 1 birds, under identical conditions to those experienced by their parent, until they have reached twelve months of age, and then we measure their approach scores. Let's label the mean approach score of these generation 2 individuals as x_2, and imagine this value to be 70 seconds. The difference between this mean (x_2) and the mean of the entire population in the first generation (x_0) is referred to as the response to selection, or R. It is a measure of how much truncation selection has changed approach scores across generations 1 and 2. In our case, $R = 10$ seconds. Heritability is defined as R/S, so in our population of birds, the heritability of approach is 10/30 or about 0.33. In other words, one-third of all the variance in approach is due to genetic variance upon which natural selection can act.

Work in evolutionary biology, population genetics, and animal behavior all suggest that many traits—ranging from morphological to behavioral—show low (0.0 to 0.1) to moderate (0.1 to 0.4) heritability (Hoffmann, 1999; Mousseau and Roff, 1987; T. Price and Schulter, 1991; Weigensberg and Roff, 1996).

The same protocol we have employed for measuring variability, fitness, and heritability of approach scores can be used for any number of behaviors. This is not to say that obtaining concrete experimental evidence that natural selection operates on animal behavior *in the wild* is an easy task—it isn't, and it often takes years and years of effort to do so (Endler, 1986; Mousseau et al., 1999). Many studies *infer* how natural selection has operated on behavior in the wild, but experimental studies that measure natural selection in the field are very difficult to design and implement.

PARENT-OFFSPRING REGRESSION. In addition to the truncation selection method, narrow-sense heritability can also be measured through **parent-offspring regression**. The idea here is simple. Parents pass on genes to their offspring, so when narrow-sense heritability is high, the behavioral variation in the offspring should map onto the behavioral variation observed in parents. The greater the role environmental variance—differences between the environments experienced by parents and offspring in relation to diet, location, and so forth—plays in determining variance in behavior, the lower the heritability of that behavior. To see how parent-offspring analysis can help us understand behavioral variance, consider Charles and Mary Brown's work on behavior in cliff swallows (*Petrochelidon pyrrhonota;* Figure 2.9). The Browns used parent-offspring regression to dissect the behavioral variance in an individual's preference for living in larger or smaller groups.

For the past twenty-five years, the Browns have conducted field studies of cliff swallow birds—a species in which group size affects survival, for example, by affecting rates of parasitic infection (C.R. Brown and Brown, 2004a, 2004b). Using five clusters of cliff swallow colonies, and a sample of 2,581 birds, Brown and Brown found that the group size in which individual swallows lived was statistically similar to the group size in which their parents lived (Brown and Brown, 2000). This was true for birds that bred at the same site as their parents,

A   **B**

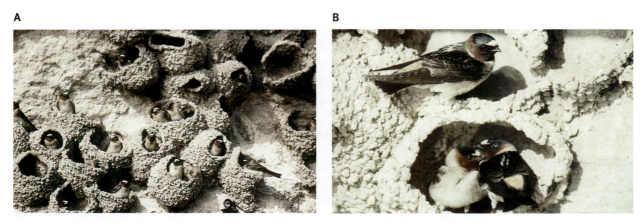

FIGURE 2.9. Cliff swallows in their nests. In cliff swallows, preference for group size is a heritable trait. **(A)** Cliff swallow nests are often clustered together. **(B)** A closeup of one nest with chicks, and the mother standing next to the nest. *(Photo credits: Charles Brown)*

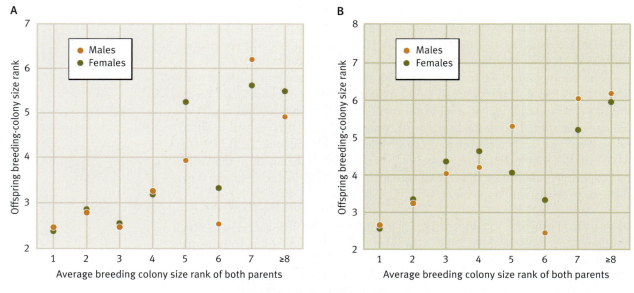

FIGURE 2.10. Breeding colony size in parents and offspring. There is a correlation between breeding colony size in parents and offspring in cliff swallows. Charles and Mary Brown sampled thousands of birds in their native habitat and found a strong correlation between parent and offspring colony size. This held true for all offspring **(A)**, as well as for offspring that bred away from their natal site **(B)**. To calculate heritability for a parent-offspring graph like this, the "line of best fit" for a set of data points is drawn (not shown here). If the line of best fit has a slope significantly greater than zero, the narrow-sense heritability is significant. *(From C. Brown and Brown, 2000)*

as well as for offspring that emigrated elsewhere, suggesting that the correlation between parent and offspring was not a function of living in the same place per se (Figure 2.10). If the correlation in group-size preference was due to parents and their offspring living in the same environment—that is, if this variance in group size was due to environmental variation—we would expect that correlation to disappear when offspring emigrated and lived in different environments from their parents. But Brown and Brown found that the correlation held, suggesting that preference for group size might be a heritable trait.

In a second experiment, Brown and Brown used nestlings from two *large* and five *small* colonies. Half the young from the nests in the large colonies were removed (the other half remained in the nest), and the young from small colonies were placed in their stead. Likewise, half the young in small colonies were replaced by young from large colonies. As such, in all their manipulated nests, Brown and Brown had offspring from both large and small colonies. This is referred to as a **cross-fostering experiment**, and of the almost 2,000 birds in this experiment, 721 were recaptured and used for a parent-offspring regression on preference for group size.

When examining the preference for group size in cross-fostered offspring, the Browns found a positive correlation with the group-size preferences of their genetic parents; a negative correlation was uncovered when group-size preferences of offspring and foster parents were compared. In other words, even after the cross-fostering, the young displayed the same group-size preference as their genetic parents (not their foster parents), suggesting a significant heritability for this complex behavioral trait.

Sociobiology,
Selfish Genes, and Adaptation

Sociobiology is the study of the evolution of social behavior (Wilson, 1975). The sociobiological notion that genes—in our case, genes associated with behavior—are the units upon which natural selection acts, is often referred to as the "selfish gene" approach to ethology. The phrase "selfish genes" was first coined by Richard Dawkins in 1976. As Dawkins himself makes clear, genes aren't "selfish" in any emotional or moral sense. But, genes can sometimes be treated as though they were selfish in that the process of natural selection favors those alleles that increase the expected relative reproductive success of their bearers (Dawkins, 2006; Grafen and Ridley, 2006). (The Conservation Connection box shows how the adaptation approach can inform conservation biology studies.)

Any allele that codes for a trait that increases the fitness of its bearer above and beyond that of others in the population will increase in frequency. So natural selection often, but not always, produces genes that *appear* to be selfish. Apply this approach to animal behavior, particularly animal social behavior, and you have one of the main ways in which ethologists think about genes and animal behavior (Alexander and Tinkle, 1981; J. L. Brown, 1975; Grafen and Ridley, 2006; E. O. Wilson, 1975). But it is worth emphasizing again that genes have no inherent qualities such as selfishness. Thinking of a gene as "selfish" just provides a convenient means to conceptualize some problems in animal behavior.

We need a term for traits that natural selection molds, and that often match organism to environment so exquisitely. Such traits are typically referred to as **adaptations**, a term that is defined in many ways in the literature (Mayr, 1982; Mitchell and Valone, 1990; Reeve and Sherman, 1993; Sober, 1987). For our purposes, the most important point is that natural selection is the primary process generating adaptations—traits associated with the highest relative fitness in a given environment (Reeve and Sherman, 1993).

Kern Reeve and Paul Sherman argue that a trait is an adaptation when that trait results in its bearers having the highest fitness among a *specified set of behaviors in a specified environment* (Reeve and Sherman, 1993). One critical aspect of this definition is that it shows that while natural selection produces adaptations, it is not necessarily true that adaptations are only the product of natural selection. If adaptations are traits that provide their bearers with the highest relative fitness in an environment, then any number of forces, including chance, could produce an adaptation. As long as the environment being studied is the same environment in which natural selection operated, it's likely that most adaptations *are* the product of natural selection; nonetheless, under Reeve and Sherman's definition, adaptations *don't have to be* the product of natural selection (S. J. Gould and Lewontin, 1979).

Let's examine three case studies of adaptation: (1) antipredator behavior in guppies, (2) cooperative behavior in naked mole rats, and (3) mate choice in humans.

ANTIPREDATOR BEHAVIOR IN GUPPIES

Ethologists and behavioral ecologists have often used the guppy (*Poecilia reticulata*) to study the evolution of behavior, and dozens of papers a year are published on some aspect of guppy ethology (Houde, 1997; Magurran, 2005).

Conservation Biology and Symmetry as an Indicator of Risk

One of the goals of conservation biologists is to detect populations under risk. But it is often difficult for them to identify which environments are stressful, and in which environments populations are at especially high risk. In some cases, of course, changes to an environment may be so extreme that it clearly can no longer support a population. But ideally, conservation biologists want to identify these stressful environments before so much damage is done that it might be irreversible. What tools does "natural selection thinking" offer conservationists in such situations?

One approach is to look for certain generalizable traits that can be used as indicators of the genetic quality of an individual. Evolutionary biologists and ethologists have long searched for traits that indicate the ability to respond to the changing (and often adverse) conditions that animals face throughout their development. If such indicator traits exist and are heritable, animals might use such indicator traits to select their mates.

A number of studies suggest that one trait that may serve as an indicator of genetic quality is the symmetry of the left and right side on an individual (Figure 2.11). Research suggests that symmetry is a cue that an individual has fared well in responding to the changing conditions that it faces during its development (Leamy and Klingenberg, 2005; Swaddle, 2003).

What are the implications of this fascinating discovery on conservation biology? If symmetry is an indicator of the ability to handle development stress, then as we move from less stressful to more stressful environments we would expect to see more asymmetry in traits. Luc Lens and his colleagues have proposed that conservation biologists can use asymmetry as an early warning system for detecting populations under risk (Lens et al., 2002).

To test this idea, Lens and his colleagues examined symmetry and survival in three

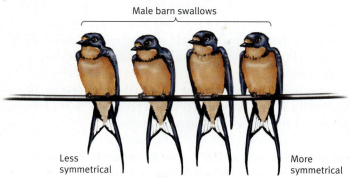

Male barn swallows

Less symmetrical

More symmetrical

FIGURE 2.11. Symmetry. Male barn swallows differ in how symmetric their tail feathers are.

populations of Taita thrushes (*Turdus helleri*). This species lives in forests in Kenya, and their forests have been fragmented by human activity since the 1960s. Fragments in this forest have been classified by conservation biologists as "low degeneration" (least damage, most similar to undisturbed forest), moderate degeneration, and high degeneration. Lens and his team studied one population of thrushes in each type of fragment. Not surprisingly, they found decreased rates of survival as they moved from the low-degeneration to high-degeneration sites. But as the researchers captured and marked each bird, they also measured the symmetry of the left and right tarsus. They found that symmetry decreased as forest degeneration increased:

individuals in the high-degeneration areas had higher levels of asymmetry than those in low-degeneration areas (Figure 2.12). Symmetry of the left and right tarsus bone in these thrushes appears to be an early warning signal that conservation biologists could use to identify populations at high risk. Rather than wait for the grim differences in survival between high- and low-degeneration forests to manifest themselves, conservation biologists could proactively take measures in populations showing high measures of asymmetry. For example, researchers could pay special attention to populations showing high measures of asymmetry, conducting more periodic inspections in such populations by surveying population size and growth rate.

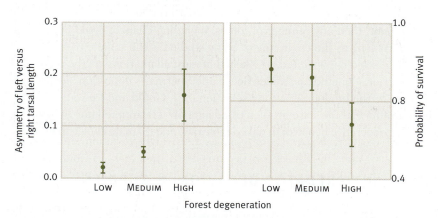

FIGURE 2.12. Symmetry as an early warning cue. In the Taita thrushes (*Turdus helleri*), symmetry of the left and right tarsus bone may be an early warning signal to identify populations at high risk. As degeneration increases, survival decreases and asymmetry (in the length of left and right tarsus bones) increases. *(From Lens et al., 2002)*

AB

FIGURE 2.13. Different guppy environments. (A) An upstream, low-predation stream in Trinidad, and **(B)** a downstream, high-predation stream. Guppies in these streams have been subject to different natural selection pressures. *(Photo credits: Lee Dugatkin)*

One reason for this interest in guppies is not only that they breed quickly, facilitating multigenerational studies, but that their population structure is ideal for studies of natural selection and behavior. Most guppy studies use fish from the Northern Mountains of Trinidad and Tobago. In many of these streams, guppies can be found both upstream and downstream of a series of waterfalls (Houde, 1997; Magurran, 2005; Seghers, 1973; Figure 2.13). These waterfalls act as a barrier to many of the guppies' predators. Upstream of waterfalls, guppies are typically under only slight predation pressure from one small species of fish; in contrast, downstream of the waterfalls, populations of guppies are often under severe predation pressure from numerous piscine (fish) predators.

High-predation and low-predation sites in the same streams are often only kilometers apart, but there is little gene flow between high- and low-predation sites, so if the type of predators differ dramatically based on site, then natural selection should favor different traits in upstream and downstream guppy populations. And indeed, between-population comparisons in guppies have found differences with respect to many traits, including antipredator behavior, color, number of offspring in each clutch, size of offspring, aging patterns, and age at reproduction (Endler, 1995; Houde, 1997; Magurran, 2005; Reznick, 1996; Figure 2.14). Recent genomewide genetic scans of guppies from many sites have found "genetic signatures" indicative of natural selection favoring different sets of alleles linked to behavioral traits at high- and low-predation sites (Willing et al., 2010).

Using a combination of field and laboratory experiments, David Reznick and his colleagues have found that guppies from high-predation sites mature faster, produce more broods of smaller offspring, and tend to channel their resources to reproduction when compared to guppies from low-predation sites (Figure 2.15). Why? At high-predation sites, guppy predators tend to be much larger (Figure 2.16A) and can eat a guppy no matter how large the guppy gets. At such sites, producing many smaller fish should be favored in guppies, as this is akin to buying lots of lottery tickets and hoping that one is a winner. At low-predation sites, only a single small fish predator (*Rivulus hartii*) of guppies

FIGURE 2.14. Male color patterns. Males at low-predation sites are more colorful than males at high-predation sites. Here we see a colorful male (right) and a female (left) from the Paria River in Trinidad. *(Photo credit: Anne Magurran)*

Predator (*Rivulus hartii*)

Prey (Guppies)

Low-predation site
Females produce fewer, but bigger offspring

High-predation site
Females produce many small offspring

Predator (*Crenicichla alta*)

Prey (Guppies)

FIGURE 2.15. Natural selection and predation. Natural selection acts differently on guppy populations from high-predation sites (with *Crenicichla alta*) and low-predation sites (with *Rivulus hartii*). At high-predation sites, natural selection favors guppies producing many small young, but at low-predation sites, natural selection pressure is flip-flopped, favoring fewer, but larger, offspring. Offspring are pictured below female guppies in left and right circles above.

exists (Figure 2.16B). If guppies can get past a certain size threshold, they are safe from *R. hartii*. As such, natural selection favors females producing fewer, but larger, offspring that can quickly grow large enough to be out of the zone of the danger associated with *R. hartii*, and this is precisely what we see (Reznick, 1996).

Natural selection has also operated on various aspects of guppy *behavior* (Endler, 1995; Houde, 1997; Magurran, 2005; Reznick, 1996). One suite of behaviors that has been studied extensively in natural populations of guppies is their antipredator activities (Magurran, 2005; Magurran et al., 1995; Seghers,

FIGURE 2.16. Guppy predators. (A) A pike cichlid, *Crenicichla alta*. This predator is common in downstream sites native to guppies. **(B)** A killifish, *Rivulus hartii*. This small, fairly innocuous predators can eat only tiny guppies and is found in upstream (low-predation) sites in the rivers of Trinidad. These pictures are not to scale. *C. alta* is much larger than *R. hartii*. (*Photo credits: Pete Oxford/naturepl.com; © Jesús Salas y Carlos Garrido*)

TABLE 2.2. The effects of predation. An abbreviated list of some behaviors that differ across populations as a function of predation pressure in guppies (*P. reticulata*), sticklebacks (*G. aculeatus*), and minnows (*P. phoxinus*). *(Adapted from Magurran et al., 1993)*

BEHAVIOR	AT AREAS OF HIGH PREDATION PRESSURE	SPECIES
Schooling	Larger and more cohesive schools	*P. reticulata* *P. phoxinus*
Evasion tactics	More effectively integrated in high-risk populations	*P. reticulata* *G. aculeatus* *P. phoxinus*
Inspection and predator assessment	Increase in inspection frequency	*G. aculeatus* *P. phoxinus*
	Increase in inspection group size	*P. phoxinus*
Habitat selection	Remain near surface and seek cover at edge of river	*P. reticulata*
Foraging	Increased feeding tenacity	*P. reticulata*
Female mating choice	Preference for less brightly colored males	*P. reticulata*
	Avoidance of sneaky mating attempts	*P. reticulata*
Male mating tactics	Increased use of sneaky mating tactics in high-risk populations	*P. reticulata*

1973). Depending upon whether they evolved in populations with heavy or light predation pressure (guppies, as well as sticklebacks and minnows) have a very different suite of antipredator behaviors (Table 2.2).

Two components of antipredator behavior—called shoaling and predator inspection—have been studied in great detail by animal behaviorists. Shoaling, or swimming together in a group (also referred to as schooling), is a measure of group cohesiveness (Keenleyside, 1955; Pitcher, 1986), whereas **predator inspection** behavior refers to the tendency for individuals to move toward a predator to gain various types of information (Dugatkin and Godin, 1992; George, 1960; Pitcher, 1992; Pitcher et al., 1986).

Because research from many fish species has found that swimming in large groups provides more protection from predators than swimming in small groups, many ethologists have hypothesized that guppies from high-predation sites would shoal more tightly, and in greater numbers, than guppies from low-predation sites (Houde, 1997; Magurran, 2005; Magurran et al., 1995). The data from the field are in line with this prediction (Figure 2.17). Furthermore, guppies from high-predation sites inspect a predator more cautiously, but more often, than their low-predation counterparts. This difference, too, is likely the result of contrasting natural selection pressures at high- and low-predation sites—inspecting a threat cautiously, but frequently, should be more strongly favored in areas of high versus low predation.

In the early 1990s, Anne Magurran and her colleagues discovered a unique opportunity to examine a "natural experiment" on the evolution of antipredator behavior in guppies. It seems that back in 1957, one of the original researchers on guppy population biology, C. P. Haskins, transferred 200 guppies from a high-predation site (in the Arima River) to a low-predation site (in the Turure River) that had been unoccupied to that point by guppies. Magurran realized that this was a golden opportunity to examine natural selection on

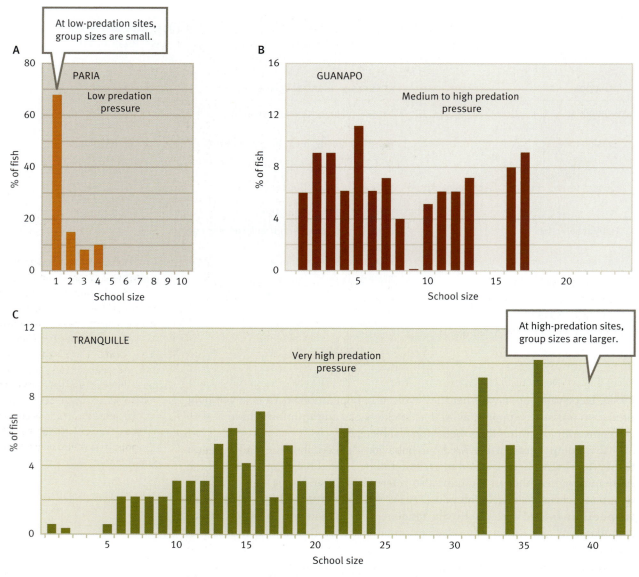

FIGURE 2.17. Predation pressure and group size. As a result of different natural selection pressures, group size differs across guppy populations: **(A)** a low-predation site, **(B)** a medium-to high-predation site, **(C)** a very-high-predation site. *(From Magurran and Seghers, 1991)*

antipredator behavior. If natural selection, via predation, shapes antipredator responses, then the lack of predation pressure in the Turure should have led to selection for weakened antipredator behavior in guppy descendants studied in the 1990s.

Magurran and her colleagues sampled numerous sites in the Turure (Magurran et al., 1992; Shaw et al., 1992). Genetic analysis suggested that the high-predation fish transferred from the Arima River back in 1957 had indeed spread all around the previously guppy-free site in the Turure River. More to the point, as a result of strong natural selection pressures, the descendants of the Arima River fish showed shoaling and predator inspection behaviors that were more similar to those of guppies at low-predation sites than they were to the behaviors of their ancestors from the dangerous sites in the Arima River.

FIGURE 2.18. **Transplants and natural selection.** One way to examine natural selection in the field is to do reciprocal transplant experiments, which in effect switch selection pressures. Here, some guppies from high-predation sites (with *C. alta*) are transplanted to low-predation sites, and vice versa. Over the course of several generations, as a result of new selection pressures transplanted populations converge on the characteristics of the fish in the populations into which they were transplanted.

This sort of result has also been found in other transplant experiments that had examined shifts in morphology and life history (Reznick, 1996; Reznick et al., 1990; Figure 2.18).

Magurran's group uncovered another curious finding. In addition to colonizing the low-predation areas of the Turure River (located upstream, where there are few predators), over the course of time the descendants of the Arima River fish moved downstream in the Turure River, back into areas of greater predation pressure. When tested, these fish showed antipredator behavior similar to that of their ancestors from the original high-predation site in the Arima River. One possible explanation is that the original colonizers spread fast, and since their antipredator behavior was beneficial when they reached high-predation sites in the Turure River, natural selection simply maintained such behavior. A more tantalizing, but to date untested hypothesis, is that the Arima River fish and their descendants colonized their new habitat at a much slower rate. If this were the case, natural selection may have shifted the colonizers and early descendants one way—toward the norm for low-predation sites—and then, later on, shifted the late descendants back in the opposite direction—toward the norm for fish from high-predation sites.

In addition to nicely illustrating how ethologists study the evolution of behavior, the guppy example illustrates that while natural selection may take hundreds of thousands of years to shape some traits, it can act much more quickly. Upstream and downstream guppy populations have been separated from one another for less than 10,000 years, yet as a result largely of differences in predation pressure, natural selection has produced significant differences in behavior in guppy populations over this fairly brief evolutionary time period (Endler, 1995). Indeed, both Magurran and Reznick's work on transfer experiments demonstrates that natural selection can act on antipredator behavior in wild populations on the time scale of years to decades.

KINSHIP AND NAKED MOLE RAT BEHAVIOR

By definition, genetic relatives (siblings, parents, offspring, and so on) share many of the same alleles. Because of this genetic similarity between kin, we expect more cooperative and altruistic behavior among kin than among unrelated individuals. In fact, as discussed in Chapter 9, according to theory, the more related individuals are, the more we expect to see cooperative and altruistic behaviors (Hamilton, 1964). To see how genetic relatedness can effect the evolution of cooperative and altruistic behaviors, we shall look into the fascinating, but bizarre life of naked mole rats (*Heterocephalus glaber*; Figure 2.19).

Few mammals have captured the fancy of both scientists and laypersons to the extent of the naked mole rat (Jarvis, 1981; Sherman et al., 1991). These small, hairless rodents of tropical Africa display **eusociality**, an extreme form of sociality that is present in many social insect groups (Table 2.3). Naked mole rats were the first vertebrates discovered to display the three characteristics associated with this extreme form of sociality, namely:

▶ A reproductive division of labor in which individuals in certain castes reproduce and individuals in other castes do not.
▶ Overlapping generations, such that individuals of different generations are alive at the same time.
▶ Communal care of young.

Naked mole rats are the longest-living rodent species, and their genome has recently been sequenced, shedding light on their longevity (Kim et al., 2011). They live in large groups in which a *single queen* and usually somewhere between one and three males are the only individuals that mate and reproduce in the entire colony. Although within-colony aggression certainly exists in this species, various forms of cooperative behavior are much more common (Reeve, 1992; Reeve and Sherman, 1991; Stankowich and Sherman, 2002). Nonreproductive male and female naked mole rats, which live much shorter lives than reproductive naked mole rats, undertake a wide variety of cooperative behaviors, such as digging new tunnels for the colony, sweeping debris, grooming the queen, and defending against predators (Lacey and Sherman, 1991; Pepper et al., 1991). But why do individuals yield exclusive reproduction to a single queen and a few males, as well as work for the colony? The answer in part centers on the high genetic relatedness between individuals within naked mole rat colonies.

Kinship theory suggests that the more genetically related individuals are, the more cooperation they will show with each other. Using this logic, Kern Reeve and his colleagues hypothesized that naked mole rats are very cooperative with

FIGURE 2.19. Naked mole rats. Naked mole rats show very high within-colony relatedness. This maps nicely onto numerous cooperative and altruistic behaviors common to this species, such as digging tunnels, sweeping dirt or debris out of the tunnels, grooming the queen, or defending against predators. Here, workers are in the process of digging a tunnel. Naked mole rats use their sharp teeth to break up the dirt and then move it back through the tunnel to a worker that throws it out of the tunnel. *(Photo credit: Gregory G. Dimijian/Photo Researchers, Inc.)*

TABLE 2.3. The relationship between social behavior in naked mole rats and social insects. Comparisons between naked mole rats and eusocial insects indicate basic similarities in such characteristics as reproductive division of labor, overlapping generations, and communal care of young. *(From Lacey and Sherman, 1991)*

SPECIES	SIMILARITIES TO NAKED MOLE RATS	DIFFERENCES FROM NAKED MOLE RATS
Paper wasp (*Polistes fuscatus*)	Single breeding female per colony Aggressive domination of other colony members by reproductive female No permanent sterility in subordinate foundresses Size-based subdivision of nonbreeding caste Similar colony sizes (20–100 workers) Slightly larger size in queens than in workers	All female workers Haplodiploid genetics New nests each spring Outcrossing promoted by dispersal of reproductives from nest site Carnivore
Honeybee (*Apis mellifera*)	Single breeding female per colony Colony reproduction by fissioning (swarming) Change in behavior with age in nonreproductives Slightly larger size in queens than in workers	All female workers Haplodiploid genetics Primarily chemical, not behavioral, reproductive suppression of workers Permanently "sterile" nonreproductive females Vastly larger colonies in honeybees Outcrossing promoted by aerial mating aggregations Nectarivore
Wood termite (*Kalotermes flavicollis*)	Single breeding female per colony Diploid genetics Male and female workers Delayed caste determination Division of labor among nonbreeders Extremely long life of reproductives Diet of plant material; breakdown of cellulose by gut endosymbionts Eating of feces Opportunity for workers to become reproductives if they escape the breeder's influence Working behavior begins before adulthood	Chemical, not behavioral, reproductive suppression Vastly larger colonies in termites Queens many times larger than workers Outcrossing promoted by aerial mating aggregations

others in their colony in part because they are close genetic relatives (Reeve et al., 1990).

To test this idea, Reeve's group sampled the DNA of fifty naked mole rats and then employed a molecular technique called DNA fingerprinting to examine the genetic relatedness among colony members (DeSalle and Schierwater, 1998; Jeffreys et al., 1985). Using DNA from liver, muscle, and brain samples, they created three DNA "probes" to isolate three distinct DNA fragments for each mole rat. These probes produced a series of prominent black bands, which consisted of between 1.6 and 1.8 kilobytes of DNA. Individuals had an average of about twenty-nine such distinct bands, and these bands together represented the **DNA fingerprint** of that mole rat. DNA fingerprints of different individuals could then be compared—the better the bands matched up, the more closely related were the individuals (Reeve et al., 1992).

Reeve and his colleagues found that because of very high levels of inbreeding—a result of mating between relatives—the average genetic relatedness in their colonies of naked mole rats was extremely high, as predicted by kinship theory (see Braude, 2000, for an alternative view). The exact value they came up with was an average relatedness of 0.81 within naked mole rat colonies. To put

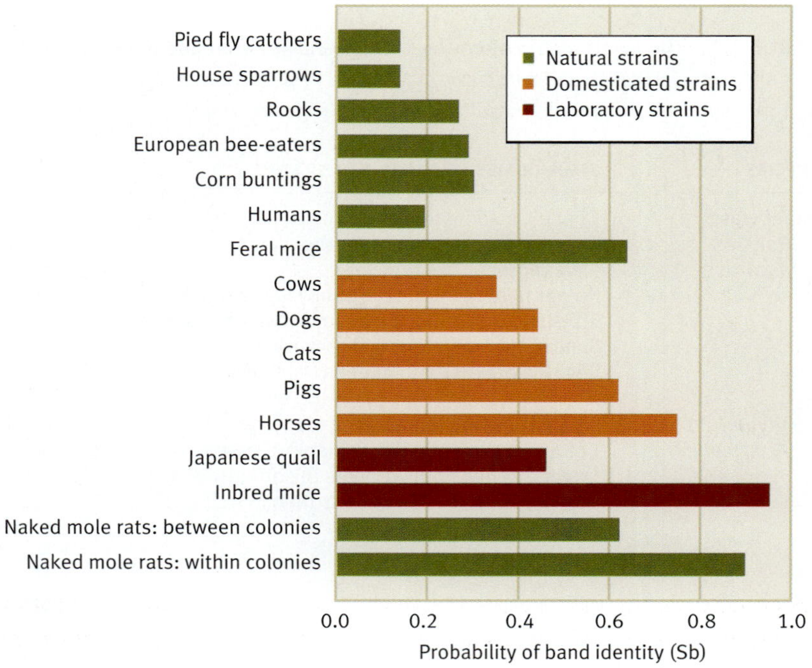

FIGURE 2.20. Naked mole rats show high levels of genetic relatedness. Reeve and his colleagues found that naked mole rats in a colony were more genetically related to one another than any non-inbred strain of animal known. *(From Reeve et al., 1990)*

this number in some perspective, unrelated individuals have a value of 0 for this indicator, brothers score (on average) 0.5, and the most related of all individuals, identical twins, score 1.0 (Figure 2.20). Thus, naked mole rat individuals, on average, fall between normal siblings and identical twins on a relatedness scale, and they even lean toward the identical twins' side of the equation. The cooperative and altruistic behavioral adaptations seen in naked mole rats are then largely driven by the high degree of genetic relatedness seen within colonies.

MATE CHOICE IN HUMANS

Natural selection acts on human behavior, as well as on the behavior of nonhuman animals. For example, in a now-classic study of human mate choice, David Buss (1989) used evolutionary theory to test predictions about patterns of human mate choice around the world. All in all, Buss examined the mate choice of more than 10,000 males and females in 37 different cultures (Figure 2.21). Buss tested many predictions first proposed by Darwin in his 1871 book *The Descent of Man and Selection in Relation to Sex* and subsequently expanded and formalized by behavioral ecologists like Robert Trivers (Trivers, 1972).

Here, we focus on predictions regarding the importance of resource acquisition abilities in a potential mate. Theory that we will delve into in much greater detail in Chapters 7 and 8 predicts that the sex that invests more in offspring should be the sex that values resource acquisition abilities in a potential mate more strongly. Usually, but not always, the female sex invests more in offspring. Females produce larger gametes—eggs—that are more expensive per unit to produce in the sense that the amount of resources allocated to each gamete is greater. In mammals that reproduce by internal gestation, including humans, the difference in female versus male parental investment is even greater, and females invest dramatically more. As such, females should be "choosier" than

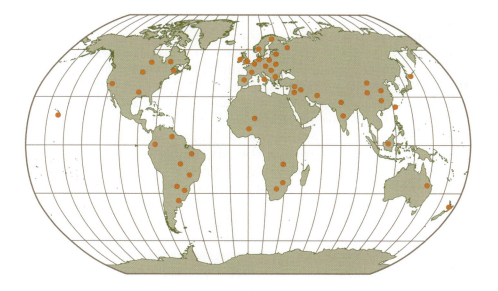

FIGURE 2.21. Human mate choice. A map of locations where Buss (1989) studied human mate choice around the world.

males about selecting mates that provide resources for them and their offspring. The argument is an economic, as well as evolutionary, one. The cost of expensive gametes and internal gestation generates stronger natural selection pressure on females to choose mates that will provide resources for their offspring.

Buss surveyed males and females in the 37 cultures that he and his colleagues were testing. They asked people to rate the importance of a mate being a "good financial prospect" on a scale of 0 (unimportant) to 3 (indispensable). In 36 of the 37 cultures he studied, women valued "good financial prospect" as a more important trait in a potential mate than did men (this difference was statistically significant in all 36 cultures; in the 37th culture [Spain], no sex differences were found).

Phylogeny and the Study of Animal Behavior

In *On the Origin of Species*, Darwin summarized his ideas on evolution in "two great laws" that centered on (1) conditions of existence, and (2) common ancestry. When Darwin spoke of the "conditions of existence," he was describing the living (biotic) and nonliving (abiotic) environment that sets the stage on which natural selection operates. The effect that a given variant of a trait has on reproductive success is not absolute, but depends on the environment in which an organism finds itself—it depends on the conditions of existence. The first part of this chapter has been devoted to examining animal behavior and natural selection. Here we turn to Darwin's other great law—common ancestry—as it applies to the study of ethology.

PHYLOGENETIC TREES

To study common ancestry, evolutionary biologists construct **phylogenetic trees**, which depict the evolutionary history of a group of species, genera, families, and so forth. These trees graphically depict the phylogeny of the groups of organisms in question. The term "phylogeny" was first introduced by a German contemporary of Darwin, Ernst Haeckel, who was deeply involved in disseminating ideas on evolution throughout Europe. Although Darwin himself did not use the word *phylogeny* until the fifth edition of *On the Origin of Species*, he was clearly thinking about phylogenetic trees as early as 1837, twenty-two years before the publication of the first edition (Dayrat, 2005). Figure 2.22A depicts a hypothetical tree that Darwin included as the only figure in the first edition of *On the Origin of Species*, whereas Figure 2.22B shows one of the first sketches of a phylogenetic tree from Darwin's early notebooks.

Species that share a recent common ancestor tend to have many traits in common for the very reason that they share a common ancestor. For Darwin, common ancestry was important because it helped explain "that fundamental agreement in structure which we see in organic beings in the same class, and which is quite independent of their habits of life" (Darwin, 1859). For example, from an ethological perspective, suppose that Species 1 and Species 2—let's say two species of squirrels—descended from a recent common ancestor, labeled Species 3. If Species 1 and 2 display the same set of antipredator behaviors, one possible reason for this similarity is that both Species 1 and 2 possess this suite of traits because their common ancestor, Species 3, possessed it.

READING A PHYLOGENETIC TREE. The tree shown in Figure 2.23 depicts the evolutionary relationships among the vertebrates. In this figure, each branch tip represents a group of related organisms or taxa—birds, crocodilians, mammals, and so on. Figure 2.23 illustrates two different ways of showing the same

A   **B**

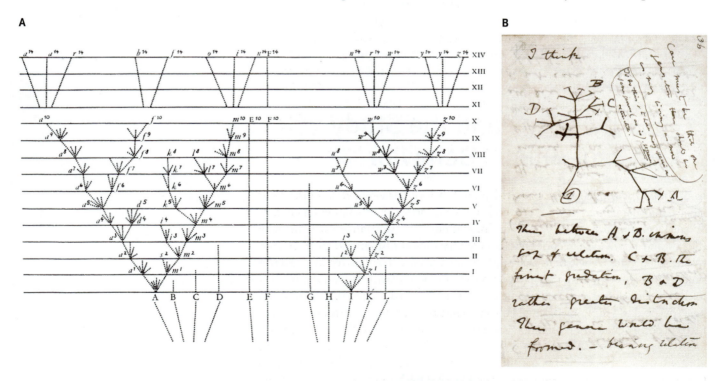

FIGURE 2.22. Darwin and phylogenetic trees. (A) Hypothetical phylogenetic tree from *On the Origin of Species*. Ancestral species A–L are on the bottom, and time is along the y-axis *(From Darwin, 1859).* **(B)** First known sketch of an evolutionary tree by Charles Darwin, who drew it in an early notebook in 1837. Notice the "I think" in the top left corner. Both A and B show divergence over time. *(Reproduced by permission of Syndics of Cambridge University Library)*

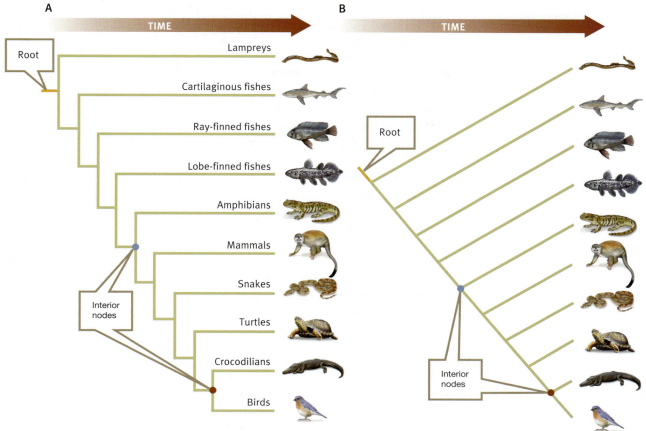

FIGURE 2.23. Two ways of drawing a phylogeny. The two phylogenies of the vertebrates shown illustrate exactly the same information. The phylogeny on the left **(A)** is sometimes referred to as a *tree* representation; the one on the right **(B)** is termed a *ladder* representation. In each, time flows from left to right, so that the branch tips at the right represent current groups, whereas the *interior nodes* (nodes on the inner section of the tree) represent ancestral populations. The red dot indicates the common ancestor of birds and crocodilians, whereas the blue dot indicates the common ancestor to all tetrapods. The yellow line segment is the root of the tree. *(From Bergstrom and Dugatkin, 2012)*

information: on the left side, the phylogeny is drawn in *tree format*; on the right side, the same phylogeny is illustrated in a slanting structure known as *ladder format* (Novick and Catley, 2007). These two ways of drawing a phylogeny are interchangeable. The orientation of the tree *does not matter*: phylogenetic trees can be drawn with the root at the left and the branch tips at the right or, equivalently, with the root at the bottom and the branch tips at the top—it makes no difference to the meaning of the tree.

The points where the tree splits—the **nodes**—represent common ancestors to the species that come after the splitting or branching point. All branch tips arising from a given branching point are descendants of the common ancestor at that branching point. For example, a red dot highlights the node representing the common ancestor to birds and crocodilians, and a blue dot indicates the common ancestor to all tetrapods. Notice that the hypothesized common ancestor to tetrapods would not have been identical to any currently living tetrapod. Rather, evolutionary change has occurred along every branch leading from this ancestor to the species we currently observe on earth.

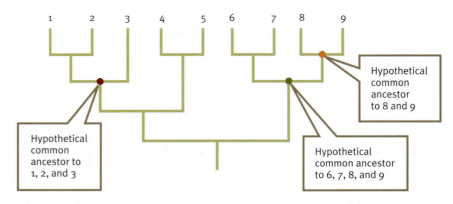

FIGURE 2.24. Finding common ancestors on a tree. Finding the common ancestor for a group involves tracing backward in time. Follow the dashed lines to see the common ancestors of different groups in this phylogeny. *(From Bergstrom and Dugatkin, 2012)*

At the base of the tree is the **root**—the common lineage from which all species indicated on the tree are derived—indicated in yellow. To find the most recent common ancestor of two or more species, trace backward along the tree until the branches leading to these species converge. Figure 2.24 illustrates this process.

BUILDING A PHYLOGENETIC TREE. The first step in the process of building phylogenetic trees is to measure some set of traits, often referred to as characters, in the organisms under study. Traits can, for example, be structural, developmental (embryological), molecular, or most relevant here, behavioral. Until the molecular revolution of the 1970s, almost all trait measurements were morphological or anatomical—for example, bone length, tooth shape, and so on—many of them were gathered from fossil evidence. In this context, a trait need not correspond to a single structural measurement, such as bone length. It can instead be a measure of a number of related structures, such as jaw shape, the relative position of a group of bones, and so forth. With the advent of molecular genetics, many "traits" measured today are DNA sequences, and evolutionary biologists can use these molecular genetic maps to build phylogenies by comparing and contrasting sequences across different species.

Not all traits are equally valuable in constructing phylogenies. To understand why, we need to distinguish between two basic types of traits—homologies and homoplasies. A **homology** is a trait shared by two or more species because those species shared a common ancestor. For example, all female mammals produce milk for their young, and they all possess this homologous trait because mammals share a common ancestor that produced milk.

A **homoplasy** is a trait that is not due to descent from a common ancestor shared by two or more species but instead is the result of natural selection acting independently on each species. A classic example of a homoplasy is shown by the wings of birds, mammals (for example, bats), and insects (Figure 2.25). Superficially, the wings found in these different groups are constructed in a similar manner, but not because wings were present in a common ancestor of birds, mammals, and insects. In the case of wings, natural selection pressures in the three taxa were similar enough that an appendage associated with flight would have some of the same basic characteristics. Such homoplasies—the result of shared natural selection

Bird   Bat   Insect

FIGURE 2.25. Convergent evolution in wing structure. Convergent evolution has led to wings in birds, bats, and insects. Wings in these groups are analogous traits.

pressure—are referred to as **analogies**, and the process leading to the production of analogous traits is called **convergent evolution**.

Homologous traits are used in phylogeny building because they reflect shared evolutionary histories. Homoplasies, on the other hand, while they do shed light on natural selection, do not reflect the historical relationships between species and, in fact, distort and misrepresent these relationships when used in phylogenetic tree building. Ideally, evolutionary biologists would like to be able to build phylogenetic trees by including homologies but eliminating homoplasies and they have developed many techniques for distinguishing between these two types of traits (Bergstrom and Dugatkin, 2012).

When attempting to construct phylogenetic trees, we need a method to ascertain which varieties of a trait appeared first—that is, which are ancestral—and which varieties are derived from these ancestral states. When we study the historic order in which different varieties of a trait appear, we are examining what is referred to as **polarity**, or the direction of historical change in a trait.

Evolutionary biologists have many ways for determining polarity, including using the fossil record in conjunction with techniques for dating fossils. Indeed, for most of the history of evolutionary biology, the fossil record was the primary source for adding a temporal component to phylogenetic trees. The oldest state of a trait found in the fossil record is assumed to represent the oldest true state—what is called the ancestral or primitive state of the trait.

Today, molecular genetic data can be used to help date major changes that occur on phylogenetic trees. To do so, DNA sequences from different species are compared against one another at the nucleotide level. Complex mathematical models that include an estimate of the rate of molecular genetic change can then be used to calculate the amount of time that would be required to produce the nucleotide differences we have measured. In essence, even in the absence of information from the fossil record, these molecular genetic data allow us to work backward from extant (present-day) species to build and date phylogenetic trees.

When evolutionary biologists use the techniques described above, they often end up with many possible phylogenetic trees, each of which *might* represent the true phylogenetic history of the taxa they are studying. But, of course, there is only one actual phylogenetic history for any taxa, so we need a method for distinguishing between candidate trees. Researchers have developed a number of techniques to handle the problem of distinguishing between possible phylogenetic trees (Felsenstein, 2004; Hillis et al., 1992). The most common of these techniques is **parsimony analysis**. The concept of parsimony is most often associated with the ideas of a fourteenth-century philosopher named William of Ockam (sometimes spelled Occam), who argued that "entities are not to be multiplied beyond necessity." This concept is now referred to as Ockam's razor, and within evolutionary biology it has been taken to mean that the phylogenetic tree that requires the least number of evolutionary changes is the most likely to be correct.

Every phylogenetic tree is a *hypothesis* of the evolutionary history of the groups under study. When morphological and molecular genetic analyses produce similar phylogenetic trees, our confidence increases that our phylogenetic tree is correct. In addition, as more morphological, molecular genetic, and fossil evidence is uncovered in the future, we can examine whether the new data are consistent with our phylogenetic tree. If they are, we gain confidence that our tree accurately reflects the evolutionary history of the taxa we are studying.

PHYLOGENY AND INDEPENDENT CONTRASTS. Ethologists use phylogenetic trees in many ways. In some cases, behavioral data can be the primary source on which a phylogeny is constructed, but such cases are not very common. A second, more common way for ethologists to use phylogenetic trees is to find an already established phylogenetic tree for the taxa they are studying (for example, a tree based on the methods discussed above). Then, ethologists could take the behavioral data they have—let's say data on mating systems—and superimpose them onto the phylogenetic tree to examine the evolutionary history of mating systems. That is, they take the mating systems data they have collected on each species and place them on a phylogenetic tree, so that mating systems can be viewed in the context of evolutionary history. Then, animal behaviorists can ask questions such as: Which of the mating systems seen today are derived? Which are ancestral? They can use statistical techniques to calculate which ancestral state (of mating system) would require the fewest number of evolutionary changes to produce the mating systems that we see in our extant species. These calculations help them make inferences about the mating systems of species that predated those they are studying, even if they are extinct.

Ethologists can also use existing phylogenetic trees to make inferences about natural selection. Once we know something about the phylogenetic history of the group we are studying, we can ask whether certain selection pressures consistently favor one combination of traits or another *independent of phylogenetic history*, or whether the co-occurrence of the traits in certain species is the result of a common ancestry for those species.

As an example, suppose we are interested in understanding whether natural selection consistently favors organisms that display both nocturnal activity and an arboreal (tree-based) lifestyle (Felsenstein, 2004). We can address this question using the **comparative method**, by collecting information on both of these traits in a number of species. Suppose we find the pattern of characters shown in Figure 2.26. At first glance, this figure appears to offer extremely

FIGURE 2.26. Relationship between nocturnal and arboreal behavior in ten species. Nocturnal is shaded in red, diurnal in orange, arboreal in green, and terrestrial in yellow. Nocturnal and arboreal co-occur often, as do diurnal and terrestrial. *(Adapted from Feselstein, 2004)*

strong support for the hypothesis that nocturnal and arboreal lifestyles go hand in hand. One interpretation here would be that natural selection has independently favored this combination of traits, over and over.

But there is a problem here, in that we are not accounting for any shared evolutionary history among these species. Suppose that we discover that the phylogenetic history of these species is as depicted in Figure 2.27A. From this phylogeny, we can infer the evolutionary changes that most likely gave rise to the characters we observe. This is shown in Figure 2.27B. With this information in hand, we might take a different view of the character pattern that we have observed. Rather than natural selection favoring a pairing of nocturnal and arboreal habits in ten independent cases, the entire pattern has likely arisen from a *single pair* of evolutionary changes, one for the diurnal/nocturnal trait and one for the arboreal/terrestrial trait.

Now that we have walked through a brief primer on phylogeny, let's look at a few more detailed examples in which ethologists have used phylogenetic analyses to shed light on animal behavior.

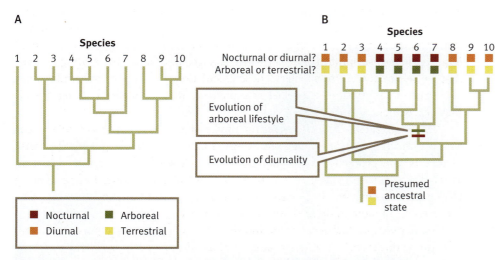

FIGURE 2.27. Relationship between nocturnal and arboreal behavior is not independently favored in many lineages. (A) The phylogenetic relationships among the ten species. **(B)** Nocturnal behavior and arboreal behavior evolved just one time each along one branch of our tree. *(Adapted from Feselstein, 2004)*

Ethologists often use phylogenetic analysis to try to reconstruct the order in which a suite of behaviors has appeared over evolutionary time. We shall use this approach here to focus on the evolution of parental care behavior. At the most basic level, there are four "states" that capture parental care systems in animals: no parental care, maternal care, paternal care, and biparental (maternal and paternal) care. We can then ask: In what historical order, if any, do these forms of parental care most often appear in some taxa we are studying? For example, if we assume that no parental care is the primitive, or original, state in the taxa we are studying, do we most often see maternal care appear next? Paternal care? Biparental care? This phylogenetic analysis could then be followed up with additional studies on why natural selection favored the ordering uncovered in the phylogenetic work, if such an ordering was discovered.

Early theories on the order in which parental care systems evolve focused on fish and anuran frogs (Gross and Sargent, 1985). These analyses suggested a "stepping-stone" model in which parental care appeared in the following order: no parental care, paternal care, biparental care, and finally maternal care (as a result of male desertion). The sample size for these analyses, however, was relatively small, and subsequent studies in the same groups produced contradictory results (N. B. Goodwin et al., 1998; Summers et al., 1999).

To better address the order in which parental care systems typically evolve— if such an order even exists—we need to work with a large group of species. Ideally we would use a group in which we see each variety of parental care (many times), and one in which we have good information on the phylogenetic relationship between species. One such group is the ray-finned fish (the *Actinopterygii*), which contain on the order of 400 families of fish, made up of over 20,000 species (Figure 2.28). Ray-finned fish make up more than one-half of all known vertebrate species, and include the group most often studied by ethologists—the bony fish, or teleosts.

To begin their phylogenetic study of parental care in ray-finned fish, Judith Mank and her colleagues first used morphological data and molecular genetic

FIGURE 2.28. Parental care in fish. Various forms of parental care (maternal care, paternal care, both maternal and paternal care, and so on) have been uncovered in ray-finned fish. Here we see a male clown anemonefish tending the eggs laid by its mate. This fish shows the bright coloration and paternal care often found in species with external fertilization. *(Photo credit: Fred Bavendam/Minden Pictures)*

data, and then assessed trait polarity. They used this information to build a phylogenetic tree of 224 families of ray-finned fish and then examined the evidence for parental care in these various families (Mank et al., 2005). That is, their phylogenetic tree was not built using data on parental care behavior—instead, inferences about the evolution of parental care were made based on the tree Mank and her colleagues built using morphological and molecular genetic data. They began by categorizing the species in their established phylogenetic tree by using already published data on parental care—no parental care, maternal care, paternal care, or biparental care—and then they examined the order in which these systems appeared over evolutionary time. Second, because ray-finned fish have two fairly distinct ways of reproducing—external or internal fertilization—they asked whether particular parental care systems were associated with either external or internal fertilization. Finally, because male color pattern and nest-building behavior have been found to play an important role in fish mating systems, they examined whether the presence or absence of male coloration and/or the presence or absence of nest-building behavior was associated with specific parental care systems.

Parental care was found in approximately 30 percent of the families of ray-finned fish analyzed by Mank and her colleagues. In external breeders, parental care was found in 25 percent of the families studied. For internally fertilizing fish, parental care was found in 90 percent of the families analyzed. Mank and her colleagues found that in ray-finned fish paternal care evolved on twenty-two independent occasions—in other words, there were twenty-two independent occurrences of a shift from "no parental care" in an ancestral species to paternal care in a descendant species. Maternal care and biparental care also evolved independently numerous times within the ray-finned fish (Figure 2.29). No

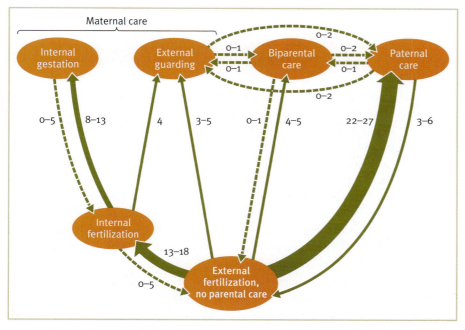

FIGURE 2.29. The evolution of parental care in ray-finned fish. Thickness of the arrows reflects relative numbers of evolutionary transitions (numerals near arrows indicate the estimated range of the number of transitions). Dashed arrows indicate cases in which transitions are unclear. *(From Mank et al., 2005)*

evidence, however, was found for the stepping-stone model described earlier. Indeed, aside from the fact that paternal care was the most likely evolutionary state to emerge from no parental care, no clear ordering of parental care systems was found.

Other findings from Mank and her colleagues included:

1. Maternal care tended to evolve after species had moved from external to internal fertilization.
2. In species with external fertilization, paternal care was often found alongside intense male coloration and nest construction.
3. Male coloration was also associated with systems in which females fertilized their eggs internally.

Overall, then, Mank and her colleagues' ambitious phylogenetic analysis did not find evidence for a sequential stepping-stone model of the development of parental care. Instead, what phylogenetic analyses found was evidence for multiple origins of all parental care systems (particularly paternal care), as well as some fascinating evolutionary relationships between parental care, nesting behavior, mode of fertilization (external or internal), and male coloration pattern.

PHYLOGENY AND COURTSHIP BEHAVIOR

Though adaptationist and phylogenetic studies have historically been conducted by different groups of researchers, the two types of studies can complement each other, and together they provide a deep understanding of the evolution of behavior. For example, in the discussion of adaptation earlier in this chapter, considerable space was devoted to explaining various behavioral adaptations in guppies. The behavior of guppies, in particular their mating behavior, has also been examined from a phylogenetic perspective (Bisazza et al., 1997).

Guppies are live-bearers (females give birth to live young) and they are part of the family Poeciliidae, a family composed of only live-bearers. Because live-bearers are usually small and often adapt well to the laboratory, ethologists have studied their behavior for decades. Much of this work has been focused on mate choice, especially female mate choice (which is examined in greater detail in Chapter 7). But male mating behavior has also been studied in this family.

Males use two very different types of mating strategies. In some species, males have bright coloration (a type of sexual ornament) and display vigorously to females. If a female is receptive, the male and female mate, which involves the male inserting his gonopodium into a female and inseminating her. In other species, males tend to have drabber coloration and rely primarily on what is known as gonopodial thrusting, which involves a male approaching a female from behind and forcing a mating by inserting his gonopodium into the female.

From a phylogenetic perspective, which male strategy is the ancestral (primitive) mating strategy in poeciliid males, and which is the derived mating strategy? That is, which came first? To address this question, Angelo Bisazza and his colleagues built a phylogeny of a part of the family Poeciliidae using sequences from a mitochondrial rRNA gene. They then mapped male mating strategy onto their phylogeny.

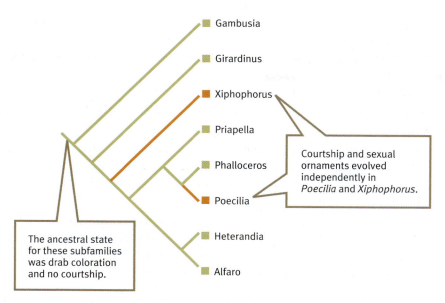

FIGURE 2.30. Phylogenetic history of male courting strategies. The evolutionary history of male mating fertilization strategies in Poeciliinae fish. Orange branches are species that use courtship and sexual ornaments, including color. Green branches lead to species that employ gonopodial thrusting and drab coloration. *(Adapted from Bisazza et al., 1997)*

What they found was that in the groups they examined, gonadapodial thrusting was the ancestral (primitive) state in poeciliids, including the guppy. In poecillids, courtship and sexual ornamentation are derived from the primitive gonopodial thrusting state (Figure 2.30). Part of the power of this sort of phylogenetic analysis is that adaptationist methodology can then be employed to address what sort of selective pressures might have favored courtship and sexual ornamentation in guppies. Much research has been done on this question, and in conjunction with the phylogenetic studies it is providing us with a rich and detailed picture of the evolution of mating in live-bearing fish (Houde, 1997; Magurran, 2005).

This chapter has focused entirely on ultimate or "why" questions in animal behavior. It has examined the process of natural selection and how selection pressures shape animal behavior, as well as how evolutionary history—phylogeny—can help shed light on questions in ethology. It is important to recognize that just because we have discussed natural selection and phylogeny in different sections does not imply that they are unconnected to one another. Along with mutation, recombination, and the other sources of new genetic variation, the phylogenetic history of a species affects the variation available for natural selection to act upon at any given time. Conversely, when natural selection shapes traits in Species 1 it is, to some extent, affecting the way that *descendants* from Species 1 will look, act, and so on.

Ultimate questions are often contrasted with proximate questions, which focus on immediate causes of behavior (release of a hormone, firing of a neuron, and so on). Chapters 3 and 4 are primer chapters that examine proximate factors. For the remainder of the book, both proximate and ultimate perspectives are integrated into each and every chapter.

Dr. Alan Grafen

If you had to explain natural selection to an alien who knew nothing of life on earth, how would you do it?

Richard Dawkins has convincingly argued that all adaptive complexity must arise through natural selection. It follows that an alien capable of communicating with me would themselves be the product of natural selection, and furthermore would be likely to come from a civilization that had understood how adaptive complexity arose. Thus, I would look forward to hearing what the alien had to tell me about natural selection.

Some have argued that the process of natural selection is a tautology—something that is circular, and true by definition. Where have they gone wrong?

They've almost certainly read the wrong explanation. You can't read Darwin's *Origin of Species*, or Dawkins's *The Selfish Gene*, and pay attention and come away thinking natural selection is a tautology. It is one of the features of natural selection that nearly everyone thinks they intuit the idea without being exposed to it! Indeed, Darwin's correspondence suggests he suspected that even Thomas Henry Huxley, who came to be known as "Darwin's bulldog," did not understand natural selection very well.

There are surely people who claim that special relativity is a tautology, or that Newtonian mechanics is fatally flawed, but they are a tiny green-inkish minority. The anti-intellectuals' assault on natural selection is conducted on a broader front and,

more worryingly, seems to be much more acceptable in respectable company.

The idea of natural selection needs to be treated with respect, so don't waste your time arguing with someone who thinks he's worked it all out for himself and claims to have

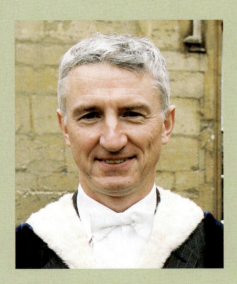

discovered a mistake in it! Insist that he should have a proper explanation in front of him, make him point to the line where he doesn't follow the argument, and do your honest best to help him over his obstacle.

Can it really be true that the same process—natural selection—operates on everything from viruses to humans?

That is what science is like. Gravity too operates on everything from viruses to humans and, just like natural selection, it operates in

some ways just the same, and in other ways very differently, because of the different context.

Why is there such reluctance in the social science community to accept that natural selection has shaped much of our own behavior?

This is an interesting question, but there is another foot that may have a boot on. Might a social scientist not ask "Why is there such reluctance in the biological community to accept that human behavior has unique aspects and essentially cultural aspects for which biology provides no framework?"

It is a primary duty of an academic to respect other subjects, and the seriousness of their intellectual work, so we should not try to maneuver social scientists into giving up their lines of inquiry or their conceptual frameworks.

It is actually quite hard to understand another subject's basic ideas, and this leads to trouble whenever subjects grow into new areas and acquire new neighbors. To have a dialogue with people absolutely requires mutual respect and a serious attempt at understanding each others' positions.

One of my own papers, "Fertility and Labour Supply in *Femina economica*" was a consciously constructed attempt to build a model in which the same behavior could be understood biologically and economically. The resolution of the conflict between the points of view is all done formally and mathematically in that paper. There is no simple "one side is right, the other is wrong"—rather the explanatory forces are partly overlapping and partly intertwined.

It seems uncontroversial that at some level, biology will have something important to say about human behavior, and equally, that many of the patterns of behavior and explanations already discovered by social scientists will remain unaffected by biology. But in between these two easily accepted propositions lie many areas in which only detailed study and analysis can determine the role, if any, biology should play. Useful study of those areas can be done only by people with a good understanding both of the social sciences and of biology.

In what way has our fundamental understanding of the process of natural selection been changed by the revolution in molecular genetics?

Molecular biology has revolutionized the detailed facts we can discover about the organisms we're studying. Do female birds have offspring by more than one male in the same nest? How long ago did humans and chimps diverge? The range of useful facts will no doubt grow and grow.

There is one, terribly important, aspect of natural selection that is completely unaffected by molecular biology, and it is in many ways the very core of Darwin's argument: that the only known process capable of producing complex adaptations is natural selection. This argument was developed mathematically by the great biologist (and statistician and geneticist) R. A. Fisher, and Bill Hamilton wrote in his encomium on the variorum edition of Fisher's *Genetical Theory of Natural Selection*: "And little modified even by molecular genetics, Fisher's logic and ideas still underpin most of the ever broadening paths by which Darwinism continues its invasion of human thought."

The only known process capable of producing complex adaptations is natural selection.

There is a tendency, particularly prevalent for obvious reasons among some practitioners, to believe that new technologies require a revisiting of all old knowledge. As if the discovery of the telescope might revise our view of how frequently the moon circles the earth. Read the *Origin of Species*, and be persuaded of the reality and biological significance of natural selection. Read G. C. Williams's *Natural Selection and Adaptation* (1966) and understand how to reason about natural selection. Read Dawkins's *The Selfish Gene* and see how Darwin's logic, rendered logically explicit, unifies the range

of evolutionary theories biologists apply today. Biologists know a lot about natural selection, and molecular genetics will not make any substantial difference to their arguments.

Molecular genetics is nevertheless a hugely important scientific enterprise, both medically and intellectually. What *does* it have to offer to students of natural selection? For one thing, many more examples of Darwinian adaptations. To learn about the workings of the ribosomes, or DNA topoisomerases, or hundreds of other fabulous molecules, is to be filled with wonder at the achievements of natural selection, just as much as the sonar of bats or the workings of the vertebrate eye. In some ways the molecular examples are more impressive, because they reveal how life itself is possible at a very fundamental level.

The problem Darwin solved in 1859 has remained solved. While nothing in molecular genetics is likely to require the reopening of that case, molecular genetics *will* continue to supply further examples of adaptations and fuel further sophistications of our understanding of natural selection.

***Dr. Alan Grafen** is a professor at Oxford University, where he is a member of the Animal Behavior Research Group. He is the coauthor of* Modern Statistics for the Life Sciences *(Oxford University Press, 2002) and is one of the leading theoretical biologists in the area of evolution by natural selection.*

SUMMARY

1. Artificial selection is the process of humans choosing certain trait varieties over others through implementing breeding programs that cause one or more selected varieties to increase in frequency. We have been shaping animals and plants by the process of artificial selection for at least 10,000 years.

2. Much of animal behavior work revolves around the notion that behavior, just as any trait, has evolved by the process of natural selection. The process of natural selection requires limited resources, variation, fitness differences, and heritability. For example, if individuals in a population display a number of different foraging variants, and these variants translate into fitness differences and are heritable, then natural selection will act to increase the frequency of the foraging variant associated with the greatest relative fitness.

3. Even a small selective advantage—on the order of 1 percent per generation—is sufficient for natural selection to dramatically change gene frequencies over evolutionary time.

4. Natural selection produces adaptations—traits with the highest relative fitness in a given environment—but not all adaptations are necessarily produced by natural selection.

5. To study common ancestry, evolutionary biologists construct phylogenetic trees, that depict the evolutionary history of various species, genera, families, and so forth. These trees graphically depict the phylogeny, or history of common descent, between the groups of organisms under study.

6. Phylogenetic trees can be used in many ways by ethologists. In some cases, behavioral traits are used to construct phylogenies, but more often behavioral traits are mapped onto already existing phylogenetic trees to assess the order in which various traits appeared over evolutionary time.

DISCUSSION QUESTIONS

1. How was it possible for Darwin to come up with his theory of natural selection in the complete absence of a science of genetics? Many modern studies in ethology rely on genetics, particularly molecular genetics, as a critical tool. How did Darwin manage without this?

2. François Jacob and Jacques Monod have referred to natural selection as a tinkerer. Why is this a particularly appropriate analogy for how the process of natural selection operates?

3. Read Reeve and Sherman's (1993) paper, "Adaptation and the Goals of Evolutionary Research," in volume 68 (pp. 1–32) of the *Quarterly Review of Biology*. After reviewing all the definitions of *adaptation*, do you agree with Reeve and Sherman's approach to this subject? If you do not agree, explain why not by listing your reasons.

4. When considering how much variation in a behavioral trait is genetic and how much is environmental, how does the "uterine environment," in which a fetus matures, complicate matters?

5. Secure a copy of Gould and Lewontin's (1979) paper, "The Spandrels of San Marcos and the Paglossian Paradigm: A Critique of the Adaptationist Programme" in volume 205 (pp. 581–598) of the *Proceedings of the Royal Society of London, Series B*. List both the merits and flaws in Gould and Lewontin's approach. Overall, do you think their critique is a fair one?

6. Maternal care behavior has been documented in such diverse groups as insects, fish, and mammals. Why would it probably be a mistake to use maternal care as a character in any analysis that examined the phylogenetic relationship between these very different groups? Cast your answer in terms of convergent evolution.

SUGGESTED READING

Buss, D. (2012). *Evolutionary psychology.* (4th ed.). Boston: Allyn & Bacon. An excellent book on how "natural selection thinking" can shed light on the evolution of human behavior.

Darwin, C. (1859). *On the origin of species.* London: J. Murray. The starting point for both evolutionary biology and modern ethology and a joy to read.

Gardner, A. (2009). Adaptation as organism design. *Biology Letters*, 5, 861–864. A discussion of recent work on the power of the adaptationist approach to studying behavior.

Keller, L. (2009). Adaptation and the genetics of social behavior. *Philosophical Transactions of the Royal Society B,* 364, 3209–3216. A bit technical, but a nice review of how recent work in genetics sheds light on the evolution complex social behaviors.

Williams, G. (1966). *Adaptation and natural selection.* Princeton, NJ: Princeton University Press. A critical approach to adopting the adaptationist perspective. Regarded by many as a must read for those studying animal behavior.

Hormones and Neurobiology

Suppose that you are attending a roundtable discussion on visual acuity in birds. Sitting in the room with you are four scientists. The speaker on the podium is discussing visual acuity in robins, and on the screen in front of you is information on the anatomy of a robin's eye, the neuronal underpinnings of robin vision, and even information about the molecular genetics underlying visual acuity in robins.

When she has finished her presentation, the speaker asks everyone in the room to briefly explain visual acuity in robins from the perspective of their particular discipline. The first scientist proceeds to explain how natural selection produced an increase in visual acuity over time. She outlines the process of natural selection on vision, and describes experiments that have found how better vision produces a robin that is a superior forager and more adept at avoiding predators. The next speaker stands up and explains how studying the curvature of the eye leads to a better understanding of why modern-day robins have better visual acuity than some other birds. The third scientist then describes the neural circuitry underlying vision and how that helps explain the increased visual acuity in robins. Finally, the fourth scientist proceeds to detail the changes associated with robin vision at the molecular genetic level, and how these changes are critical to understanding vision in robins.

Who's right? They all are, but their explanations reflect different levels of analysis. The first scientist has provided an ultimate explanation—one based on evolutionary processes—whereas the last three scientists have provided answers at the proximate level. Ever since Tinbergen discussed the four sorts of questions ethologists ask (Chapter 1), animal behaviorists have understood the importance of studying behavior from both an ultimate and a proximate perspective. One goal of animal behaviorists is to integrate these different perspectives, to the greatest extent possible, into a comprehensive view of animal behavior.

Ultimate and Proximate Perspectives

One common error made by those not trained in the field of ethology is to confuse the level at which a behavior is being analyzed. For example, we can answer the question "Why do songbirds sing?" in terms of physiology, neurobiology, survival value, phylogenetic relationship to other species of birds, and so on (Sherman, 1988). Confusion arises when you ask a question about behavior at one level of analysis, and I answer your question at a different level. My answer to why songbirds sing might be based on the bird's voice box (syrinx), its musculature, and the nervous system connecting the syrinx to the brain, when in fact you were looking for a reply that explains why natural selection seems to have favored birdsong in many species, or perhaps you were searching for an answer cast in terms of the role that phylogeny plays in explaining birdsong. When we talk past each other this way, it is because we are approaching the question from different perspectives—one dealing with immediate causation (proximate), and one centering on forces that have shaped a trait over evolutionary time (ultimate). To solve this "levels of analysis" problem in behavior, we need to understand both ultimate and proximate causation, so as not to confuse them (Alcock and Sherman, 1994; Mayr, 1961; Reeve and Sherman, 1993; Sherman, 1988).

One way to think of the distinction between proximate and ultimate perspectives is that proximate questions include "How is it that . . . ?" and "What is it that . . . ?" sorts of questions, while ultimate questions are typically in the form "Why is it that . . . ?" (Alcock, 2001, 2003). In that sense, proximate and ultimate explanations complement one another. Sometimes proximate causes are defined as those that are not evolutionary in nature. But this is probably not the most productive definition for two reasons. First, it can often be very useful to understand evolutionary (ultimate) forces when asking "how" and "what" questions, and that utility can be lost if proximate causation is defined as everything that isn't evolutionary. Second, this definition makes it seem as if the proximate level of analysis is less important than analysis at the ultimate level, whereas both levels of analysis are equally important.

Conceptually, we need to think in different ways when working with proximate and ultimate questions. In a proximate analysis, we are working with factors that operate within the lifetime of an organism, in the here and now, rather than inferring (or testing) adaptationist or phylogenetic arguments as we do in an ultimate analysis about what evolutionary forces have operated in the past, how they are operating now, or how such evolutionary forces might operate in the future. But it is also important to recognize that there are fundamental links between proximate and ultimate analyses. Suppose we are undertaking a proximate analysis of some behavior—let's say male aggressiveness. If we know something about the natural selection pressures that have acted on male aggression, that knowledge can help us design a better way to do our proximate analysis. For example, if we suspect that males direct aggression toward other males to gain access to food, we might focus on different proximate factors than if we suspect that males direct aggression toward other males during the breeding season to secure access to females. We would approach the question differently because the proximate underpinnings of foraging and mating behaviors are often not the same, and our understanding of natural selection might help provide focus for our proximate analysis.

Conversely, a proximate analysis might provide information that will help us better understand *how* natural selection has shaped male aggression. We might find that the hormones involved in male aggression also lower a male's resistance to disease, suggesting that we need to incorporate this disease effect into our natural selection model of male aggression. What's more, and just as important, a proximate analysis, by shedding light on neurobiology, endocrinology, molecular genetics, and so on, will help us understand the raw material that natural selection may operate on in the future. That is, proximate analysis will shed light on the variation available for natural selection to act on in the future.

Before we turn our attention to proximate causation per se, let us look at an in-depth example of how animal behaviorists employ both proximate and ultimate perspectives to address a single trait—and how analysis from one perspective often leads to new questions being asked from *both* perspectives. The example we will examine is Geoff Hill's work on plumage, or feather coloration, in the house finch (*Carpodacus mexicanus*).

In the house finch, male plumage coloration is brighter than female plumage coloration. Hill was interested in understanding this difference in plumage coloration, so he designed experiments to study plumage coloration at both a proximate and an ultimate level (Figure 3.1). He wanted to know *what* causes males and females to differ in plumage coloration (a proximate question) and

FIGURE 3.1. Natural variation. Significant natural variation exists in house finch coloration. This variation set the stage for Hill's work on proximate and ultimate questions related to plumage coloration. *(Photo credit: Geoff Hill)*

why such color differences persist over evolutionary time (an ultimate question). To answer these questions, Hill looked at foraging behavior, mate choice, and parental care, all of which play a part in understanding plumage coloration in male and female house finches. We begin by discussing Hill's work on proximate causation, and then move on to his work on ultimate causation. But note that proximate studies do not necessarily precede ultimate studies—just as often the converse is true.

To examine the proximate basis for plumage, Hill employed within- and between-population comparisons and controlled feeding experiments. When he began this work, Hill already knew that plumage coloration in the male house finch was due to carotenoid color pigments (primarily red) that the birds ingested (A. Brush and Power, 1976; G. Hill, 1992, 1993a; G. Hill et al., 2002; Inouye et al., 2001). House finches are unable to synthesize their own carotenoid pigments, and they rely completely on diet for this substance (A. Brush, 1990; A. Brush and Power, 1976; T. Goodwin, 1950). Hill's prior work and that of others had demonstrated that at the proximate level, differences in *male* plumage brightness within and between populations of finches were correlated with the amount of carotenoids in their diet. That is, differences in plumage were due to developmental differences in what carotenoid-based foods males ate as they matured, as well as which of these foods males added to their diets as adults. Hill next wanted to understand what was responsible for differences in plumage in females (A. Brush and Power, 1976; Butcher and Rohwer, 1989; G. Hill, 1993c).

Hill undertook a series of controlled feeding experiments on two groups of females that were fed a fixed diet while living in aviaries at the University of Michigan. Both groups were fed a commercially made finch food, and both groups had their diets supplemented with water and apples. In one of these groups, however, the water and apples were treated with canthaxanthin—a red carotenoid pigment. Females in the canthaxanthin treatment developed much brighter plumage after their diet was supplemented, but females fed just normal apples and water maintained a drab plumage pattern.

From a proximate perspective, Hill could now address two questions: (1) What causes between-population differences in female coloration? and (2) What causes differences in plumage coloration between males and females? These sorts of "What causes . . . ?" questions are often where a proximate analysis begins. With respect to between-population differences, Hill found that female plumage coloration differed among females from Michigan, New York, and Hawaii (G. Hill, 1993a, 1993c). These differences appear to be a function of the amount of carotenoid-based food sources in these localities: The more such food is present in the environment, the brighter the average female is in a population. At the proximate level, the *differential availability* of carotenoid (primarily red) pigments in food across populations appears to explain the difference in plumage coloration among females across populations.

Differences in plumage coloration *between* males and females *within* a population seem to be due to differences in the way in which males and females forage, rather than to the availability of carotenoid-based foods for males and females. Males actively search for and ingest carotenoid-based foods. Females will eat cartenoid-based foods, but they do not actively search for such food. A proximate analysis focuses attention not solely on the amount of carotenoid-based food present in the environment, but on the differences in foraging strategies in males and females.

FIGURE 3.2. Plumage manipulation. As part of the study on plumage coloration, Geoff Hill artificially brightened (top left photo to top right photo) or lightened (bottom left photo to bottom right photo) the plumage coloration of male house finches. *(Photo credit: Geoff Hill)*

This proximate explanation for differences in male and female coloration leads to an ultimate question: Why do males, but not females, actively search for carotenoid-based foods? Hill hypothesized that males receive significant benefits for having colorful plumage, but females do not. But what exactly were these benefits that males received for having bright plumage? To find out, Hill conducted a number of experiments using hair dyes to either experimentally brighten or lighten the plumage coloration of a group of wild eastern male house finches (G. Hill, 1990, 1991, 1993b; Figure 3.2).

Males that had their color experimentally brightened were much more likely to get a mate than males whose colors had been experimentally lightened (Table 3.1). One of the ultimate reasons that males search for carotenoid-rich food is that females are attracted to males with bright plumage as potential mates. With an understanding of the benefits males received for bright plumage coloration, Hill went on to examine *why* females prefer males

TABLE 3.1. Plumage manipulation. Hill examined how brightening and lightening plumage coloration affected a suite of variables in male house finches. There were forty males in the brightened condition, twenty males in the sham control, and forty males in the lightened condition. *(From G. Hill, 1991)*

MALE CHARACTERISTICS	BRIGHTENED	SHAM CONTROL	LIGHTENED	STATISTICAL SIGNIFICANCE (P)
Original plumage score	140.7	139.9	141.00	0.95
Manipulated plumage score	161.6	139.9	129.40	0.0001
Proportion paired with a mate	1.0	0.6	0.27	0.0001
Time to pair (days)	12.1	20.2	27.80	0.07

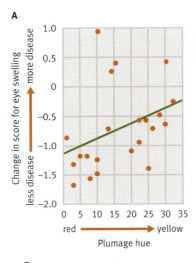

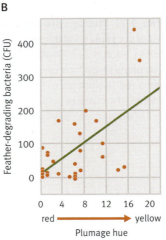

FIGURE 3.3. Plumage coloration and disease. (A) The rate at which birds recovered from *Mycoplasma gallicepticum*, which causes eye swelling, was linked with their plumage coloration. Birds with more red coloration recovered more quickly (in later phases of the experiment) than birds with more yellow coloration. **(B)** Individuals with redder plumage carried fewer feather-degrading bacteria. *(From G. Hill and Farmer, 2005; Shawkey et al., 2009)*

with bright plumage. This is an ultimate question regarding mate choice in females—namely, are there benefits for females that choose males with brighter plumage coloration?

One benefit females receive for mating with redder-colored males is that such males appear to be better able to fight off pathogens. When Hill and Farmer inoculated twenty-four male house finches with *Mycoplasma gallicepticum*, a bacterial pathogen, they found that males with more elaborate plumage coloration were able to rid themselves of this phatogen more quickly than drabber-colored males (Figure 3.3A). Follow-up work also discovered that colorful males that had been selected as mates had lower levels of bacteria that degrade the quality of feathers (Shawkey et al., 2009; Figure 3.3B). Such bacteria are not lethal in and of themselves, but degradation of feathers can lead to problems with thermoregulation and flight. Females expose themselves to fewer pathogens by selecting males that are more resistant to disease over males that are less resistant (we explore this benefit in more depth in Chapter 7). And if such disease resistance is heritable, females that choose the more disease-resistant males as mates will produce offspring that are better able to stave off disease (Duckworth et al., 2001; G. Hill and Farmer, 2005).

A second benefit that females may receive by choosing brighter-colored males over drabber-colored males centers on parental care and food provisioning. When examining the relationship between male color and male parental care in eastern populations of house finches, Hill found that the mean number of times a male fed a chick at his nest was positively correlated with the intensity of his plumage coloration: brighter males fed chicks more than twice as often as drabber males (G. Hill, 1991; Figure 3.4A). Females in this population likely prefer redder males because such males make good fathers with respect to feeding young. In other populations of house finches, however, more colorful males provide less parental care than drabber males, but females still prefer the more colorful males as mates (Badyaev and Hill, 2002; Duckworth et al., 2004; K. McGraw et al., 2001). Why such differences exist between population remains an interesting, but open, question.

There is a third benefit that females may receive by choosing more colorful over less colorful males as mates. In eastern populations, brightly colored males appear to be better foragers (and not only with respect to feeding chicks at their nest), and hence they survive with a relatively high probability. If the traits responsible for this increased survival are passed on to offspring, then mating with colorful males may lead to the production of healthy, long-lived offspring. Hill found that males with more red plumage produced sons with colorful plumage (Figure 3.4B). But remember, red plumage coloration can't be inherited—such coloration is diet dependent. This suggests that males that are good foragers, and thus find and ingest the carotenoid-based foods needed to produce colorful plumage, produce sons that are also good foragers.

Hill's work is an excellent example of how a combination of ultimate and proximate perspectives can provide an in-depth picture of animal behavior. Among other things, it shows how proximate explanations of what causes differences in coloration can also point the way to ultimate reasons for such differences.

In the remainder of this chapter, we examine the hormonal and neurobiological proximate factors affecting behavior. Chapter 4 tackles molecular genetic and developmental proximate factors. Keep in mind that dividing proximate causation into four components—hormonal, neurobiological, molecular genetic, and developmental—is artificial. For example, neurobiological differences often

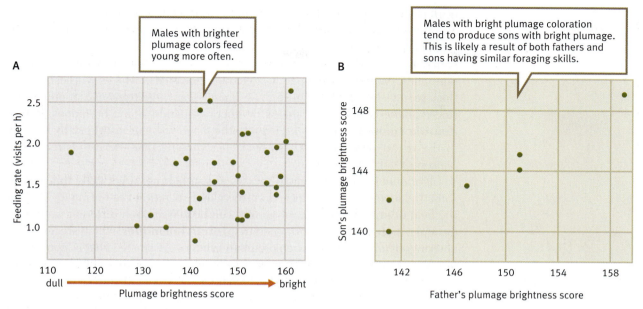

FIGURE 3.4. Plumage, feeding, and between generation correlation. (A) The relationship between male plumage and the rate of feeding offspring. **(B)** A significant positive relationship exists between father and son plumage brightness scores in house finches. *(From G. Hill, 1991)*

underlie hormonal changes in behavior (and vice versa), and one could argue that all hormonal and neurobiological differences associated with behavior can be understood at the molecular genetic level. As another example, hormonal differences can lead to behavioral differences that translate into animals having very different experiences during development. Conversely, developmental differences can lead to different hormonal responses to some stressor in the environment, as well as affect various properties of neurons associated with behavioral responses. All of this is to say that the relation among hormonal, neurobiological, molecular genetic, and developmental proximate causation is complex, and that boundaries between them can be difficult to define. The justification for separating them here is that before we can understand the complicated *interactions* among these four types of causation, we need a basic understanding of how each one works.

After the "primer" chapters (Chapters 2–6) of this book, Chapters 7–17 will deal with a number of major behavioral venues (foraging, predation, cooperation, aggression, and so on), weaving in discussion of both proximate and ultimate causation.

Hormones and Proximate Causation

A bit of basic background will help set the stage for our discussion of proximate causation, hormones, and behavior. The **endocrine system** is a communication network that influences many aspects of animal behavior. This system is primarily composed of a group of ductless glands that secrete hormones (derived from the Greek word for "excite") directly into the bloodstream (in vertebrates) or into fluid surrounding tissue (in invertebrates). In vertebrates, major glands producing hormones include the adrenal gland, pituitary gland, thyroid gland, pancreas, the gonads, and hypothalamus, but many other organs also produce hormones. Within these glands are endocrine cells, which synthesize and then secrete hormones (Figure 3.5). There is an important exception to the rule that hormones are produced by ductless glands. Some hormones, called **neurohormones**, can be released into the blood via neurons (typically located in the brain) that secrete these hormones directly into the bloodstream (R. Nelson, 2011).

Hormones act as chemical messengers, affecting target cells that reside some (relatively long) distance from the gland secreting the hormone. Within the field of **endocrinology**, hormones can be classified in different ways. For example, most vertebrate hormones are **protein hormones**, which are made up of strings of amino acids. When protein hormones are small—that is, composed of only a small number of amino acids—they are called **peptide hormones** (for example, prolactin). Protein hormones can be stored in endocrine cells and do not have to be released immediately into the bloodstream. Protein hormones are soluble in water and blood (and therefore are referred to as hydrophilic, or "water-loving") and do not require any other "carrier" chemicals to travel through blood. All other things being equal, the larger a protein hormone, the greater its half-life—that is, the longer it takes for half of the hormone to be removed from the blood.

Steroid hormones such as testosterone are different from protein hormones in many ways. The lag time between when a stimulus is sensed and when a hormone is produced can be much longer for steroid hormones than for protein hormones. But steroid hormones cannot be stored in cells, so once they are produced by an endocrine gland such as the adrenal gland, they are immediately released into the bloodstream. Because they are fat soluble but

FIGURE 3.5. Endocrine cells and target cells. A schematic of how endocrine cells work and how hormones eventually affect target cells. Enzymes in the Golgi apparatus process proteins into hormone molecules and package them inside secretory vesicles. These vesicles fuse with the cell membrane and release the hormone molecules into the bloodstream. The hormone molecules travel through the bloodstream until they reach the receptor sites of the target cell, where they bond and initiate a series of interactions. Here we see the schematic for a membrane-bound receptor, which is generally part of a peptide hormone system. Steroid hormones often pass right through a membrane to bind to receptors. *(Adapted from R. Nelson, 2005)*

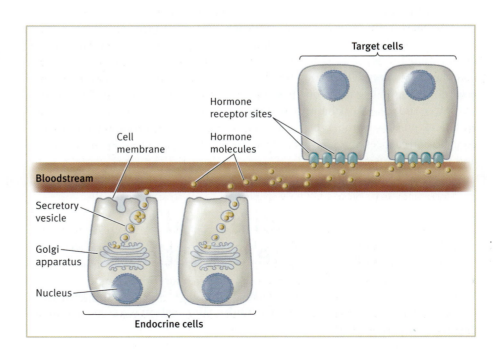

not water soluble, steroid hormones are called hydrophobic ("water-hating") and usually require a chemical "chaperone" (often a protein) to move them through the bloodstream to their target organ (R. Nelson, 2011).

Many glands can secrete more than one hormone, and the same hormone can have different effects on different target cells. Hormones affect many traits, including various behaviors, both directly and indirectly through, for example, changes in cell metabolism or DNA expression. Hormones also affect intracellular processes that promote cell division, induce the ion channels leading to neuronal signal paths to open, cause muscle contractions, and lead to the production of other hormones. Malfunctioning of the endocrine system, either through diminished secretion (hyposecretion) or excessive secretion (hypersecretion), affects functions such as growth, metabolism, reactions to stress, aggression, and reproduction.

The endocrine system is not a disconnected amalgam of ductless glands. Hormones secreted by one endocrine gland can stimulate the production and secretion of hormones from another gland. These chains may have important behavioral consequences. One example is the activation of the **hypothalamic-pituitary-adrenal (HPA) axis**, which leads to changes in behavior. Let's walk through this chain of events by focusing on hormonal cascades after a subordinate animal interacts with a more dominant animal in its vicinity. In response to the production and secretion of one hormone called CRH (corticotropin-releasing hormone) from the hypothalamus, the anterior pituitary gland of the subordinate animal secretes another hormone called ACTH (adrenocorticotropic hormone), which in turn stimulates the adrenal glands to produce glucocorticoid hormones like cortisol (see also the Conservation Connection Box). These glucocorticoids play a critical role in such behavior as reduced aggression in the presence of dominant individuals. Conversely, the secretion of some hormones *inhibits* the production of hormones in other ductless glands. As such, we need to think of the endocrine system as an interconnected and complex communication network that has both direct and indirect effects on animal behavior.

Hormones are almost always transported through the bloodstream, and with one exception (the cells in the lens of the eye), all cells in the body have a direct blood supply. But a hormone will have an effect on a cell only if the cell has the proper receptor site for that hormone. Cells with the receptor site for a particular hormone are referred to as **target cells** for that hormone. The receptor site, which is often located on the surface of a cell, and the hormone itself act as a lock-and-key system that is known as the hormone receptor complex. A receptor (lock) is not activated until the correct hormone (key) reaches it, and a hormone has no effect on a cell unless the cell has the correct receptor. For our purposes, what this means is that the relationship between animal behavior and endocrinology is put into play only when hormones reach target cells.

In this lock-and-key system, once a hormone reaches a cell and bonds with the receptor site, a series of interactions occur that affect the expression of genes and the synthesis of proteins. Precisely how these changes occur depends on the specific hormone, but changes in gene expression and protein synthesis can either directly or indirectly affect an animal's behavior. For example, increased gene expression or protein synthesis could, through a series of intermediary steps, lead to an increase in foraging behavior or an increase in aggressive behavior, or perhaps both.

Community-Based Ecotourism: Using Hormones to Measure Effects on Animal Well-Being

Ecotourism has become an increasingly popular vacation choice for environmentally conscious travelers. Ecotourism destinations typically involve a joint partnership between private enterprise and conservation biologists. In principle, ecotourism is designed to draw tourists to ecologically beautiful and often ecologically imperiled areas of the world; to use the funds generated from this tourism to protect the wildlife in these areas; and to promote the local, often indigenous, human culture that lives around the ecotourism site. It is not yet clear if the economics of ecotourism will in the end benefit conservation initiatives or prove a net positive for human populations in the area (Brightsmith et al., 2008; Coria and Calfucura, 2012; Kiss, 2004). But there are hundreds, perhaps thousands, of ecotourist programs in place. Ethologists and conservation biologists therefore seek to understand the effects of these programs on various aspects of animal behavior. One way that these have been measured is through the use of techniques developed in **field endocrinology**—measuring hormone levels in populations in nature (B. Walker

et al., 2005). As an example, let's look at the ethological implications of a large ecotourism program built around a colony of hundreds of thousands of Magellanic penguins in Punta Tombo, Argentina (B. Walker et al., 2005; Figure 3.6).

At the turn of the twenty-first century, Punta Tombo was home to a large colony of breeding Magellanic penguins (*Spheniscus magellanicus*)—two hundred thousand breeding pairs of birds. In an ecotourism program designed to help sustain this population, several hundred penguin nests have been fenced off, and tourists can walk freely in this area. This program draws seventy thousand ecotourists per year. Observational work suggests that adult penguins exposed to tourists have habituated to their presence and show reduced defensive responses in the presence of humans compared with adult penguins from nests not exposed to tourists. Conservation biologists describe adults from areas exposed to tourists as "calm and accepting of humans with penguins often walking directly among the tourists" (B. Walker et al., 2005).

But has this habituation had a positive, negative, or neutral effect for the animals' well-being? At first glance, it appeared that ecotourism was not having adverse effects on the animals or causing them to display dangerous behavior. In fact, when field endocrinologists measured corticosterone "stress" hormone levels in penguin adults, they found that those from areas with tourists showed lower stress responses to handling and release (Fowler, 1999).

More detailed follow-up work with chicks exposed to ecotourists, however, shows a much different, more disconcerting picture (B. Walker et al., 2005). Within days of hatching, chicks from nests exposed to ecotourism showed very high

levels of corticosterone compared with chicks from areas where ecotourists were excluded (Figure 3.7). Researchers are not certain exactly what causes this rise in corticosterone levels, but they believe that it is linked to the altered behavioral responses of adults at the chicks' nests. In nests from unexposed areas, parents spend much of their time brooding on the nest. But in ecotourist areas, parents spend more time standing up and even walking among the tourists. This constant change in temperature for the chicks, in conjunction with the increased distance between parent and chick, may be the cause of the corticosterone stress response in the chicks from ecotourist areas. In many other species, including numerous Adele penguins, increased corticosterone during the chick stage has been shown to have detrimental effects on individuals when they mature (Bonier et al., 2009, Ninnes et al., 2011). Whether such detrimental effects will emerge in the Magellanic penguins remains to be tested, but the work to date from field endocrinology and ethology certainly flags these populations as at risk.

FIGURE 3.6. Magellanic Penguins. Field endocrinology experiments on Magellanic penguin behavior have shed light on issues of conservation biology. *(Photo credit: Copyright © Chappell, Mark/Animals Animals—All rights reserved.)*

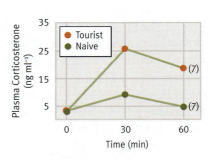

FIGURE 3.7. Stress and ecotourism. Very young penguins exposed to humans from ecotourist groups showed much higher levels of stress hormones, such as plasma corticosterone, than did individuals from control groups. *(From B. Walker et al., 2005)*

The lock-and-key system is put into operation only when a hormone is released into the bloodstream. But what causes the release in the first place? The answer to this question often lies in the complex ways that animals respond to stimuli in their environment. To see this, let's look at one component of the endocrinology of reproductive behavior in birds.

Many bird species breed during the spring and summer, when temperatures are warmer and food is more plentiful. Changes in day length are typically a reliable cue for seasonal change, and in many birds, levels of the hormones gonadotropin (which stimulates sperm production) and testosterone increase in males as day length increases. Some testosterone binds to receptor cells in the brain and is associated with a suite of behaviors related to mating and paternal care. For example, males are more likely to be aggressive toward one another to gain access to females, guard their mates, build nests, and defend their brood when testosterone is high. Some testosterone is converted into estradiol or dihydrotestosterone, and these, too, bind to receptors and lead to behaviors linked to mating and parental care. Thus, environment change (day length) is associated with increased levels of testosterone and related substances, and increased levels of these chemicals affect how animals respond to environmental stimuli like the presence of other males and the appearance of nest predators (G. Ball, 1993; Dawson, 2002; Wingfield et al., 2001; Figure 3.8). When ethologists experimentally manipulated day length in the laboratory, they discovered that delaying the onset of increased day length can have profound effects. Because the cues for the onset of spring and

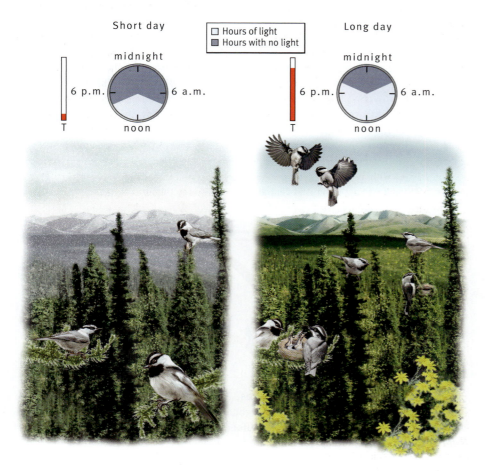

FIGURE 3.8. Day length, hormones, and behavior. Many bird species breed during the spring and summer, and changes in day length are an excellent cue for seasonal change. As day length changes, it affects circulating levels of testosterone (T). Increases in T increase the probability that males are aggressive toward one another (to gain access to females), guard their mates, build nests, and defend their broods. *(From B. Walker et al., 2005)*

summer breeding seasons were lacking, male testosterone levels remained low, and all of the testosterone-mediated behaviors just described were not set in motion. Similar experimental work has shown that manipulations of day length also affect hormone levels and the mating and breeding behavior of females as well as males.

HOW THE ENDOCRINE SYSTEM INTEGRATES SENSORY INPUT AND OUTPUT

If we think of animals as an engineer might, we might describe them as possessing three interactive systems (Figure 3.9): (1) an input system made up of all the sensory systems (smell, sight, and so on), (2) a central processor made up of integrators that process and integrate the sensory information received, and (3) output systems—effectors such as muscles that move when stimulated (R. Nelson, 2005). What hormones do is change the probability that a specific sensory input leads to a specific output.

Hormones can affect the probability that a specific sensory input leads to a specific behavioral response in many ways. Hormonal changes might *modify* some ongoing behavior by increasing or decreasing the frequency or duration of that behavior, or they might *trigger* the onset or end of a behavior or behavioral sequence. In addition, hormones might *prime* animals so that they are more or less likely to behave in a specific way in a specific environment. For example, when testosterone levels are high, males are more likely to engage in aggressive behavior than when testosterone levels are low. That is, when baseline levels of testosterone are high, males are primed for aggressive behavior, in the sense that they are more likely to display aggression when encountering another male than they are otherwise. With testosterone, as with many hormones, there is a hormonal-behavioral feedback loop in play. If an animal wins a fight, partly as a result of behaviors resulting from high baseline levels of testosterone, the act of winning may in turn increase the

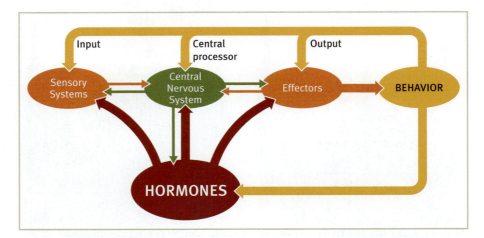

FIGURE 3.9. Complex effects of hormones. Hormones can affect input systems (sensory systems such as those for smell, sight, or hearing), central nervous system functions (processing), and output systems (for example, effectors such as muscles controlling movement). *(Adapted from R. Nelson, 2005)*

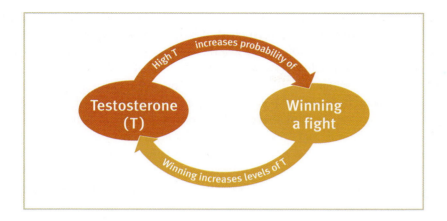

FIGURE 3.10. Testosterone (T) and aggression feedback loop. A positive feedback loop exists between levels of T and probability of winning a fight. High levels of T increase the probability of winning, whereas winning further increases circulating levels of T.

probability of winning future fights by further increasing testosterone levels or by lowering the level of stress hormones such as cortisol (Figure 3.10).

Hormones also affect the *organization* of behavior systems during early development. For example, female mice gestate many fetuses at the same time. If a developing male mouse fetus is surrounded by female fetuses, it is often exposed to lower levels of circulating testosterone (Figure 3.11). When such males mature, they tend to be less aggressive and less sexually active than males that were surrounded by male fetuses in utero. Here we see a case in which

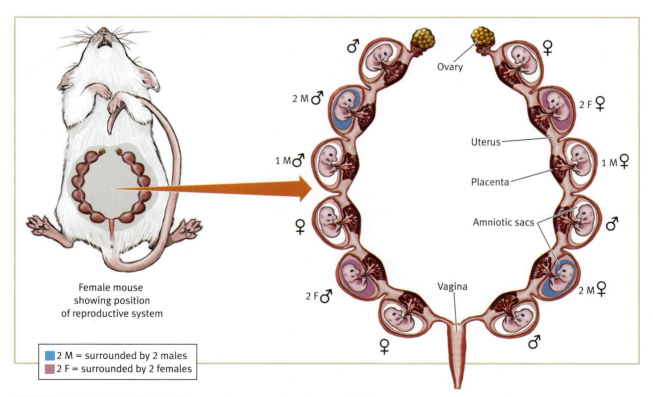

FIGURE 3.11. Intrauterine position. Males surrounded by two females in utero act relatively "feminized," whereas females surrounded by two males act relatively "masculinized." These behavioral differences are typically a result of differential exposure to hormones in utero. 1M = adjacent to one male; 2M = surrounded by two males; 2F = surrounded by two females. *(From vom Saal, 1989)*

the entire behavioral repertoire associated with aggressive interactions could be fundamentally altered by hormonal effects early on in development. We will delve into testosterone and in utero effects like this in more detail in the next section.

Although the specific relationship between behavior and the endocrine system varies in different species, early work by Hans Selye found that the behavioral/endocrinological response to danger—sometimes called the "fight or flight" response—is very similar across different types of stress. Subsequent work found that this fight or flight response was similar across many vertebrate species, including humans (R. Nelson, 2005; Selye, 1936; Stratakis and Chrousos, 1955). To understand the complex interaction between the endocrine system and behavior in a bit more detail, let's examine some of the specifics of the fight or flight response.

When an individual senses a stressor—a predator, for example—the hypothalamus initiates a response, which works along two pathways. In the first pathway, the adrenal glands begin to secrete epinephrine (also known as adrenaline). Epinephrine, which is secreted by both the adrenal glands and certain nerves in the central nervous system, binds to receptors on the smooth muscles around blood vessels and causes constriction or dilation, depending on which type of receptor is involved. Epinephrine also acts directly on receptors in the heart and lungs to increase cardiopulmonary activity. This burst of epinephrine (and norepinephrine) leads to a quick and very large increase in blood sugar that, along with oxygen, is delivered quickly to vital organs. In particular, the brain, skeletal muscles, and heart receive both increased blood sugar and oxygen. As a result of the increased sugar and oxygen, circulation increases and "nonessential" systems—for example, the digestive and reproductive systems—are shut down. All of these effects allow for an appropriate behavioral response to whatever the stressor may be—in this case, a predator. For example, the increase in sugar and oxygen might enable the animal to quickly flee from a predator or perhaps to fight off this danger.

In a second pathway, another "responsive reaction" chain is also put into motion when a stressor is sensed. Here corticotropin-releasing hormone (CRH), growth hormone-releasing hormone (GHRH), and thyrotropin-releasing hormone (TRH) are secreted by the hypothalamus. In turn, CRH stimulates the anterior pituitary gland to increase the production of ACTH, which then stimulates the adrenal gland to secrete cortisol. This hormonal cascade converts noncarbohydrates into sugars—energy that can be used to handle the stressor in question by, for example, fleeing quickly from danger or standing put and fighting. The adrenal gland also increases its production of aldosterone, which increases water retention and reduces bleeding if the stressor causes injury. These two parallel hormonal/behavioral responses to stressors illustrate the intricate and complex ways that organisms respond to dangers in their environment. It is only when stress is chronic that the endocrine system fails to respond in the appropriate manner.

With this brief overview of the endocrine system and behavior complete, let's examine a few case studies from the animal behavior literature.

THE LONG-TERM EFFECTS OF IN UTERO EXPOSURE TO HORMONES

Let us return to long-term behavioral effects of in utero exposure to testosterone. In a number of rodent species, the sex of the two siblings surrounding an individual in utero can have dramatic effects on testosterone level and an individual's behavior after birth (B. Ryan and Vandenbergh, 2002; vom Saal, 1989). Males that were surrounded by two other males in utero (labeled 2M males) are more aggressive than males that were surrounded by two females (labeled 2F males). In addition, 2M males mark their territories by scent and mount more females than 2F males, both in the laboratory and the field (Drickamer, 1996; B. Ryan and Vandenbergh, 2002; vom Saal and Bronson, 1978). To better understand the behavioral endocrinology of testosterone and intrauterine position, we begin by examining mating and parental care in male gerbils (*Meriones unguiculatus*; Figure 3.12).

In 1991, Mertice Clark and her colleagues measured in utero testosterone levels of 2M and 2F male gerbils (M. Clark et al., 1991). What they found was that testosterone levels in 2M males were significantly higher than in 2F males (Figure 3.13). It is still not clear exactly what causes this difference, but it is likely, in part, a result of 2M males being exposed to higher levels of testosterone in their amniotic sacs because they are surrounded by two other males (Clemens et al., 1978). These in utero differences between 2M and 2F males have long-term consequences—adult 2M males have twice the level of circulating testosterone levels as adult 2F males (M. Clark et al., 1991, 1992a; Even et al., 1992; vom Saal, 1989). Clark and her colleagues hypothesized that testosterone differences that had been initiated in utero and continued through life would affect male reproductive behavior, in terms of both obtaining mates and providing parental care (M. Clark and Galef, 2000; Clark et al., 1992b, 1998).

Clark and her team began their comparison of the mating behavior of 2M and 2F males by housing a single male (either 2M or 2F) and a single female together in a cage. They then measured how long it took before the male sexually mounted the female, and the time it took between mounting the female and ejaculation. The 2M males mounted females more quickly, and they ejaculated sooner (similar results have been obtained in comparisons of 2M and 2F mice; vom Saal, 1989; vom Saal et al., 1983). Follow-up experiments indicated that 2M males also sired more offspring after being paired with a female than did 2F males.

The work described above demonstrates increased mating effort on the part of 2M males. Clark also hypothesized that as a result of high testosterone levels 2M males would dispense less parental care (M. Clark and Galef, 1999; M. Clark et al., 1998). As a test of this hypothesis, twenty-three 2F males and twenty-one 2M males were each housed individually with a female, until the female was pregnant and gave birth. Once a female had given birth, the behavior of males toward pups was examined. Both when their mates were present and when their mates were absent, 2M males spent significantly less time in contact with pups than 2F males did (Figure 3.14). To be more confident that the differences between 2M and 2F males were linked to relatively high testosterone levels in 2M males, Clark and Galef undertook a follow-up experiment in which they castrated adolescent males to experimentally reduce testosterone levels. Such castrated males increased the time they spent caring for pups. When castrated males had their testosterone levels brought back to normal, through silicon implants, their parental behavior decreased, suggesting a causal link between testosterone and parental care (M. Clark and Galef, 1999; M. Clark et al., 2004; Figure 3.15).

In gerbils, the position of *females* while in utero also affects the level of testosterone to which they are exposed. When 2M females (females surrounded by two males in utero) mature, they have relatively high levels of circulating

FIGURE 3.12. Helping in male Mongolian gerbils. Male helping behavior as a function of prior intrauterine position has been examined in Mongolian gerbils. *(Photo credit: Julian Barker)*

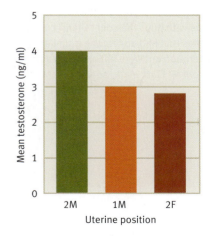

FIGURE 3.13. Testosterone measurements. Circulating testosterone levels for males that were adjacent to two (2M), one (1M), or zero (2F) other males in utero. *(Adapted from M. Clark et al., 1991)*

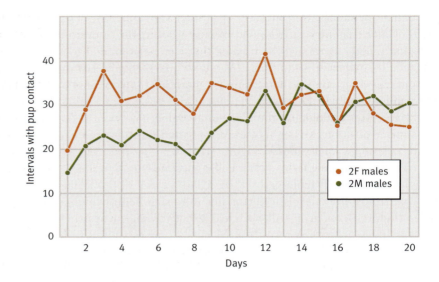

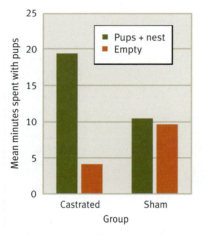

FIGURE 3.14. In utero position and subsequent parental behavior. Males that were surrounded by two males in utero (2M males) provided less parental care when they matured than did males that were surrounded by two females in utero (2F males). *(From M. Clark et al., 1998)*

FIGURE 3.15. Testosterone and male parental care. When male Mongolian gerbils were castrated, they spent more time with pups than did "sham" castrated males that had undergone a similar operation but were not actually castrated. *(From M. Clark et al., 2004)*

testosterone and lower levels of circulating estradiol. These 2M females are less attractive to males, in part because they are more aggressive than other females. They also have fewer litters and begin reproducing later than other females.

Researchers have a good understanding of how exposure to testosterone produces the behavioral variation between 2M females and other females. The key here seems to be how in utero exposure to testosterone affects metabolic activity in certain brain areas of 2M females. In gerbils, the preoptic area of the hypothalamus portion of the brain controls copulatory behavior, and this area of the brain is sexually dimorphic, meaning that males and females show different activity patterns in the preoptic area; 2M females have preoptic metabolic activity that resembles that of males more than that of 2F females. More specifically, 2M females have 20 percent greater metabolic activity in the preoptic area of the hypothalmus than do 2F females (D. Jones et al. 1997). One long-term effect of in utero exposure to high levels of testosterone in gerbils is to "masculinize" the preoptic area of the brain of 2M females. This relationship between in utero testosterone exposure and subsequent effects on brain activity shows nicely the tight connection between the hormonal and neurobiological underpinnings of behavior.

VASOPRESSIN AND SOCIALITY IN VOLES

Two closely related hormones, vasopressin and oxytocin, play a central role in reproduction and parental care in mammals. Vasopressin and oxytocin are neurohormones (although they can also act as neurotransmittors), and each is nine amino acids long. They are produced in the hypothalamus and then transported to various brain regions or projected to the pituitary gland for release. Homologs of vasopressin and oxytocin—that is, hormones that share the same evolutionary history—originated about 700 million years ago and are found in a diverse array of animals (Figure 3.16). Vasopressin and oxytocin themselves appear to have arisen from a gene duplication that occurred before the emergence of vertebrates. Just as vasopressin and oxytocin are produced and often expressed in the hypothalamus of vertebrates, the homologous hormones are produced and expressed in the equivalent brain regions in worms and fish (Donaldson and Young, 2008).

Vasopressin and oxytocin have been studied extensively in the context of parental care—or the lack of it—in two vole species: the prairie vole (*Microtus ochrogaster*) and the meadow vole (*Microtus pennsylvanicus*). Here, we focus on studies of vasopressin. (In Chapter 4, we will examine the role that oxytocin plays in these vole systems by discussing molecular genetic work centering on oxytocin in these species.)

Prairie voles are monogamous—both males and females have a single mate for a given breeding season (Chapter 8)—and males often display parental care and guard their mates. Meadow voles have a polygynous mating system, wherein males mate with multiple females during a breeding season, and males

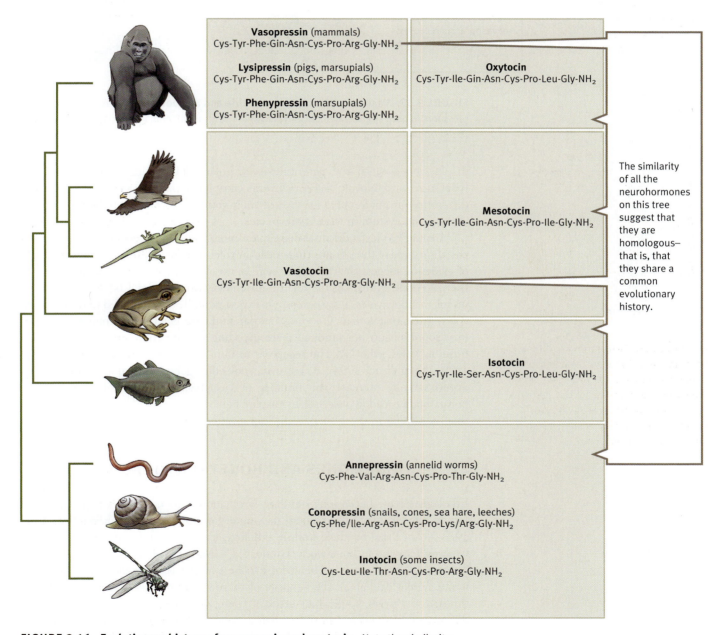

FIGURE 3.16. Evolutionary history of vasopressin and oxytocin. Note the similarity in amino acid structure between all the neurohormones on this phylogenetic tree. *(From Donaldson and Young, 2008)*

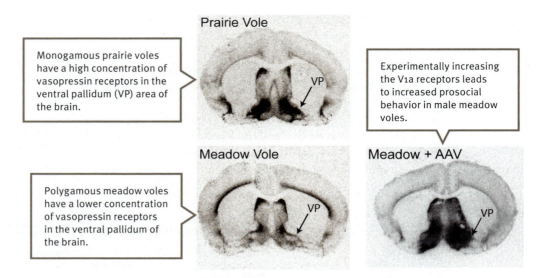

FIGURE 3.17. Vasopressin receptors in prairie and meadow voles. Vasopressin receptors in both species are concentrated in the ventral pallidum (VP) area of the brain. *(From Donaldson and Young, 2008)*

display very little, if any, parental care or prosocial behavior toward their mates (McGuire and Novak, 1984; Oliveras and Novak, 1986). One of the major differences in male behavior toward their young and their mate centers on the vasopressin system in these two species.

In prairie voles, individuals have many more vasopressin receptors in the ventral pallidum area of their brains than meadow voles do (Figure 3.17). A number of lines of evidence suggest that this difference in the number of vasopressin receptors in the vole brain is responsible for the difference in male social behavior in prairie versus meadow voles. If vasopressin is experimentally administered to male prairie voles, it stimulates mate guarding and parental care. Such experimental increases in vasopressin do not produce mate guarding and parental care in polygamous meadow voles, who lack the receptors to bind the extra vasopressin (Cho et al., 1999; Wang et al., 1994). If, however, molecular genetic techniques are used to experimentally increase the number of vasopressin *receptors* in the typically polygamous meadow vole, males display behaviors toward their mating partners that are similar to those seen in the monogamous prairie vole (Lim et al., 2004).

HORMONES AND HONEYBEE FORAGING

Ethologists have had a long-standing fascination with honeybees (*Apis mellifera*; Figure 3.18). When Niko Tinbergen, Konrad Lorenz, and Karl von Frisch won the 1973 Nobel Prize for their work in ethology, it was based in part on von Frisch's work on foraging in honeybees (Hinde, 1973; Marler and Griffin, 1973). But this fascination with honeybee behavior can be traced further back to Darwin, who wrote a great deal about the behavior of these social insects. Indeed, by Victorian times, the British public had fallen in love with bees, particularly when they discovered that worker bees that were fed "royal jelly" developed into queens. What's more, they were spellbound by the idea that the same bee egg develops into a male if it remains unfertilized, but develops into a female if it is fertilized by a drone's sperm (Prete, 1990, 1991; Richards, 1987).

FIGURE 3.18. Honeybee foraging.
The proximate underpinnings of foraging behavior in honeybees have been studied in depth. *(Photo credit: George D. Lepp/ Corbis)*

In later chapters, we explore one complex communication system—the famous waggle dance—that has evolved in the context of foraging behavior in honeybees. Much work has been done on honeybee foraging from this ultimate perspective. One of the aspects of honeybee foraging that makes it so attractive as a system to study, though, is that this work on ultimate questions has been linked to work on the proximate underpinnings of foraging in bees. This chapter and Chapter 4 use honeybee foraging behavior as a case study of an integrated proximate approach toward animal behavior, examining the effects of hormones, neural structures, genes, and development on honeybee foraging behavior. We begin with hormones and honeybee foraging.

Honeybees play different roles within their hive after they emerge from their pupal envelopes. Depending on the specific needs of a colony, during the first three weeks of life, young bees will clean the hive, feed larvae, make wax, process honey, guard a hive, or fan the hive for thermoregulation. At approximately twenty-one days of age, many bees become foragers that leave the nest, search for pollen and nectar, and return to the nest with what they find. The shift to forager is associated with an increase in levels of what is called juvenile hormone III (JH III), which is produced in a gland called the corpus allatum (Hagenguth and Rembold, 1978; Wyatt and Davey, 1996). Gene Robinson and his colleagues hypothesized that an increase in JH III was not only *correlated* with the shift to the forager stage but that changes in levels of JH III *caused* changes to the rate at which bees curtailed within-hive activites like cleaning and hive thermoregulation and became foragers. To test their ideas, Joseph Sullivan, Robinson, and their team surgically removed the corpus allatum (in a procedure known an allatectomy) in experimental groups of bees in four colonies (J. Sullivan et al., 2000, 2003; Figure 3.19). The behavior of bees in

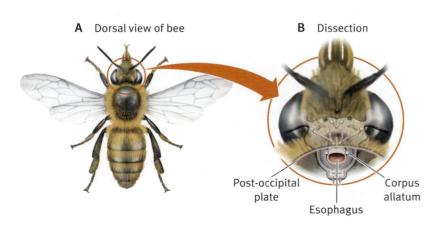

A Dorsal view of bee   **B** Dissection

Post-occipital plate

Esophagus

Corpus allatum

FIGURE 3.19. Bee surgery. To examine the effect of juvenile hormone (JH III) in honeybee foraging, Sullivan removed the corpus allatum—the gland that produces this hormone. The inset shows a view through the incision. *(Adapted from J. Sullivan et al., 2000)*

the allatectomized group was then compared with that of two control groups, one in which the bees went through a similar surgical procedure but did not have their corpus allatum removed, and the other in which the bees were only anesthetized.

Sullivan and his colleagues found that although both allatectomized bees and control bees eventually commenced foraging, in three of the four colonies observed, bees that had been allatectomized (and thus had no JH III) began foraging significantly later than did bees in the control groups. When allatectomized bees eventually did begin foraging, they returned to the nest much less often than did bees in the control treatments, and some work suggests that allatectomized bees are poor navigators in their environments. Sullivan and his team uncovered additional support for the link between JH III and the shift to foraging when they found that no other major behavior changes resulted from removing the corpus allatum—just changes related to foraging (J. Sullivan et al., 2000, 2003).

The researchers hypothesized that if increased JH III levels were causing increased foraging behavior, experimentally increasing the JH III levels in allatectomized bees would result in a return to normal foraging behavior by the bees. Experiments designed to test this idea have found that when allatectomized bees are given a dose of a chemical called methoprene—a chemical that is similar to JH III—they show no differences in age-related foraging when compared with control bees, providing strong evidence for a causal role of JH III in foraging behavior in honeybees (G. Robinson, 1985, 1987; G. Robinson et al., 1989).

A neurohormone called octopamine has also been linked to increased foraging activity in honeybees. Octopamine is found in invertebrates, but it is structurally similar to noradrenaline found in vertebrates—indeed, octopamine and noradrenaline appear to be homologous (Farooqui, 2007; Roeder, 1999; Verlinden et al., 2010; Figure 3.20). Early work had shown that octopamine modulates learning and memory in honeybees and also affects their visual, olfactory, and gustatory senses (Erber et al., 1993; Hammer and Menzel, 1998; Scheiner et al., 2002). Foraging bees have higher concentrations of octopamine in their brains than do nurse bees that stay at the hive, and this neurohormone reaches its highest concentration when a bee switches from nest-bound activities to foraging activities, including foraging-related flight behavior (Schulz et al., 2003; Wagener-Hulme et al., 1999). But does an increase in octopamine affect other activities in newly foraging bees, or does it just target flight activities related to foraging? How specific are the effects?

Barron and Robinson found two lines of evidence that the effects of octopamine were specific to foraging activities (Barron and Robinson, 2005): (1) Although experimental treatment with octopamine increased flight activity related to foraging, it did not increase a second flight-related behavior—the removal of corpses from the colony; and (2) When bees treated with octopamine

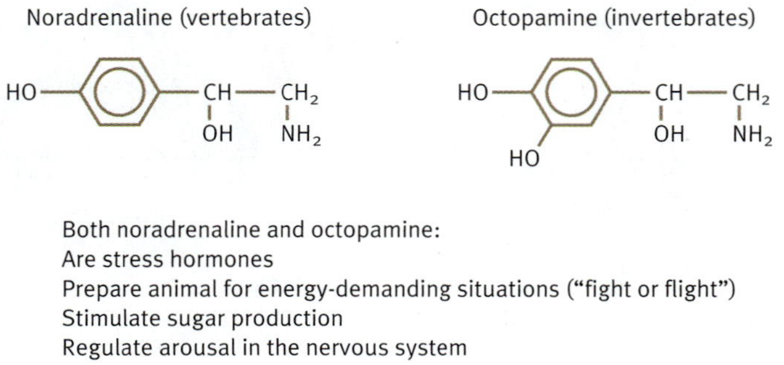

FIGURE 3.20. Comparing vertebrate and invertebrate systems. A comparison of the vertebrate adrenal system with the invertebrate octopamine system. *(Adapted from Roeder, 1999)*

Noradrenaline (vertebrates)   Octopamine (invertebrates)

Both noradrenaline and octopamine:
Are stress hormones
Prepare animal for energy-demanding situations ("fight or flight")
Stimulate sugar production
Regulate arousal in the nervous system

were exposed to other hormones associated with the production of a new brood of offspring, they increased their foraging behavior (to feed a now-larger hive), but they did not increase other activities associated with caring for the new brood.

Controlled experimental work on octopamine, along with JH III, illustrates the ways in which a proximate perspective of the honeybee endocrine system can shed light on a fundamentally important behavior in the life of a bee—foraging.

Neurobiological Underpinnings of Behavior

The endocrinological system is based on chemical communication. We have seen the power of such a communication system, but chemical communication takes *time*—on the order of minutes to hours. At the proximate level, there are other ways that information relevant to behavioral decisions can be transferred. A second communication network—the nervous system—relies on electrical impulses that allow for *much quicker* responses, including behavioral responses. This section of the chapter will serve as a primer on the nervous system and behavior, providing an overview of the neurobiological underpinnings of behavior. This field of work is often called **neuroethology**.

Animals possess specialized nerve cells, called neurons, which share certain similarities regardless of what message they conduct. Each neuron has a cell body that contains a nucleus and one or more nerve fibers. These nerve fibers, called **axons**, transmit electrical information from one cell to another. Axons can range in size from less than a millimeter to over a meter long. Axons also differ in terms of diameter. This variation is important because the speed of the nervous impulse affects the speed at which animals respond behaviorally, and the thicker the diameter of an axon, the faster the nervous impulse travels along it.

Each neuron has only a single axon. The first section is called the axon hillock, and the last section consists of the axon terminals (sometimes called synaptic terminals). There may be many branches of axon terminals, and it is from these branches that information leaves a neuron as it passes along the nervous system (Figure 3.21). Neurons receive impulses from other cells via fibers called **dendrites**. A neuron may have thousands of dendrites, forming what is called a dendritic tree. In addition, in certain types of neurons, each "branch" on this dendritic tree may have many *dendritic spines* that receive input from other neurons.

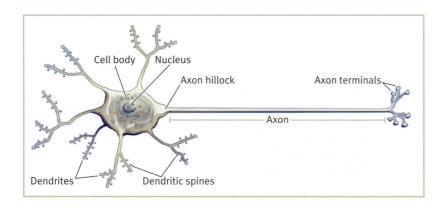

FIGURE 3.21. Nerve cell. Information is collected by dendrites (which often have dendritic spines projecting off their surfaces), conducted along an axon, and transmitted from the axon terminals across the synaptic gap to the dendrites of neighboring cells.

THE NERVOUS IMPULSE

Let's trace what happens from the point at which the nervous system of an animal responds to something in the environment to the point at which some sort of behavioral response can be measured. Suppose the external stimulus is tactile—that is, the animal is touched by something. In response to this stimulus, a wave of electrical activity sweeps down along the axons of sensory nerve cells that are in contact with the skin. Not all stimuli produce such a response. For this process to begin, the stimulus—the external touch in our example—must exceed the nerve cell's "threshold." Technically, the threshold is a function of the amount of change in the voltage across a neuron's membrane, but what is most important is that stimuli that don't meet this threshold fail to cause the nerve cell to fire, and stimuli above the threshold always cause the nerve cell to fire.

To use an analogy from electrical engineering, nerve cell thresholds are the equivalent of on-off switches, rather than a dimmer switch. Any stimulus greater than the threshold, regardless of how much greater, causes the nerve cell to fire. The nerve cell fires in exactly the same manner whether the stimulus is 1 percent greater than the threshold or 1,000 percent greater. But even though the nerve cell fires the same way each time its threshold is reached, organisms are still able to use their nervous systems to gauge the strength of a stimulus in at least two ways: (1) the number of times a neuron fires increases with the strength of the stimulus, and (2) the number of neurons that fire in response to a stimulus also increases as a function of stimulus strength.

Once an impulse has reached the end of an axon, it is transmitted to other neurons. This transmission may involve an electrical impulse jumping across the synaptic gap between neurons or, more commonly, the release of a neurotransmitter—for example, acetylcholine—from thousands of synaptic vesicles located on the tips of the branches at the end of the axon terminals. Once neurotransmitters migrate across the synaptic gap, they are absorbed by the membrane of the next neuron, or they bind to receptors that open an ion channel and/or initiate intracellular signaling cascades. Neurotransmitters add a chemical component to the nervous system—that is, some nervous impulses requires both electrical and chemical transmission.

If the next neuron is sufficiently stimulated—that is, if its threshold has been met—it will fire. This process continues over and over as the nerve impulse migrates along its neural path. The neural pathway that was initiated by an animal being touched may end when the terminal neurons in the pathway *innervate* (stimulate) some *effector*, such as a muscle, to take action; in this case, terminal neurons might stimulate foraging behavior by opening or closing the jaw muscle (if the animal has encountered potential prey) or initiate flight behavior by innervating muscles associated with such flight (if the touch was that of an aggressive, dominant group member). In other instances, depending on the stimulus, nerve pathways might end with the secretion of a chemical that causes an endocrine gland to secrete a hormone. This cause-and-effect relationship between the terminal neurons and the secretion of hormones that affect behavior shows nicely that the neural and endocrinological communication systems are not independent of one another but are part of a communication "web" that provides an organism with information about its external and internal environments.

This communication web can work in many ways. Our example involving the neuroethological response to touch focused on a stimulus that causes an excitatory response along the nervous system, eventually translating into an animal taking some behavioral action (for example, foraging, fleeing). But some stimuli cause inhibitory effects, whereby a neuron sends a signal that does not meet the threshold of the cell that receives the signal, eventually resulting in the inhibition of behavior. Such inhibitory responses might be lifesaving—as a result of this process an animal might, for example, freeze in the presence of predators that hone in on the motion of prey.

At a very general level, nervous systems across many animals show some consistent evolutionary trends: Nerve cells that served specific functions—including functions linked to behavior—became clustered. Over evolutionary time, nervous systems became centralized, and longitudinal nerve cords became the major highways across which nervous impulses traveled. The front end of the longitudinal nerve cord became dominant, leading to the evolution of the brain.

Many new technologies are now in place that allow neuroscientists to scan activity in the brains of animals, providing researchers with a deeper understanding of the way that the nervous system shapes the immediate, as well as the long-term, behavioral responses of animals to stimuli in their environment. For example, functional magnetic resonance imaging (fMRI) allows scientists to measure neuronal activity across large sections of animals' brains, and to do so with high resolution. This procedure allows ethologists to address important questions related to behavior: Do different salient stimuli—a potential mate, a competitor, a predator, a food item—translate into different neural activity at the brain level, and if so, how (Ferris et al., 2008)? Do we see similar patterns in closely related species? Do we see similar patterns in distantly related species? Does the same general behavioral pattern—for example, aggressive behavior—involve different brain activity patterns in different contexts? At the brain level, does aggression toward a potential predator look similar to aggression toward a conspecific? If so, why? Do behaviors associated with a positive reward—food, access to mates, and so on—generally look similar at the level of the brain? Why or why not? Though researchers are still trying to understand exactly how to interpret patterns across large areas of the brain, early work in whole-brain neuroimaging is already providing useful insight into how important stimuli in the environment "map" onto brain activity (Figure 3.22).

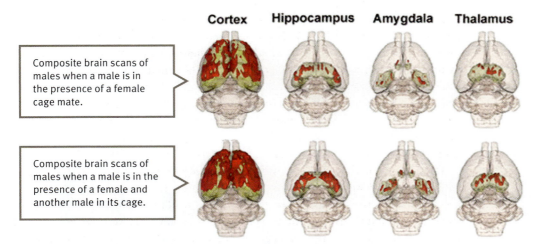

Cortex Hippocampus Amygdala Thalamus

Composite brain scans of males when a male is in the presence of a female cage mate.

Composite brain scans of males when a male is in the presence of a female and another male in its cage.

FIGURE 3.22. Brain scans. Ten male rats were presented with either their female cage mate or their female cage mate and a male intruder. Different activity patterns in different parts of the brain were detected (red indicates activity). *(From Ferris et al., 2008)*

With this brief review of the animal nervous system in hand, we can now move on to examine how animal behaviorists might address proximate questions using a neuroethological approach.

NEUROBIOLOGY AND LEARNING IN RODENTS

Let's return to the prairie and meadow voles discussed earlier in the chapter. Recall that prairie voles are monogamous, but meadow voles are polygynous. Polygynous male meadow voles have very large home ranges—home ranges that are sometimes ten times the size of that of a female, and that often encompass the home ranges of many females (Gaulin and Fitzgerald, 1986, 1989). Given that males have larger home ranges than females and must keep track of females within those home ranges, animal behaviorists hypothesized that males would have better navigational skills than females.

When Steven Gaulin and Randall Fitzgerald ran male and female meadow voles through a series of mazes in the laboratory, they found that males showed superior spatial learning abilities. Gaulin and Fitzgerald also examined spatial learning skills in the more monogamous prairie vole, where males and females have home ranges that are approximately equal in size. Because males do not have larger home ranges and do not have to navigate across many female home ranges, the researchers hypothesized that the difference in spatial learning in males and females that was seen in the polygamous meadow vole would be absent in the monogamous prairie vole. When Gaulin and Fitzgerald tested prairie voles in mazes, they found no difference between the spatial abilities of males and females of this species (Figure 3.23).

The hippocampus is an area of the brain known to be important in spatial navigation (Eichenbaum et al., 1990; R. Morris et al., 1982; O'Keefe and Nadel, 1978). Based on the spatial learning results in voles, Lucia Jacobs and her colleagues hypothesized that males in polygamous species would have larger hippocampuses than females, but that no such sex difference in hippocampal size would be seen in monogamous species of voles. For their experiments, the researchers used meadow voles as the polygamous species (as above), but they

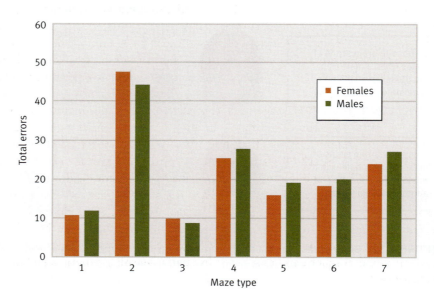

FIGURE 3.23. Learning in voles. No significant differences in learning mazes (as shown here by the number of errors on seven maze types) were found in monogamous prairie voles, indicating no differences in spatial learning between male and female prairie voles. *(Adapted from Gaulin and Fitzgerald, 1989)*

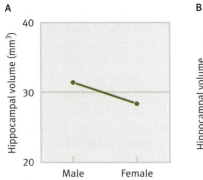

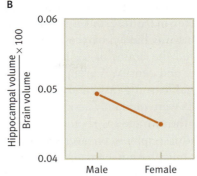

FIGURE 3.24. Sex difference in size of hippocampus. In meadow voles, males have a larger hippocampus, in both **(A)** absolute size and **(B)** relative size, than do females. *(From L. Jacobs et al., 1990)*

chose the pine vole (*Microtus pinetorum*), rather than the prairie vole tested by Gaulin and Fitzgerald, as their monogamous species.

When Jacobs and her team compared the size of the hippocampus in relation to overall brain size, they found sex differences in meadow voles, but they did not find sex differences in pine voles (L. Jacobs et al., 1990; Figure 3.24). This sort of comparative approach is a powerful neuroethological tool. Behavioral differences between species can lead to predictions about neurobiological differences regarding specific parts of the brain—in this case, the hippocampus—and these predictions can then be tested under controlled conditions.

Work on spatial learning and neurobiology in rodents is not restricted to studies on the hippocampus. Researchers have found that the parietal and prefrontal cortex areas of the brain also play an important role in the spatial learning process, and in particular that the number of dendritic spines in these areas of the brain is correlated with learning ability.

Martin Kavaliers and his colleagues have examined spatial abilities and the number of dendritic spines in the prefrontal and parietal cortices of male and female meadow voles (Kavaliers et al., 1998). To do so, they tested both male and female meadow voles in a water maze, and they determined spatial learning by measuring how quickly the voles learned to find the hidden platform in the maze over the course of eight trials. As in Gaulin and Fitzgerald's study, males were consistently superior to females at the spatial learning task, in terms of both the speed of learning and the ability to retain the information they had learned (Figure 3.25). Once the spatial learning trials were completed, the brains of a subset of the males and females were dissected and the number of dendritic spines in their prefrontal and parietal cortices was measured. Kavaliers and his team found that males had more dendritic spines than females in both areas of the cortex (Figure 3.26).

The Kavaliers study demonstrates that learning and neuronal density are correlated, but it does not tell us about causation. It might be that male meadow voles had more dendritic spines than females before the water maze trials and hence had superior spatial learning abilities. Or it could have been that males and females had the same number of dendritic spines before the water maze test, but that learning during the water maze trials might have increased spine density in males, but not in females. Or perhaps both are true. No studies to date on voles allow us to distinguish between such possibilities, but May-Britt Moser and her colleagues have provided evidence from rats suggesting that dendritic spine number can be changed by spatial learning experience (Moser et al., 1994).

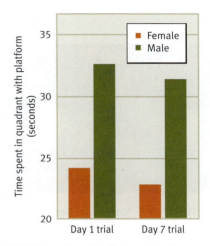

FIGURE 3.25. Sex differences in water maze trials. Male meadow voles spent more time near the platform in water maze trials than did female meadow voles. *(Adapted from Kavaliers et al., 1998)*

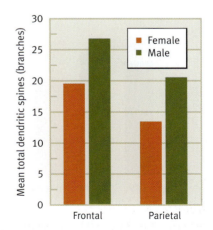

FIGURE 3.26. Sex difference in number of dendritic spines. In meadow voles, males have more dendritic spines (branches) in both the frontal and the parietal cortex of the brain than do females. *(From Kavaliers et al., 1998)*

Moser and her team measured spatial learning skills in rats using both the water maze test and a second spatial learning task that involved finding food that was hidden in a cage. After the spatial learning trials, the researchers dissected the brains of a subset of these animals to measure dendritic spine number and density. A group of control animals was not exposed to either type of spatial learning task, and the researchers also dissected their brains. Given that rats were randomly assigned to the control and spatial learning groups, the rats in the two groups should have started out with roughly the same number of dendritic spines, on average. Therefore, any difference between the two groups must have been caused by the spatial learning activity. The researchers predicted that rats that went through the spatial learning trials would have more dendritic spines than the control animals. Moser and her team found just such a relationship in rats. This work by Moser and her colleagues is a good example of what is called **neural plasticity**—defined as "the ability of the brain, and the nervous system more generally, to alter its structure and physiology to mediate both stability and changes in behavior" (Teskey et al., 2010).

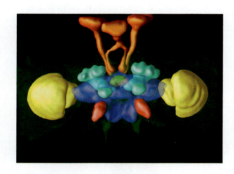

FIGURE 3.27. Mushroom bodies.
The mushroom bodies—shown in light blue—are clusters of neurons located at the front of the bee brain and are involved in spatial learning. The yellow denotes the optic lobes of the bee brain.
(Photo credit: Birgit Ehmer)

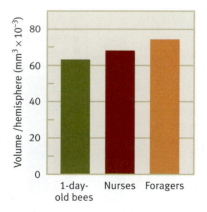

FIGURE 3.28. Mushroom bodies and foraging. Mushroom bodies were larger in foragers than in bees that remained in a colony (one-day-old bees and nurses who care for the larvae in the hive).
(From Withers et al., 1993)

MUSHROOM BODIES AND HONEYBEE FORAGING

We now return to the proximate analysis of foraging behavior in bees that we earlier discussed from a hormonal perspective, but here we approach this topic from the standpoint of neuroethology. Honeybee foragers must navigate outside their hives or nests in search of food, and the ability to remember and retrieve information from the environment is critical. As we have seen, in vertebrates this spatial navigation is often associated with the hippocampus. In invertebrates, however, it is most often linked with a cluster of small neurons located at the front of the brain. This cluster, known technically as the *corpora pedunculata*, is often referred to as the **mushroom bodies** (Capaldi et al., 1999; Fahrbach, 2006; Figure 3.27).

Mushroom bodies play a central role in spatial navigation and foraging behavior in honeybees—a species that often travels many kilometers in search of food (Visscher and Seeley, 1982). As discussed earlier in the chapter, younger bees usually reside within the hive and later switch to foraging outside the hive (G. Robinson, 1992). Such foragers use both visual and olfactory cues in their search for food.

When they first leave their home hive, rather than immediately beginning a search for food in the nearby environment, would-be foragers often turn back toward the nest and hover up and down for several minutes, in what is referred to as an "orientation flight," orienting the foragers to the relative position of their nest in the environment (Willmer and Stone, 2004). Bees often start undertaking orientation flights when they are about one week old, though they don't actually begin foraging until they are about three weeks of age. Ginger Withers, Susan Fahrbach, and Gene Robinson examined bees of different ages and bees that undertake different tasks within a colony (foraging versus caring for the larvae in the hive) to see whether the relative size of their mushroom bodies differed as a function of task allocation. The researchers found that the mushroom bodies of foragers were 14.8 percent larger than those of the other groups that they measured, suggesting a link between mushroom bodies and foraging (Withers et al., 1993; Figure 3.28). The volume of other nerve

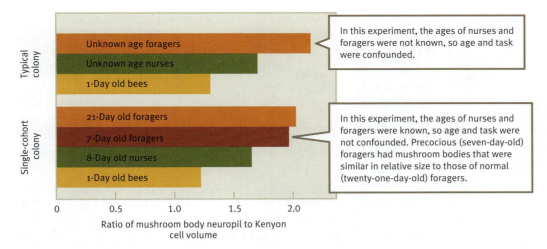

FIGURE 3.29. Disentangling age and task in the division of labor. Researchers could disentangle age and task in relation to honeybee foraging. They found that the relative size of mushroom bodies of precocious foragers was about equal to that of normal foragers. *(From Withers et al., 1993)*

clusters *relative* to total brain size in the honeybee brains remained unchanged as a function of task allocation; only the mushroom bodies increased in relative size. But there is a confound here: The exact age of the foragers was unknown, and the age of the "nurse" bees from the hive was also unknown. The relationship between mushroom body size and foraging might be the result of age differences, rather than differences in task.

To separate the effect of age and task on mushroom body size, Withers and her team gathered together one-day-old bees from numerous groups and formed a colony composed only of one-day-old bees. Creating a colony of young, same-aged individuals induces early foraging behavior in bees. Whereas normal bees begin this task at about twenty days of age, bees in the experimental treatment began foraging at around four to seven days old. The configuration of the mushroom bodies in these precocious foragers resembled that of normal-aged foragers, suggesting that activities related to foraging triggers a series of neural-based changes in mushroom body volume, illustrating yet another instance of neural plasticity (Sigg et al., 1997; Withers et al., 1993; Figure 3.29). In the next chapter, we will examine the molecular genetic underpinnings of mushroom body growth and neural plasticity (Lutz et al., 2012).

VOCALIZATIONS IN PLAINFIN MIDSHIPMAN FISH

The neurobiological underpinnings of communication have been investigated by Andrew Bass and his colleagues in their long-term work on the plainfin midshipman fish (*Porichthys notatus*). Bass and his team have distinguished what they call type I and type II male plainfin midshipman (Bass, 1998; Bass and Grober, 2009; Bass and Zakon, 2005; Figure 3.30). Behaviorally, type I and type II males act very differently. Type I males build nests, are four times larger than type II males, have a very high gonad-to-body size ratio, and produce

FIGURE 3.30. Vocal fish. In plainfin midshipman, some male types produce vocalizations while others do not. **(A)** The two smaller fish on the ends are type II sneaker males (that do not sing), whereas the fish that is second from the left is a "singing" type I parental male. **(B)** A type I male in his nest with his brood attached to the rocks. *(Photo credits: Andrew Bass)*

sounds in a number of behavioral contexts (Table 3.2). Type I males generate short-duration grunts when engaged in aggressive contests with other males and generate long-duration "hums" when courting females (Bass and McKibben, 2003; Bass and Zakon, 2005; Brantley and Bass, 1994). If females select a type I male as a mate, they remain on the nest of that male and lay their eggs there.

In contrast, type II males are small and have low gonad-to-body size ratios. They are often referred to as "sneakers" because they do not build nests, but rather stay around the nests of type I males, where they dart in and shed sperm in an attempt to fertilize the nesting female (see Chapter 7 for more on this behavioral strategy in bluegill sunfish). In terms of communication, unlike type

TABLE 3.2. Traits of type I and type II males. A summary of the differences between type I and type II plainfin midshipman, and a comparison with plainfin midshipman females.

SEXUALLY POLYMORPHIC TRAITS	TYPE I MALE	TYPE II MALE	FEMALE
Nest building	yes	no	no
Egg guarding	yes	no	no
Body size	large	small	intermediate
Gonad-size/body-size ratio	small	large	large (gravid), small (spent)*
Ventral coloration	olive-gray	mottled yellow	bronze (gravid), mottled (spent)*
Circulating steroids	testosterone, 11-Ketotestosterone	testosterone	testosterone, estradiol
Vocal behavior	hums, grunt trains	isolated grunts	isolated grunts
Vocal muscle	large	small	small
Vocal neurons	large	small	small
Vocal discharge frequency	high	low	low

*Gravid connotes pregnant; spent connotes postpregnant.

I males, type II males do *not* hum to attract females, and they only occasionally produce grunt sounds.

What are the neurobiological underpinnings of the difference in the way that type I and II males communicate with others? The vocal organ of the midshipman is a set of paired sonic muscles attached to its swim bladders (Figure 3.31). *Both* type I and type II males have such sonic muscles, so the presence of the vocal organ per se does not explain the different ways that type I and II males produce sounds. But type I males have larger sonic muscles with more muscle fibers, and this difference in size affects sound production (Bass, 1992; Bass and Marchaterre, 1989). However, the difference in size of the sonic muscles is only part of the reason that type I and II males produce such different vocalizations. Each of the sonic muscles is innervated by—that is, receives nervous impulses from—a sonic motor nucleus that runs from the hindbrain of the fish down along the spine to the sonic muscles. The two sonic motor nuclei themselves are innervated by "pacemaker" neurons that generate impulses in a regular cyclic, rhythmic pattern. Bass and his colleagues have mapped out much of this pathway and have found differences in the rate at which pacemaker neurons generate impulses. The pacemaker neurons in type I males fire at a rate that is 15 to 20 percent higher than that of type II males, helping explain the differences in the vocalizations of type I and II males, and therefore the differences relating to reproductive behavior and aggression.

Bass and his team have demonstrated a tight linkage between neurobiological and endocrinological approaches to sound production in type I and type II males (Bass and Zakon, 2005; Remage-Healey and Bass, 2007). On the one hand, the hormone arginine vasotocin (AVT) inhibits activity in the neurobiological circuitry associated with the production of sounds in type I males, but AVT does not affect vocal motor activity and sound production in type II males. On the other hand, the hormone isotocin (IT) has the exact opposite effect, inhibiting activity in the neurobiological circuitry associated with sound production in type II males, but having no effect on sound production in type I males (Bass and Baker, 1990, 1991). The areas of the brain involved in sound production in the midshipman— more specifically, the preoptic and anterior hypothalamus—are similar to brain

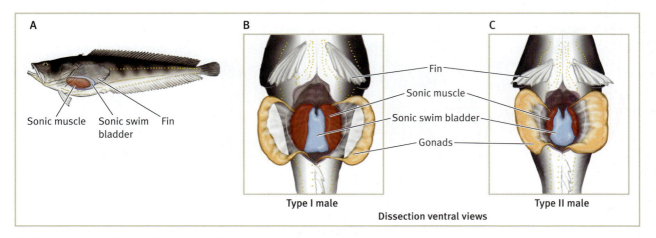

A

Sonic muscle   Sonic swim   Fin
                bladder

B

Fin
Sonic muscle
Sonic swim bladder
Gonads

Type I male

C

Type II male

Dissection ventral views

FIGURE 3.31. Sonic muscles and vocalization. (A) The vocal organ of the plainfin midshipman is made up of a pair of sonic muscles attached to the walls of the swim bladder. **(B)** Sonic muscles of type I males are well developed in comparison with muscles from **(C)** type II males. *(From Bass, 1996, p. 357)*

sections involved in sound production in numerous other species. The linkage among hormones, neurobiology, and sound production in the midshipman thus may have some important general applications for other vertebrate groups. In that sense, midshipman fish are a "model species" for examining the interplay of the neuronal and endocrinological systems involved in the production of sound. Work on plainfin midshipman also shows how *internal* communication systems— endocrinological and neuronal—help us understand an *external* communication system, namely, sound production (Goodson and Bass, 2001).

SLEEP AND PREDATION IN MALLARD DUCKS

Ethologists have studied sleep behavior from both a proximate and an ultimate perspective (Capellini, 2008; Lesku et al., 2008, 2011; Rattenborg et al., 2011; Roth et al., 2010). While it is clear that sleep is necessary for normal functioning in animals, sleep nonetheless poses a problem, in that sleeping individuals are more susceptible to being attacked by predators. Here we will look at the proximate underpinnings of an incredible antipredator behavior seen in a number of creatures—sleeping with one eye open and one eye shut. This type of sleep was first recorded in chickens (Spooner, 1964), but here we focus on John Rattenborg and Steve Lima's studies of sleep and antipredator behavior in mallard ducks (*Anas platyrhynchos*; Lima et al., 2005; Rattenborg et al., 1999a, 1999b, 2000).

Mallards are able to sleep with one eye open and one hemisphere of the brain awake (N. Ball et al., 1988). These birds sleep with half their brain awake and half their brain asleep, in what is referred to as unihemispheric sleep. In the laboratory, Rattenberg and Lima examined unihemispheric sleep in mallards and found that not only could they sleep with one eye open but mallards on the periphery of the group, where they are more susceptible to predation, relied on unihemispheric sleep more than birds in the center of a group (Figure 3.32). Indeed, birds on the periphery of the group slept with the open eye outward, away from their group and toward areas of potential predation (Figure 3.33).

Infrared cameras

FIGURE 3.32. Sleeping apparatus. This experimental housing unit was employed to record eye state and electrophysiology of four mallard ducks. Eight infrared cameras were used to allow the movement of each eye of each mallard to be recorded. Birds on the extreme left and right were considered to be on the edges of the group. *(From Rattenborg et al., 1999a)*

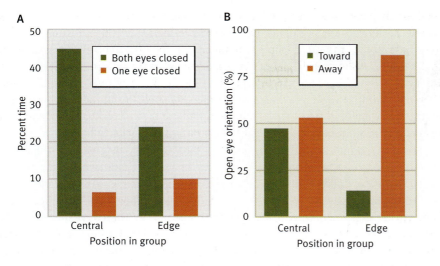

A (chart)
Percent time (y-axis: 0, 10, 20, 30, 40, 50)
Position in group (x-axis: Central, Edge)
Legend: Both eyes closed (green), One eye closed (orange)

B (chart)
Open eye orientation (%) (y-axis: 0, 25, 50, 75, 100)
Position in group (x-axis: Central, Edge)
Legend: Toward (green), Away (orange)

FIGURE 3.33. Unihemispheric sleep.
(A) Percentage of time ducks spent with one eye closed or both eyes closed as a function of position in the group (at the group's center or on its edge). **(B)** When ducks were at the edge of a group and had one eye open, they spent much more time looking away from the group's center than when they had one eye open and were at the center of the group. *(From Rattenborg et al., 1999a)*

How do mallards manage this split-brained sleep? It seems that they are capable of putting one hemisphere of the brain—the hemisphere active during sleep—into what is called slow-wave sleep (Rattenborg et al., 1999a, 1999b). *Slow-wave* here simply refers to the frequency of the brain waves that investigators record using a device called an electroencephalogram (EEG). Slow-wave sleep in birds has a signature (in terms of wave frequency and amplitude) that is quite different from other states of sleep or wakefulness. This slow-wave state allows quick responses to predators, but it does not interfere with the sleeping half of the bird's brain until danger is present. EEG recordings indicate that the part of the brain controlling the open eye during unihemispheric sleep showed the low-frequency range characteristic of slow-wave sleep, while the other half of the mallard's brain showed EEG patterns that were similar to those of true sleep.

Much of the work on unihemispheric sleep has been undertaken in birds, but they are not the only group in which this form of sleep has been documented. Although most mammals sleep with both halves of the brain asleep (bihemispherically), aquatic mammals are an exception (Figure 3.34). In aquatic mammals like dolphins, whales, fur seals, and sea lions, unihemispheric sleep is thought to allow individuals to swim to the surface and breathe during sleep

FIGURE 3.34. Sleep in aquatic mammals. In some aquatic mammals, like the fur seal, unihemispheric sleep is thought to allow individuals to swim to the surface and breathe during sleep. Here the fur seal is shown sleeping on its left side as its left hemisphere sleeps. The right hemisphere remains awake, controlling paddling of the left flipper and keeping its nostrils above water. When the right hemisphere sleeps, the left hemisphere controls paddling of the right flipper and breathing. *(From Rattenborg et al., 2000)*

INTERVIEW WITH

Dr. Geoffrey Hill

Why do you think so few ethologists design experiments to test both proximate and ultimate questions? Why do researchers tend to specialize in one or the other?

Increasing specialization has been a trend in Western culture for hundreds of years, and specialization in all endeavors is accelerating in the twenty-first century. Science is a reflection of society at large, and in modern science all researchers are specialists. Consider a scientist during the time of Darwin or Newton. Scientific meetings in the nineteenth century included researchers from all branches of science and every scientist was conversant and up to date on the latest discoveries in all fields of science. Part of the inspiration for Darwin's theory of natural selection was a book that he read that presented a new theory for geological processes. Today, scientists cannot realistically stay abreast of discoveries even within a single discipline like biology or chemistry. With the explosion of knowledge in past decades, a scientist has to spend all of his or her time just keeping up with the latest research in a very narrow subdiscipline like behavioral ecology. So, it is not surprising that many researchers who are focused on ultimate questions related to the function and evolution of behaviors do not engage in parallel lines of research investigating the proximate neural or hormonal mechanisms underlying the behaviors that they study. The techniques for one scientific approach are essentially unknown to individuals trained in another scientific approach.

A trend in modern science that helps to counteract the move toward increasing specialization is collaboration. Collaborative, multidisciplinary studies are drawing in more top scientists, and many funding agencies are actively encouraging multidisciplinary studies by providing special funding for such projects. Collaborative projects allow a topic to be studied from both ultimate and proximate perspectives by combining the skills and knowledge of scientists with different training. Such collaborative efforts can lead to breakthroughs in our understanding of behavioral processes.

What drew you to focus your research program on bird behavior?

Since I was a preteen I've been a bird nut. I got my first pair of binoculars when I was eleven and I've been chasing birds ever since. I'm very fortunate in being able to turn my hobby into my profession. I never seriously considered studying anything but bird behavior. As much as I enjoyed finding and identifying

different species of birds as a young bird-watcher, my greatest exhilaration came when I captured birds, put bands on their legs, and followed and chronicled their individual successes and failures. There is really nothing like walking onto a university campus where every individual house finch is wearing colored bands that I placed on their legs and slowly gaining an understanding of why the birds behave as they do.

Early on in my career, I made the conscious decision to give up some travel opportunities and focus on common local bird species that are logistically easy to study. At times I envy my colleagues who get to travel to exotic locations for their studies of bird behavior, but the exhilaration of having a tractable population of birds where I can conduct convincing tests of key behavioral hypotheses is more than compensation for travel opportunities lost.

Generally speaking, how do you think proximate and ultimate perspectives interact in helping us to better understand animal behavior?

I think we clearly need both ultimate and proximate explanations of all behavioral phenomena if we want to really understand them. The two levels of explanation do not work in opposition to each other—they reinforce one another. Evolution of behaviors can only proceed within the constraints of the underlying proximate mechanisms. Proximate mechanisms are shaped by evolution. An understanding of proximate control of behavioral traits inevitably leads to new insights about how they function and why they evolved. Conversely, an evolutionary prospective can guide studies about proximate mechanisms and make such mechanisms more comprehensible.

As an example of how ultimate and proximate studies reinforce each other, consider studies of red coloration in fish and birds. Explaining such red coloration was a focus of both physiologists and ethologists in the mid-twentieth century, but through several decades these two groups of scientists paid little attention to each other's work. During this period, ethologists showed that red coloration often functioned in mate attraction, and physiologists determined that red coloration of both birds and fish was commonly created by deposition of carotenoid pigments. Moreover, physiologists discovered that carotenoid pigments could not be synthesized by birds or fish but had to be ingested as macromolecules. Coming out of this era, scientists had both a proximate and ultimate understanding of red coloration, but a comprehensive understanding of red coloration eluded both groups.

A breakthrough in the understanding of red coloration (and ornamental traits in general) occurred when pigment physiology was united with observations of mating preferences and sexual selection theory. Observations and bits of information that seemed disconnected and inexplicable suddenly made perfect sense. Carotenoid pigmentation was a useful criterion in mate choice because carotenoid pigments cannot be synthesized. These pigments have to be ingested, and acquisition and utilization of carotenoid pigments could be interrupted by environmental conditions. Red pigmentation was a signal of success in dealing with environmental challenges. This conclusion could only have been reached by working from and uniting the foundations laid by initial proximate and ultimate studies.

Suppose, for the sake of argument, that a hypothetical "house finch genome project" was now complete. How would this sort of information affect the way you studied behavior in this species?

A finch genome would lead to breakthroughs in our understanding of many aspects of the behavioral ecology of house finches, such as the genetic basis for behaviors, the co-evolution of display traits and preferences for such traits, and the link between genetically based immunity and ornamentation. There is no house finch genome project in

> **I think we clearly need both ultimate and proximate explanations of all behavioral phenomena if we want to really understand them. The two levels of explanation do not work in opposition to each other—they reinforce one another.**

the works, but the genome of the domestic chicken has already been sequenced, and the first songbird genome—from the zebra finch—is currently in the works. The chicken genome is already very useful for interpreting genetic data from house finches, and a zebra finch genome will be even more useful, even though zebra finches are in a different songbird family than house finches.

Even without a house finch genome, my colleagues and I have seen great opportunities for testing behavioral hypotheses with new genetic tools, and I've been increasingly involved in genetic work with house finches. I do not ever actually get into the lab and do DNA work. My genetic studies are a great example of the sort of collaborative

research that I mention above. I provide data sets on behavior and reproductive history and blood samples as a source of DNA from my field studies, and my collaborators sequence the DNA and interpret the genetic information.

Over the next five years, what sort of advances in the proximate study of behavior do you envision?

We live in an exciting age of discovery. Breakthroughs in DNA sequencing and gene expression will revolutionize our understanding of the genetic control of many traits, including behavioral traits. In addition, equipment for all sorts of analytical analyses—from hormone assays to action potentials along individual neurons—will allow behaviorists to gain a firmer understanding of the mechanistic basis of behaviors. Truly exciting, but perhaps more than five years away, is the delivery of specific doses of hormones or neurotransmitters to specific tissues of the body or the instantaneous measurement of circulating hormones or the release of neurotransmitters. Such ability to manipulate the chemistry of the body or to measure short-term changes in body chemistry will rapidly advance our understanding of the proximate bases for behaviors.

As great a tool as these new analytical methods present, it will always be the creative application of technology by insightful researchers that leads to breakthroughs in our understanding of how nature works.

Dr. Geoffrey Hill is a professor at Auburn University. His work on the house finch is a classic example of employing both proximate and ultimate perspectives when studying animal behavior. This work is summarized in his wonderful book A Red Bird in a Brown Bag: The Function and Evolution of Colorful Plumage in the House Finch *(Oxford University Press, 2002).*

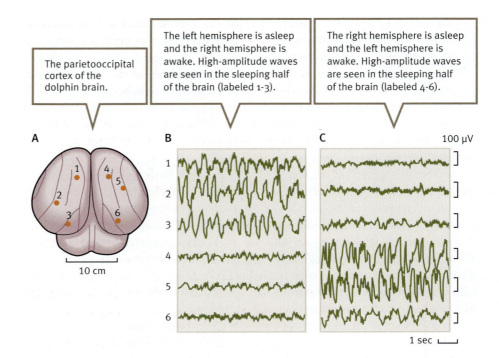

The parietooccipital cortex of the dolphin brain.

The left hemisphere is asleep and the right hemisphere is awake. High-amplitude waves are seen in the sleeping half of the brain (labeled 1-3).

The right hemisphere is asleep and the left hemisphere is awake. High-amplitude waves are seen in the sleeping half of the brain (labeled 4-6).

10 cm

100 μV

1 sec

FIGURE 3.35. Sleep in dolphins. EEG activity was measured in dolphins during unihemispheric sleep. *(From Mukhametov et al., 1988)*

(Lyamin and Chetyrbok, 1992; Mukhametov et al., 1988; Figure 3.35). Various techniques have been used across these studies to measure activity in the two sides of the brain, including brain temperature measurements, in which temperature was found to be higher in the "awake" side of an animal's brain than in the "sleeping" side (R. Berger and Phillips, 1995; Kovalzan and Mukhametov, 1982).

SUMMARY

1. One way to distinguish between proximate and ultimate perspectives is to remember that the former tends to address "How is it that . . . ?" and "What is it that . . . ?" questions, while the latter addresses "Why is it that . . . ?" questions. If we were, for example, examining vision in birds, a proximate perspective might lead us to ask how the visual system is set up or what the neurobiological and molecular underpinnings of vision are. An ultimate, or evolutionary, approach would focus on why the visual system is designed as it is. One of the most frequent errors made by those new to ethology is to confuse proximate and ultimate perspectives.

2. Proximate and ultimate approaches complement one another and together provide a comprehensive picture of the ethological trait under study.

3. The endocrine system, a powerful communication system that strongly influences many aspects of animal behavior, is composed of a group of ductless glands that secrete chemical messengers, in the form of hormones, directly into an animal's bloodstream. Correct functioning of the endocrine system is of fundamental importance to behavioral functions and to the modification of rates and directions of various cellular functions.

4. The nervous system provides an electrical impulse designed for instantaneous communication. Animals possess specialized nerve cells called neurons

that share certain similarities, regardless of what message(s) they conduct. Understanding how neurobiology affects behavior is an important component of proximate analyses of behavior.

5. The example of honeybee foraging behavior allows us to bring together numerous approaches to the proximate study of behavior—hormonal, neuronal, and other approaches—in a single system, and in so doing, provides us with a comprehensive understanding of the proximate underpinnings of a behavior that has fascinated ethologists since the time of Darwin.

DISCUSSION QUESTIONS

1. Go back and reexamine one of the behaviors discussed in Chapter 2, on ultimate causation. Suggest how you might go about studying this behavior from a proximate perspective, and how the ultimate and proximate perspectives together provide a richer understanding of whatever behavior you have chosen.

2. Find the 1998 special issue of *American Zoologist* (vol. 38), which was devoted to proximate and ultimate causation. Choose two papers in this issue, and compare and contrast how they try to integrate proximate and ultimate causation.

3. If you could design some behavioral endocrinology experiments to add to the integrative work on plumage and carotenoid food in finches discussed in this chapter, what sort of experiments would you design? How would these experiments complement what is already known?

4. How does the work on the neuroethology of learning in voles illustrate the way that proximate studies can use between-population and between-sex comparisons to help us better understand animal behavior?

5. The Conservation Connection box in this chapter describes how an endocrinological approach to animal behavior might inform questions in conservation biology. What might be an equivalent example regarding neuroethology and conservation biology?

SUGGESTED READING

Adkins-Regan, E. (1998). Hormonal control of mate choice. *American Zoologist*, 38, 166–178. A review of the endocrinology of mate choice.

Alcock, J., & Sherman, P. (1994). The utility of the proximate-ultimate dichotomy in ethology. *Ethology, 96*, 58–62. A short defense of the proximate/ultimate classification.

Bass, A. H., & McKibben, J. R. (2003). Neural mechanisms and behaviors for acoustic communication in teleost fish. *Progress in Neurobiology, 69*, 1–26. A nice illustration of how animal nervous systems play a critical role in behavioral decision making.

Capaldi, E., Robinson, G., & Fahrbach, S. (1999). Neuroethology of spatial learning: The birds and the bees. *Annual Review of Psychology, 50*, 651–682. A review of the neural underpinnings of spatial learning.

Donaldson, Z. R., & Young, L. J. (2008). Oxytocin, vasopressin, and the neurogenetics of sociality. *Science, 322*, 900–904. An overview of the neuroendocrinology of social behavior, with an emphasis on the meadow and prairie vole systems.

4

Molecular Genetics and Development

In Chapter 3, we delved into proximate analyses of animal behavior by focusing on endocrinology and neuroethology. In this chapter, we extend our examination of the proximate causes of behavior and discuss the ways that molecular genetics and development shed light on ethological questions. Studying genes from a proximate perspective may strike you as strange. Why are studies on molecular genetics considered proximate analyses? After all, most of the primer chapter on evolutionary approaches to behavior (Chapter 2) was devoted to discussing how natural selection favors one behavioral *genetic* variant over another. Genes, however, can also be used in the proximate explanation of a trait. If, rather than expounding on what *selective* forces are involved in changing allele frequencies, we study which specific allele or set of alleles code for a trait, then genes serve as a proximate causative factor. If we find that allele 1 is associated with a variant of behavior Y, and allele 2 with another variant of the behavior, we are casting genes in a proximate, rather than an ultimate, light.

Work on termite sociality is a good example of the way developmental and molecular genetic proximate approaches to behavior can complement each other. In most colonies of social insect species—bees, ants, wasps, and termites—the queen or queens in the colony produce virtually all the offspring in that colony. Some of these colonies can have thousands, even millions, of workers that do not produce offspring even though they are physiologically capable of doing so. Why? What causes such a division of labor (Smith et al., 2008)? We can address that question from both a proximate and an ultimate perspective. From the ultimate perspective, the question is why natural selection would ever favor this sort of reproductive caste system. We return to that question in Chapter 9. For our purposes in this chapter, the question is "What is the molecular genetic, proximate basis for workers that, during development, do not even attempt to reproduce when they are capable of doing so (Korb et al., 2009; Weil et al., 2007)?"

One clue to answering this question comes from what happens when the queen in a termite colony dies. Upon the queen's death, some workers become aggressive, butting one another often. After a series of such aggressive interactions, one of the workers becomes queen. Judith Korb and her team hypothesized that termite workers not reproducing was the product of queen-worker *chemical signaling* (Korb et al., 2009). When this signaling ceases, the butting behavior among the workers begins. A gene homologous with the termite gene *Neofem2* is known to be involved in queen-worker communication in other insect species (Cornette et al., 2003; Weil et al., 2007). Korb and her colleagues therefore focused their proximate analysis on this gene, which in termites appears to be involved in the production of chemical signals called pheromones that the queen produces.

Korb and her team used sophisticated molecular genetic techniques involving RNAi (RNA interference) to silence the expression of the *Neofem2* gene in queens. Silencing this gene had no effect on the behavior of the queens themselves—they behaved like their counterparts in control colonies in which *Neofem2* was not silenced in queens. But silencing the expression of this gene in the queen, and suppressing the chemical signals it codes for, did dramatically affect the behavior of workers, increasing their butting behavior toward one another (Figure 4.1). In addition, when the queen in a colony failed to produce the pheromones coded by the *Neofem2* gene, workers began a series of aggressive interactions that, in nature, would eventually lead to one of them reproducing. These results suggested that the suppression of worker reproduction by the

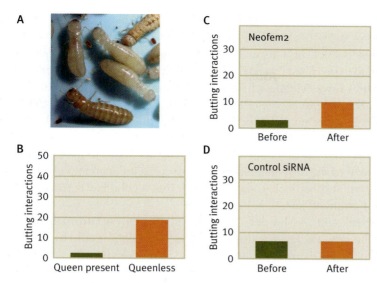

FIGURE 4.1. Genes, termite workers, and queens. (A) Queen *Cryptotermes secundus* (bottom left) with male (top right) and workers. **(B)** Butting among workers in colonies with queen and without queens. **(C)** and **(D)** Butting before and after treatment of the queen with *Neofem2* small interfering RNA (siRNA) and control siRNA. *(Photo credit: Korb, J., et al (2009). Science, 324, 758–758)*

queen was directly related to the pheromones associated with the *Neofem2* gene. This study is then a good example, from a proximate perspective, of the way the *developmental pathway* in termites is affected by the presence or absence (in our case, the silencing) of a *particular gene* in the queen of the colony.

Molecular Genetics and Animal Behavior

Identifying the genes associated with behavior is a large-scale endeavor going on in hundreds of labs around the world today. Researchers involved in this area have examined the molecular genetic underpinnings of traits such as foraging, mate choice, aggression, division of labor, and so on. Below we first look at the field known as **behavioral genetics** and subsequently examine four case studies involving proximate explanations at the molecular genetic level: (1) ultraviolet vision in birds, (2) song acquisition in birds, (3) foraging in birds, and (4) within-family interactions in voles.

We have already looked at one aspect of behavioral genetics in the heritability analyses discussed in Chapter 2. Here we will look at two other important ways that behavioral genetics contributes to the study of ethology:

▶ Using Mendel's laws of genetics to predict the distribution of behavioral phenotypes.
▶ Using quantitative trait loci (QTL) analysis to map the location of clusters of genes linked to behavioral traits.

As a consequence of Gregor Mendel's simple but brilliant experiments on inheritance in pea plants, behavioral geneticists have formulated what are known as **Mendel's laws**. Mendel's first law—the principle of segregation—states that individuals have two copies of each gene (or "factors," as Mendel called them), that such genes remain distinct entities, and that these genes segregate (that is, they are distributed) fairly during the formation of eggs or sperm. Mendel's second law—the law of independent assortment—states that whichever allele is passed down to the next generation at one locus is independent of which allele is passed down at other loci. Today, we know this second law is true only for what are called unlinked loci.

With respect to Mendel's first law, alleles can be **dominant**, meaning that at the genetic level a single copy of the allele is all that is necessary for a trait to be expressed, or they can be **recessive**, in which case two copies of an allele are necessary for the expression of a trait. To understand behavioral genetics and dominant and recessive alleles as they apply to animal behavior, let's look at the mating behavior of male ruff birds (*Philomachus pugnax*).

In ruffs, males display one of two behavioral strategies during the mating season (we shall return to these mating strategies in later chapters). "Independent males," which make up the majority of individuals in most populations, guard small mating territories. "Satellite males," in contrast, do not defend their own territories but instead temporarily share an independent male's mating arena and form a kind of alliance with independents, in which both individuals court simultaneously to attract the attention of females. Independent and satellite males differ not only in mating strategies but also in coloration and body mass; satellites are smaller and have lighter plumage than independents (Bachman and Widemo, 1999; Hogan-Warburg, 1966; Lank et al., 1995; Figure 4.2).

A

B

FIGURE 4.2. Satellite and independent ruff males. Some of the differences in the mating behavior of **(A)** satellite males and **(B)** independent males are controlled by a single gene with two alleles labeled S and s. *(Photo credits: BS Thurner Hof).*

David Lank found that these two alternative mating types—independent and satellite—are primarily controlled by a single gene with two alleles labeled S and s. The S allele is dominant and codes for satellite male behavior, whereas the s allele is recessive, and two copies of this allele are necessary for the development of an independent male. As such, SS and Ss males are satellites, and ss males are independents. It is remarkable that a single gene with two alleles codes for such a complex set of traits—mating strategy, size, and body mass. If we take the knowledge that mating behaviors in ruffs are coded by one gene with two alleles, and we combine that with a natural history–based understanding of mating preferences (which males with which genotypes are preferred as mates), we can make predictions about the distribution of SS, Ss, and ss genotypes over time. But keep in mind that the relationship between genotype and behavioral phenotype is often not nearly as simple as that seen in the case of ruff mating behavior. Many, if not most, behaviors have a much more complex underlying genetic structure.

LOCATING GENES FOR POLYGENIC TRAITS

In many cases, variation in more than a single gene is responsible for variation in behavior. Using a number of experimental and quantitative techniques, behavioral genetics can shed light on the basis of **polygenic** behavioral traits—traits associated with variation at more than one locus. For polygenic behavioral traits, behavioral geneticists often search for a *set* of genes, each of which contributes a small amount to the expression of the trait of interest. When researchers conduct such searches, they are looking for what are called **quantitative trait loci** (**QTLs**; Flint and Mackey, 2009).

First, let's walk through an overview of how QTL experiments are designed.

It can be difficult to identify the precise loci that are responsible for quantitative traits (so-called quantitative trait loci, or QTLs), but **QTL mapping** is a powerful way of finding at least the general region of the genome in which quantitative trait loci reside. The idea is that we can use *marker loci* that are easily assayed, but causally unrelated to the trait in question, in order to identify the approximate locations of the unknown alleles that affect the behavioral trait of interest. Figure 4.3 illustrates the basic concept behind the QTL mapping procedure.

Step 1. We typically begin the process by selecting two parental strains that (1) differ considerably in their values of the quantitative trait and (2) differ at a set of marker alleles. Parental strain 1 has a lower distribution of trait values than does strain 2; strain 1 is homozygous for the *A, B,* and *C* marker alleles, while strain 2 is homozygous for the *a, b,* and *c* marker alleles.

Step 2. The next step is to cross these two strains to produce a set of progeny. This new generation is referred to as the F_1 generation. If the parents are homozygous at the marker loci, these F_1 progeny will be heterozygous at each marker locus, and typically they will manifest intermediate values of the quantitative trait.

Step 3. The F_1 individuals are then mated to produce an F_2 generation. For the F_2 individuals, we measure (1) the genotypes at the marker loci, and (2) the value of the quantitative trait. From this information, we can infer which marker loci are most closely associated with QTLs for the behavioral trait in question. The F$_2$ generation in Figure 4.3 illustrates the basic logic behind this inference. In each frame, the quantitative trait values are plotted with the genotypes sorted according to one of the marker loci.

In Figure 4.3, we see a large difference in the quantitative trait values associated with the *AA*, *Aa*, and *aa* genotypes in the F$_2$ generation. This does not mean that the *A* marker locus is itself influencing the quantitative trait value, but it does imply that this locus is linked to an important quantitative trait locus.

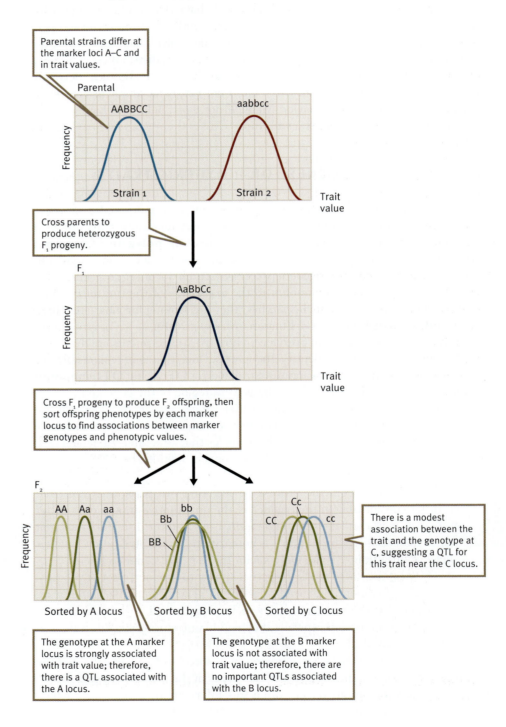

FIGURE 4.3. Quantitative Trait Loci Mapping. Quantitative Trait Loci (QTL) mapping allows researchers to find the general region of the genome in which quantitative trait loci reside by using *marker loci* that are easily assayed, but unrelated to the trait in question, in order to identify the approximate locations of the unknown alleles that affect the trait of interest. *(From Bergstrom and Dugatkin, 2012)*

Parental strains differ at the marker loci A–C and in trait values.

Cross parents to produce heterozygous F$_1$ progeny.

Cross F$_1$ progeny to produce F$_2$ offspring, then sort offspring phenotypes by each marker locus to find associations between marker genotypes and phenotypic values.

There is a modest association between the trait and the genotype at C, suggesting a QTL for this trait near the C locus.

The genotype at the A marker locus is strongly associated with trait value; therefore, there is a QTL associated with the A locus.

The genotype at the B marker locus is not associated with trait value; therefore, there are no important QTLs associated with the B locus.

Let's examine the work of Jonathan Flint and his colleagues on QTLs and fear/fearlessness in mice (Flint and Mackay, 2009; Flint and Mott, 2008; Flint et al., 1995; Solberg et al., 2006; Talbot et al., 1999; Yalcin et al., 2004). Flint's group first studied fear and anxiety by recording "open-field behavior" in mice. Open-field behavior measures fear expression when animals are placed in large, open, well-lit environments. Flint's group used two genetic lines of mice that had been bred under artificial selection for many generations: one line of mice had been selected for high open-field activity, and the other for low open-field activity. The behavior of mice from both lines was measured in open-field tests, and their fear/anxiety response was measured when they were placed in two different types of mazes. Fear was measured in a number of ways, including measuring the mice's activity level and the rates at which they defecated, because low activity and high defecation rates are associated with fear in rodents.

After all the behavioral tests were complete and the more fearful mice were identified, Flint and his colleagues collected DNA from the animals' spleens. Using a more complicated, and slightly different, version of the QTL analyses discussed above, Flint and his colleagues were able to identify QTLs for fear on six mouse chromosomes—chromosomes 1, 4, 12, 15, 17, and 18. QTLs for fear during open-field trials were found on these six chromosomes, while QTLs for fear displayed in mazes were found on only a subset of these chromosomes (1, 12, and 15). In follow-up studies, Flint's group studied 1,636 laboratory-bred mice and examined the fear response shown by mice in five different laboratory environments (open field, mazes, mirrored chambers, and so on). Across these five behavioral measures, researchers found evidence of QTLs associated with fear on fourteen chromosomes (Turri et al., 2004). More recent work has found QTLs linked to "emotionality," including fear and anxiety, on twenty chromosomes in mice (Willis-Owen and Flint 2006; Figure 4.4). Those sorts of analyses give us a much more detailed and in depth picture of how molecular genetic variation effects behavioral variation.

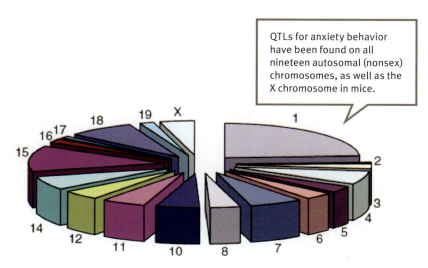

QTLs for anxiety behavior have been found on all nineteen autosomal (nonsex) chromosomes, as well as the X chromosome in mice.

FIGURE 4.4. QTLs for mouse behavior. Chromosomal distribution of QTLs linked to anxiety in mice. *(From Wills-Owen and Flint, 2006)*

GENES, MRNA, AND HONEYBEE FORAGING

Let's return to the proximate underpinnings of foraging in honeybee colonies that we discussed in Chapter 3. To examine the possible molecular genetic underpinnings of the developmental shift from working in the hive by younger bees to foraging outside the hive by older bees, Alberto Toma and his colleagues built on earlier studies indicating that the *period* (*per*) gene influences circadian rhythms and development time in fruit flies (Konopka and Benzer, 1971; Kyriacou et al., 1990). Because of these findings, they focused on the *per* gene as a "candidate" gene involved in control of the developmental changes in foraging that occur in honeybees (M. Fitzpatrick et al., 2005; G. Robinson et al., 2005). More specifically, they examined how levels of *per* messenger RNA (mRNA)—a single-stranded RNA that is critical for protein synthesis—might influence the developmental changes associated with the transition to forager (Toma et al., 2000).

Toma and his colleagues measured mRNA levels in the brains of three groups of laboratory-raised honeybees. The three groups were composed of bees that were four to six days old, seven to nine days old, and twenty to twenty-two days old. In addition, individuals in a group of one-day-old bees were marked and added to a natural colony of bees in the field. Researchers recaptured marked individuals at day 7 and day 24 of their experiment, and they measured *per* mRNA levels. In both laboratory and natural populations, *per* mRNA was significantly greater in older individuals that foraged for food and brought such food to their colony, when compared with younger bees that remained at the hive (Figure 4.5).

It is possible that increased *per* mRNA levels could be due to age differences alone, rather than to the age-related shift to forager. As in the Withers experiment examined in Chapter 3, Toma studied "precocious" foragers—foragers that begin

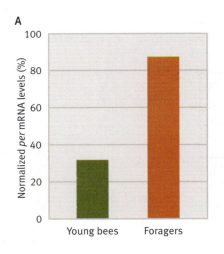
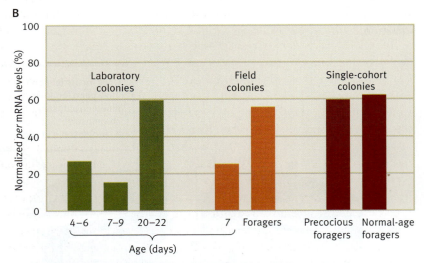

FIGURE 4.5. Foraging, age, and mRNA. (A) Foraging bees have significantly higher levels of *per* mRNA than younger, nonforaging bees. This difference could be due to age, behavior caste (forager versus nonforager), or both. **(B)** Some individual bees developed into precocious foragers that began searching for food much earlier than usual. When ten-day-old precocious foragers were compared with normal foragers (twenty-two-day-old foragers), Toma and his colleagues found no statistical differences in *per* mRNA levels. *(From Toma et al., 2000)*

searching for food outside of the nest at about seven days old. These precocious foragers provided Toma and his colleagues with the chance to remove age effects from the *per* mRNA/foraging connection, as they now had a sample of bees that foraged but were much younger than typical food gatherers. Precocious foragers had *per* mRNA levels that did not differ from those of typical (older) foragers, suggesting a link between *per* mRNA and foraging, as opposed to a more general connection between *per* mRNA and development. We still do not know the direction of the causality here: Is increased *per* mRNA level a cause or a result of increased foraging? But the correlational link is well established (Bloch et al., 2001, 2004).

Over the last few years, with the explosion of molecular genetic technology, a much more detailed picture of genes, mRNA abundance in the brain, and foraging in honeybees has emerged. For example, in a large-scale study of 5,500 genes, Charles Whitfield, Anne-Maire Cziko, and Gene Robinson found that changes in mRNA levels associated with 39 percent of these genes were involved in the transition from hive work to foraging behavior in honeybees (Whitfield et al., 2003, 2006). In addition, QTL analysis has been used to localize genes associated with age at first foraging to chromosomes 4 and 5 (Rueppell, 2009).

These sorts of large-scale genomic approaches to animal behavior will become common as the genomes of more species are sequenced (G. Robinson et al., 2005; S. Sumner, 2006). The challenge will be figuring out how to go from the massive amount of data gathered when a genome is sequenced to understanding any specific animal behavior. For example, although the field of genomics and behavior is growing quickly, at present we can't examine each of the thousands of genes and mRNA products associated with the transition to honeybee forager. Nor would we want to, as little is known about exactly how these genes and their mRNA products operate in the transition to honeybee forager.

We can, however, examine one more well-studied gene associated with the transition to foraging in honeybees to give us a finer understanding of the proximate underpinnings of social insect foraging. Here we discuss the work of Yehuda Ben-Shahar, Nichole Dudek, and Gene Robinson, who examined the effect of the gene *malvolio* (*mvl*) on manganese transport to the honeybee brain, and its implications for foraging in this species (Ben-Shahar et al., 2004). The gene *mvl* is an excellent candidate gene for this sort of work, as earlier experiments have shown that it affects the way fruit flies respond to sucrose (an important component of honeybee food and drink).

In honeybees, foragers that specialize in collecting pollen have a higher responsiveness to sucrose than those that specialize in nectar foraging, and both types of foragers have a stronger response to sucrose than younger "nurse" bees (that feed the larvae) in the hive. Ben-Shahar and his colleagues examined whether such differences were linked to differences in *mvl* transport of manganese to the honeybee brain. They found that both the amount of manganese in the head of a honeybee and the amount of *mvl* mRNA in the honeybee brain were high in pollen foragers and nectar foragers, and low in nurses.

On a finer scale, when Ben-Shahar and his team looked at pollen foragers versus nectar foragers, they found even more evidence suggesting a link between *mvl* and foraging. Recall that pollen foragers show a stronger response to sucrose than do nectar foragers. If *mvl* plays an important role in foraging, we would expect to see more manganese in the heads of pollen foragers versus nectar foragers, and the evidence suggests that this is the case (Figure 4.6). Further evidence for the

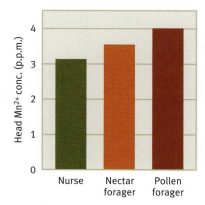

FIGURE 4.6. Pollen foragers, nectar foragers, and manganese. Pollen foragers have more manganese in their heads than both nectar foragers and nurses. *(From Ben-Shahar et al., 2004)*

proximate link between *mvl* and foraging can be found in the fact that when bees were experimentally treated with manganese, their response to sucrose increased, and they made the transition from nurse to forager at an early age.

ULTRAVIOLET VISION IN BIRDS

Animals, including humans, experience the world through sensory input. Humans rely heavily on visual input to make sense of the environment, but we only use some of the visual spectrum for input. We do not pick up the ultraviolet (UV) part of the spectrum, whereas many other species do, including some fish, amphibians, reptiles, birds, insects, and other mammals (G. Jacobs, 1992; Salcedo et al., 2003; Shi et al., 2001; Yokoyama, 2002). Ultraviolet vision in vertebrates is determined by retinal visual pigments and is used in the context of mating, foraging, hunting, and social signaling (Bennett et al., 1996; Church et al., 1998; Fleishman et al., 1993; Tada et al., 2009; Viitala et al., 1995). For example, female zebra finches use ultraviolet wavelength in determining which males they select as their mates (Church et al., 1998). Until recently, however, little was known about UV vision's proximate underpinnings. What allows some organisms, but not others, to hunt, signal, and attract mates in this wavelength?

Using molecular genetic cloning techniques, Yokoyama and his colleagues examined UV vision in zebra finches (*Taeniopygia guttata*; Yokoyama et al., 2000; Figure 4.7). Zebra finches are normally able to see in UV, but Yokoyama and his team transformed the ultraviolet pigment—the pigment that allows the birds to see in UV—into a violet pigment that doesn't allow for UV vision. With the change of a single amino acid, ultraviolet light perception was gone and violet perception was present. But that is only half of the story.

Even more remarkably, Yokoyama and his colleagues performed the reverse experiment using chickens and pigeons, neither of which possesses ultraviolet pigments (Kawamura et al., 1999). Using sections of DNA that were very similar to those found in the finch, Yokoyama and his colleagues were able to generate ultraviolet pigments from violet pigments, again by changing a single amino acid. One change in amino acid, and pigeons and chickens were able to see in UV wavelength via ultraviolet pigments.

FIGURE 4.7. A small change goes a long way. A single amino acid change is all that separates the coding for ultraviolet versus violet perception in the zebra finch. *(Photo credit: Ann and Rob Simpson)*

Following the work of Yokoyama and his associates, a number of studies have further elucidated the molecular genetic, proximate underpinnings of ultraviolet vision in animals. For example, evidence now suggests that the molecular genetic underpinnings of UV vision in invertebrates such as fruit flies can also be tied to changes in a single amino acid (Salcedo et al., 2003), and that in some birds in which the ancestral species had lost the UV pigment, some descendant bird lineages regained UV pigment by a single amino acid change (Odeen et al., 2011, 2012; Shi and Yokoyama, 2003).

As a whole, this set of work not only demonstrates the power of molecular genetics to help unravel proximate questions in animal behavior but also illustrates that what appears to be a complex trait like ultraviolet vision—a trait that functions in the context of mating, foraging, and so on—can originate from relatively simple changes at the molecular level.

SONG ACQUISITION IN BIRDS

Songbirds use their songs in many contexts, including attracting potential mates and fending off intruders from their territory. A molecular genetic approach to birdsong highlights the role of gene expression and gene regulation—the "on and off switches" that control the production of proteins by a specific allele or set of alleles. For example, researchers have found that the expression of the *FOXP2* gene in brain regions is associated with both song perception in birds and language acquisition in humans (Enard et al., 2002; Haesler et al., 2004; Teramitsu et al., 2004). Experimental work in young zebra finches has found that when the *FOXP2* gene is "knocked out," that is, when it is deactivated, their ability to copy the song of adults is severely impaired (Haesler et al., 2007; Figure 4.8).

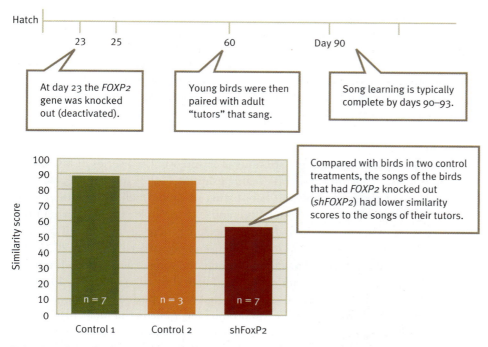

FIGURE 4.8. *FOXP2* **and song learning.** When the *FOXP2* gene was knocked out in young zebra finches, their ability to learn other finches' songs was diminished. *(From Haesler et al., 2007)*

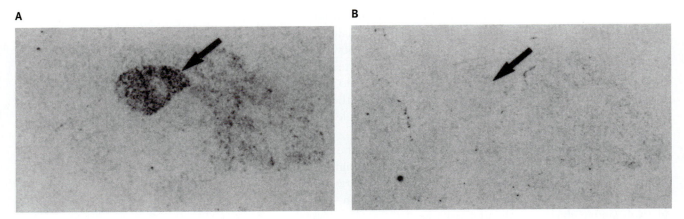

FIGURE 4.9. ZENK and exposure to song. (A) Induction of *zenk* mRNA in the forebrain of a male zebra finch that has been exposed to zebra finch song for forty-five minutes. **(B)** Same area in male that was not exposed to song. *(From Mello et al., 1992, p. 6819; courtesy Claudio Mello)*

The zebra finch (*Taeniopygia guttata*) has become a model system to illustrate the relationship between gene expression and birdsong. Early work in this species by David Clayton, Claudio Mello, and their colleagues involved exposing zebra finches to birdsong, and then measuring mRNA levels in the brains of these birds (Mello et al., 1992). Clayton and his colleagues focused on mRNA levels in a section of the forebrain called the neostriatum, because this section of the brain is associated with song pattern recognition, song discrimination, and the processing of auditory cues in birds. They found that mRNA levels associated with a gene called *zenk* increased after the birds heard zebra finch songs, and that increase was associated with an increase in the number of neurons in the neostriatum (Figure 4.9). Other research also supports the role of *zenk* in zebra finch birdsong: (1) zebra finches exposed to the song of another species (the canary) showed a much-reduced *zenk* mRNA response to the other species' birdsong compared with their responses to a conspecific's song, and (2) no increase in *zenk* expression was discovered either after the birds were placed in a "no song" experimental treatment, or when they heard "tone bursts" that did not correspond with the song of any bird species.

If a zebra finch is exposed to the same conspecific song repeatedly, all of the strong responses described above begin to decrease and slowly return to their baseline levels—that is, to the levels seen before exposure to the song for the first time (Figure 4.10A). This sort of **habituation** is often seen when an animal is exposed to the same stimulus over and over. In a series of fascinating experiments, Clayton and his team examined the interaction between expression of the *zenk* gene and habituation of zebra finches to familiar song. To do so, Mello, Fernando Nottebohm, and Clayton first exposed a zebra finch to the song of a conspecific and found the same increase in *zenk* mRNA and the number of neurons in the forebrain discussed earlier (Mello et al., 1995). Once a bird had habituated to a familiar song, the song of a new zebra finch was played, and an increase in *zenk* mRNA and neural recruitment to the forebrain was again recorded. These results suggest that after repeated exposure, a song is categorized by a bird as "familiar," and it no longer elicits the molecular genetic and neural changes associated with increased *zenk* expression (Figure 4.10B), but a new song will generate new molecular genetic and neural changes.

It is still not clear precisely how the increased *zenk* mRNA levels and neuron development associated with *zenk* expression are tied to song learning.

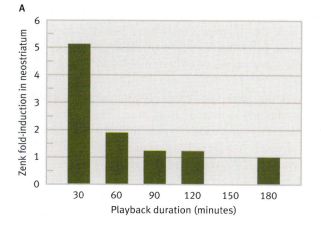

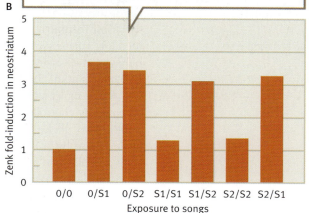

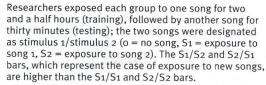

Researchers exposed each group to one song for two and a half hours (training), followed by another song for thirty minutes (testing); the two songs were designated as stimulus 1/stimulus 2 (0 = no song, S1 = exposure to song 1, S2 = exposure to song 2). The S1/S2 and S2/S1 bars, which represent the case of exposure to new songs, are higher than the S1/S1 and S2/S2 bars.

The protein produced by the *zenk* mRNA may affect the auditory neurons, which are connected to song recognition. More generally, there is some evidence that zenk may be part of a complex genetic pathway leading to the neural plasticity discussed earlier in the chapter—neural plasticity that is critical to song learning in birds (Mello et al., 2004). Indeed, the entire genome of the zebra finch has been sequenced, and researchers have discovered changes in gene expression in at least 807 genes in the zebra finch brain when males sing (Warren et al., 2010; Figure 4.11).

FIGURE 4.10. *Zenk* levels and habituation. (A) Induced *zenk* mRNA levels decreased with increased exposure to the same song in male zebra finches. **(B)** Induced *zenk* mRNA levels increased when a male zebra finch was exposed to a new song. *(Adapted from Mello et al., 1995)*

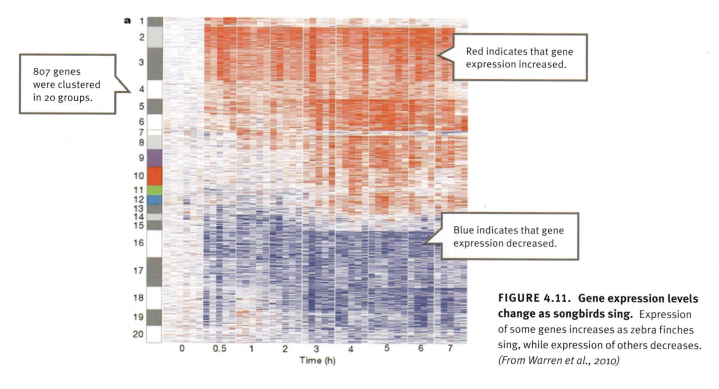

807 genes were clustered in 20 groups.

Red indicates that gene expression increased.

Blue indicates that gene expression decreased.

FIGURE 4.11. Gene expression levels change as songbirds sing. Expression of some genes increases as zebra finches sing, while expression of others decreases. *(From Warren et al., 2010)*

AVPR1A, VASOPRESSIN, AND SOCIALITY IN VOLES

Chapter 3 discussed the critical role that the hormone vasopressin plays in prairie voles (*M. ochrogaster*), a species in which males often display parental care and guard their mates, and meadow voles (*M. pennsylvanicus*), a species in which males display very little, if any, parental care or prosocial behavior toward their mates. Here, we return to this system but focus on *within*-species variation in prairie voles. Though they display much more parental care than meadow voles, some male prairie voles display more of this type of prosocial behavior than other males in the same population (Hammock and Young, 2002; R. Roberts et al., 1998).

At the molecular genetic level, the expression of vasopressin and of vasopressin receptors is controlled by a gene known as *avpr1a* (Insel et al., 1994; Lim et al., 2004). Two alleles of this gene—the long-version allele and the short-version allele—have been the subject of much research. Early work, both within prairie voles and between prairie and meadow voles, suggested that the long version of the *avpr1a* allele was associated with prosocial behaviors like parental care and affiliative interactions with mates (Phelps and Young, 2003). To examine this connection in more detail, Elizabeth Hammock and Larry Young bred two lines of prairie voles— one line was homozygous for the long version of the *avpr1a* allele, and one line was homozygous for the short version of the *avpr1a* allele (Hammock and Young, 2005).

If the length of the *avpr1a* allele was primarily responsible for male behaviors toward mates and offspring, then males from these two lines should display different suites of social behaviors. That is in fact what Hammock and Young found. Males homozygous for the longer version of *avpr1a* displayed more pup licking and grooming of pups, and responded more positively toward familiar females than males that were homozygous for the shorter version of *avpr1a* (Figure 4.12). These studies complement those discussed in Chapter 3 and provide another piece to the puzzle regarding what proximate factors are responsible for key behavioral aspects of sociality in voles. So far, we have evidence on these behaviors from a hormonal, neuroethological, and molecular genetic perspective. In the next section, we examine the role of development in shaping vole sociality.

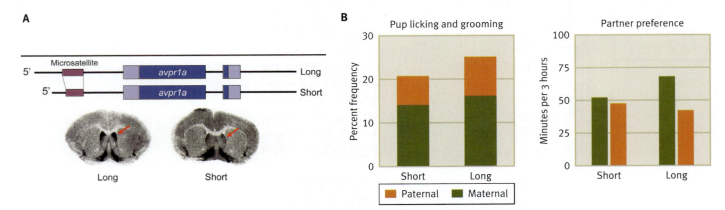

FIGURE 4.12. *Avpr1a* **length and behavior.** **(A)** A schematic of the long and short *avpr1a* alleles and the expression of vasopressin receptors in the brain of prairie voles (differences in expression can be seen at the two red arrows). **(B)** Both male care of offspring and the strength of partner preference were greater in males who were homozygous for the longer *avpr1a* allele. *(From Donaldson and Young, 2008; Hammock and Young, 2005)*

Development and Animal Behavior

As discussed in Chapter 1, in his classic paper "On the Aims and Methods of Ethology," Niko Tinbergen emphasized the importance of developmental factors in shaping an animal's behavior (N. Tinbergen, 1963). Tinbergen and those who followed in his footsteps were using the term *development* in a broad sense to encompass everything from the in utero effects in rodents discussed in Chapter 3 to the specific effects that environmental factors might have throughout an individual's life (Stamps, 2003; M. West et al., 2003). Environmental factors relevant to development and behavior encompass both abiotic (nonliving) and biotic (living) factors.

Throughout this book, we will explore myriad ways that development affects behavior. Here, we focus on four case studies, examining (1) family structure and the development of vole social behavior; (2) development of cichlid fish in the nest; (3) early development and its effect on parental behavior in the oldfield mouse; and (4) development, temperature, and ovipositing behavior in wasps.

DEVELOPMENT, TEMPERATURE, AND OVIPOSITING BEHAVIOR IN WASPS

One abiotic factor that often has profound developmental consequences for animals is temperature. Animals that are native to colder climates often show developmental differences when raised in warmer environments, and vice versa. Temperature, for example, is known to have strong effects on olfactory senses in insects (Herard et al., 1988), and because olfaction is a primary way in which insects interact with both abiotic and biotic factors in their environment, developmental differences caused by temperature have become an area of interest in ethology (see the Conservation Connection box for more).

Joan van Baaren and her colleagues examined temperature-related developmental changes that occur in *Anaphes victus*, a parasitoid wasp species. Parasitoids usually lay their eggs inside a host species, and adult females learn how to avoid hosts that have already been parasitized (Papaj and Lewis, 1993; Papaj et al., 1989). Van Baaren and her colleagues tested whether the temperature that female larvae were exposed to during development would affect their ability to learn how to find suitable hosts when they themselves were ready to lay eggs (van Baaren et al., 2005). Conceptually, here we are considering the manner in which the environment interacts with development via an animal's ability to learn, which then affects how parasitoids infect their hosts.

To examine the effect of temperature on learning and host choice behavior, van Baaren's team raised larvae at 4° C in two experimental treatments— one of which lasted three weeks and one of which lasted twelve weeks. In addition, they ran a control treatment in which larvae were not exposed at all to a temperature of 4° C. Afterward, female larvae were raised at a more typical temperature for this species (24° C). Their ability to find suitable hosts was later examined when these females were ovipositing (laying) their own eggs (Figure 4.14). When tested, females were presented with three different

Development, Dispersal, and Climate Change

For many years, ethologists assumed that dispersal behavior—behavior associated with moving from one habitat to another—was fixed, in the sense that animals did not use environmental cues to change the type of dispersal behavior they used. More and more evidence from insects, however, suggests that animals have more than one technique for dispersal, and that they sometimes use environmental cues they experience during development to choose which dispersal strategy to adopt (Ronce, 2007).

Two dispersal strategies seen in many species of spiders are rappeling and ballooning behavior. The spider *Erigone atra* uses silk threads for both rappeling and ballooning dispersal behavior. In ballooning behavior, the spiders rely on the silk threads to sail long distances, often hundreds of yards. When rappeling, spiders use silk thread to create a bridge that they can move along for short dispersals (Figure 4.13).

The life cycle of *E. atra* involves colonizing and then breeding in crop fields to which they migrate in the early spring. Dispersal in the spring is not especially risky, in the sense that crop habitats are often large with many available areas for breeding. In the fall, these spiders migrate to noncrop areas, where they may breed again and then spend the winter (Toft, 1995). Mortality risks are higher during cooler fall migrations to noncrop areas (Bonte et al., 2003, 2006). Bonte and his colleagues predicted that spiders would use short-distance rappeling behaviors more often during the spring migrations, when potential territories are abundant, and then switch to ballooning dispersal—a riskier strategy, in which

FIGURE 4.13. Dispersal strategies in *Erigone altra*. These spiders use temperature as a development cue for when to use risky (ballooning) versus less risky (rappeling) dispersal strategies. *(Photo credit: TDP Invertebrate Surveyors and Consultants, Trevor Pendleton)*

spiders have little control where they land—during the cooler fall migration when fewer suitable spider habitats are available. During the fall migration, the chances of getting a suitable habitat close by others is small enough that ballooning to distances far away may be worth the costs of having little control where a landing will occur. Bonte and his colleagues further hypothesized that spiders would employ temperature as the environmental cue they used during development to decide which strategy to employ. When the researchers examined the dispersal choices of spiders raised under controlled temperature conditions, they found that, as predicted, spiders that were raised at higher temperatures (similar to those they would experience in the spring) were most likely to rappel; spiders that were raised at lower temperatures

(similar to those they would experience in the fall) were most likely to balloon (Bonte et al., 2008).

If the choice of dispersal strategy can differ depending on various temperature cues during development, think for a moment how spiders' dispersals might be affected by climate change. If animals use temperature as reliable cues for selecting how to disperse in both the spring and the fall, then climate change induced by humans—change that is unlikely to reflect other natural environmental changes—might induce the wrong dispersal choice (during either fall, spring, or both), in the sense of leading an individual to disperse using a strategy that is not beneficial in that environment. On a larger scale, Bonte and colleagues argue that if enough animals made such choices, population sizes might decline, potentially causing species-level extinction.

patches, which differed in the number of good hosts (unparasitized) and "bad" hosts (already parasitized). Van Baaren and her colleagues then measured the number of eggs the females laid and the females' ability to learn to avoid already parasitized hosts.

The results indicated that exposure to cold temperature during development had significant effects on (1) the number of eggs a female laid inside a host and (2) the female's ability to discriminate among hosts of different quality. Equally interesting is the finding that exposure to cold temperature had a strong negative effect on the speed at which females *learned* to avoid already parasitized hosts. Females from low-temperature regimes would usually reject parasitized hosts after injecting their ovipositor into such hosts, but compared with females raised at higher temperatures, they fared poorer at discriminating hosts through use of the external cues on the host after it had been parasitized. Such external cues were learned more quickly by females raised at higher temperatures.

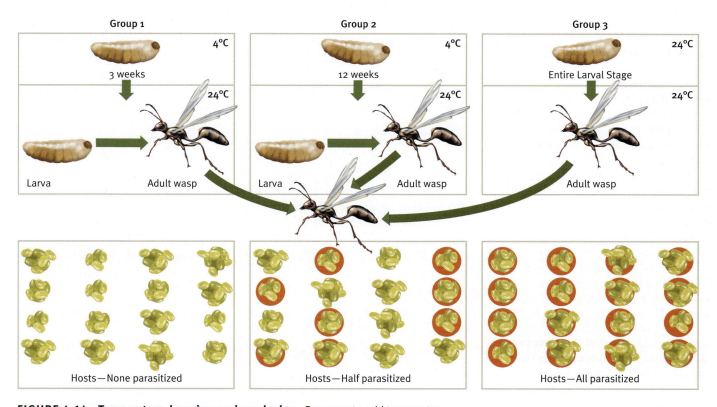

FIGURE 4.14. Temperature, learning, and egg laying. Exposure to cold temperature (4° C for three weeks for larvae in group 1, 4° C for twelve weeks for larvae in group 2) during development had significant effects on a female wasp's ability as an adult to discriminate between hosts of different quality. Larvae in group 3, which were continuously raised at normal temperatures of 24° C, did not experience these effects. Exposure to cold temperature also had a strong negative effect on the speed at which females learned to avoid already parasitized hosts, as they had to choose to lay their eggs in patches in which none of the hosts had been parasitized (left), half had been parasitized (center), or all had been parasitized (right). Orange circles indicate a parasitized host. *(Based on van Baaren et al., 2005)*

FAMILY STRUCTURE, DEVELOPMENT, AND BEHAVIOR IN PRAIRIE VOLES

One of the most salient features of an animal's development is the family social environment in which it is raised. In particular, the type of parental care received and the frequency with which it is obtained can have important consequences later in life. Here we look at two case studies that center on development and interactions (or lack of interactions) with parents and other adults. We first examine this aspect of development in prairie voles, and then in cichlid fish.

In some species, such as prairie voles, there is significant variation in family structure. For example, Getz and Carter (1996) found that in the prairie vole population they were studying, about one-third of pups were raised by only their mother, one-third were raised by their mother and father, and one-third were raised in communal nests where they received care from their mother, their father, and other adult nest mates.

Ahem and Young (2009) wanted to understand how this variation in parental care affected the behavioral development of voles. In particular, they examined the amount of care a pup received when raised by a mother alone (labeled SM for single mother) versus when raised by a mother and father (labeled BP for biparental), and tracked whether the care an individual received affected its own parental behavior later in life. Pups raised in the SM group were left in their nest alone (no parent present) more often than pups in the BP care group, and BP pups received significantly more grooming and licking behavior than SM pups (Figure 4.15A). The amount of grooming and licking provided by mothers in the SM group was approximately the same as that provided by mothers in the BP group; group differences were due to the absence of a male and any licking or grooming he might have provided. When pups from the SM and BP groups matured, Ahem and Young found significant differences in their social behaviors. SM females licked and groomed their pups less than BP individuals (Figure 4.15B). Both males and females from the SM group took longer to find a mate and bond with that individual than did males and females from the SM group. Differences in early development—being raised by a single parent or by two parents—had long-term effects on the ontogeny of vole social behavior.

FIGURE 4.15. Amount of licking and grooming differed in one- versus two-parent nests. (A) In prairie voles, pups in the biparental (BP) treatment received more licking and grooming than pups in the single-mother (SM) treatment. **(B)** After they matured, females that were raised by a male and female (BP-reared) displayed more licking/grooming and pup care than females raised by only a female (SM-reared). These females also spent less time away from their pups than females raised by only a female. *(From Ahem and Young, 2009)*

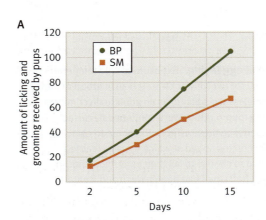

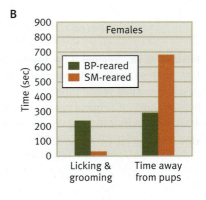

EARLY NEST DEVELOPMENT AND
BEHAVIOR IN CICHLID FISH

Similar sorts of experiments to those just described in prairie voles have been undertaken in other species. For example, to those just described in Prairie Voles Cornelia Arnold and Barbara Taborsky studied the early nest environment and its long-term behavioral effects in the cooperatively breeding cichlid, *Neolamprologus pulcher* (C. Arnold and Taborksy, 2010). In this species, the young in nests are raised not only by adults but also by older siblings that act as "helpers-at-the-nest," aiding adults in rearing the latest clutch of offspring (Chapter 9). Adults and helpers defend the nest against predators and remove parasites from eggs and developing fry, but they rarely interact directly with developing young. Arnold and Taborsky hypothesized that the presence of parents and helpers-at-the-nest may provide younger individuals with behavioral skills that are beneficial in a species that lives in complex social groups, by (1) freeing up time for developing offspring to interact with others, rather than to be vigilant for predators; and/or (2) serving as role models for the developing offspring, which could copy their actions.

To test their hypothesis, the researchers raised newly born fish in groups together. These groups differed in terms of what other fish were present at the nest. In one treatment, no adults were present, in a second treatment an adult male and female were present, and in a third treatment an adult male and an adult female plus helpers-at-the-nest were present. In all cases, when the subjects matured, individuals from different treatments were tested in a series of competition experiments with one another. For example, an individual might be given time to establish a territory and then have ownership of that territory challenged by another individual. What Arnold and Taborsky found was that fish raised with either adults or adults and helpers displayed behaviors that were less costly in terms of energy but were nevertheless very effective in defending their territories compared with fish that were raised in the absence of adults or adults and helpers. It remains unclear whether this difference was due to young copying the behavior of older individuals at the nest or simply having more time to interact socially with one another when there were adults present, but in either case it is clear that the early environment experienced by these fish had important consequences for behavioral decisions later in life.

EARLY DEVELOPMENT AND ITS EFFECT ON
PARENTAL BEHAVIOR IN THE OLDFIELD MOUSE

Animals often become better parents as they produce more and more offspring. Direct experience as a parent, however, is only one way to learn how to become a successful mother or father. Developmental factors early in life can also potentially affect future parental behavior. For example, in many species of birds and mammals, some individuals remain in their natal group even after they are capable of reproduction, and they often help their parents raise additional broods of offspring (as in the cichlid example above). Is it possible that such developmental experience may affect subsequent parenting success in helpers that eventually leave their natal territory? Susan Margulis and her colleagues examined this possibility in oldfield mice (*Peromyscus polionotus*).

Data on helping behavior among oldfield mice in natural settings are difficult to gather, but a significant body of indirect evidence suggests that some females

Dr. Gene Robinson

Why did you decide to work on honeybees?

I first became fascinated with honeybees at the age of eighteen, when I took time off from college to work on a kibbutz in Israel. One day I was asked to help the kibbutz beekeeper with the bees, and from that very first day I was struck by the combination of order and chaos in their behavior. Because of that experience, I decided to major in biology, with a focus in entomology. After college, I worked as an apiary inspector, a queen breeder, and a bee-keeping trainer overseas. It was in the course of my doctoral research that I started to study the division of labor in the hive, first from an endocrinological perspective and then, as a postdoc, from a genetic perspective. Since then, one of the main goals of my research has been to determine the various mechanisms that control the highly ordered yet highly plastic division of labor among honeybees.

Why sequence the genome of the honeybee?

Honeybees have proved to be an ideal model system for studying complex animal society. The honeybee system is at once both organized and flexible, depending on the needs of the colony. Research has already shown how the division of labor is influenced by various factors including hormones, pheromones, and environmental changes. It is also clear from my research that bees of different genotypes have different behavioral tendencies, and this also contributes to the structuring of the division of labor.

To further these studies, I decided in 1998 to begin laying the groundwork to sequence the honeybee genome. A sequenced genome provides many forms of knowledge to help in mechanistic and evolutionary analyses of social behavior. We have

used several, including transcriptomic analyses of all the genes active in the brain during the performance of specific behaviors. For example, genetic research has allowed us to discover that hive bees and foragers differ in approximately 40 percent of their brain gene expression and that these differences arise from hereditary, environmental, and physiological effects. We also have partially sequenced the genomes of ten other species of bees, some social

and some with more solitary lifestyles, and employed molecular evolution analyses. We found a particularly strong signature of selection for genes related to metabolism. In other words, genes involved in regulating basic metabolic processes appear to have been shaped by natural selection during the evolution of complex social life in the bees.

What is the most difficult part of taking what you learn at the genomic level and applying it to the study of behavior?

There are many levels of biological organization in the brain that stand between the genome and behavior. Understanding how changes at the genomic level give rise to changes

in the brain that ultimately cause changes in behavior is a key challenge.

What has been your basic strategy when trying to understand the complex developmental pathways we see in honeybees?

There are two key elements to our strategy. On the behavioral side, our strategy has been to start with the behavior as it occurs naturally in the field and then use robust assays to capture the natural behavior in such a way that it is amenable to genomic analysis. On the genomic side, we have generally applied the candidate gene approach to either developing hypotheses about individual genes or interpreting results from large-scale transcriptomic analyses. That is, we have used knowledge of gene function from other species or knowledge of bee physiology to guide our study of genes. However, we also maintain an open mind and are occasionally surprised by results that implicate a family of genes or a biological process that we had not previously targeted. For example, we have implicated changes in brain metabolism, especially a down-regulation of genes in the oxidative phosphorylation pathway, as being critical to increases in arousal and aggression, and we would not have predicted that.

What is the next big question you and your lab are going to tackle?

We are trying to understand the transcriptional architecture underlying socially regulated gene expression in the brain and how it has evolved from the architecture underlying solitary behavior.

Dr. Gene Robinson is a professor at the University of Illinois at Urbana-Champaign and a member of the National Academy of Sciences. He is one of the world's foremost experts on the honeybee.

remain at the nest and help their mothers raise the next litter of young. For example, natural history data suggest that a mother can be both pregnant and nursing a brood of mice, while an older litter still remains at the nest, providing ample opportunity for potential helpers to aid in rearing their younger siblings (Foltz, 1981). Margulis tested whether "experienced" females—that is, females that remained at their parents' nests during the rearing of a litter of their younger siblings—were subsequently better mothers to their own offspring than "inexperienced" females that did not remain at the nest while a younger litter of siblings was being reared (Margulis et al., 2005). That is, they examined whether a developmental trajectory involving helping one's mother affected the helper's own parenting behavior.

Margulis and her colleagues experimentally created inexperienced and experienced females by removing (or not removing) females from their natal nests. They began their work using a large colony of oldfield mice housed at the Brookfield (Illinois) Zoo, and they used mice that were ten to fifteen generations removed from wild-caught individuals. Margulis and her team formed a series of male-female pairs. Quickly thereafter, mating occurred, and pups were born. In the "inexperienced female" (IF) treatment, IF females were removed from the nest at twenty days of age, and they were raised in all-female groups until the experiment began. In the "experienced female" (EF) treatment, when EF females were twenty days old, they were *not* removed from the nest but rather remained at the nest with their pregnant mother until she gave birth again, and nursed and weaned a second brood. At that point, EF females were removed from the nest and reared in an all-female group until the experiment began.

At the start of the experiment, IF and EF females were paired with inexperienced males, with whom they mated and produced offspring, and the females' parental activities and offspring survival were recorded. The results suggested that all females—both EF and IF—became better parents as they produced more and more litters over time, but the key comparison was between inexperienced and experienced females at any given point in time. Here, Margulis and her colleagues found that the litters of experienced females had a higher probability of survival than those of inexperienced females, in part because of the superior nest-building behavior displayed by experienced females (Figure 4.16).

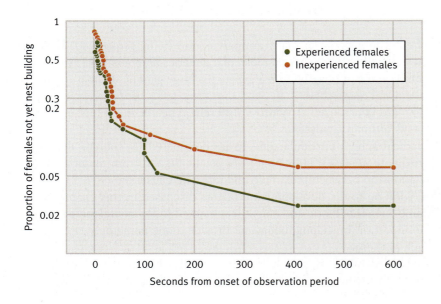

FIGURE 4.16. Nest building and experience in oldfield mice. The proportion of females that did not start building nests was lower in experienced females (green) versus inexperienced females (orange). Experienced females began to build nests sooner and built superior nests than inexperienced. *(From Margulis et al., 2005)*

The results indicate that the developmental experience of being present when one's mother raises a subsequent clutch of offspring has long-term consequences for parenting abilities.

SUMMARY

1. Molecular genetics can be a powerful tool in proximate analyses of behavior. Researchers can use genes in the proximate explanation of a trait by studying which specific gene or set of genes codes for a trait. Studies of animal behavior and molecular genetics, often identifying (and sometimes sequencing) the genes behind behavioral traits, are large-scale endeavors going on in many labs around the world today.

2. Gene expression and gene regulation—whether, when, and to what extent a gene is switched on—are also the subject of more and more ethological work focusing on proximate causation.

3. Development in a broad sense encompasses everything from in utero effects to the specific effects that environmental factors might have after birth on the behavior of a developing organism. Environmental factors relevant to development and behavior include both abiotic (nonliving) and biotic (living) components of the environment.

4. Developmental pathways affect dispersal strategies and can be influenced by temperature change. More and more evidence suggests that animals choose which dispersal strategy to adopt as a function of the temperature they experience during development.

5. Two examples—honeybee foraging behavior and comparative social behavior in two closely related species of voles—have run through Chapters 3 and this one on proximate approaches to behavior. These examples allow us to bring together numerous approaches to the proximate study of behavior—molecular genetic, hormonal, developmental, and neurobiological approaches—in a single system, and provide us with a comprehensive understanding of the proximate underpinnings of behavior.

DISCUSSION QUESTIONS

1. The entire genome of many animals is now being sequenced. How might these studies affect work on molecular genetics and behavior? Are there any pitfalls you can imagine to having such a large genetic database available to behavioral researchers?

2. How are genes used to address different types of questions in proximate versus ultimate analyses?

3. Medical school curricula tend to be dominated by courses dealing with proximate analyses. How might physicians benefit from understanding the relationship between proximate and ultimate causation?

4. In addition to the developmental factors discussed in this chapter, can you name one other biotic and one other abiotic developmental factor that might be important in shaping behavior? Provide a one-paragraph explanation for each of these.

5. Make a detailed argument that no matter which of the four proximate approaches you employ, you can always complement your study with one

or more of the other approaches. You may use the honeybee foraging and/ or sociality in voles examples discussed in this chapter as a fulcrum for your discussion.

SUGGESTED READINGS

Flint, J., & Mackay, T. F. C. (2009). Genetic architecture of quantitative traits in mice, flies, and humans. *Genome Research, 19,* 723–733. An overview of molecular genetic approaches to complex behavioral traits.

Korb, J., Weil, T., Hoffmann, K., Foster, K. R., & Rehli, M. (2009). A gene necessary for reproductive suppression in termites. *Science, 324,* 758. A short but excellent paper on the termite study discussed at the start of this chapter.

Smith, C. R., Toth, A. L., Suarez, A. V., & Robinson, G. E. (2008). Genetic and genomic analyses of the division of labour in insect societies. *Nature Reviews Genetics, 9,* 735–748. A review that brings together all four proximate approaches to behavior to study social behavior in insects, especially honeybees.

Veenema, A. H. (2012). Toward understanding how early life social experiences alter oxytocin- and vasopressin-regulated social behaviors. *Hormones and Behavior, 61,* 310–312. A good review of how development regulates key hormones in rodents.

West, M. J., King, A. P., & White, D. J. (2003). The case for developmental ecology. *Animal Behaviour, 66,* 617–622. Also Stamps, J. (2003). Behavioural processes affecting development: Tinbergen's fourth question comes of age. *Animal Behaviour, 66,* 1–13. Two reviews of developmental approaches to animal behavior.

5

Learning

P aper wasps (*Polistes fuscatus*) live in well-lit colonies in which individuals are constantly interacting with one another. Reproduction in paper wasp colonies is tightly linked to the position a wasp holds in a dominance hierarchy, so knowing who is who has important consequences for reproductive success. Research has shown that paper wasps recognize their hive mates, but how do they do so? Because these wasps have facial marks that might allow for such recognition, researchers have examined whether *facial learning* occurs in the species. This type of specialized learning, which is especially prominent in humans, has been demonstrated in other mammals, but until Michael Sheehan and Elizabeth Tibbets's work on paper wasps, had not been found in other species (Sheehan and Tibbetts, 2011).

To test for facial learning, Sheehan and Tibbets took individual wasps and exposed them to the facial images of two other (stimulus) wasps. One of the pictures was paired with an electric shock, but the other was not. The researchers examined whether the wasps learned to avoid the facial image associated with the electric shock, and how quickly they learned. Results indicated that wasps were indeed able to pair a specific facial image with the electric shock. One clue that the wasps were truly exhibiting facial learning, and not some general ability to learn, was that when the pictures of the stimulus wasps lacked antennae, or the faces on these pictures had been artificially rearranged, wasps were not capable of pairing one image with an electric shock—only intact faces produced learning. What's more, wasps were not capable of pairing geometric patterns with an electric shock, again strongly suggesting facial learning (Figure 5.1).

Given that paper wasps displayed facial learning, Sheehan and Tibbets wanted to understand the selective forces that might have shaped such an ability. They reasoned as follows: if this sort of facial learning had been selected in paper wasps because it allows them to recognize individuals in their colony, and if such recognition has effects on reproduction, then in other wasps that live a

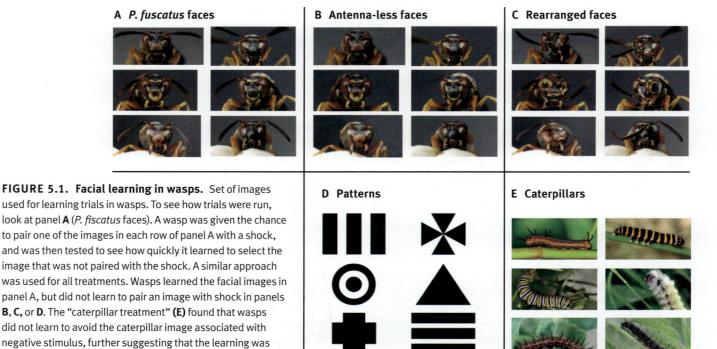

A *P. fuscatus* faces

B Antenna-less faces

C Rearranged faces

D Patterns

E Caterpillars

FIGURE 5.1. Facial learning in wasps. Set of images used for learning trials in wasps. To see how trials were run, look at panel **A** (*P. fiscatus* faces). A wasp was given the chance to pair one of the images in each row of panel A with a shock, and was then tested to see how quickly it learned to select the image that was not paired with the shock. A similar approach was used for all treatments. Wasps learned the facial images in panel A, but did not learn to pair an image with shock in panels **B, C,** or **D**. The "caterpillar treatment" **(E)** found that wasps did not learn to avoid the caterpillar image associated with negative stimulus, further suggesting that the learning was specific to conspecific facial learning. *(Photo credit: Sheehan et al., 2011, Science, 334, 1272–1275)*

more solitary lifestyle and lack specific markings on their face, facial recognition may be absent. To test this idea, Sheehan and Tibbets ran experiments like those described above, but this time using *Polistes metricus*, a species in which individuals typically nest alone, and in which individuals have much less facial pattern variability. No facial learning was observed in this species.

What Is Individual Learning?

This chapter examines the role that **individual learning** plays in animal behavior. The role of social learning—that is, learning from other individuals—is the subject of Chapter 6. In our analysis of individual learning—henceforth referred to as learning in the rest of this chapter—we start by addressing three interrelated questions: How do animals learn? Why do animals learn? What do animals learn?

Before tackling these overarching questions, it is important to address a few baseline issues, first and foremost among them being: What do we mean when we speak of learning? This is a complicated question, but the definition of learning we adopt here is straightforward, and fairly widely accepted within psychology. *Learning* refers to a relatively permanent change in behavior as a result of experience (Shettleworth, 1998). This definition does have a downside, in that it is not clear how long a time period is encompassed by the words *relatively permanent*. That being said, this is a working definition, already adopted implicitly by most ethologists, and it will serve our purposes.

It is interesting to note that the phrase *relatively permanent* was added to an older definition of learning, which was something akin to "a change in behavior as a result of experience." The insertion of *relatively permanent* was meant to address a particularly difficult problem regarding what constitutes learning. Sara Shettleworth describes the problem as follows: A rat that experiences no food for twenty-four hours is more likely to eat than a rat that has just been fed. Most people would say that hunger per se, not any learning by the rat, explains its increased proclivity to forage, even though the experience of *not being fed* did affect the rat's behavior when it was presented with food (Shettleworth, 1998). Insertion of the phrase "relatively permanent" into the definition of learning eliminates this problem.

Our definition of learning—"a relatively permanent change in behavior as a result of experience"—also reveals an interesting relationship between learning and what evolutionary ecologists refer to as "phenotypic plasticity" (Gianoli and Valladores, 2012; Harvell, 1994; Levins, 1968; Pfennig et al., 2010; West-Eberhard, 1989, 2003). A **phenotype** is typically defined as the observable characteristics of an organism (P. Walker, 1989), and **phenotypic plasticity** is broadly defined as the ability of an organism to *produce different phenotypes* depending on environmental conditions. For example, many invertebrates such as the bryozoan *Membranipora membranacea* live in colonies (Harvell, 1998). When living in such colonies, individuals typically lack the spines that are used as an antipredator defense in related species. These spines are simply not grown when a *Membranipora membranacea* colony develops in the absence of predators. Yet individuals will grow spines relatively quickly when exposed to predatory cues (Harvell, 1991, 1994; Tollrian and Harvell, 1998; Figure 5.2). The resultant change, from spineless to spined, constitutes a case of phenotypic plasticity. The phenotype of this bryozoan shifts dramatically as a result of environmental changes—in this case, the addition of a predator—and hence is thought of as "plastic" (Figure 5.3).

A

B

FIGURE 5.2. Inducible defenses.
In some bryozoans, like *Membranipora membranacea*, colonies produce spines when predators are present. **(A)** Spines are shown protruding from a colony as a defense against predators (red arrows point to spines), and **(B)** of a colony of *Membranipora membranacea*. (*Photo credits: Ken Lucas/Visuals Unlimited; © Sue Daly/naturepl.com*)

If learning is "a relatively permanent change in behavior as a result of experience," it then becomes one type of phenotypic plasticity if we think of behavior as part of a phenotype (Dukas, 1998a). That is, if we replace *behavior* with *phenotype* in our definition of learning, phenotypic plasticity becomes the broader category under which learning is subsumed. So all learning is a type of phenotypic plasticity, but not all phenotypic plasticity involves learning. To better understand this distinction, consider the "flushing" behavior often seen in foraging birds. While searching for food, birds may move their tails and wings in a way that flushes insects out from cover—insects that the bird then eats. In the painted redstart (*Myioborus pictus*), for example, when individuals are under branches, they increase their wing and feather motion and flush insects from the overhanging branches. One hypothesis to explain this flushing behavior is that the birds learn that when they are under branches and flap their wings, they will get food. However, this response could be based on a relatively fixed genetic response rather than learned.

Piotr Jablonski and his colleagues tried to distinguish between the two hypotheses. What the researchers found is that, while it is true that birds in nature increase their wing-flapping behavior when they are under branches, the same increase in wing flapping also occurs in the laboratory, even when the birds

FIGURE 5.3. PHENOTYPIC plasticity. When colonies of the bryozoan *Membranipora membranacea* are exposed to chemical stimuli from a predator, individuals in these colonies grow spines. This graph shows the response to a single "dose" of water conditioned with bryozoan predators. Large colonies produce more spines. (*From Harvell, 1991, p. 4*)

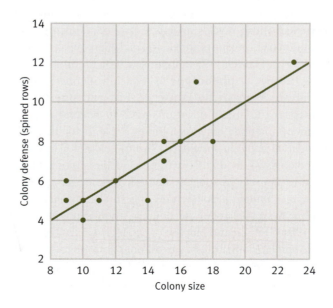

are not rewarded for such behavior. That is, when Jablonski's team put naive birds under branches, they started flapping more as well, even when they got no food for doing so (Jablonski et al., 2006). These results suggest that flushing insects under branches does represent a case of phenotypic plasticity—the ability of an organism to *produce different phenotypes* depending on environmental conditions (whether the birds are under trees or not)—but it is not a case of learning.

How Animals Learn

In this section, we delve into *how* animals learn what they learn. There is a huge psychological literature on this, both theoretical and empirical, but here we will just review some basic ideas on how animals learn, or what psychologists often refer to as the processes underlying learning. This discussion of how animals learn follows an outline developed by Cecilia Heyes, both because of its conciseness and its attempt to tie *how* animals learn to *why* they learn (Heyes, 1994). Heyes notes that there are three commonly recognized types of experience that can lead to learning—namely, single stimulus, stimulus-stimulus, and response-reinforcer—each of which facilitates certain forms of learning.

LEARNING FROM A SINGLE-STIMULUS EXPERIENCE

The simplest experience that can lead to learning involves a single stimulus—a stimulus that can take almost any form. For example, let's imagine that we are interested in studying learning in rats and that numerous times throughout the day we simply place an arbitrary cue in a rat's cage. We will make the cue a blue-colored stick, which is considered an arbitrary cue because it had no relevance to the rat before we began our experiment (Figure 5.4). Rats will often take note of such a disturbance and turn their heads in the direction of the blue stick. If, *over time*, the rats become more likely to turn their heads in the direction of the blue stick—that is, if they become more sensitive to the stimulus with time, **sensitization** has occurred. Conversely, if, over time, the animals become less likely to turn their head, **habituation** is said to have taken place. Sensitization and habituation are two simple single-stimulus forms of learning.

The process of habituation can be problematic for experimental ethologists, particularly to those who work in the laboratory because it is sometimes difficult

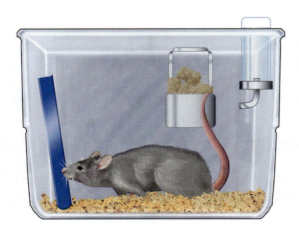

FIGURE 5.4. Habituation and sensitization. Numerous times each day, a blue stick is placed in a rat's cage. If the rat takes less and less notice of the stick, habituation has occurred. If the rat pays more attention to the blue stick over time, sensitization has taken place.

Predator fish

If the experimental setup involves keeping the potential predator in one tank...

... and the prey species in an adjacent tank, the prey may learn that the predator is not dangerous and habituate to its presence.

Prey fish

FIGURE 5.5. Habituation as a problem. In controlled laboratory experiments prey may habituate to the presence of a predator over time.

to examine behavior if animals habituate quickly to stimuli. For example, in many ethology experiments involving antipredator behavior, predators may be housed such that visual interactions between predator and prey are possible, but the predator can't actually harm the prey (Figure 5.5). This ethical compromise spares the life of the potential prey, but also it creates a scenario in which the prey may now habituate to the predator, having learned that the predator cannot in fact move close enough to present any real danger (Huntingford, 1984). Because of these sorts of issues, ethologists often need to go to great lengths to be certain that habituation has *not* occurred in their study system (Rowland and Sevenster, 1985).

Conservation biologists also worry about habituation in animals that may be trained in captivity and then released into the wild. In the safety of the lab, such animals may have habituated to certain stimuli that might prove very dangerous upon the animals' release into the wild (Bauer, 2005). Animal trainers, however, particularly those working with economically valuable animals such as horses, often try to habituate the animals they work with to various stimuli, so that the animals become less frightened by stimuli that elicit innate fear responses (Christensen et al., 2006).

A single stimulus that results in habituation or sensitization may have many consequences for learning processes. An animal's habituation to a stimulus may interfere with later attempts to get the animal to associate that stimulus with some other cue. For example, if rats habituate to the blue stick, it might prove more difficult for them to subsequently learn that the blue stick signals the arrival of food. In contrast, if sensitization to a single cue has occurred, it may facilitate the association of the sensitized stimulus with other cues.

PAVLOVIAN (CLASSICAL) CONDITIONING

Suppose that rather than giving a rat a single stimulus like the blue stick, from the start we *pair* this stimulus with a second stimulus, let's say the odor of a cat—an odor that rats fear, even when they are exposed to this odor for the

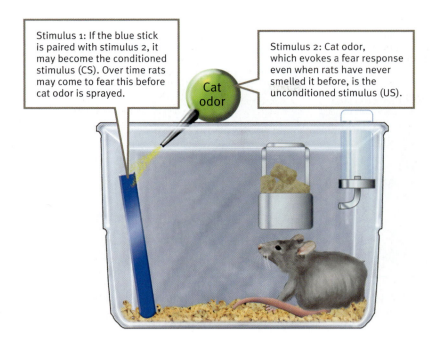

Stimulus 1: If the blue stick is paired with stimulus 2, it may become the conditioned stimulus (CS). Over time rats may come to fear this before cat odor is sprayed.

Cat odor

Stimulus 2: Cat odor, which evokes a fear response even when rats have never smelled it before, is the unconditioned stimulus (US).

FIGURE 5.6. Paired stimuli. Five seconds after a blue stick (stimulus 1) is placed in a rat's cage, the odor of a cat (stimulus 2) is sprayed in as well. The question then becomes: Will the rat pair the blue stick with danger (cat odor)?

first time. Now, five seconds after the blue stick is in place, we spray the odor of a cat into one corner of the cage (Figure 5.6). If the rat subsequently learns to associate stimulus 1 (blue stick) with stimulus 2 (cat odor) and responds to stimulus 1 by climbing under the chip shavings in its cage as soon as the blue stick appears, but *before* the odor is sprayed in, we have designed an experiment in **Pavlovian** or **classical conditioning** (Kim and Jung, 2006).

This form of conditioned learning was first developed by Ivan Pavlov in the late 1800s (Pavlov, 1927; Figure 5.7). Pavlovian conditioning experiments involve two stimuli—the conditioned stimulus and the unconditioned stimulus

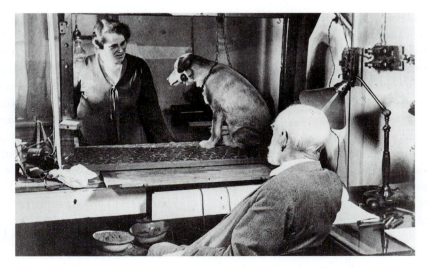

FIGURE 5.7. Ivan Pavlov and classical conditioning. Pavlov watches a classical conditioning experiment as it is being conducted in his laboratory. In the experiment, a device to measure salivation has been attached to the dog's cheek, the unconditioned stimulus is a dish containing meat powder, and the conditioned stimulus is a light. *(Photo credit: Sovfoto/Eastfoto)*

Only the blue stick (CS) is present.

The rat hides, showing the conditioned response (CR).

FIGURE 5.8. Conditioned response. If the rat pairs the blue stick (CS) and the cat odor (US), it will hide under the chips when the blue stick alone is presented. Such hiding represents a conditioned response (CR).

(Domjan, 2005, 2006). A **conditioned stimulus (CS)** is often defined as a stimulus that initially fails to elicit a particular response but comes to do so when it becomes associated with a second (unconditioned) stimulus. In our rat example, the blue stick is the conditioned stimulus, as initially the rat will probably have no fear of this object. The **unconditioned stimulus (US)** is a stimulus that elicits a vigorous response in the absence of training. In our rat example, the US would be the cat odor, which inherently causes a fear response in rats. Once the rat has learned to hide after the blue stick (CS) alone is in place, we can speak of its hiding as being a **conditioned response (CR)** to the presence of the blue stick (Figure 5.8).

Before examining Pavlovian conditioning in a bit more detail, we must define a few more terms. In the learning literature, any stimulus that is considered positive, pleasant, or rewarding is referred to as an **appetitive stimulus**. Appetitive stimuli include food, the presence of a potential mate, a safe habitat, and so on. Conversely, any stimulus that is unpleasant—shock, noxious odors, and so forth—is labeled an **aversive stimulus**. Another important distinction made in the learning literature is between positive and negative relationships. When the first event (placement of the blue stick) in a conditioning experiment predicts the occurrence of the second event (cat odor), there is a positive relation between events. Conversely, if the first event predicts that the second event will *not* occur—imagine that a blue stick is always followed by *not* feeding an animal at its normal feeding time—there is a negative relationship. Positive relationships—for example, blue stick predicts cat odor—produce **excitatory conditioning**. Negative relationships—for example, blue stick leads to no food at a time when food is usually presented—produce **inhibitory conditioning**.

Pavlovian conditioning experiments can become very complicated when second-order conditioning is added to the experimental equation (Figure 5.9). In second-order conditioning, once a conditioned response (CR) has been

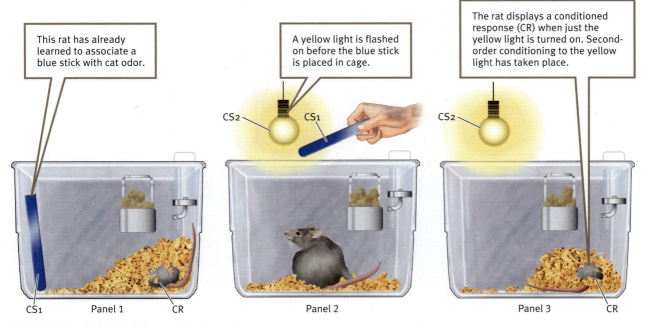

FIGURE 5.9. Second-order conditioning. The rat learns to respond to a second CS—the yellow light—with the conditioned response.

learned by pairing US and CS1, a new stimulus is presented before CS1, and if the new stimulus itself eventually elicits the conditioned response, then the new stimulus has become a conditioned stimulus (CS2). In our case, any rat that has learned to pair the blue stick (CS1) with danger might now see a yellow light (CS2) preceding the appearance of the blue stick. Once the rat has learned to pair the yellow light (US) with the danger associated with cat odor, second-order Pavlovian conditioning has occurred.

BLOCKING, OVERSHADOWING, AND LATENT INHIBITION. Pavlovian conditioning affects not only behavior per se but also what is referred to as *learnability*, that is, the ability to learn under certain conditions. We will explore three types of learnability: overshadowing, blocking, and latent inhibition. To see how these operate, consider an experiment with four groups of rats. Suppose that group 1 individuals undergo a standard Pavlovian paradigm with two stimuli: the blue stick (CS1) and a cat odor (US). In group 2, a second conditioned stimulus (CS2), a yellow light, is always presented *simultaneously* with the blue stick, just before the cat odor is sprayed (Figure 5.10). Subjects from both groups are then tested in response to the blue stick alone. If the yellow light is **overshadowing** the blue stick, rats in group 2 will respond less strongly to the blue stick (when it is presented alone) than will rats in group 1. The CS2—the yellow light—has made it more difficult for the rats to pair the blue stick and the cat odor—it has retarded the rats' ability to learn.

Now let us add a third group to our experiment. In group 3, individuals are first trained to associate the blue stick with the cat odor, but after this training, the yellow light is presented at the same time as the blue stick, and this compound stimulus is paired with the cat odor. Recall that in group 2, the blue stick and the yellow light were always presented together. Rats in group 3 differ

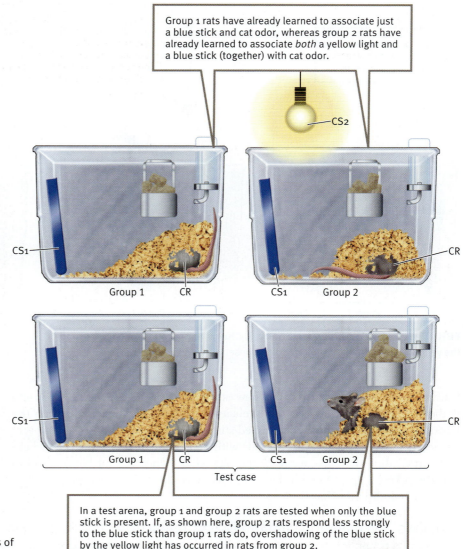

Group 1 rats have already learned to associate just a blue stick and cat odor, whereas group 2 rats have already learned to associate *both* a yellow light and a blue stick (together) with cat odor.

CS2

CS1

CR

Group 1   CR

CS1   Group 2   CR

CS1

CS1

Group 1   CR

CS1   Group 2   CR

Test case

In a test arena, group 1 and group 2 rats are tested when only the blue stick is present. If, as shown here, group 2 rats respond less strongly to the blue stick than group 1 rats do, overshadowing of the blue stick by the yellow light has occurred in rats from group 2.

FIGURE 5.10. Overshadowing. The process of overshadowing is shown in two groups of rats.

from those in group 2 in that they first learned to associate the blue stick and cat odor before any yellow light is added to the protocol. Now, after the animals are trained, let us compare the reaction of rats in groups 2 and 3 when they are presented with the stimulus of a yellow light. **Blocking** occurs when those in group 3 respond less strongly to the yellow light (when it is presented alone) than individuals in group 2 do (Kamin, 1968, 1969). It is as if initially learning to associate the blue stick alone with the cat smell *blocked* group 3 subjects' ability to pair the yellow light with the cat odor (Figure 5.11).

Finally, let's consider a fourth group of rats that is initially exposed to a blue stick, but no cat odor, for a long period of time. We then try to pair the blue stick with cat odor at some subsequent point in time. If we find that the rats in group 4 have more difficulty learning than the rats in group 1 (where standard Pavlovian pairing has occurred), then we would say that **latent inhibition** is responsible for this lack of learning in group 4.

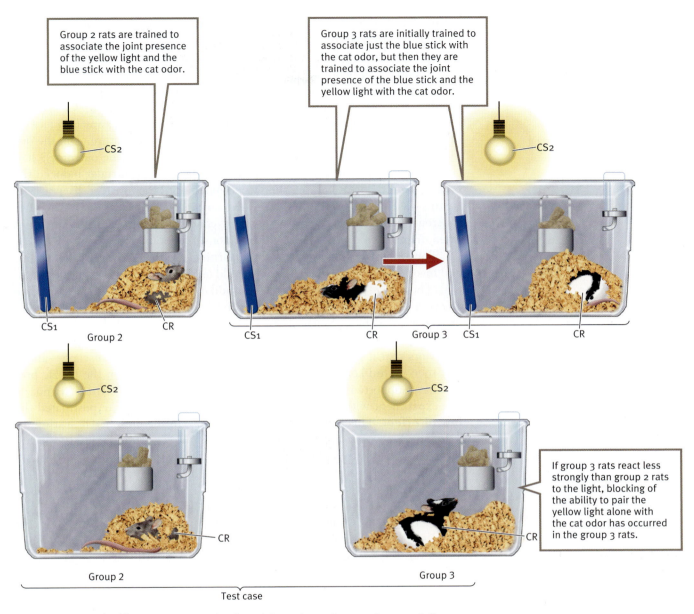

Group 2 rats are trained to associate the joint presence of the yellow light and the blue stick with the cat odor.

Group 3 rats are initially trained to associate just the blue stick with the cat odor, but then they are trained to associate the joint presence of the blue stick and the yellow light with the cat odor.

CS2

CS1   CR
Group 2

CS1   CR   Group 3   CS1   CR

CS2

CS2

If group 3 rats react less strongly than group 2 rats to the light, blocking of the ability to pair the yellow light alone with the cat odor has occurred in the group 3 rats.

CR

CR

Group 2

Group 3

Test case

FIGURE 5.11. Blocking. Learning can be slowed down depending on prior association or lack of association between stimuli.

INSTRUMENTAL (OPERANT) CONDITIONING

Instrumental conditioning, also known as **operant** or **goal-directed learning**, occurs when the response that is made by an animal is reinforced (increased) by the presentation of a reward or the termination of an aversive stimulus, or when the response is suppressed (decreased) by the presentation of aversive stimulus or the termination of a reward. One of the most fundamental differences between Pavlovian and instrumental learning is that, in instrumental learning, the animal must undertake some action or response in order for the conditioning process to produce learning. The classic example of instrumental learning is a rat pressing some sort of lever (that is, taking an action) to get food to drop into its cage. Rats associate pressing on the lever (response) with some probability of getting food (outcome) and learn this task.

Not all instrumental learning, however, is as easy to quantify as in the case of a rat pushing a lever. Michael Domjan illustrates this by considering a dog that barks at intruders until the intruders go away (Domjan, 1998). Clearly, the dog has to bark (undertake an action) to make the intruder go away (produce an outcome), but does this illustrate instrumental *learning* in nature? Did the dog learn that barking (the response) made the intruder leave (the outcome)? It is certainly possible. But it is equally plausible that a novel intruder induced the barking, and that this barking simply continued until the intruder disappeared. In this case, the dog may have learned nothing. Because of the difficulties inherent in these sorts of cases, operant conditioning experiments are most often undertaken under controlled conditions in the laboratory.

The earliest work on instrumental learning was that of Edward Thorndike and involved testing how quickly cats could learn to escape from "puzzle boxes" that Thorndike had constructed (Thorndike, 1898, 1911). When Thorndike placed a cat in a locked box, the cat initially tried all sorts of things to get out of its confined space. Some of these behaviors, by chance, led to a successful escape from the box. Thorndike postulated that the cat began to pair certain behaviors that it undertook in the box with a positive effect—escape—and it was then more likely to use such behaviors when confined in the puzzle box. His data supported this contention. Combining the findings from his puzzle box experiment with other results he had obtained, Thorndike postulated the **law of effect**. This law states that if a response in the presence of a stimulus is followed by a positive event, the association between the stimulus and the response will be strengthened. Conversely, if the response is followed by an aversive event, the association will be weakened.

Thinking on the subject of instrumental learning was revolutionized by the work of B. F. Skinner, who devised what is now known as a Skinner box (Skinner, 1938). His idea was to create a continuous measure of behavior that could somehow be divided into meaningful units. When a rat pushes down on a lever, it is making an **operant response** because the action changes the rat's environment by adding food to it (Figure 5.12). Because "lever pushing" is a relatively unambiguous event that is easily measurable, and because it occurs

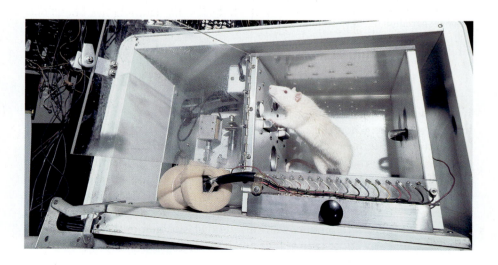

FIGURE 5.12. Rats in a Skinner box. To test various theories of animal learning, rats are often placed in "Skinner boxes," where they have to take an action (here, pressing a button) to get a reward of food or water. *(Photo credit: Walter Dawn/Science Photo Library/Photo Researchers, Inc.)*

in an environment over which the rat has control, the Skinner box has greatly facilitated the work of psychologists doing research within the instrumental learning paradigm.

There is still some debate among psychologists over the relative merits of Pavlovian and instrumental learning techniques. From our perspective, what matters most is that over the years both instrumental and Pavlovian conditioning techniques have become fine-tuned and have provided much in the way of understanding how animals learn what they learn.

Why Animals Learn

With an understanding of *how* animals learn, we can now take a more evolutionary perspective and ask *why* animals learn. If the ability to learn is under some sort of complex genetic control, we can ask whether this ability is favored by natural selection, and if so, under what circumstances. In so doing, we will address three related questions: (1) How can within-species studies help us understand natural selection and learning? (2) How can population comparisons be used to shed light on the evolution of learning? and (3) What theories examine how learning evolves in different environments?

WITHIN-SPECIES STUDIES AND THE EVOLUTION OF LEARNING

Psychologists have done thousands, perhaps tens of thousands, of studies of learning in animals, particularly in rodents and pigeons. For the most part, until recently, these experiments have not been designed to take into consideration the natural environment of the organism or the evolutionary forces acting on the organism. Both Edward Thorndike and Ivan Pavlov argued that, aside from the details, the qualitative features of learning are the same in all animals, including human beings—that is, all animals learn in a fundamentally similar fashion (Bitterman, 1975; Pavlov, 1927; Thorndike, 1911). This view became widely accepted, and was promoted by such psychologists as B. F. Skinner and Harry Harlow (Harlow, 1959; Skinner, 1959). M. E. (Jeff) Bitterman, a leader in the field of comparative animal learning, notes that "work on learning has been dominated from the outset by a powerful theory which denies that learning has undergone any fundamental evolutionary change" (Bitterman, 1975, p. 699). If Thorndike and Pavlov were right that the particular environment an organism evolved in has no effect on learning, the same sort of learning should be seen in all creatures that learn, regardless of the sort of learning tasks with which they are presented.

From an adaptationist approach (Chapter 2), the above claim seems hard to fathom. Could it really be the case that natural selection on learning does not lead to differences in learning across population and in different species? This seems very unlikely. The ability to learn should be under strong selection pressure, such that individuals that learn appropriate cues that are useful in their particular environment should be strongly favored by natural selection. This is the "ecological learning" model, and its influence is getting stronger as our knowledge about evolution and learning increases (Dukas, 1998a; Johnston, 1985; Shettleworth, 1998). Such ecological learning first emerged in the psychological literature with the work of John Garcia and his colleagues.

GARCIA'S RATS. In the mid-1960s, Garcia and his team ran a series of learning experiments using rats. Experiments on rat learning were hardly news in either the psychological or biological community, but Garcia's results made both groups rethink their approach to studies of learning (Garcia et al., 1972; Seligman and Hager, 1972). What is particularly amazing about Garcia's work is that in many ways the protocol he used was very similar to that already being used in psychology learning experiments. In essence, Garcia attempted to have rats form an association between a series of cues (Garcia and Koelling, 1966). One of those cues was either "bright-noisy" water (that is, water associated with a noise and an incandescent light, an audiovisual cue) or "tasty" water (that is, water with a particular taste, a gustatory cue). The researchers then paired the bright-noisy water or the tasty water with one of the following negative stimuli: radiation, a toxin, immediate shock, or delayed shock. The radiation and the toxin made the rats physically ill, whereas the shocks were painful. For example, in the bright-noisy water/radiation treatment, bright-noisy water would be presented and the rats that were being trained to this cue would drink the water and then be exposed to radiation that would make them ill.

Garcia and Koelling found a fascinating interaction between the type of punishment a rat received for drinking water and the type of water the rat consumed. X-ray and toxin treatments, each of which made the rat physically ill, were easily associated with the tasty water (gustatory) cues, but not with the bright-noisy (audiovisual) cues. That is, the rats quickly learned that tasty water was to be avoided after this cue was paired with X-rays or toxins, but they did not learn to avoid bright-noisy water after it was paired with X-rays or toxins. In contrast, when Garcia and Koelling examined the rats that were given shock treatments, they found that shock was easily paired with the bright-noisy water (audiovisual cue), but not with tasty water (gustatory cue).

Garcia and Koelling explained their results in terms of adaptation—something quite unusual for psychologists of the 1960s. They argued that, on the one hand, natural selection would favor the ability to pair gustatory cues (tasty water) with internal discomfort (getting ill). After all, many instances of internal discomfort in nature are likely to be caused by what an animal has consumed, and rarely are food cues associated with audiovisual cues (as in the bright-noisy water treatment). On the other hand, peripheral pain, like that caused by a shock, might be more commonly associated with some audiovisual cue like hearing or seeing a conspecific or a predator, so again natural selection should favor the ability to pair these cues together.

In another apparent blow to orthodoxy in psychological learning circles of the period, Garcia found that learning in rats will occur without immediate reinforcement (Garcia et al., 1966). Most psychologists believed that delays in reinforcement on the order of seconds can cripple animal learning, yet Garcia found that learning occurred even after delays of seventy-five minutes, when injections of noxious substances were paired with drinking saccharin-flavored water. From an adaptationist perspective, one would expect a delay between the time that a rat consumed a substance and any subsequent negative effect of such consumption. As such, natural selection would have favored rats that were able to associate what they ate with becoming ill, even if the events were separated by significant time intervals.

OPTIMAL FORGETTING IN STOMATOPODS. The ability to remember events plays an obvious role in animal learning. In fact, depending on your precise definition of *memory*, learning is impossible without some form of memory.

Psychologists have long studied memory and learning by, for example, looking at **extinction curves** (graphical representations of the weakening and then ending of paired associations) in learning experiments. Such experiments typically test how long an animal will remember a paired association once the pairing itself has stopped. These tests are critical to numerous aspects of psychological learning theory, but they fail to ask a question that jumps to mind when an adaptationist approach to the subject of memory is invoked: Given the ecology of the species, might natural selection shape an optimal memory span? That is, while psychologists tend to focus on how long associations last, an evolutionary animal behaviorist would ask why we see variation in the length of time that stimuli stay paired, and how natural selection may have favored certain extinction curves over others.

It may be that certain events in an animal's life are timed such that it would be beneficial to remember them for a certain amount of time, but after this time period, recalling and acting on such memories would be either of no benefit or perhaps even detrimental. For example, imagine a forager that is able to remember the location and amount of food in patches spread across their environment. If patches are replenished after they are depleted, then while it may pay to remember the location of a food patch, it isn't beneficial for the forager to remember how much food was there the last time it visited, as this information might lead to erroneous decisions based on outdated information (Figure 5.13).

Before proceeding to an example of how behavioral ecologists and ethologists might test the idea of optimal memory, it is important to realize that no natural selection model of any behavior predicts that an animal will reach some global

Bee

Nectar-producing flower

FIGURE 5.13. How long to remember? Imagine a bee foraging at a nectar-producing flower. While it might be beneficial for the bee to remember the flower's location, it might not be to remember specific nectar content, as that shifts within and between days.

optimum. Rather, natural selection models predict that, given the constraints that an animal faces, selection will favor moving as close to an optimum as possible (Alcock, 1998; Mayr, 1982). In other words, ethologists don't search for the best of all possible solutions to a problem and assume that selection will take a population there, but rather they recognize that many constraints prevent selection from ever reaching such a global optimum. With respect to learning there are, in all likelihood, neurobiological constraints on memory and other cognitive functions, and these serve as guideposts for the limits of natural selection's ability to shape memory capacity (Dukas, 1998a, 1998b; Shettleworth, 1998).

To see how ethologists might tackle the question of natural selection and memory, consider Roy Caldwell's 1992 study of aggressive behavior in the stomatopod crustacean *Gonodactylus bredini* (R. Caldwell, 1992; Healey, 1992). Male and female *G. bredini* share a nest cavity for a few days before they breed, and they actively repel all intruders from this area (R. Caldwell, 1986; Dingle and Caldwell, 1972; Shuster and Caldwell, 1989). Shortly after mating, however, the male leaves the breeding cavity and goes out in search of a new mate. Caldwell set up an experimental design that allowed him to examine whether the males that left the brood cavity and the females they left behind recognize each other after the male's desertion, and if so, for how long and why (R. Caldwell, 1992).

It turns out that both male and female *G. bredini* remember each other after the male has left. Males are less likely to be aggressive toward their former mates, while females put up weaker resistance to such males. With respect to memory, males and females recognize each another for at least two weeks—the duration of the experiment (R. Caldwell, 1992). What makes this study fascinating is that, based on his understanding of stomatopod biology, Caldwell hypothesized that former mates might actually be able to recognize each other for up to four weeks. But why four weeks?

Females guard their brood for four weeks, after which the brood leaves its cavity for good. At the four-week point, then, males do not pay significant costs for being aggressive toward former mates, as such aggression will not harm their own brood (Figure 5.14). Caldwell predicted that male and female stomatopods will no longer recognize each other after this four-week time period, or at the very least, they will act as if they had forgotten each other.

Unfortunately, Caldwell's experiment did not run for four weeks. But suppose it had, and suppose he had found that, up to four weeks, males behaved less aggressively toward former mates, but no longer did so after this time. It would be difficult, if not impossible, to determine whether the stomatopods actually forgot information or simply ignored information that they still retained. This is a general problem in animal learning studies, however, and is beyond the scope of what we are addressing here. That is, since we measure learning by what an animal does, in the absence of language, it is hard to be specific about what is going on cognitively when an animal doesn't do something. Did it forget? Did it not act on the information it possessed? There are some ways around this problem, but they are often circuitous (K. Roper et al., 1995). For example, if an animal is placed in an obvious life-and-death situation and fails to perform a learned behavior that might save its life, we can infer that it simply forgot the appropriate life-saving behavior, rather than that it recalled such behavior but ignored it. In any case, the stomatopod study raises some fascinating questions about how natural selection might be acting on learning and memory in animals.

FIGURE 5.14. Stomatopod threat. Male and female stomatopod crustaceans share a cavity for a few days before they breed. Although the males leave the breeding cavity soon after mating, they tend to remember their former mates and to be less aggressive toward them during the weeks that their brood remains in the cavity. This is a photo of a male *Gonodactylus smithii* in a threat position. *(Photo credit: Roy Caldwell)*

POPULATION COMPARISONS AND THE EVOLUTION OF LEARNING

Recall from Chapter 2 that one technique that ethologists employ to understand the evolution of behavior is to compare behavior across different populations of the same species. Often, such studies compare two or more populations to one another and make predictions about behavioral similarities and differences between these populations, based on both a knowledge of their ecology and some hypotheses about how natural selection has operated in each (Balda et al., 1998; Bitterman, 1975; Dukas, 1998a; Roth et al., 2010; Shettleworth, 1998). Here we will adopt this approach with respect to how natural selection has shaped learning in doves and sticklebacks.

LEARNING, FORAGING, AND GROUP LIVING IN DOVES. Animals in groups often find food faster and have more time available for foraging than solitary foragers (Krebs and Davies, 1993). With respect to learning, Pascal Carlier and Louis Lefebvre predicted that individuals that live in groups, who must compete with others in their group and hence gain much from learning about their food sources, should learn more quickly than territorial (and hence more isolated) individuals (Carlier and Lefebvre, 1996). Ideally, one would like to test this hypothesis in a single species, where natural selection has favored group living in some populations but solitary living in others. Zenaida dove populations from Barbados (*Zenaida aurita*; Figure 5.15) fit the bill perfectly, as solitary living (via territoriality) is the norm in one dove population studied by Carlier and Lefebvre, while in another population only nine miles away, group living is the norm. Because these populations are geographically close to one another, differences in other environmental variables, above and beyond those associated with territoriality, were likely minimal.

Sixteen doves from each of the group-living and territorial populations were captured and brought into the laboratory. All subjects were then individually presented with the challenge of learning how to operate an experimental apparatus that required the birds to pull on a metal ring, which then opened a drawer containing food. Carlier and Lefebvre found evidence that group-living doves learned this task more quickly than did birds from the territorial population (Figure 5.16; see also

FIGURE 5.15. Zenaida doves. Zenaida doves from populations where individuals live in groups appear to be better at learning foraging tasks than individuals from populations where doves are territorial. *(Photo credit: Jean-Philippe Paris)*

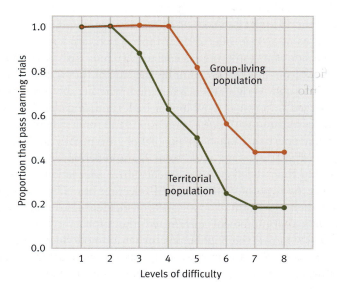

FIGURE 5.16. Group living and learning. More birds from the group-living population surpassed the "learning criteria" for foraging tasks than did birds that had lived alone (territorial population). *(From Carlier and Lefebvre, 1996, p. 1203)*

Sasvari, 1985). Furthermore, they found that the more difficult the learning task the birds had to solve, the more pronounced the between-population differences.

Let's examine two possible explanations for these differences (Carlier and Lefebvre, 1996). First, the animals may have already differed in foraging experience before the experiment, and hence some of the differences the researchers uncovered may have been due to what individuals had experienced, and potentially learned, prior to being brought into the laboratory. Second, above and beyond what experiences any given set of birds took into Carlier and Lefebvre's experiment, natural selection may have operated on learning ability across these populations. The verdict is still out as to which of the possibilities best explains differences between dove populations. But in principle, one could obtain a much clearer idea of which of these explanations is more powerful by raising both territorial and group-living doves in the laboratory under controlled conditions, and testing whether differences in learning still remained. If they did, this finding would suggest that natural selection had favored learning more strongly in the group-living population of doves.

LEARNING AND ANTIPREDATOR BEHAVIOR IN STICKLEBACKS. As we have discussed, one way to partially circumvent the confounding effects of learning per se and natural selection acting on the ability to learn is through the use of controlled laboratory experiments. In the lab, it is possible to raise individuals from two very different populations in a similar environment, and in so doing minimize experience differences. Huntingford and Wright used this approach in their study of avoidance learning in two populations of three-spined stickleback (*Gasterosteus aculeatus*; Huntingford and Wright, 1992). Some sticklebacks live in locales that contain a variety of predators, and some populations live in lakes with virtually no predators.

Huntingford and Wright raised individuals derived from predator-rich and predator-free streams in the laboratory, and during their development none of the individuals had any interaction with predators. If natural selection has acted more strongly on antipredator strategies—including learning about danger—in the sticklebacks from the predator-rich population, then we should see such differences in our experiment.

Huntingford and Wright began by training eight sticklebacks from each of their population groups to associate one side of their home tank with food. They found no differences in learning across populations in the context of foraging alone—individuals from both populations were equally adept at learning that food would come to one side of their tank. Then, after a stickleback had learned that one side of its tank was associated with food, fish were subjected to a simulated attack from a heron predator on the side of the tank that contained food. Huntingford and Wright then examined whether between-population differences emerged in terms of how long it took the fish to learn to avoid the side of the tank associated with heron predation (and food; Figure 5.17).

While all but one fish from both populations eventually learned to avoid the dangerous end of their tank, fish from high-predation areas learned this task more quickly than did fish from predator-free populations. Two lines of evidence support the hypothesis that natural selection has operated on learning and antipredator behavior in these populations of sticklebacks. First, Huntingford and Wright's laboratory protocol minimized the probability that individual experiences differed across the populations they examined. Second, and equally important, Huntingford and Wright did not find between-population differences in all learning contexts; when the task was a simple association of food and place,

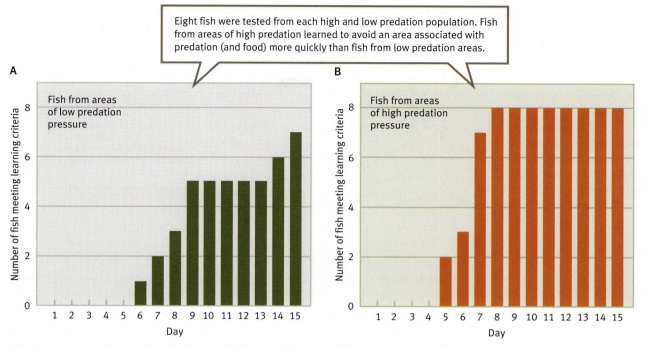

FIGURE 5.17. **Learning differences across populations.** The number of fish that learned to avoid areas associated with predation. **(A)** Fish descended from individuals from low-predation sites. **(B)** Fish descended from individuals from high-predation sites. *(From Huntingford and Wright, 1992)*

no population differences in learning emerged. The researchers only discovered interpopulational differences with respect to learning and antipredator behaviors in the behavioral scenario in which they hypothesized natural selection was operating—in learning to avoid areas associated with danger.

A MODEL OF THE EVOLUTION OF LEARNING

A number of mathematical models have been developed to examine when natural selection might favor the ability to learn (Bergman and Feldman, 1995; Borenstein, et al., 2008; Boyd and Richerson, 1985; Irwin and Price, 1999; Nakahashi, 2010; Odling-Smee et al., 2003; Stephens, 1991, 1993). Theoreticians typically begin the process of modeling the evolution of learning by taking a cost-benefit perspective. Imagine a behavioral scenario in which the options are to respond to some stimuli with a fixed genetically programmed response, or to respond to stimuli based on prior experience in the same situation—that is, by learning. Are the net benefits associated with learning greater or less than the net benefits that might be associated with a fixed genetic response to some stimuli?

Ethologists, behavioral ecologists, and psychologists have argued that natural selection should favor the ability to learn over the genetic transmission of a fixed trait when the environment an animal lives in changes often, but not too often. To see why, we must recognize two assumptions that underlie many of these models: (1) most models assume that there is some cost to learning, even if it is only a very small cost; and (2) when we speak of learning in such models we are referring to the ability to learn being a trait that has an underlying genetic

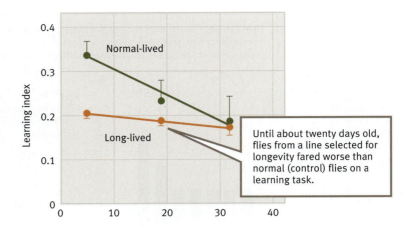

FIGURE 5.18. A trade-off between learning and life span. "Learning index scores" for normal (control) fruit flies and fruit flies from a line selected for artificially prolonged life spans. The difference between these groups suggests a trade-off between long life and the ability to learn. *(From Burger et al., 2008)*

Text within figure:
Normal-lived

Long-lived

Until about twenty days old, flies from a line selected for longevity fared worse than normal (control) flies on a learning task.

basis. Though difficult to measure, evidence for both these assumptions has been found in fruit flies and butterflies (Burger et al., 2008; Burns et al., 2011; Kawecki, 2010; Figure 5.18).

When the environment *rarely changes*—and hence the environment that offspring encounter is similar to that of their parents—information is best passed on by a fixed genetic rule, since such a means of transmission avoids the costs of learning. On the other end of the spectrum, if the environment is *constantly changing*, there is nothing worth learning because what is learned is completely irrelevant in the next situation. When the environment is constantly changing, acting on past experience is worthless, as past experience is of no predictive value. Hence, genetic transmission of a fixed response, rather than a costly learned response, is again favored. Somewhere in the middle, in between an environment that never changes and one that always changes, learning is favored over the genetic transmission of a fixed response. Here, it is worth paying the cost of learning. The environment is stable enough to favor learning, but not so stable as to favor genetic transmission.

David Stephens has challenged the way that environmental stability has been represented in these models, saying that many models conflate two types of stability that need to be separated (Stephens, 1991, 1993). The model that Stephens developed breaks environmental predictability into two types: (1) predictability within the lifetime of an individual, and (2) predictability between the environment of parents and offspring. These two types of predictability can be very different, and conflating them may hinder our understanding of the evolution of learning (Figure 5.19). For example, consider a case in which early in life the offspring of species A move to environments that are far removed from those of their parents, and the environment to which they migrate is stable over the course of their lifetime. In such a case, between-generation environmental predictability in species A is quite low, while within-lifetime environmental predictability is much higher. This distinction is lost when models lump together within- and between-generation environmental predictability.

Learning is favored in Stephens's model when predictability within the lifetime of an individual is high, but environmental predictability between generations is low. To see why, let's walk through each of the possible scenarios in Figure 5.19. In boxes 1 and 2, predictability within generations is low, so

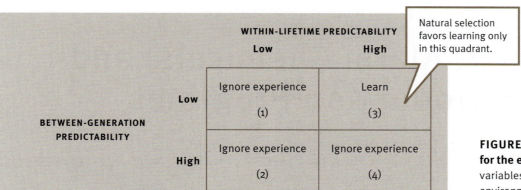

FIGURE 5.19 Stephens's model for the evolution of learning. The key variables in this model are within-lifetime environmental predictability and between-generation environmental predictability.

neither strategy does particularly well. But since learning has a cost associated with it, genetic transmission of a fixed response is favored. Fixed genetic transmission is again favored in box 4, because with high predictability at all levels, the cost of learning is never worth the investment. Only in box 3, with high within-generation predictability but low between-generation predictability, is learning favored. Learning is favored here because once an organism learns what to do, it can repeat the appropriate behaviors during its lifetime. But isn't this the sort of predictability that usually favors a fixed genetic transmission? Yes, but now the environment changes so much between generations that fixed genetic transmission, per se, would be less advantageous than learning.

What Animals Learn

In order to provide a broad (but brief) overview of animal learning, in this section we examine evidence for learning with respect to predation, mating, familial relationships, and aggression. The Conservation Connection box discusses learning with respect to habitat and conservation biology. Learning about food sources is covered in Chapter 11.

LEARNING ABOUT PREDATORS

Prey often live in areas that contain both predatory and nonpredatory species, and learning which species is which has fitness consequences. Even encounters with predators are not always the same, as at any given time some predators are in hunting mode while others are not actively hunting prey (Chivers et al., 1996). If prey can learn to distinguish between dangerous and benign encounters with potential predators, they may free up time for other activities such as foraging or mating—in other words, learning about possible predation pressure may allow animals to handle the trade-offs they constantly face.

In aquatic systems, recognition of predators often involves processing chemical stimuli (Ferrari, Capitania-Kwok et al., 2006; Ferrari and Chivers, 2006; Ferrari and Messier, 2006; Ferrari et al. 2005). If the food that a potential

Learning, Alarm Chemicals, and Reintroduction Programs

One way in which conservation biologists try to protect threatened or endangered species is through **reintroduction programs** (Ewen et al., 2012). These programs typically involve managers raising individuals of a threatened or endangered species in captivity and then releasing them into an area that the species formerly occupied. Reintroduction programs have had mixed success. One problem is that reintroduced individuals are often especially susceptible to predation, in part because they experience no threats while being raised in captivity. A similar issue arises in translocation programs, when individuals are moved from one natural habitat to another, and in fisheries, when fish are released into the wild (Olson et al., 2012).

As you have learned in this chapter, animals can learn about many different aspects of their environment, and what they learn can affect survival and reproduction. Conservation biologists understand this very well, and many programs now try to present individual animals with the opportunity to learn something about one aspect or another about the environment into which they will be released or transferred, *before* the release or introduction occurs. For example, hellbenders

(*Cryptobrancus alleganeinsis*), large aquatic salamanders, whose natural range has declined dramatically, have an innate fear response when exposed to an alarm chemical—a white mucus—that is produced by other hellbenders. To better understand how to design reintroduction programs for hellbenders, Crane and Mathis used a classic conditioning protocol in which one group of hellbenders was given the opportunity to pair the alarm chemical with the scent of brown trout—a predator of hellbenders. For a second (control) group, the alarm chemical was paired with just water (Crane and Mathis, 2011). In follow-up experiments, the hellbenders that were given the chance to pair the trout odor and the alarm chemical showed more fine-tuned antipredator behaviors in response to trout than the hellbenders in the control group (Figure 5.20). In nature, presumably, hellbenders also learn to associate these alarm chemicals with indicators of the presence of a predator. This classical conditioning protocol could be used in reintroduction programs for hellbenders. Individuals trained to show fear responses when exposed to real predator cues would be more likely to take action to avoid the danger and therefore survive when encountering predators in nature for the first time.

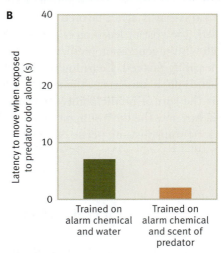

FIGURE 5.20. Learning and response to predators. (A) A hellbender salamander. **(B)** Hellbenders that were given the opportunity to pair the alarm secretion and the odor of a predator moved quickly when exposed to the odor of the predator alone. *(Photo credit: Robert J. Erwin/Photo Researchers)*

predator eats produces a chemical cue that is recognizable to its prey, then this chemical cue acts as "the scent of a predator" and may provide an opportunity for such prey to learn what is dangerous and what isn't (Crowl and Covich, 1990; Gelowitz et al., 1993; Howe and Harris, 1978; Keefe, 1992; Mathis and Smith, 1993a, 1993b; D. J. Wilson and Lefcort, 1993). Douglas Chivers and his colleagues examined this possibility by studying learning and antipredator behavior in the damselfly (*Enallagma spp*; Chivers et al., 1996; Figure 5.21).

What makes Chivers's study especially interesting is that it shows how to distinguish *learning* about predators from *innate behavioral responses* that prey may display toward predators.

Damselfly larvae are found in ponds with minnows, and both species are often attacked and eaten by pike (*Esox lucius*). Chivers and his colleagues hypothesized that damselfly larvae might learn about the potential dangers associated with pike encounters by using chemical cues. To test this hypothesis, the researchers fed pike predators either minnows, damselflies, or mealworms. Minnows and damselflies are prey for pike, whereas mealworms are not and served as a control. After four days on one of these three diets, a pike was removed from its tank, and damselflies that had never before had any contact with a pike were exposed to the water from the pike's tank.

When damselfly larvae were exposed to the water containing chemical cues from a pike that had eaten damselflies or a pike that had eaten minnows, Chivers and his colleagues found that the damselfly larvae significantly reduced their foraging behavior. They did not reduce their foraging behavior, however, when they were exposed to the water treated with pike that had eaten mealworms (Figure 5.22). Because damselflies are found in the same ponds as minnows, but the damselflies tested by Chivers had never before experienced a pike, these results strongly suggest that damselflies innately associate the scent of pike plus damselfly or pike plus minnow with danger, but they make no such association between pike, mealworm, and danger. That is, the damselflies here hadn't learned anything; they simply were predisposed to respond to the smell of pike and prey (minnows and damselflies) as dangerous.

Chivers's team followed up their initial experiment by examining the role of learning and antipredator behavior in damselflies. Here they took the damselflies

FIGURE 5.21. Damselfly. Larval damselflies learn about predation threat through chemical cues. An adult damselfly is shown here. *(Photo credit: Kim Taylor/ natureplace.com)*

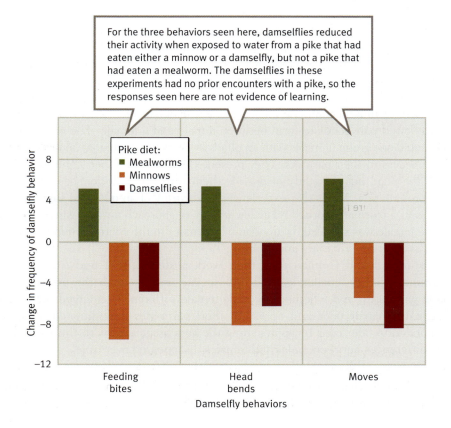

For the three behaviors seen here, damselflies reduced their activity when exposed to water from a pike that had eaten either a minnow or a damselfly, but not a pike that had eaten a mealworm. The damselflies in these experiments had no prior encounters with a pike, so the responses seen here are not evidence of learning.

Pike diet:
- Mealworms
- Minnows
- Damselflies

Damselfly behaviors

Feeding bites   Head bends   Moves

Change in frequency of damselfly behavior

FIGURE 5.22. Chemically mediated changes. Numerous aspects of damselfly behavior, including the frequency of feeding bites, head bends, and moves, changed as a function of whether the damselflies were exposed to chemical stimuli from a pike predator that had eaten mealworms, minnows, or damselflies. *(From Chivers et al., 1996)*

that had been exposed to the three treatments above (water from pike plus damselfly, water from pike plus minnow, and water from pike plus mealworm) and isolated them for two days. Then each damselfly was exposed to water from a pike that had been fed mealworms. With respect to learning and antipredator behavior, there were three groups of damselflies; group 1: damselflies that were initially exposed to water from pike plus damselfly, but that were subsequently exposed to pike plus mealworm water; group 2: damselflies that were initially exposed to water from pike plus minnow, but that were subsequently exposed to pike plus mealworm water; and group 3: damselflies exposed to water from pike plus mealworms twice. Damselflies from groups 1 and 2 had responded with antipredator behaviors in the first experiments, but damselflies in group 3 had not.

Damselflies in groups 1 and 2 responded to the scent of pike plus mealworm by decreasing their foraging activities. Damselflies in group 3 did not decrease their foraging. Recall that in the first experiment, damselflies did *not* curtail feeding when they encountered the smell of pike plus mealworm for the first time, but they did curtail feeding in this second experiment. These results suggest that, based on their earlier experience in the first experiment, damselflies in groups 1 and 2 in the second experiment had learned to associate pike plus *the scent of any potential prey* with danger, and this association translated into a reduced foraging rate even when they encountered the scent of pike and mealworm.

LEARNING ABOUT YOUR MATE

If an individual could learn to associate certain cues with mating opportunities, this ability should translate into increased reproductive opportunities (Graham and Desjardins, 1980; H. West et al., 1992; Zamble et al., 1985, 1986). Michael Domjan and his colleagues have been examining Pavlovian learning and mate choice in a number of different species. Here we discuss their work with the Mongolian gerbil (*Meriones unguiculatus*; Domjan et al., 2000).

Mongolian gerbils are burrowing desert rodents that rely on chemical communication during many forms of social exchange, including the formation of pair bonds (Ågren, 1984; Thiessen and Yahr, 1977). Given this reliance on chemical communication, Ron Villarreal and Domjan ran an experiment that first allowed pair bonds to form between a male and a female. The researchers presented one group of males with an olfactory cue (mint or lemon) and then gave them access to their partners. They presented males in another (control) group with the odor, but they did not provide them with access to females following this presentation. Males that experienced the pairing of an odor and subsequent access to a pairmate learned relatively quickly to approach an area where access to the female was signaled by an odor, while males in the control group made no such association (Villarreal and Domjan, 1998).

Villarreal and Domjan took their work a step further and went on to examine whether females learned to associate an odor with the presence of their pairmates, and whether any male/female differences emerged. Females did learn to pair odor and access to their pairmates—conditioned females responded to odor cues by approaching the area associated with this cue, and any differences between males and females disappeared over time.

Domjan and Karen Hollis hypothesized that differences between males and females in their learning abilities should be positively correlated with differences in male and female parental investment (Domjan and Hollis, 1988). The more equally parental investment is shared, the more the sexes should be similar in terms of learning the location of partners. One way to think about Domjan and Hollis's hypothesis is in terms of how much each sex is willing to invest in future offspring (Trivers, 1974). If both males and females provide resources for offspring, selection pressure for learning ability about partners should be strong on both sexes. Assuming the appropriate cognitive infrastructure, in such a case we expect males and females to have the ability to learn where their mate—the co-provider of resources for their offspring—is at any given time. This "learning the location of mates" ability may also be demonstrated during courtship.

In many mating systems, only a single sex provides resources for offspring, and that sex is predominantly the female sex. There are many reasons that females are more likely to take this role (see Chapters 7 and 8), but for the purposes of Domjan and Hollis's hypothesis, what this translates into is that males in such systems should be better at learning about the location of mates than females. To see why, think about it like this: In terms of parental care for subsequent offspring, females are a valuable resource for males, as they alone provide food for offspring. From a parental care perspective, males are much less valuable a resource to females. Males then are under strong selection pressure to find receptive females, while females can almost always find males that are willing to mate. Hence, in such mating systems, selection for learning locations associated with potential mates should be stronger in males than in females. More experimental work needs to be done in this area, but some support for this hypothesis has been found. First, in both Mongolian gerbils and gourami fish, parental investment is shared, and differences in learning about mates between the sexes is small. Second, in contrast to Mongolian gerbils and gourami fish, in Japanese quail, where there is no parental investment on the part of males, males show greater learning abilities than do females, though the extent of differences in learning between males and females may be specific to certain mating contexts (Gutierrez and Domjan, 1996, 2011; Hollis et al., 1989; Villarreal and Domjan, 1998; Figure 5.23).

A **B**

FIGURE 5.23. Parental investment and learning ability. (A) Parental investment is shared in blue gourami, and differences in learning about pairmates between the sexes is small. **(B)** In Japanese quail, the females care for the young, as shown by this female at the nest with her eggs, and there is no parental investment by males. In this species, males show greater learning about mates abilities than females. *(Photo credits: © Wil Meinderts/ FotoNatura/Minden Pictures; public domain)*

LEARNING ABOUT FAMILIAL RELATIONSHIPS

How animals recognize kin is still a subject of contention, but in some species learning may play a role (Fletcher and Michener, 1987; Hepper, 1991). If animals can learn how they themselves are related to others, as well as how different individuals in their group are related to one another, natural selection might favor altruistic and cooperative behavior being preferentially allocated to close genetic kin (Chapter 9). Here we examine learning and kin recognition in the context of helpers-at-the-nest.

As discussed in Chapters 2 and 3, in some species of birds and mammals, individuals forgo direct reproduction and instead help their relatives raise their offspring (Brown, 1987; Solomon and French, 1996; Stacey and Koenig, 1990). Often, these offspring remain at their natal nest and help their parents raise a subsequent brood (the helper's siblings). But this is only one way in which helping may emerge. For example, young, reproductively active long-tailed tits (*Aegithalos caudatus*) breed independently as soon as they can, but most nests fall victim to predation on the young (Figure 5.24). At that point, breeders often become helpers at the nests of their close genetic relatives, and such helpers accrue indirect fitness benefits by helping raise their kin (Hatchwell et al., 2004). How do the birds know who are kin? Do they learn who is kin, and who isn't, and if so, how?

To address these questions, Stuart Sharp and his colleagues ran a series of experiments that focused on the "churr" call made by long-tailed tits. This call develops before young birds fledge and leave the nest, and it remains very consistent throughout the lifetime of an individual (Sharp and Hatchwell, 2005). Churrs are given by males and females in the context of short-range communications, such as those regarding nest-building and aggression (Gaston, 1973; Hatchwell et al., 2001; Sharp and Hatchwell, 2005).

Sharp and his team used a "playback" experiment, in which an individual would hear the taped call of either a close genetic relative or a nonrelative. Individual birds showed a strong preference for the calls given by their kin (Figure 5.25). Following the playback experiment, Sharp and his colleagues designed an experiment to assess whether the birds learned the churr calls of their relatives, or whether their preference was based on genetic predispositions for certain types of churr calls. They ran a cross-fostering experiment (see Chapter 2), in which chicks were either raised with their biological parents or were switched to another nest and raised by foster parents.

A number of lines of evidence suggest that the churr call is learned: (1) The calls of foster siblings raised together were about as similar as the calls of biological siblings raised together; (2) the calls of biological siblings raised apart were as dissimilar as the calls of unrelated individuals in nature; (3) the songs of foster parents and their foster offspring were very similar, whereas the songs of biological parents and their offspring were different when those offspring were raised by foster parents. These results all suggest an important role for learning in the development of churr calls: calls that are subsequently used to distinguish kin from nonkin.

LEARNING ABOUT AGGRESSION

Ethologists and behavioral ecologists have had a long-standing interest in aggressive behavior (see Chapter 15; Archer, 1988; Dugatkin, 1997a; Gadagkar,

FIGURE 5.24. Learning who is kin. Young long-tailed tits (*Aegithalos caudatus*) often become helpers at the nests of their close genetic relatives, helping build nests and forage for food to feed the chicks. As helpers, they accrue indirect fitness benefits by contributing to the survival of their close genetic kin. *(Photo credit: Andrew MacColl)*

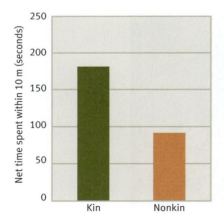

FIGURE 5.25. Playback calls and kin. Individual birds showed a strong preference for the calls given by their close genetic kin, staying for a longer time near the speakers that gave off the calls of their kin. *(From Sharp and Hatchwell, 2005)*

1997; Huntingford and Turner, 1987). In the animal behavior literature, aggression is often partitioned into two components: intrinsic and extrinsic factors (Landau, 1951a, 1951b). Intrinsic factors usually refer to traits that correlate with an animal's inherent fighting ability (G. A. Parker, 1974a). The most common of these factors is some measure of size (Archer, 1988; Huntingford and Turner, 1987; Hurd, 2006).

Extrinsic factors include what have come to be known as "winner" and "loser" effects (see Chapter 15). Winner and loser effects are defined as an increased probability of winning an aggressive interaction based on past victories, and an increased probability of losing an aggressive interaction based on past losses, respectively. But how can we study whether learning is involved with either winner or loser effects? To address this question experimentally, ideally we would work with a species in which aggression per se was well understood, so that we could have a strong baseline from which to launch into studying learning and winner and loser effects in the context of such aggression. For this reason, Karen Hollis and her colleagues worked on learning and winner and loser effects in blue gourami fish (*Trichogaster trichopterus*), a species in which males are territorial and aggression has been studied in numerous contexts (Hollis et al., 1995).

Hollis and her team had previously demonstrated that male blue gouramis could be trained to associate a light with either the presence (+) or absence (−) of an intruder male in their territory. They found that after the presentation of the light cue, individuals that had learned that a light meant the presence of another fish were much more aggressive toward an intruder than fish that saw a light that was not paired with an intruder, presumably because the former were prepared for a fight (Figure 5.26). This work, in and of itself, does not touch

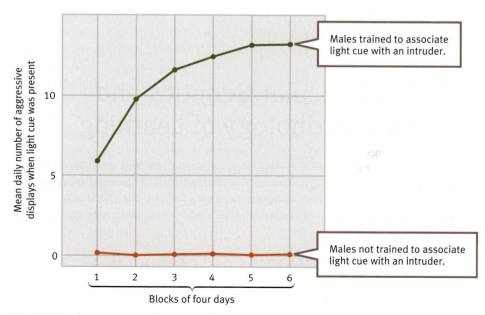

FIGURE 5.26. Pavlovian learning in fish. Males that had learned to associate a light with the presence of another male were more aggressive when the light cue was present than were males that did not associate the light with the presence of another male. *(From Hollis, 1984)*

A

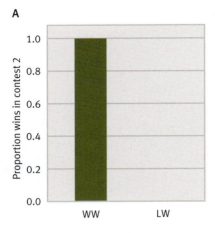

B

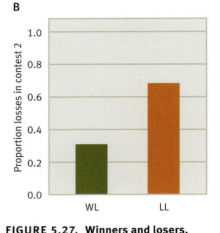

FIGURE 5.27. Winners and losers.
(A) Males that won in contest 1 were more likely to win in contest 2 (WW) than were males that had lost in contest 1 (LW).
(B) Males that lost in contest 1 were more likely to lose in contest 2 (LL). *(From Hollis et al., 1995, p. 129)*

on the issue of winner and loser effects per se, but rather sets the stage for experimental work that does.

Hollis used a similar protocol to examine the effect of conditioned learning on winner and loser effects (Hollis et al., 1995). The idea was to examine whether conditioning (Pavlovian learning) could have long-term effects on territorial males by creating the start of a series of encounters that would produce a winner or loser effect. Half the gouramis in the study were trained to associate a red light with the presence of a territorial intruder. The other half were control males that were presented with the red light six hours before they saw a territorial intruder, and as a result of the time delay did not pair the red light with an intruder. Males were randomly selected to be in the intruder/light treatment or the control, so there should have been no difference between males in fighting ability in the treatment versus the control groups. Hollis's experiment involved two contests. In the first contest, after a red light was turned on, trained males and controls were pitted against each other until a clear winner and loser emerged. As expected from the earlier work, trained males were much more likely to win such encounters.

In the second contest, winners and losers from the first contest were paired with new intruders three days later. In these pairings, no red light was shown to either fish. In this experiment, both winner and loser effects were found. All individuals that won their fights in the second contest had been winners in the first contest, and these individuals were the trained males that had seen the signal light before their first encounter (Figure 5.27). That is, males trained through conditioning were not only more likely to win initial encounters, but they also won subsequent encounters in which the stimulus they had been trained with was now absent. A similar effect was found with respect to losing. Those males that were trained to associate the red light and an intruder, but lost their fight in the first contest, were likely to lose to new intruders even in the absence of the stimulus. These results suggest that learning may be a powerful force in shaping aggressive interactions.

Molecular Genetics and Endocrinology of Learning

Modern techniques in both molecular genetic analysis and endocrinology are shedding light on proximate aspects of learning. Both of the studies discussed in this section focus on fear avoidance learning in rats, but the lessons drawn from these studies may apply to learning in a wide variety of contexts.

MOLECULAR GENETICS OF LEARNING IN RATS

Psychologists have long used rats as a model system for studying learning. Recent work on learning has attempted to tie together long-term breeding experiments in rats with the molecular underpinnings of various types of learning. For example, Shumin Zhang and his colleagues examined the molecular genetics of avoidance learning in two lines of rats that have been selectively bred for over forty years (F. R. Brush, 2003; F. R. Brush et al., 1999; Zhang et al., 2005).

The two lines of rats—known as the Syracuse High Avoidance (SHA) line and the Syracuse Low Avoidance (SLA) line—are descended from a single, large population of rats in 1965. Each generation of rats was tested on its tendency to avoid auditory and visual cues associated with a foot shock. Although the protocol varied slightly over time, in the basic protocol a rat was placed in a cage with two compartments and could move freely between these compartments. Just before a series of foot shocks was to be delivered to the compartment that the rat was in, a light and a tone were set off. The rat underwent ten training sessions in such a protocol. Rats were then tested in sixty "avoidance" trials to see how often they would move to the other compartment once the sound and tone were presented—that is, the rats were tested as to whether they paired these signals with shock and would attempt to avoid such shock. In each generation, those rats that were best at avoiding shock (SHA) were bred with one another, and those that were poorest at such an avoidance task (SLA) were bred with one another.

Over the course of more than forty years of selective breeding, SHA animals eventually avoided shocks in forty of sixty trials (on average), while the SLA typically displayed such avoidance learning in none of the sixty trials. Other work had shown that the SHA and SLA rats were equally active in their normal daily routines and that they did not differ in their ability to detect shock or the visual and auditory cues used during the experiments. But these two strains of rats did differ in "fearlessness," such that the SLA rats showed much higher levels of anxiety in a series of experiments that were separate from the avoidance learning trials described above. It seems that SLA rats were both anxious and poor at learning to avoid unpleasant cues (shocks), while the converse held true for SHA rats. Further support for this association comes from studies that demonstrate that rats that were administered drugs that are known to reduce anxiety became better at avoidance learning (Fernandez-Teruel et al., 2002; Pereira et al., 1989; Sansone, 1975).

To better understand the molecular underpinnings of the learning differences between the SHA and SLA lines of rats, Zhang and his colleagues ran a series of tests to determine whether these lines showed differences in gene expression in the hippocampus—an area of the brain known to be important in avoidance learning as well as anxiety (Zhang et al., 2005). To begin their work, Zhang's team ran rats from both lines through ten training sessions and then through sixty avoidance learning trials. Then they selected SHA rats that showed avoidance learning in 70 percent (or more) of their trials and SLA rats that displayed such avoidance learning in less than 10 percent of their trials.

After the learning trials, the researchers removed and measured gene expression in the hippocampus of each rat. Initially sifting through gene expression in 7,500 genes, and correcting for statistical problems associated with sampling expression patterns in so many genes, Zhang and his team were able to distill their system down to eight candidate genes that were differentially expressed in the SLA and SHA rat lines. Four of these genes—*Veli1, SLC3a1, Ptpro,* and *Ykt6p*—showed greater expression in the hippocampus of SHA rats, while four others—*SLC6A4/5HTT, Aldh1a4, Id3a,* and *Cd74*—were expressed in greater quantities in the brains of SLA rats. From these results, Zhang and his colleagues argue that complex traits like avoidance learning may be controlled by many genes, each of which contributes only a small amount to phenotypic expression. Exactly how the differences in gene expression that Zhang and his team found translate into different behavioral phenotypes is not yet understood.

ENDOCRINOLOGY OF LEARNING IN RATS

The molecular genetic studies we have just discussed elucidate one proximate factor underlying learning in rats. Let's now review a case study in which another proximate factor associated with learning in rats—hormones, and in particular stress hormones—has been examined.

Glucocorticoids such as corticosterone are hormones that play a large role in the stress responses and learning of many animals (de Kloet et al., 1999). In fact, experimental work shows not only that glucocorticoids have an effect on learning and memory in adult animals but also that when pregnant female rats are stressed and glucocorticoid levels rise, the offspring of such females show high levels of anxiety and perform suboptimally in learning tests (Lemaire et al., 2000; Weinstock, 1997). In both mature individuals and juveniles, glucocorticoids easily cross the "blood-brain" boundary and enter the brain, where they can affect emotional state and cognitive abilities. This may explain why stressing pregnant mothers affects anxiety in their offspring.

The hippocampal section of the brain contains receptors, including the mineralocorticoid receptor (MR) to which glucocorticoids bind (de Kloet et al., 1998). To better understand the relationship between glucocorticoids, stress, and learning, Ana Herrero and her colleagues administered a series of behavioral tests to a group of rats and then measured the level of various hormones in the animals. Their results are correlational rather than causal, but they shed light on learning and stress (Herrero et al., 2006). Herrero and her team first exposed a group of rats to tests designed to examine the fear response (one test involved rats moving through a maze, and the second involved measuring how rats respond to large open fields). Rats appear to fear open environments—that is, environments lacking cover—and both of these tests involve placing a rat in an open environment and examining its stress response to that environment. The researchers assessed the rat's stress by measuring such variables as the amount of time the rat spent frozen (unmoving) and the rate at which the rat defecated. The higher these values, the greater the stress and anxiety attributed to the animal. Some rats were only exposed to one fear test, and others were exposed to both fear tests. Rats exposed to both tests were consistent in that they were either relatively anxious in both or not anxious in either.

Once Herrero and her colleagues had established that rats are consistent in their response to open environments, they ran 140 rats through one fear test, and they classified the rats as "high-anxiety" or "low-anxiety" animals. After this fear test, rats were tested on their spatial learning skills. This involved placing the rats in a water maze and measuring their abilities to find and remember the location of a submerged escape platform on which they could rest. At various points in their experiments, Herrero's team measured either the rats' plasma corticosterone levels—a rough measure of the amount of corticosterone circulating in their blood—or the number of mineralocorticoid receptors in their hippocampus.

Although all animals—those classified as high and low anxiety— eventually learned to swim to the submerged platforms, high-anxiety individuals took significantly longer than low-anxiety animals to learn to do so, mostly because high-anxiety animals spent more time swimming close to the edge of the water tank (Figure 5.28). When blood corticosteroid levels were measured in rats that had been run through the water maze, Herrero found that high-anxiety animals had higher corticosterone levels than did low-anxiety animals (see Chapter 3). In addition, high-anxiety animals had *fewer* mineralocorticoid receptors in their hippocampus. This finding makes some sense, as having fewer mineralocorticoid receptors results in a reduced ability to bind corticosterone, and indirectly leads to an increase in circulating stress hormones.

It is not clear what the cause-and-effect relationship is in Herrero's studies. It may be, for example, that rats with few mineralocorticoid receptors and high-circulating corticosterone became anxious in the open-environment tests, and then scored poorly in the water maze test. Or it could be that the open environment tests caused a change in the availability of mineralocorticoid receptors and circulating corticosterone, and animals with increased circulating corticosterone and decreased availability of mineralocorticoid receptors then did poorly on the water maze test. Further experiments are needed to decipher a cause-and-effect relationship, but the work of Herrero and her colleagues clearly demonstrates a link between stress hormones, stress hormone receptors, anxiety, and learning.

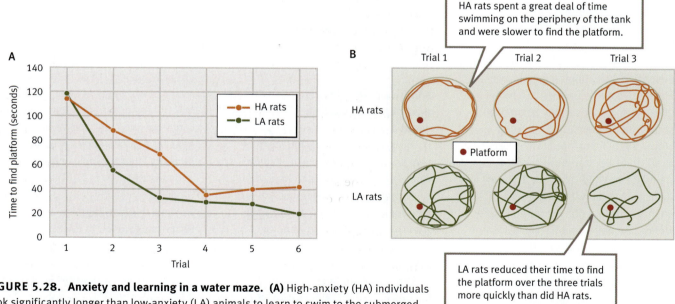

FIGURE 5.28. Anxiety and learning in a water maze. (A) High-anxiety (HA) individuals took significantly longer than low-anxiety (LA) animals to learn to swim to the submerged platform, mostly because high-anxiety animals spent more time swimming close to the edge of the water tank, as shown in **(B)** diagrams of the rats' swimming paths in their first three training trials. *(From Herrero et al., 2006)*

INTERVIEW WITH
Dr. Sara Shettleworth

How did you become interested in studying learning? Were you trained as a psychologist or a biologist?

As an undergraduate, I was drawn to psychology by the introductory course taught by Henry Gleitman. He was a wonderful lecturer who went on to write a popular introductory textbook and win awards for his teaching. Much of the course dealt with the then-current controversies in learning theory, and I found the interplay of theory and experiments fascinating.

In graduate school I was exposed to biological approaches to behavior, which piqued my interest in what role learning played in the natural lives of animals. As it happened, it was at about that time that conditioned taste aversion was discovered by Garcia and his colleagues, and not long afterward Brown and Jenkins first described autoshaping in pigeons. These discoveries were mind-blowing at the time because they seemed to show that animals learn some things with very minimal input and others not at all. To understand these patterns, we might need to take into account the role that learning might play in the animal's natural life. Although I have worked on quite a number of different species and learning problems since then, it has always been with a commitment to this point of view. As for biological training, I have been fortunate to collaborate or otherwise be associated with biologists who are interested in psychological questions, and I think we have learned a lot from one another.

Do we know of any species of animal that can't learn something?

I don't know of any. Considering that even the single-celled organism *Stentor* can modify its behavior in simple ways, as shown by Jennings 100 years ago, it might be hard to find one that did not at least show habituation or sensitization.

Is there any reluctance in mainstream psychology to study learning from an evolutionary perspective? If so, why?

Historically, learning and memory were studied pretty much in isolation from anything about the biology of the species doing the learning. People have suggested that the prominence of learning in American psychology and its abiological emphasis are attributable to the American faith in the importance of the environment as opposed to hereditary factors in individual development and adult achievement. The relationship between psychology and evolutionary thinking has been changing, partly due to discoveries within psychology like

those I mentioned in answering the first question. The increasingly important role of neuroscience and genetics in psychology probably also play a role in making people think more and more of psychology as, in effect, part of biology. At the same time, behavioral ecologists have brought the evolutionarily based study of behavior full circle, back to an interest in causal mechanisms and development, which tended to be neglected in the early days of that field. Indeed, the term *cognitive ecology* was invented to refer to the study of the role of cognitive mechanisms—perception, learning, memory, and the like—in solving ecological problems. These developments have led to more cross-disciplinary communication, collaboration, and training. This integrated approach is especially evident in the burgeoning subfield of comparative cognition, which deals not only with processes of learning but other aspects of cognition, such as social and physical understanding. Many of the questions addressed in contemporary comparative cognition arise from observations of natural behavior, such as tool use by corvids and primates.

How much of a quagmire is the terminology used in the study of learning? To the outside reader it seems as if there are endless definitions and subdefinitions. Is that a fair statement?

No, I don't think it's fair. Fields of science all have their own specialized terminologies, which are necessary to convey specialized ideas and distinctions that can't be expressed concisely in ordinary language. When I ask psychology students to read literature from behavioral ecology, they find terms like *ESS, MVT, conspecific, homology versus analogy, phylogeny,* and the like, pretty baffling.

What do you think ethologists can glean from work on human learning?

In his famous paper "On Aims and Methods of Ethology," Tinbergen pointed out that a complete understanding of behavior includes answers to four questions. Two of them are "What is the current (or proximate) cause of this behavior?" and "How does it develop in the individual?" (The others are "How did it evolve?" and "How does it function?" or "What is its survival value?") Causation and development are essentially what psychologists study. So it seems obvious that ethologists and psychologists (whether they study humans or other animals) should have a lot to learn from each other.

Could you weigh in on the "modular mind" debate? Do you think learning is better viewed as one all-purpose algorithm or as a series of smaller programs, each designed by selection to allow animals to cope with particular sorts of problems (foraging, mating, etc.)?

I am definitely of the "modular mind" school because I think it makes more functional sense and because it also makes better sense of some data. It is also a more sensible way to approach a broad comparative psychology than the old idea that species differ in some single dimension like "intelligence." Along with other psychologists of this persuasion, I tend to think of modules more as cognitive subunits that perform distinct information-processing operations rather than as mechanisms for separate biological functions like mating and food-finding. For example,

learning and using information from landmarks might be identified as a distinct (modular) part of spatial cognition, but it could be employed in finding food or mates or a nest. Of course, there have been rather few tests of whether this is correct. In laboratory tests, in rats, spatial learning seems to proceed similarly whether the animal is rewarded with food (i.e., foraging) or escaping from a water tank or some other aversive situation. As a key concept in evolutionary developmental biology, modularity also provides a

> The term *cognitive ecology* has been invented to refer to the study of the role of cognitive mechanisms— perception, learning, memory, and the like— in solving ecological problems.

framework for comparing cognition across species. For instance, young human children share basic processes of spatial, numerical, social, and other aspects of cognition with other animals, whereas later-developing abilities such as counting and using maps may be uniquely human.

What's the next breakthrough to look for in the study of animal learning?

Although there are still many unanswered questions about animal learning and cognition that need to be studied entirely at the level of behavior, at present there is probably more research on learning being done by behavioral neuroscientists and geneticists than by people like me who focus on behavior of normal

intact animals. Thus, statistically it is most likely that the biggest breakthroughs will be made in studies of the neural and molecular basis of learning. However, while much work of this kind simply uses traditional behavioral tests like maze learning or Pavlovian conditioning, some behaviorally sophisticated researchers in this area are making novel observations about how animals learn and remember. For example, in a quest to develop simple tests for rats in which the animals would remember many items of information, Howard Eichenbaum and his colleagues discovered that—not surprisingly given their nocturnal way of life— rats are extraordinarily good at olfactory learning. A rat can learn and remember the significance of many different odors and at the same time. As well, when odors are used, rats can easily learn kinds of tasks that they would learn only with great difficulty, if at all, with visual stimuli, which have traditionally been used. This example illustrates how an integrated approach to learning can lead to new findings: by using ethologically relevant stimuli the researchers have revealed new facts about animal learning and memory, as well as making possible investigations of the neural basis of learning that would not be possible otherwise. An important future frontier will be to use knowledge of the neural and molecular basis of cognition with information about species differences in cognition. For example, can we relate cognitive differences among humans, chimpanzees, and bonobos to differences in their brains and ultimately their genomes?

Dr. Sara Shettleworth is a professor at the University of Toronto, Canada. Her work integrating biological and psychological approaches to the study of animal cognition and learning has made her a leader in that field.

SUMMARY

1. Here, we define *learning* as "a relatively permanent change in behavior as a result of experience." This definition does, however, have a downside, in that it is often difficult to determine how long a time period constitutes "relatively permanent."

2. The ability of organisms to learn provides them with the opportunity to respond in a very flexible fashion to environmental change. Learning, in the most general sense, is considered a form of phenotypic plasticity.

3. The simplest form of learning involves a single stimulus. Sensitization and habituation are two simple, single-stimulus forms of learning. Habituation in particular is often of concern to experimental ethologists, whose interpretation of results can be difficult to interpret if animals habituate too quickly to stimuli.

4. Pavlovian, or classical, conditioning experiments involve two stimuli: the conditioned stimulus and the unconditioned stimulus. A *conditioned stimulus* (CS) is defined as a stimulus that fails initially to elicit a particular response, but comes to do so when it becomes associated with a second (unconditioned) stimulus. Second-order conditioning, excitatory conditioning, inhibitory conditioning, learnability, blocking, overshadowing, and latent inhibition are issues often addressed in Pavlovian conditioning experiments.

5. Instrumental conditioning, also known as operant or goal-directed learning, occurs when a response made by an animal is somehow reinforced. One fundamental difference between Pavlovian and instrumental learning centers on the fact that, in the latter, the animal must undertake some "action" or "response" in order for the conditioning process to produce learning.

6. Interpopulation comparisons in learning are a powerful tool employed by ethologists interested in learning. By making comparisons across populations that differ in their abilities to learn, ethologists can address both proximate questions (What are the differences?) and ultimate questions (Why do such differences occur?).

7. Ethologists, behavioral ecologists, and psychologists have long argued that learning is favored over the genetic transmission of a fixed trait when the environment in which an animal lives changes often, but not too often. Work that expands on this idea has investigated within- and between-generation variability. Theory suggest that learning is favored when predictability within the lifetime of an individual is high but predictability between generations is low.

8. Animals learn in many different contexts, including but not limited to, foraging (what to eat?), habitat selection (where to live?), predators (what's dangerous?), mates (what constitutes a mate, and what constitutes a good mate?), and familial relationships (who is genetic kin?).

9. Modern techniques in both molecular genetic analysis and endocrinology are shedding light on learning. Researchers, for example, have studied the molecular genetics of avoidance learning in rats. Evidence suggests that at the proximate level differences in avoidance learning are associated with a small number of candidate genes that are differentially expressed in these two lines of rats.

DISCUSSION QUESTIONS

1. Following up on the "optimal forgetting" study in stomatopods that we discussed in this chapter, can you think of other situations in which it might pay for animals to forget, or at least not act on, information they have obtained? Try and come up with three cases, and write a paragraph on each case justifying its selection.

2. Obtain a copy of parts I and II of Tooby and Cosmides's 1989 article "Evolutionary psychology and the generation of culture," in volume 10 (pp. 29–97) of the journal *Ethology and Sociobiology*. After reading this article, explain how "Darwinian algorithms" work and how they relate to our discussion of animal learning.

3. Read Domjan and Hollis's 1988 chapter "Reproductive behavior: A potential model system for adaptive specializations in learning," which appeared in Bolles and Beecher's book *Evolution and Learning* (pp. 213–237). Then outline how classic psychological models of learning can be productively merged with evolutionary approaches to learning.

4. Design an experiment that can distinguish between the two alternative explanations for interpopulational differences in dove foraging, as described in Carlier and Lefebvre's 1996 "Differences in individual learning between group-foraging and territorial Zenaida doves," which appeared in volume 133 (pp. 1197–1207) of the journal *Behaviour*.

SUGGESTED READING

Domjan, M. (2009). *The principles of learning and behavior* (6th ed.). Belmont, CA: Wadsworth Publishing. An excellent primer on the psychology and biology of learning.

Kim, J. J., & Jung, M. W. (2006). Neural circuits and mechanisms involved in Pavlovian fear conditioning: A critical review. *Neuroscience and Biobehavioral Reviews, 30,* 188–202. A review of Pavlovian conditioning and the role that neurobiology plays in this form of learning.

Krebs J. R., & Inman, A. J. (1992). Learning and foraging: Individuals, groups, and populations. *American Naturalist, 140,* S63–S84. A review of learning and its role in foraging studies.

Mery, F., & Burns, J. G. (2010). Behavioral plasticity: An interaction between evolution and experience. *Evolution Ecology, 24,* 571–583. A review of learning and phenotypic plasticity.

Shettleworth, S. (2009). *Cognition, evolution and behavior.* (2nd ed.). New York: Oxford University Press. A very detailed treatment of behavior and evolution, with much on learning.

6

Cultural Transmission

FIGURE 6.1. Imo the monkey. Imo, a Japanese macaque, introduced a number of new behaviors (for example, potato washing) that spread through her population via cultural transmission. *(Photo credit: Umeyo Mori)*

In the discussion of genetics and heritability in Chapter 2, we noted that for natural selection to act on a behavior, a mechanism for transmitting that behavior across generations is required. When Mendel's work on genetics was rediscovered in the early 1900s, it became obvious that genes are a means of transmitting traits across generations, enabling natural selection to act on genetically encoded traits. In fact, until recently, evolutionary biologists and ethologists operated under the assumption that genes were not only *one* way to transmit information across generations, they were the *only* way. Slowly, this view is beginning to change. Recall, for example, our discussion in Chapter 1 of Jeff Galef's work on foraging and information transfer across generations of rats. More generally, there is a growing recognition of the importance of the **cultural transmission** of behavior—often defined as the transfer of information from individual to individual through social learning or teaching—both within and between generations of animals (see below for more on the definition of *cultural transmission*; Bonner, 1980; Boyd and Richerson, 1985; Danchin et al., 2004; Heyes and Galef, 1996; Laland and Janik, 2006; Odling-Smee et al., 2003; Reader and Laland, 2003; Zentall and Galef, 1988).

To get a glimpse of how cultural transmission operates in nonhumans, let's consider the case of Imo, a Japanese macaque monkey that lived on Koshima Islet, Japan, in the 1950s (Kawai, 1965; Kawamura, 1959; Figure 6.1). Imo's story begins when ethologists studying her troop of macaques threw sweet potatoes on the sandy beach for the monkeys to gather and eat. When Imo was a year old, she began to wash the sweet potatoes in water before she ate them. This novel and creative behavior, which was never before seen in Imo's population, allowed her to remove all the sand from the sweet potatoes before she ingested them. But what made this new behavior remarkable is not that Imo found a novel solution to cleaning her food; rather, the key feature of this system is that many of Imo's peers and relatives learned the skill of potato washing from Imo via **social learning**, which is usually defined as the process of learning by watching others. By 1959, most infant macaques in Imo's troop intently watched their mothers, many of whom had acquired Imo's habit, and they learned to wash their own sweet potatoes at early ages. Because a novel behavior—potato washing—had spread through Imo's troop by social learning, we can speak of the spread of potato washing as an example of cultural transmission (Figure 6.2).

When Imo was four, she introduced an even more complicated new behavior into her group. In addition to the sweet potatoes that researchers gave the monkeys on Koshima Islet, they also occasionally treated the monkeys to a food item to which the macaques were partial—wheat. The introduction of this new source, however, caused a problem, which was that the wheat was provisioned to the monkeys on a sandy beach, and the wheat and sand mixed together. Imo came up with a novel solution—she tossed her wheat and sand mixture into the water, where the sand sank and the wheat floated. As with the sweet potatoes, her groupmates soon learned this trick from her. It took a bit longer for this trait to spread through the population, however, as monkeys aren't used to letting go of food once they get it, so it was hard to learn to throw the sand-covered wheat into the water. But eventually this new behavioral trait spread to many group members via cultural transmission.

Imo's actions were not the only ones to attract attention about the cultural transmission of behavioral traits in non-humans and many similar cases are

FIGURE 6.2. Potato washing in monkeys. In Japanese macaques living on Koshima Islet, Japan, the skill of potato washing appears to be transmitted culturally. *(Photo credit: Frans de Waal)*

now on record. Michael Huffman, for example, found another instance of cultural transmission in macaques (Huffman, 1996; Nahallage and Huffman, 2006). Huffman's study revolved around decades of work on the Japanese macaques of the Iwatayama National Park in Kyoto, Japan. Early in his work, Huffman began to observe a behavior never before noted in macaques—individuals would play with stones, particularly right after eating (Hiraiwa, 1975; Figure 6.3).

In the Iwatayama National Park where Huffman studies macaques, stone play behavior was first observed in 1979 when Glance-6476, a three-year-old female macaque, brought rocks in from the forest and started stacking them up and knocking them down. Not only that, Glance-6476 was very territorial about her stones and took them away when approached by other monkeys. When Huffman returned to study Glance's troop four years later, stone play (also referred to as stone handling) was common and was being transmitted from older to younger individuals. Interestingly, cultural transmission in this system seems to work down the age ladder, but not up. While many individuals

FIGURE 6.3. Stone play in monkeys. In one population of Japanese macaques in the Iwatayama National Park in Kyoto a tradition of "stone play," in which individuals stack up stones and then knock them down, has been observed. *(Photo credit: Michael Huffman)*

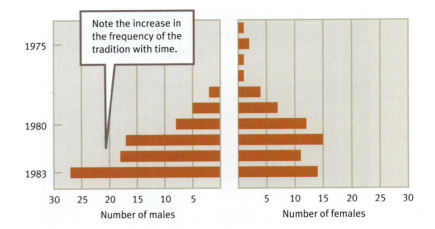

FIGURE 6.4. Stone play tradition spreads. Orange bars represent verified stone handlers. *(From Huffman, 1996, p. 276)*

younger than Glance-6476 acquired her stone play habit, no one older than her added stone play to their behavioral repertoire (Figure 6.4).

Charmalie Nahallage and Huffman later examined stone play in a captive troop of macaques in the Primate Research Institute in Kyoto, Japan (Nahallage and Huffman, 2006). When they did a detailed study of the use of stone play in juvenile and adult macaques, they uncovered age-specific differences in the use of culturally acquired behavior. Juveniles tended to engage in many short bouts of stone play that involved vigorous body actions (for example, shaking, running, and jumping). This pattern of behavior is consistent with what is called the "motor training" hypothesis of play behavior (see Chapter 16), which suggests that stone play may facilitate the development of perceptual and cognitive skills (Bekoff and Byers, 1981; Leca et al., 2011). Adults, on the other hand, engaged in fewer, but longer bouts of stone play, which may slow down the deterioration of cognitive processes often seen in aging primates (Figure 6.5). More work is needed to fully test these hypothesized benefits of stone play.

Stone play has now been uncovered in other captive troops of macaques like the monkeys that Nahallage and Huffman study. But this behavior has never been documented in wild populations that are not provisioned by humans.

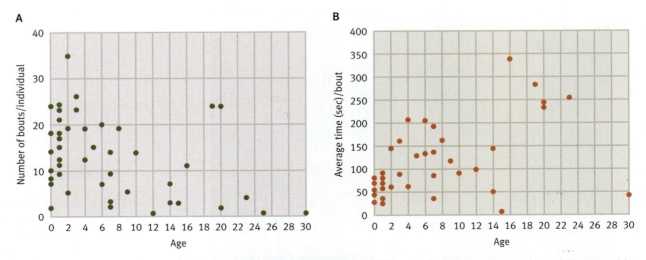

FIGURE 6.5. Stone play and age. (A) The number of stone play sessions decreases with age. **(B)** The average time per stone play session increases with age. *(From Nahallage and Huffman, 2007)*

Stone play appears to occur only in populations that have significant "leisure time"—that is, in populations that are freed up because their food is provided for them (Nahallage and Huffman, 2007).

Cultural transmission has also been extensively documented in other primates, including chimpanzees (Whiten, 2005, 2011; Whiten and Boesch, 2001; Whiten et al., 1999, 2005). Over the course of the last thirty years, six long-term studies of natural populations of chimpanzees have documented an incredible array of culturally transmitted traits, ranging from "swatting flies" and opening nuts with stones, to squashing parasites with leaves. In some cases, researchers have clear evidence of young chimps watching and learning these behaviors from adults; in other cases, the evidence is more indirect (see Chapter 16).

When looking across these chimpanzee populations, we see a hallmark signature of cultural transmission: numerous traits are culturally transmitted in every population, but populations differ from one another in terms of *which* traits are transmitted culturally. For example, hammering open nuts with stones is seen in two chimp populations, but it is absent in four. Hammering nuts, once it arose by such trial-and-error learning, could quickly spread by cultural transmission, but only in the populations in which one of the members stumbles on the solution in the first place (or a population into which such an individual migrates).

The reason for opening a discussion of cultural transmission with primate examples is not that they provide the most convincing experimental evidence of cultural transmission recorded in animals. Rather, these studies are observational, and they often lack the controlled experimental manipulations that underlie other work on nonhuman cultural transmission (Galef, 2004). But these examples of food washing, stone handling, hammering open nuts, and swatting away flies dramatically illustrate that cultural transmission in animals can be a powerful force.

What Is Cultural Transmission?

In order to understand the role of cultural transmission in animal societies, we need to begin with a definition. What exactly do we mean when we speak of cultural transmission in animals? This turns out to be a more difficult question than it might appear (Figure 6.6). In fact, even back in 1952, anthropologists already had more than 150 definitions of *culture* and *cultural transmission* (Kroeber and Kluckhohn, 1952). Evolutionary and behavioral biologists, however, have tended to use the following definition. Cultural transmission is a system of information transfer that affects an individual's phenotype by means of either teaching or some form of social learning (Boyd and Richerson, 1985). Recall the Norway rats discussed in Chapter 1. As scavengers, rats often encounter potential new foods. The cost to scavenging is that many novel foods may in fact be dangerous to eat. Scavenging is an ideal behavior in which to examine the cultural transmission of information, as new items are always being encountered, and information on their potential danger constantly needs to be updated. Obtaining some sort of information about food from other rats is one way to update knowledge about the suitability of novel foods. In fact, Norway rats learn what new foods to try by smelling

FIGURE 6.6. What constitutes cultural transmission? The child (left) is learning to use utensils by watching others (that is, through cultural transmission). The young chimp on the right, also has learned its nut-cracking skills from watching others.

their nestmates and subsequently trying the new food items that those nestmates have recently ingested (Galef, 1996, 2009; Galef and Wigmore, 1983). Returning to our definition of what constitutes cultural transmission, one aspect of a rat's phenotype—in this case, what sorts of food it eats—is modified by information that it has learned from other individuals.

WHAT'S SO IMPORTANT ABOUT CULTURAL TRANSMISSION?

Before moving on to a more detailed discussion of social learning and teaching in animals, let us briefly touch on an issue that ethologists, psychologists, and anthropologists often raise when discussing cultural transmission in animals. "Why," the question goes, "is there so much focus on cultural transmission?" After all, we spent a great deal of time on the topic of learning in Chapter 5, and if cultural transmission is just one form of learning—that is, learning from other individuals—why not simply consider it a special kind of learning and move on? In other words, as Boyd and Richerson ask, "Why not simply treat culture as a . . . response to environmental variation in which the 'environment' is the behavior of conspecifics?" (Boyd and Richerson, 1985). Why not think of cultural transmission as just another means by which organisms adapt to the environment?

The answer to these questions is that learning from other individuals in a group differs dramatically from individual trial-and-error learning in two important ways. To begin with, cultural transmission involves the spread of information *from individual to individual* (Figures 6.7 and 6.8). That is, if you learn from others and they learn from you, information can be spread through a population. This potentially translates into the behavior of a single individual in a population dramatically shifting the behavior patterns seen in an entire group—recall how Imo's novel food-washing techniques spread via cultural transmission. This is not the case for other types of learning. What an

FIGURE 6.7. Imitation in infants.
Imitation is a form of social learning that begins early in humans. Here an infant claps his hands in imitation of his mother clapping her hands. *(Photo credit: Jerry Tobias/Corbis)*

individual learns via individual learning—the sort of learning we discussed in Chapter 5—disappears when that individual dies (and perhaps earlier). When cultural transmission is in play, however, what is learned by one individual may be passed down through generations. If Imo had learned to clean her sweet potatoes by washing them in water in a population in which social learning was absent, this foraging innovation would have vanished when she died. Instead, decades later, one can still go to Koshima Islet and see monkeys washing their sweet potatoes.

Cultural transmission can occur within a population very quickly, which makes it a particularly potent form of information transfer. When natural selection acts to change the frequency of genes that code for behavior, the time scale can range from a few dozen generations (as in the guppy case we explored in Chapter 2) to much longer time scales (thousands of generations). And when natural selection acts on major morphological change, the time scale may be even longer (Figure 6.9). Cultural transmission of information, on the other hand, operates much faster, and can cause important changes in the behavior seen in populations in just a few generations. In fact, cultural transmission can have a dramatic impact within a single generation (Boyd and Richerson, 1985, 2004, 2005; Henrich et al., 2005; Odling-Smee et al., 2003; Reader and Laland, 2003).

FIGURE 6.8. Cultural transmission via teaching. Teaching is a form of cultural transmission in which the teacher imparts some information to a student faster than the student could learn it on her own. This piano teacher is teaching her young student how to place her hands to produce particular chords on the piano. *(Photo credit: fotosearch)*

EFFECTS OF OTHERS ON BEHAVIOR

Cultural transmission involves a "model" individual—sometimes called a demonstrator or tutor and an "observer,"—who learns a specific behavior or response from the model. But it is important to recognize that there are situations that involve an interaction between observers and models, but that do *not* constitute social learning or teaching. In these cases—labeled local enhancement and social facilitation—the observer is drawn to an area by a model or by the action of a model, or is simply in the presence of models, but the observer does not learn a particular behavior or response from the model, so cultural transmission is not occurring. We will first examine local enhancement and social facilitation before moving on to instances of cultural transmission via social learning and teaching.

FIGURE 6.9. Speed of change. In some cases, natural selection may take millions of years to produce major changes, as in horse evolution. Cultural evolution works on a much faster time scale. Natural selection can work much quicker than in the horse case, but it is generally much slower than cultural evolution. *(From Ridley, 1996)*

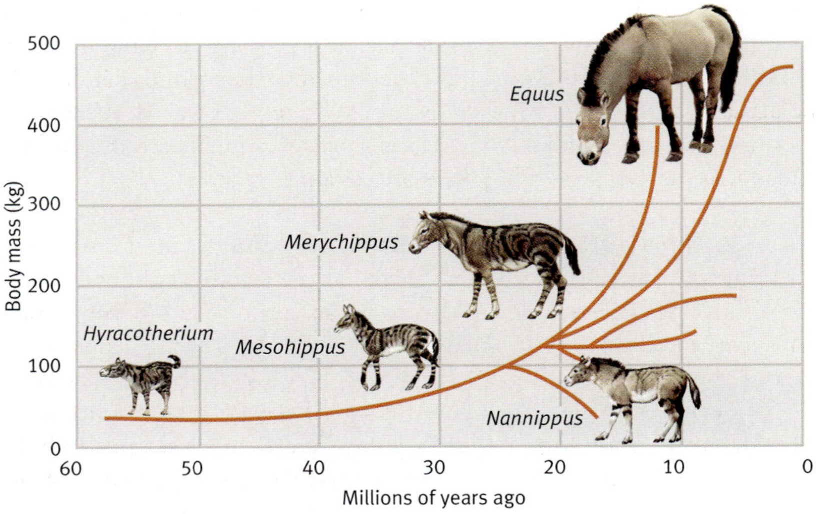

LOCAL ENHANCEMENT. William Thorpe coined the term **local enhancement** to describe the case in which individuals learn from others, not so much by doing what they observe, as by being drawn to a particular area because another individual—a model—was in that location (Heyes, 1994; Thorpe, 1956, 1963; Figure 6.10). In other words, when local enhancement is in play, a model simply draws attention to some aspect of the environment by the action it undertakes there (for example, digging for worms). Once the observer is drawn to the area, the observer may learn on its own, that is, via individual learning.

FIGURE 6.10. Local enhancement. If fish 1 is drawn to where fish 2 is foraging (near a stone) local enhancement is at work.

As an example of local enhancement, consider foraging behavior in colonially nesting cliff swallows, birds that feed in groups ranging in size from 2 to 1,000 individuals (see Chapter 2). Charles Brown has found that, in addition to the actual transfer of information between groupmates (C. R. Brown, 1986), local enhancement facilitates foraging, as some individuals are drawn to good foraging areas just because other birds are foraging there.

SOCIAL FACILITATION. **Social facilitation** differs in a subtle, but important, way from local enhancement. During local enhancement, the *action* of a model draws attention to some aspect of the environment. But under social facilitation, the *mere presence* of a model, regardless of what it does, is thought to facilitate learning on the part of an observer (Zajonc, 1965). For example, there are many instances in the foraging literature in which increased group size caused increased foraging rates per individual, perhaps because individuals had learned that the mere presence of others made them safer (Giraldeau and Caraco, 2000; Stephens and Krebs, 1986; Figure 6.11).

FIGURE 6.11. Social facilitation. In social facilitation, the mere presence of one or more models draws in an observer. Here a lone starling is attracted to a group, not because of what group members are doing or where they are, but simply because it is drawn to the presence of others. "Safety in numbers" might be a benefit to such facilitation.

FIGURE 6.12. Capuchin foraging.
Foraging in capuchins (*Cebus paella*) has been examined with respect to both local enhancement and social facilitation. These capuchin monkeys are foraging on palm nuts. *(Photo credit: Tui De Roy/Minden Pictures/Getty Images)*

Social facilitation and local enhancement can be experimentally separated from one another. To see this, consider a series of experiments that Elisabetta Visalberghi and Elsa Addessi ran on foraging behavior in capuchin monkeys (Visalberghi and Addessi, 2000; Figure 6.12). These researchers examined what factors affected a capuchin monkey's probability of eating a novel food. In one treatment, a *lone* capuchin was tested on its tendency to try a new food type (vegetables that had been color dyed). In a second treatment, a capuchin and the novel food type were on one side of a test cage, and a group of capuchins was on the other side of the cage. No food was placed on the side of the cage with the group. The third treatment was identical to the second, except that a familiar food type was placed on the side of the cage with the group, which made it likely that they would eat the food. Now the lone capuchin saw not only a group, but a group that was eating food, but not eating the novel food type (Figure 6.13). These treatments were carefully designed to create potential for social facilitation (treatment 2: mere presence of others) and local enhancement (treatment 3: presence of others that are eating). Treatment 1 served as the control condition.

When comparing across their treatments, Visalberghi and Addessi found evidence of local enhancement, but not of social facilitation. Evidence for local enhancement was uncovered in that during any given observation period, the test capuchin in treatment 3 (test capuchin + group with food) was more likely to be eating food than capuchins in treatment 1 (test capuchin alone). Local enhancement was occurring as the test capuchin's attention was drawn to the novel food when it saw the other capuchins eating food, and then the test capuchin proceeded to eat more itself. However, no definitive evidence

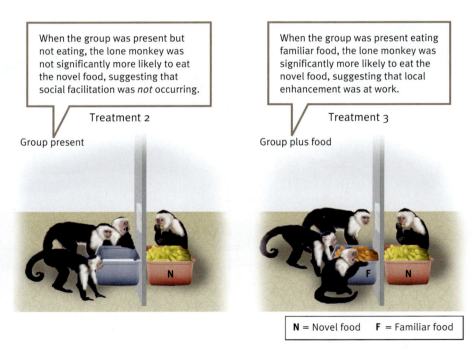

A lone capuchin is tested with a novel food.

Treatment 1

Alone

When the group was present but not eating, the lone monkey was not significantly more likely to eat the novel food, suggesting that social facilitation was *not* occurring.

Treatment 2

Group present

When the group was present eating familiar food, the lone monkey was significantly more likely to eat the novel food, suggesting that local enhancement was at work.

Treatment 3

Group plus food

N = Novel food   F = Familiar food

FIGURE 6.13. Social facilitation and local enhancement in capuchin foraging. The experimental design for the three treatments used in Visalberghi and Addessi's study. *(From Visalberghi and Addessi, 2000)*

for social facilitation was uncovered. While a comparison of treatments 1 and 2 (test capuchin versus test capuchin + group without food) did show more foraging in treatment 2, this difference was not statistically significant—that is, the mere presence of others who were not eating did not lead the lone capuchin to eat significantly more of the novel food.

SOCIAL LEARNING

Examples of social learning—observers learning specific behaviors from models—abound in the psychological literature on humans, as well as in the ethological literature on animals (Bandura, 1977, 1986; Dugatkin, 2000; Galef, 2009; Heyes and Galef, 1996; Rosenthal and Zimmerman, 1978; Zentall and Galef, 1988; Figure 6.14; see the Conservation Connection box). Social learning is sometimes referred to as "observational learning" in the psychology literature. A classic example of an experiment on social learning in humans is Albert Bandura's "bobo-the-clown" doll study (Bandura et al., 1961; Giancola and Chermack, 1998). In this experiment, children were exposed to one of two treatments. In one treatment, children were working on an art project when an adult in the room began punching and kicking a "bobo" doll. This doll was inflatable and weighted at the bottom, so it snapped back up when knocked down. In the other treatment, children working on an art project saw a calm adult. In this treatment, the bobo doll was still in the room; it just wasn't being assaulted by an adult. Children from both groups were then given the choice of playing with aggressive toys or nonaggressive toys around the bobo doll.

FIGURE 6.14. Watch, learn, and decide. Chimps learn how to "fish" for termites by watching others. Chimps may judge how effective a foraging technique is, and choose whether or not to add it to their behavioral repertoire.

Crop Raiding, Elephants, and Social Learning

In many areas around the world, humans are cultivating crops in areas that have not historically been used for agriculture. This activity can cause conflict between indigenous people and native wildlife that attempt to forage on such crops. For example, around the Amboseli National Park in Kenya, approximately one of out three adult male elephants that have dispersed from their natal groups raid crop fields (Figure 6.15). Raiding crop fields is a dangerous behavior for male elephants, as many are injured or killed by farmers during such raids (Obanda et al., 2008). To better understand how to protect elephants from such injury, and to better protect staple crops grown by indigenous people who may rely on these crops as a primary food source, we need to better understand crop-raiding behavior.

Elephants live in complex social networks, in which both individual learning and cultural transmission play important roles (Lee et al., 2011; Plotnik et al., 2011). Patrick Chiyo and his colleagues tested the hypothesis that male elephants learn how to raid crops—and especially how to be vigilant for farmers when doing so—through some form of social learning (Chiyo et al., 2012). In particular, they predicted that males that associated with others that raided crops would raid crops at a higher rate than males that associated with others that did not raid crops. In addition, they predicted that this effect would be strongest when males associated with *older associates* that raided crops, because such associates would be the best models.

Chiyo and his team worked with a population of 1,400 elephants in the Amboseli National Park. This population has been studied intensively since 1972, and individual elephants are recognized by unique tusk and body markings. The researchers observed fifty-eight male elephants often enough with others to rank these individuals in terms of their association patterns. Chiyo and his team were also able to gather information independently on the crop-raiding behavior of these fifty-eight males.

Chiyo and colleagues found evidence that social learning was important in crop raiding. As predicted, a male was more likely to raid crops if the individual it associated with most often was a crop raider. (This was also true when its second-closest associate was a crop raider, but not if only a third, fourth, or more distant associate was a crop raider.) Also, as predicted by Chiyo and colleagues, this effect was most pronounced when associates were older.

One of the long-term goals for ethologists involved in such projects is to guide policy makers in the difficult task of developing management practices that simultaneously minimize harm to animals and maximize crop productivity. Much work is being done to develop ideas on this front, but one application from the study on social learning in elephants might be something like this: *If* a plan were developed for providing elephants an alternative food source so that crop raiding is decreased, then this plan should first target older males that serve as models for others in their group.

FIGURE 6.15. Crop raiding behavior in elephants. The role that social learning plays in crop raiding has been studied in elephant populations in the Amboseli National Park in Kenya. *(Photo credit: Martin Harvey/Alamy)*

FIGURE 6.16. The bobo doll experiment.
In Bandura's classic bobo-the-clown doll experiments, the power of social learning is frighteningly evident: Young children treated the bobo doll, as well as other toys, as aggressively as the adult they observed. *(Photo credit: Albert Bandura)*

The children exposed to the "violent adult" treatment not only chose to play with aggressive toys, but they often beat up the doll and yelled the same things that the adult had yelled, while children in the other treatment played much more calmly (Figure 6.16). No doubt, some readers will look at the bobo doll experiment and think "Of course Bandura got the results he did; it's obvious to anyone that this is what would happen." But two things should be kept in mind when looking at these results: First, Bandura didn't "have to" obtain the results he did. The kids could have ignored the adult punching the bobo doll, or they could even have been particularly nice in response to the adult's aggressive behavior. In fact, some social psychology theories predicted that watching aggression would somehow reduce the observers' aggressive tendencies by providing an outlet. It could have worked that way in Bandura's experiment; it just didn't. Second, science is full of examples in which researchers actually did the experiment to test "something that was obviously true," and they found out that it wasn't. The only way to know for sure is to do the experiment. The same thing can be said of work on social learning in animals. In retrospect, it might seem obvious that rats learn what to eat and what not to eat by social learning. But it didn't have to turn out that way. Rats could have ignored any information they obtained from smelling their nestmates, but they didn't.

Social learning can take many different forms in both humans and nonhumans. Below we examine two forms of social learning: imitation and copying.

IMITATION. George Romanes was one of the first scientists to suggest that cultural transmission plays an important role in animal societies. In a series of books, Romanes argued that imitation is integral to a wide variety of animal

This bird has already been trained to push on the blue lever to get food.

Observer watches and may learn through imitation.

Trained bird

Observer bird

FIGURE 6.17. Imitation. Here an observer bird watches a trained pigeon that must lift its foot and push on a lever to open a small circular entrance to a food source. Imitation occurs when the observer learns this new task by watching the model lift its leg and push the lever down.

actions (Romanes, 1884, 1889, 1898). Since Romanes wrote his books more than a century ago, the term **imitation** has been used in many different ways in the psychological literature (R. W. Byrne, 2002; Heyes and Galef, 1996; Miklosi, 1999; Whiten, 1992). Cecilia Heyes defines *imitation* as the "acquisition of a topographically novel response through observation of a demonstrator making that response . . ." (Heyes, 1994). To demonstrate imitation, there must be some *new* behavior learned from others, and that behavior must involve some sort of *new spatial (topographic) manipulation* as well as lead to the achievement of some goal (Figure 6.17). We have already touched on a good case of imitation at the start of this chapter. When Imo washed her sweet potatoes in water before she ate them, and others in her group observed and then learned this novel behavior, which requires a new sequence of spatial actions, imitation was taking place.

Another instance of imitation can be seen in the behavior of blue tit birds (*Parus caeruleus*) in Britain in the 1940s. At the time, glass milk bottles covered with foil caps were delivered to people's homes and left on their doorsteps by milkmen in the morning. Some customers found that the foil caps on the top of the milk bottles were being torn off before they could retrieve the freshly delivered bottles from their doorsteps (Figure 6.18). J. Fisher and Robert Hinde suggested that this new behavior had been accidentally stumbled upon by a

FIGURE 6.18. Blue tit birds opening milk bottles. Blue tit birds learned to peck open the top of milk jugs decades ago. This behavior may have spread via cultural transmission. *(Photo credit: Roger Wilmhurst/Foto Natura/Minden Pictures)*

lucky blue tit and that others learned this nifty trick, at least in part, from watching the original milk thief (J. Fisher and Hinde, 1949). That is, birds observed a demonstrator taking action to manipulate the foil by pecking at it and tearing through to the milk. This was a novel response that was learned by the observers, that then also undertook the spatial actions necessary to achieve the goal of obtaining milk from the bottle.

Jeff Galef and his colleagues have uncovered another interesting example of imitation in birds. In this study, budgerigars (parakeets) observed others lifting a flat cover off a dish, using either their beak or their foot (Galef et al., 1986). Those budgerigars observing a model that used its beak to gain access to food were themselves likely to use their beaks when placed in the same situation, while those budgerigars observing a model that used its foot to unlock hidden food also used their feet. That is, the birds learned a topographically novel response from watching others.

Several interesting questions arise when we think critically about imitation. For example, when individual 1 attempts to imitate individual 2, it can only see individual 2's movements, but not the muscle activation underlying such movements. So how does individual 1 know what to do to make such movements itself? This is referred to as the correspondence problem (Brass and Heyes, 2005). Another issue associated with imitation is that of "perspective taking." Suppose, for example, that you and I are facing one another, and I raise my right hand and move it in circles. If you wish to imitate this action, you need to take into account our positions relative to one another. If you simply raise the hand that was on the same side as the hand I raised, you would be raising your left hand, and not precisely imitating my action.

Behavioral neuroscience is beginning to shed light on both correspondence and perspective-taking problems (Rizzolatti et al., 2006). Studies on these questions in humans often involve subjects who are placed in a magnetic resonance imaging (MRI) device and given some problem to solve while activity in their brain is scanned by the MRI machine. Typically, the problem is very simple—for example, tapping a specific finger a set number of times, or repeating a particular musical chord on the guitar. These studies indicate that certain sections of the brain—the inferior frontal gyrus, the dorsal and ventral premotor cortex, as well as other areas—are consistently active during imitation (Decety et al., 2002; Grezes and Decety, 2001; Grezes et al., 2003; Iacoboni et al., 2001; Koski et al., 2002; Muhlau et al., 2005). Interestingly, our brains respond more strongly to opportunities to imitate an action when we see another human doing it than when we see the same action being performed by a robot (Kilner et al., 2003; Tai et al., 2004).

There is less behavioral neuroscience work on these issues in nonhumans, but that is beginning to change. For example, work on imitation in monkeys has found that a set of "mirror" neurons in what is referred to as the F5 area of the premotor area of the monkey brain is very active when a monkey observes an action—such as grasping a piece of food on a tray—and then repeats that action (Molenberghs et al., 2009; Rizzolatti and Craighero, 2004; Rizzolatti et al., 2001). What makes the mirror neurons in the F5 area of the brain particularly relevant is that some of these neurons are motor neurons (neurons that are needed to repeat an act during imitation) and some are visual neurons (neurons that are necessary for watching a model). An action must first be observed before mirror neurons will fire, again suggesting a connection to imitation. Indeed, these neurons will fire if an individual sees a hand manipulating an object, but not if it sees the object alone or if it sees the object being manipulated by a tool (Gallese et al., 1996; Molenberghs et al., 2012; Rizzolatti et al., 1996). Some evidence suggests that mirror neurons in humans also reside in the equivalent of the F5 section of our brain, and that such neurons are involved in human imitation as well, but much work in this area remains to be done.

COPYING. When animals **copy** one another, an observer repeats what it has seen a model do. Often, the copier is then rewarded for whatever behavior it has copied. In the psychological literature, the rewards associated with copying can be extrinsic (the food items in the above case) or intrinsic (related to animal emotions and feelings). Copying differs from imitation in that what is copied need not be novel and need not involve learning some new topographical action. That is, an individual can copy the action of another, even if it already knows how to do what the model is doing, and even if it does not involve learning some new spatial orientation to do what the model does. As a case in point, in the absence of copying opportunities, animals will select a mate, but they may still copy the mate choice of a model.

Some evidence to date of copying comes from mate-choice copying, during which an individual—usually a female—copies the mate choice of those around her. I examined female mate-choice copying in guppies (*Poecilia reticulata*) using a ten-gallon aquarium situated between two separate end chambers constructed of clear Plexiglas (Dugatkin, 1992a). A single male was put into each of these end chambers. The "observer" female—the individual that potentially copies the behavior of other females—was placed in a clear canister in the middle of the central aquarium. At the start of a trial, one male was put into each end chamber (Figure 6.19). Removable glass partitions created left or right sections of the aquarium into which another female—the "model" female—was placed. A key component of this experiment was that the placement of the model—either near the male in the left end chamber or near the male in the right end chamber—was determined by the flip of a coin. This was done to rule out the possibility that one male may have been more

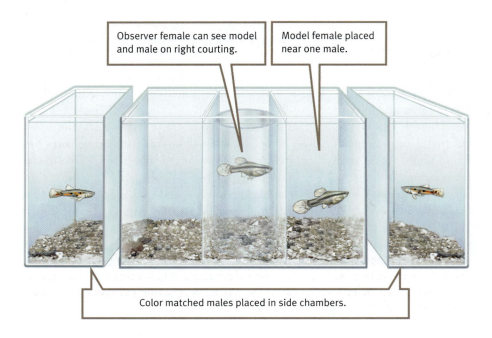

Observer female can see model and male on right courting.

Model female placed near one male.

Color matched males placed in side chambers.

FIGURE 6.19. Mate-choice copying.
The experimental apparatus used in the guppy mate-choice copying experiments. Whether the model is placed near the male on the left or on the right is determined by the flip of a coin.

attractive than the other and that the model and the observer independently each chose that male.

Once the model, the observer, and the two males were in place, fish were given ten minutes during which the observer female could watch the model female near one of the two males. The model female and the glass partitions were then removed, and the observer female was released from her canister and given ten minutes to swim freely and choose whichever male she preferred (as indicated by which end chamber she swam near). In these trials, the observer female chose the male that had been chosen by the model female seventeen out of twenty times.

While the results of this experiment are consistent with the hypothesis that females copy the mate choice of others, there are several alternative explanations. For example, since guppies live in social groups (schools of fish), the observer female might simply be choosing the area that had recently contained the largest number of fish (in this case, two). A control treatment was conducted to test this schooling hypothesis. It was identical to the above protocol, except that females were placed in the end chambers. In this case, the observer female chose the female in the end chamber closest to the model in only ten out of twenty trials. As such, the tendency to school (stay near two fish rather than one fish) per se does not explain the results of the first experiment; if it had, observer females would have consistently chosen the end chamber closest to where the model had been placed. Results from other control experiments were also consistent with mate-choice copying. Using a protocol similar to that of the guppy experiments, mate-choice copying has also been observed in a number of different species of fish, as well as in some birds and mammals (see Chapter 7).

Another example of copying can be found in the fear response of mice to stable flies (*Stomoxys calcitrans*; Kavaliers et al., 1999, 2001). Typically mice that have been exposed to stable flies before do not show any immediate behavioral response to the presence of a stable fly. After an individual is bitten by a stable fly, one of its defensive responses is to bury itself under whatever debris it can find (Figure 6.20). When a naive mouse observes another mouse being bitten by a fly and then burying itself, the observer quickly buries itself when it is exposed to a fly—that is, the

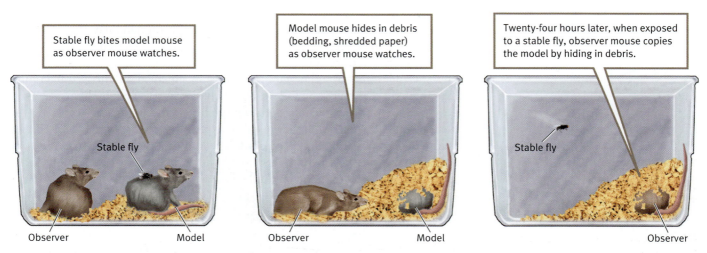

Stable fly bites model mouse as observer mouse watches.

Model mouse hides in debris (bedding, shredded paper) as observer mouse watches.

Twenty-four hours later, when exposed to a stable fly, observer mouse copies the model by hiding in debris.

Stable fly

Observer   Model

Observer   Model

Stable fly

Observer

FIGURE 6.20. Copying a defensive response. After a mouse is bitten by a stable fly, one of its defensive responses is to bury itself under debris. A mouse that observes another mouse being bitten by a stable fly and then hiding will copy the hiding behavior of the model mouse as soon as it is exposed to a fly.

observer does not wait until it is bitten to adopt a defensive position, but does so as soon as it is exposed to the fly—it copies the defensive action of the model.

In order to understand the underlying molecular genetics of copying in these mice, Martin Kavaliers and his colleagues focused on what is known as the NMDA receptor (Kavaliers et al., 2001). At the cellular level, this receptor plays an important role in neural plasticity—that is, neural flexibility—which itself is an important component of learning and memory (see Chapter 3). When Kavaliers and his colleagues blocked the NMDA receptor of an observer mouse, using an NMDA antagonist chemical, they found that the observer did not learn to bury itself as soon as it was exposed to a fly—that is, the NMDA antagonist blocked copying in these mice. A similar experiment used rats that could copy a model's foraging behavior and also found that blocking the NMDA receptor impeded social learning by means of copying (M. Roberts and Shapiro, 2002).

THE RISE AND FALL OF A TRADITION

In the literature on social learning and cultural transmission, when a new preference emerges and then becomes common within a group, it is referred to as a **tradition**. Traditions have been documented in animal populations in the lab, but rarely studied in a controlled manner in nature (Thornton & Clutton-Brock, 2011). In 2009, Alex Thornton and Aurore Malapert ran one of the first controlled experiments on traditions in wild populations, using nine groups of meerkats (*Suricata suricatta*) in the South African Kalahari (Figure 6.21A; Thornton and Malapert, 2009b; see Thornton and Malapert, 2009a and Thornton et al., 2010).

To experimentally examine traditions, Thornton and Malapert created two arbitrary landmarks in the area of each group of meerkats. These landmarks were combinations of plastic geometric objects that differed in color and pattern. Both landmarks were placed near a reliable new water or food source for the meerkats. In seven of the nine meerkat groups, a demonstrator individual was trained by the researchers to show a preference for *one* of the two color- and shape-coded landmarks (Figure 6.21B). Which landmark an individual was trained to prefer in a specific group was determined randomly. In two groups, no demonstrators were trained.

What Thornton and Malapert found was the rise, and subsequent fall, of a tradition. Initially, after seeing a demonstrator at the landmark to which the demonstrator had been trained, meerkats in a group preferred that landmark. A completely arbitrary tradition—both landmarks were equally profitable, but one was preferred by other group members—emerged in the seven groups. This preference lasted a few days but then slowly dissipated. What seems to have occurred is that meerkats initially preferred the landmark visited by the demonstrator, but over time they began to explore the other landmark. Once they learned *themselves* that this other landmark was just as good as the one they had been frequenting, they spent time at both landmarks, eroding the tradition. Social learning produced a tradition, but individual learning led to its demise.

TEACHING IN ANIMALS

The hypothesis that animals teach one another is one of the more contentious in the literature on animal cultural transmission (Caro and Hauser, 1992; Franks and Richardson, 2006; M. Hauser, 1997, 2000; M. Hauser and Konishi, 1999; Leadbeater et al., 2006; Thornton and McAuliffe, 2006). While there are many

A

B

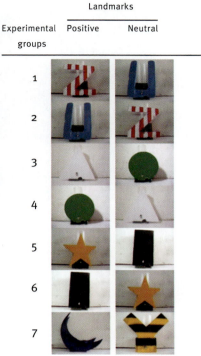

FIGURE 6.21. Meerkat traditions come and go. (A) Meerkats foraging in a group. **(B)** The set of visual cues used to create landmarks. Such arbitrary landmarks became attractive to meerkat group members if they observed others feeding there. "Positive" means a trained model preferred the area near this landmark. "Neutral" means no model was associated with this landmark. *(Photo credit: Martin Harvey/Getty Images; Courtesy of Alex Thorton; from Thornton and Malapert, 2009b)*

definitions of **teaching**, most include the idea that one individual serves as an instructor or teacher, and at least one other individual acts as a student who learns from the teacher (Caro and Hauser, 1992; Ewer, 1969; Fogarty et al., 2011; Fragasy and Perry, 2003; Galef et al., 2005; Maestripieri, 1995).

Teachers have a much more active complicated role than just being a model that another individual mimics, as in social learning. In a review on teaching in animals, Tim Caro and Marc Hauser suggest that for a behavior to be labeled as "teaching," a teacher must provide an immediate benefit to students but not to him- or herself, he or she must teach only naive "students," and he or she must impart some new information to students faster than they would otherwise receive it. This definition is interesting, not only because of the emphasis on what must take place for teaching to occur but also for what kinds of behaviors are excluded from the realm of teaching. For example, in the case of the blue tits opening the foil caps of milk bottles in Britain in the 1940s, imitation rather than teaching was taking place. While blue tits learned how to open foil caps by observing others, those that opened the caps did so regardless of who watched them. According to Caro and Hauser's definition, they weren't teaching other birds anything since they opened the milk bottles in the same way even if they were alone and thus they weren't modifying their behavior only in the presence of naive observers. They were indeed obtaining immediate benefits for themselves in that they obtained the milk after they opened the bottles.

What sort of examples might fall under the Caro and Hauser definition of *teaching*? Consider a female cat that captures live prey and allows its young to interact with this prey, making sure that the prey doesn't escape along the way. If mother cats only engage in this behavior when in the presence of young cats,

A

B

FIGURE 6.22. Cheetah teaching? **(A)** A mother cheetah brings a Thomson's gazelle to her cubs and allows them to "kill it," even though it was already dead. **(B)** Until the young cheetah is taught how to hunt, it can only kill small items like the hare shown here. *(Photo credits: Tim Caro)*

we may have a true case of teaching. Anecdotal examples of this kind of teaching have been documented in domestic cats (Baerends-van Roon and Baerends, 1979; Caro, 1980; Ewer, 1969), lions (Schenkel, 1966), tigers (Schaller, 1967), and otters (Liers, 1951). Teaching has been examined in more detail in both cheetahs (Caro, 1994a) and meerkats (Thornton and McAuliffe, 2006). Consider Caro's description of three ways that mother cheetahs used "maternal encouragement" to facilitate hunting skills in their offspring (Figure 6.22):

> Firstly, they pursued and knocked down the quarry but instead of suffocating the victim allowed it to stand and run off. By the time the prey had risen, the cubs had normally arrived. Second, mothers carried live animals back to their cubs before releasing them, repeatedly calling (churring) to their cubs. Third, and less often, mothers ran slowly during their initial chase of a prey and allowed their cubs to overtake them and thus be the first to knock down the prey themselves. (Caro, 1994a, pp. 136–137)

While this sort of behavior is consistent with teaching, it is not sufficient to demonstrate teaching, as it is unclear whether young cheetahs accelerated their hunting skills as a result of these interactions with their mother or for other reasons (Caro and Hauser, 1992; Galef et al., 2005; Thornton and Clutter-Brock, 2011).

Another example of teaching has been documented in the meerkats discussed earlier in the chapter. In meerkats, young pups are incapable of catching their own prey. At about a month old, these pups begin following groups of foragers, and the pups are assisted in their own foraging attempts by older groupmates called "helpers" (Figure 6.23). Many of the prey that meerkats ingest are difficult to catch and, in the case of scorpions, are also quite dangerous. Helpers will often incapacitate scorpions by removing their stingers and present them to pups as food. Alex Thornton and Katherine McAuliffe observed that very young pups were either fed dead or incapacitated scorpions, but as the pups got older, the helpers presented them more and more often with live scorpions (Thornton and McAuliffe, 2006).

To experimentally examine whether helpers were teaching pups how to forage on dangerous prey, the researchers took advantage of the fact that helpers

FIGURE 6.23. Meerkat foraging and teaching. Older groupmates assist younger pups in their foraging attempts and also teach the younger pups to catch prey, including dangerous scorpions, as shown here. *(Photo credit: Alex Thornton)*

respond to the begging calls of pups even when pups cannot be seen, and that such begging calls change in predictable ways as pups age. When Thornton and McAuliffe had a group of young pups, but played the calls of older pups, helpers were more likely to bring live prey over to the pups. When the group contained older pups, but the researchers broadcast the begging calls of younger pups, helpers were more likely to bring over dead or incapacitated scorpions. Helpers were changing what type of prey they delivered in a manner that would help and might even teach the pups.

Thornton and McAuliffe found additional evidence for teaching by helpers in that helpers: (1) spent significant time monitoring pups after presenting them with food; (2) retrieved prey when pups lost their food; (3) on occasion, further modified a scorpion (removing the stinger, killing the scorpion, and so on) after it was lost but later retrieved by the pups; and (4) nudged pups that were reluctant to eat scorpions, increasing the probability that the pups would eat the scorpion that they had initially rejected. All in all, the evidence that helpers modified their behavior in costly ways—spending time that they could have used to forage for themselves but instead spent with pups—and that such modifications helped pups learn how to forage on dangerous prey suggests teaching on the part of meerkat helpers.

COMMON THEMES IN EXAMPLES OF ANIMAL TEACHING. When Caro and Hauser searched for common themes that underlie cases of teaching, they found two. First, except for the meerkat case (which was undertaken after Caro and Hauser's analysis), almost all instances of animal teaching focus on the parent/offspring relationship. This may strike you as intuitive, but remember that in many animal societies, young can learn from others besides their parents. Furthermore, in principle, adult animals could presumably teach other adults, but there is little, if any, evidence that they do. These common themes certainly suggest something special about the costs and benefits of teaching in the parent/teacher, offspring/student relationship. The genetic kinship that bonds teacher and student—that is, parent and offspring or perhaps between siblings—may be one of the only benefits large enough to make up for the costs of teaching (Caro and Hauser, 1992; Galef et al., 2005).

Second, Caro and Hauser argue that cases of teaching tend to fall into one of two categories: "opportunity teaching" and "coaching." In opportunity teaching, teachers actively place students in a "situation conducive to learning a new skill or acquiring knowledge." In contrast, coaching involves a teacher who directly alters the behavior of students by encouragement or punishment. The majority of examples of animal teaching fall under opportunity teaching, presumably because this type of teaching is the simpler of the two. The meerkat example, however, shows nicely how both forms of teaching can be in play in the same system. Meerkats use opportunity teaching by manipulating prey for young pups, while at the same time coaching pups by nudging them and thereby encouraging them to try new, potentially dangerous, food items.

Modes of Cultural Transmission

With a better understanding of what constitutes teaching and social learning, we are now ready to examine three different modes of cultural transmission: vertical, oblique, and horizontal transmission (Cavalli-Sforza and Feldman, 1981).

FIGURE 6.24. Cultural transmission in finches. Birdsong in some species of finches is learned culturally. *(Photo credit: Greg Lasley Nature Photography)*

VERTICAL CULTURAL TRANSMISSION

Vertical cultural transmission occurs when information is transmitted across generations from parent(s) to offspring. This type of cultural transmission might take place through either teaching or social learning—offspring might simply learn from their parents by observation, or parents might teach a behavior to their offspring. For example, in some finch species, vertical transmission occurs when males learn the song that they will sing from their fathers, as well as when females develop song preferences in potential mates based on the songs their father sang (B. R. Grant and Grant, 1996; Figure 6.24).

Vertical cultural transmission has been studied in bottlenose dolphins (*Tursiops truncatus*) that imitate one another, both in the laboratory and in the wild, and that live in groups that are structured in a way that allows for transmission of behavioral traits from parents to offspring (Kuczaj et al., 1998; Xitco and Roitblat, 1996). In particular, researchers have focused on vertical transmission of foraging skills in bottlenose dolphins, as these dolphins display a complex and rich assortment of foraging strategies (Patterson and Mann, 2011, Sargeant and Mann, 2009). For example, one fascinating foraging strategy called "beaching" or "beach hunting" involves a dolphin surging out of the water and onto a beach to catch a single fish, and then quickly returning to the water with its prey (Mann and Sargeant, 2003). This form of beaching of individual fish is a fairly rare behavior—in a long-term study of one bottlenose dolphin population in Shark Bay, Australia, only four adult females and their calves were observed beaching with any consistency (Sargeant et al., 2005). Another form of beaching among other dolphins, however, has been observed in salt marshes in the southeastern United States, where one to six dolphins isolate a school of fish and herd the fish toward land, creating a wave to send them onto the land and then surging out of the water to capture the stranded fish (Hoese, 1971; Figure 6.25). Although beaching can be profitable when successful, it can also be quite dangerous, as beaching dolphins may get stranded.

FIGURE 6.25. Beaching behavior. Dolphins may trap fish by stranding them on a beach and then surging out of the water to catch them. The dolphins then quickly return to the water with their prey. *(Photo credit: Neil Lucas/Nature Picture Library)*

Janet Mann and her colleagues found that only calves born to mothers that themselves were "beachers" displayed this specialized foraging strategy. The behavioral evidence suggests that young calves learn beaching from their mothers, likely because the young calves spend a great deal of time with their mothers and learn this dangerous foraging tactic via vertical cultural transmission (Krutzen et al., 2005).

A second fascinating case of vertical transmission of foraging strategies has also been observed in the bottlenose dolphins of Shark Bay, and is the first known case of tool use in dolphins (Krutzen et al., 2005). Here the foraging behavior, called "sponging," involves female bottlenose dolphins breaking a marine sponge off the seafloor and placing it over their mouth, and then using this tool to probe the seafloor for fish prey (Smolker et al., 1997; Figure 6.26). As with beaching behavior, sponging is seen almost exclusively in females, and young females learn sponging from their mothers.

FIGURE 6.26. Sponging behavior in dolphins. Female bottlenose dolphins break a marine sponge off the seafloor and place it over their mouth. This tool is used to probe the seafloor for fish prey and to protect them from scrapes and stings as they forage. *(Photo credit: Janet Mann)*

OBLIQUE CULTURAL TRANSMISSION

Oblique cultural transmission refers to the transfer of information across generations, but not via parent/offspring interactions. In this form of cultural transmission, young animals get information from adults that are not their parents. This sort of transmission might be particularly common in animal species where there is no parental care, and hence where most interactions between younger and older individuals would be between nonrelatives.

Oblique transmission has been uncovered in many scenarios, such as learned snake aversion in rhesus monkeys. Early work comparing wild- and laboratory-raised primates found that the lab-raised animals that had never seen a snake before did not respond to snakes in the same manner as wild-raised individuals that had the chance to experience snakes in nature (K. R. Hall and Devore, 1965; Seyfarth et al., 1980; Struhsaker, 1967). Susan Mineka and her colleagues designed an experiment to examine whether oblique cultural transmission played a role in the development of the fear of snakes (Cook et al., 1985; Mineka and Cook, 1988; Mineka et al., 1984).

Mineka and colleagues began their laboratory-based experiment with juvenile rhesus monkeys that were not afraid of snakes. Shortly after observing an adult model respond to snakes with typical fear gestures and actions (Figure 6.27), juveniles themselves adopted these same gestures (for at least three months). It made no difference whether the individuals observed were the subjects' parents (vertical transmission) or unrelated adult monkeys (oblique transmission)—exposure to adult models showing fear in the presence of snakes led to observers showing similar fear responses. Mineka and colleagues also found that when observers saw a model that had been trained to display fear in the presence of a neutral object—a flower—the observers did not display fear when they were exposed to flowers, suggesting that a predisposition to fear snakes is interacting with oblique cultural transmission (Cook and Mineka, 1989). Further evidence of this interaction can be seen in monkeys that had first seen an adult model interact with a snake in a nonfearful manner, but then saw a second model respond with fear to snakes. Observers then displayed less fear than individuals exposed only to a model that had displayed fear. These monkeys were culturally "immunized" against associating snakes and fear (Mineka and Cook, 1986).

FIGURE 6.27. Monkeys and fear of snakes. Rhesus monkeys in the field often fear snakes after watching others respond to such potential danger. Monkeys raised in the lab that do not normally fear snakes can be made to fear them through observing older monkeys reacting fearfully to snakes. Young monkeys were shown this video of an older monkey that fled to the back of the cage and cringed in fear at the sight of two snakes. *(Photo credit: Sue Mineka)*

HORIZONTAL CULTURAL TRANSMISSION

Cultural transmission is not limited to the transfer of information from older to younger individuals. In humans, for example, most information comes from peers—people who are of the same approximate age group. This type of transmission of information, called **horizontal cultural transmission**, operates not only for adults but for children as well. In fact, in humans, horizontal transmission of information is so powerful that adults spend much of their time trying to subdue its effects on their offspring. We want our children to learn some things from their peers, but not everything.

Horizontal cultural transmission plays a role in animal as well as human behavior. Consider the case of horizontal transmission of foraging-related information in guppies, a species in which fish segregate into schools based on age. Kevin Laland and Kerry Williams trained guppies of approximately the same age to learn different paths to a food source—a long path and a short path (Laland and Williams, 1998). Not surprisingly, it was more difficult to train fish to take the longer path when both paths were present, but Laland and Williams found a clever way (involving what amounts to trapdoors) to do so.

Once the "long-path" and "short-path" groups were trained, Laland and Williams slowly removed the original members of each group and replaced them with new "naive" individuals that didn't know either of the paths to the food source. Initially, groups contained five trained guppies, then four trained individuals and one untrained, then three trained individuals and two untrained, and so on, until none of the original trained group members remained. The question, then, was whether the fish remaining at the end of the experiment—none of which were trained to a particular path—maintained the path taken by the original fish in their group.

Laland and Williams found that in both the short-path and long-path groups, guppies at the end of the experiment still followed the path to which the original fish had been trained. Horizontal transmission of information was operating here, as the only "models" from which to learn were same-age individuals. What makes this experiment particularly interesting is that it demonstrates that cultural transmission can produce "maladaptive" (long-path use), as well as adaptive (short-path use) behavior. In fact, horizontal transmission in the long-path groups not only resulted in guppies acquiring the "wrong" information, it actually made it more difficult for the fish in that treatment to subsequently learn to use the shorter path (C. Brown and Laland, 2002; Reader et al., 2003).

The Interaction of Genetic and Cultural Transmission

Because cultural and genetic transmission operate on animal behavior, we can study cases in which both forms of transmission may both be operating in the same system. To better understand how such interactions work, let's consider two different case studies that examine the interaction of genes and culture in two animal systems: birds and fish.

THE GRANTS' FINCHES

Peter and Rosemary Grant have been studying finches in the Galápagos Islands for more than three decades. Among the many problems the Grants have tackled is the role of cultural transmission in the evolution of finch song. In Galápagos finches, cultural transmission not only shapes birdsong but it interacts in an unexpected manner with the genetics of reproductive isolation and speciation in these birds (P. R. Grant and Grant, 1994, 1997).

The medium ground finch (*Geospiza fortis*) and the cactus finch (*G. scandens*) both live on the Galápagos island of Daphne Major. Although these are considered to be different species, some cases of interbreeding between these two finch species have been uncovered, and the hybrids do not appear to suffer a decrease in reproductive success as compared to the two pure species—that is, there is no fitness penalty for hybridization across ground and cactus finches. Yet although there seems to be no cost for hybridization, medium ground finches and cactus finches rarely interbreed. Why? Does cultural transmission play a role in inhibiting such interbreeding (Freeberg, 2004; Lachlan and Servedio, 2004; D. A. Nelson et al., 2001; Slabbekoorn and Smith, 2002)?

In finches, male birdsong plays a prominent role in the mating process, and, as in all songbirds, male finches learn the songs they sing. When the Grants studied the songs sung by ground and cactus finches during the mating season, they found that these songs were transmitted across generations via cultural transmission (B. R. Grant and Grant, 1996). The evidence for the cultural transmission of birdsong comes from comparing the songs of sons, fathers, and grandfathers. To see why, consider the case in which fathers and sons have very similar songs. This could be due to the song being a genetic trait passed from father to son, or it could be due to the cultural transmission of the song from father to son. Now, compare the songs of sons to those of their paternal and maternal grandfathers. If songs are genetically controlled, we should see that the songs of sons are similar to the songs of both their paternal and maternal grandfathers, since they inherit genes from both grandfathers. But if the songs are culturally transmitted from father to son, then the songs of the sons should resemble those of their paternal grandfather, but not those of their maternal grandfather. This is because the paternal grandfather would have transmitted the song to the father, who would then have transmitted the song to the son. The evidence suggests that the songs of sons resemble the songs of their paternal grandfathers, but not the songs of their maternal grandfathers (Figure 6.28). Birdsong appears to be culturally transmitted.

The Grants also found that the songs of these two species were quite different from one another. For example, the song of the cactus finch has shorter components that are repeated more often than the components of the song of the ground finch. These differences in their songs—a culturally transmitted trait—have a dramatic impact on gene flow across species. Of 482 females Grant and Grant sampled, the vast majority (over 95 percent) mated with males who sang the song appropriate to their own species. It seems that cultural transmission of song allows females to recognize individuals of their own species. In addition, females tend to avoid males who sing songs that are similar to the songs that their own fathers sang, which suggests that song also plays a role in preventing inbreeding. Because song is culturally transmitted from father to son, females may decrease the probability of mating with genetic relatives when they avoid mating with males that sing like their fathers.

In their long-term study, the Grants uncovered eleven cases in which the male of one species sang the song of another species. In most of these cases,

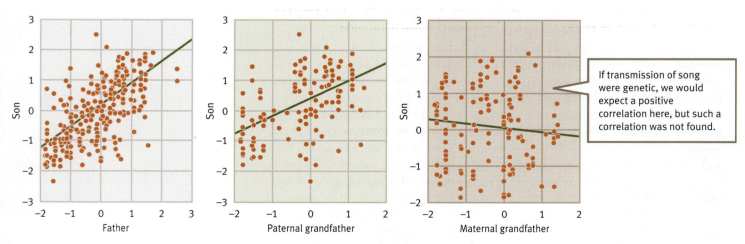

If transmission of song were genetic, we would expect a positive correlation here, but such a correlation was not found.

FIGURE 6.28. Finch songs across generations. Components of male finch song are positively correlated with those of their father and their paternal grandfather, but not their maternal grandfather. This is consistent with cultural transmission of male song. *(From B.R. Grant and Grant, 1996)*

cross-species breeding would then occur, resulting in hybrid offspring. In other words, remove the normal pattern of cultural transmission and the barrier to breeding across species disappears, suggesting a new and exciting avenue of research in the interaction of cultural transmission and genetics. In this example, then, we see how powerful cultural transmission of traits can be in that it can have significant effects on such evolutionary issues as cross-species hybridizations resulting from interbreeding.

GUPPY MATE CHOICE

As we have discussed earlier in the chapter, female guppies copy the mate choice of other females. Observer females that viewed a model female choose one male over another were much more likely to choose that male themselves. In addition to this type of cultural transmission of information, genetic transmission of traits also plays an important role in guppy mate-choice behavior (Houde, 1997; Magurran, 2005).

Guppies from some populations—including fish from the Paria River in Trinidad and Tobago—prefer to mate with males that have lots of orange body color. Interpopulational comparisons suggest that this preference of Paria River females for males with more orange body color is heritable (Houde, 1988; Houde and Endler, 1990). In addition, orange body color itself is a heritable trait in Paria River males (Houde, 1992, 1994).

To examine how genetic and cultural transmission interact in shaping mate choice in females from the Paria River, I set up an experiment with four different treatments (Dugatkin, 1996b). In each treatment, a female was exposed to a pair of males. In treatment I, males were matched for orange body color (the mean difference in orange color between the males was less than 4 percent). In treatments II, III, and IV, males differed in total orange body color by an average of 12 percent (II), 24 percent (III), and 40 percent (IV). In each treatment, the experiment was designed such that observer females always saw a model female near the male with less orange body color. That is, the information being culturally transmitted to models was always in direct opposition to a female's genetic predisposition to choose males with more orange body color as mates.

Results of the experiment suggest that females in treatment I—in which males were matched for orange body color—copied each other's mate choice. When males differed in orange body color by an average of 12 or 24 percent (treatments II and III, respectively), females consistently preferred the less orange of the two males, again copying the mate choice of the model female; in these treatments, culturally transmitted information overrode a female's genetic predisposition to mate with males with lots of orange body color. But when male orange body color differed by an average of 40 percent (treatment IV), females consistently preferred the more orange of the two males, thus overriding any effects of mate-choice copying (Figure 6.29).

These results suggest that it may be possible to experimentally examine the relative strength of both cultural transmission and genetic predisposition. In the guppy system, it appears that whether or not females copy a model's mate choice is affected by a threshold difference in the amount of orange body color in the male. If the male's orange body color is beneath this threshold, the effects of copying are predominant. But if the orange body color is above this

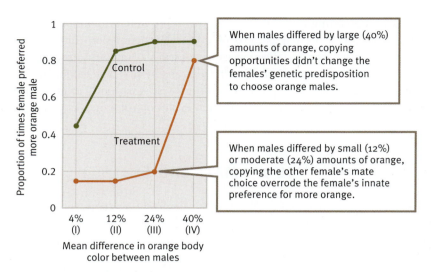

FIGURE 6.29. Mate-choice copying in guppies. In the control trials a female chose between males that differed in orange body color and no model female was present. In these trials, females show a strong preference for more orange males. In the treatment trials, a model female was always placed near the less orange male. *(Based on Dugatkin, 1996b)*

threshold, genetic preferences mask any cultural effects. This sort of finding reinforces the idea that both genetic and cultural transmission can shed light on important issues in the study of animal behavior, and it demonstrates that a comprehensive understanding of ethology will often, but certainly not always, require that both forms of transmission be taken into account.

Cultural Transmission and Brain Size

In this chapter, we have seen examples of cultural transmission in everything from fish to primates (B'shary et al., 2002). Clearly, at least some forms of cultural transmission are possible even in animals with very small brains. That being said, it may be that a positive relationship between large brain size and cultural transmission exists.

E. O. Wilson suggested that in a population of large-brained animals, new innovations—the discovery of novel solutions to problems—might arise and spread more often than would happen in a population of small-brained animals. The data on this question are mixed (Dunbar, 1992; Sawaguchi and Kudo, 1990). But Simon Reader and Kevin Laland, in the most comprehensive study to date, found that across more than 100 species of nonhuman primates, there was a significant positive correlation between brain size and both innovation and tool-use frequency, confirming the predicted trends (Dunbar, 1992; Reader and Laland, 2002; Sawaguchi and Kudo, 1990).

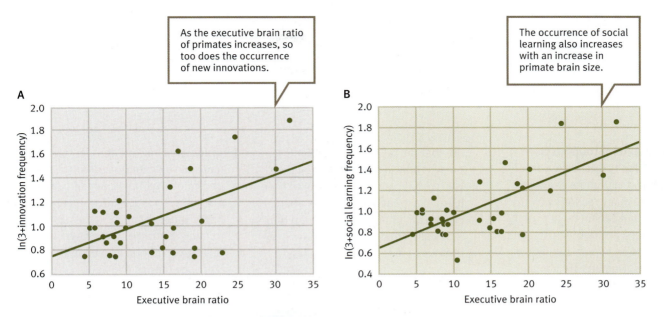

FIGURE 6.30. Effects of increasing brain size in primates. Reader and Laland determined the executive brain ratio (the executive brain size divided by the brain stem size) to examine the relationship between **(A)** innovations and **(B)** social learning and brain size. Both of the relationships shown in the graphs held true for absolute executive brain size as well as executive brain ratio. *(From Reader and Laland, 2002)*

Reader and Laland defined *innovations* as "novel solutions to environmental or social problems." Using this definition, they uncovered 533 recorded instances of innovations, 445 observations of social learning, and 607 episodes of tool use that covered 116 of the 203 known species of primates. They then mapped out these behaviors against something called "executive brain" volume, a measure that includes both the neocortex and striatum sections of the brain (Jolicoeur et al., 1984; Keverne et al., 1996). Innovation, social learning, and tool use all had a positive correlation with the absolute value of executive brain volume (Figure 6.30).

A similar trend between large brain size and an increased propensity toward innovation was found for birds in North America, Britain, and Australia (Lefebvre et al., 1997a; Lefebvre et al., 1997b; Lefebvre et al., 2004; Sol, Lefebvre et al., 2005). This relationship has interesting implications for questions relating to the conservation and ecology of birds (Overington et al., 2011; Sol, Duncan et al., 2005; Sol, Lefebvre et al., 2005; Sol et al., 2010). For example, Daniel Sol and his team examined whether the relationship between brain size and innovation affected bird species when they were moved to novel environments through large-scale, human "introduction" programs (where humans introduce a new species to a novel habitat). Using global data on more than 600 such introduction programs, Sol and his colleagues found that bird species in which individuals had a high brain size/body ratio were more likely to survive and thrive (that is, to have greater "invasion potential") after introduction to novel environments than were species with lower brain size/body ratios. In addition, the researchers found that when large-brained species were introduced to novel environments, they increased their rate of innovation—for example, by using a new foraging technique—which in turn increased their probability of success (their "invasion potential") in their new habitats.

INTERVIEW WITH

Dr. Cecilia Heyes

Why study cultural transmission in animals? Isn't culture the sort of subject that sociologists examine?

There are many reasons why the study of culture cannot be left solely to sociologists, anthropologists, and other experts on human cultures and societies. Humans are, among other things, animals, so even if culture were a distinctively human phenomenon, it would be essential to understand it at the biological and psychological levels—to find out where it comes from in evolutionary terms, and to identify the proximate biological and psychological processes that make it possible. Comparing human and nonhuman cultures (or, if you prefer, human cultures with animal traditions) helps us to understand what Thomas Huxley, "Darwin's bulldog," called "Man's place in nature" (1863). Each of us is part of at least one human culture, and usually we belong to many. For example, I am soaked in English culture, but I'm also influenced by Western culture, the culture of science, and, as a result of my upbringing, by Roman Catholic culture. Because each of us is embedded in the matrix of one or more human cultures, it can be difficult for us to get an aerial view of what culture really is and how it works. Comparing ourselves with other animals can give us this aerial view. It can reveal the fundamental processes that make culture possible, the ingredients that are and are not needed to make cultural transmission "take off" to become a system of cultural inheritance, analogous to that of genetic inheritance. It might turn

out that the full set of ingredients is present only in humans, but on the way to that conclusion we would have learned a great deal, not only about culture but about a type of learning —social learning—that can have a profound influence on behavioral adaptation.

There is a good deal of contention about how to define *culture* in animals. What is your take on this issue?

I think it may be too soon to be using the term *culture* to describe examples of social transmission or traditional behavior in nonhuman animals. The risk is that, if we categorize them as examples of culture too soon, we'll forget that the critical ingredients of culture have not yet been identified, and then we won't prioritize the research that would enable us to answer this important question. This kind of maladaptive shortcut is made surprisingly often in the study of animal behavior. For example, field primatologists in the

1980s were so confident that they had seen examples of intentional deception that they claimed quite emphatically that nonhuman apes have "theory of mind"—the capacity to think about the thoughts and feelings of others. As a result, it has been hard in the last twenty years for researchers such as Daniel Povinelli and Michael Tomasello to get the animal behavior community to recognize that more careful research was needed to find out whether animals can really attribute mental states to themselves and others.

So, I'm uneasy about calling socially transmitted behavior "culture." Having said that, I should also point out that I don't think the exact content of a definition is terribly important. What's essential is that a definition should be explicit and reasoned. If it isn't clearly stated, then there will be cross-talk and confusion. I might say "I believe that fruit bats have culture," and, if I haven't said what I mean by culture, you might think I'm claiming that fruit bats go to the opera. When definitions aren't reasoned—for example, if I describe the behavior of my favorite species as "cultural" without thinking, or just because it sounds more exciting that way—then it's hard to make real scientific progress. Much of pre-Darwinian biology was about giving arbitrary names to things; it involved the kind of cataloging that goes on in a library. But contemporary biology is about understanding structures and processes. A definition without a reason is a throwback to cataloging science, but a reasoned definition usually represents or leads to a testable hypothesis. For example, if a researcher chooses to call all social transmission in animals "culture" because he believes that the nongenetic transmission process is the most important ingredient of culture, then his definition represents a testable hypothesis, and therefore contributes to progress in the field.

If you had to make an educated guess, how common would you say that cultural transmission is in nonprimates?

There are many kinds of social learning and, although most of them do not require the learner to be especially "clever," they all have the potential to yield very substantial adaptive advantages. Therefore, if *culture* is understood liberally to mean behavior affected by social learning, then I think it's very common in nonprimate as well as primate taxa. The phenotype of any animal that lives in a social group, and in an environment that is sufficiently variable to warrant adaptation through learning, is likely to be influenced by social learning.

When I was a graduate student, I had the bright idea that adult Syrian hamsters may have poor social learning ability because they are "solitary"; except when mating, they are highly intolerant of even the presence of mature conspecifics. However, my attempts to show this experimentally, by comparing the social learning ability of adult and predispersal juvenile hamsters, were a rank failure. In each experimental task, the adult hamsters cheerfully learned from their "demonstrators," and thereby heartlessly proved me wrong.

If *culture* is defined more strictly—for example, as a system of inheritance that allows cumulative,

selection-based evolution—then I'm not yet convinced that it is present in any nonhuman species. To answer this intriguing question we would need much more information about the dynamics of social transmission in animals, and particularly about the fidelity with which behaviorial variants are copied across successive links in a transmission chain. If adaptive innovations are not transmitted faithfully, then an inheritance system based on social learning cannot have effects comparable to those of gene-based selection.

How do you see advances in neurobiology affecting the study of animal culture over the next five years?

The discovery of "mirror neurons" in the premotor cortex of macaque monkeys has caused a huge stir among those who study imitation and related processes in humans. These cells fire when the animal executes an action—for example, grasping an object—but also when the animal passively observes a human performing the same action. Curiously, the discovery of mirror neurons does not seem to have had such a dramatic impact on those studying social learning in animals. Part of the reason for this may be that there are still many gaps between research on the functions and mechanisms of animal behavior. The work of neurobiologists, psychologists, and ethologists is becoming ever more integrated, but these are still, to some extent, three separate scientific cultures.

However, there may be another, more specific reason why those studying social learning and culture in animals have not run with the lead provided by mirror neurons.

At first glance, it looks like mirror neurons could mediate imitation or copying of very specific features of behavior—the shape or topography of body movements. But, due to doubts that monkeys are able to copy at this level of specificity, many people have rejected this idea, and instead focused on the possibility that mirror neurons are involved in action understanding, theory of mind, or empathy, processes that are not so closely related to culture. I think that was a mistake. Looking into the crystal ball, I predict that the connection between imitation and mirror neurons—in the premotor cortex and elsewhere in the brain, in monkeys and in other taxa—will be "rediscovered," and that well-integrated behavioral and neurobiological studies will begin to answer important questions about mirror neurons. This will be greatly assisted by more widespread availability of scanners for functional magnetic resonance imaging (fMRI) of primates and rodents. The most important question, in my mind, is: Where do mirror neurons come from? Are some animals born with mirror neurons for a range of behaviors, or are mirror neurons made through associative learning? The answer to this question will tell us a great deal about the origins of the capacity to imitate and, insofar as imitation is involved in cultural transmission, about the origins of culture.

Dr. Cecilia Heyes is a professor at University College London in England. Her research focuses on the evolution of cognition. She is a co-editor of Social Learning in Animals: The Roots of Culture *(Academic Press, 1996) and* Evolution of Cognition *(MIT Press, 2000).*

SUMMARY

1. Cultural transmission involves the acquisition and transfer of information via social learning and teaching and may represent a powerful force for the acquisition and spread of behaviors both within and between generations of animals.

2. What is learned via cultural transmission is passed on from individual to individual. This can translate into the behavior of a single individual shifting the behavior patterns seen in an entire group. When cultural transmission is operating, what is learned by one individual may be passed down through many generations.

3. Cultural transmission involves a "model" individual and an "observer" that learns a specific behavior or response from the model. There are situations, however, that involve an interaction between observers and models, but that do *not* constitute cultural transmission because observers do not learn any particular behavior or response from models. Two examples of this are local enhancement and social facilitation.

4. Cultural transmission can occur through learning from other individuals via social learning (copying or imitation) and/or teaching.

5. To demonstrate imitation, there must be some new behavior learned from others, and that behavior must involve some sort of new spatial (topographic) manipulation as well as lead to the achievement of some goal. Copying differs from imitation in that what is copied need not be novel and need not involve learning some new topographical action.

6. Traditions can be experimentally examined in animal societies. Traditions can rise and fall over time.

7. Teaching, when rigorously defined, implies that one individual serves as an instructor and at least one other individual acts as a student who learns from the teacher. Teaching entails providing an immediate benefit to the student but not to the teacher, instructing naive "students," and imparting some new information to students faster than they would otherwise receive it.

8. Cultural transmission may occur via vertical, oblique, or horizontal transmission. Vertical cultural transmission involves the transfer of information from parent to child; oblique cultural transmission is the transfer of information from older to younger individuals, excluding transfers from parents to offspring; and horizontal cultural transmission occurs when information comes from same-aged peers.

9. Genetic and cultural transmission can operate independently on animals, but often they interact in interesting ways, as exhibited in the cultural transmission of birdsong in finches and the evolution of mate choice in guppies.

10. Data on more than 100 species of nonhuman primates have shown that there is a significant positive correlation between brain size and both innovation and tool-use frequency. A similar trend between large brain size and an increased propensity toward innovation was found for bird species in North America, Britain, and Australia.

DISCUSSION QUESTIONS

1. Why do you suppose it took so long for ethologists to focus on the possibility that cultural transmission was an important force in animals? Can you imagine any biases—scientific, ideological, and so on—that might be responsible for this?

2. Suppose I run an experiment in which I take a bird (the observer) and let it view another bird (the demonstrator) opening a sealed cup by pecking at a circle on the cover of the cup. I then test the observer and see that it now opens the cup by pecking at the circle. What can I infer about social learning here? What other critical treatment is missing from this experiment?

3. Imagine that adults in some population of monkeys appeared to pick up new innovations (for example, potato washing, stone play) from observing others. How might you disentangle vertical, oblique, and horizontal cultural transmission as possible explanatory forces?

4. List the pros and cons of Hauser and Caro's definition of *teaching*. How might you modify this definition to address what you listed on the "cons" side of your ledger?

5. Suppose that after extensive observations, you determine that certain animals in a population appear to rely on social learning much more often than other individuals, and that such differences are due to genetic variation. How might you use the truncation selection technique described in Chapter 2 to examine the heritability of the tendency to employ social learning?

SUGGESTED READING

Bonner, J. T. (1980). *The evolution of culture in animals*. Princeton, NJ: Princeton University Press. One of the earliest books looking at culture from an evolutionary perspective.

Boyd, R., & Richerson P. J. (1985). *Culture and the evolutionary process*. Chicago: University of Chicago Press. This book can get very technical, but it is worth the effort to work through it.

Brass, M., & Heyes, C. (2005). Imitation: Is cognitive neuroscience solving the correspondence problem? *Trends in Cognitive Sciences*, 9, 489–495. An interesting paper on the role of neuroscience in explaining a long-standing "perspective-taking" problem in the study of imitation.

Galef, B. G. (2009). Strategies for social learning: Testing predictions from formal theory. *Advances in the Study of Behavior*, 39, 117–151. An overview of ethological theories of social learning and how these theories have been tested.

Lefebvre, L. (2011). Taxonomic counts of cognition in the wild. *Biology Letters*, 7, 631–633. A review of the evidence for innovations in natural populations of birds and mammals.

7

Sexual Selection

E ach of the first six primer chapters focused on a fundamental issue in animal behavior (natural selection, learning, and so on), identified basic theory, and then examined the issue across a whole suite of behaviors. In the remaining chapters, we continue to examine theoretical, empirical, and conceptual questions, but each of these chapters focuses on a specific behavior—or more generally, a specific set of related behavioral issues (sexual selection, mating systems, foraging, predation, cooperation, and so forth). We begin with the topic of **sexual selection**.

In *The Descent of Man and Selection in Relation to Sex*, Charles Darwin provided the first detailed evolutionary theory of sexual selection, defining this process as one that "depends on the advantage which certain individuals have over other individuals of the same sex and species in exclusive relation to reproduction" (Darwin, 1871). Darwin's work on sexual selection focuses on: (1) **intrasexual selection**, in which members of one sex compete with each other for access to the other sex, and (2) **intersexual selection**, in which individuals of one sex choose which individuals of the other sex to take as mates. Because differential reproductive success drives the process of natural selection, intersexual and intrasexual selection are among the most studied topics in ethology. (For a discussion of mate choice and conservation biology, see the Conservation Connection box.)

In this chapter, we shall examine sexual selection from both proximate and ultimate perspectives, with particular emphasis on (1) the evolution of mating preferences, (2) learning and sexual selection, (3) cultural transmission and female mate choice, and (4) male-male competition.

Intersexual and Intrasexual Selection

Darwin proposed that one key factor leading to differences in reproductive success is access to mating opportunities—intrasexual selection. In most species, it is males that compete for females as mates and not vice versa. This fundamental difference between the sexes is due, in part, to the different type and number of gametes produced by males (sperm) and females (eggs). By definition, females produce fewer, but larger gametes (Figure 7.1). Each egg is extremely valuable, because of both its size and its relative scarcity. Each sperm, on the other hand, requires much less energy to produce, and sperm are usually found in prolific quantities. Male reproductive success is limited by the much lower rate of gamete production of females compared with that of males. Thus, while males often produce millions of sperm, creating the possibility that some males will have extraordinary reproductive success, females' eggs are often, but not always, few and far between, causing intense competition for this scarce resource (Trivers, 1985). These ideas were summed up by the geneticist A. J. Bateman, who studied sexual selection in fruit flies. According to what is known as **Bateman's principle**: (1) females should be the choosier sex because eggs are expensive to produce and because a female's potential reproductive success is limited compared with that of a male, and (2) females' greater choosiness in

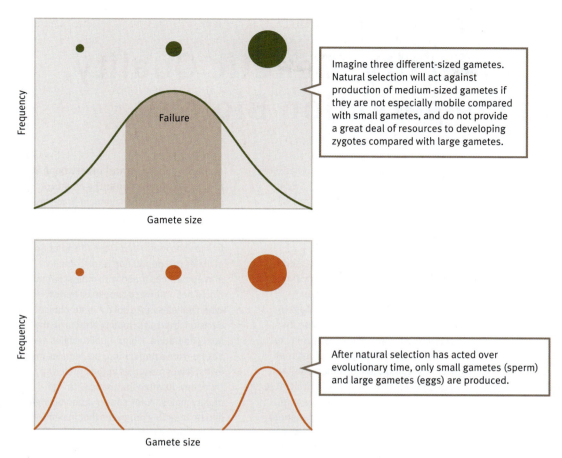

Imagine three different-sized gametes. Natural selection will act against production of medium-sized gametes if they are not especially mobile compared with small gametes, and do not provide a great deal of resources to developing zygotes compared with large gametes.

After natural selection has acted over evolutionary time, only small gametes (sperm) and large gametes (eggs) are produced.

FIGURE 7.1. Natural selection and gamete size. Natural selection favors large and small gametes over medium-sized gametes. *(From Low, 2000)*

mate selection should translate into greater variance in the reproductive success of males (Bateman, 1948).

Darwin argued that, all else being equal, any male trait that confers mating and fertilization advantages and is passed down across generations will, over time, increase in frequency in a population, because males with such traits will produce more offspring than their competitors. Darwin's idea about the struggle among males for mating opportunities forms one basic foundation of our current understanding of sexual selection. Of course, from a proximate perspective, the ways in which hormones, neurobiology, development, environment, and many other factors operate on any particular behavior play a very important role in the sexual selection process as well.

Competition for mates can take many forms, depending on ecology, demography, and cognitive ability. For example, males may fight among themselves, occasionally in dramatic "battles to the death," but often in less dangerous bouts, to gain mating opportunities with females (Figure 7.2). This latter form of male-male sexual competition is illustrated by male stag beetles and red deer (*Cervus elaphus*), which use their "horns" (enlarged jaws) and antlers, respectively, in physical fights over females; the winners of such contests mate more often than the losers.

Genetic Diversity, Genetic Quality, and Conservation Biology

Conservation biologists have long understood the negative consequences of low genetic diversity in populations, but only recently have they come to realize the importance of mate choice per se for conservation issues. Suhel Quader has suggested that human disturbances, as well as actions that conservation biologists and managers undertake to protect species, may have a broad spectrum of effects on mate choice, particularly female mate choice (Quader, 2005). Such effects on mate choice may then have implications for genetic diversity as well as the genetic quality of individuals in natural and managed populations of animals. A few examples will help illustrate the sometimes subtle ways in which such effects can occur.

1. Increasing hybridization: Many natural environments are being heavily polluted by human activities, from dumping commercial waste products to washing laundry in otherwise pristine water sources. Above and beyond the obvious effects on animal mortality, such pollution also affects mate choice. For example, Lake Victoria is home to many closely related species of cichlid. Females often distinguish between males of their species and those of other species by male color patterns. But with pollution increasing in Lake Victoria, the turbidity of the water has also increased. Females cannot see as well in this turbid water, and research has found that cichlid females mate with males from other species because of their inability to distinguish male color patterns (Seehausen et al., 1997). Pollution, through its effect on increased turbidity, has led to higher rates of hybridization in Lake Victoria cichlid species. The long-term implications of such hybridization on these cichlid species are still not well understood.

2. Sexual imprinting: As discussed later in this chapter, sexual imprinting is important in establishing mate preference in some species. Managed animal populations, such as those on conservation refuges or in zoos, may be forced to live under circumstances that reduce, or completely eliminate, suitable models on which to sexually imprint (P. R. Grant and Grant, 1997; Slagsvold et al., 2002). When closely related endangered species are reared together, individuals may even imprint on adults from the wrong species, leading to potentially maladaptive behaviors (M. Wallace, 1997).

3. Danger and decreased mating: Individuals from some species may interpret human disturbances from censusing, habitat manipulation, and so on as cues of increased danger. These cues may then cause females to spend more time being vigilant for danger and less time choosing their mates. As a result, low-quality males may have mating opportunities they would not normally (without human disturbance) obtain. Under extreme circumstances, this increased time spent on vigilance could cause females to skip breeding altogether. For example, the female blackbuck antelope is in estrus for only twenty-four hours per breeding season. If females should sense increased danger as a result of human disturbance, they may forestall breeding for an entire season (Mungall, 1978).

4. Mating and parental investment in managed populations: The effects of conservation and management practices on female mate choice in animals can be quite subtle. For example, work on zebra finches has found that when the number of males in a population is low, females may sometimes breed with males that they would not otherwise choose as mates. When females make such a mate-choice decision, there are rippling effects on the next generation. If high-quality males are absent from a population, the females may invest fewer resources in their offspring than they do when they mate with high-quality males. When conservation biologists initiate breeding programs that maximize the number of individuals mating (and hence genetic diversity), they may inadvertently force females to mate with suboptimal males, and the result may be that females devote fewer resources to their offspring. Offspring from such matings may be less healthy and suffer other adverse affects, such as weakened immune systems (Burley, 1986; Burley et al., 1982).

An understanding of mating systems and issues such as increased hybridization, sexual imprinting, the interaction of predation pressure and mating, and parental investment will provide conservation biologists and managers additional tools to foster natural populations, as well as develop reserves, wildlife parks, and managed populations of all sorts.

A    B

FIGURE 7.2. Competition for mates. (A) Male deer battle with their antlers. **(B)** Male horses compete for females. *(Photo credits: Manfred Danegger/Photo Researchers, Inc.; Rich Pomerantz)*

Until the 1970s, much of the work on mate selection focused almost exclusively on intrasexual competition, rather than mate choice (intersexual selection; Andersson, 1994; Andersson and Simmons, 2006). One reason for this focus may be that male-male competition is very easy to observe in nature. In addition, some prominent evolutionary biologists of the 1930s dismissed mate choice as unimportant and directed research toward male-male competition (J. Huxley, 1938).

Just as intrasexual selection, in principle, could involve either males fighting for females or females fighting for males, so too does intersexual selection encompass female choice of a male and male choice of a female. Both occur in nature, but female mate choice is more prevalent, likely because females stand to lose much more than males by making a bad choice of mates. In the first place, as mentioned above, females invest much more energy in each gamete they produce, so they should be choosier than males in terms of who has access to their gametes. In addition, in species with internal gestation, females typically devote much energy to offspring before they are born, and hence they should be under strong selection pressure to choose good mates that will produce healthy offspring.

Above and beyond the physical genitalia that are necessary for the act of mating, males often possess other traits that play an important role in attracting mates. These traits are referred to as secondary, or epigamic, sexual characteristics. Many conspicuous secondary sexual traits in males such as ornamental plumage, bright colors, and courtship displays have been favored by sexual selection because of their effect on female mate choice. What's more, the underlying genetics of such secondary sexual characteristics are the subject of much experimental work. For example, male fruit flies "sing" to females during courtship by vibrating their wings. This courtship song, which is an epigamic trait, not only influences female mate choice but may also be important in the process whereby new species are formed (speciation) in fruit flies (Spieth and Ringo, 1983; Tomaru and Oguma, 1994). One particular form of song in fruit flies, called "pulse song," is quite conspicuous during courtship, and the interval between pulses (the interpulse interval, or IPI) appears to be critical in terms of female fruit flies' mate choices (Ewing and Bennet-Clark, 1968; Ritchie et al., 1999; Schilcher, 1976a, 1976b). Early work on the genetics of courtship song in *Drosophila* suggested pulse song might be controlled by a large number of different genes, each of which would contribute a small amount to the expression of the song. More recent work, however, suggests that the genetics of song appear to involve three loci that account for much of variance in courtship song (Gleason et al., 2002).

Both intersexual and intrasexual selection play a role in virtually all mating systems (see Chapter 8). Whether the system is *monogamous* (where a single male pairs up with a single female), *polygamous* (some males mate with many females), or *polyandrous* (some females mate with numerous males), sexual selection plays a role. Generally speaking, sexual selection will be stronger in polygamous and polyandrous systems than in monogamous systems. In polygamous and polyandrous systems, some individuals obtain many mating opportunities and some obtain no mating opportunities, while there is generally less variation in reproductive success in monogamous systems (see Chapter 8). This greater variation in reproductive success in polygamous and polyandrous species creates stronger sexual selection pressure.

Evolutionary Models of Mate Choice

As mentioned earlier, until the 1970s male-male competition was the most studied type of sexual selection. The tables have clearly turned of late, and female mate choice is now the most actively studied form of sexual selection (Andersson, 1994; Ord et al., 2005). Until recently, one unifying theme running through virtually all such studies was that they assumed that a female's choice of mates is under some sort of genetic control (R. A. Fisher, 1958). That is, genetic variation in mate-choice behavior in females was assumed to be responsible for variation in mate choice seen in nature. Of course, the environment played a huge role in determining *which* genes were favored over evolutionary time, and proximate factors played a role in determining *how* genes affected mate choice, but at the bottom of it all there was assumed to be a set of genes that controlled the process of mate choice. We will begin by working within this framework and in so doing discuss four evolutionary models of sexual selection that examine female preferences, as well as the traits that females prefer in their mates. Then, we will discuss the idea that learning and cultural transmission also play an important role in sexual selection.

Evolutionary models of female mate choice can be broken down into four classes: "direct benefits," "good genes," "runaway selection," and "sensory exploitation" models. We will examine the logic of these models, and then look at case studies for each. For all of these models, ethologists are attempting to understand how evolutionary change has shaped the process of mate choice. It is important to note that we will focus on case studies in which the evolution of a sexually selected trait is best explained by one of our four models. In many species, of course, the evolution of sexually selected traits might best be explained by a combination of two or more models. The logic of focusing on the more clear-cut cases, where a single model best explains the evolution of a sexually selected trait, is to provide a basic understanding of the dynamics of female mate choice, so that we do not put the cart before the horse. Once each model is understood on its own, students of animal behavior are better prepared to understand the more complex ways that sexual selection operates in nature.

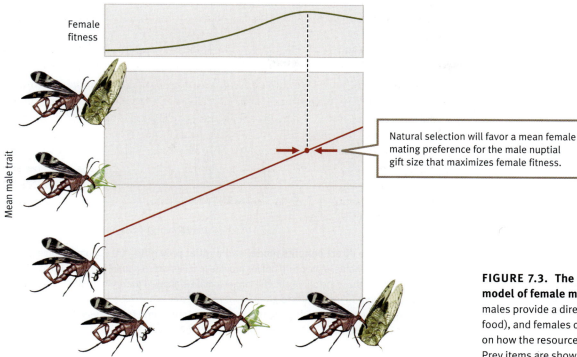

Natural selection will favor a mean female mating preference for the male nuptial gift size that maximizes female fitness.

FIGURE 7.3. The direct benefits model of female mate choice. Here males provide a direct benefit (in this case, food), and females choose males based on how the resource affects their fitness. Prey items are shown in green. *(Based on Kirkpatrick and Ryan, 1991)*

DIRECT BENEFITS AND MATE CHOICE

In the direct benefits model, sexual selection favors females that have a genetic predisposition to choose mates that provide them with tangible resources—above and beyond sperm—that increase their fecundity (Andersson, 1994; Kirkpatrick and Ryan, 1991; Møller and Jennions, 2001; T. Price et al., 1993; Figure 7.3). That is, any female that tends to choose males that provide her with some important resource—food, safe shelter, assistance with parental care, and so on—will do better than her counterparts that are less choosy, and through time we should see more picky females. One well-studied case of direct benefits centers on nuptial gifts in scorpionflies.

DIRECT BENEFITS AND NUPTIAL GIFTS. Randy Thornhill and his colleagues have tested the hypothesis that direct benefits provided by males influence female mate choice in the scorpionfly (*Hylobittacus apicalis*) in both the laboratory and the field (Thornhill, 1976, 1980a). They found that female scorpionflies choose their mates using a simple rule: choose males that bring relatively large prey items—primarily aphids, flies, beetles, and so on—during the courtship process (Figure 7.4). These **nuptial gifts**, which are consumed during courtship, provide females with a direct tangible benefit—food.

Males that bring no prey are immediately rejected as potential mates by females (Cordero, 1996). But females do more than simply choose a mate based on nuptial gifts: they also determine how long they will mate with a male based on the size of his gift—that is, based on the direct benefit they receive. When the nuptial gift is small (between 3 and 19 mm²), there is a

FIGURE 7.4. Scorpionflies and nuptial gifts. To obtain mates, male scorpionflies present females with nuptial gifts, which are prey items that the females consume during courtship or mating. The male (top) provided the female (bottom) with a blowfly (at arrow), which she eats as they copulate. *(Photo credit: Randy Thornhill)*

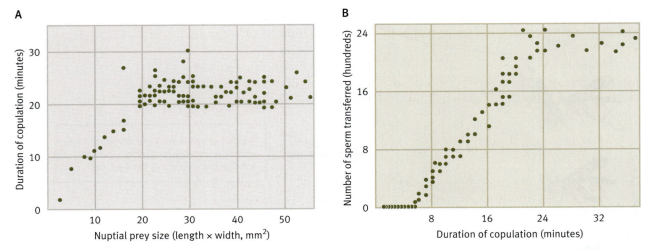

FIGURE 7.5. The direct benefits model and nuptial prey gifts. (A) Copulation time as a function of direct benefits. Male scorpionflies provide females with a nuptial prey gift. Up to about 19mm², the greater the value of the gift, the longer a male is allowed to copulate with the recipient. **(B)** Longer copulation time leads to greater sperm transfer in scorpionflies. When males provide large nuptial gifts to females, the added copulation time translates into more sperm transferred to the female. *(From Thornhill, 1976, 1980b)*

positive relationship between prey size and copulation time. When nuptial gifts are in this size range, it is the female that always terminates copulation, by pulling her abdominal tip away from the male. When the nuptial gift size is larger (greater than 19 mm²), copulations last much longer (Figure 7.5A). Copulation time is important, because there is a strong positive relationship between it and the number of sperm transferred during such matings—indeed, copulation times of less than approximately seven minutes often involve no sperm transfer (Figure 7.5B).

Given that females can clearly distinguish males with large gifts from other males, what benefits do they receive by doing so? A female that actively chooses males that bring large nuptial gifts produces *more* eggs and, in all likelihood, has a longer life span, both because of the nutrition she receives directly from the nuptial gift and the decreased amount of time she must allocate to hunting (Thornhill, 1976, 1979b, 1980a, 1980b; Thornhill and Alcock, 1983).

The fact that nuptial prey size has such a strong effect on mating opportunities (and sperm transfer) creates strong sexual selection pressures on males to bring large nuptial gifts to females. Finding prey that are large enough (greater than 19 mm²) to result in long copulations with a female is both time consuming (many prey are much smaller) and quite dangerous (increased foraging time exposes males to greater risks of predation). Indeed, at any given time, only about 10 percent of the males in a population are in possession of such prey. These constraints have led to some rather remarkable adaptations in male foraging behavior. Males will often sample, but then discard, prey that are too small to result in long copulation opportunities with females. In contrast, females hunting on their own almost never discard small prey.

Selection pressure on males to bring large nuptial gifts has also resulted in males stealing large prey from one another. In one study, Thornhill recorded such prey theft on 345 occasions; a follow-up study discovered that one way that males manage such thefts is to mimic the behavior of a female and then subsequently steal prey brought to them by other males (Thornhill, 1976, 1979a).

Thornhill's work not only provides evidence that females select among males based on resources that provide direct benefits but it also illustrates the way that such choice shapes male foraging and mating behavior.

GOOD GENES AND MATE CHOICE

Females receive more than direct resources such as food and shelter from their mates; they also get sperm, and with it, obviously, they receive genes that are passed on to offspring. Much theory, and some empirical work, now suggests that sexual selection favors choosing females mates that possess "good genes" (Andersson, 1994; R. A. Fisher, 1915; Kodric-Brown and Brown, 1984; Kokko et al., 2003; Mays and Hill, 2004). Good genes are those that code for some suite of favorable traits—traits that can be inherited by offspring of the appropriately choosy female. In **good genes models**, females choose the males with genes best suited to their particular environment—for example, genes associated with superior foraging skills or the ability to fend off predators. In doing so, the females receive "indirect" benefits, in the sense that their offspring receive some of the good genes that led their mother to choose a particular male as a mate in the first place (Cameron et al., 2003). Good genes models are therefore sometimes called "indirect benefit models" of sexual selection.

Good genes models of sexual selection apply to mating systems in which the primary benefits received by females lie in the good genes their offspring receive as a result of their mate choice. For example, in pronghorn antelopes (*Antilocapra americana*), males provide females with no direct benefits, and they do not appear to actively coerce females into mating in any measurable way. Instead, female pronghorns undertake a long and energetic search to find mates by visiting different males that have harems of females. Early work had shown that females select males as mates based on a male's ability to defend his harem (J. A. Byers et al., 1994). Eventually, most females end up mating with a small subset of males in their population. This leads to high variance in harem size among males and to males with large harems in prior years siring a disproportionate number of offspring. John Byers and Lissete Waits hypothesized that such males possessed good genes (and so they labeled them as "attractive" males). But how could a female know about a male's prior reproductive success? Here, the researchers assumed that females were using current harem defense as an indicator of good genes in males. If so, males with large harems (attractive males) should have offspring that are more likely to survive than offspring from other (nonattractive) males (J. Byers and Waits, 2006).

Byers and Waits tested their hypothesis in a population of individually marked pronghorns that live in the National Bison Range in Montana. They followed females during the mating season and noted which males were selected as mates. When offspring were born, they too were marked (usually the first day they were born), and their survival was measured as a surrogate for fitness. Byers and Waits found that, in accordance with predictions from most good genes models, offspring from attractive males had higher survival rates than offspring from other

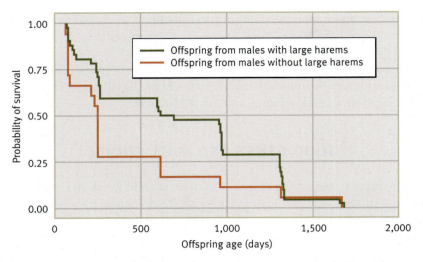

FIGURE 7.6. Indirect benefits in pronghorns. Offspring males that had large harems (green line) had higher survival rates than offspring from other males (orange line), suggesting that females were selecting males based on some measure of a male's genetic quality. *(From J. Byers and Waits, 2006, p. 16344)*

males. These findings suggested that females were selecting males based on some measure of a male's genetic quality (Figure 7.6).

In pronghorns, females use harem defense as an indicator of good genes in males, but this is one of the few instances where ethologists and evolutionary biologists have good evidence of what male traits females use to infer good genes in potential mates. But how can females gauge which males have good genes in other systems? In addition, how can females determine whether males have cheated in indicating that they possess good genes, given that sexual selection would be likely to favor males that cheat in such a manner? That is, wouldn't selection favor males that attempt to give the impression that they possess appropriately good genes, even if they don't? The answer is yes—sexual selection should favor males that do just this. This, in turn, creates new sexual selection pressures on females. Now only traits that are true and **honest indicators** of male genetic quality should be used by females when choosing mates. If females focus on such honest indicators, they can overcome the male cheater problem—at least temporarily.

PARASITE RESISTANCE AND GOOD GENES. Which male traits should females accept as honest indicators of a male's genetic quality? One theory is that honest indicator traits should be generally "costly" to produce—the costlier the trait, the more difficult it is to fake, and hence the more likely it is that this trait is a true indicator of good genes (Zahavi, 1975; Zahavi and Zahavi, 1997; Figure 7.7). One costly trait that has been studied in the context of honest advertising is parasite resistance. Females that choose males with strong resistance to parasites—a trait that would be difficult to fake—may receive indirect benefits in that they may be mating with individuals with good genes—here, genes conferring parasite resistance. The predictions of honest advertising models, when applied to parasites, go under the name of the Hamilton-Zuk hypothesis, after Bill Hamilton and Marlene Zuk, who first applied the idea of good genes to parasite resistance (Hamilton and Zuk, 1982).

FIGURE 7.7. Peacock's tail. An example of an elaborate, costly trait in males. *(Photo credit: Corbis)*

Consider the case of endoparasites that reside within their hosts and can't be seen. How do females know which males have good genes with respect to endoparasite resistance? The answer to this dilemma must in some way lie in the female's ability to use some other male trait that correlates with the ability of a male to avoid being parasitized. If possessing trait T also means that males are good at fighting endoparasites, females can use trait T as a proxy for judging what they really need to know. That is, if information on internal parasitization is unavailable, using some other trait (T) that correlates well with the one of interest should be favored by sexual selection. One such proxy cue appears to be body coloration (Figure 7.8), which has been studied extensively in birds and fish. Healthy males tend to be very colorful, while infected males have much duller colors (Milinski and Bakker, 1990; Figure 7.9). Although still the subject of some heated debate, numerous studies in birds and fish have found results that are consistent with the predictions of the Hamilton-Zuk hypothesis, in that females often choose the most colorful (and least parasitized) males (Figure 7.10).

MHC AND GOOD GENES. If females are searching for honest indicators of good genes in males, honing in on traits associated with disease resistance should be favored by sexual selection. One set of genes that is involved in disease resistance in many animals is known as the Major histocompatibility complex (MHC; Milinski, 2006; Penn and Potts, 1998, 1999). Proteins produced by MHC genes guide the body in identifying "self" versus "foreign" cells (Wills, 1991).

A unique aspect of the MHC is that it is the most variable set of genes ever uncovered. Very few (if any) individuals have exactly the same MHC. Given this incredible genetic variability, biologists have predicted that animals may prefer mating with others that have a dissimilar MHC (S. V. Edwards and Hedrick, 1998; Ziegler et al., 2005). Such a preference could arise because the resulting offspring may have particularly strong immune systems. Why? Because diseases evolve so rapidly that it is as if animals are trying to hit a moving target in trying

FIGURE 7.8. Elaborate coloration in males. In many birds, such as the red-knobbed hornbill (*Rhyticeros cassidix*), males are much more colorful than females as a result of different sexual selection pressures. *(Photo credit: Mark Jones/Minden Pictures)*

FIGURE 7.9. Color, parasites, and good genes. One reason stickleback females may prefer the most colorful (red) males is that color indicates resistance to parasites.

to combat them. In this case, their weapon against such a moving target is a set of MHC genes that is constantly changing across generations.

How can females determine which males have MHCs that differ from their own? Work using mice and rats suggests that females use odors to determine if another individual is a good MHC match—that is, whether he or she has sufficiently different MHC genes (Penn and Potts, 1998). This work in rodents has spurred experiments on odor, MHC, and mate choice in humans. To test the hypothesis that humans use MHC when choosing mates, and that odor plays a role in the process, male and female undergraduate students in Switzerland were tested by Claus Wedekind (Wedekind and Furi, 1997; Wedekind et al.,

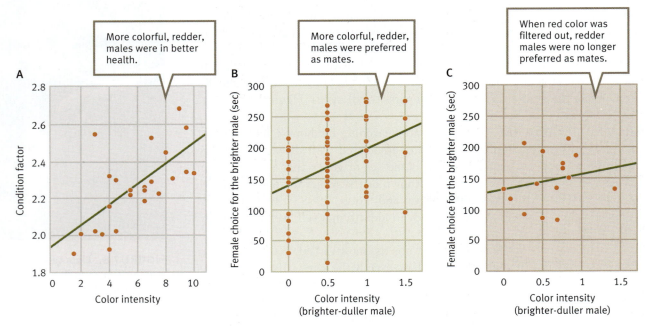

FIGURE 7.10. Female sticklebacks prefer brighter males. Brighter (redder) males were healthier and were preferred as mates, but not when red coloration was filtered out. (*From Milinski and Bakker, 1990, p. 331*)

1995). In Wedekind and Furi's (1997) study, men were instructed to wear a cotton T-shirt for two nights. Blood samples from each of these males were then analyzed to determine MHC. Females also had blood samples taken for MHC analysis. These women were then given T-shirts from males with MHCs similar or dissimilar to their own. Women not on oral contraceptives consistently found the odors of the T-shirts from males with dissimilar MHCs sexier, suggesting that MHC has a significant effect on female mate choice in humans. Indeed, it seems that not only do females choose males with dissimilar MHC alleles but they themselves use perfumes that specifically magnify their own MHC-mediated odors (Milinski and Wedekind, 2001).

Thorston Reusch and his colleagues have hypothesized that, in order to produce disease-resistant offspring, individuals should, all else being equal, prefer mates with many MHC alleles, rather than choosing a mate that is dissimilar in MHC-related genes (S. V. Edwards and Hedrick, 1998; Reusch et al., 2001). They tested their "MHC allele counting" hypothesis using wild-caught stickleback fish from three populations of fish that reside in interconnected lakes. Reusch and his team used molecular genetic tools to sequence genes at a number of MHC loci in their sticklebacks. Sticklebacks from these three lakes varied in the number of MHC (class IIB) alleles they possessed, ranging from two to eight such alleles. When females were given a choice between males—some of whom had few MHC alleles and some of whom had many such alleles—they consistently preferred males with a greater number of MHC alleles (up to eight; Figure 7.11).

In a fascinating series of experiments, Manfred Milinski and his colleagues (including Reusch) examined the proximate cues that female sticklebacks may be using to assess the MHC qualities of potential mates (Milinski et al., 2005). These researchers knew from prior work that females can assess the number of MHC alleles a male has using chemical cues alone. The question, then, was how do female sticklebacks make this assessment? The answer may be related to the fact that the greater the number of MHC alleles found in an individual, the greater was the number of MHC peptides, which are short strings of amino acids displayed at the cell surface (Rammensee et al., 1997). Milinski and his collaborators hypothesized that female sticklebacks were able to use odor to assess the diversity of MHC peptides for a particular male. More specifically, they proposed that MHC peptide ligands—that is, molecules that bind to proteins—were the key underlying proximate mechanism that females were assessing during mate choice.

Milinski and his team tested their hypothesis by simultaneously exposing a female to two different water columns. One column had water drawn from a tank that had male X swimming in it, and the other column also had water from male X's tank, but the water in this column was supplemented with MHC peptide ligands. The MHC peptide ligand diversity of both the male and the female were known, and a given pair together had either the optimal number of peptide ligands (producing offspring with better disease resistance) or a suboptimal number of MHC peptides (Kurtz et al., 2004; Wegner, Kalbe, et al., 2003; Wegner, Reusch, et al., 2003).

What Milinski and his colleagues found was that when a pair had a less-than-optimal number of MHC peptide ligands, the addition of synthetic ligands to one side of the water column made the odor on that side more attractive, suggesting that the number of peptide ligands was the proximate cue being used by females to select males with good genes. Conversely, when a male and a female had the optimal number of MHC peptide ligands, the addition of ligands

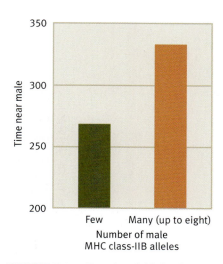

FIGURE 7.11. Female sticklebacks prefer males with more MHC alleles. Female sticklebacks spent more time on the side of the tank with males that had many different MHC alleles than with males that had few MHC alleles. *(From Reusch et al., 2001)*

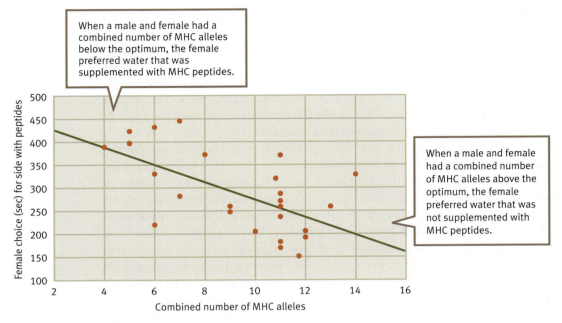

FIGURE 7.12. Female sticklebacks use peptides to assess MHC. Female sticklebacks were given a choice between water from a tank containing a lone male, or water from a tank with that same male plus added MHC peptides. *(From Milinski et al., 2005)*

to one of the water columns made the odor associated with that column less attractive (Figure 7.12).

In a powerful follow-up experiment, Milinski's group tested another prediction regarding odor and MHC peptide ligands and good genes. Soon after females give birth, they avidly forage, often raiding nests of other sticklebacks and eating their eggs. Males guard nests from such female attacks. As such the researchers hypothesized that a female should be repelled by the very male MHC peptide ligand odor she was attracted to prior to mating, because such a repulsion would reduce the chance that she accidentally raids the nest of her mate and cannibalizes her own eggs. When Milinski and his team tested whether foraging females were repelled by the MHC peptide ligand odor of their former mates, they found strong evidence that this was indeed the case.

RUNAWAY SEXUAL SELECTION

The process of runaway sexual selection was first proposed by Sir Ronald Fisher. Models of runaway sexual selection center on the relationship between alleles at two loci. In runaway sexual selection models, one locus houses alleles that code for female preference, and the other houses alleles associated with the male trait that females prefer. Over evolutionary time specific alleles from the two genes become associated with each other—when one allele is present in male offspring in a clutch, the other allele is likely to be present in female offspring from that clutch (Andersson, 1994; R. A. Fisher, 1958; Kirkpatrick, 1982; Mead and Arnold, 2004).

To see how this runaway selection process works, imagine a population in which some proportion of females have a heritable preference for brightly colored males, and the remainder of females choose males randomly with respect to color.

Further suppose that in this population, the degree of male coloration is also a heritable trait—some males are more colorful than others. So, we have a group of females, some of whom prefer brightly colored males and some of whom don't, and a group of males, some of whom are more colorful than others. Now, we will make an important assumption that is critical to runaway models—the loci for both male color and female preference are found in both males and females, but the alleles at these loci are only expressed, that is, "switched on," in the appropriate sex (preference genes in females, color genes in males).

Females that mate with colorful males should not only produce colorful sons but also daughters that possess their mother's genetically coded *preference* for colorful males. Over time, the allele in females that codes for the preference for colorful males and the allele in males that codes for color in males become linked in the sense that, as the frequency of one changes, the frequency of the other changes as well. Once this positive feedback loop is set in motion, it can, under certain conditions, "run away," like a snowball rolling down a snowy mountain. Across generations, selection may produce increasingly exaggerated male traits (for example, male color pattern) and stronger and stronger female preferences for such exaggerated traits.

STALK-EYED FLIES AND RUNAWAY SELECTION. Empirical evidence of runaway sexual selection comes from Jerry Wilkinson and his colleagues' work on the stalk-eyed fly (Wilkinson, 1993; Wilkinson, Kahler et al., 1998; Wilkinson and Reillo, 1994). In stalk-eyed flies, females prefer to mate with males possessing eyes that are at the end of long eye "stalks" (Figure 7.13). In one treatment of an experiment that spanned over thirteen generations of flies, males with the largest eye stalks were selected and allowed to breed with females. In a second treatment, males with the shortest eye stalks were allowed to breed. In neither case were females chosen for any particular trait; that is, females were selected randomly in these two treatments. Not surprisingly, in the treatment where long eye-stalk length was selected for thirteen generations, the average male eye stalk increased in length, producing a more exaggerated version of this trait. Conversely, in the case where individuals with short eye stalks were selected, the size of the average male eye stalk decreased in length.

The critical finding in this study, however, did not center on the length of male eye stalk per se. At the heart of the runaway sexual selection model was the discovery of a positive link between the length of the male eye stalk and the *female preference for this trait*. Recall that in both treatments—the long eye-stalk and the short eye-stalk male treatments—the females that mated with males were selected at random. All of the selection pressures were with respect to male eye-stalk length. Yet, after just a few generations, when females were given the choice between mating with males with short and long eye stalks, females from the short eye-stalk line preferred males with short eye stalks. This preference for short male eye-stalk length by females in the short eye-stalk line changed in response to selection pressure on *males*, in accordance with the prediction of the runaway selection model. In the long eye-stalk line, females preferred males with long eye stalks, but then so did females that had not been subject to any selection treatment (that is, females in a control line of flies). In other words, contrary to predictions of the runaway selection model, Wilkinson and his colleagues did not find evidence that selection on males in the long eye-stalk treatment produced significant changes in female preference for eye-stalk length. While there were a number of possible reasons why results were similar

FIGURE 7.13. Stalk-eyed flies and runaway selection. Male stalk-eyed flies show variation in the length of their eye stalks. Wilkinson bred lines of flies with long and short eye stalks to test the runaway model of sexual selection. *(Photo credit: Mark Moffett/National Geographic Stock)*

in the long eye-stalk length treatment and the control, Wilkinson and Reillo speculate that female preference might very well have been detected had the experiment gone on for a longer period of time. Whether or not that is the case can only be determined by further experimentation.

Before we leave the subject of female preferences and male traits co-evolving to shape sexual selection, it is worth noting that recent work in two closely related species of Hawaiian crickets suggests that female preference and male trait expression may be controlled by the *same* alleles. In this case, complex co-evolutionary runaway selection is not required to explain how male trait and female preference may change together over time (Shaw and Lesnick, 2009).

SENSORY BIAS AND THE EMERGENCE OF MATE CHOICE

Before we complete our discussion of evolutionary models of mate choice, we need to consider one last model known as the **sensory exploitation**, **sensory bias**, or **preexisting bias** model of mate choice (Endler and McLellan, 1988; M. J. Ryan, 1990; West-Eberhard, 1979, 1981). Sensory bias models hypothesize that as a male trait first emerges it may be preferred by females because it elicits a neurobiolocial response that is already in place in females, and that such a response initially has nothing to do with mating preferences.

As a hypothetical example, suppose that red berries are the most nutritious food source available to a fruit-eating, blue-feathered songbird species. Females that are best able to search out and subsequently eat red berries survive and reproduce better (Rodd et al., 2002; C. Smith et al., 2004). Natural selection should then favor neurobiological circuitry in females that allows them best to hone in on red things in their environment (Kirkpatrick and Ryan, 1991).

Once a preference for all things red is in place, if red feathers should suddenly arise in males of this normally blue-feathered species, birds with these red feathers may be chosen as mates because the female's nervous system is already designed to respond preferentially to red objects. Males with red feathers, then, are exploiting the preexisting neurobiologically based preferences of females—preferences that evolved as a result of other selection pressures.

The sensory bias model leads to a clear prediction regarding phylogenetic history. When we look at a phylogeny that includes information on both female preference and the male trait preferred by females, the female preference trait should predate the appearance of the male trait (Figure 7.14). In our example of red berries, the preference for red should be in place before red feathers are present. In a moment, we shall examine two case studies that test this sort of prediction.

One thing to keep in mind as we discuss the sensory bias hypothesis is that it was designed to elucidate the *origin*, not the long-term maintenance, of female preference. That is, the sensory bias hypothesis centers on how a female preference initially arose in a population, not how it was maintained in a population by natural selection over evolutionary time. So as we walk through the sensory bias model, remember that the three other models we have discussed (direct benefits, good genes, and runaway selection) focus on how natural selection acts on mate choice, but the sensory bias model focuses on the emergence of mate choice.

TRICOLOR VISION, FRUITS, AND SENSORY BIASES IN PRIMATES. Let's continue with our discussion of red food and a possible sensory bias for red

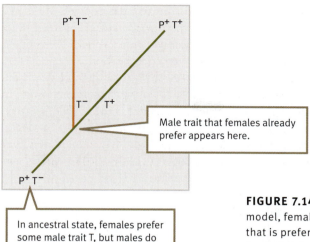

FIGURE 7.14. Sensory exploitation model. In the sensory exploitation model, female preference (P+) is assumed to predate the male trait (T) that is preferred. In one lineage here (shown in orange), the male trait is immediately preferred once it appears (T+). *(From M. J. Ryan and Rand, 1993, p. 189)*

coloration, but now move from a hypothetical bird example to a real study involving primates. There is growing evidence from work on sexual selection in primates that females prefer males with red fur coloration and/or red facial skin (P. Sumner and Mollon, 2003; Figure 7.15). The question with respect to the sensory bias model is whether female preference for red coloration in males arose because of a sensory system that favored such coloration (Fernandez and Morris, 2007). To answer that question, we begin with a discussion of the primate visual system, particularly with regard to foraging.

Many mammals have two types of photoreceptor cones in the eye (dichromatic vision), but some primate species have three types of cones (trichromatic vision). Possessing a third type of cone allows some primates to detect red-orange colors, whereas many other mammals are color-blind in this spectrum. Ethologists and others have suggested that trichromatic vision allows primates to more easily pick out red-orange fruits and edible ripe red leaves in their natural environment (Allen, 1879; P. W. Lucas et al., 1998; Mollon, 1989).

Peter Sumner and J. D. Mollon examined this hypothesis by measuring the visual (spectral) properties of the fruit and leaves eaten by six primate species (Sumner and Mollon, 2000). When they compared the visual systems of these primates with the spectral properties of both the fruit and leaves these animals ate, they found that trichromatic vision provided primates with an enhanced ability to detect red foods (fruits and leaves) in a visually complex environment. Indeed, individuals in many (but not all) primate species with trichromatic color vision prefer red fruits and leaves as food (P. W. Lucas et al., 2003).

Returning to the sensory bias hypothesis, we can now ask whether the female mating preference for red coloration in males might have originated as a sensory bias associated with trichromatic vision and its effect on foraging for red food items. To address this question, Andre Fernandez and Molly Morris used already published data on both color vision and red fur/face color in primates and superimposed this data on a phylogenetic tree of more than 200 primate taxa (Fernandez and Morris, 2007). This phylogenetic data allowed them to test the sensory bias hypothesis by examining whether trichromatic vision that allows for the detection of red-orange colors predated a preference for red fur/face coloration. If trichromatic vision appeared before red fur/face coloration, this is

FIGURE 7.15. Primate preference for red. There is growing evidence from work on sexual selection in primates that females prefer males with red fur coloration and/or red facial skin, as seen in this red uakari. *(Photo credit: Claus Meyer/Minden Pictures)*

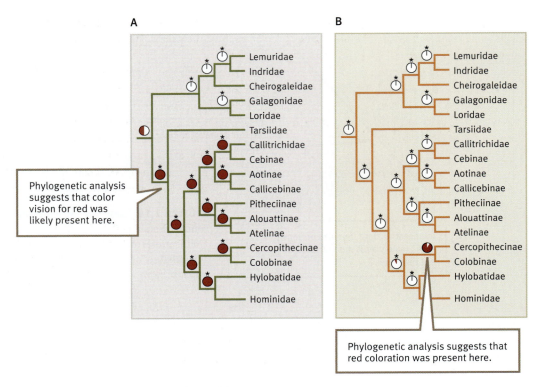

FIGURE 7.16. Color vision and red skin in primates. (A) A phylogenetic tree in which the presence of trichromatic color vision is shown to be ancestral for most modern lineages. The degree to which a pie graph is filled with red indicates the probability that red color vision was present. **(B)** A phylogenetic tree for red skin color. The degree to which a pie graph is filled with red indicates the probability that red fur was present; white indicates the absence of red skin. In both phylogenetic trees, asterisks indicate that there is statistical confidence for reconstruction of the trait in that branch of the phylogeny. A comparison of these two phylogenies shows that trichromatic vision evolved earlier than red coloration in males. *(From Fernandez and Morris, 2007, p. 14)*

FIGURE 7.17. Frog calls and sensory bias. *Physalaemus pustulosus* males (shown here) add a unique "chuck" sound to the end of their calls. The calls of *Physalaemus coloradorum* males lack a chuck. The calls of these two species have been used to test models of the sensory bias hypothesis. *(Photo credit: Michael Ryan)*

consistent with the sensory bias hypothesis. If trichromatic vision appeared after red fur/face coloration, the sensory bias hypothesis should be rejected in this case.

Results from Fernandez and Morris's phylogenetic analysis is consistent with the sensory bias hypothesis for the origination of female mate preference. They found that trichromatic vision evolved earlier than red coloration in males (Figure 7.16). This result suggests that the *origin* of the female preference for red-colored males was linked to a sensory bias in females for foraging on red food items in their environment.

FROGS AND SENSORY BIASES. One of the first studies of sensory exploitation involved two closely related species of frogs—*Physalaemus pustulosus* and *Physalaemus coloradorum* (M. J. Ryan et al., 1990; Figure 7.17). Males in these two species use calls to attract females. As is the case for most species in the genus *Physalaemus*, males of both the *pustulosus* and *coloradorum* species begin their call with what is referred to as a high-frequency "whine." Females pick up the whine part of a male's call through the basilar papilla in their inner ear.

Pustulosus males also add a low-frequency "chuck" sound to the end of their call. When *pustulosus* females choose between *pustulosus* males that chuck and those that don't, they prefer to mate with the former. Females detect the chuck through the amphibian papilla section of the inner ear, which is more sensitive to low-frequency sounds. Michael Ryan and his colleagues hypothesized that the preference for chucks in female *pustulosus* was the result of an already in-place sensory bias in favor of such low-frequency sounds.

Phylogenetic and behavioral evidence support the contention that the preference for chucks is due to a sensory bias. To understand why, recall that the chuck part of the call is absent in the call of *coloradorum* males. When Ryan and his colleagues used molecular and morphological data to reconstruct the evolutionary history of the genus *Physalaemus*, they found that the common ancestor of *coloradorum* and *pustulosus* did not use a chuck call. That is, not only do male *coloradorum* not use the chuck call, but the chuck call appears never to have been used in any of the species that make up the evolutionary lineage leading to *coloradorum* (Figure 7.18). Yet when modern computer technology is used to add a chuck call to the end of prerecorded *coloradorum* male calls,

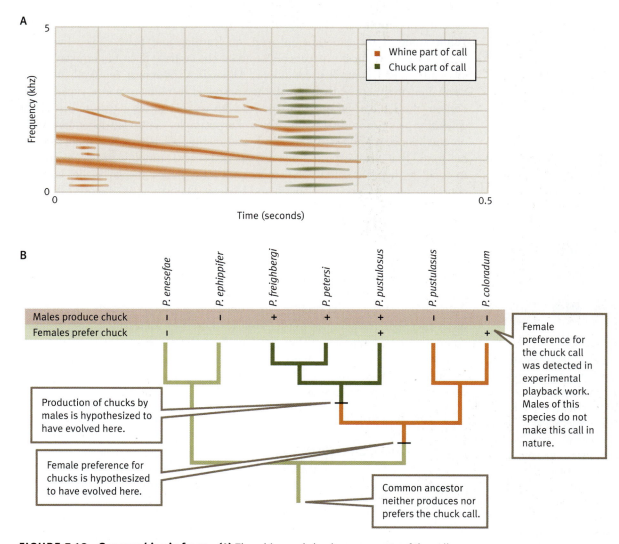

FIGURE 7.18. Sensory bias in frogs. (A) The whine and chuck components of the call made by *Physalaemus pustulosis*. **(B)** Preference for the chuck calls arose before the chuck call itself. *(Adapted from Ryan and Rand, 2003)*

coloradorum females show a preference for calls that include a chuck. In other words, Ryan's experimental manipulation of calls suggests that as soon as chucks appear in a *coloradorum* population, females prefer males that produce such calls. These studies suggest that the auditory circuitry in *Physalaemus* frogs has been designed in such a way as to prefer a certain class of sounds—low-frequency calls like chucks. The *coloradorum* studies provide evidence that the preference for chucks predated the actual appearance of chucks in the *Physalaemus* species, in accordance with the sensory bias hypothesis.

Learning and Mate Choice

While evolutionary models of female mate choice have greatly expanded the understanding of sexual selection, these models do not usually consider that individuals may *learn* various aspects of mate selection. To get a more complete understanding of mate choice, psychologists and ethologists have conducted many experiments examining the role of learning in selecting mates (Domjan, 2009). While some of the experimental work on learning and mate choice has been done in the field, most experiments on mate choice have been undertaken in the laboratory under controlled conditions. Although they appear removed from the normal environment of the animal, the controlled stimulus-response experiments done in the lab can be quite instructive with respect to what they tell us about general aspects of animal mate choice. For example, controlled laboratory work on conditioned stimuli (see Chapter 5) and mating behavior across many species has found that, after exposure to a conditioned sexual stimulus, males are quicker to copulate, become better competitors with other males, display higher levels of courtship, and produce more sperm and progeny (Domjan, 2006; Figure 7.19).

SEXUAL IMPRINTING

One form of learning studied in the context of mate choice is **sexual imprinting** (Bateson, 1978; Lorenz, 1935; ten Cate and Vos, 1999). In sexual imprinting, young individuals learn mating preferences from their interactions with adults (most often their parents). Imprinting is often restricted to some small time window during normal development, but the length of this window varies dramatically across species (including humans) and behavioral contexts (Bereczkei et al., 2004).

Ethologists have developed numerous ways to experimentally examine sexual imprinting, including:

1. Using the cross-fostering approach discussed in Chapter 2. In the case of imprinting, one can test the hypothesis that offspring raised by adoptive parents show different mating preferences than those raised by biological parents, and one can examine whether such preferences can be linked to something about the behavior or morphology of adopted parents (Cooke and McNally, 1975; Cooke et al., 1972, 1976).
2. Employing the "novel trait" approach, in which offspring are raised by parents that have some novel trait that an experimenter has introduced.

Let's examine the novel trait approach to studying sexual imprinting by focusing on Klaudia Witte and her colleagues' work on the mannikin bird, *Lonchura*

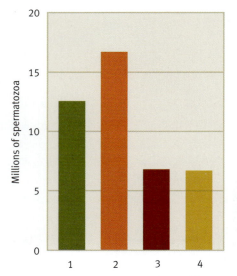

FIGURE 7.19. Pavlovian conditioning and number of sperm. Two different lines of male quail learned to pair a distinctive experimental chamber with a chance to mate with a female. Males were then placed in the chamber with a model female. Afterward, spermatozoa samples were taken. Males from both lines that had learned to pair the chamber with the chance to mate produced significantly more spermatozoa than control males (treatments 3 and 4, which consisted of males from the two lines that had not been given the chance to associate these two cues). *(From Domjan et al., 1998)*

leucogastroides (Witte et al., 2000). To study the role of sexual imprinting in this species, Witte added a novel trait to some adults in her population. The novel trait was a red feather that the researchers attached to the forehead of adult mannikins, so that the feather stood up like a crest. Witte and her colleagues hypothesized that offspring that were raised in the presence of such adults adorned with a red head feather would display a sexual preference for individuals with red feathers when they matured. To examine this, Witte raised juvenile mannikins in one of four groups. Group 1 served as a control in which offspring were raised with their mother and father, both of whom lacked the red feather. In group 2, offspring were raised with their mother and father, each of whom had a red head feather experimentally attached to their forehead by Witte and her colleagues. In group 3, offspring were raised with a mother with no feather, but a father with a red feather; in group 4, offspring were raised with a father with no feather, but a mother with a red feather.

When offspring reached sixty days of age, they were separated from their parents and their mate preferences were examined. These birds were given the choice between two members of the opposite sex—one that had a red head feather added, and one that did not. Witte and her team found evidence that young mannikins imprinted on the red head feather of their parents, and expressed a preference for such birds when they themselves matured. Compared with the control groups (in which individuals often preferred unadorned mates), males raised with a mother with a red feather preferred females with red head feathers as mates, and females raised with a father with a red feather preferred males with red head feathers as mates.

Ethologists also examine the neurobiological underpinnings of imprinting. As an example, let's return to the dendritic spines (on neurons) first discussed in Chapter 3. Recall that neuroethologists have argued that an increase in dendritic spines is often associated with learning. If this is correct, it might also be true that when animals rely *less* on learning—including learning about potential mates— we might see a decrease in dendritic spine density. In the context of sexual imprinting, this might manifest itself as a decrease in dendritic spine density soon after an individual has imprinted on the phenotype of those of the opposite sex, since it has now completed one stage of learning about what traits to search for in a mate, and rarely changes such learned preferences after imprinting. Evidence has been found for such a decrease in spine density in the zebra finch.

Following up on early work by Klause Immelmann, behavioral studies have shown that when a male zebra finch is raised by its parents for the first month of life, but then kept in isolation for two months, he often imprints on the phenotype of the first female he encounters after isolation (Bischof and Clayton, 1991; Immelmann, 1972; Immelmann et al., 1991). When Hans Bischof and his colleagues measured dendritic spine density in one area of the zebra finch brain (the media neo/hypostriatum or MNH area), they found a decrease in the spine density after exposure to a female (Bischof and Rollenhagen, 1999; Rollenhagen and Bischof, 1991, 1994). Compared with spine density when males were in isolation, Bischof found fewer dendritic spines in males that had completed the process of sexual imprinting and hence were less reliant on learning about whom to choose as a mate.

SEXUAL IMPRINTING ON FACES IN HUMANS. Sexual imprinting occurs in humans as well as in other animals. One hypothesis for how imprinting occurs in humans is that during some critical time in childhood, a young person creates a mental template of his or her opposite-sex parent, and then uses that template as a guide for finding mates when he or she matures. If men imprint on their mother's features when they are children, then we would expect a similarity between a man's mother and his subsequent wife.

Bereczkei and colleagues found that when subjects were shown pictures of women and asked to pick out pictures of a woman's mother-in-law—the mother of her husband—they were able to do so with some accuracy (Bereczkei et al., 2002). One limitation with this type of study is that people may use some of their *own* facial features—"matching-to-self"—when searching for mates. Because of the similarity between parent and offspring, matching-to-self rules and sexual imprinting would both result in the sort of results found by Bereczkei and his colleagues. How might we distinguish between these two hypotheses?

To distinguish between the matching-to-self and sexual imprinting hypotheses, Bereczkei and colleagues looked at the similarity between the mate choice of females who had been raised as adopted daughters (Bereczkei et al., 2004). If these women selected mates who resembled *themselves*, this would support the matching-to-self hypothesis; if they selected mates who resembled their *adopted fathers*, this would support the sexual imprinting hypothesis. Results indicated a much greater similarity between a daughter's adopted father and her subsequent husband, as predicted by the sexual imprinting hypothesis. In addition, the stronger the emotional bonds that had existed between a daughter and her adopted father, the greater the similarity between adopted father and future husband, suggesting that sexual imprinting is influenced by emotional bonds between parent and child.

LEARNING AND MATE CHOICE IN JAPANESE QUAIL

By studying learning and mate choice in the Japanese quail, Michael Domjan and his colleagues found evidence that classical conditioning (see Chapter 5) in adults affects mate choice and that adult male Japanese quail will quickly learn to stay in areas in which they have the opportunity to mate with a female (Domjan et al., 1986; Mahometa and Domjan, 2005; Mills et al., 1997). While earlier work had demonstrated that sexual imprinting affects Japanese quail mate choice (Gallagher, 1976), Nash and Domjan found that learning can also play another role in mate choice. They tested the hypothesis that classical conditioning at the adult stage might override the effects of sexual imprinting as a juvenile (Nash and Domjan, 1991).

Nash and Domjan began by using different strains, or varieties, of Japanese quail. Each trial in their experiment had three subjects: an adult male quail from the standard "brown" colored strain, an adult female quail from the brown strain, and an adult female quail from a lighter-colored "blond" strain of quail. The brown males used in this study were raised with other brown quail, and these males had already sexually imprinted on the phenotype of brown females. All else being equal, when such males matured, they typically showed a strong preference for brown females as mates. But all else was not equal in Nash and Domjan's experiment.

In phase 1 of a trial, a brown male was allowed to see a blond female and was then given the opportunity to copulate with her. In phase 2 of a trial, the same male could see a brown female quail, but was never in physical contact with her. In other words, a male learned that, in the laboratory, the presence of a blond female meant a mating opportunity, but the presence of a brown female did not. In the last phase of the experiment, males were tested to see how much time they spent near brown and blond females. The researchers found that blond female quails elicited a much stronger response (Figure 7.20). That is, adult-stage learning about who was likely to be a receptive mate overrode the effects of early sexual imprinting.

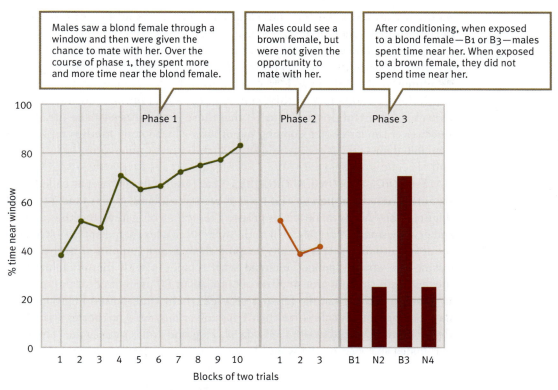

FIGURE 7.20. **Overriding an imprinted sexual preference.** Male Japanese quail were raised in an environment with normal brown-colored quail and based on prior work done by reaserchers, these males were expected to have imprinted on brown females as future mates. These males were then put through a battery of tests in which experimenters measured the percentage of time they spent near a window. Conditioning—in which males learned that they would have the chance to mate with blond, but not brown, females—overrode sexual imprinting. B = blond, N = brown strain. *(Based on Domjan, 1992, p. 53)*

Cultural Transmission and Mate Choice

In addition to individual learning, the role of cultural transmission has also been examined with respect to sexual selection. One area of sexual selection that has been explored is the extent to which **mate-choice copying** occurs—that is, when a female's mate-choice preference is affected by the preference of other females in her population (Dugatkin, 2000; Pruett-Jones, 1992; Westneat et al., 2000; see Chapter 6). Cultural transmission and sexual selection have also been studied with respect to song learning and mate choice in birds (Freeberg, 1998, 2004).

MATE-CHOICE COPYING

We will begin our discussion of cultural transmission and mate choice with a definition of mate-choice copying. Following Stephen Pruett-Jones's definition, we will say that female mate-choice copying has occurred when a male's probability of being preferred as a mate increases as the result of having been preferred by a female in the past (Pruett-Jones, 1992). In other words, if a male has an X percent chance of mating if he has not recently mated, and a Y percent chance if he has recently mated, the effect of mate-choice copying is defined as the difference between Y and X. Mate-choice copying occurs if Y − X > 0; the greater Y − X, the stronger the effect of mate-choice copying. One nice feature of this definition is that these probabilities can be measured, and hence a mate-choice copying hypothesis can be supported or refuted by available data.

MATE-CHOICE COPYING IN GROUSE. Black grouse mating arenas called **leks** are interspersed throughout the bogs of central Finland, Wales, and Scotland (Figure 7.21). Many male black grouse (*Tetrao tetrix*) gather together at leks and occupy small territories, which they defend. Males then display— strutting about with their tail spread out, flapping their wings, jumping in the air, and hissing—to attract females to mate with them. As with many lek-breeding species, a single male grouse often obtains about 80 percent of all the matings at an arena. Before mating, females visit leks many times, often in groups that stay together and synchronize their trips to various male territories within an arena (Höglund and Alatalo, 1995). Jacob Höglund and his colleagues observed that males that had mated were likely to mate again fairly quickly, suggesting a possible role for mate-choice copying among females (Höglund et al., 1990). In addition, older females mated, on average, three days earlier than younger females, suggesting that mate-choice copying, if it occurred, was most common among younger females.

Höglund and his colleagues placed stuffed "dummy" females placed on male territories within a lek (Höglund et al., 1995). In their experiment, each of seven randomly chosen males on a particular lek had stuffed black grouse females (model females) placed on their respective territories early in the morning before real females arrived. Males courted these model females and even mounted them and attempted copulations. After observing the males on the lek, females were more interested in mating with a male that had copulated with other females on his territory, even if they were model females. Because models were placed on the territories of randomly chosen males, this finding suggests that mate-choice copying, rather than some set of physical traits possessed by males, explains the skew in male reproductive success among black grouse.

FIGURE 7.21. Black grouse and mate-choice copying. Males try to attract females with which to mate by displaying on leks. This involves strutting, flapping their wings, jumping in the air, and hissing while females observe. *(Photo credit: Jan Smit/Foto Natura/Minden Pictures)*

Mate-choice copying likely plays a role in another species of grouse as well. In the early 1990s, Robert Gibson and his colleagues studied female mate choice in sage grouse (Gibson et al., 1991). They examined the mating behavior of sage grouse females (*Centrocercus urophasianus*) in two different leks over a four-year period. One of their hypotheses regarding mate-choice copying behavior was that the unanimity of female mate choice would increase as more females mated on a given day, because more opportunities to observe and copy mate choice would exist on such days.

To test this hypothesis, each day Gibson and his team arrived early in the morning before a lek formed and then, using telescopes, observed all interactions on that lek—when females arrived, which males mated, which males courted, and so on. Gibson and his colleagues tested their mate-choice copying hypothesis using data from fifty-six days of observation. They began by using a computer simulation to estimate how often females would have chosen the same male on a lek if they had *not* been copying each other. Then they compared the results of their computer simulation with the data they had gathered on the relationship between the number of females visiting the lek and the unanimity of female mate choice. This comparison uncovered support for the mate-copying hypothesis, as the unanimity of female mate choice increased as more females appeared, and this increase occurred more quickly than expected by chance (Figure 7.22).

MATE-CHOICE COPYING IN MICE. Recent work has examined the underlying endocrinological and genetic basis of mate-choice copying. Elena Choleris, Martin Kavaliers, and their colleagues studied mate-choice copying in mice, examining the role of oxytocin in facilitating such copying. Oxytocin (OT) is a hormone secreted by the pituitary gland, and it appears to play an important role in social behaviors, including mate choice, maternal bonds, and individual recognition (Choleris et al., 2004; Ferguson et al., 2001; Pedersen and Boccia, 2002). Mice with deletions of the OT gene—so-called OT "knockout" mice—appear to learn normally, except in the context of social learning and mate choice (Choleris et al., 2003; Clipperton-Allen et al., 2012; Ferguson et al., 2000; Kavaliers et al., 2003; Gabor et al., 2012).

Choleris, Kavaliers, and their team studied mate-choice copying in three lines of mice, two of which were normal—that is, these mice had the OT gene—

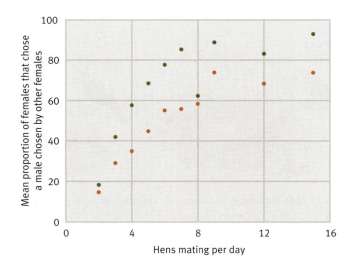

FIGURE 7.22. Female mate-choice copying in sage grouse. As predicted by models of mate-choice copying, when the number of hens mating per day increased, so too did the proportion of females choosing to mate with a male chosen by other females. The green points are the observed matings, and the orange points are the expected matings if the females had not been copying the mate choices of other females. *(From Gibson et al., 1991)*

and one of which was an OT knockout line of mice. They allowed a female to choose between two healthy males, one of which had recently associated with an estrous female and had traces of the odor of that estrous female in his area, and one of which had not. Females from the two normal lines of mice copied the mate choice of other females—they were attracted to males that had recently associated with estrous females. Females from the OT knockout line, however, did not copy the other females' mate choice. In conjunction with a number of control experiments run by Choleris and her team, these results suggest that knocking out the gene for oxytocin has the specific effect of inhibiting mate-choice copying behavior in mice.

SONG LEARNING AND MATE CHOICE IN COWBIRDS

Mate-choice copying is not the only way that cultural transmission can affect mate choice in animals. Song learning in birds provides another good example of cultural transmission acting on sexually selected traits (see Chapter 6). One thing that is common to almost all songbirds is that they learn the songs they sing. In particular, much of the song-learning process involves learning songs from others that are often referred to as tutors. Songs, in other words, are culturally transmitted.

Todd Freeberg undertook a series of experiments with cowbirds to learn more about the cultural transmission of birdsong and its long-term consequences for mate choice in this species (Freeberg, 2004). He collected juvenile and adult birds from two populations of cowbirds—one population from South Dakota (SD) and one population from Indiana (IN; Freeberg, 1998). Freeberg chose these particular populations because cowbirds display different social behaviors and sing different songs across populations.

Freeberg used the IN and SD birds in a modified cross-fostering experiment (Chapter 2) designed to examine cultural transmission and mate choice. He raised juvenile SD birds with either SD adults or IN adults (Figure 7.23). In Freeberg's experiment, if cowbirds behaved like the individuals they were raised with, regardless of whether these individuals were from their native populations, then this would suggest a role for cultural transmission of birdsong.

When juvenile cowbirds matured, they were observed in a large aviary that contained *unfamiliar* birds from both the SD and IN populations. Freeberg

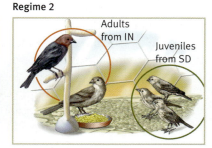

FIGURE 7.23. Cross-fostering. In Freeberg's cross-fostering experiments, juvenile cowbirds from South Dakota were raised with adults from South Dakota (regime 1) or adults from Indiana (regime 2).

found that when placed into such groups, birds paired up and mated with others that came from the same rearing regime (IN or SD) in which they were raised. That is, SD birds raised with SD adults preferred SD birds as mates, and SD birds raised with IN birds preferred IN birds as mates. The mating preferences of the SD birds were strongly dependent on the social environment in which the birds were raised.

Freeberg and his colleagues explored what it was about the social environment during development that might be responsible for SD birds' preference for mates that were like the birds with which they were raised. The answer appears to involve copying and perhaps sexual imprinting. Males copied the songs of the adults with which they were raised, regardless of whether those adults were from South Dakota or Indiana (Freeberg et al., 2001). Furthermore, female SD birds preferred songs that were like those of the males with which they were raised (M. J. West et al., 1998). This preference might either be an example of females copying the song preferences of adult females in their population, or it might be a consequence of females imprinting on the songs of the males they were exposed to during development. Further work is needed to distinguish between these possibilities.

Male-Male Competition and Sexual Selection

As mentioned at the start of the chapter, once Darwin introduced the notion of sexual selection, much of the work in this area focused on intrasexual selection in the form of male-male competition for females. Intrasexual selection is often more dramatic than female mate choice, as it typically involves some sort of direct competition between males, as in the classic case of various males bashing horns to determine access to a female. But intrasexual selection need not be as dramatic as fierce fights between males. Competition may be less direct, as in the case of male cuckoldry or sperm competition, which are examined in detail in Chapter 8.

RED DEER ROARS AND MALE-MALE COMPETITION

Tim Clutton-Brock and his colleagues have been studying male-male competition in roaring behavior in red deer (*Cervus elaphus*; Clutton-Brock and Albon, 1979; Clutton-Brock et al., 1979). Male red deer stags on Rum Island, Scotland, form harems during the mating season (the rut) and fight off other males to defend their harems. While serious fighting is not common, 23 percent of harem holders show some sign of a fighting injury and 6 percent are permanently injured, suggesting strong selection pressure for accurately assessing an opponent's strength to avoid injury and possible death at the hands of a much stronger rival. Clutton-Brock and Stuart Albon hypothesized that male red deer should use honest indicators of each other's fighting ability to determine how much time and energy to invest when competing with other males for access to females (Clutton-Brock and Albon, 1979).

Clutton-Brock and Albon proposed that the main indicator of strength and fighting ability in males was roaring. Evidence suggesting that roaring was being used to assess an opponent's fighting ability in the context of harem holding included:

- Roaring and associated activities such as "parallel walks," in which males walk alongside one another to assess their size and fighting ability relative to others, were almost exclusively seen during the mating season (Figure 7.24).
- During the mating season, harem holders roared more than those without a harem (Figure 7.25).
- Roaring rates increased when a harem holder was approached and his ownership of his harem was threatened by another male.
- The more often a male roared, the more likely he was to win a subsequent fight.
- Roaring contests were much more common between mature stags of approximately equal fighting ability than between stags of unequal fighting ability, and roaring rarely escalated (via the "parallel walk" stage) to fighting except among the most closely matched stags, which suggested that roaring was a good indicator of fighting ability and one upon which stags often relied.

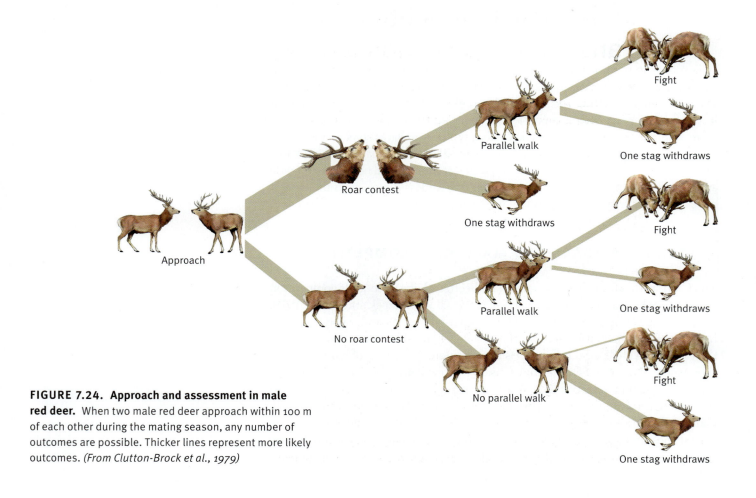

FIGURE 7.24. Approach and assessment in male red deer. When two male red deer approach within 100 m of each other during the mating season, any number of outcomes are possible. Thicker lines represent more likely outcomes. *(From Clutton-Brock et al., 1979)*

Clutton-Brock and his team experimentally manipulated red deer roars to examine the hypothesis that males use roar rate to assess opponents. Using stags of different sizes and abilities, the researchers played tapes for red deer males of other males roaring at different rates. They then measured the responses of the stags to the different roars on the tapes. When playback tapes included sounds from a larger opponent, male red deer responded by being more attentive to such calls and increasing their own rate of roaring (Reby et al., 2005).

Together, all of the data above suggest that male-male competition for females has led to a social system involving fine-tuned communication designed to maximize access to females while minimizing injuries that result from serious fights.

MALE-MALE COMPETITION BY INTERFERENCE

Although many examples of male-male competition involve males fighting *before* a female is present, one interesting subset of male-male interactions consists of males interfering with another male while that male attempts to mate with a female. This sort of behavior is common in amphibians and insects (Duellman and Trueb, 1994). For example, in the European earwig (*Forficula auricularia*), heavier males often succeed in interrupting copulation between lighter males and females by displacing lighter males to gain access to females (Forslund, 2000).

In at least some cases of male-male interference, females appear to solicit males to try to remove a rival during the actual mating event. Cox and Le Bouef hypothesized that females incite male-male competition because it increases the probability of mating with the highest-ranking male in a group (Cox and Le Boeuf, 1977). They tested their hypothesis using the elephant seal (*Mirounga angustirostris*), a species in which males form large harems of females and in which the males are much larger than the females (Figure 7.26). Harems of

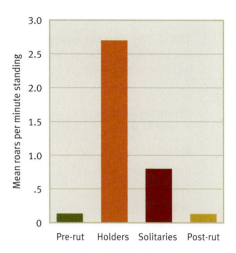

FIGURE 7.25. Roaring rates in red deer. Clutton-Brock and Albon found that males in possession of a harem (holders) roared much more than males not holding a harem (solitaries), as well as more often than males in the pre-rut or post-rut periods. *(Based on Clutton-Brock and Albon, 1979)*

FIGURE 7.26. Elephant seal fights. Male elephant seals are much larger than females and fight for access to females during the breeding season. *(Photo credit: François Gohier)*

TABLE 7.1. Female elephant seals protest when mounted by males that were not top-ranked in dominance hierarchy. Cox and Le Boeuf found not only that alpha (dominant) males mounted females more often but that the females protested their mounts less often than mounts by nondominant adults or subadults (Subadult [1], Subadult [2]). *(From Cox and Le Boeuf, 1977, p. 324)*

AGE OR SOCIAL RANK OF MOUNTING MALE	MOUNTS OBSERVED* (N)	TOTALLY PROTESTED (%)	PARTIALLY PROTESTED (%)	NOT PROTESTED (%)
Alpha	74	37	43	20
Non-dominant adult	70	49	34	17
Subadult (1)	9	78	22	0
Subadult (2)	4	100	0	0

Interrupted mounts are excluded.

up to forty females are exclusively defended by a single dominant (large) bull seal, and there is consequently huge variation in reproductive success among males. In one long-term study, only 8.3 percent of males mated, but some males inseminated 121 females (Le Boeuf and Reiter, 1988).

Cox and Le Bouef marked 271 estrous female elephant seals and tracked attempted matings (mounts) with these females. Approximately 87 percent of the attempted mounts were "protested" to some degree or another by the females. Protesting included many behaviors, most prominently loud calls and constant back-and-forth movement to prevent the male from copulating. These protests were quite effective, as 61.4 percent of protested mounts were interrupted to some extent, as opposed to 25 percent of mounts that went unprotested. Females rarely protested dominant male mounts, but they often protested mounts attempted by subordinates (Table 7.1). This led to an increased probability of copulation between the highest-ranking males in the vicinity and the females that had protested mounts with subordinate males.

The elephant seal example shows how difficult it can be to completely disentangle female mate choice from male-male competition. Clearly, interruption of the copulation attempts of males by other males involves male-male competition, but in the elephant seal example, this competition is initiated by a female to increase her probability of mating with the highest-ranking male in the area.

SEXUAL SIZE DIMORPHISM AND MALE-MALE COMPETITION FROM A PHYLOGENETIC PERSPECTIVE. Sexual selection via intense male-male competition has been examined from a phylogenetic perspective. For example, Patrik Lindenfors and his colleagues have studied the relationship between sexual size dimorphism and male-male competition in the pinnipeds (seals, walruses, and sea lions), including the elephant seals described above (Lindenfors et al., 2002). The researchers hypothesized that in species in which harem size is large, and hence where male-male competition is usually most intense, sexual size dimorphism (males larger than females) should be greatest. In addition, since sexual selection for large body size in pinnipeds should act strongly on males but

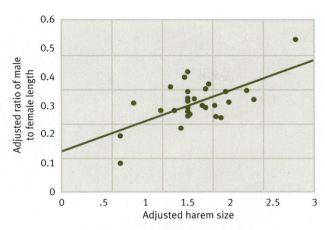

FIGURE 7.27. Harem size and sexual selection. In pinnipeds, there is a positive relationship between harem size and the relative difference in size between males and females. Each point represents a comparison between two species of pinnipeds. The x-axis and y-axis are adjusted independent contrast measures. *(From Lindenfors et al., 2002, p. 189)*

not females, Lindenfors and his team predicted that, as harem size increased, female body size would remain fairly constant. The researchers wanted to make certain, however, that if this relationship was uncovered, it was a result of sexual selection pressures, rather than the fact that pinnipeds share a recent common ancestor, and hence share behavioral and morphological attributes because of common descent. To distinguish between these possibilities, they used independent contrasts (Chapter 2), which remove the effects of phylogeny from patterns found among related species. To test their hypotheses, they first gathered published data on harem size, male size, and female size for thirty-eight species of pinnipeds. They incorporated this data into a phylogenetic analysis, using a phylogeny of pinnipeds as their starting point. Their analysis uncovered a significant positive correlation between harem size and sexual size dimorphism: the larger the average harem size in a species, the larger was the relative size of males compared with females. This analysis also found that the relative size of males increased as harem size increased, but the relative size of females stayed fairly constant (Figure 7.27).

MALE-MALE COMPETITION BY CUCKOLDRY

In many species of fish, ethologists have found different male reproductive types, or morphs, that are distinct in structural, physiological, endocrinological, and behavioral traits (Gross, 1985; Gross and Charnov, 1980). In bluegill sunfish (*Lepomis macrochirus*), three male morphs—known as parental, sneaker, and satellite morphs—coexist within populations (Gross, 1982; Neff et al., 2003). Parental males are light-bodied with dark yellow-orange breasts, build nests, and are highly territorial, chasing off any other males that come near their territory. It takes between six and thirteen days for their eggs to hatch, and males exert significant amounts of energy fanning the eggs (to oxygenate them) and defending the nest against predators during this time (R. Coleman et al., 1985; Figure 7.28).

FIGURE 7.28. Parental male bluegill sunfish tending his eggs. Parental male sunfish build nests and care for their eggs by fanning them and defending them against predators. *(Photo credit: Dr. Bryan D. Neff)*

Sneaker males look and act very different from parental males. These males are smaller, less aggressive, and do not hold territories. Instead, they spend time in hiding places near a parental male and swim quickly into a territory while the parental male and female are spawning. Sneakers then shed their sperm and swim away; the whole process takes less than ten seconds (Gross, 1982). Male-male competition here then involves sneakers trying to outcompete others by a "hit-and-run" type strategy in which they sneak into the parental male's territory and cuckold the parental male (Figure 7.29). Using molecular genetic analysis, researchers have found that, depending on their relative numbers in a population, sneaker males fertilize between 0 and 58.7 percent of all bluegill eggs laid in Lake Opinicon, Canada (Philipp and Gross, 1994).

A third male morph, labeled satellite males, can also be found in some bluegill populations. Satellite males tend to look like females, and they position themselves between a spawning pair. If the parental male is tricked and treats the satellite male as another female, he will attempt to spawn with both the female and the impostor satellite male, at which point the satellite male will release his own sperm.

Because of the very different reproductive behaviors seen among these three bluegill morphs, Bryan Neff and his colleagues tested whether there are also differences in sperm production and sperm quality across morphs (Neff et al., 2003). They predicted that because of their hit-and-run mating strategy, sneaker males might invest most heavily in sperm production. Results are consistent with this prediction. Although parental males are much larger than sneakers and have much larger testes, when Neff and his colleagues examined the ratio of testes size to body size—the relative

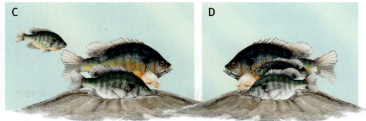

FIGURE 7.29. Bluegill morphs.
(A) Bluegill parental male preparing a nest.
(B) Sneaker males hiding behind plants awaiting a chance to quickly sweep into a parental nest. **(C)** A satellite male swimming over a nest containing a male and female.
(D) A satellite male swimming between a parental male and a female. **(E)** A composite of A–D. *(Based on Gross, 1982)*

investment in testes—sneaker males had the highest ratio, followed by satellites and then parentals.

The relative investment in testes size is an indirect measure of sperm production. A more direct measure would be the number of sperm produced per ejaculate. Given that the sperm produced by sneakers are always competing with parental sperm (sneakers are always in competition with the parental male they are attempting to cuckold), but parental sperm are not always competing directly with sneaker sperm (not all parentals are cuckolded), a high density of sperm per ejaculate should be more strongly favored in sneaker males. When Neff and his team looked directly at the density of sperm per ejaculate, they found that sneakers produced more sperm per ejaculate. Sneakers do, however, pay costs for investing so heavily in sperm production. First, their sperm are shorter lived than parental sperm. Second, when Neff and his colleagues stripped sperm from both sneakers and parentals, and then released the same number of parental and sneaker sperm over eggs, parental sperm were more likely to fertilize eggs than were sneaker sperm. If we focus on the sneaker and parental morphs, the work on testes and sperm production show two very different reproductive strategies: sneakers invest in producing many short-lived, lower-quality sperm, whereas parentals invest in producing fewer but higher-quality sperm.

INTERVIEW WITH

Dr. Anne Houde

You've done extensive research on sexual selection in guppies. Why have you studied that species? How did you first get interested in working with these small fish?

An aspiring scientist can choose a study system in two different ways. First is to sit at your desk and read the literature for a while, find a research question, and then find a study system to help answer it. Second is to start by getting involved with organisms that interest you, either in the field or in the lab. Once you know your system, you can start asking questions. This is what I did: started with the system and then thought of the questions. Of course, to think of good research questions, I needed to have a good grounding in my field, animal behavior, and its literature.

I started out working on birds—terns—but found them to be so long-lived that it was hard to answer questions that interested me in a reasonable amount of time. Then I went in the opposite direction and started looking at the sexual behavior of *Drosophila*, but I did not find flies very exciting (I did not know about their songs and pheromones). When David Reznick came to study life history evolution in my department and gave me some guppies, I started looking at their sexual behavior. I soon realized that I could address some important questions about sexual selection simply by recording male color patterns and observing behavior. I turned out to be correct about this, and have been working on guppies for the last twenty-five years!

Back in 1871 Darwin proposed that females might exert some choice in choosing their mates. Why do you think it took so much longer to study this form of sexual selection compared to studying male-male combat?

The study of mate choice by females did not really get going until the late 1970s, more than 100 years

after Darwin's book about sexual selection. I think there are a few reasons for this. People did not take the idea of natural selection seriously until decades later, people had difficulty imagining that nonhuman animals could do things like choose their mates, and science itself has had a long history of being biased toward a male perspective.

When most people think of male-male competition they envision two rams butting heads. Is it possible for male-male competition to be more subtle?

Just as "survival of the fittest" does not only mean that the strong survive by beating up the weak, male-male

competition can take many forms. Darwin's classic distinction says that sexual selection (differences in mating success) can occur if female behavior results in greater mating success of some males than others (female choice), or if interactions between males lead to differences in mating success (male-male competition). So yes, two rams might butt heads until one emerges the victor and the other leaves the area. Presumably the winner then has greater access to reproductive females while the loser risks being beaten up if he comes near the females again. But the combat does not need to be overtly physical.

What else can males do to increase their own success at the expense of others? In many species, male-male interactions are mediated by signals such as coloration, visual displays, or songs and calls rather than fighting, which can be very costly. Males are able to communicate their prowess and likelihood of winning in physical combat through signals like these. Amazingly, these signaling systems seem to work in lieu of combat, perhaps because the signals evolve to be "honest." Animal behaviorists have been fascinated by the seeming paradox of how signals can be "honest"—evidently, males who "bluff" still face the risk of getting beaten up.

The effect of symmetry on mate choice has proven to be quite a contentious issue. Where do you come down on the importance of symmetry when it comes to choosing mates?

There certainly seems to be something to the idea that mate

choice can be based on symmetry—the question is Why? There are plenty of empirical studies in which the researchers have related mate choice to some measure of symmetry, including studies of humans. The implication is that there must be some benefit to choosing a symmetrical mate if this preference has evolved. The argument goes that an individual with "good genes" will be developmentally stable and be more symmetrical than an individual with lower genetic quality. Thus, an animal choosing a mate should be able to use symmetry as a signal of good genes, and a preference for symmetry should be favored. Perhaps certain sexual ornaments like tail plumes or bright color patterns have evolved because they show off symmetry particularly well.

The problem I see is that there can be circularity in this reasoning. Let's suppose that "good genes" do result in greater symmetry (a point that needs more study). But individuals with good genes might show other signs of this too. Perhaps they are especially vigorous and athletic and potential mates can detect this. So, in studies that find a correlation between symmetry and attractiveness, choice might just be based on a different, unrelated correlate of good genes so that the correlation with symmetry is spurious. A case in point is a study of scorpionflies by Randy Thornhill showing that females prefer the odor of more symmetrical males, even with no chance to assess male symmetry. There is a way out of this problem, and there are studies that convince me that there are preferences for symmetry per se.

How far can we extrapolate from animal studies of mate choice to human mate choice? Are there warning signs one can look for that tell us we have crossed some line we shouldn't cross?

We can learn a great deal from animal studies to understand mate choice in humans. The now-famous "sweaty T-shirt" studies, in which mate choice based on odors has been detected in humans, are a good example. These are adaptations of studies done on rodents, and they reach similar conclusions about, for example, the role of relatedness or MHC genotype in mate choice, despite humans' notoriously poor sense of smell. So mate choice, narrowly defined, has some clear similarities between humans and other species. What are the pitfalls?

First of all, it is too easy to forget that the role of culture in human behavior is vastly more complex and important than in any other species. Studies of classified ads reveal that men prefer youth and beauty in women, while women prefer wealth in men. Do these preferences maximize reproductive success by identifying women with high reproductive value and men who are good resource providers, similar to many animal species? Maybe, but we need to think harder about the interplay of biological and cultural evolution in how men and women construct classified ads.

Second, there is a danger of falling into the "naturalistic fallacy," especially when thinking more generally about patterns of sexual behavior in humans. Yes, most animal species show the stereotype of macho males and choosy females. The same stereotype in humans probably has an evolutionary origin. Does this mean oppression of women by men is morally justified and inescapable? No, of course not. The answer to this paradox is to think more deeply about human moral systems (themselves the product of natural selection) and also the interplay of natural selection and cultural evolution in shaping the roles of men and women in human society.

How do you think the current focus on brain function will affect our understanding of the evolution of mate choice over the next five years?

It is always hard to see into the future, so let me start with two areas I think are especially promising right now. For several years, we have been seeing studies using various kinds of brain scans (e.g., PET scans), in which "active" parts of the brain are visualized. Mostly these are studies of humans, comparing brain activity in different contexts or between different people. I think this approach could be potentially useful in studies of sexual behavior and mate choice. For example, we might be able to look at the interplay between attraction, fear, and aggression when an individual responds to different potential mates. This kind of "whole-brain" imaging approach can lead to more detailed, neuron-level studies of specific brain circuits involved in producing or modulating behavior.

An even more powerful genomic approach to understanding brain function during mate choice involves directly assessing gene expression in the brain. Scientists are beginning to use this method to understand which genes are involved in a variety of behavioral activities and contexts, including mate choice. Ultimately, perhaps, we will be able to look at how local and global gene expression patterns and neural activity interact in the context of sexual behavior.

Dr. Anne Houde is a professor at Lake Forest College. Her work on sexual selection in guppies is summarized in her book, Sex, Color and Mate Choice in Guppies *(Princeton University Press, 1997).*

SUMMARY

1. When developing his theory of sexual selection, Darwin outlined two important processes: intersexual selection and intrasexual selection.
2. Intrasexual selection involves competition among one sex, usually males, for mating access to the other sex.
3. Intersexual selection involves mate choice in which individuals from one sex, usually the female sex, choose their mates from among members of the opposite sex.
4. There are four types of evolutionary models of female mate choice: "direct benefits," "good genes," "runaway sexual selection," and "sensory exploitation" models.
5. Females may learn how to select mates through sexual imprinting and classical conditioning in which they are rewarded with the opportunity to mate.
6. Mate choice is one of the more active areas of ethology in terms of studying the cultural transmission models of behavior. Work in this area includes studies of mate-choice copying, as well as song learning and mate choice in birds.
7. Male-male competition for access to females with which to mate can occur in many ways, including, but not limited to, fighting, roaring, interfering with another male as he attempts to mate a female, and cuckolding another male by fertilizing his mate's eggs.

DISCUSSION QUESTIONS

1. Suppose that a group of males was engaged in a series of fights, and male A emerged as the dominant individual. Now suppose that a female assessed all the males involved in fights and chose male A. Why might this example blur the distinction between intersexual selection and intrasexual selection?
2. Find a copy of Kirkpatrick and Ryan's 1991 paper "The Evolution of Mating Preferences and the Paradox of the Lek," in *Nature* (vol. 350, pp. 33–38). Drawing from this paper, list the similarities and differences between sexual selection models in terms of both assumptions and predictions.
3. Why do you suppose it is so difficult to demonstrate mate-choice copying? Choose a species and design an experiment that would examine whether mate-choice copying is present in that species. How many controls did you need to construct to rule out alternative hypotheses to mate-choice copying?
4. Pick any of the numerous "reality" shows on television that focus on dating, and watch a series of episodes. As you watch, imagine yourself as an ethologist studying mate choice. What traits do males prefer in females? What traits do females prefer in males? Can you say anything about how your observations match up against current models of sexual selection?
5. Read over Lindenfors, Tullberg, and Biuw's 2002 paper entitled "Phylogenetic Analyses of Sexual Selection and Sexual Size Dimorphism in Pinnipeds" in *Behavioral Ecology and Sociobiology* (vol. 52, pp. 188–193). Explain why pinnipeds are a good group for a phylogenetic analysis that examines the relationship between sexual dimorphism and sexual selection.

SUGGESTED READING

Andersson, M. (1994). *Sexual selection*. Princeton, NJ: Princeton University Press. This book is considered the modern reference guide for work on sexual selection.

Darwin, C. (1871). *The descent of man and selection in relation to sex*. London: J. Murray. The book in which Darwin outlined his theory of sexual selection.

Fisher, R. A. (1915). The evolution of sexual preference. *Eugenics Review, 7,* 184–192. One of the earliest models of the process of sexual selection.

Houde, A. E. (1997). *Sex, color and mate choice in guppies*. Princeton, NJ: Princeton University Press. A book-length case study of sexual selection in a model system: the guppy (*Poecilia reticulata*).

Kirkpatrick, M., & Ryan, M. (1991). The evolution of mating preferences and the paradox of the lek. *Nature, 350,* 33–38. This article provides a nice view of the contrasting models in the sexual selection literature.

Kraaijeveld, K., Kraaijeveld-Smit, F. J. L., & Maan, M. E. (2011). Sexual selection and speciation: The comparative evidence revisited. *Biological Reviews, 86,* 367–377. A review of the relationship between the strength of sexual selection and the rate of speciation.

8

Mating Systems

In Chapter 7, we examined intersexual and intrasexual selection, and worked through a number of theories on how and why individuals end up with particular types of mate(s). In this chapter, we will look at a related question: How can we understand the *mating systems* that we observe in nature? Why, for example, do some animals choose a single mate for life, while others mate with a single partner each breeding season, but switch partners across breeding seasons? And, why, for that matter, do some mating systems involve a male (or female) having two (or more) opposite-sex partners during a single breeding season?

Mating systems occur in many forms and gradations. For example, we often think of our own species as being monogamous. In general, *monogamy* means "having one mate," but this definition sometimes fails to capture important aspects of mating systems. For example, people in most modern industrial societies are better characterized as serially monogamous, since, with the high divorce rate, many individuals will be married to more than one person during their lifetime (serial monogamy), just not at the same time.

Before we examine different mating systems in animals and the underlying proximate and ultimate factors that shape these systems, it is important to understand that mating systems are not static and change over time, as a result of selection pressures. And change can occur quickly, if selection pressures are strong. For example, over the course of a ten-generation experiment, when female *Drosophila pseudoobscura* fruit flies were placed with some males that carry a deleterious allele, which decreases the reproductive success of the male and the reproductive success of females who mate with him, and some males who did not carry this deleterious allele, the tendency for females to mate with more than a single male increased in frequency, compared with control populations (T. A. R. Price et al., 2008). In ten generations, the mating system itself evolved in these populations.

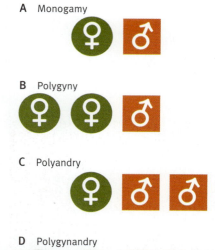

A Monogamy

B Polygyny

C Polyandry

D Polygynandry

FIGURE 8.1. Four mating systems.
These are **(A)** monogamy (1 male, 1 female), **(B)** polygyny (1 male, more than 1 female), **(C)** polyandry (1 female, more than 1 male), and **(D)** polygynandry (more than 1 male, more than 1 female). Each mating system can be further subdivided in a number of ways.

Different Mating Systems

Let's start with a quick survey of different forms of mating systems. We begin by tackling a definitional question. There has been some debate in the mating systems literature as to whether "pair bonds" between males and females— that is, long-term relationships that often involve rearing offspring—should be part of how we classify mating systems (Davies, 1991; S. T. Emlen and Oring, 1977; Thornhill and Alcock, 1983). Here we sidestep that question and instead use the number of mating partners as the critical variable in defining our mating systems (Figure 8.1). Occasionally, however, it will be helpful to examine the role of pair bonding, and we will do so when needed.

A general classification of mating systems is shown in Table 8.1.

MONOGAMOUS MATING SYSTEMS

A **monogamous mating system** is defined as a mating system in which a male and female mate with each other, and only each other, *during a given breeding season.* As such, we can have animal societies in which pairs mate only with one another during season 1, but in subsequent years find new mates (serial monogamy)— indeed, this sort of mating system is very common in territorial animals.

TABLE 8.1. Mating combinations. Various mating combinations and how they map onto potential mating success for males and females. *(From Davies, 1992, p. 29)*

	MATING SUCCESS	
MATING COMBINATION	**FOR A MALE**	**FOR A FEMALE**
Polyandry (e.g., 2 ♂ 1 ♀)	Share one female	Sole access to several males
Monogamy (1 ♂ 1 ♀)	Sole access to one female	Sole access to one male
Polygynandry (e.g., 2 ♂ 2 ♀)	Share several females	Share several males
Polygyny (e.g., 1 ♂ 2 ♀)	Sole access to several females	Share one male

In some animal populations, a male and female will mate only with one another during their entire life span (or more precisely, the life span of the individual that dies earliest). As an example of this type of lifetime monogamy, let us look at the oldfield mouse, *Peromyscus polionotus* (Dewey and Dawson, 2001; Foltz, 1981; Figure 8.2).

Early studies suggested that monogamy was rare in mammals (Eisenberg, 1966; Kleiman, 1977), but there was some suggestion that this might reflect a lack of work on the subject. In the early 1980s, David Foltz hypothesized that while monogamy was uncommon in large, conspicuous, diurnal species (those active in the daytime), it was more common in smaller diurnal groups, particularly rodents (Foltz, 1981). Behavioral work suggested that Foltz might be correct, even though the genetic evidence for monogamy in rodents at the time was much weaker (Kleiman, 1977). Foltz studied the breeding system of the oldfield mouse, in part because these rodents were relatively easy to study in both the laboratory and the field, and in part because other work suggested that a great deal of genetic variation existed in this species (Selander et al., 1971).

Foltz excavated more than 500 oldfield mouse burrows, captured the individuals in each burrow, and then brought them into the laboratory. One hundred and seventy-eight families were collected from these burrows, and a genetic analysis was conducted. From a subset of these families, Foltz calculated that 90 percent of the offspring found in a family group were fathered by the male in their burrow (Figure 8.3). Even higher rates were found in the closely related species *Peromyscus californicus* (Gubernic and Nordby, 1993). Furthermore, behavioral observations of both males and females found that most females remained with the same mates across litters, suggesting long-term, perhaps lifetime, monogamy to be the rule, rather than the exception, in this species.

Although long-term monogamy appears to be the most common mating behavior seen in oldfield mice, Foltz found that approximately 13 percent of the burrows excavated contained only an adult female, rather than the typical male and female. When this small subset of females without live-in mates was analyzed, they were found to be more likely to switch mates across litters, compared with females that were already paired with a mate in the field. Whether these females were choosing a more polygamous mating strategy, or whether their mates had simply died, and hence they were polygamous by necessity, remains to be tested.

FIGURE 8.2. Long-term monogamy in the oldfield mouse. From data on 500 oldfield mouse burrows, Foltz found that 90 percent of the offspring in a family group were fathered by the male in their burrow. Behavioral observations also suggest that many females remained with the same mates across litters, suggesting long-term monogamy. *(Photo credit: © James F. Parnell)*

FIGURE 8.3. Who fathers whom?
In the oldfield mouse (*Peromyscus polionotus*), the male found in the burrow, on average, fathers 90 percent of the pups in that burrow.

MONOGAMY AND FITNESS CONSEQUENCES. In mating systems like that of the oldfield mouse, where most individuals pair with a mate across multiple breeding seasons, there should be significant fitness consequences associated with choosing a high-quality mate—that is, a mate that would give its partners high relative reproductive success by siring attractive offspring, offspring with strong immune systems, and so on. Karen Ryan and Jeanne Altmann tested this idea in a series of fitness-related mate-choice trials involving oldfield mice (K. K. Ryan and Altmann, 2001). A male was given the choice between two virgin, but sexually mature, females, and the male's preference was recorded (Edward and Chapman, 2011). Then, in one treatment, males were paired with the female for which they had expressed a preference, and in a second treatment, males were paired with the female that they had not preferred.

When Ryan and Altmann examined the number of pups that survived in each treatment, they found significantly more pups were born to pairs made up of a male and his preferred mate than to males who mated with females they did not prefer (Figure 8.4A). The researchers then ran a fascinating set of follow-up experiments. They again let a male choose between two females, and they recorded which female the male preferred—let's call this part I of the experiment. Then, in part II, a naive male—a male that had no experience with either of the females—was paired with either the female that had been preferred in part 1 (in one treatment) or the female that had been rejected in part 1 (in a second treatment). The researchers found that the number of offspring raised in these two treatments was approximately equal, suggesting that what constituted a good mate was different for each individual oldfield mouse—otherwise we would have expected to see more offspring from matings involving females that been preferred in part I of Ryan and Altmann's experiment (Figure 8.4B). The exact underlying proximate mechanism responsible for this difference is not yet understood. It may be due in part to genetic incompatibility, and in part to differences in male or female behavior toward pups across the two experiments (Gowaty, 2008; Spoon et al., 2006).

Animal behaviorists continue to examine the costs and benefits associated with monogamy. It is difficult to understand why males would ever choose to engage in monogamous pairings when they could potentially inseminate many females over a short period of time. From an adaptationist perspective, ethologists predict that monogamy should occur in ecological situations that create significant benefits to a male that opts to remain with a single female. Later in the chapter,

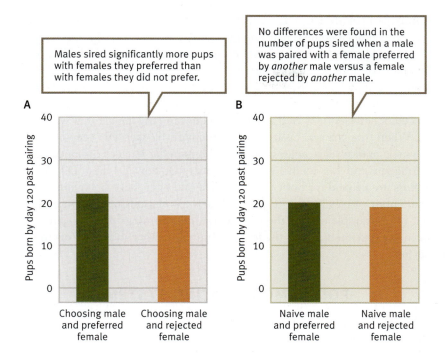

A

Males sired significantly more pups with females they preferred than with females they did not prefer.

Pups born by day 120 past pairing

Choosing male and preferred female | Choosing male and rejected female

B

No differences were found in the number of pups sired when a male was paired with a female preferred by *another* male versus a female rejected by *another* male.

Pups born by day 120 past pairing

Naive male and preferred female | Naive male and rejected female

FIGURE 8.4. Fitness consequences of choice in monogamous systems. Researchers compared the number of pups sired by a male in two different treatments in each of two experiments (as shown in **A** and **B**). *(Based on K. K. Ryan and Altmann, 2001, pp. 438, 439)*

we will see some research that suggests what those circumstances may be. In general, when resources are relatively scarce, a male's reproductive success may be highest when he is part of a monogamous pair and provides some sort of care for the offspring. In such situations, males may help raise offspring by bringing food to the nest and defending the offspring from potential predators. This is often referred to as the mate-assistance hypothesis of monogamy (S. T. Emlen and Oring, 1977; Kenter et al., 2011; Woodroffe and Vincent, 1994).

PROXIMATE UNDERPINNINGS OF MONOGAMY. Animal behaviorists have also studied the proximate underpinnings of monogamous mating systems (Donaldson and Young, 2008; McGraw and Young 2010; L. J. Young et al., 2005). One species that has proved quite useful for understanding the neurobiology of monogamy and pair bonding is the prairie vole (*Microtus ochrogaster*) that we have discussed at length in Chapters 3 and 4. In this species, males and females that are courting approach one another, and this affiliative—"friendly"— behavior is a prerequisite to partner choice. Once individuals mate and form a strong pair bond, they are actively aggressive to other members of the opposite sex (Carter et al., 1995). Brandon Aragona and his colleagues studied how this change from affiliative to aggressive behavior in a monogamous species takes place (Aragona et al., 2006; McGraw and Young, 2010).

Because earlier work had found that dopamine transmission in the nucleus accumbens (n.a.) section of the brain was linked to both affiliative and aggressive behavior in the prairie vole, Aragona and his team focused on changes in dopamine in the n.a. (Aragona et al., 2003; Liu and Wang, 2003). They first established that a specific area of the n.a. called the rostral shell was most associated with affiliative, mating-related behavior. Within the rostral shell, activation of two dopamine-related receptors—labeled D1 and D2—was critical to the formation of long-term monogamous relationships. When D2 receptors were activated by dopamine, pair bonding was facilitated, whereas when D1 receptors were activated, pair bonding was inhibited (Hostetler et al., 2011). Aragona and his colleagues hypothesized that D2 receptors mediated pair-bond formation, while D1 receptors were critical to the aggression seen toward unfamiliar, opposite-sex individuals after two individuals had formed a pair bond.

To test their hypothesis, Aragona and his team examined D1 receptors in males that had recently formed pair bonds with females. These males showed a surge in D1 receptor activation, as well as aggression toward any female that was not their mate. When Aragona and his team experimentally blocked D1 receptor activation in the male prairie voles, aggression toward unfamiliar females disappeared.

Aragona and his team's work demonstrates how D1 and D2 dopamine receptors are critical for both the affiliative and aggressive behaviors exhibited by male prairie voles in long-term monogamous relationships. Their studies are also a good example of how the proximate underpinnings of monogamy can be examined experimentally, at the level of brain receptors and their role in mating behaviors.

POLYGAMOUS MATING SYSTEMS

Polygamy is defined as a mating system in which either males or females have more than one mate during a given breeding season/cycle. Polygamy includes **polygyny**, in which males mate with more than one female per breeding season, and **polyandry**, in which females mate with more than one male per breeding season. Polygamy can be also be subdivided temporally, in that polygamy can be simultaneous or sequential. Simultaneous polygamy refers to the case in which an individual maintains numerous mating partners in the same general time frame, whereas sequential polygamy involves individuals forming many short-term pair bonds in sequence during a given breeding season.

One important thing to keep in mind with respect to polygamy is that this type of breeding system increases the variance in reproductive success in the sex that has more than one mate per season (see Chapter 7). For example, in polygynous systems, there is often intense competition among males to mate with as many females as possible, and this typically produces a distribution of mating success in which a few individuals do extraordinarily well, but many males obtain no mates whatsoever (Figure 8.5). The converse holds true for the case of polyandry, although variance in reproductive success is usually less dramatic in this situation.

Polyandry has been particularly well studied in jacanas—a group of shorebirds in which the males incubate the eggs and care for the young, and the females compete aggressively for multiple mates. Stephen Emlen and his colleagues studied one of these species, the wattled jacana (*Jacana jacana*) in Panama (S. T. Emlen and Wrege, 2004a, 2004b; S. T. Emlen et al., 1998; Figure 8.6). While polyandry in some species of jacanas is simultaneous, and females nest with numerous males at one time, wattled jacana females are usually sequentially polyandrous. In this species, males have small territories (40 meters in diameter) that abut one another, but female territories are considerably larger and contain anywhere from one to four male territories.

In the wattled jacana, a female lays clutches of eggs sequentially, after mating with males on her territory, with intervals of less than two weeks often separating the production of sequential clutches. Because a female's territory encompasses a number of male territories, it was rather surprising that Emlen

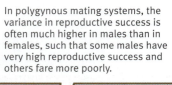

In polygynous mating systems, the variance in reproductive success is often much higher in males than in females, such that some males have very high reproductive success and others fare more poorly.

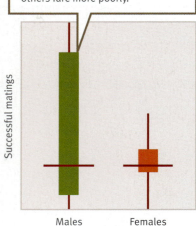

Successful matings

Males   Females

FIGURE 8.5. Variance and polygyny. Since every successful mating involves a male and a female, the average reproductive success must be equal in both sexes. But the varience in reproductive success differs across these mating systems. (Horizontal lines represent average reproductive success.) *(Based on Low, 2000, p. 55)*

and his team found a very high probability that a chick in a nest on a particular male's territory was his genetic offspring. Given the possibility that a female was also engaging in copulations with other males in her territory, the fact that there was a 92.5 percent probability that a given chick in a nest was the offspring of the male guarding that nest was rather remarkable. This figure decreased when males in surrounding nests were not incubating eggs or tending young chicks and had more time available for mating attempts.

FEMALE DEFENSE POLYGYNY. We will return to the evolution of polygamous mating systems in general later in this chapter, but for the time being, let's examine one species characterized by a polygynous mating system—a very small Australian wasp in genus *Epsilon* (A. P. Smith and Alcock, 1980). In this species of wasps, males mature earlier than females. When they mature, males search for unopened brood cells from which females will emerge. Such cells are often clustered together, and males are very territorial once they uncover such a cluster of females they can defend. Once a virgin female wasp emerges from her cell, the closest male present climbs on her back and mates with her. The benefits to males that guard against intruders may be mating opportunities with as many as two dozen virgin females (Figure 8.7). Because males mate with many females and they defend females from mating attempts by other males, the mating system of these wasps is referred to as **female defense polygyny**.

Randy Thornhill and John Alcock have argued that three characteristics are often found in insects displaying female defense polygyny: (1) females are short-lived and have low fecundity, receiving all the sperm they will ever use from a single male; (2) females mate shortly after becoming adults; and (3) females are grouped close together in space (Thornhill and Alcock, 1983). These three characteristics make female defense polygyny beneficial for males. This mating system's effects on female fitness remains less well understood.

LEKS. One type of polygyny is **lekking**, or **arena mating** (see Chapter 7). This type of polygyny has been studied in birds, mammals, amphibians, fish, and insects. It occurs when males set up and defend small arenas called leks—temporary territories specifically for mating—that contain no apparent resources (food or shelter, for example). Females come to these leks and select mates from among the males present. Often a single male will obtain a very large proportion (sometimes more than 80 percent) of all the matings at that

FIGURE 8.7. **Epsilon wasps and female defense polygyny.** Male wasps wait at the nest for females to emerge from their brood cells. Pictured here is one male (left) on the nest and a female (at arrow) that is starting to emerge from a hole in the mud nest.

lek, leading to a great deal of variation in reproductive success across males in a population (Hoglund and Alatalo, 1995; Mackenzie et al., 1995).

Leks have long fascinated ethologists, and they have been studied from many different perspectives. Here we shall focus on two questions: (1) What benefit(s) do females obtain from this form of polygyny? and (2) What benefit(s) do males obtain?

The benefit(s) that females receive from choosing among males on a lek has been a contentious issue in ethology, in part because females appear to receive sperm, and only sperm, from males—that is, females appear to receive few direct, material benefits (see Chapter 7). One possibility is that females select among males using indicators of male condition—for example, size, number of parasites, and other indicators of health and vigor—and that such choice will lead to offspring with higher survival probabilities because they possess good genes with respect to health. An alternative explanation is that females may use these same indicators to select among mates because such choice will lead to the production of male offspring that are attractive to the next generation of females—an idea sometimes referred to as the **sexy-son hypothesis**. Through a series of experiments using the sandfly (*Lutzomyia longipalpi*), Theresa Jones and her colleagues attempted to distinguish between these two hypotheses regarding the benefits that females receive in polygynous lek mating systems (T. M. Jones, 2001; T. M. Jones and Quinnell, 2002; T. M. Jones et al., 1998).

Sandflies form leks in which males defend small arenas that measure about 4 cm in diameter. Males emit a chemical attractant called a pheromone, and females can choose freely among courting males. As in many lekking species, a single male on a lek will often obtain all matings with visiting females, and neither males nor females will provide parental care to the young.

Jones and her team set up experimental sandfly leks in the laboratory, and they ran a two-part experiment. In the first part of the experiment, five randomly selected males were put together in an arena, and once a lek was established, a single, virgin female sandfly was allowed to choose among these males. Once that female mated with a male, she was removed from the lek, and a second female was allowed to select a mate. This process was repeated until ten females had selected mates and had been inseminated in a given lek. Then, for the second part of their experiment, Jones and her colleagues took the males that had rarely been selected by females in the first part of their experiment, and they put them in new leks. They then allowed a sequence of ten females to select among males at these new leks—that is, in the second part of the experiment, females were forced to select among males that other females had rejected as potential mates. By comparing the results obtained in the two halves of the experiment, the experimenters could begin to decipher what benefits females might receive when selecting males in a highly polygynous lek mating system.

To distinguish between the good genes and sexy-son hypotheses, the researchers first compared the survival of offspring that were the product of females choosing among a random sample of males (part I) and the offspring of females that were forced to choose between males that had been chosen by other females (part II). No evidence for the good genes model was uncovered, as offspring from both part I and part II survived with approximately the same probabilities. Jones and her team, however, did find support for the sexy-son hypothesis. When male offspring from part I and part II were placed in a lek, females showed a strong mating preference for the former, suggesting that first-generation females were receiving sexy-son benefits as a result of their choice among lekking males.

The benefits that males might obtain from a lek mating system have been investigated in birds (also see Chapter 7, where we discuss lekking in grouse). From the perspective of a male on a lek, the fitness benefits are huge *if* you are the individual chosen as a mate by most females. Consider an extreme case of polygyny in which one male—let's call him male 1—out of all the males on a lek obtains all the matings at that lek. The benefits associated with being male 1 might compensate for the low probability that male 1 is the one that obtains all mating. But there are other potential benefits that males on a lek might receive—for example, a male may receive benefits from helping his genetic relatives on a lek (Hoglund, 2003; Hoglund et al., 1999; Kokko and Lindstrom, 1996; Petrie et al., 1999).

Peacocks (*Pavo cristatus*) generally have many different leks that females can visit, and the females are often drawn to leks that contain the most males (Alatalo et al., 1992; Kokko et al., 1998; Figure 8.8). If only one or a few males at a given lek obtain matings, and the lek is composed of many males that are genetic relatives, these relatives may receive indirect benefits from the matings that their kin receive.

Marion Petrie and her colleagues studied a group of 200 peacocks living in Whipsnade Park in England (Petrie et al., 1999). These peacocks divide up into many different leks, and males defend their temporary territories within the leks all day during the breeding season. Petrie and her team did a molecular genetic analysis to determine whether individuals in a lek were genetic kin, and they found that indeed they were, with the average genetic relatedness within the groups being equivalent to that of half-siblings. Petrie and her colleagues then ran an experiment in which they released into Whipsnade Park a group of peacocks raised elsewhere. These males were raised in such a way that they did not interact with their genetic relatives any more than they interacted with strangers during their development. Petrie's group found that when the birds that had been raised without interactions with their genetic relatives formed their own leks, genetic relatives set up their temporary territories much closer to one another than one would expect by chance. That is, even without the opportunity to learn about who is kin and who isn't during their development,

FIGURE 8.8. Peacocks on leks. The benefits to mating on leks have been measured in the peacock. **(A)** Four males gather on a lek before any females arrive (one of the males is under the brush in the top left corner). **(B)** When a female appears on the lek, the male displays his tail in an attempt to get her to choose him as her mate. *(Photo credits: Marion Petrie; Frans Lanting/Minden Pictures)*

the peacocks were able to gauge genetic relatedness and clustered near genetic kin within their leks. And again, in such leks of genetic kin, even males that did not mate received indirect benefits from their genetic relatives that did mate.

PHYLOGENETIC HISTORY OF POLYGYNY IN WARBLERS. Animal behaviorists have also undertaken phylogenetic studies on the origin of polygyny using Acrocephaline warblers, a group that includes such species as the Seychelle warbler, the moustached warbler, and three species of reed warblers. Mating systems in these species vary from monogamous to polygynous, and warblers in monogamous systems show much higher levels of parental care than do warblers in polygynous systems, with monogamous males providing much more food to chicks than do polygynous males. Warbler species in this family of birds also differ in terms of the quality of the habitats they inhabit, ranging from poor habitats with little food to much better habitats that contain significantly more and better food types for warblers.

To examine the evolution of mating systems in warblers, Bernd Leisler and his colleagues used molecular genetic data to build a phylogeny of seventeen warbler species (Leisler et al., 2002). They also gathered published data on habitat quality, parental care, and mating system (monogamous versus polygynous) in warblers (Figure 8.9). Habitat quality was scored from poor (few, poor-quality food items) to medium (more, larger prey) to good (highly productive areas with quickly renewing food sources), and paternal care was scored as "full paternal care," "reduced paternal care," or "no paternal care." Finally, the researchers took the data on habitat quality, male parental care, and mating system and superimposed them onto their phylogenetic tree.

From an ecological and behavioral perspective, Leisler's work found a strong correlation between mating system and habitat quality. Most monogamous systems were found in poor habitats, and most polygynous systems were found in better habitats. In addition, males were much more likely to provide significant parental care to developing chicks in poor habitats. These two findings together suggest that monogamy is associated with poor habitat quality because in such habitats food is scarce enough that it takes two parents to gather enough food to provision developing chicks.

Leisler and his team's phylogenetic analysis also examined which type of mating system—monogamy with male parental care or polygamy

FIGURE 8.9. Warbler mating systems. The phylogeny of mating systems has been studied in warblers, including **(A)** the polygynous great reed warbler (*Acrocephalus arundinaceous*) and **(B)** the monogamous Seychelle warbler (*Acrocephalus sechellensis*). (Photo credits: John Hawkins/FLPA; M. D. England/www.ardea.com)

A

B

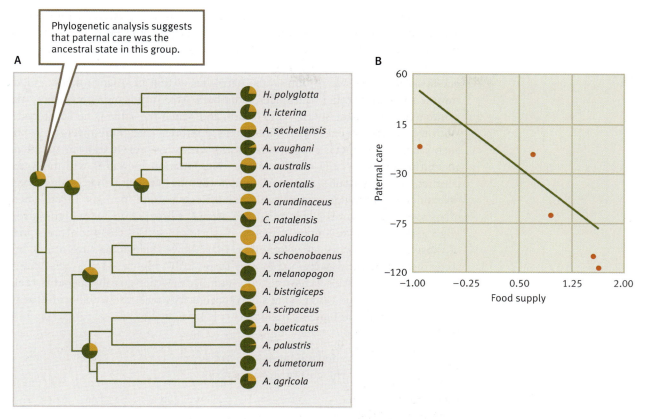

A

Phylogenetic analysis suggests that paternal care was the ancestral state in this group.

H. polyglotta
H. icterina
A. sechellensis
A. vaughani
A. australis
A. orientalis
A. arundinaceus
C. natalensis
A. paludicola
A. schoenobaenus
A. melanopogon
A. bistrigiceps
A. scirpaceus
A. baeticatus
A. palustris
A. dumetorum
A. agricola

FIGURE 8.10. Phylogeny of warbler mating systems. (A) Paternal effort is represented as the proportion of a pie chart that is green (when the pie chart was more than 60 percent filled, researchers classified the system as full paternal care). This paternal effort was mapped onto the warbler phylogeny. **(B)** The relationship between paternal care and food supply in several species of warblers is shown by the data points. Where there was less food, there was greater paternal care; where there was more food, there was less paternal care. Values on the x-axis and the y-axis can be negative because of statistical transformations associated with this analysis. *(From Leisler et al., 2002, p. 384)*

with reduced male parental care—was ancestral, and which was derived (see Chapter 2). To do so, they used a statistical technique called maximum likelihood analysis, which allowed them to take data from their phylogenetic tree and calculate probable ancestral states of various traits. They found evidence that the ancestral state in warblers in this family was a monogamous system in which males displayed parental care and the birds lived in poor habitats. Polygynous systems with reduced care are derived from this ancestral state. What this phylogenetic analysis allows us to do is reconstruct the evolution of polygyny. In warblers, it appears that, through evolutionary time, some species began inhabiting better-quality habitats. Once this occurred, and it was possible for chicks to receive enough food from a single parent, males were freed from parental care duties, and the evolution of polygyny was favored (Figure 8.10).

POLYANDRY IN SOCIAL INSECTS. Polyandry refers to a mating system in which females mate with more than one male per breeding season. This type of mating system has been well studied in social insects, where a single queen will often mate with many worker males. Compared with nests where a single male mates with a queen, polyandrous nests often have greater levels of within-group conflict.

The reason that within-group conflict is higher in the nest of polyandrous queens is the presence of numerous "patrilines"—that is, offspring descended from a common mother, but different fathers. When polyandry is absent, all workers have the same mother and father, and hence the same genetic interest. With the establishment of patrilines, genetic interests are more divergent, and each patriline competes with the others for greater representation in the next generation (Seeley, 1995, 1997). While this situation increases within-group conflict, the decreased genetic relatedness among offspring that comes from polyandry does provide queens with at least one benefit (perhaps more)—her offspring, as a group, are likely to survive when the hive is infected with disease. High genetic diversity among a queen's offspring increases the odds that some of those offspring will have a genotype that allows them to survive attack by some disease-causing agent (Seeley and Tarpy, 2007; Figure 8.11).

Thornhill and Alcock have listed a number of other possible benefits that female insects accrue when using a polyandrous mating strategy (Thornhill and Alcock, 1983). These benefits include:

1. Sperm replenishment
   ▸ Female adds to depleted or low sperm supply.
   ▸ Female avoids the cost of storing sperm.
2. Material benefits
   ▸ Nutrients
   ▸ Reduced predation
   ▸ Protection from other males
3. "Genetic benefits"
   ▸ Replacement of "inferior" sperm
   ▸ Increased genetic variance in offspring
4. Convenience
   ▸ Female avoids the costs of fending off copulation attempts by male.

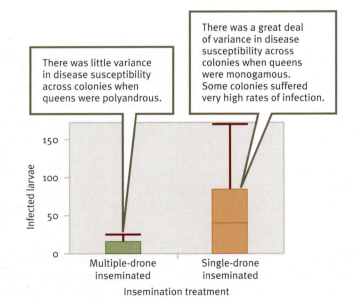

FIGURE 8.11. Polyandry and disease resistance in honeybees. One benefit of polyandry is resistance to disease. Honeybee colonies were inoculated with spores of *Paenibacillus larvae*, a bacterium that causes a highly virulent disease called American foulbrood. The mean number of brood infected did not differ among treatments, but the variance was significantly greater in colonies in which queens were inseminated by only one male. *(Adapted from Seeley and Tarpy, 2007)*

There was little variance in disease susceptibility across colonies when queens were polyandrous.

There was a great deal of variance in disease susceptibility across colonies when queens were monogamous. Some colonies suffered very high rates of infection.

Of these, Thornhill and Alcock hypothesize that "material benefits" account for most of the polyandry seen in insects. These material benefits include (1) seminal fluid that sometimes contains chemicals that a female can sequester and use; (2) nutritious spermatophores—sperm-encompassing packets filled with nutrients and produced by male insects such as moths and butterflies; (3) "nuptial gifts" such as prey items presented by courting males; (4) access to superior feeding sites and oviposition (egg-depositing) sites; and (5) male parental care (as in some waterbugs).

PROMISCUOUS MATING SYSTEMS

When polyandry and polygyny are occurring in the same population at the same time a breeding system is said to be promiscuous. There are two very different kinds of **promiscuity**, which vary dramatically as a function of the presence or absence of pair bonds between mating individuals. In one form of promiscuity, both males and females mate with many partners and no pair bonds are formed. For example, a male may defend a territory that contains food, and females may visit such territories, obtain food, mate with a male, and then repeat this sequence many times (Davies, 1991). Promiscuity need not be tied to male territoriality. In many primate species, when females are in estrous, both males and females mate repeatedly, often in rapid succession, with many opposite-sex partners. For example, during their estrous period, female Barbary macaques (*Macaca sylvanus*) search for new male mating partners after each copulation, and males use female signals to gauge sexual receptivity, which has the effect of minimizing the chances that a female will mate with just one male (Semple and McComb, 2000; Taub, 1980, 1984; Whitten, 1987). At the same time, this pattern of female behavior results in males having numerous female sexual partners, producing a promiscuous breeding system.

In the second type of promiscuous breeding system, **polygynandry**, several males form pair bonds with several females simultaneously. For example, in the dunnock (*Prunella modularis*), pairs of males will often jointly defend the territory(ies) of a pair of females (Figure 8.12). Nick Davies has studied the dunnock mating system and the help provided by males to females (Davies,

FIGURE 8.12. Dunnock mating system. The dunnock mating system is extremely versatile and includes monogamy, polyandry, polygyny, and polygynandry. Here we see a female with a newly hatched offspring (just visible in egg). *(Photo credit: Nick Davies)*

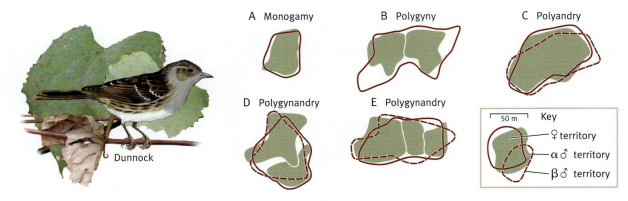

FIGURE 8.13. Incredible variation in dunnock breeding systems. Female territories are shown in green, while alpha male territories are depicted by solid red lines and beta male territories are shown by dashed red lines. In a single dunnock population, we can find mating systems ranging from monogamy to polygamy, polyandry, and polygynandry. *(From Davies, 1992, p. 27)*

1986). He observed cases in which one or two males were resident on a female's territory, and he found that females received help from one of the males, from both, or from neither of them, but that, on average, polygynandrous females received the equivalent help that a female with only a single mate would receive (Davies, 1992; Figure 8.13). The more help a polygynandrous female received, the higher the mean nestling weight of the chicks in the brood and the lower the chick mortality rate due to starvation. We will examine Davies's work on polygynandry in the dunnock in more detail later in the chapter.

DISEASE AND PROMISCUITY. Margie Profet has added an interesting dimension to the study of promiscuous breeding by examining this mating system in relation to female reproductive health (Profet, 1993). Her basic argument goes as follows: Sperm are known to be vectors—that is, carriers—of sexually transmitted diseases, some of which are quite dangerous to females (Ewald, 1994). During insemination, bacteria and other disease-causing agents from both male and female genitalia move to sperm tails and attach to the uterus (Profet, 1993).

Menstruation, Profet argues, is a defense mechanism that has evolved in females to rid the female reproductive tract of pathogens carried in by sperm. If this hypothesis is correct, menstruation should be most common in promiscuous breeding systems, as the probability of disease transfer via sperm should be highest in systems where both males and females engage in sexual activity with numerous partners. Promiscuous breeding systems not only expose females to greater quantities of sperm but to a greater diversity of sperm, and hence to a greater diversity of diseases that use sperm as vectors. And both of these factors increase a female's risk of infection (Figure 8.14). While Profet suggests that a qualitative test of the data on breeding systems and menstruation supports her hypothesis (Profet, 1993), a more quantitative analysis by Beverley Strassmann did not find any link between promiscuity and copiousness of menstruation in primates (Strassmann, 1996). Clearly, further work needs to be undertaken to better elucidate the evolutionary relationship between promiscuity and menstrual bleeding.

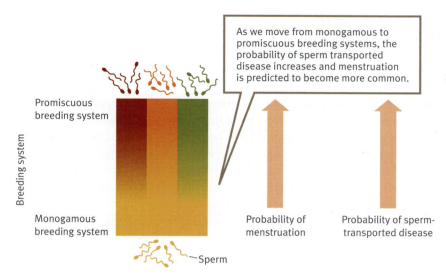

As we move from monogamous to promiscuous breeding systems, the probability of sperm transported disease increases and menstruation is predicted to become more common.

Promiscuous breeding system

Breeding system

Monogamous breeding system

Probability of menstruation

Probability of sperm-transported disease

Sperm

FIGURE 8.14. Sperm, breeding systems, and menstruation. Promiscuous breeding systems deliver not only more sperm to females but a greater diversity of sperm (denoted by different colors). As the probability of sperm-transported disease increases, menstruation should become more common.

The Ecology and Evolution of Polygynous Mating Systems

With a survey of different mating systems in hand, let's now address questions about the ecology and evolution of these different mating systems. We will first look at the role of resource dispersion on female mating decisions and then at a model to predict when polygyny is favored by natural selection.

POLYGYNY AND RESOURCES

The study of animal mating systems has long been tied to female dispersion patterns—that is, how females move about in relation to one another and their environment (Bradbury and Vehrencamp, 1977; S. T. Emlen and Oring, 1977). The basic argument goes like this: A female can often fertilize all her available eggs by mating with one or a very few males. So, female fecundity is not so much tied to the availability of mates as it is to the availability of resources—food, defense, and so forth (Krebs and Davies, 1993). The more resources that are available, the more offspring females can produce. Males, in contrast, can potentially fertilize large numbers of females, so male reproductive success is more tightly associated with access to females than with access to resources. The predictor is then that, female dispersion patterns should track the distribution of resources, whereas male dispersion patterns should track the dispersion of females.

If females track resources and males track females, then the mating system in a population is intimately tied to the distribution of resources because resource distribution will affect whether males can defend more than one female at a time. If resources are dispersed fairly homogeneously and/or if females must cover large areas to obtain enough resources to survive, it may not be economically feasible, in terms of costs and benefits, for males to mate with and potentially defend more than one female.

When resources are clumped—for example, when food is located in discrete groupings—the economics of mating systems may shift, and males may be able to mate with and defend several females at once (see the Conservation Connection box). For example, seal populations can be found aggregating on both ice packs and beaches. On ice packs, females are widely dispersed, and males typically guard and mate with anywhere from one to a few females. In contrast, when the seals are on land, females cluster tightly in particularly safe areas on the beach, and males are able to defend herds of females against the approach of other males, leading to a more polygynous mating system (Le Boeuf, 1978).

The effect of female dispersion on male mating patterns has been examined experimentally. Rolf Ims manipulated the distribution of the food eaten by the grey-sided vole (*Clethrionomys rufocanus*) and found that female dispersion patterns changed in the predicted manner—when resources became more clumped, females clustered themselves together in the area of their resources (Ims, 1987). Ims then introduced the grey-sided vole to a small island in Norway. In one treatment, he used caged female voles to simulate an environment in which females moved about a home range and were fairly spaced out. In a second treatment, caged females were clustered together. As a control, Ims ran a reverse experiment in which males were in cages and their distribution was manipulated. As predicted, Ims found that males tracked the distribution of caged females across treatments, but female dispersion was unaffected by the distribution of caged males.

THE POLYGYNY THRESHOLD MODEL

Even though females often track resources and males track females, decisions regarding polygyny are not made only by males. For example, in systems where males have territories that can sustain numerous mates simultaneously, females need to decide which territory they settle on by choosing between alternative male territories. In 1969, Gordon Orians created the **polygyny threshold model** (PTM) to predict the behavior of females in such a scenario (Orians, 1969).

HOW THE PTM WORKS. To see how the PTM works, imagine that ten males each have their own territory. Let's refer to these territories as territories 1–10 (T1–T10), and let's call the males on these territories males 1–10 (M1–M10). These territories vary with respect to some resource that is valuable to females— for example, food. Before females begin selecting mates, T1 has the most food, T2 the second most, down to T10, which has the least amount of food. The first female arriving to choose a male territory on which to settle can base her choice simply on what territory is optimal with respect to food intake, so she should choose male 1's territory—that is, she should choose T1. A second female choosing between male territories now faces the same decision as the first, except that T1 is occupied by a male and another female. If this second female chooses T1, she will only get some fraction of the food available there. If the food still available on T1, even though it is already occupied by another female, is greater than the food available on any of the other nine male territories, our second female should

Anthropogenic Effects on Animal Mating Systems

Jeffrey Lane and his colleagues (2011) review four anthropogenic effects— effects caused by humans—that have been shown to have important consequences on animal mating systems: (1) habitat fragmentation, (2) climate change, (3) pollution, and (4) sport hunting (Colborn et al., 1993; Lurling and Scheffer, 2007; E. M. Olsen et al., 2004; Parmesan, 2006). Let's examine two of these: habitat fragmentation and climate change.

In an influential paper, S. T. Emlen and Oring (1977) hypothesized that habitat fragmentation produces clumped resources. Emlen and Oring then argue that clumped resources create scenarios in which females, that track such resources, also clump together in space. Under such conditions, it becomes possible for males to guard numerous mates simultaneously, producing a polygynous mating system. Emlen and Oring were originally considering environmental factors that might lead to clumped resources, but their hypothesis applies just as well when anthropogenic effects lead to such clumping.

Emlen and Oring's hypothesis has been indirectly tested in two populations of the brushtail possum (Trichosurus cunninghami), that live adjacent to one another (J. K. Martin and Martin, 2007). In the population residing in an unfragmented forest, females have large home ranges, which are necessary to obtain sufficient food in such an environment. Monogamy, the most common mating system found in the brushtail possum, was found in this forest population. An adjacent population of brushtail possums lived alongside a road in a more fragmented, piecemeal habitat. In this fragmented population,

food was clustered and females had smaller home ranges, allowing males the potential to defend numerous females during breeding season. Polygyny was found in this fragmented population with more clustered resources and more clustered females. In the case of the brushtail possum, anthropogenic road building led to a fragmented population in which the typical monogamous mating system seen in brushtail possums was replaced by polygyny (García-Peña et al., 2009). The implications of such changes to an evolved mating system are not yet understood.

Climate change—including anthropogenic-caused climate change— may affect the evolution of mating systems, including such behaviors as paternal care of offspring. Using the phylogenetic analyses discussed in Chapter 2, researchers have examined the relationship between migration distance and mating systems in shorebirds (García-Peña et al., 2009; Figure 8.15). As in previous studies on migratory

shorebirds, García-Peña et al. found that longer migration routes were correlated with reduced paternal care, presumably because the energy used in migration was not available to care for the young. Subsequent phylogenetic analysis found that changes in migration distances tended to precede changes in paternal care patterns, suggesting a link between migration distance and paternal care. In terms of anthropogenic climate change, global warming may affect the distance that migratory species must travel to reach an appropriate end point. In some instances, human-caused climate change may lead to increased migratory distances (and reduced paternal care); in others, it may lead to decreased distances (and increased paternal care)—the specifics will depend on the species and whether migration is linked to changes in temperature, rain level, or other factors. But the overarching point is that climate change may affect animal breeding systems in ways that we are only just beginning to understand.

FIGURE 8.15. Migration, climate change, and mating systems. The relation among migration patterns, climate change, and mating systems has been studied across many migratory species of birds. Pictured here are black-bellied plovers (*Pluvialis squatarola*) during their spring migration. *(Photo credit: Jason Stone/NHPA)*

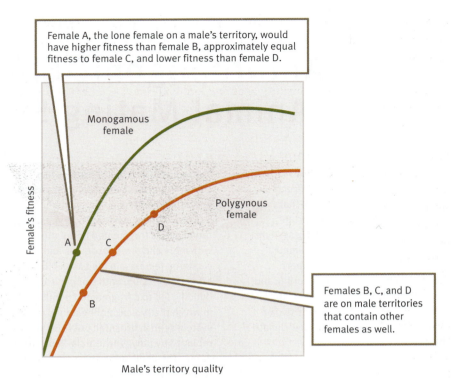

Female A, the lone female on a male's territory, would have higher fitness than female B, approximately equal fitness to female C, and lower fitness than female D.

Monogamous female

Polygynous female

D

A   C

B

Female's fitness

Male's territory quality

Females B, C, and D are on male territories that contain other females as well.

FIGURE 8.16. Female choice of territories. Imagine a female deciding among territories. She should choose the territory with the highest quality, that is, with the most food available or the most shade and so on, regardless of whether she would be the lone female (monogamous territory) or one of several females (polygynous territory), because such a choice would provide her with the greatest fitness.

choose to settle with M1 on his territory, T1, anyway. In so doing, she has passed the "polygyny threshold"—that is, she opts to stay on a territory with another female and thus to be part of M1's polygynous relationships.

Suppose that the second female does not pass the polygyny threshold for staying on T1 with M1, and she settles on T2, that is, M2's territory. Our third female now comes in and chooses between T1 (another female present), T2 (another female present), and the territories of males 3–10 (no other females present). Her situation is slightly different from that of female 2 when she arrived, however, because the best open territory is now T3—M3's territory— and that isn't as profitable (in terms of fitness effects) as male 2's territory. If female 3 chooses to settle on T1 or T2, then the polygyny threshold has been crossed for whichever of these she settles on. This sort of logic can be applied to all subsequent females, and it allows us to make some clear-cut predictions with respect to territory settlement (J. L. Brown, 1982). In figure 8.16, you can see a graphical representation of the PTM with four females deciding where to settle.

One prediction from the PTM is that polygyny should often occur in "patchy" environments, where resources such as food are clumped in discrete patches that are spread out over the landscape. In such patchy environments, males can defend territories that have these valuable resources, and the quality of a male's territory will affect his mating success—males with better territories (more food, better safety, and so on) should have higher reproductive success than males with poorer territories. The PTM also makes a prediction about female fitness. When two females settle at about the same time and hence face the same sort of economic decisions regarding the availability of resources and where to settle, a female that opts to settle on a territory on which she is the lone mate of the male (a territory that then has monogamous mating pair) and a female that decides to settle on a territory that already contains a male and one or more females (a territory in which we see polygamy) *should have about equal fitness* (Krebs and Davies, 1987). It is this equivalency of fitness among monogamous and polygamous females that makes the PTM stable. Because monogamous females and females in polygamous relationships settle in such a way as to produce approximately equal fitness, there is no temptation for females to move from territory to territory once this state has been reached, as any such move would in fact lower an individual's reproductive success (see Borgerhoff-Mulder, 1990, for some limitations of the PTM).

THE PTM AND MATE CHOICE IN FEMALE BIRDS. Wanda Pleszczynska used lark buntings (*Calamospiza melanocorys*) in one of the first experimental tests of the PTM. In lark buntings, the resource that primarily determines female settlement onto male territories is shade cover, as the main cause of nestling mortality in lark buntings is overheating. The more shade on a territory, the better the territory, and this shade effect can be shown experimentally by artificially increasing the shade in a given territory, which leads to an increase in nestling survival (Pleszczynska and Hansell, 1980).

Because shade protection is such a critical resource, it will often be the case that the male territories that are best suited to provide shade have already been settled by other females (Figure 8.17). Settling on a territory in which there already is a male and a female and becoming a "secondary" female allows a female access to shading. The cost of becoming such a secondary female is that a male only provides paternal care to the nestlings of his "primary" female, the mate that arrived first (Krebs and Davies, 1987). Pleszczynska and Hansell found that increasing shade availability on a territory not only made it much more sought after by females, but that as the PTM predicts, secondary females that bred in areas with shade cover (but that did not receive male aid) had about the same reproductive success as did monogamous females that bred on territories with less shade cover (but that did receive male aid).

Territory 1 (more shade)

Male

Female

Female

Female

?   ?

Territory 2 (less shade)

Male

FIGURE 8.17. The polygyny threshold model. In the lark bunting, shade is a limiting resource, as it affects nestling survival. Females may choose the territory of a male with good shade cover (territory 1) over a territory with less shade cover (territory 2), even if this decision means entering a polygynous relationship rather than a monogamous one.

EXTRAPAIR COPULATIONS

Male and female birds often form pair bonds at their nest during breeding season. It was once thought that such pairings were monogamous, in the sense that members of a given pair only mated with their nesting partner during a breeding season. Starting in the early 1980s, however, ornithologists began to uncover more and more instances of what are now referred to as **extrapair copulations**, or EPCs (R. Ford and McLauglin, 1983; McKinney et al., 1984). Ethologists were finding that males and females were leaving their territories during the mating season and mating with other individuals, usually those in nearby territories. The occurrence of such EPCs prompted some animal behaviorists to make a distinction between social monogamy and genetic monogamy. Most bird species in which EPCs were recorded formed pair bonds with just a single partner during a mating season, and as such displayed what is referred to as social pair bonding, or social monogamy. Yet, genetically, these systems resembled promiscuity more than monogamy, as mating occurred both with the social partner and with other individuals during the mating season.

The increased reproductive success of males that leave territories and engage in EPCs seems clear, as they can fertilize more females. But why would a female be involved in EPCs? The answer depends on the particular species and its ecology and demographics, but in general, females engaging in EPCs may (1) increase the probability that all their eggs are fertilized (the fertility insurance hypothesis); (2) maximize genetic diversity in their offspring, thereby increasing the chances that some of the offspring fare well in the environment in which they mature (Blomqvist et al., 2002, 2003; Griffith and Montgomerie, 2003); (3) use EPCs to select males that have good genes (see Chapter 7) but that might not be willing to form a pair bond and provide direct benefits to their offspring (Griffith et al., 2002; Neudorf, 2004); and (4) increase the amount of direct benefits—food, protection, and so on—that they receive from males.

David Westneat was among the first researchers to uncover the extent to which EPCs are occurring in nature. His behavioral observations of indigo buntings suggested that about 13 percent of all matings were EPCs (Westneat, 1987b; Figure 8.18). Westneat was concerned, however, that such observations might underestimate the actual percentage of all offspring that were sired via extrapair copulations. Because buntings were hard to follow for long periods at

FIGURE 8.18. Indigo buntings.
(A) A female indigo bunting. Although socially monogamous, female indigo buntings are often involved in extrapair copulations. (B) A male indigo bunting. Males defend their territories against intruders. *(Photo credits: David Westneat)*

A

B

a time, a significant number of EPCs may have been missed. Furthermore, it was not clear how many extrapair *matings* translated into extrapair *fertilizations*. For example, indigo bunting females generally resisted EPCs to a greater extent than mating with their nesting partner: Females resisted EPCs in 34 out of 43 attempts, but only resisted mating with their pairmates in 72 out of 320 attempts (Westneat et al., 1987). Not all EPCs result in offspring.

To examine what impact EPCs had on mating dynamics in buntings, Westneat ran a genetic analyses of parentage done in conjunction with a detailed behavioral study (Westneat, 1987a; Figure 8.19). This study was conducted before DNA fingerprinting techniques were widely available, and it relied on a technique called electrophoresis, which, although less powerful than DNA fingerprinting, does allow ruling out a particular adult individual as the parent of a particular offspring.

Over the course of two years, Westneat obtained DNA samples from hundreds of individually recognizable buntings. Using electrophoretic comparisons, and plugging that data into existing mathematical models, he found that, of the 257 young that were examined, 37 had genotypes that were *not* consistent with the genotype of one of their presumed parents (Westneat, 1987a), so at least 14 percent (37 out of 257) of all young were sired via an extrapair copulation— right in line with the 13 percent Westneat had predicted based on his behavioral observations. After that analysis, however, updated mathematical models showed that electrophoretic estimates of the percentage of young fathered by extrapair fertilizations in buntings were underestimates. Plugging Westneat's numbers into these newer models uncovered extrapair fertilization rates between 27 to 42 percent (depending on the year) in buntings (Westneat et al., 1987).

Since Westneat's work, many studies have documented EPC frequency in birds (for example, Adler, 2010; C. E. Hill et al., 2011; Schmoll, 2011), and have found that EPCs account for 76 percent of all young in one population of the superb fairy wren (*Malurus cyaneus*; Mulder et al., 1994; M. S. Webster

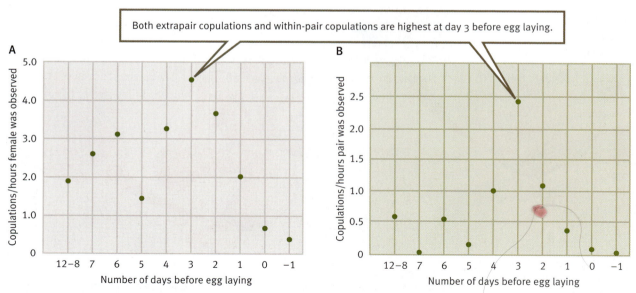

FIGURE 8.19. Copulations in indigo buntings. Occurance of copulations in **(A)** extrapair and **(B)** within-pair matings. The day the egg was laid is shown as day 0 on the *x*-axis. *(Based on Westneat, 1987b)*

and Westneat, 1998; Westneat et al., 1990). Given that monogamy was long considered to be the norm in birds, these are staggering numbers—numbers that are available only because of the revolution in molecular genetics that is still underway today.

SPERM COMPETITION

In many polyandrous and promiscuous breeding systems, there is a great deal of variation in reproductive success between individuals. With respect to males, a few individuals may obtain the vast majority of matings in a population, and many males may fail to obtain even a single mating opportunity. In Chapter 7, we saw how both male-male competition and female mate choice can affect which males are on the upper and lower ends of this mating curve distribution. Here, we will look at the effect of **sperm competition**—that is, the direct competition between the sperm of different males to fertilize a female's eggs—on mating success and the evolution of mating systems. (Birkhead and Møller, 1992, 1998; Birkhead and Parker, 1997; G. A. Parker and Pizzari, 2010; Tourmente et al., 2011).

In some promiscuous (as well as some polyandrous) mating systems, males compete not only for access to mating opportunities with females, but directly for access to eggs. In these systems, competition also occurs *after* a female has mated with numerous males. If females store sperm from numerous matings, sperm from different males may compete with one another over access to fertilizable eggs (Figure 8.20). When sperm competition exists, selection can operate directly on various attributes of sperm, such as sperm size and shape.

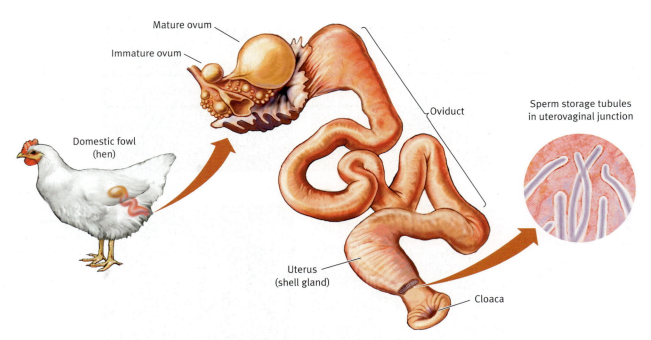

FIGURE 8.20. Sperm competition. The reproductive system of domestic fowl. Sperm storage occurs in sperm storage tubules (SST) at the uterovaginal junction. Females in many species can store sperm from multiple males, setting the stage for sperm competition. Only a small proportion of sperm makes it into the SST. *(Based on Birkhead and Møller, 1992)*

SPERM COMPETITION IN DUNGFLIES. Sperm competition has been extensively documented in many groups of animals, and a similar sort of competition, known as "pollen competition," is known to occur in plants (Delph and Havens, 1998). One of the leaders in this field, in terms of both empirical and theoretical aspects of sperm competition, is Geoff Parker, whose initial work on sperm competition in dungflies is regarded as the origins of modern research in this area (Immler et al., 2011; G. Parker, 1970b, 2001; G. A. Parker and Pizzari, 2010).

Dungflies use the droppings of large, often domestic, animals for breeding sites. While in most insects, copulations often last a matter of seconds, in dungflies they can last on the order of thirty minutes or more. After a detailed analysis of the natural history of mating in dungflies in pastures, Parker was faced with a number of unresolved issues regarding dungfly mating.

When a new dung pat is created and females begin to arrive, there ensues intense male competition for mating opportunities. A thousand or so males can descend on a single dung pat, all in search of females. Males that find a female and begin copulating are under constant physical attack from other males trying to break up their pairing and start their own round of copulating (Figure 8.21).

To test for the role of sperm competition in this mating system, Parker relied on a technique that entomologists of the day had been using in various biocontrol programs (G. Parker, 1970a). He irradiated the sperm of certain males, creating males whose sperm would then fail to produce eggs that hatched. As such, Parker could take pairs of males, irradiate one of them, and then examine the relative success of each male by simply determining the proportion of eggs that failed to hatch—that is, the proportion of fertilizations attributable to the irradiated male.

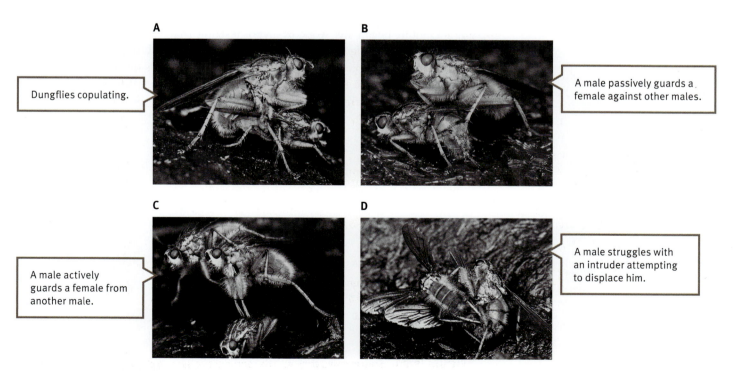

A Dungflies copulating.

B A male passively guards a female against other males.

C A male actively guards a female from another male.

D A male struggles with an intruder attempting to displace him.

FIGURE 8.21. Dungfly mating. In dungflies, sperm competition can be intense, with the last male copulating with a female fathering up to 80 percent of her offspring. *(Photo credits: Geoff Parker)*

Parker found that the number of eggs that were fertilized by the last male to mate with a female was proportional to how long such a mating lasted. The longer the last mating, the greater the reproductive success of the male. More specifically, the longer such a copulation, the greater the extent to which the last male's sperm displaced the sperm of males that had copulated with the female earlier (Figure 8.22). Such "last male precedence" is common when sperm competition is in play, but it is not ubiquitous. (In some mating systems, sperm competition appears to favor the first, rather than the last, male to mate with a female.)

In the dungfly system, the last male to mate with a female copulated on average for thirty-six minutes, and as a result he fathered approximately 80 percent of the young in the clutch of eggs deposited by a female (G. Parker, 1970a). If copulation time correlates with greater displacement of a competitor's sperm, why don't males copulate for even longer periods and thereby attempt to displace 100 percent of a competitor's sperm? The answer appears to be that males must weigh such an option against what else could be done with the time in question. While increasing the time spent with female A will increase the displacement effect, it is also time that the male could have used to find another female with whom to mate. Because the rate of sperm precedence slows down with time, it will often benefit a male to use such additional time to find other potential mates (G. A. Parker, 1974b; G. A. Parker and Stuart, 1976).

SPERM COMPETITION IN SEA URCHINS. While sperm competition is often thought of in terms of its effect in utero, it can also play an important role in species that do not have internal fertilization. To get a better sense of the importance of sperm competition in such systems, let's consider Don Levitan's work on sperm velocity and fertilization rates in the sea urchin (Levitan, 2000).

Levitan began his work by hypothesizing that variation in the speed at which sperm traveled correlated with fertilization rate. Using sea urchin sperm makes this task a bit easier than using sperm from birds or mammals, because sea urchins secrete their sperm and eggs into seawater. With a video camera that can tape sperm swimming along and a microscope to see which eggs

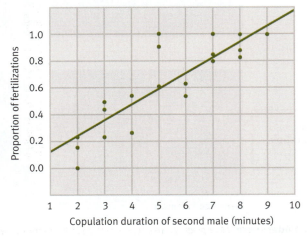

FIGURE 8.22. Sperm competition in dungflies. The longer a male dungfly mates with a female dungfly, the greater his fertilization success. *(From Simmons, 2001)*

are fertilized, Levitan was able to measure the sperm's swimming speed and fertilization success.

Levitan found that to fertilize the same number of eggs, males that produced slow-moving sperm needed to release up to 100 times more sperm than males that produced fast-moving sperm (Figure 8.23). Levitan then examined what happens to sperm as they age. In so doing, he was examining a predicted trade-off between the speed at which a sperm moves and how long that sperm survives. Because swimming fast and swimming for a long time both require energy, fast-moving sperm shouldn't live as long as slow-moving sperm.

When examining the expected trade-off between speed and longevity, Levitan first found that *all* sperm slow down as they get older (Figure 8.24). Not only did sperm decrease their swimming speed with age, but as they aged, they were much less likely to fertilize an egg, even when they encountered one. For example, sperm that were only an hour old could be up to 100 times less likely to fertilize an egg as were newly released sperm. After two hours, sperm fertilized no eggs at all.

With data on longevity and speed in hand, Levitan could return to the question raised earlier: Is there a trade-off between sperm speed and sperm life span for individual sea urchins? The answer appears to be yes, as Levitan found a negative correlation between velocity and endurance. Individuals that produced fast-moving sperm had their sperm become ineffective at much quicker rates than other individuals. The energy used up in swimming fast resulted in less energy for swimming for a long time, as well as a shorter life span for sperm.

OTHER EFFECTS OF SPERM COMPETITION. Sperm competition not only affects the speed at which sperm swim but also affects the various shapes and functions that sperm can take (Bellis et al., 1990; Gomendio et al., 1998; Holman and Snook, 2006; H. Moore et al., 1999, Tourmente et al., 2011). For example, Roger Baker and Mike Bellis's **kamikaze sperm hypothesis** suggests that natural selection might favor the production of some sperm types that are

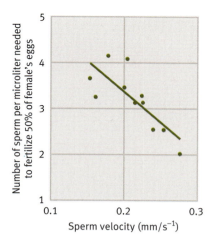

FIGURE 8.23. Sperm velocity and fertilization. In sea urchins, slower sperm fare poorly. The slower the sperm, the more sperm needed to fertilize a female's eggs. *(Based on Levitan, 2000)*

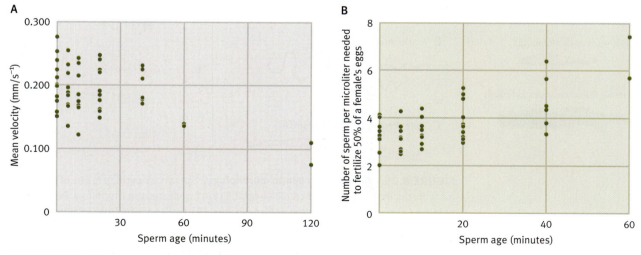

FIGURE 8.24. Older sperm fare poorly. (A) In sea urchins, older sperm swim slower, and **(B)** a greater quantity of such sperm is needed to achieve high fertilization rates. *(Based on Levitan, 2000)*

designed to kill other males' sperm rather than fertilize eggs (Baker and Bellis, 1988).

While the evidence is equivocal in humans, sperm competition has had clear effects on sperm morphology in insects, frogs, mammals, birds, fish, and worms (Baer et al., 2009; Birkhead and Møller 1992, 1998; Eberhard, 1996; Firman and Simmons, 2010; Stockley et al., 1997; Tourmente et al., 2009, 2011; Figure 8.25). For example, in a phylogenetic analysis of 100 species of Australian (myobatrachid) frogs, Philip Byrne and his colleagues found that males in species with intense sperm competition produced sperm with relatively long tails (P. G. Byrne et al., 2003). While the exact advantage of longer-tailed sperm in these frogs is not yet known, other studies have found that sperm with longer tails swim faster than their shorter-tailed competitors, and hence have an increased probability of fertilizing an ovum (Oppliger et al., 2003).

Sperm competition also has effects on the number of sperm produced per ejaculate. One prediction from sperm competition theory is that the number of sperm per ejaculate should be a function of the probability that a female has recently mated with other males (Baker and Bellis, 1993). To see this, consider two males that we will call M1 and M2. Suppose that M1 is about to copulate with a female. The greater the chance that such a female has mated with M2 in the recent past and that his sperm are still present, the greater the chance that M1's and M2's sperm will be in direct competition to fertilize the eggs of the female. Under this scenario, sperm competition theory predicts that M1 will ejaculate more sperm in an attempt to increase the chances that he will fertilize the female's eggs.

Baker and Bellis tested this hypothesis in humans. They obtained data on the interval between copulations in a given pair of individuals, and assumed that the longer this interval, the greater the chances that a partner would have had a sexual encounter with someone besides their partner. Then they obtained

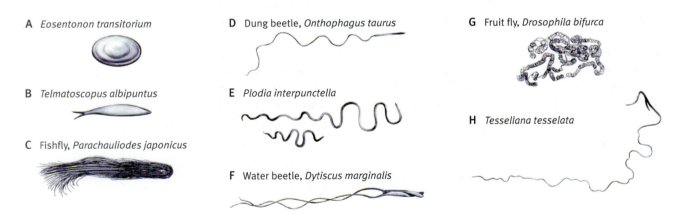

A *Eosentonon transitorium*

B *Telmatoscopus albipuntus*

C Fishfly, *Parachauliodes japonicus*

D Dung beetle, *Onthophagus taurus*

E *Plodia interpunctella*

F Water beetle, *Dytiscus marginalis*

G Fruit fly, *Drosophila bifurca*

H *Tessellana tesselata*

FIGURE 8.25. Variability in sperm morphology. Sperm competition is one of the many forces that have led to incredible variability in insect sperm morphology. Pictured here are **(A)** sperm from *Eosentonon transitorium*, **(B)** fishlike sperm from *Telmatoscopus albipuntus*, **(C)** sperm bundle from the fishfly *Parachauliodes japonicus*, **(D)** 1 mm sperm from the dung beetle *Onthophagus taurus*, **(E)** short and long sperm from *Plodia interpunctella*, **(F)** paired sperm from the water beetle *Dytiscus marginalis*, **(G)** giant 58 mm sperm from *Drosophila bifurca*, and **(H)** hook-headed sperm from *Tessellana tesselata*. *(Based on Simmons, 2001)*

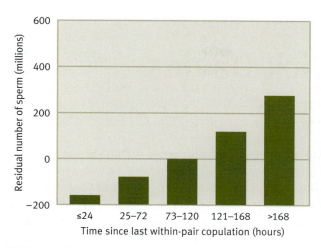

FIGURE 8.26. Sperm number in humans. In humans, the number of sperm ejaculated during a copulation is a function of the time since a pair last copulated. Note that the y-axis is a measure of "residuals"; hence negative values are possible. *(Based on Baker and Bellis, 1993)*

sperm samples from individuals the next time they copulated with their partner. Baker and Bellis found that not only did sperm number increase as a function of time since last copulation (which could be due to many different factors), but that even when absolute time was statistically removed from the equation, the relative amount of time couples spent together predicted sperm volume as well (Figure 8.26). When couples spent more time together, and hence the risk of extrapair copulations was low, sperm count was significantly lower than when couples spent less time together.

With respect to sperm competition, it is important to keep in mind that females are not simply "inert environments" that serve as receptacles of male sperm (G. Parker, 1970b). Rather, females themselves may play an active role in sperm competition via **cryptic mate choice**—that is, female mate-choice behavior that is not obvious to males. William Eberhard, for example, suggests that, among other things, cryptic choice may affect how much sperm a female allows a copulating male to inseminate her with, how she goes about transferring such sperm to the organs where sperm are stored, and which sperm she may select for actual fertilization (Eberhard, 1996). Precisely how females select among sperm remains unknown in most cases, but this is an active area of study within animal behavior research (Birkhead and Pizzari, 2002; Holman and Snook, 2006).

Multiple Mating Systems in a Single Population?

In the case of dunnocks, which we discussed earlier in the chapter, we see monogamous, polygynous, polyandrous, and polygynandrous mating groups all in the same population. Let's return to this mating system, and examine it in a bit more detail.

Nick Davies and his colleagues have an ongoing, long-term study on a population of about eighty dunnocks that reside in the Botanical Gardens of

Dr. Nick Davies

The dunnocks you work with have proved to be a model system for so many questions in behavioral ecology and animal behavior. Why did you choose to work with this species?

In 1979, when I started as a young university lecturer at Cambridge, I was excited by the new theoretical ideas of Bill Hamilton, Robert Trivers, John Maynard Smith, and Geoff Parker. They were beginning to explore the evolutionary consequences of individual conflicts of interest, not only the long-recognized conflicts among rival males but also those between the male and female of a breeding pair and between parents and their offspring. These ideas led to a profound change in how we interpret individual adaptations. For example, the classical work on clutch size (pioneered by David Lack) had considered what would be optimal from a pair's point of view. Once genetic conflicts within families were recognized, a whole new world of possibilities was opened up, involving deception, manipulation, cheating, and compromise. Likewise, previous studies of mating systems had emphasized ecological pressures (food, nest sites, predation) that might lead to one system rather than another. But now we began to consider social conflicts as important selection pressures too.

So I was on the lookout for a field study which would allow me to take a fresh look at bird mating systems. While wandering in the University Botanic Garden, I noticed the dunnocks chasing around the bushes in pairs, threes, and fours and decided to color-band them for individual recognition to see what was going on. Once I had this initial excuse to begin the study, I then discovered more interesting new questions from the bird watching rather than from the theoretical literature.

Your work on the dunnock suggests these birds have an incredibly plastic mating system. Why do you think that is?

It is clear from watching behavior that males and females have different mating preferences. These conflicts make good sense in relation to individual reproductive payoffs from the different mating combinations. For a female, polygyny (one male plus two females) was the least successful option, because she had to share the help of one male with another female, and her chicks often starved. In polygyny, females often chased and fought each other. If one of them could drive the other away, then she could claim the male's full-time help and so enjoy the greater success of monogamy. A polyandrous female (one female with two males) was even more successful: if she shared copulations between both her males, then both helped to feed her brood. With this extra help, more chicks survived. This explained why a female encouraged both alpha and beta males to copulate with her. Our experiments showed that she maximized total male help by sharing copulations equally between the two males.

By contrast, a male fared least well in polyandry. Although more young were raised, there was the cost of shared paternity. Our DNA fingerprinting results showed that an alpha male did better with full paternity of a pair-fed brood. This explains why alpha males acted against the female's wishes, and tried to drive beta males away, or at least prevent them from copulating. A male did best of all in polygyny, the system in which a female did worst. Although each female was less productive, the combined output of two females in polygyny exceeded that of a monogamous female. So once again there was conflict—the male intervened in the squabbles between his two females in an attempt to retain them both.

The variable mating system thus reflects the different outcomes of sexual conflict. Where a male can prevent the conflicts between two females, we find polygyny. Where a female can escape the close guarding of her alpha male, and give a mating share to the beta male, we observe polyandry. Monogamy occurs when neither sex can gain a second mate. Polygynandry—for example, two males with two females—can be viewed as a "stalemate": the alpha male is unable to drive the beta male off and so claim both females for himself, and neither female can evict the other and so claim both males for herself.

So the interesting question is What determines whether particular individuals can get their best option despite the conflicting preferences of others? This is influenced by individual competitive ability and various practical considerations, such as vegetation density (which determines how easily an alpha male can follow a female). The main point is that the social conflicts are played on an ecological stage, which may affect the likely outcomes.

Sexual selection and mating systems are often presented as two distinct topics. Is it possible to understand one without the other?

Darwin's theory of sexual selection aims to explain the evolution of traits that increase an individual's mating success: size, weapons, ornaments, mate choice, and so on. Mating systems are the outcomes of this competition for mates. So you need to think about both process and outcome for a full understanding.

Why do you think polyandry is rare compared to polygyny?

When you look at social groups, it is certainly true that you more often see one male defending a group of females rather than the converse—namely, one female defending a group of males. In theory, this is exactly what you would expect, because a male usually has a greater potential reproductive rate than a female. This means that an increase in the number

of mates leads to a much greater increase in male reproductive success than it does in female reproductive success. Thus, males often go for quantity, while females seek quality, when they search for mates. Nevertheless, until recent years we underestimated the frequency with which females seek matings with multiple males. They often do this surreptitiously, sometimes outside their social groups, to increase their access to resources, to increase male care, to reduce male harassment, or to improve the genetic quality of their offspring.

> **Males often go for quantity, while females seek quality, when they search for mates. Nevertheless, until recent years, we underestimated the frequency with which females seek matings with multiple males. They often do this surreptitiously. . . .**

There has been some debate as to who calls the shots when it comes to mating systems—males or females? Clearly any given mating requires both a male and female, but are there certain situations in which you expect males to be controlling the mating system, and others where mating systems are primarily under the control of females?

Darwin's theory of sexual selection aimed to explain how individuals competed for mates. He proposed two processes: people readily accepted the first, namely, male-male competition (Darwin's

"law of battle"), but were reluctant to believe the second, namely, female choice. We now have good evidence for this, of course. It is interesting to see reactions to Parker's theory of sperm competition (sexual selection after the act of mating) follow the same history. At first, everyone focused on males—mate guarding, frequent copulation, mating plugs, sperm removal, testis size, and so on. Consideration of female roles ("cryptic female choice" of sperm) came later, and it is still controversial. My guess is that female roles will be crucial, simply because females should be better able to control events inside their own bodies.

In some cases, the act of mating seems to be under male control; for example, where female insects lay their eggs in localized patches (e.g., cow pats), powerful males can monopolize the laying sites, and a female is forced to mate in order to gain access. In other cases, females can more easily escape males; for example, insects that lay their eggs in more dispersed sites can often refuse matings, and in birds (where females can often easily escape males by flying off) mating rarely occurs without female consent. It will be interesting to see whether cryptic female choice of sperm is more likely in cases where females have less control over the act of mating.

Dr. Nick Davies is a professor at Cambridge University in England. His long-term work on dunnocks, summarized in Dunnock Behaviour and Social Evolution *(Oxford University Press, 1992), elegantly shows the complexity of animal mating systems and how one goes about testing fundamentally important hypotheses in such a system.*

Cambridge University, and their work has provided a rare detailed portal into a complex mating system (Davies, 1992). What makes the dunnock breeding biology so fascinating from a mating systems perspective is the long-term persistence of monogamy, polygyny, polyandry, and polygynandry *in the same population*. For Davies, unraveling the story behind the dunnock mating system is akin to detective work: "The puzzle is the dunnock's extraordinary breeding behavior and variable mating system. The job of the nature detective is to understand alternative options facing individuals, to assess their reproductive payoffs, and then to discover whether different mating strategies might emerge as a result of individuals competing to maximize their reproductive success" (Davies, 1992, p. 21).

Underlying much of the variance in mating systems, including that of the dunnock, is the fact that the fitness of males and females is affected in different ways by the mating system. Reproductive success of the most successful males will often be lowest when they share access with other males to a single female (polyandry), and then increase in the following order: sole access to a single female (monogamy), joint access to two females (polygynandry), and sole access to numerous females (polygyny). In other words, male reproductive success increases as a function of both the number of mates and the degree to which a male has sole reproductive access to such mates.

The reproductive success of the most successful females increases in precisely the opposite direction, with polyandrous and polygynandrous females having the highest reproductive success. As such, a conflict of interest between the sexes—sometimes referred to as a "battle of the sexes"—exists with respect to what constitutes the optimal breeding system (Arnqvist and Rowe, 2005; Hosken et al., 2009). In dunnocks, females appear to be winning this battle of the sexes (at least for now), as over the course of ten years, 75 percent of females and 68 percent of males observed by Davies and his team were involved in either polyandrous or polygynandrous mating groups (Davies, 1992).

The battle of the sexes, in conjunction with the dispersal patterns of dunnocks, helps us better understand the complex breeding system found in this bird. Early in the breeding season, females compete with one another to establish territories, and such female territories are chosen independently of the position of males. Males then attempt to build their own territories so that they overlay as many female territories as possible. Given these conditions, Davies argues that the difference between monogamy and polygyny is a function of *male* territory size. Polygynous males had larger territories than monogamous males, but when these two breeding systems were compared, the territory size of females remained constant. In contrast, the difference between polyandry and polygynandry was a function of female territory size. Male territory size remained constant across these systems, while female territory size was significantly larger in the former.

Because all individuals in this population were marked (with color rings) and rarely moved more than two miles from the Botanical Gardens, Davies was able to undertake experiments designed to gauge more precisely how resource defense and territoriality influenced the dunnock mating system. Davies and Arne Lundberg hypothesized that because females were in strong competition with one another for territories with the best resources, if the resources available on a territory were experimentally supplemented, territory size should shrink, as females would then be able to obtain the same amount of resources without having to defend as large an area (Davies and Lundberg, 1984).

TABLE 8.2. Food and territory size. Supplementing the food on dunnock territories led to a decrease in female territory size but not male territory size. *(From Davies, 1992, p. 63)*

	TERRITORY SIZE (m²)						
	WITH FEEDER			CONTROL			SIGNIFICANCE OF DIFFERENCE
	MEAN	(SE)	*n*	MEAN	(SE)	*n*	
Females	2,776	(379)	28	4,572	(456)	39	$P < 0.01$
Males							
Territory defended by one male	2,864	(340)	11	2,642	(416)	13	NS
Territory defended by two males	5,276	(797)	14	6,614	(674)	17	NS

To test their hypothesis, Davies and Lundberg placed artificial feeders on a randomly selected set of female territories, and they did indeed find that the female territories shrank as predicted. Moreover, they found that the male territories that overlay the manipulated female territories did not change in size (Table 8.2). What did occur, however, was a shift in the distribution of mating systems, in such a manner as to favor males. When female territory size shrank as a result of supplemental resources, males were better able to monopolize more than one female, and a shift away from polyandry and toward polygynandry occurred. This finding supports Davies's argument that, in the dunnock, females track resources, males track females, and the resulting interaction helps us better understand the incredible variation in mating systems in this small brown bird.

SUMMARY

1. Animal mating systems are classified as monogamous, polygynous, polyandrous, polygynandrous, or promiscuous, depending on the number of mates that males and females take and the timing of such mating in relation to breeding season.
2. A female can often fertilize all her available eggs by mating with one or a very few males, so female fecundity is not so much tied to the availability of mates as it is to the availability of resources, such as food, defense, and so on. Up to a point, the more resources available, the more offspring females can produce.
3. Males can potentially fertilize many females, so their reproductive success is tied more to access to females than to access to resources. Male dispersion patterns should track the dispersion of females.
4. Recent work in neurobiology and endocrinology has provided animal behaviorists with a better understanding of the proximate underpinnings of both monogamy and polygamy.
5. Phylogenetic work has shed light on the relationship between mating systems and habitat quality. Monogamous systems are often found in poor habitats, and polygynous systems are found in better habitats.

6. The polygyny threshold model predicts under what conditions polygyny should occur in nature. In this model, females weigh the costs and benefits associated with being in a polygynous relationship on a good territory versus a monogamous (or more precisely, a less polygynous) relationship on a poorer territory.

7. While extrapair copulations and extrapair matings were once thought to be rare in birds, genetic evidence suggests that this is far from the case in many species.

8. In some mating systems, males compete not only for access to mates but directly for access to eggs. If females store sperm from numerous matings, sperm from different males compete with one another over access to fertilizable eggs in what is known as sperm competition.

DISCUSSION QUESTIONS

1. Define and distinguish among serial monogamy, serial polygyny, simultaneous polygyny, promiscuity with pair bonds, and promiscuity without pair bonds.

2. Read Jenni and Colliers's 1972 article "Polyandry in the American Jacana (*Jacana spinosa*)" in *Auk* (vol. 89, pp. 743–765). What selective forces favored polyandry in jacanas?

3. Why do you think that polygamous mating systems more strongly favor the evolution of virulent diseases in animals and humans than do monogamous breeding systems? Think about this from the perspective of the disease-causing agent.

4. Define an EPC. How does this differ from an extrapair mating? Why did it take ethologists so long to recognize the extent of EPCs in nature? How has molecular genetics revolutionized the way we think of mating systems in birds?

5. How has natural selection via sperm competition shaped both sperm morphology and male behavior? Create a list of potential ways in which females may affect sperm competition and its outcome.

6. What is a lek, and why is that form of polygyny especially interesting to ethologists? How has knowledge of kinship bonds contributed to an understanding of why males form leks?

SUGGESTED READING

Arnqvist, G., & Rowe, L. (2005). *Sexual conflict*. Princeton, NJ: Princeton University Press. A book-length treatment of sexual conflict and the way that it shapes mating systems.

Davies, N. B. (1992). *Dunnock behaviour and social evolution*. Oxford: Oxford University Press. A delightful book about Davies's long-term work on dunnock behavior, with an emphasis on dunnock mating behavior.

Lane, J. E., Forrest, M. N. K., & Willis, C. K. R. (2011). Anthropogenic influences on natural animal mating systems. *Animal Behaviour, 81,* 909–917. An overview of how four anthropogenic factors may affect animal mating systems.

Orians, G. (1969). On the evolution of mating systems in birds and mammals. *American Naturalist, 103*, 589–603. This is the paper in which Orians presents the polygyny threshold model.

Parker, G. (1970a). Sperm competition and its evolutionary consequences in insects. *Biological Reviews of the Cambridge Philosophical Society, 45*, 525–567. The seminal paper (no pun intended) of sperm competition and animal behavior.

Shuster, S. M. (2009). Sexual selection and mating systems. *Proceedings of the National Academy of Sciences, U.S.A., 106*, 10009–10016. A review of mating systems in a special issue devoted to the 150th anniversary of the publication of Darwin's *On the Origin of Species.*

9

Kinship

In an open field somewhere, a group of ground squirrels feed. Seemingly out of nowhere, a long-tailed weasel (*Mustela frenata*) appears, targeting the squirrels in the field as its prey. Suddenly an alarm call given by one squirrel alerts others of the impending danger. The field comes to life with squirrels making mad dashes everywhere, doing whatever they can to reach their burrow, or at least some safe haven. Later, when the predator has departed, the squirrels reemerge.

In terms of costs and benefits, this type of alarm seems counterintuitive. Why should an individual squirrel give off an alarm call? Emitting alarm calls as loud as possible, if nothing else, should make the alarm caller the single most obvious thing in the entire field. Why would the alarm caller do anything to attract a predator in its direction and make itself the predator's most likely next meal? Why not let another squirrel take the risks?

Paul Sherman has been addressing these sorts of questions in long-term studies of alarm calls in Belding's ground squirrels (*Spermophilus beldingi*; Sherman, 1977, 1980, 1981, 1985; Figure 9.1). Sherman has found that genetic relatedness affects animal behavior in important ways, playing a large role in whether or not natural selection favors squirrels emitting alarm calls when a predator is detected.

In this chapter, after an introductory section demonstrating the power of genetic kinship to affect animal behavior, we will examine:

▸ the theoretical foundation underlying "inclusive fitness," or kin selection models of social behavior;
▸ the evolution of the family unit;
▸ parent/offspring conflict and sibling rivalry; and
▸ how and why animals recognize kin.

A

B

FIGURE 9.1. Alarm calling in squirrels. In Belding's ground squirrels, females **(A)** are much more likely than males to emit alarm calls when predators are sighted. Such alarm calls warn others, including female relatives and their pups **(B)**. *(Photo credits: George D. Lepp; Paul W. Sherman)*

Kinship and Animal Behavior

Belding's ground squirrels, like many other species, such as prairie dogs, give alarm calls when a predator is spotted (Hoogland, 1983, 1995). These calls signal that a predator is in the vicinity and others respond to this signal by moving toward places of safety. To begin to answer why Belding's ground squirrels give alarm calls at the risk of their own lives, we need to recognize that alarm calls in these squirrels are most often emitted by females. That is, female squirrels give alarm calls when a predator is in the vicinity more often than expected by chance, whereas males give fewer alarm calls than expected by chance (Figure 9.2). The question of interest then is not "Why are alarm calls emitted?" but "Why do females give alarm calls so often?" The answer lies in gender differences in where the squirrels live and in their proximity to their genetic kin.

In Belding's ground squirrels, males emigrate from their group to find mates, but females mature in their natal area (that is, their place of birth). This male-biased dispersal creates an imbalance in the way males and females are related to the individuals that live around them—females find themselves surrounded by genetic relatives, while adult males are generally in groups that do not contain many genetic relatives (Figure 9.3). When females give alarm calls, they are warning genetic kin. Any alarm calls given by adult males, however, primarily warn unrelated individuals. Kinship, then, lies at the heart of female alarm calling. Further support for the kinship-based alarm-calling hypothesis includes Sherman's finding that, in the rare instances in which adult females do move away from their natal groups and into groups with fewer relatives, they emit alarm calls less frequently than do native females.

Kinship not only promotes prosocial behavior but also acts as a force in deterring antisocial behavior as well. As an extreme case, consider homicide in humans. Martin Daly and Margo Wilson examined 512 homicide cases

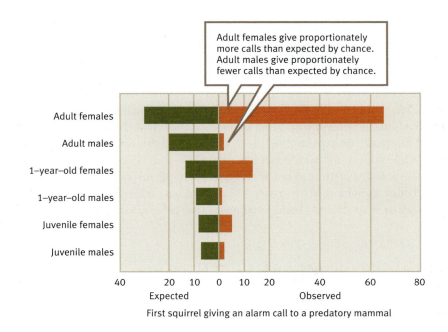

FIGURE 9.2. Ground squirrel alarm calls. When comparing the observed (orange bars) versus the expected (green bars) frequencies of alarm calls in Belding's ground squirrels, females emit such calls at a rate greater than that expected by chance (p < .001). As a result of dispersal differences across sexes, females, but not males, are often in kin-based groups. *(From Sherman, 1977)*

FIGURE 9.3. Kin selection and ground squirrels. Belding's ground squirrel groups are typically made up of mothers, daughters, and sisters that cooperate with one another in a variety of contexts. Males that emigrate into such groups cooperate to a much smaller degree. *(Based on Pfennig and Sherman, 1995)*

occurring in 1972 in Detroit, Michigan (Daly and Wilson, 1988). In the police records, 127—a full 25 percent—of these murders were committed by what the police records denote as "relatives." The police, however, classify in-laws, and even boyfriend-girlfriend pairs, as relatives, rather than limiting this category to genetic kin. When Daly and Wilson considered only genetic kin, rather than these other categories, only 6 percent of the murders involved relatives. Genetic kin don't kill each other all that often because harming genetic relatives is selected against for the very same reason that dispensing altruism to relatives is favored—they both have indirect consequences on those who share the same alleles.

With respect to Daly and Wilson's homicide data from Detroit, it might be argued that the reason that homicide rates among genetic kin are low is that, in modern society, people encounter unrelated individuals much more often than genetic kin. For example, if killers spent 94 percent of their time with unrelated individuals and 6 percent with genetic kin, then the 6 percent murder rate among genetic kin would be expected simply by chance, and this would not indicate that genetic relatedness reduces homicide. Yet, Daly and Wilson found that, even when the amount of time spent with genetic kin versus everyone else is taken into account, genetic relatives rarely kill each other (Table 9.1). Few forces have the power to shape animal behavior the way that genetic kinship can.

Kinship Theory

The modern study of animal behavior and evolution began in the early 1960s, when W. D. Hamilton, one of the leading evolutionary biologists of the twentieth century, published his now famous papers on genetic kinship and the evolution of social

TABLE 9.1. Risk of homicide in cases where the victim and offender were cohabitants in Detroit in 1972. Observed values indicate the number of homicides that were actually committed. Expected values indicate the number of homicides in each category that we would expect if genetic kinship were not playing a role. Relative risk rates were much higher for individuals who were not genetic relatives. These numbers are underestimates since the "parent" and "offspring" categories include some stepfamily members and some in-laws. *(From Daly and Wilson, 1988)*

THE AVERAGE DETROITER ≥ 14 YEARS OLD IN 1972 LIVED WITH 3.0 PEOPLE	NUMBER OF VICTIMS		RELATIVE RISK (OBSERVED/EXPECTED)
	OBSERVED	EXPECTED	
0.6 Spouses	65	20	3.32
0.1 Nonrelatives	11	3	3.33
0.9 "Offspring"	8	29	0.27
0.4 "Parents"	9	13	0.69
1.0 Other "relatives"	5	33	0.15

behavior (Hamilton, 1963, 1964). These papers formalized the theory of "inclusive fitness" or "kinship" theory and revolutionized the way scientists understood the evolution of behavior. Recall from Chapter 1 that inclusive fitness is a measure of an individual's total fitness based both on the number of its own offspring and the contribution it makes to the reproductive success of its genetic relatives.

But *why* is kinship so powerful an evolutionary force in promoting social behaviors like cooperation and altruism (in Chapter 10 we will discuss other paths leading to such behaviors)? Hamilton had this to say in his seminal paper tying together genetic kinship and the evolution of altruism:

> In the hope that it may provide a useful summary we therefore hazard the following generalized unrigourous statement of the main principle that has emerged from the model. *The social behavior of a species evolves in such a way that in each distinct behavior-evoking situation the individual will seem to value his neighbors' fitness against his own according to the coefficients of relationship appropriate to that situation* [Hamilton's italics]. (Hamilton, 1964, p. 19)

Although rightly credited with being the founder of modern kinship theory, Hamilton was not the first to recognize the power of kinship to shape behavior (Dugatkin, 2006). Before Hamilton, Charles Darwin suggested that the suicidally altruistic defense behavior that he observed in social insects like bees may have evolved as a result of bees defending hives filled with their kin—that is, under certain conditions, natural selection could favor such extreme altruism if the recipients of the altruistic act were genetic relatives (Figure 9.4). About seventy-five years later, population geneticist J. B. S. Haldane discussed altruism and genetic kinship (Haldane, 1932). It is rumored that Haldane once said that he would risk his life to save two of his brothers or eight of his cousins. Haldane, a brilliant mathematician,

FIGURE 9.4. Helping offspring. One classic case of helping genetic relatives is that of mothers feeding their young. In bank swallows, young chicks remain at the nest, and mothers remember the location of their nests and return after foraging to feed youngsters there. When chicks learn to fly, mothers learn to recognize their offspring's voices. *(Based on Pfennig and Sherman, 1995)*

made this rather surprising statement by counting copies of an allele that might code for cooperative and altruistic behavior. Such a gene-counting approach to kinship and the evolution of cooperation has been formalized by theoreticians, but in its most elementary form, it is at the core of inclusive fitness theory. Let's see how it works.

RELATEDNESS AND INCLUSIVE FITNESS

The *Random House Dictionary* defines kinship as "family relationship," but the evolutionary definition is much more restrictive. In evolutionary terms, relatedness centers on the probability that individuals share copies of alleles that they have inherited from common ancestors—parents, grandparents, and so on. Alleles that are shared because of common ancestry are referred to as "identical by descent." For example, you and your brother are kin because you share some of the same alleles and you inherited them from common ancestors—in this case, your mother and father. In a similar vein, you and your cousins are kin because you share alleles in common; only now your most recent common ancestors are your grandparents. In general, most recent common ancestors are those individuals through which two (or more) organisms can trace alleles that they share by descent.

Once we know how to find the common ancestor of two or more individuals, we can calculate their genetic relatedness, labeled r, which is equal to the probability that they share alleles that are identical by descent. For example, two siblings are related to one another by an r value of 0.5. To see why, recall that all of the alleles that siblings share come from one of two individuals—their mother or father. As such, there are two ways, *and only two ways*, that siblings can share a copy of allele X—via mother or father. If sibling 1 has allele X, then there is a 50 percent chance she received it from her mother; if sibling 2 has allele X, there is again a 50 percent chance that her mother passed this allele to sibling 2. Thus there is a 1 in 4 chance that the siblings share allele X through their mother. The same argument can be made to demonstrate that there is a 1 in 4 probability that the father is the reason that the siblings share allele X. To calculate the chances that the siblings share allele X through *either* their mother or their father, we add the probabilities for each and obtain $1/4 + 1/4 = 1/2$, or 0.5. This value—labeled *r*—can be calculated for any set of genetic relatives, no matter how distant. For example, the genetic relatedness between cousins is 1/8 (that is, $r = 0.125$), between grandparent and grandchild is 1/4 (that is, $r = 0.25$), and between aunts/uncles and their genetic nieces and nephews is also 1/4 (that is, $r = 0.25$; Figure 9.5).

Let us work through a few more examples of calculating genetic relatedness. In Figure 9.6A, individuals X and Y are half siblings, with the same mother but different fathers. To compute the coefficient of relatedness (*r*) between X and Y, we first must find the most recent common ancestor or ancestors. In this case, there is one: their mother. Second, we compute the probability that a given allele copy in the mother is passed to both offspring. The probability is 0.5 that the allele will be passed to X, and the probability is 0.5 that it will be passed to Y, so the probability that it will be passed to *both* is $0.5 \times 0.5 = 0.25$. Because the mother is the sole most recent common ancestor, this is the total coefficient of relatedness (*r*).

In Figure 9.6B, X and Y have a single most recent common ancestor who is X's maternal grandmother and Y's mother. The chance that a given allele copy in this ancestor reaches X is 0.25, because there is a 0.5 chance that it will reach X's mother, and if it does, there is an additional 0.5 chance that it will go on to reach X, for a net chance of 0.25. The chance that a given allele will reach Y is 0.5. Thus, the chance that the given allele copy will reach *both* X and Y is $0.25 \times 0.5 = 0.125$. The coefficient of relatedness between X and Y is therefore 0.125. (If B had been a full sibling to X's mother, the coefficient of relatedness between X and Y would have instead been 0.25.) Similar calculations allow us to compute the genetic relatedness between any pair of individuals with a known pedigree.

To this point, we have been thinking about an allele in terms of the effect it has on the *individual* in which it resides, but kinship calculations suggest that this is an overly restricted view. Given that genetic relatives, by definition, have a higher probability of sharing allele X through common descent than do nonrelatives, then allele X may increase its chances of getting copies of itself into the next generation by how it affects not just the individual in which it resides, but that individual's genetic relatives as well.

Think about it like this: When an individual reproduces and its offspring survive, copies of that individual's alleles make it into the next generation. But that is not the *only* way that alleles can increase their representation in future generations. If an allele—let's call it allele X—codes for preferentially aiding genetic kin, then that allele can increase its representation in the next generation because it is coding for aid to individuals who are likely to have X as well (Hamilton, 1963). How likely a recipient is to have a copy of X is equal to the genetic relatedness of the donor and recipient (50 percent probability for

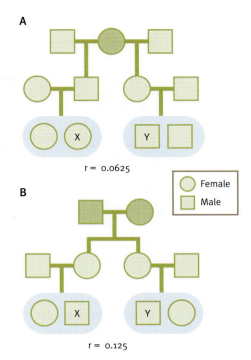

FIGURE 9.5. Pedigrees for calculating relatedness. Individuals X and Y may have one or two most recent common ancestors (dark shading). **(A)** X and Y have the same grandmother but different grandfathers. Thus, their grandmother is their sole most recent common ancestor. **(B)** X and Y have the same maternal grandmother and the same maternal grandfather. Thus, maternal grandparents are the most recent common ancestors. *(From Bergstrom and Dugatkin, 2012)*

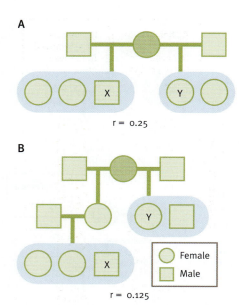

A

r = 0.25

B

Female

Male

r = 0.125

FIGURE 9.6. Example pedigrees for computing coefficients of relatedness. (A) X and Y are half siblings. (B) A more complicated scenario, in which X and Y come from different generations. Here, Y is X's aunt. *(From Bergstrom and Dugatkin, 2012)*

siblings, 25 percent probability for uncle and nephew, and so on). When we depict fitness in this manner, and consider both direct and indirect components to fitness, we are talking about **inclusive fitness**.

With an understanding of how r is calculated, we can now examine inclusive fitness theory in more detail. Hamilton tackled the question of kinship and animal behavior in a pair of papers, "The Genetical Evolution of Social Behavior, I and II" (Hamilton, 1964). The essence of inclusive fitness models is that they add on to "classical" models of natural selection by considering the effect of an allele, not only on the individual in which it resides, but on individuals (genetic kin) carrying alleles that are identical by descent. The equations in some of Hamilton's papers on kinship can be daunting, even to those with a mathematical background. Fortunately, these equations can be captured in what is now referred to as "Hamilton's Rule" (Hamilton, 1963). This rule states that an allele associated with some trait being studied increases in frequency whenever:

$$\left(\sum_{1}^{A} rb \right) - c > 0$$

where b = the benefit that others receive from trait under study (recall the benefit that squirrels received when they heard one of their groupmates give an alarm call), c = the cost accrued to the individual expressing the trait (think of the alarm caller and its risk of being taken by a predator), r is our measure of relatedness ($r = 0.5$ for siblings, $r = 0.125$ for cousins, and so on), and A is a count of the individuals affected by the trait of interest (e.g., those that hear the alarm call and head to safety; Grafen, 1984). In other words, the decision to aid family members is a function of how related individuals are, and how high or low the costs and benefits associated with the trait turn out to be. When genetic relatedness is high, then r times b is more likely to be greater than c than when genetic relatedness is low. What this means is that natural selection more strongly favors kin helping one another when r is high. In addition, as the benefit that recipients obtain (b) increases, and/or the cost (c) to the donor decreases, the probability that r times b is greater than c increases—in other words, natural selection should strongly favor kin helping one another when b is high and/or c is low. Finally, as A—the number of relatives helped by an act of altruism—increases, selection more strongly favors altruism.

Inclusive fitness theory has had a profound impact on the work of ethologists, behavioral ecologists, and comparative psychologists. Moreover, the impact of these ideas has been even greater as a result of Jerram Brown's reformulation of Hamilton's equation. Fieldworkers in animal behavior had found the b and c terms of Hamilton's model difficult to measure in nature, but Brown solved the problem by coming up with the "offspring rule," which used the number of offspring that were born and survived as the currency of measure (J.L. Brown, 1975). This formulation set up the possibility of field manipulations in which Hamilton's and Brown's ideas could be tested by counting the number of offspring across different experimental treatments. For example, if an ethologist wanted to know the positive effects that young "helpers-at-the-nest" might have on raising their siblings, she could examine the difference in the average number of chicks that survive in the presence and absence of such helpers (J.L. Brown et al., 1982; Figure 9.7). In terms of measuring the costs to the helper of helping, ideally ethologists would measure the number of offspring produced by individuals that did not help versus those that did help. All else being equal, the difference between these values would allow for an estimation of the cost of helping.

FAMILY DYNAMICS

While Hamilton's Rule makes some very general predictions about animal social behavior, subsequent work by animal behaviorists and behavioral ecologists has generated more specific predictions about what can be called "family dynamics" (S. Emlen, 1995b). In particular, Stephen Emlen has developed an "evolutionary theory of family" that aims to test specific predictions regarding "the formation, the stability, and the social dynamics of biological families" (S. Emlen, 1995b, p. 8092).

The building blocks for Emlen's work on family dynamics are (1) inclusive fitness theory; (2) ecological constraints theory, which examines dispersal options of mature offspring, and specifically the conditions that favor dispersal from home rather than remaining on a natal territory (J.L. Brown, 1987; S. Emlen, 1982a, 1982b; Koenig and Pitelka, 1981; Koenig et al., 1992); and (3) reproductive skew theory, which examines how reproductive opportunities are divided among potential breeders by predicting conditions that should favor conflict or cooperation with respect to breeding decisions (R. Johnstone, 2008; Nonacs and Hager, 2011; Shen-Feng et al., 2011; Figure 9.8).

Emlen has made fifteen specific predictions about animal family dynamics, and for each of these, he reviewed the evidence from the animal literature, both for and against his predictions (S. Emlen 1995b; Table 9.2). Two years after publication of Emlen's paper, Jennifer Davis and Martin Daly tested Emlen's fifteen predictions as they relate to human families (J. Davis and Daly, 1997).

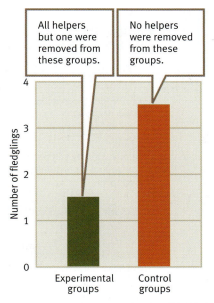

FIGURE 9.7. The effects of helping kin. In grey-crowned babblers (*Pomatostomus temporalis*), reproductive success, as measured by the number of fledglings, was significantly lower in the experimental groups because they had fewer helpers. Helpers increased the reproductive success of others—their kin—in their group (*Based on Brown et al., 1982*)

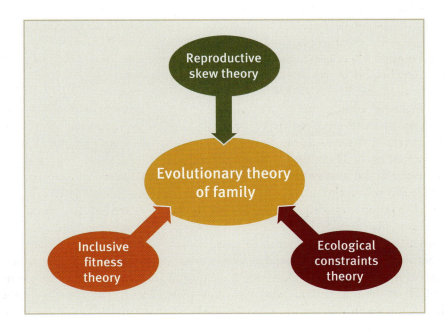

FIGURE 9.8. Evolutionary theory of family. Emlen's evolutionary theory of family is generated by combining inclusive fitness, reproductive skew, and ecological constraints theory.

TABLE 9.2. Predictions generated by the evolutionary theory of the family model. The table lists the fifteen hypotheses associated with Emlen's evolutionary theory of the family. *(From Emlen, 1995b, p. 8093)*

NO.	ABBREVIATED PREDICTION	EVIDENCE
1	Family groupings will be unstable, disintegrating when acceptable reproductive opportunities materialize elsewhere.	**Supportive:** 7 avian species; 2 mammalian species
2	Family stability will be greatest in those groups controlling high-quality resources. Dynasties may form.	**Supportive:** 5 avian species
3	Help with rearing offspring will be the norm.	**Supportive:** 107 avian species; 57 mammalian species **Counter:** 5 avian species; 6 mammalian species
4	Help will be expressed to the greatest extent between closest genetic relatives.	**Supportive:** 5 avian species; 3 mammalian species **Counter:** 1 avian species
5	Sexually related aggression will be reduced because incestuous matings will be avoided.	**Supportive:** 18 avian species; 17 mammalian species **Counter:** 1 avian species; 3 mammalian species
6	Breeding males will invest less in offspring as their certainty of paternity decreases.	**Supportive:** 1 avian species (many additional studies, some supportive, others counter, have been conducted on nonfamilial species)
7	Family conflict will surface over filling the reproductive vacancy created by the loss of a breeder.	**Supportive:** 6 avian species
8	In stepfamilies, sexually related aggression will increase because incest restrictions do not apply to replacement mates. Offspring may mate with a stepparent.	**Supportive:** 4 avian species
9	Replacement mates (stepparents) will invest less in existing offspring than will biological parents. Infanticide may occur.	**Supportive:** 2 avian species; 2 studies summarizing mammalian data
10	Family members will reduce their investment in future offspring after a parent finds a new mate.	**Supportive:** 2 avian species **Counter:** 2 avian species
11	Stepfamilies will be less stable than biologically intact families.	No data available
12	Decreasing ecological constraints will lead to increased sharing of reproduction.	**Supportive:** 2 avian species
13	Decreasing asymmetry in dominance will lead to increased sharing of reproduction.	**Supportive:** 2 avian species; 1 mammalian species
14	Increasing symmetry of kinship will lead to increased sharing of reproduction.	**Supportive:** 4 avian species
15	Decreasing genetic relatedness will lead to increased sharing of reproduction. Reproductive suppression will be greatest among closest kin.	**Supportive:** 5 avian species; 2 mammalian species

Nonbreeding Groups and Inclusive Fitness Benefits in Gorillas

The Louke population of western lowland gorillas (*Gorilla gorilla gorilla*) in the Congo is made up of approximately 400 individuals. Over the course of their lives males occupy three social positions: (1) a solitary male; (2) a member of a nonbreeding group (NBG), which contains younger individuals (usually males) and often one older, "silverback" male; and (3) a member of a breeding group (BG), in which they are the lone mature male and the remainder of the group are adult females and sexually immature males.

The gorillas in this Louke population are individually recognizable (primarily by fur patterns), and many have been genotyped (from dung samples). This population of gorillas, along with others, has been studied for decades, and although the social dynamics of breeding groups in the wild is fairly well understood, little is known about NBGs (Fossey, 1983; A. Harcourt, 1978; Robbins, 1996). What is known is that males shift from being solitary to being part of NBGs and BGs (Figure 9.9). Florence Levrero and her team were interested in whether there might be inclusive fitness benefits associated with being part of an NBG, both for immature males in the group and for the single male silverback in an NBG (Levrero et al., 2006).

To examine whether inclusive fitness benefits were important in the formation and stability of NBGs, Levrero and her colleagues first determined levels of genetic relatedness between immature males in NBGs. They found no evidence that males preferentially joined or remained in groups in which other non-breeding males were their relatives. Males did, however, show a strong preference for joining NBGs that contained a silverback male, as such groups tend to be safer and associated with more food than NBGs with no silverback.

Among NBGs that contained a silverback, immature males preferred to join groups in which they were related to the silverback. Silverbacks in NBGs, then, receive indirect benefits by providing food and protection to their genetic relatives, many of whom will go on to form their own breeding groups later in life. Indeed, in some populations of gorillas, there is evidence that the silverback in an NBG preferentially provides support to relatives in his NBG when such relatives are in aggressive interactions with those not related to the silverback in that group (D. P. Watts, 1990).

While Levrero and her colleagues were able to study the inclusive fitness benefits to silverbacks in NBGs, the inclusive benefits to young males joining an NBG group that contains a related silverback are unclear. It may be that the benefits are passive, in that such emigrating young males find that NBGs with a related silverback are simply easier to join. For example, young males may encounter less resistance when attempting to join these groups than when trying to join NBGs that contain silverbacks to whom they are not related.

Gorilla populations are dwindling quickly and are severely endangered. Understanding the role of the indirect fitness benefits that silverback males in NBGs receive may help provide some guidance when developing conservation plans for these populations, both in the wild and in captivity. When managing such populations, attempts to manipulate NBGs in any way that undermines their social structure may interfere with the inclusive fitness benefits that the silverback in such groups normally receives.

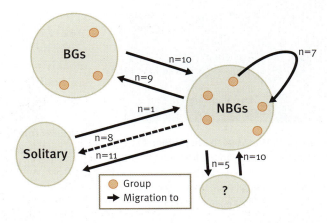

FIGURE 9.9. Group structure of lowland gorillas. (A) A group of lowland gorillas. **(B)** BG = breeding group, NBG = nonbreeding group, ? = unknown group structure. Over the course of their lives, most males will be part of all three group structures. *(Photo credit: Christophe Courteau/naturepl.com; from Levrero et al., 2006)*

Whereas Emlen's data are from a wide variety of animals, Davis and Daly's analysis is necessarily restricted to one species—*Homo sapiens*. Most of their data came from the Canadian General Social Survey, a telephone survey that amassed information on family dynamics in 13,495 households. Such a survey is probably reflective of modern Western society, but it is important to recognize that it does not necessarily represent all societies.

A review of the papers by Emlen and by Davis and Daly provides us with a unique opportunity to examine and test evolutionary theories of family in both humans and nonhumans. We will examine a subset of three of Emlen's predictions (his predictions numbered 1, 2, and 4) in more detail. These three predictions were chosen to show the diversity of issues that kinship touches upon within animal and human behavior.

PREDICTION 1. "Family groupings will be unstable, disintegrating when acceptable reproductive opportunities materialize elsewhere."

This prediction focuses on costs and benefits associated with family life. Broadly speaking, individuals who have a higher inclusive fitness when remaining with their family should stay as part of the family unit, while those who have opportunities for increasing their inclusive fitness elsewhere should depart (see Conservation Connection box). Evidence in support of this prediction in animals comes from many studies of birds and mammals.

One technique for experimentally examining prediction 1 is to create new, unoccupied territories and examine whether mature offspring leave their natal area to live in such newly created areas (Komdeur, 1992; Pruett-Jones and Lewis, 1990; Walters et al., 1992). To see how such an experiment is undertaken, consider Stephen Pruett-Jones's work with superb fairy wrens (*Malurus cyaneus*), an insectivorous (insect-eating) Australian bird species (Pruett-Jones and Lewis, 1990; Figure 9.10). In superb fairy wrens, a breeding pair is often helped by its nonbreeding young male offspring, which provide their siblings with such resources as additional food and protection. In contrast, female superb fairy wrens emigrate from their natal territory and do not help raise siblings at their parents' nest. To test the prediction that families will break down when suitable territories emerge for young helper males, Pruett-Jones and Lewis removed the breeding males from twenty-nine superb fairy wren territories.

By removing breeding males from their natal territories, new breeding opportunities arose for male helpers in nests in new areas that were near the area of the removals. All but one of the thirty-two potential male helpers that could have dispersed to the newly opened territories did so, and they did so quickly—new territories were usually occupied by former male helpers within six hours. But why did males immediately leave home when reproductive opportunities emerged? A shortage of females and breeding territories created a scenario in which reproductive opportunities were exceedingly rare, so male helpers quickly seized the opportunity for a breeding territory and thus disbanded family life when the chance arose (Figure 9.11). Pruett-Jones and Lewis's work suggests that helping-at-the-nest may raise the inclusive fitness of young males when territories are limited, but not otherwise.

The picture is not as clear-cut when it comes to testing prediction 1 in humans. In their analysis of the data from the Canadian General Social Survey, Davis and Daly found that married individuals were much more likely to live away from their parents than were single individuals in the same age/sex

FIGURE 9.10. Superb fairy wren. In superb fairy wrens, young males often act as helpers-at-the-nest. When breeding males are removed from their territories, almost all potential male helpers that could have dispersed to newly opened territories did so. *(Photo credit: Graeme Chapman)*

FIGURE 9.11. Family breakup. In the superb fairy wren, male helpers often assist their parents. If a vacant territory opens up, however, male helpers are quick to leave the family unit and attempt to start their own family.

category. This suggests that new marriages—that is, new opportunities for reproductive success—cause existing family units to dissolve. It is important to understand that it need not have turned out that way. Davis and Daly *might* have found that married individuals were *more* likely than single individuals to live with one set of parents, but instead the data supported Emlen's first prediction.

While the above data on dispersal and residence patterns suggest that marriage causes the dissolution of existing family units, while creating other new family units, prediction 1 was not supported when Davis and Daly used another set of data to test this prediction. When they examined whether married and single individuals living away from their parents differed in terms of contact with parents or grandparents—differences we might expect if marriage did break up already existing families—very few differences were uncovered. For most age/sex categories, married individuals living apart from either set of parents were just as likely to stay in contact via phone, visits, and letters with parents and grandparents as were single individuals living away from home, in clear contrast to prediction 1.

Davis and Daly tested prediction 1 in other ways as well, and they argue that as a whole, the data from the Canadian GSS do not support Emlen's first prediction. Rather, Davis and Daly believe that, with some exceptions, it

FIGURE 9.12. Dynasty building in acorn woodpeckers. In cooperatively breeding acorn woodpeckers (*Melanerpes formicivorus*), young birds not only survive better on territories with more storage holes but are also more likely to remain on their natal territories throughout their life, creating a "family dynasty." *(Photo credit: Steve and Dave Maslowski/Photo Researchers, Inc.)*

appears that human parents act as post-reproductive helpers to their own offspring, which may select for strong family bonds that do not easily dissolve when offspring get married.

PREDICTION 2. "Families that control high-quality resources will be more stable than those with lower-quality resources. Some resource-rich areas will support dynasties in which one genetic lineage continuously occupies the same area over many successive generations."

Inclusive fitness theory predicts that individuals may remain in their natal territory if there are enough resources for them to mate and provide for their own offspring. That is, if the benefits associated with remaining on a natal territory are sufficiently great—lots of food and the space to attract a mate and breed, for example—then those benefits, in conjunction with the indirect benefits of helping relatives, create incentive for keeping families intact. However, individuals will tend to leave their families if there are not enough resources at their natal territories.

Emlen argues that offspring from families that control high-quality resources are likely to be much more reluctant to vacate the natal territory, as few alternative territories provide the resources that are available at home. Over the long run, this will create dynasties in families that occupy the very highest-quality territories (see Chapter 14). Not only are the offspring that remain on high-quality territories receiving a benefit, but their parents are as well, since they then pass down the best-quality territories to their genetic kin (J.L. Brown, 1974).

Data from six species of birds support the dynasty-building hypothesis in that birds from high-quality family territories are indeed less likely to disperse from the natal territory than their counterparts from families with inferior territories. For example, in cooperatively breeding acorn woodpeckers (*Melanerpes formicivorus*), the critical measure of territory quality is the number of storage holes (Koenig et al., 2011; Figure 9.12). In a New Mexican population of acorn woodpeckers studied by Peter Stacey and David Ligon, territories varied from less than 1,000 to greater than 3,000 storage holes for acorns.

Individuals on territories with many storage holes produced a greater average number of offspring (Stacey and Ligon, 1987; Figure 9.13). More critical to testing Emlen's prediction, in areas with more than 3,000 storage holes, 27 percent of the young remained on their natal territories and helped their relatives, while only 2 percent of the young on territories with fewer than 1,000 holes stayed and helped. The benefits of remaining on a high-quality territory appear to be real, as (male) birds that served as helpers had a relatively high probability of eventually entering the breeding population, often breeding in turn on their natal territory, either at the same time as their parents or after their parents had died (Stacey and Ligon, 1987).

In terms of human family dynamics, prediction 2 translates into the hypothesis that well-to-do families will be more stable than poorer families. Davis and Daly found that if a stable family is defined in terms of co-residence (as in the nonhuman case), then this prediction is not supported. To cite just one of Davis and Daly's examples, young adults from wealthy families tend to be *less* likely to be living with their parents than are same-age individuals

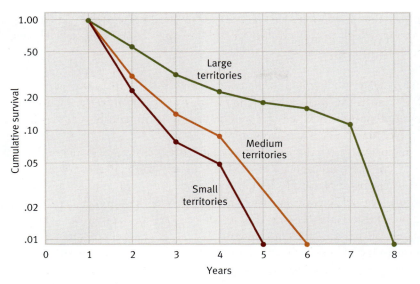

FIGURE 9.13. Territory quality and survival in acorn woodpeckers. Increasing territory size, and hence increasing number of storage holes, led to increased rates of survival. *(Based on Stacey and Ligon, 1987, p. 663)*

from poorer families (White, 1994). Nonetheless, since resources are much more mobile than ever in today's Western economies, it might be argued that familial co-residence is an inappropriate yardstick for measuring family stability. If the measure of stability is defined in terms of maintaining family contacts and providing social support during adulthood, the data are more supportive of prediction 2.

At the most general level, data suggest that contact and support are indeed found more often in wealthy families (Eggebeen and Hogan, 1990; Taylor, 1986; White and Reidmann, 1992). Davis and Daly used GSS data to address the more detailed question of whether contact with kin is not only more likely but more frequent, as a function of wealth. Using letter, phone, or face-to-face conversations as a measure of contact, they examined whether individuals in wealthier families kept in contact more often with parents, grandparents, and siblings than did individuals in poorer families. The GSS data suggest that for most age/sex cohorts, wealthier individuals did keep in touch with relatives more often than did lower-income individuals.

PREDICTION 4. "Assistance in rearing offspring (cooperative breeding) will be expressed to the greatest extent between those family members that are the closest genetic relatives."

Inclusive fitness theory suggests that, all else being equal, when given the choice between helping individuals that differ with respect to r (the coefficient of relatedness), more aid should be dispensed to the closest genetic kin than to more distantly related kin.

Most studies published on cooperation in birds or mammals that live in extended families find that individuals do extend aid as a function of genetic relatedness. For example, in white-fronted bee-eaters (*Merops bullockoides*;

FIGURE 9.14. White-fronted bee-eater kinship. Inclusive fitness models of behavior have been tested extensively in white-fronted bee-eaters. *(Photo credits: N. J. Demong)*

Figure 9.14), helpers chose to aid individuals they were most closely related to in 108 of 115 opportunities (Figure 9.15).

In addition to supporting a basic prediction of kinship theory, results from the study on bee-eaters helped resolve a thorny issue surrounding Hamilton's Rule. Beginning in 1975, a number of researchers had suggested that individuals should dispense altruistic aid to relatives in *direct proportion* to their genetic relatedness (Barash, 1975; West-Eberhard, 1975). Let's label this the "proportional altruism" model. For example, imagine that an individual has nine units worth of aid that it can dispense to relatives. Suppose then that this individual interacts with one sibling ($r = 0.5$) and one uncle ($r = 0.25$). Since siblings share an r value twice as great as that between uncle and nephew, the proportional altruism model predicts that six units of aid should be dispensed to the sibling and three units of aid should be dispensed to the uncle.

Stuart Altmann argued that the proportional model rested on faulty logic, because an individual always increases its inclusive fitness most when it is altruistic toward its closest genetic relative (Altmann, 1979). Instead, Altmann predicted that an individual should dispense *all* of its aid to the recipient that is its closest genetic relative (let's call this the "all-or-nothing" model). In our hypothetical case, Altmann's model predicts that all nine units should be dispensed toward the donor's sibling. In principle, Altmann is right, but the question is whether animals actually do behave in accordance with Altmann's predictions. Emlen's work on white-fronted bee-eaters enables us to answer this question, for it allows behavioral and evolutionary biologists to determine which of these two models better fits data gathered in the wild. In support of Altmann's model, Emlen found that helpers not only overwhelmingly chose to help their closest genetic relative, but that once a helper made a choice, it dispensed all of its aid toward the chosen individual (S. Emlen 1995b).

Many studies of kin-based cooperation and altruism have been done in eusocial (Chapter 2) insects like bees, ants, and wasps, which are part of the insect order hymenoptera. Hymenoptera have an odd genetic architecture that creates sisters that are "super relatives." These super relatives come about

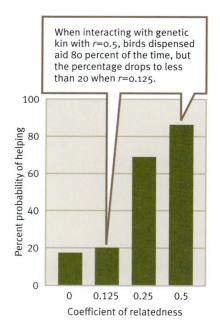

When interacting with genetic kin with $r=0.5$, birds dispensed aid 80 percent of the time, but the percentage drops to less than 20 when $r=0.125$.

FIGURE 9.15. Helping close relatives. In white-fronted bee-eaters, individuals are more likely to help those to whom they are more closely related (as indicated by r, the coefficient of relatedness). *(Based on S. Emlen, 1995a)*

because many social insects have a haplodiploid genetic system. Normally, we think of all individuals in a species as being either diploid (possessing two copies of each chromosome) or haploid (possessing only one copy of each chromosome). Haplodiploid species defy this convention in that males are haploid, while females are diploid.

As a result of the genetics underlying haplodiploidy, sisters are related to one another on average by a coefficient of relatedness of 0.75, which has the effect of making females more related to their sisters than to their own offspring. This value differs from the standard average relatedness of sisters in diploid species ($r = 0.5$), because in haplodiploids, full sisters inherit exactly the same alleles from their father, while in diploid species, females have only a 50 percent chance that an allele that they inherited from their father is identical to an allele that their sister inherited from their father. Not only are female social insects highly related to one another, but social insect colonies tend to have very high female : male sex ratios, leading to many females potentially interacting and helping one another (Trivers and Hare, 1976).

With an r of 0.75 between sisters, one would expect high levels of aid giving— just the sort of thing for which social insects are well known—and it is the highly related female workers in many species that go to suicidal lengths to defend a hive full of their sisters (for more on kinship and altruism in social insects, see Abbot et al., 2010; Herbers, 2009; Nowak et al., 2010; Ratnieks et al., 2011). A bee's stinger is designed for maximal efficiency, to the extent that the stinger is often ripped from the body of the stinging bee, causing it to die. Kinship need not, however, produce such ultra-altruism. If individuals are able to gauge their relatedness to others, then social insects may be influenced by kinship in any number of ways.

In the social insects, eusociality has evolved on at least nine separate occasions (W. Hughes et al., 2008). Eusociality in social insects is not *completely* explained by the high genetic relatedness that comes about because of their haplodiploid genetics. All hymenopteran species are haplodiploid, but only *some* hymenopteran species are eusocial, and there are also examples of eusociality in diploid species such as naked mole rats and termites. While haplodiploidy alone does not explain the evolution of eusociality, it does help explain, in part, why eusociality is overrepresented in social hymenopterans.

The hypothesis that high genetic relatedness is important to the evolution of eusociality in at least some hymenoptera can also be tested using phylogenetic analyses. Genetic relatedness is highest in social insect groups when queens are *monandrous*—that is, when they have a single mate. When females are *polyandrous* (see Chapter 8), the average genetic relatedness in groups goes down, as not all individuals in a group share the same father, so ethologists have predicted that eusociality in bees should often be associated with a monogamous mating system.

To test this prediction, William Hughes and his colleagues began by recognizing that eusociality has independently evolved five times in bees, three times in wasps, and once in ants (W. Hughes et al., 2008; Ratnieks and Helantera, 2009). Today we see both monandry and polyandry in these eusocial lineages. But Hughes and his colleagues hypothesized that for eusociality to have taken hold in these groups to begin with, their evolutionary histories should indicate that the ancestral mating system in most of these lineages was monandrous. A phylogenetic analysis of eight of the nine lineages (data were not available to test one lineage of bees) indicates that, as predicted by inclusive fitness theory, monandry was the ancestral state in *all* eusocial lineages examined (Figure 9.16).

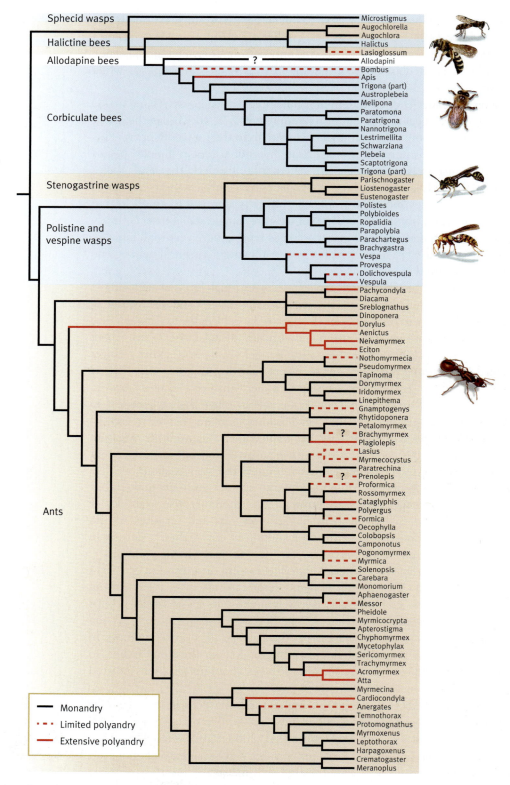

FIGURE 9.16. Phylogeny of ant, bee, and wasp species. Ethologists have predicted that eusociality in bees should often be associated with a monandrous mating system. The phylogeny shown here is for ants, bees, and wasps for which data on female mating frequency are available. Each independent origin of eusociality is indicated by alternately colored—blue or orange—clades. (A clade is a taxonomic grouping including an ancestral group and its descendants.) Cases of high polyandry are depicted by red branches, and completely monandrous groups are shown with black branches. All eight clades here have monandry as the ancestral state. *(Adapted from Hughes et al., 2008)*

A second example of how genetic kinship can influence behavior in eusocial insects can be seen in **worker policing** in honeybees (*Apis mellifera*), in which sterile worker bees use information associated with genetic relatedness to "police" their hive, and destroy eggs that are less related to them resulting in an increase to their inclusive fitness (Ratnieks and Visscher, 1989).

In honeybee hives, queens produce most of the offspring, but workers can also produce unfertilized eggs that always develop into males. Using the mathematics of inclusive fitness theory, Francis Ratnieks and P. Kirk Visscher found that in honeybee colonies with a single queen that mates one time, female workers are more related to their nephews (their sisters' sons, $r = 0.375$) than to their brothers (the queen's sons, $r = 0.25$; Ratnieks and Visscher, 1989). But this inequality switches when the queen mates multiple times. And indeed, honeybee queens typically mate with ten to twenty different males. When multiple mating takes place, workers may be more closely related to brothers (males produced by the queen) than to nephews (males produced by their sister workers), with the exact values of relatedness depending on the number of different males with whom a queen mates. Under such conditions— when female workers are more related to brothers than to nephews—Ratnieks has hypothesized that worker policing of honeybee reproduction may evolve (Ratnieks and Visscher, 1988, 1989). Such policing, for example, may take the form of workers favoring those eggs to which they are most highly related (Figure 9.17).

Ratnieks and Visscher examined the possibility that honeybee workers may favor brothers over nephews. They found that honeybee workers showed remarkable abilities to discriminate between worker-laid eggs, which produce nephews, and haploid queen-laid eggs, which produce brothers. After twenty-four hours, only 2 percent of the worker-laid eggs remained alive, while 61 percent of the haploid queen-laid eggs remained alive (Figure 9.18). Workers appear to use a specific egg-marking pheromone produced only by queens to distinguish which eggs to destroy and which eggs to leave unharmed, and in so doing, they police the hive in a manner that increases their inclusive fitness (Ratnieks, 1995; Ratnieks and Visscher, 1989).

A **B**

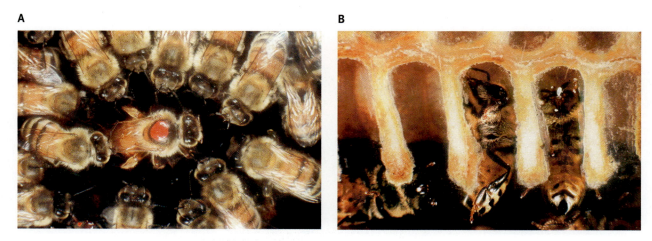

FIGURE 9.17. Honeybee policing. (A) While the queen (designated by the red dot on her back) typically lays the eggs in a honeybee colony, workers also attempt to lay unfertilized eggs. **(B)** When an egg laid by a worker is detected by worker police, it is eaten or destroyed. Workers are much more likely to destroy eggs produced by other workers than eggs produced by the queen. Such "policing" has inclusive fitness benefits associated with it. *(Photo credits: Francis Ratnieks)*

FIGURE 9.18. Worker policing in honeybees. In honeybees, where queens often mate with ten to twenty males, workers are more related to the male offspring of the queen (their brothers) than to offspring of other workers (their nephews). Workers police the hive and search out and eat the eggs of other workers. *(From Ratnieks and Visscher, 1989)*

Tom Wenseleers and Ratnieks extended the logic of policing behavior to further explore the relationship between kinship and reproduction in insects (Wenseleers and Ratnieks, 2006; Figure 9.19). If policing was effective at removing the eggs laid by workers, they hypothesized that it should create strong selection pressures against worker reproduction in the first place. They tested this idea by examining policing behavior in ten species—nine species of wasps and the honeybee. They found that the more effective policing was at removing worker eggs, the less often workers attempted to reproduce in the first place (Figure 9.20).

Unfortunately, in Davis and Daly's examination of prediction 4 in humans, the GSS data were not collected in a way to address this question. For the most part, individuals in the GSS study were either related by an r value of 0.5 or 0.0, and therefore the distinction between how different relatives—that is, individuals with different positive values of r—are treated could not be

FIGURE 9.19. Wasp policing. In the wasp *Dolichovespula saxonica*, workers often lay (haploid) eggs, in nests with both single-mated and multiply mated queens. Such eggs are often eaten when detected by other workers. **(A)** The wasp in the middle of the photo is a worker that has just laid an egg. **(B)** Here a worker is eating another worker's egg. Policing is much more common in wasp colonies where the queen has mated with many males. *(Photo credits: Kevin Foster)*

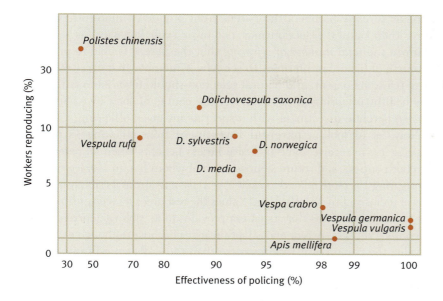

addressed. Indirect evidence for prediction 4, however, can be found in studies of divorce. When males believe they have a low probability of being the genetic father of their ex-spouse's children, they decrease the amount of resources they invest in those children (Anderson et al., 2007).

Conflict within Families

Most often, inclusive fitness theory is used to understand why relatives so often cooperate with one another. But inclusive fitness theory can also be used to study conflicts within families. To examine this phenomenon more closely, we now turn to the subjects of parent-offspring conflict and sibling-sibling conflict.

PARENT-OFFSPRING CONFLICT

Inclusive fitness theory predicts that parents should go to great lengths to help their offspring because parents and offspring have an average *r* of 0.5. Furthermore, parents are almost always in a better position to help offspring than vice versa. As such, parental aid should be seen in many contexts. And indeed it is. Hundreds of studies have shown that parents, mothers in particular, provide all sorts of aid to their offspring.

Yet there are limits to this aid, as first conceptualized by Robert Trivers in his parent-offspring conflict theory (Trivers, 1974). This theory recognizes that **parent-offspring conflict** arises with respect to a parent's decisions about how much aid to give to any particular offspring. From the perspective of the parent, these decisions are affected by how much energy is available for helping current offspring, and by how many offspring it is likely to have in the future.

In principle, a parent could dispense every ounce of energy it has to provide offspring 1 with all the benefits at its disposal. But if such an effort kills the parent or severely hampers the parent from producing more offspring in the future, then natural selection may not favor such behavior, as it might not maximize the *total number of offspring* that the parent is able to produce over the course of his or her lifetime. To see why, remember that every offspring has an *r* of 0.5 to its parent, and natural selection should favor parents that raise as many healthy offspring as possible over the course of their lives. So, there are limits on parental investment with respect to any given offspring.

Now, let us look at **parental investment** from offspring 1's perspective. Offspring 1 will receive some inclusive fitness benefits when its parent provides aid to both current and future siblings, each of whom has an average *r* of 0.5 to it. Yet, offspring 1 is more related to itself ($r = 1$) than to any of its siblings. As such, in terms of inclusive fitness, offspring 1 values the resources it receives from its parent more than the resources that its parent provides to its siblings (current or future). The conflict between parent and offspring arises because, although each offspring will value the resources it receives more than those dispensed to its siblings, all offspring are equally valuable to a parent, in terms of the parent's own inclusive fitness. This then sets up a zone of conflict between how much offspring 1 want, and how much a parent is willing to give (the former always being greater than the latter). This zone is where parent-offspring conflict takes place (Figure 9.21).

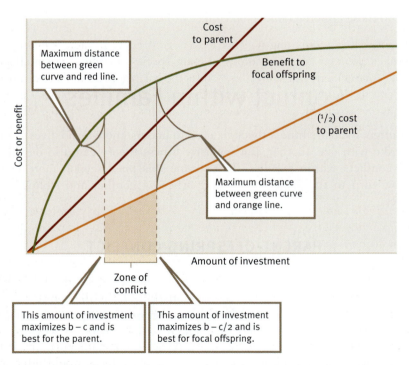

FIGURE 9.21. Parent/offspring conflict. Parents can provide resources to a "focal" offspring or use those resources on other current or future offspring. The *x*-axis shows the resources invested in the focal offspring, and the *y*-axis shows fitness costs (c) or benefits (b). The parent is equally related to all of its offspring, but the focal offspring is only half as related to its full siblings as it is to itself. As a result, parent and offspring prefer different amounts of resource allocation. This zone of conflict is shaded in the figure. To the left of the zone, parents and offspring alike benefit from increasing allocation to the offspring. To the right of this zone, parents and offspring alike benefit from decreasing allocation to the offspring.

PARENT-OFFSPRING CONFLICT AND MATING SYSTEMS IN PRIMATES. The degree of parent-offspring conflict predicted is in part a function of the mating system (see Chapter 8) that exists in a population (Hain and Neff, 2006; Long, 2005). To see why, recall that natural selection favors offspring that weigh (1) the inclusive fitness benefits associated with receiving continued parental assistance versus (2) the inclusive fitness benefits of curtailing the degree of parental assistance received, and leaving a parent with more resources to produce future offspring.

The degree of relatedness between current offspring and future offspring is not fixed, but rather is a function of the mating system. In long-term monogamous species, current offspring and future offspring will have an average genetic relatedness of $r = 0.5$, because they are likely to have the same mother and the same father. But suppose the mating system is polyandrous (see Chapter 8), so that a female mates with many males. Then the genetic relatedness between current and future offspring will be somewhere between 0.5 (for full siblings) and 0.25 (for half-brothers or half-sisters). Compared with the case of monogamous mating systems, in polyandrous mating systems, natural selection will favor offspring that attempt to extract more in the way of parental assistance. Parent-offspring conflict should then be more intense in polyandrous versus monogamous mating systems (Macnair and Parker, 1978; Mock and Parker, 1997; G. Parker and Macnair, 1979; Trivers, 1974).

Tristan Long hypothesized that offspring will attempt to extract more resources from parents in polyandrous systems than in monogamous systems. He tested his hypothesis by examining whether fetuses grew faster in utero—taking more maternal resources—in polyandrous primate species. In utero parent-offspring conflict is particularly fascinating, as it shifts the balance of power between parent and offspring. In most cases of parent-offspring conflict, a mother has the upper hand, as she is almost always behaviorally dominant to her offspring. When the offspring is still in utero, however, it is more difficult (but not impossible) for mothers to deprive offspring of resources without depriving themselves too, thus shifting the balance of power away from the mother and toward the developing fetus.

To examine this possible in utero parent-offspring conflict, Long used the independent contrast phylogenetic method discussed in Chapter 2 (Felsenstein, 1985, 2004). Long asked whether, if he controlled for phylogenetic effects, strong parent-offspring conflict would be more likely to occur in polyandrous or in monogamous primate species (Mastripieri, 2002).

Long began by using a well-established phylogenetic tree for primates. From this tree, he was able to find sixteen pairs of primates to use in his independent contrast analysis. Each pair was made up of species that had diverged from a recent common ancestor—one of these species was monogamous, and the other was polyandrous. Long then compared already published data on fetal growth rates for each of the species in his pairwise comparison (Long, 2005). He predicted that in polyandrous mating systems, a fetus would attempt to sequester more resources during development, and would show faster rates of growth than fetuses in species that were monogamous. Long's independent contrast analysis found just such a relationship.

Long also examined how mating systems were connected to parent-offspring conflict in a slightly different way. Because sperm competition (see Chapter 8) is more intense in polyandrous species, males in such species tend to have larger testes. Testes size, then, can often be used as a proxy for the degree of polyandry. When Long examined the relationship between testes size and parent-offspring

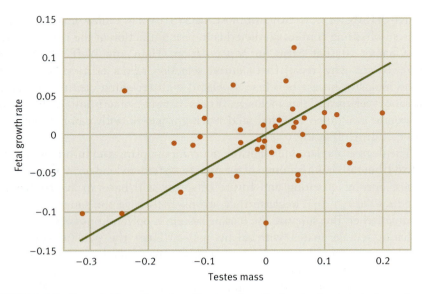

FIGURE 9.22. **Parental investment and testes size.** Testes size tends to be larger in males from polyandrous versus monogamous species. The relationship between testes size (a proxy measure of polyandry) and parent-offspring conflict (measured by fetal growth rate) was positive in Long's analysis of primates. The *x*- and *y*-axes measure residual log values of testes size and fetal growth rate, respectively. *(Based on Long, 2005)*

FIGURE 9.23. Mothers and babies.
While the parent-offspring relationship is usually cooperative **(A)**, parent-offspring conflict can occur, even in utero **(B)**.
(Photo credits: Ariel Skelley/Corbis; Science VU/Visuals Unlimited)

conflict (measured by fetal growth rate), his phylogenetic analysis again found a positive relationship, demonstrating how parent-offspring conflict can be mediated by the type of mating system in place (Figure 9.22).

IN UTERO CONFLICTS IN HUMANS. Parent-offspring conflict may also occur in humans (Geary, 2000; Haig, 1993; Schlomer et al, 2011; Figure 9.23). Parent-offspring conflict in pregnant women occurs because mother and fetus do not have identical interests in terms of how to maximize inclusive fitness. Using published medical literature, Haig argues that, in humans, fetal cells have invaded the maternal endometrium—the membrane lining the mother's

A

B

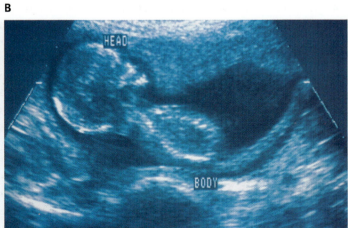

uterus—during implantation, and that such cells manipulate maternal spiral arteries in such a way as to make constriction of the arteries, which would make fewer resources available to the fetus, much more difficult. Such an action benefits the fetus in two ways: (1) by providing the fetus with direct access to maternal arterial blood and allowing the fetus to release hormones and other substances directly into the maternal bloodstream; and (2) by putting the volume of blood—and the nutrients it contains—under fetal, rather than maternal, control.

Haig suggests that placentally produced hormones, such as human placental lactogen and human chorionic gonadotropin, change the in utero environment in a manner that benefits the fetus at the cost of the mother. For example, a fetus may use human placental lactogen to manipulate insulin in such a manner that sugar would remain in the blood a longer time than normal. This manipulation would provide the fetus with more time to access such sugar for itself. The maternal counterresponse is to increase the production of insulin. If this countermeasure is unsuccessful, the fetus obtains extra sugar, but the mother suffers from gestational diabetes (Wells, 2007).

Gestational diabetes may thus be a possible outcome of parent-offspring conflict, and how it is viewed may have serious implications for the treatment of pregnancy-related medical conditions. If, for example, medical doctors viewed gestational diabetes as a "disease" that needed to be cured, they might act differently than if they viewed gestational diabetes as an evolutionary measure selected by fetal genes to increase sugar flow to the fetus (T. Moore, 2012). These sorts of issues are being studied by researchers in the field of evolutionary medicine (Ewald, 2000; Nesse and Williams, 1995; Nesse et al., 2010, Stearns et al., 2010).

SIBLING RIVALRY

Most readers are probably familiar with **sibling rivalry** from basic psychology classes, but animal behaviorists have been fascinated with such rivalries as well, and they have developed a substantial empirical and theoretical literature on this subject (Mock and Parker, 1997). The logic underlying the models of sibling rivalry is similar to that of parent-offspring conflict.

Mathematical models of sibling rivalry often consider sibling rivalry among many siblings, but here we focus on the evolution of this sort of behavior when only pairs of siblings are involved. Consider two siblings that we will label sib 1 and sib 2, who share an r of 0.5. Such genetic relatedness means that what is good for sib 1 is *usually* good for sib 2, but it also means that in a situation in which resources are limited and the siblings must compete for these resources, each individual will act as if it is more important to receive resources for itself than to have the same resources go to its sibling (Figure 9.24). We have adopted this sort of logic in our discussion of parent-offspring conflict, but here the competition is directly between siblings in a clutch of offspring, rather than between parent and offspring per se.

Imagine an extreme environment in which there is only enough food for one sibling to survive. Because of their genetic relatedness to one another, each sibling values the other at a level that is half of that at which it values itself. In such a resource-poor environment, we would expect intense, perhaps even lethal, competition to emerge among siblings. In a less harsh environment, we would expect sib-sib interactions to be less competitive, but because each values

Abundant resources

Scarce resources

FIGURE 9.24. Sib-sib conflict. Kin selection theory predicts that individuals should not be very aggressive toward kin such as sibs, especially when there are abundant resources. But if there are limited resources, conflict over the resources will increase, because each individual is more related to itself ($r = 1$) than to its sib ($r = 0.5$).

FIGURE 9.25. Sib-sib competition in birds. In nests of egrets, sib-sib competition can be intense and can result in the death of smaller, less dominant chicks. Sibling rivalry can be seen in the fights between siblings in the nest, as shown here, where the chick on the left is preparing to bite its sibling on the back of its head. *(Photo credit: Millard H. Sharp/ Photo Researchers, Inc.)*

itself more than the other, some level of competition should still be the norm, rather than the exception. We simply expect the rivalry to emerge in less lethal ways when resources are not as limited.

Sibling rivalry is illustrated nicely in studies of egrets by Douglas Mock and his colleagues (Mock, 2004; Mock and Parker, 1997). As idyllic as the interactions of downy chicks in a bird nest may seem to the casual observer, the interactions among egret chicks actually resemble prizefights more closely than some picturesque scene of nature taken from a Disney film (Figure 9.25). Consider the following summary of work on sib-sib interactions:

Sibling fights take many forms, depending mainly on how the loser concedes and how quickly it does so. The simplest fights, which usually occur while the participating dyad has had a series of increasingly one-sided battles, are those in which the attack inspires no retaliation. At the next level, return fire is brief until the loser is tagged with several unanswered shots and crouches low. From there, the severity of the beating is left largely to the victor's discretion. Sometimes it continues to jab at its opponent, causing the latter to screech and hide its face. As an alternative to jabbing, a dominant chick may seize the cowering victim by its head or neck, lift that part a few centimeters and then slam it down forcefully against the nest cup. If the attack persists for more than a few extra blows, the loser is likely to flee, sometimes squawking loudly and racing about the nest dodging behind the other nest occupants while being hit. During such chases, the primary target is the back of the head. Frequently bullied chicks soon develop a characteristic baldness, dotted with fresh and crusted blood, where the nape feathers have been plucked forcibly during fights. (Mock and Parker, 1997, pp. 103–104)

When egret chicks first hatch, parents bring back enough food to fill all the chicks' guts, minimizing sib-sib aggressive interactions. As the chicks grow, however, a point is quickly reached at which the food brought to the nest is insufficient to feed everyone to satiation, and intense competition among siblings emerges when a parent returns to the nest with food and regurgitates it in the nest. The key to obtaining the food is positioning within the nest, and specifically vertical positioning (the higher the better when mom returns). Even a chick tilting its head up above horizontal is a cue likely to spark aggression in egrets' nests.

Egrets, like most birds, hatch eggs asynchronously—that is, they lay their eggs in sequence, rather than all at one time. Thus, hatching order produces chicks that can differ in age by many days. Such age differences play a critical role in determining who emerges as the victor in sib-sib interactions, since chicks that hatch first start to feed sooner and hence receive more food, which leads to a weight advantage over chicks that hatch later. As a result, a very clear age-related dominance hierarchy exists among chicks. First-hatched chicks are often much larger than second-hatched chicks, who are often much larger than chicks hatched still later (Figure 9.26A). In sib-sib interactions, large size means better fighting ability, which translates into significantly more food (Figure 9.26B).

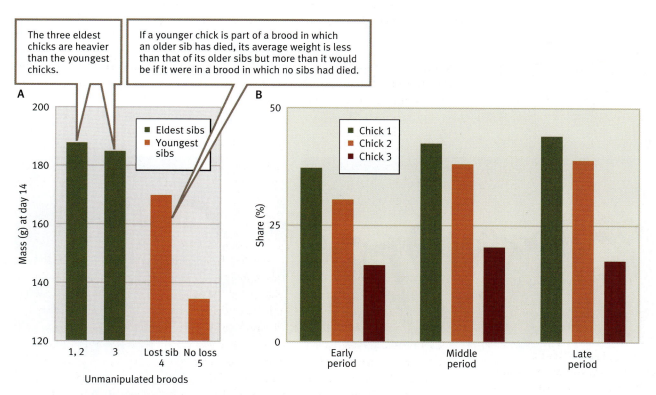

FIGURE 9.26. **Birth order and food intake.** **(A)** Normal broods of little blue herons include four to five chicks that are hatched asynchronously. **(B)** In egret broods, the oldest, dominant chick (1) receives more food than the middle chick (2), who in turn gets more than the youngest chick (3). This holds for the early period after hatching (1–13 days), the middle period (14–21 days), and the late period (21–30 days). *(Based on Mock and Parker, 1997)*

Kin Recognition

Given the power of kinship to affect social interactions, ethologists and behavioral ecologists have a long-standing interest in how kin recognize one another (Fletcher and Michener, 1987; Hepper, 1991; Hepper and Cleland, 1998; Holmes, 2004; Pfennig and Sherman, 1995). Early in this chapter, we went through a general procedure to show how to calculate relatedness (*r*) (for discussions of kin recognition in humans, see Bressan and Zucchi, 2009; Kaminski et al., 2009; Lieberman et al., 2007; Lundstrom et al., 2009). Of course, animal behaviorists don't assume that nonhumans are able to calculate genetic relatedness in that manner. We need only assume that natural selection favors individuals that act in a manner that makes it appear as though they are making such calculations.

Kin recognition in animals has often been studied in situations in which there are important fitness benefits for recognizing kin, but in which the task of kin recognition is difficult—for example, when individuals live in very large groups. Consider the remarkable kin recognition abilities seen in some species of penguins. Parents travel long distances to the sea to obtain food to take back to the inland areas where their chicks have hatched. When parents return from their journey, they must find their young among scores—sometimes thousands—of screaming, hungry chicks in a colony (Aubin and Jouventin, 2002; Brumm and Slabbekoorn, 2005). How do parents returning from a foraging bout (that might last weeks) reunite with their own young? For species like the emperor penguin (*Aptenodytes forsteri*) and the king penguin (*Aptenodytes patagonicus*) (Figure 9.27), the answer appears to center on complex vocal cues that allow for kin recognition via "vocal signatures" emitted by the young (Aubin and Jouventin, 1998; Aubin et al., 2000; Jouventin et al., 1999; Lengagne et al., 2000).

A

B

FIGURE 9.27. Kin recognition in penguins. Kin recognition via vocal signatures has been examined in **(A)** the emperor penguin (*Aptenodytes forsteri*) and **(B)** the king penguin (*Aptenodytes patagonicus*). Both species of penguins live in large colonies, and parents returning from foraging with food for their chicks use vocal cues to find their offspring in the middle of many other chicks. *(Photo credits: Hans Reinhard/Photo Researchers, Inc.)*

Yet not all penguin species are as proficient as the king and emperor penguins at recognizing the vocal signatures of their offspring. Studies indicate that penguins that build nests are not as adept at recognizing the vocal calls of their young as are individuals that live in dense colonies and do not nest (Jouventin and Aubin, 2002; Searby et al., 2004). But why? Parents in nest-building species can find their offspring by remembering the location of their nests, presumably because any chick in their nest is their offspring (see below), and hence natural selection to recognize offspring by vocal cues in these species is weak. When the problem of kin recognition is more difficult—in dense colonies with no nests—natural selection favors the evolution of more complex vocal recognition systems.

MATCHING MODELS

Many models of kin recognition center on individuals having some "internal template" against which they match others and gauge relatedness (Reeve, 1989). These **kin recognition matching models** differ in their specifics, but the basic idea is that individual 1 attempts to assess whether individual 2 is kin or nonkin, depending on how closely individual 2 matches the internal template of individual 1. The internal template may be generated genetically, via learning, or via social learning, but in all cases, the animal estimates the degree of kinship as some function of the extent to which others match its own template (Alexander, 1979, 1991; Boyse et al., 1991; Crozier, 1987; S. Robinson and Smotherman, 1991). Templates can range from dichotomous "kin/nonkin" classification systems to more graded systems of kinship, in which individuals can distinguish among kin at a finer level (sibling, cousin, and so on).

TEMPLATE MATCHING IN TADPOLES. David Pfennig and his colleagues have studied template matching in the cannibalistic behavior of spadefoot toad tadpoles (*Scaphiopus bombifrons*; Elgar and Crespi, 1992; Pfennig 1999; Pfennig et al., 1993, 1999; Figure 9.28). Two feeding morphs of spadefoot toads exist: Juveniles that feed on detritus (small, often drifting vegetative clumps of food)

FIGURE 9.28. Tadpole cannibals.
As in the spadefoot toad, two different tadpole morphs—a carnivorous cannibal and an herbivorous omnivore—exist in a number of amphibian species. Here a tiger salamander (*Ambystoma tigrinum*) cannibal morph (right) is eating an omnivore morph (left). (*Photo credit: David Pfennig*)

typically develop into herbivorous omnivores, while those that feed on shrimp tend to mature into carnivorous cannibals.

Pfennig and his team examined kin recognition abilities in the herbivore and cannibal spadefoot morphs by testing both morphs in the presence of either *unfamiliar* siblings or *unfamiliar* nonrelatives. When visual cues (behavior and morphology, for example) and chemical cues (odors, for example) were both in play, herbivores preferred associating with their siblings over unrelated individuals, presumably because of the inclusive fitness benefits associated with interactions with genetic kin. Carnivorous individuals that cannibalize other tadpoles were taken from the same sibship as the omnivores that were being tested. When these cannibals were tested by Pfennig and his colleagues, they spent more time near unrelated individuals, presumably to avoid the costs of killing their genetic kin (Figure 9.29).

Pfennig and his colleagues also offered carnivores a choice between unfamiliar siblings and unfamiliar nonrelatives in a protocol that allowed carnivores to actually eat other tadpoles. Carnivores were not only more likely to eat unrelated individuals, but they were able to distinguish between relatives and nonrelatives by taste cues. That is, carnivores were equally likely to suck relatives and nonrelatives into their mouths, but they released their relatives much more frequently than they released unrelated individuals. Being able to recognize kin has clear advantages, as ingesting kin would generally be selected against whenever alternative food sources were available. But as the costs and benefits of eating kin change, Pfennig and his team predicted that tadpoles' behavior would change. And indeed, the researchers found that cannibalistic toads were much less picky when they had been starved for twenty-four hours or more—that is, when they were very hungry, they would occasionally eat even genetic kin (Figure 9.30).

MHC, KINSHIP, AND TEMPLATES. Recall from Chapter 7 that animals sometimes use potential partners' major histocompatibility complex (MHC) genes, which they identify by odor, to determine which mate to choose. MHC also plays a role in kin recognition (J. L. Brown and Eklund, 1994; Frommen

FIGURE 9.29. Kin recognition in spadefoot toads. Spadefoot toad tadpoles come in two morphs: carnivorous and herbivorous. Individuals from each tadpole morph were placed between two groups of tadpoles, one of which contained sixteen unfamiliar siblings, the other of which was composed of unfamiliar nonsiblings. *(Based on Pfennig et al., 1993)*

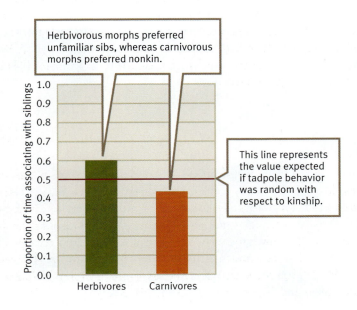

Herbivorous morphs preferred unfamiliar sibs, whereas carnivorous morphs preferred nonkin.

This line represents the value expected if tadpole behavior was random with respect to kinship.

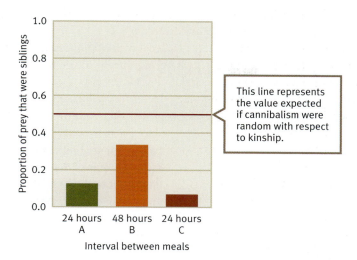

FIGURE 9.30. Hunger and carnivorous toads. The carnivorous morph of spadefoot toads prefers to eat nonkin over kin. When carnivorous morphs were starved for 24 hours (A), only a little more than 10 percent of individuals eaten were kin. If they were starved for 48 hours, this figure rises. As a control, toads were again starved for 24 hours (C), and results were similar to the original 24-hour deprivation treatment (A). *(Based on Pfennig et al., 1993)*

et al., 2007; Manning et al., 1992). Jo Manning and her colleagues examined MHC in house mice (*Mus musculus domesticus*), a species in which females nest together and nurse all offspring at their nest (Manning et al., 1992). When female mice nest together, they all receive a benefit, which is protection from infanticidal males that sometimes attack and kill offspring that are not their own (Manning et al., 1995). At the same time, communal nesting creates a situation in which females can be "cheated"—this occurs when other females at their nest are protected from danger but do not nurse all pups present. One way to minimize the cheater problem and to maximize inclusive fitness benefits would be for females to form communal nests with their genetic relatives. And because MHC differences are correlated with differences in odor, one way that females may discriminate among kin and nonkin is through odors associated with the MHC (Brennan and Kendrick, 2006; Packer et al., 1992).

Manning and her colleagues worked with six wild populations of house mice, individually marking each mouse and determining its MHC "haplotype" (similar to a genotype). They observed pregnant females and examined whether females that had just given birth opted to nest alone or in a communal nest. Ninety percent of the females chose to nest communally. With respect to kin recognition, when females selected which communal nests to join, they chose nests with individuals that had an MHC similar to their own. While these results do not definitively show that females use MHC as a cue for kinship, they are consistent with such a hypothesis.

RULE-OF-THUMB MODELS OF KIN RECOGNITION

As we discovered in the penguin studies discussed earlier, in species in which individuals do not raise their offspring in dense colonies, but instead live in discrete nests that are physically separated, a second form of kin recognition may evolve. In such scenarios, natural selection might favor a kin recognition rule of the form "if it lives in your nest/cave/territory, then treat it like kin" (Blaustein, 1983; Holmes and Sherman, 1983; Sherman and Holmes, 1985; Waldman, 1987). Such rules are referred to as "rules of thumb" (Houston et al., 2007; Hutchinson and Gigerenzes, 2005; Rands, 2011).

INTERVIEW WITH

Dr. Francis Ratnieks

Why has so much work on animal behavior and kinship been done in the social insects? How did you decide to work with this group?

Becoming a social insect biologist was literally plan B. On completing my BSc in Ecology at the University of Ulster in Northern Ireland, I decided to study pest management. I felt this would be useful and would combine my interests in insects and ecology. I applied to do a Ph.D. in the Department of Entomology at Cornell University, as this seemed the best place for studying insects. The professor of apiculture, the late Roger Morse, offered me a studentship. Although studying honeybees was something I had never thought of, I accepted. I quickly became enthusiastic about honeybees and beekeeping. More gradually, I developed interests and ideas about social evolution and behavior, and taught myself how to model social evolution. I also expanded into wasps and ants.

Social insects have been the most important group of organisms for testing predictions arising from William Hamilton's inclusive fitness theory. It has been a two-way street. Social insects have played a key role in validating the theory, and the theory has revolutionized our understanding of social insects, especially how eusociality evolved and the reproductive behavior of insect societies.

Social insects can be used to study practically any question. They are literally a gateway to biology. An immense number of important discoveries have been made with the

honeybee alone, which is only one of 20,000 social insect species. Social insects have been very useful for testing Hamilton's theory because relatedness, the theory's central parameter, varies both within and between species because queens can be mated to one or more males,

because there can be different numbers of queens per colony, and because haplodiploidy (in ants, bees, and wasps) causes relatedness to differ between the sexes. Insect colonies are also very practical to study. A colony of ants can be kept in a plastic box. A colony of honeybees can be kept in a hive. Honeybees are easy to study once you know how. Some honeybee studies are even based on decoding waggle dances by viewing the dancing bees through the glass walls of an observation hive. This is the only case in animal behavior where the animals "talk to" the researchers.

I often tell students that Hamilton's inclusive fitness theory is as important to the field of evolution and behavior as Einstein's e=mc² is to physicists. Is that an overstatement?

It is not an overstatement. Both are elegant and concise mathematical representations of a fundamental underlying relationship. Hamilton's Rule tells us the condition under which any behavior or trait that affects other individuals of the same species will be favored or disfavored by natural selection. Einstein's equation tells us the relationship between mass and energy. The relationship in Einstein's equation is inviolable. Technically speaking, Hamilton's Rule is not inviolable because gene frequencies, and therefore the traits they code for, can also be affected by genetic drift as well as by natural selection. Does that make Einstein's equation more important? I don't think so. It just reflects a difference between physics and biology.

How is it that a simple mathematical rule can be so important? The reason is simple. Many biological processes can be represented mathematically. For example, population growth can be represented by multiplication. The relatedness term in Hamilton's Rule comes from the simple fact that each gene in an organism has a precise probability of being passed on to an offspring. The probability is usually 0.5, but can be 1, for example, from a haploid male bee to his daughter.

Hamilton's Rule is also a good example of the importance of mathematics in biology. The mathematics is not hard. High school algebra is enough to understand it. The hard part for the student is combining the mathematics with the biology. The best way is to jump in and have a go!

Kinship isn't the only factor that promotes social behavior and altruism in animals—is it?

Kinship is very important. Consider the evolution of eusociality. The problem is to explain altruism—how can natural selection select for individuals to forgo reproduction to help others? Eusociality evolved within families, with offspring helping their parents rear more brothers and sisters. Helpers are as closely related to the individuals reared (brothers and sisters) as to their own sons and daughters. (Although many queen bees, wasps, and ants mate to several males, which diminishes relatedness, this evolved after eusociality.)

But if all that is needed for eusociality to evolve is high relatedness, why is eusociality not more common? Two other things are needed. First, a nest or some way of keeping the family together, so that help is directed to kin. Second, some way of helping, such as by providing food or by defending. Eusociality arose many times in the *Hymenoptera* because many species have nests to which the mother brings food for her offspring. Helpers can help simply by bringing more food. In termites food was not needed as the family was living inside its food—a log. Here, defense was the key.

In many modern-day insect societies, worker altruism is also caused by social pressure. Workers in most bee, wasp, and ant species have ovaries and can lay eggs. But in many such species, worker reproduction is rare. In the honeybee, fewer than 0.1 percent of the workers lay eggs.

Egg laying by workers is deterred by an effective policing system that kills worker-laid eggs. This means that worker honeybees are better off working rather than laying eggs, given that almost all their eggs will be killed if they lay any.

Do the terms *kinship* and *family* differ in meaning when discussed in ethology as opposed to when they are used by nonscientists in the course of normal conversation? How so?

I don't think there is much difference. Sometimes people will refer to other individuals as a brother or sister when they are not true relatives. But the people using these words probably know the difference. By referring to someone unrelated to you as brother or sister is often a way of showing that you have a common interest because you belong to the same group within society. The fact that we humans have what seems like a keen natural understanding of kinship suggests that it is important to us and is a human universal. That is, it is something innate in being human, rather than something that is purely cultural. Given the importance of relatedness in social behavior, this is not surprising.

Can you envision a day when sociologists and animal behaviorists will be using a common framework for studying kinship and behavior? What might such a framework look like?

A common interdisciplinary framework is something that is possible. However, even if this is established, the subjects and goals of the different disciplines may be sufficiently different that they may be using very different parts of a large and unwieldy framework. This is especially true when studying humans, given the vast number of disciplines involved, including history, economics, anthropology, sociology, political science, criminology, psychology, and biology. What insights would a historian studying the Tudors take from evolutionary biology, for example?

The value of a common framework can be seen when some important insight from one discipline is ignored in another discipline. For example, studies by evolutionary biologists Martin Daly and Margo Wilson have shown that kinship may influence abuse of children by parents and stepparents. This idea, which comes from Hamilton's theory (and is also part of common knowledge, given well-known stories like *Cinderella*), ran counter to the way that sociologists were trained. The question then is: Why were sociologists not trained to consider this and are they now doing so?

The debate triggered by the publication of the book *Sociobiology* by E. O. Wilson in 1975 is a good example of the friction that can be caused when disciplines and ideologies collide. It is easy for more heat than light to be generated. The value of ideas or theories originating in one discipline and exported to another can be gauged by the new insights they give and the degree that they unify previously disparate fields when tested with real data. Interdisciplinary cross-pollination is not just one way, from biology to social science. The study of animal behavior has greatly benefited from insights from game theory, which was originally developed within the social sciences.

Dr. Francis Ratnieks is a professor at Sussex University, England. His seminal work on social behavior has focused on the role of genetic relatedness in shaping insect societies.

FIGURE 9.31. Kin recognition breakdown. While adopting an "if it's in your nest, it's your offspring" rule often works for mothers, the system can be sabotaged by "nest parasites." Here a mother dunnock is feeding a baby cuckoo that has been dumped into her nest. *(Photo credit: Eric and David Hosking/ Corbis)*

To see how such a rule of thumb might work, imagine a population of animals in which family units live in a fixed area—let's call it a nest—that is set apart from other nests. In such a population, all of the machinery (cognitive, genetic, sensory) necessary to distinguish kin from nonkin may be superfluous. It may be that a rule that instructs individuals to treat all individuals in their nest as kin works just as well in terms of kin recognition as the more complicated rules associated with matching. After all, if everyone in a nest is almost always kin, selection should favor the simplest possible kin recognition rule. Of course, such kin recognition rules are subject to cheating. Cowbirds and cuckoo birds, for example, lay their eggs in the nests of other species, which then raise those chicks as if they were their own (Ortega, 1998; Figure 9.31). Such "nest parasites" are in an evolutionary arms race with their hosts: Hosts are selected to detect and reject foreign eggs, and nest parasites to circumvent any detection system that might evolve in their host (Dawkins and Krebs, 1979).

Spatial cues and kin recognition rules can often change through the lifetime of an individual. For example, in bank swallows (*Riparia riparia*), parents *initially* feed any chick in their nest (Hoogland and Sherman, 1976; see Figure 9.4). Because chicks cannot fly for the first three weeks of life, it is extremely likely that any chick in a burrow is kin. At three weeks, however, chicks learn to fly, and there is consequently much more mixing among young. Michael Beecher has found that when bank swallow chicks are about twenty days old, their mothers switch from the rule of thumb (feed what is in your nest) to using distinctive vocal cues (that is, a template-based system) to recognize and feed their offspring (Beecher et al., 1981, 1986).

The study of animal behavior was revolutionized by the introduction of inclusive fitness models. Since W. D. Hamilton introduced these models in the early 1960s, almost every animal behaviorist who has studied social behavior has at one time or another thought about whether kinship plays a role in the system that he or she is studying. As we have seen, kinship theory not only allows researchers to make predictions about when animals should be cooperative and altruistic toward their kin, but it also makes predictions about when they should not be so (as in parent-offspring conflict, sibling rivalry). Inclusive fitness continues to be one of the most actively researched areas in ethology. Modern work employs molecular genetic and phylogenetic analyses to expand the frontiers of research in this area.

SUMMARY

1. "Inclusive fitness" or "kinship" theory has revolutionized the way that scientists understand the evolution of behavior.
2. In evolutionary terms, relatedness centers on the *probability* that individuals share alleles that they have inherited from some *common ancestor*—parents, grandparents, and so on. The essence of inclusive fitness models is that they add on to "classical" models of natural selection by considering the effect of an allele not only on the individual that bears it but also on those sharing alleles that are identical by descent (that is, genetic kin).
3. The decision to aid family members is a function of how related individuals are, and the costs and benefits associated with the trait. For example, when individuals are highly related and an allele codes for an action that provides a huge benefit at a small cost, selection strongly favors this trait.

4. Parents should be willing to go to great lengths to help their offspring. But a zone of parent-offspring conflict is also predicted under basic kinship theory.
5. While sibling rivalry is most often associated with the field of psychology, basic kinship theory also defines the conditions under which sibling rivalry should be favored.
6. Many models of kin recognition center on individuals having some "internal template" against which they match others and gauge relatedness.
7. In a species in which kin groups are spatially segregated from one another over relatively long periods of time, a second, simpler, form of kin recognition may evolve. In such scenarios, natural selection favors a kin recognition rule of the form "if it lives in your nest/cave/territory, then treat it like kin."

DISCUSSION QUESTIONS

1. What might be some of the benefits to gauging very small differences in genetic kinship relationships? Why, for example, would it be better able to distinguish relatives at the level of cousin ($r = 0.125$) than simply sibling ($r = 0.5$)? What sorts of benefits might be possible when small differences in relatedness could be gauged?
2. Build a family tree and use it to calculate the genetic relatedness between paternal first cousins. Then expand the tree to the case of paternal second cousins.
3. Based on the parent-offspring conflict model, what differences in weaning behavior would you expect to see between younger and older mammalian mothers?
4. How might both kin selection and kin recognition rules be useful in understanding cases of "adoption" in animals?
5. How does a phylogenetic comparison of mating systems in primates shed light on kinship and parent-offspring conflict?

SUGGESTED READING

Emlen, S. T. (1995b). An evolutionary theory of the family. *Proceedings of the National Academy of Sciences, U.S.A., 92,* 8092–8099. Emlen lays out all fifteen predictions derived from his "evolutionary theory of family."

Gardner, A., West, S. A., & Wild, G. (2011). The genetical theory of kin selection. *Journal of Evolutionary Biology, 24,* 1020–1043.

Hamilton, W. D. (1963). The evolution of altruistic behavior. *American Naturalist, 97,* 354–356. Hamilton's ideas on kinship, boiled down to their main ingredients.

Lieberman, D., Tooby, J., & Cosmides, L. (2007). The architecture of human kin detection. *Nature, 445,* 727–731. A provocative paper on how humans recognize kin.

Sherman, P. W. (1977). Nepotism and the evolution of alarm calls. *Science, 197,* 1246–1253. This paper on kin selection and alarm calls in Belding's squirrels is perhaps the most well-cited empirical study in all the inclusive fitness literature.

10

Cooperation

The previous chapter focused on the strong role that genetic relatedness plays in promoting prosocial behavior. Animals that are not genetic relatives, however, cooperate with each other in many contexts. Here we shall examine such cooperation among unrelated individuals.

For example, female elephants help raise the offspring of others in their group, protect their calves and the calves of others from predators, and even push other elephants out of the path of danger (Douglas-Hamilton et al., 2006; Lee, 1987). And perhaps most remarkably, female elephants know when providing assistance to another member of their group would be futile.

But how do ethologists know what elephants know about cooperation? Joshua Plotnik and his colleagues (2011) addressed this question in an ingenious experiment. They adapted a protocol that had been originally devised to test cooperation and cognition in chimpanzees, and modified it to examine elephant cooperation (Bates et al., 2008; Melis et al., 2006). The experiment began with an elephant learning to pull a rope that was attached to a table that was otherwise outside of its reach. Between the table and the elephant was a bowl of food. When the rope was pulled, it moved the table toward the elephant, and the table moved the bowl of food within the elephant's reach. Twelve elephants learned this task, and this group was then split into six pairs. Each pair was then given a new task to learn. Now the pair had to work in a coordinated way and simultaneously pull on a rope to move the table and food toward them. If they succeeded, they could reach the food (Figure 10.1).

In one treatment of this experiment, the two elephants were released at the same time and allowed to move toward the rope. Elephants in this treatment quickly learned how to pull the rope simultaneously and obtain the food as a reward. But this coordinated action may not represent cooperation at all.

If both elephants pulled their ropes at the same time, the the table moved toward them, pushing the food bowls along with it.

Ropes

Table attached to ropes

FIGURE 10.1. Elephant cooperation. A multiview perspective of the apparatus and the elephants. The inset shows the setup from above. *(Adapted from Plotnik et. al., 2011)*

An elephant might have been using a very simple rule: "Pull the rope, and food comes."

In a second treatment, the release time of elephants in a pair was staggered. To obtain food, each elephant released into the experimental setup now had to wait until its partner was allowed in, and then the pair could simultaneously pull on the rope. Elephants learned this social coordination task—that is, they learned to wait for their partner and then simultaneously pull on a rope with that partner to obtain food. This finding *suggests* that the elephants were cooperating with one another to get food. At the very least, this treatment shows elephants were not just using a "pull the rope" rule; if they were, they would not have waited for their partners.

A third treatment provides more evidence that rope pulling was cooperative. In this treatment, a pair was released simultaneously, but one end of the rope was tied up so that the elephant near it could not pull on the rope. This elephant often remained idle, and its partner was much less likely to pull on the rope than in the first treatment (in which both partners had the rope available). When the partner's inaction made it apparent that pulling on the rope would have been futile in terms of getting food, an elephant did not expend time and energy pulling on its end of the rope. When cooperation would have yielded no reward, elephants did not cooperate.

Defining Cooperation

The word **cooperation** typically refers to an *outcome* in which two or more interacting individuals each receives a net benefit from their joint actions, despite the costs they may have to pay for undertaking such actions. For example, jointly hunting prey may provide each of two hunters with food, even though there are costs (possible injury, energy expended) associated with hunting. In addition to looking at outcomes (that is, successfully capturing prey), it is also important to examine cooperation in terms of individual action. Suppose, for example, that to successfully hunt prey, a pair of hunters needs to both (1) flush the prey into an open area and (2) pounce on the prey when it is flushed into the open. A successful hunting strategy might be for hunter 1 to flush out the prey, and for hunter 2 to follow up by pouncing on the prey. If hunter 1 flushes the prey into the open, then it acted cooperatively, in that hunter 1's behavior made a successful capture possible.

An *individual* can cooperate by acting in a way that would potentially benefit itself and its partner, even if its partner didn't cooperate. In our case, hunter 1 could flush out the prey, but hunter 2 might not pounce. What this means is that, in addition to defining cooperation as an out-come, it is necessary to define what it means "to cooperate." Here *to cooperate* means to behave in such a way as to make the benefits that could be obtained from joint action possible, even though they may not necessarily be achieved (Dugatkin et al., 1992; Mesterton-Gibbons and Dugatkin, 1992).

Cooperation occurs in many species and in a wide variety of behavioral contexts. To better understand the origins and the costs and benefits of cooperation among unrelated individuals, the following questions are addressed in this chapter:

► What paths leading to cooperation have been identified by ethologists? What theory lies behind each? What empirical evidence supports each of these paths?

- What do we know about both the ultimate and proximate underpinnings of cooperation?
- What role does phylogeny play in explaining the distribution of animal cooperation?
- What role does cooperation play in coalition formation?
- How can we explain *interspecific* cooperation—that is, cooperation among animals from different species?

The Range of Cooperative Behaviors

Although cooperation is far from ubiquitous in the animal kingdom, it does occur in many contexts: foraging, predation, antipredator behaviors, mating, play, aggressive interaction, and so forth. To get a sense of the sorts of cooperation seen in animals, we will discuss: (1) helping in the birthing process, and (2) social grooming in primates.

HELPING IN THE BIRTHING PROCESS

Thomas Kunz and his colleagues describe an interesting case of cooperation in the Rodriques fruit bat (*Pteropus rodricensis*). In this bat, unrelated female "helpers" assist pregnant individuals in the birthing process (Kunz et al., 1994). To study this form of cooperation, Kunz and his team observed females for three hours before the start of the birthing process to its completion, and found that helpers were integrally involved at almost every stage (Figure 10.2). To begin with, pregnant Rodriques

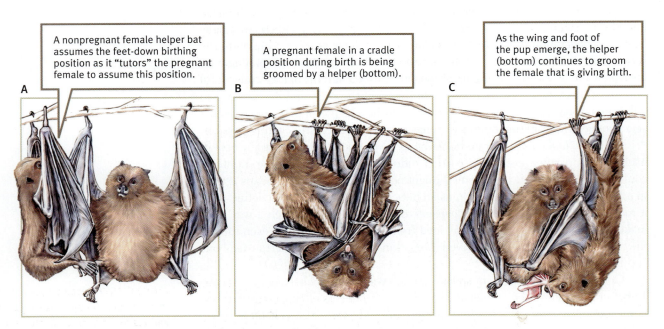

A nonpregnant female helper bat assumes the feet-down birthing position as it "tutors" the pregnant female to assume this position.

A pregnant female in a cradle position during birth is being groomed by a helper (bottom).

As the wing and foot of the pup emerge, the helper (bottom) continues to groom the female that is giving birth.

FIGURE 10.2. Bat midwives. Nonpregnant female helpers sometimes tutor pregnant females in the birthing process. The three stages of the process are shown here **(A–C)**. *(Based on Kunz et al., 1994)*

fruit bats must assume a "feet-down" position in order to give birth. Yet, expectant mothers do not take the feet-down position until they observe a helper female assume this position. This sort of tutoring on the part of helpers is remarkable—helpers themselves are not pregnant and hence would normally not assume such a position.

During the birthing process, the helper continues providing assistance by grasping the wings of the pregnant female, providing both protection and warmth, and subsequently cleaning (licking) newborn pups upon their emergence. Once pups are born, helpers guide the newborns into a suckling position, where they can obtain milk from their mother. These bat "midwives" cooperate with pregnant mothers during virtually every stage of the birthing process. Although not nearly as well documented, other anecdotal evidence for assistance during birth has been recorded in marmosets (*Callithrix jacchus*), Indian elephants (*Elephas maximus*), African hunting dogs (*Lycaon pictus*), raccoon dogs (*Nyctereutes procyonoides*), and bottle-nosed dolphins (*Tursiops truncatus*; Essapian, 1963; Gaffrey, 1957; N. Lucas et al., 1937; Poppleton, 1957; Yamamoto, 1987).

SOCIAL GROOMING

Many animals groom, or clean, themselves in an attempt to remove external parasites. This sort of grooming might involve scratching an area of skin or licking it. **Social grooming**, or **allogrooming**, in which one individual grooms another, is one of the most obvious and frequently observed cooperative behaviors recorded by primatologists. From the time of the earliest experimental work on primates, primatologists have recognized that, while social grooming often serves the function of removing parasites from the body of a partner, it may also have many other functions as well—indeed, social grooming has been considered a large part of the glue that holds primate troops together (Carpenter, 1942; Schino and Aureli, 2007, 2009; S. Zuckerman, 1932; Figure 10.3).

One benefit linked to social grooming in primate groups is "tension reduction" (Schino et al., 1988; Sparks, 1967; Terry, 1970). Aggression, especially aggression by top-ranking individuals, can be very dangerous. In addition to the serious costs associated with fighting-related injuries, aggression among top-ranking members can often disrupt normal everyday group living, causing all

FIGURE 10.3. Primate grooming. Primates of many species engage in various forms of grooming behavior. Such behavior may serve numerous functions simultaneously. *(Photo credit: Chris Crowley/Visuals Unlimited)*

group members to suffer. One way in which primates reduce the chances of escalating violence in a group is through social grooming, which has the effect of lowering the level of tension between putative combatants. From a proximate perspective, one means by which social grooming lowers tension levels is through increasing the circulating levels of hormones such as endorphins and opioids (Keverne et al., 1989; Martel et al., 1993). This sort of tension reduction through social grooming may play a role in nonprimate species as well, as research has found that social grooming on the part of subordinate meerkat females placates dominant meerkats (Kutsukake and Clutton-Brock, 2006, 2010).

Many studies have suggested that primates are capable of exchanging one sort of resource—for example, social grooming—for another resource in what amounts to a "biological marketplace" (de Waal, 1997; Fruteau et al., 2011; Noe and Hammerstein, 1995; Noe et al., 2002; Newton-Fisher & Lee, 2011). For example, consider the ability to groom another individual as one resource, and providing aid to an individual that is involved in some sort of aggressive interaction as another resource. In terms of biological markets, social grooming can be exchanged for aid during aggressive interactions. If individual 1 groomed individual 2, individual 2 might "pay back" the act of grooming by aiding individual 1 when it is involved in an aggressive interaction with individual 3 (we shall explore these sorts of coalitions in more depth later in this chapter).

Numerous experiments in primates have examined whether individuals cooperate with one another by exchanging social grooming for aid during aggressive interactions. To test for such an exchange system, Gabriele Schino ran a statistical analysis of thirty-six published studies that examined both aid giving and social grooming (Schino, 2007). The data from these thirty-six studies, that involved both wild and captive primate populations, show a positive correlation between aid dispensed during combat and social grooming, suggesting an exchange-based system of cooperation.

Other work suggests that grooming can be exchanged for currencies besides aid during combat. Primates also appear to exchange social grooming for access to scarce resources such as water or food, entrance into new groups, aid in chasing potential predators away, and future association with individuals who possess "special skills" that they themselves do not (de Waal, 1989; Stammbach, 1988). Grooming others, however, can also have associated costs, as evident in both captive and wild populations. In both rhesus macaques (*Macaca mulatta*) and blue monkeys (*Cercopithecus mitis*), individuals involved in social grooming were less vigilant, subjecting themselves or their offspring to increased rates of aggression from other troop members and possibly increased predation rates (Cords, 1995; Maestripieri, 1993).

Paths to Cooperation

Within the fields animal behavior and evolutionary biology, work on the evolution of cooperation can be traced at least as far back as Darwin's interest in social insects and other groups (Darwin, 1859, 1871). Darwin was well aware that social insects displayed an array of cooperative behaviors. Although he was unable to completely explain why, Darwin did come close to explaining part of the "riddle" of such behavior via kinship. After Darwin, the study of cooperation was primarily (but not exclusively) kept alive first by Petr Kropotkin, a Russian prince and naturalist, and later by such researchers as W. C. Allee, A. E. Emerson, and others who were part of what was

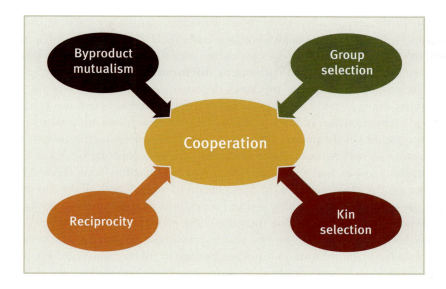

FIGURE 10.4. Four paths to cooperation. Reciprocity, byproduct mutualism, kin selection, and group selection can all lead to cooperative behavior.

called "the Chicago School of Animal Behavior" (Dugatkin, 2006, 2012; Mitman, 1992). Although these natural historians, ethologists, and population biologists amassed a good deal of data that documented cooperation among animals, they did little in the way of advancing any theory on why cooperation evolved (Allee, 1951).

A theoretical foundation for studying the evolution of cooperation has emerged over the last forty years (B'hary and Bergmueller, 2008; Clutton-Brock, 2009; Dugatkin, 1997a; Lehmann and Keller, 2006; Nowak, 2006b; Sachs et al., 2004; S. West et al., 2011). From Hamilton's innovative work in 1963 (see Chapter 9) to the present, four paths to the evolution and maintenance of cooperation in animals have been developed—kin-selected cooperation, reciprocity, byproduct mutualism, and group selection (Axelrod, 1984; J. L. Brown, 1983; Hamilton, 1964; D. S. Wilson, 1980; Figure 10.4). There is now ample empirical evidence available to put some of the models underlying these paths to the test. Since kin selection was discussed in Chapter 9, here we consider the remaining three paths to cooperation.

PATH 1: RECIPROCITY

Robert Trivers tackled the evolution of cooperation between unrelated individuals in an important paper entitled "The Evolution of Reciprocal Altruism" (Trivers, 1971). Trivers argued that if individuals benefited from exchanging acts of cooperative and altruistic behavior—what Trivers referred to as **reciprocal altruism**—then this sort of reciprocal exchange system might be favored by natural selection. That is, if individual A pays some cost to help individual B, but this cost is made up for at some point in the future, when B helps A, then natural selection might favor behaviors that lead to this type of reciprocity. Reciprocal altruism will be more likely to be favored by natural selection when individuals that live in groups interact with the same partners, as well as when individuals have the ability to recognize in the future those that had helped them in the past, as both facilitate long-term partnerships.

Trivers examined how reciprocity evolves using a theoretical framework called "game theory." Game theory is a mathematical tool that is used when the "payoff"—some resource such as food, mating opportunities, and so on—that an individual receives for undertaking an action is dependent on what behavior others adopt. In the evolutionary game theory literature, payoffs are used as indirect proxies for fitness. Trivers, with some help from W. D. Hamilton (see Chapter 9), suggested that the evolution of cooperation could best be understood using a mathematical game called the **prisoner's dilemma**.

To understand how the prisoner's dilemma game works, let's begin with a human example and then examine this game in terms of animal behavior. Imagine the following scenario: While in separate rooms, two criminal suspects are interrogated by the police in the hope of getting a confession from one or both of them. The options available to both suspects are cooperate or defect (don't cooperate). Cooperation and defection (also called "cheating") are defined from the perspective of the suspects. To defect means to "squeal" and tell the authorities that the other suspect is guilty, and to cooperate is to stay quiet and not betray the other suspect. Imagine that the police have enough circumstantial evidence to put both suspects in jail for one year, even without a confession from either. But if each suspect informs on the other, both go to jail for three years. Finally, if only one suspect informs on the other suspect, such behavior allows the defector to walk away a free man, but results in the cooperator going to jail for five years.

Table 10.1 depicts the payoffs to each suspect as a function of how he behaves and how the other suspect behaves. If both suspects cooperate, they both receive a payoff of R (R is short for the reward for mutual cooperation; in Table 10.1, R = 1 year in jail), and if they both defect, each one receives P (the punishment for mutual defection; 3 years in jail). If suspect 1 defects, but suspect 2 cooperates, the former receives a payoff of T (the temptation to cheat payoff; 0 years in jail), and the latter receives S (the sucker's payoff; 5 years in jail). If we order the payoffs in this matrix from high to low, we see that $T > R > P > S$. It is these inequalities that defines our game as a prisoner's dilemma. That is, for a game to be a prisoner's dilemma, the payoff structure of the matrix must be $T > R > P > S$ (a second more technical requirement is sometimes added to the model).

TABLE 10.1. The prisoner's dilemma game. In this game, each player can either cooperate or defect. In the matrix, T = "Temptation to cheat" payoff, R = "Reward for mutual cooperation" payoff, P = "Punishment for mutual defection" payoff, and S = "Suckers" payoff. For the matrix to qualify as a prisoner's dilemma game, it must be true that $T > R > P > S$. Each cell shows the payoff to suspect 1 (in the top left corner) and the payoff to suspect 2 (in the bottom right corner). For example, in the lower left cell, when suspect 1 defects and suspect 2 cooperates, the former gets 0 years in jail, while the latter gets 5 years in jail. There are other technical requirements for a matrix to be a prisoner's dilemma, but we will ignore them here.

		Suspect 2	
		Cooperate	Defect/Cheat
Suspect 1	Cooperate	Suspect 1: R = 1 year in jail Suspect 2: R = 1 year in jail	Suspect 1: S = 5 years in jail Suspect 2: T = 0 years in jail
	Defect/ Cheat	Suspect 1: T = 0 years in jail Suspect 2: S = 5 years in jail	Suspect 1: P = 3 years in jail Suspect 2: P = 3 years in jail

Suspect 1 will receive a better payoff individually if he defects, regardless of what suspect 2 does—if his partner defects, he does better to defect (3 versus 5 years in prison), and if his partner cooperates he does better to defect (0 versus 1 year in prison). As such, suspect 1 should always defect. The same holds true individually for suspect 2, and he should also always defect. So both suspects should defect. The *dilemma* in the prisoner's dilemma is that while each suspect receives P (3 years in prison) when they both defect, both suspects would receive a higher payoff (R, 1 year in prison) if they had both cooperated (Poundstone, 1992).

The prisoner's dilemma game can be used to model certain types of animal, as well as human, cooperation. As long as each animal does better individually to defect regardless of the action of the other player, but both do better when there is mutual cooperation rather than mutual defection, the prisoner's dilemma game can be used to model animal cooperation. Before we look at animal examples, let us explore the prisoner's dilemma in more depth and examine what sort of predictions about behavior emerge from this model.

Robert Axelrod and W. D. Hamilton used both analytical mathematics and computer simulations to examine what sorts of behavioral strategies fared well—obtained high payoffs—when individuals were playing the prisoner's dilemma game. They examined the success of an array of behavioral strategies in an *iterated*, or repeated, prisoner's dilemma game (Axelrod and Hamilton, 1981). In the iterated version of the prisoner's dilemma game, two individuals play the prisoner's dilemma game with one another many times, and in the iterated game we focus on, the exact end point is unknown. This type of game tends to mimic certain natural situations, since social, group-living animals generally encounter each other more than once, but the number of possible future encounters is uncertain. The iterated version of the game allows for the possible emergence of very complex strategies. In the iterated prisoner's dilemma game when players encounter each other many times, complex rules, including "if-then" rules of the form "if the other individual does X, then I will do Y," can be employed—for example, "if she cooperates, I will cooperate; otherwise I will defect."

Axelrod and Hamilton searched for the **evolutionarily stable strategy (ESS)** to the iterated prisoner's dilemma game (Maynard Smith, 1982; see Math Box 10.1 at the end of the chapter for an explanation of ESS). They demonstrated that if the probability of meeting a given partner in the future was sufficiently high, then in addition to the success of a simple strategy of "always be noncooperative" (labeled "always defect" or ALLD), a *reciprocity-based* strategy called **tit for tat (TFT)** was one solution to the iterated prisoner's dilemma (see Math Box 10.2 at the end of the chapter).

TFT is a strategy in which an individual cooperates on the initial encounter with a partner and subsequently copies its partner's previous move. Thus, after the first move, it operates under an if-then rule: if the partner cooperates, then cooperate; if the partner defects, then defect. It thus should reciprocate acts of cooperation, as well as acts of defection. This behavioral rule also nicely ties together work on cooperation and work on social learning (see Chapter 6), as individuals copy what their partners do, and such behavior can potentially ripple through an entire population.

Axelrod hypothesized that the TFT strategy's success is attributable to its three defining characteristics: (1) "niceness"—a player using TFT is never the first to defect, as it initially cooperates with a new partner, and then cooperates as long as its partner cooperates; (2) swift "retaliation"—an individual playing TFT immediately defects on a defecting partner since it copies its partner's previous move, so if its partner defects, it defects; and (3) "forgiving"—because

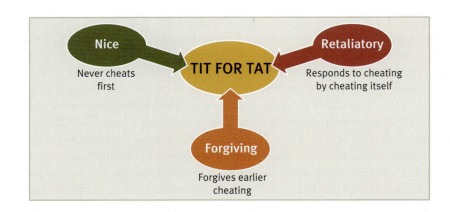

FIGURE 10.5. Tit for tat. The tit-for-tat strategy has three fundamental characteristics. The individual using TFT is (1) nice—it never cheats first; (2) retaliatory—it always responds to a partner that is cheating by cheating itself; and (3) forgiving—it only remembers one move back in time, and hence is capable of "forgiving" cheating that is done early in a sequence.

TFT instructs players to do what their partner did on the last move, those using a TFT strategy have a memory window that is *only one move back in time* (Axelrod, 1984). As such, the player using TFT forgives prior defection if a partner is currently cooperating—those using a TFT strategy do not hold grudges (Figure 10.5). Since the original models of TFT have appeared, dozens of variants of this strategy have been examined, but most share the essential characteristics we have just discussed (Dugatkin, 1997a).

Numerous studies have examined reciprocity in animals; here we will focus on three of these studies. The first examines the prisoner's dilemma and the use of TFT during antipredator behavior in guppies, the second addresses reciprocity in the context of vampire bat food sharing, and the third looks at the proximate underpinnings of human reciprocity.

PREDATOR INSPECTION AND TFT IN GUPPIES. In many streams of the Northern Mountains of Trinidad, the water is clear, and during the dry season, the behavior of guppies (*Poecilia reticulata*) can be seen from the riverbank. If you sit beside such streams, you will eventually see some guppies—perhaps a pair of them—break away from their group and approach a potentially dangerous predator, such as a pike cichlid (*Crenicichla alta*). This behavior is called predator inspection (see Chapter 2). These guppies are probably not genetic relatives, and they could just as easily have headed for cover (and many others probably did just that). Instead, they approach the predator and do so in what looks like a coordinated fashion (Pitcher et al., 1986; Figure 10.6).

FIGURE 10.6. Risk taking and cooperation in guppies. Two male guppies (lower left and lower center of photo) inspect a pike cichlid predator. Guppies cooperate during such risky endeavors. (*Photo credit: Michael Alfieri*)

TABLE 10.2. The payoffs for predator inspection. When T > R > P > S, predator inspection is a prisoner's dilemma. Fish 1's payoff is shown in the top left corner of each cell, and fish 2's payoff is shown in the bottom right corner of each cell.

		Fish 2	
		Inspect	Don't inspect/ lag behind
Fish 1	Inspect	Fish 1: R Fish 2: R	Fish 1: S Fish 2: T
	Don't inspect/ lag behind	Fish 1: T Fish 2: S	Fish 1: P Fish 2: P

How can such cooperative behavior emerge? The risk-taking guppies described above are in what appears to be a no-win situation. Think about it like this: The best thing that could happen for guppy 1 is that guppy 2 takes all the risks and passes the information it receives onto guppy 1. But, of course, the same holds true for guppy 2, as its highest payoff comes when guppy 1 approaches the predator alone, and then passes the information it receives to guppy 2. Yet if both fish opt to wait for the other to inspect the predator, each may be worse off than if they had inspected the predator as a pair. In this scenario, inspecting a predator would be acting cooperatively, while failing to inspect, or just lagging behind during inspection, would be a form of cheating. Before addressing whether guppies use TFT while inspecting a predator, we first need to examine whether a pair of fish inspecting a predator are playing the prisoner's dilemma game (Dugatkin, 1997a). As mentioned earlier, for a payoff matrix to be a prisoner's dilemma, it must be true that T > R > P > S. What evidence is there that these inequalities hold true for guppies inspecting a predator?

▶ Is T > R? Milinski argued that a fish that trailed behind its partner while inspecting a possible predator would do better than its co-inspector, as it could assess how dangerous the predator was by observing whether the lead fish was attacked (Milinski, 1987). The trailing fish would receive a payoff of T if it stayed behind, but a payoff of R if it swam beside its partner (Table 10.2). Two pieces of evidence suggest that the temptation payoff T is greater than the R payoff associated with mutual inspection. First, inspectors are more likely to get eaten the closer they approach a predator (Figure 10.7), so it is more dangerous to be leading an inspection than lagging behind. Second, inspectors transfer the information that they receive during an inspection, so that any fish lagging behind would still receive the benefits associated with inspection (Dugatkin, 1992b; Magurran and Higham, 1988; Milinski et al., 1997; Figure 10.8).

▶ Is R > P? If P is greater than R—if the payoff for inspecting in a group (R) is less than the payoff when no one inspects (P)—it would not pay for any individual to inspect, so predator inspection should be rare and maladaptive when it occurred. Given that inspection occurs in many species, this is

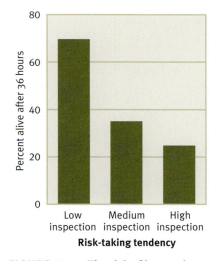

FIGURE 10.7. The risk of inspecting predators. Ten groups of six guppies each—two low inspectors, two medium inspectors, and two high inspectors—were placed with a predator in a pool that was one meter in diameter. The probability of surviving thirty-six hours was a function of inspection tendencies, with those inspecting most often suffering the highest mortality. *(From Dugatkin, 1992b)*

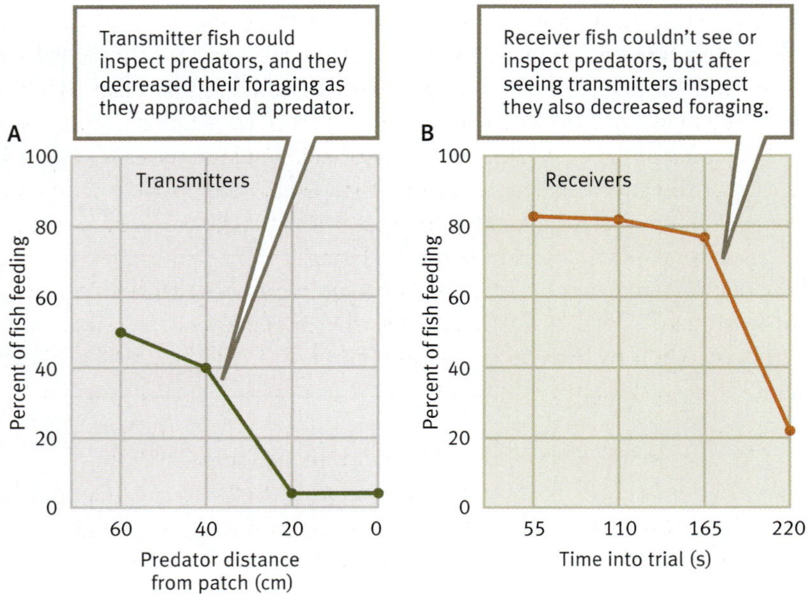

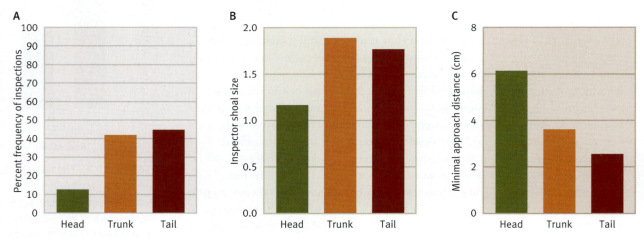

FIGURE 10.8. Information transfer in minnows. Information obtained by inspectors is somehow transferred to individuals that do not inspect. *(Based on Magurran and Higham, 1988, p. 157)*

A Transmitters

Transmitter fish could inspect predators, and they decreased their foraging as they approached a predator.

Percent of fish feeding

Predator distance from patch (cm)

B Receivers

Receiver fish couldn't see or inspect predators, but after seeing transmitters inspect they also decreased foraging.

Percent of fish feeding

Time into trial (s)

unlikely to be the case, so R, the payoff for mutual cooperation, is probably greater than P, the payoff for mutual defection (Dugatkin and Godin, 1992; Pitcher, 1992; Figure 10.9).

▶ Is P > S? If predator inspection is to qualify as a prisoner's dilemma, it must also be true that the payoff to mutual cheating (P) is greater than the payoff to inspecting alone (S). While it appears to be the case that having no inspectors in a shoal is dangerous for all group members, as no one obtains information on danger levels, the most dangerous situation for a single fish is to be the lone inspector in a group. Evidence from a number of experiments indicates that single fish (stragglers) suffer very high rates of predation, suggesting, though not definitively demonstrating, that P > S (Milinski, 1977).

FIGURE 10.9. Inspection behavior in the wild. A model predator was placed into a tributary of a river in Trinidad, and predator inspection behavior of guppies was recorded. Inspectors recognize the head region of a predator as most dangerous. **(A)** There were fewer inspections of the predator's head than of its trunk and tail. **(B)** Inspector group was smallest when inspecting the head region of a predator. **(C)** Approach distance was a function of the part of the predator's body that was being inspected; inspectors stayed farthest away when they were inspecting the predator's head. *(Based on Dugatkin and Godin, 1992)*

Measuring T, R, P, and S precisely in controlled experiments is difficult, and there is contention in the published literature over whether the payoff matrix associated with inspection meets the prisoner's dilemma requirement that T > R > P > S (see Dugatkin, 1997a, for a summary of these sometimes heated debates). This debate remains unsettled, but we can say that it is at least possible, and perhaps probable, that the payoff matrix is such that inspectors are trapped in a prisoner's dilemma. Working under the assumption that this is the case, theory suggests that fish inspecting a predator should use the TFT strategy. Do they?

The dynamic nature of inspection behavior in guppies and sticklebacks (but not mosquitofish) supports the prediction that inspectors do, in fact, use the TFT strategy when inspecting potential predators (Dugatkin, 1997a; Stephens et al., 1997). Fish that inspect predators appear to use a strategy that is:

▸ "nice," as each starts off inspecting at about the same point in time,
▸ "retaliatory," as inspectors cease inspection if their partner stops (Figure 10.10),
▸ "forgiving"—if inspector A's partner has cheated on it in the past, but resumes inspection, A then resumes inspection as well (Dugatkin, 1991).

The prisoner's dilemma game can be used to make some further predictions. If the payoffs for inspection match those of the prisoner's dilemma, then the payoffs shown in Table 10.2 suggest that each individual should prefer to associate with cooperators. This is because cooperators do better when paired with other cooperators, and defectors also do better when paired with cooperators, so whenever possible, all individuals should prefer to associate with cooperators. Evidence exists that inspectors do in fact remember the identity and behavior of their co-inspectors. When an individual fish is given the choice between associating with a fish that has been cooperative during prior encounters versus a fish that has not been cooperative, it prefers to associate with cooperators over defectors, at least when group size is small (Dugatkin and Alfieri, 1991a, 1992; Dugatkin and Wilson, 2000; Milinski et al., 1990).

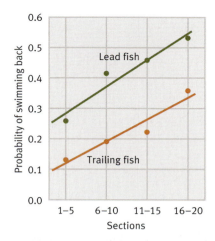

FIGURE 10.10. "Retaliation" in guppies? Pairs of guppies were given the opportunity to inspect a predator. Lead fish in a given section of an aquarium were more likely to turn back and swim to safety than were trailing fish in the same section of the aquarium. This might be interpreted as lead fish "retaliating" against trailing fish, who fail to stay by their side. *(Based on Dugatkin, 1991, p. 130)*

RECIPROCITY AND FOOD SHARING IN VAMPIRE BATS. One of the first studies of reciprocal food-sharing behavior was done in vampire bats (*Desmodus rotundus*). A group of vampire bats is composed largely of females, with a low average coefficient of relatedness (between 0.02 and 0.11; Wilkinson, 1984, 1990; Figures 10.11 and 10.12). Females in a nest of vampire bats sometimes regurgitate blood meals to other bats that have failed to obtain food in the recent past (Wilkinson, 1984, 1985). This sort of food sharing can literally be a matter of life or death, as individuals may starve to death if they don't receive a blood meal approximately every sixty hours (McNab, 1973). Jerry Wilkinson examined whether relatedness, reciprocity, or some combination of the two best explained the evolution and maintenance of sharing blood meals in this species.

Although the average relatedness in groups was low, Wilkinson found that genetic relatives were still more likely to swap blood meals with one another than with other individuals (Figure 10.13). To examine whether reciprocity per se was also important, Wilkinson created an "index of opportunity for reciprocity." When analyzing the data with this index, Wilkinson found three lines of evidence that reciprocity per se may be important in sharing blood meals in vampire bats: (1) the probability of future interaction between group members in a nest of vampire bats is high as required by TFT models; (2) blood meals provide a huge, potentially life-saving benefit for recipients, while the cost of

FIGURE 10.11. Blood-sucking reciprocators. To survive, female vampire bats need frequent blood meals. Individuals often regurgitate part of their blood meals to others, but they are much more likely to do so for those that have shared a meal with them in the past. *(Photo credit: Jerry Wilkinson)*

FIGURE 10.12. Vampire bat cooperation. If a hungry bat approaches a satiated bat, she is much more likely to get a regurgitated blood meal if she has fed the satiated bat in the past.

Hungry bat   Satiated bat

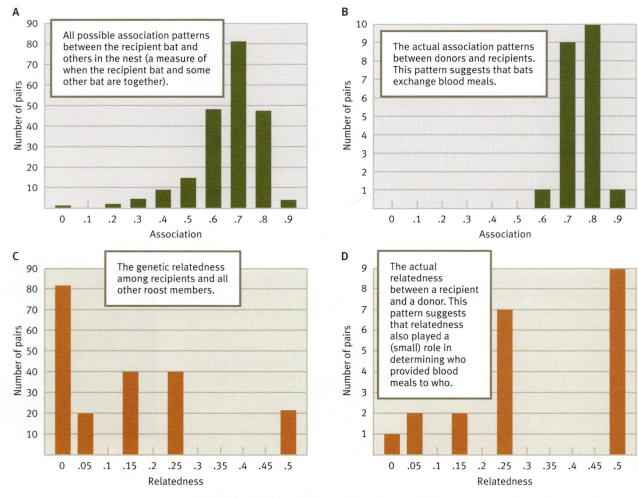

A

All possible association patterns between the recipient bat and others in the nest (a measure of when the recipient bat and some other bat are together).

B

The actual association patterns between donors and recipients. This pattern suggests that bats exchange blood meals.

C

The genetic relatedness among recipients and all other roost members.

D

The actual relatedness between a recipient and a donor. This pattern suggests that relatedness also played a (small) role in determining who provided blood meals to who.

FIGURE 10.13. Vampire bat blood meals. Wilkinson used twenty-one regurgitation events not involving mothers and their offspring to examine the role of relatedness and reciprocity in sharing blood meals. Bats were much more likely to regurgitate a meal to close kin and to those with which they associated more often. Follow-up laboratory work found that bats were capable of keeping track of those that fed them in the past and those that didn't. *(From Wilkinson, 1984)*

giving up some blood may not be as great to the donor; and (3) vampire bats are able to recognize one another allowing for the possibility of reciprocal exchange.

In addition, there is some, though not a great deal of, data that vampire bats are more likely to give blood to those that have donated blood to them in the past. While the vampire bats don't necessarily use the TFT strategy, they appear to be using some sort of reciprocal altruistic strategy.

NEUROBIOLOGICAL AND ENDOCRINOLOGICAL UNDERPINNINGS OF HUMAN RECIPROCITY. In addition to work on the evolution of cooperation, recent studies have shed light on the proximate underpinnings of cooperation, including reciprocity. As a case in point, let's examine the neurobiological and endocrinological underpinnings of reciprocity and trust in humans.

Much of this proximate work on the neurobiology of human cooperation has been undertaken by researchers in **neuroeconomics**—a collaborative research effort between economists and neurobiologists who specialize in brain science (Glimcher, 2004, 2010; Glimcher and Rustichini, 2004; Montague, 2006; Sanfey et al., 2006; Zak, 2004).

Experiments in neuroeconomics involve subjects making some economic decision—in our case, the focus will be on the economics of reciprocal altruism—while they have their brain activity patterns monitored by a magnetic resonance imaging machine (fMRI) or a positron emission tomography (PET) device.

To see how neuroeconomics can shed light on one proximate underpinning of human cooperation, let us examine a study done by James Rilling and his colleagues (Rilling et al., 2002). Rilling and his team had women subjects play the prisoner's dilemma depicted in Table 10.3. Subjects were hooked up to an fMRI machine and played the iterated prisoner's dilemma game over a networked computer with another subject who was in a different room (Figure 10.14).

Rilling and his team found that even though the highest monetary reward in this game ($3) was obtained when an individual cheated and her partner cooperated, the fMRI scan of players' brains as they made their economic decisions indicated that the most emotionally rewarding payoff was that associated with mutual cooperation (which yielded $2 to each individual).

TABLE 10.3. The monetary prisoner's dilemma game. The payoff matrix for the game played by women who were either cooperating or cheating (defecting) in an economic cooperation experiment. Subject 1's payoff in top left corner of each cell. Subject 2's payoff in bottom right corner of each cell.

		Subject 2	
		Cooperate	Defect/Cheat
Subject 1	Cooperate	Subject 1: R = $2 Subject 2: R = $2	Subject 1: S = $0 Subject 2: T = $3
	Defect/ Cheat	Subject 1: T = $3 Subject 2: S = $0	Subject 1: P = $1 Subject 2: P = $1

A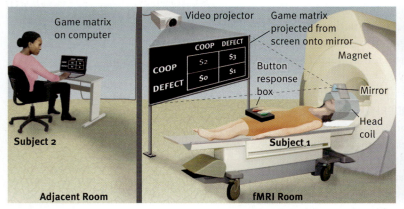

B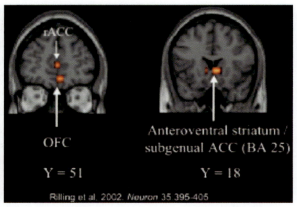

FIGURE 10.14. The prisoner's dilemma game and social cooperation.
To study the neurobiological basis of reciprocal altruism and cooperation, researchers had subjects play an iterated prisoner's dilemma game. **(A)** One of the subjects played from inside an fMRI machine that monitored her brain activity as she played, while the other subject played the game on a computer in a different room. Each subject saw the payoff matrix that represented her own payoffs. **(B)** The fMRI scans showed that, when both subjects cooperated, brain areas associated with reward processing—the ventromedial/orbitofrontal cortex (OFC), the rostral anterior cingulate cortex (rACC), the anteroventral striatum (including the caudate nucleus and the nucleus accumbens), and the subgenual anterior cingulate cortex (ACC)—were activated. *(Based on Rilling et al., 2002; photo credit: James Rilling; reprinted by permission of Cell Press)*

That is, not only did subjects say that they found mutual cooperation to be the most rewarding outcome, but it was the mutual cooperation payoff that caused the greatest activation in areas of the brain associated with reward processing in humans, namely the ventromedial/orbitofrontal cortex (OFC), the anterior cingulate cortex (ACC), and the nucleus accumbens.

Work in neuroeconomics has not only examined emotions and behavior with respect to reciprocity per se but it has also looked at what sections of the brain are active when we punish those who fail to act cooperatively during economic interactions (Izuma, 2012; Rilling and Sanfey, 2011). In particular, such behaviors often entail a *cost* to those who dole out the punishment, and a *benefit* to others not even involved in the interaction. Punishing those who violate social norms is of particular interest to those behaviorists, including animal behaviorists, who wish to understand how social norms evolved before modern legal codes came into place. Dominique de Quervain and his colleagues at the University of Zurich hypothesized that one proximate mechanism involved in maintaining punishment lies in the pleasure that we derive from enforcing social norms (de Quervain et al., 2004), and that this pleasure can indirectly be measured through neurobiological scans of the brain.

To test this hypothesis, de Quervain and his colleagues had pairs of subjects play what is called the trust game. In the trust game, two players—who are not allowed to communicate with each other—begin with 10 units of money (called monetary units, or MUs, in this experiment). Player A is hooked up to a PET scanner and starts the game by deciding whether to give his 10 MUs to player B or to keep them for himself. A is told that, if he gives his money to B, the investigator will quadruple the gift to 40 MUs, so that B now has 50 MUs, and A has no MUs. Then player B is given a choice. He can send half his MUs back to player A, or he can keep everything for himself. So if A "trusts" B to send the money back, and B does send back the money, A and B each end up with 25 MUs; while if A opts to give nothing, they both end up with 10 MUs.

De Quervain hypothesized that if A does trust B to play fairly and gives him the money, but B then decides to keep all the MUs, A should view this as a violation of trust and social norms. He hypothesized that, in such a case, A should then punish B. To allow for this possibility, one minute after B makes his decision, A can opt to "punish" B by revoking some of the MUs that he initially gave B. In one treatment of the experiment, A paid a cost in MUs for doing so, and in other treatments punishing player B did not have any cost associated with it.

If A trusts B, but B violates that trust, A will indeed punish B, even if it costs monetary units to do so. More to the point, not only did subjects say that they enjoyed punishing those who violated their trust, PET scans of their brains demonstrated that one section of the brain associated with *reward* (the dorsal striatum, which includes the caudate nucleus) is most active when A undertakes the act of retribution. That is, individuals derived *pleasure* from punishing cheaters. Indeed, results suggest that the more intense the punishment doled out to cheaters was, the more active was the dorsal striatum of the individual exacting retribution (Figure 10.15). In an interesting twist to the original experiment, A was sometimes told that B's decision was determined by the equivalent of an electronic coin toss, so it was out of B's hands. In that condition, when B did not send money back to A, A did not view this as a violation of trust and did not respond by punishing B, nor did the researchers see the increased activity in the dorsal striatum that was observed when A was punishing B.

Research has also examined the endocrinological underpinnings of reciprocity and trust in humans. As discussed in Chapters 3 and 7, oxytocin is a neurohormone that has been associated with numerous affiliative behaviors like pair bonding and parental care in nonhumans. Because there are dense accumulations of oxytocin receptors in the amygdala of the human brain, a region associated with social behavior (Loup et al., 1991), Paul Zak and his colleagues hypothesized that oxytocin would also play a role in affiliative interactions in humans (Kosfeld et al., 2005; Rilling et al., 2012; Zak 2011a, 2011b; Zak et al., 2005). In particular, they ran a series of experiments that examined the role of oxytocin in the trust game we described in the de Quervain neuroeconomics experiment.

Zak's version of the trust game was slightly different from the one described earlier. In the Zak version, player A was given $10. He was then asked how much of that $10 he was willing to give to player B, with the understanding that whatever he gave to player B would be *tripled* by the experimenter. Then player B was given a choice to send back as much money as he desired (including $0) to player A.

Rather than attaching their subjects to a brain-scanning device, Zak and his team drew blood from the subjects immediately after they had played the trust game, and they analyzed the blood sample to estimate oxytocin levels. Subjects were tested in one of two treatments. In treatment 1, player B was told that player A had made his choice voluntarily, and in treatment 2, B was told that A was forced to send a randomly selected amount of money to him. The researchers hypothesized that B would feel more trust toward A when he received money that A had voluntarily sent, and that B would then send back more money and would show an increased level of oxytocin (compared with treatment 2). Results were consistent with these hypotheses.

Zak and his colleagues found that oxytocin levels were significantly higher in treatment 1, where player B believed that the money he or she had received from player A was sent voluntarily (Figure 10.16). Further evidence for the role of oxytocin in human decisions about trust was also uncovered. Zak and his team found that the more money A sent B, the more money B sent back, and the higher B's oxytocin levels—but this was only true when B believed A had sent the money voluntarily.

The prosocial, trust-inducing effects of oxytocin also extend to groups as well as individuals. When oxytocin levels were experimentally manipulated through nasal infusions of oxytocin, subjects were more likely to donate voluntarily to charities (Barraza et al., 2011).

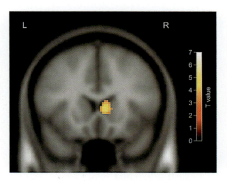

FIGURE 10.15. The trust game and punishment. Two subjects played the trust game while one of them (player A) was hooked up to a PET scanner that monitored his brain activity. The caudate nucleus, which is part of the dorsal striatum of the brain—depicted in yellow—was very active when player A punished player B for failing to return some of the money that A had provided to B. *(From de Quervain et al., 2004; reprinted by permission of the AAAS)*

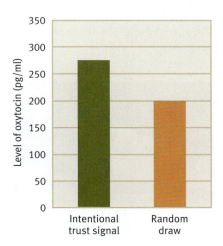

FIGURE 10.16. Oxytocin and trust. The level of oxytocin was higher when subjects believed money was sent to them voluntarily (versus sent as a function of a random draw). *(From Zak et al., 2005)*

PATH 2: BYPRODUCT MUTUALISM

A second path to the evolution of cooperation is via **byproduct mutualism** (J. L. Brown, 1983; Connor, 1995; Rothstein and Pirotti, 1987; West-Eberhard, 1975). Cooperation here is a "byproduct" of the fact that an individual would incur an immediate cost or penalty if it did not act cooperatively, such that the immediate net benefit of cooperating outweighs that of cheating. In other words, there would be no temptation to cheat, so individuals would cooperate.

Mathematical models predict that byproduct mutualism is more likely to evolve in "harsh" versus "mild" environments. What constitutes harsh and mild will depend on the system being studied (Dugatkin et al., 1992; Shen et al., 2011). For example, in Taiwanese yuhina (*Yuhina brunneiceps*), harsh environments are characterized as those with excessively high levels of rainfall, which is correlated with lower foraging success and lower hatching success in yuhinas. When Sheng-Feng Shen and his colleagues studied these communal nestings, they found significantly more cooperation in harsh than in mild environments (Shen-Feng et al., 2011; Figure 10.17). Harsh environments have also been found to favor cooperative behaviors in packs of feral dogs (Bonanni et al., 2010).

Byproduct mutualism differs from reciprocity in two ways. First, there is no temptation to cheat under byproduct mutualism, whereas such a temptation always exists in systems involving reciprocity. In addition, while most (but not all) forms of reciprocity require some form of scorekeeping, with byproduct mutualism individuals need not keep track of the past behavior of partners because in such situations it is always in the best interest of all parties to cooperate, and so past behavior is not relevant in this decision-making process.

SKINNERIAN BLUE JAYS AND BYPRODUCT MUTUALISM. One critique of the work on cooperation is that it is very difficult for investigators to measure the precise payoffs of behavioral alternatives available to individuals (Clements and Stephens, 1995). A way around this problem is to use what are called Skinner boxes, which allow precise control over the payoffs that animals encounter in the lab. Kevin Clements and David Stephens adopted this technique in their study of blue jay (*Cyanocitta cristata*) cooperation (Clements and Stephens, 1995).

Clements and Stephens tested pairs of blue jays, each of whom could peck one of two keys—a cooperate key or a defect key. After the birds made their

A

FIGURE 10.17. Harsh environments favor cooperation.
(A) A group of yuhinas (*Yuhina brunneiceps*). **(B)** During the breeding season, females cooperate with one another more in harsh than in mild environments. (*Photo credit: Shen, S., et al. 2011. Nature Communication; adapted from Shen et al., 2011*)

B

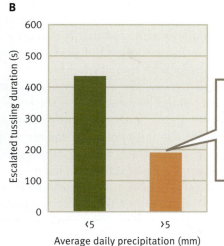

In harsh (high precipitation) environments, yuhinas cooperated with each other more often. Cooperation was measured in terms of reduced levels of aggression on the part of communally nesting females.

decisions, they were given food, the amount of which depended on what action they took (cooperate or defect), what action the other bird took, and which of two different payoff matrices the researchers had set up. The first matrix had payoffs that matched a prisoner's dilemma matrix (P matrix), while the second had payoffs that mimicked a byproduct mutualism payoff matrix (M matrix), in which there was no temptation to cheat, because the birds always received more food for cooperating. In each case, bird 1 would begin a trial by pecking one of the keys and bird 2 would end the trial (again by pecking either the cooperate or defect key). For example, if a pair was in the P matrix part of the game and bird 1 cooperated when bird 2 defected, bird 2 obtained five food items, while bird 1 received no food items (Table 10.4).

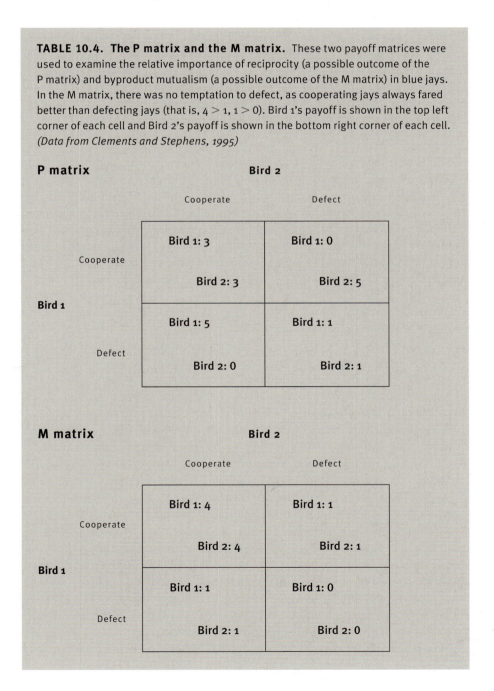

TABLE 10.4. The P matrix and the M matrix. These two payoff matrices were used to examine the relative importance of reciprocity (a possible outcome of the P matrix) and byproduct mutualism (a possible outcome of the M matrix) in blue jays. In the M matrix, there was no temptation to defect, as cooperating jays always fared better than defecting jays (that is, $4 > 1$, $1 > 0$). Bird 1's payoff is shown in the top left corner of each cell and Bird 2's payoff is shown in the bottom right corner of each cell. *(Data from Clements and Stephens, 1995)*

P matrix

Bird 2

		Cooperate	Defect
Bird 1	Cooperate	Bird 1: 3 Bird 2: 3	Bird 1: 0 Bird 2: 5
	Defect	Bird 1: 5 Bird 2: 0	Bird 1: 1 Bird 2: 1

M matrix

Bird 2

		Cooperate	Defect
Bird 1	Cooperate	Bird 1: 4 Bird 2: 4	Bird 1: 1 Bird 2: 1
	Defect	Bird 1: 1 Bird 2: 1	Bird 1: 0 Bird 2: 0

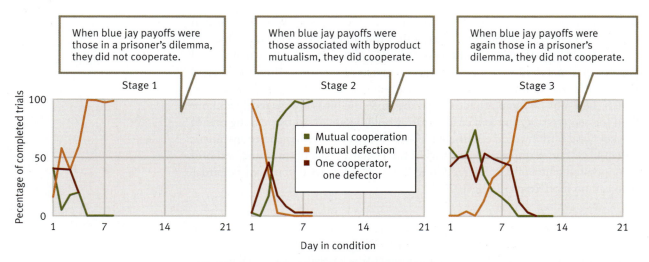

FIGURE 10.18. Byproduct mutualism and blue jays. Blue jays were tested in a three-stage experiment: stage 1 = prisoner's dilemma, stage 2 = byproduct mutualism, and stage 3 = prisoner's dilemma. Jays cooperated when the payoff matrix matched byproduct mutualism, but not when it matched the prisoner's dilemma. *(Based on Clements and Stephens, 1995)*

Birds were exposed to the P matrix, then to the M matrix, and finally to the P matrix again, and on any given day a pair of birds would play these games with each other more than 200 times. Clements and Stephens found that, regardless of whether the jays could see each other or not, birds defected in the first P matrix, cooperated in the M matrix, and reverted to defection the second time they encountered the P matrix (see Figure 10.18). Blue jays appeared to cooperate via byproduct mutualism and not reciprocity, even when a payoff matrix (the prisoner's dilemma) that should promote reciprocity was presented. Why this is so remains unknown.

BYPRODUCT MUTUALISM AND HOUSE SPARROW FOOD CALLS. House sparrows (*Passer domesticus*) produce a unique "chirrup" call when they find a food resource (Summers-Smith, 1963). These calls appear to attract other birds to a newly discovered bounty, and may involve cooperative signalling. To examine just what type of cooperation, Marc Elgar tested whether the chirrup vocalization did in fact bring conspecifics to a newly discovered food source, and if so, under what conditions (Elgar, 1986). To do so, he recorded chirrup calls at artificial feeders containing bread that was either divisible among sparrows or that was just enough food for a single bird.

Elgar did find some evidence that those sparrows arriving at a patch of food first (labeled as "pioneers") were the most likely to produce chirrup calls. Furthermore, chirrup call rates were higher when the food resource was divisible (Figure 10.19). Elgar hypothesized that if the food items were small enough that sparrows could pick them up and fly away, which is what the sparrows did in one of his treatments, they would do so without producing chirrup calls. Chirrup calls were, however, associated with larger food items—that is, those that were too big to remove from the experimental area. It may be that given that sparrows needed to remain at a feeder with large food items, and that it is safer to do so in the company of other sparrows, the benefits associated with predator detection outweighed the costs of inviting other foragers to share food at the site. In that case, chirrup calls would be an example of byproduct mutualism. They are emitted when the immediate net benefit for calling (predator detection and some food) is greater than the net benefit for not calling (more to eat).

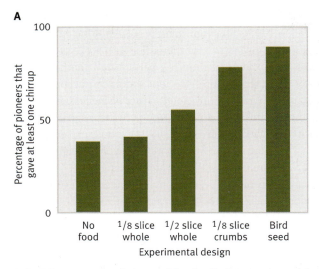

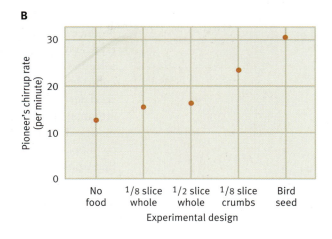

FIGURE 10.19. Food size and food calls in sparrows. (A) The first bird to arrive at a food patch (labeled the pioneer) were more likely to give "chirrup calls" that attracted other birds when resources were more divisible. **(B)** Pioneers also called more often when food was easily divisible. *(Based on Elgar, 1986, p. 171)*

PATH 3: GROUP SELECTION

A third path to cooperation is via **group selection** (Sober and Wilson, 1998; D. Wilson, 1980; D. Wilson and Wilson, 2007). Modern group selection models of cooperation are quite straightforward. These models are often called trait-group selection models, where a trait group is defined as a group in which all individuals affect one another's fitness.

The crux of trait-group selection models is that natural selection operates at two levels: within-group selection and between-group selection. In the context of cooperation and altruism, within-group selection acts *against* cooperators and altruists, because such individuals, by definition, pay some cost that others do not. Selfish types—those that do not cooperate—are always favored by within-group selection, since they receive any benefits that accrue because of the actions of cooperators and altruists, but they pay none of the costs.

As opposed to within-group selection, between-group selection favors cooperation when groups with more cooperators outproduce other groups—for example, by producing more total offspring or being able to colonize new areas faster. Consider the case of alarm calls. Alarm callers pay a cost within groups, as they may be the most obvious target of a predator honing in on such a call. But their sacrifice may benefit the group overall, as other individuals—including other alarm callers, as well as those that don't call—are able to evade predators because of the alarm call. Thus, groups with many alarm callers may outproduce groups with fewer cooperators. For such group-level benefits to be manifest, groups must differ in the frequency of cooperators within them, and groups must be able to "export" the productivity associated with cooperation (for example, more total offspring, faster colonization of newer areas, and so on).

Group selection models remain controversial. Many animal behaviorists argue that all group selection models can be mathematically translated into selfish gene models. That is, they posit that group selection models simply partition the effect of a trait into within- and between-group components, but that if you sum up the effects over all groups you get the same solution as a selfish gene model would produce (Dawkins, 1979; Queller, 1992). This is absolutely correct. One can always take a group selection model and translate the mathematics into a classic natural selection model that operates at the level of only the gene. Such mathematical equivalence, however, does not mean that group selection models do not sometimes shed new light on animal behavior, as trait-group selection models necessarily focus attention on what is happening within and between groups, whereas selfish gene models do not do so as readily (Dugatkin and Reeve, 1994). Thus, under certain conditions, group selection models may spur investigators to construct experiments that would not be obvious if they were using selfish gene models.

To get a more comprehensive view of group selection, let's focus on the evolution of specialized foraging behavior in social insect colonies (D. Wilson, 1990).

WITHIN- AND BETWEEN-GROUP SELECTION IN ANTS. Cooperative colony foundation occurs in a number of species of ants where cooperating co-foundresses are *not* closely related (Bernasconi and Strassmann, 1999; Holldobler and Wilson, 1990). Cooperative colony foundation has been especially well studied in the desert seed harvester ant *Messor pergandei*. In terms of between-group selection, in many populations of *M. pergandei*, adult ants are very territorial, and "brood raiding"—in which brood captured by ants from nearby colonies are raised within the victorious nests, and colonies that lose their brood in such interactions die—is seen among young starting colonies in the laboratory (Ryti and Case, 1984; Wheeler and Rissing, 1975). Within groups, all co-founding queens in a nest assist in excavating their living quarters, and each produces approximately the same number of offspring, so there is a positive correlation between the number of cooperating foundresses in a nest and the number of initial workers (brood raiders) produced by that colony (Rissing and Pollock, 1986, 1991; Table 10.5). Nests with more workers—that is, those with more cooperating foundresses—are more likely to win brood raids (Rissing and Pollock, 1987).

TABLE 10.5. Cooperating co-foundresses. In the ant *Messor pergandei*, unrelated queens (here W, Y, B, and O) co-found nests. The reproductive output of queens within a nest tends to be approximately equal. Differences in all three measures across the four queens were not statistically significant. *(From Rissing and Pollack, 1986)*

NEST NO.	NO. OF EGGS	PERCENTAGE LAID BY			
		W	Y	B	O
1	22	—	32	41	27
2	24	38	25	38	—
4	32	41	28	25	6
5	11	45	—	55	—
8	29	28	34	38	—
9	44	34	16	27	23
10	36	31	19	50	—
11	21	43	—	24	33
15	29	38	—	21	41

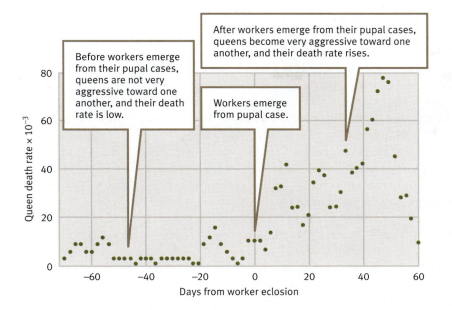

FIGURE 10.20. **From cooperation to aggression.** Queen-queen aggression and queen death rate rise as colonies move to the stage of colony development at which workers emerge from pupae and then begin helping. *(From Rissing and Pollock, 1987)*

Until workers emerge, queens within a nest do not fight, and no dominance hierarchy exists (Figure 10.20). After workers emerge and the between-group benefits of having multiple foundresses are already set in place with the presence of brood raiders, all that remains is within-group selection, which favors being noncooperative. It is at this juncture that queens within a nest often fight to the death.

The scenario depicted above is ideal for studying group selection models of cooperation, as group selection requires the differential productivity of groups based on some trait. In the case of *M. pergandei*, the trait is queen-queen cooperation. Such cooperation is selected against *within* groups because cooperators pay a cost not borne by noncooperators. At the same time, cooperation may be selected for between groups, because groups with many cooperators differentially survive brood raiding—the between-group component of group selection.

The results of these studies with respect to between-group selection have been challenged. David Pfennig constructed a field experiment contrasting single and double foundress associations in *M. pergandei*. Not only did double foundress nests not outlive single foundress nests in Pfennig's experiment, but no brood raiding at all was observed, calling into question all of the elements necessary for group selection to operate in the wild (Pfennig, 1995).

One of the strongest cases to date of group selection comes from Steve Rissing's work on another ant, *Acromyrmex versicolor* (Rissing et al., 1989). In this species, many nests are founded by multiple unrelated queens, no dominance hierarchy exists among queens, all queens produce workers, brood raiding among starting nests appears to be common, and the probability of the nest surviving the brood-raiding period is a function of the numbers of workers produced. *A. versicolor* differs from *M. pergandei*, however, in that *A. versicolor* queens forage after colony foundation (Figure 10.21).

Foraging involves bringing back materials that increase the productivity of the nest's fungus garden, which is the food source for the colony. Increased predation and parasitization makes foraging a dangerous activity for a queen. Once a queen takes on the role of forager, she remains in that role. After a queen becomes the sole forager for her nest, all queens share equally in the

FIGURE 10.21. Extreme cooperation by foraging queen. In the ant *Acromyrmex versicolor*, a single queen (shown in the blowup circle) is the forager for a nest. Such foraging is very dangerous, but all food collected is shared among (unrelated) queens.

food produced by the fungus garden—the forager assumes both the risks and the benefits of foraging, while the other queens in her nest receive the benefits but do not pay the costs (Table 10.6). But cooperation—in this case, extreme cooperation on the part of the forager—within nests appears to lead to more workers. The number of workers affects the probability that a given nest will be the one to survive the period of brood raiding, providing the between-group component necessary for cooperation to evolve (Cahan and Julian, 1999; Rissing et al., 1989; Seger, 1989). Indeed, when Rissing and his team experimentally blocked the foraging queen's opportunity to leave the nest and gather food, none of the remaining queens became the new forager, causing failure of the fungus garden and a decrease in the number of workers produced by the remaining queens (Pollock et al., 2004; Rissing et al., 1996).

TABLE 10.6. Harmony among *Acromyrmex versicolor* queens. In *A. versicolor*, unrelated queens co-found nests, and a single queen takes on the dangerous role of forager for everyone in the nest. Reproduction within nests is approximately equal between foragers and nonforagers. *(Based on Rissing et al., 1989)*

	FORAGER	NONFORAGER
Mean number of primary eggs	8.6	8.5
Mean primary egg length	0.52	0.54
Mean number of total eggs	20.37	18.94

As always, we could analyze the cooperation described in both ant examples without using group selection models, and instead rely on classic models that do not decompose selection into within- and between-group selection, and we would come to the same conclusions we arrived at from the trait-group perspective. That said, in both *M. pergandei* and *A. versicolor,* we see systems in which population biology and demographics may match those that are postulated in trait-group models. The multiple nests, intense competition between nests, and multiple unrelated foundresses in these species make them ideal for an analysis at both the within-group and between-group levels.

Coalitions

Much of the work on cooperative behavior involves pairwise, or dyadic, interactions. In these **dyadic interactions**, two individuals interact in such a way that the fitness of each is affected by both its own action and the action of its partner. Cooperation can also occur in **polyadic interactions**—that is, interactions that involve more than two individuals (see Conservation Connection box). One example of polyadic interactions involving cooperation is **coalition** behavior, defined as a cooperative action taken by at least two individuals or groups against another individual or group. When coalitions exist for long periods of time, they are often referred to as **alliances** (A. Harcourt and de Waal, 1992).

Coalitions have been documented in numerous primate species, hyenas (*Crocuta crocuta*), wolves (*Canis lupus*), lions (*Panthera leo*), cheetahs (*Acinonyx jubatus*), coatis (*Nasua narica*), and dolphins (*Tursiops truncatus*; A. Harcourt and de Waal, 1992; Mesterton-Gibbons et al. 2011; Silk, 2007; Figure 10.22). In most instances, coalitions involve an animal intervening in a dyadic, usually aggressive, interaction between other group members (A. Harcourt and de Waal, 1992). Often, the intervening individual is dominant to others involved in the interaction. In primates, interventions often take the form of the intervener coming to the aid of one of the two other individuals involved in an interaction (A. Harcourt and

FIGURE 10.22. Coalitions. (A) Three male dolphins swim together, forming a long-term coalition (or alliance). Such male coalitions "herd" females. A female is seen to the left of the three males. Occasionally different alliances join together to form superalliances that compete against other such superalliances. **(B)** Pairs of male chimps often form coalitions to act against larger, more dominant, individuals. *(Photo credits: Richard Connor; Frans de Waal)*

A

B

Cooperation, the Tragedy of the Commons, and Overharvesting

Work in fisheries conservation and management has the dual goal of protecting species and providing an economic good (food) to the public. Rules are often put into place about how much a person, a company, or even a country can fish in a given area. These rules are, in effect, attempts to make people act cooperatively, in the sense of reducing the threat of overharvesting. Yet despite such attempts, overharvesting fish populations remains a serious problem, to the extent that many species are threatened to the level of extinction. Why is this overharvesting occurring, and what can be done to solve the problem (Kraak, 2011)?

To understand why people, companies, and countries don't abide by the rules regarding harvesting fish populations, consider what Garret Hardin has famously referred to as the "tragedy of the commons." Hardin asks the reader to consider a pastoral fable: "Picture a pasture open to all. . . . As a rational being, each herdsman seeks to maximize his gain. Explicitly or implicitly, more or less consciously, he asks, 'What is the utility to me of adding one more animal to my herd?'" (Hardin, 1968; Figure 10.23).

Adding one additional animal—let's say a goat—to his own herd grazing on communal land gives the herdsman one more animal that he can use or sell. The herdsman's net benefit is one goat. But there is a cost: the overgrazing of the commons that is caused by the herdsman's extra animal. This cost is shared by everyone who grazes a herd on the commons, so even if the cost is quite large, the part that the individual herdsman pays is only a small fraction of one goat. As a result, Hardin postulates: "The rational herdsman concludes that the only sensible

FIGURE 10.23. The tragedy of the commons in grazing animals. Harden illustrated the tragedy of the commons by focusing on the decisions that people had to make about how often they should allow their herd animals (here angora sheep) to feed on a "commons" pasture that is shared by all in the community. (*Photo credit: A.N.T. Photo Library/NHPA*)

course for him to pursue is to add another animal to his herd. And another. . . . But this is the conclusion reached by each and every rational herdsman sharing a commons. Therein is the tragedy. Each man is locked into a system that compels him to increase his herd without limit—in a world that is limited."

In the case of harvesting fish, the commons area is the ocean, rather than the grazing pasture, and being altruistic means not overharvesting fish rather than not adding another grazing animal to the field. But the problem is the same as Hardin described, as is the potential end state—the tragedy when the commons falls. How to prevent this outcome is something that economists, biologists, psychologists, and many other researchers have been trying to achieve for decades. But they have done so with only moderate success, as the commons problems exists in many forms today.

In an article titled "Exploring the 'Public Goods Game' Model to Overcome the Tragedy of the Commons in Fisheries Management," Sarah Kraak (2011) suggests that future work minimizing the tragedy of the commons in fish populations should focus on (1) eliminating the options for users (people who use commons) to "buy out" of being altruistic—that is, eliminating the possibility of buying the part of the commons allocated to some other person or institution; (2) eliminating anonymity in the commons, so that fishermen are known to one another and the general community, thereby allowing for reputations to emerge; (3) providing those in the commons with detailed information about the state of the resource (in this case, fish population size); and (4) face-to-face communication among those in the commons and those managing the commons.

de Waal, 1992). This need not be the case, however, as intervening animals may break up an interaction between two others, without favoring either combatant (Dugatkin, 1998a, 1998b; R. Johnstone and Dugatkin, 2000). This type of intervention has been found in various primate species (A. Harcourt and de Waal, 1992), as well as in the cichlid fish, *Melanochromis auratus* (Nelissen, 1985).

We will examine two examples of coalitions, one in primates and one in dolphins. In each of these cases, coalitions form among males to gain access to reproductively active females.

COALITIONS IN BABOONS

Craig Packer has studied male reproductive coalitions in baboons (*Papio anubis*). In *P. anubis*, a male solicits coalition partners by rapidly turning his head between the solicited animal— the individual from which he is requesting aid—and his opponent, while at the same time threatening his opponent (Packer, 1977; Figure 10.24). In his observation of coalition formation, Packer documented ninety-seven solicitations that resulted in coalitions being formed. On twenty of these occasions, the opponent was consorting with an estrous female, and this increased the probability of a coalition being formed between the other two individuals (the enlisting male and the solicited male). On six of these twenty occasions, the estrous female deserted the opponent and went to the enlisting male, suggesting a benefit to coalition formation, at least for the enlisting individual.

What about the costs and benefits to the animal that is solicited into a coalition? Joining a coalition can be costly to solicited individuals, who rarely obtain access to the estrous female but who risk being attacked by the opponent. Packer's results suggest that solicited males may overcome such costs by having the individual that enlisted them respond when they themselves need help (reciprocal coalitions). In fact, Packer discovered that baboons had favorite partners, and that favorite partners solicited each other more often than they solicited other group members, suggesting alliance formation.

FIGURE 10.24. Baboon coalitions. A male baboon (middle) involved in an aggressive interaction (with male on left) will often solicit others to aid him by turning his head in the direction of a potential coalition partner (male on right).

Enlisting (soliciting) male

Opponent

Solicited male

Other studies of coalitions and their effect on access to reproductive females in baboons have generally reported results similar to Packer's findings (Noe, 1986; Smuts, 1985). But Bercovitch found that in olive baboons (*Papio cyanocephalus anubis*), males that enlisted others in coalitions were no more likely than solicited males to obtain mating opportunities with females, and baboons that declined to join a coalition were again solicited in the future (Bercovitch, 1988). The reason for the discrepancy between Packer's and Bercovitch's studies is as yet not clear, but it may be tied to the presence of a clear alpha male in the Bercovitch study, but not in the Packer study (Hemelrijk and Ek, 1991).

ALLIANCES AND "HERDING" BEHAVIOR IN CETACEANS

Richard Connor and his colleagues have uncovered an interesting example of alliance formation in bottlenose dolphins (*Tursiops truncatus*; Connor, 1992; Connor et al., 1992, 2001, 2011; Krutzen et al., 2003). What makes this case particularly intriguing is that dolphins are notoriously difficult to track for long periods of time, and thus while ethologists have generally thought dolphins had complex social netwoks in the wild, these have been very difficult to document. Connor and his colleagues examined pairs and trios of males forming close associations in the bottlenose dolphin population of Shark Bay, in Western Australia. He and his colleagues found not one, but two types of alliances between male dolphins, both of which were most common among males that were "herding" reproductive females to keep them close (Connor et al., 2011).

"First-order" alliances in dolphins are composed of pairs or trios of males acting in a coordinated fashion to keep females by their side, presumably for the purpose of mating. Males in first-order alliances stay very close to one another, and alliances remain stable for many years. When females herded by an alliance of male dolphins try to swim away (as they often do), the males act in a very coordinated, aggressive manner to prevent the females from leaving (Connor et al., 1992).

What makes alliance formation in dolphins unique is that different first-order alliances also join together in "second-order" superalliances and aggressively attack and "steal" females from other alliances. Connor showed that on two occasions, a defending alliance was assisted by another alliance in its attempts to maintain the female it was herding, creating a battle of second-order alliances. Second-order alliances have been documented in only one other species—humans—and Connor argues that the complex social interactions inherent in dolphin superalliances, as well as in other aspects of dolphin society, may explain the evolution of large brain size in dolphins (Connor, 1992).

A Phylogenetic Approach to Cooperation

In conjunction with inclusive fitness theory (see Chapter 9), the work we have examined so far provides the conceptual foundations for predicting when natural selection favors cooperation. We can now examine other aspects of cooperative behavior, including a phylogenetic component, which provides a window through which to examine how evolutionary history may affect cooperative behavior.

Recall that phylogenetic analysis allows us to test whether a trait may be common in a group of animal species as a result of common ancestry rather than independent selection regimes. In other words, phylogenetic analysis lets us ask whether one reason that a trait is common within a taxa is that all members of that taxa share a common ancestor that possessed this trait. Natural selection may maintain a trait in many related species, even when such a trait originated in a common ancestor, but phylogenetic analyses allow ethologists to examine such things as the number of independent origins of cooperative behavior.

PHYLOGENY AND COOPERATIVE BREEDING IN BIRDS

Numerous adaptationist hypotheses have been put forth to explain when natural selection should favor cooperative breeding behavior in birds (J. L. Brown, 1987; Stacey and Koenig, 1990; see Chapter 9). Early studies of cooperative breeding in birds, however, emphasized the role of phylogeny in cooperative breeding (J. L. Brown, 1974; D. Davis, 1942; Fry, 1977; Hardy, 1961). Ethologists now recognize that both adaptationist and phylogenetic approaches are necessary to fully understand the evolution of cooperative breeding.

Employing phylogenetic analyses made possible by more powerful computers, Scott Edwards and Shahid Naeem examined cooperative breeding in birds from a phylogenetic perspective (S. Edwards and Naeem, 1993). Using 166 species of cooperatively breeding passerine birds in ninety-seven genera, they began their work by testing whether the distribution of cooperatively breeding species was random within the genera of passerine birds on which they focused. Edwards and Naeem created a computer simulation to predict what the distribution of cooperative breeding species would be if they distributed into genera simply based on the number of species in that genera. They found that the distribution of cooperative breeding species in nature differed significantly from the random distributions generated by computer simulations, with some genera having more than the expected number of cooperatively breeding species, and others less than the expected number (Figure 10.25).

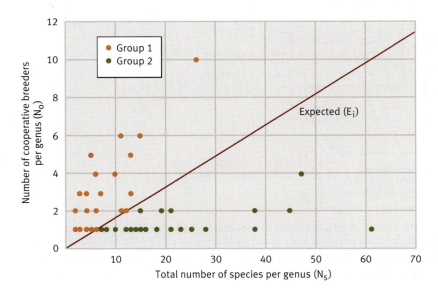

FIGURE 10.25. Phylogeny and cooperative breeding. The solid line represents the slope expected if the number of species with cooperative breeders per passerine genus (seventy-one) were proportional to genus size. In group 1, cooperative breeding was overrepresented; in group 2, cooperative breeding was underrepresented. *(Based on S. Edwards and Naeem, 1993, p. 761)*

Edwards and Naeem followed this analysis by using already published phylogenetic trees to examine the distribution of cooperative breeding. For example, their phylogenetic analysis of jays, Australian songbirds, Australian treecreepers, and New World wrens suggests that cooperative breeding may have arisen a very limited number of times in common ancestor(s) of modern-day species. Phylogenetic analysis suggests that in these groups of birds, modern species displaying cooperative breeding had a common ancestor that had this type of mating system, and modern species that do not display cooperative breeding had a common ancestor in which cooperative breeding was absent.

PHYLOGENY AND COOPERATION IN SHRIMP

Phylogenetic studies of cooperation are not limited to birds. For example, closely related species of *Synalpheus* shrimp, which live in sponge hosts, vary in the extent of cooperative behavior when defending their sponge hosts. For example, "sentinel" shrimp respond to intruder danger by recruiting others to help defend their sponge host. The sentinel and recruits then display an antipredator behavior called "snapping." Phylogenetic analysis of thirteen species discovered three independent origins of cooperative defense in these shrimp (J. E. Duffy and MacDonald, 2010; Duffy et al., 2000; Hultgren and Duffy, 2011; Figure 10.26).

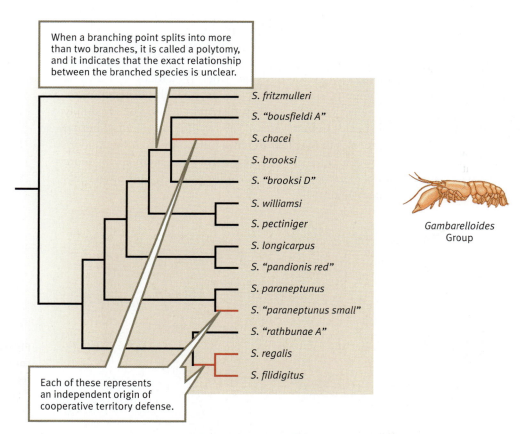

FIGURE 10.26. Multiple origins of cooperative behavior in shrimp. A phylogenetic analysis of closely related species of *Synalpheus* shrimp found three independent origins of cooperative territory defense (shown in red). The analysis was based on both morphological and molecular genetic data (Duffy et al., 2000).

PHYLOGENY AND COOPERATION IN SOCIAL SPIDERS

Most spiders do not live in groups, instead building individual webs and hunting alone. There are, however, a small number of social spider species—23 out of about 39,000 species—in which individuals display extreme levels of cooperation (Agnarsson et al., 2006). In these species, individuals build very large communal webs, jointly maintain these webs, cooperatively hunt for prey, and cooperate in raising brood born in their colonies (Figure 10.27).

Ingi Agnarsson and his colleagues examined the phylogenetic history of social spiders to estimate the number of times sociality had independently evolved in this group (Agnarsson et al., 2006; Blackledge et al., 2011). That is, of the twenty-three species in which sociality has been recorded, how many of these represented independent evolutionary events? Their phylogenetic analysis found that sociality had evolved either eighteen or nineteen different times in spiders—this was evident because, except for one instance, sociality was scattered across the spider phylogenetic tree, and not clustered in species with a common ancestor that displayed sociality. Eighteen or nineteen is a remarkably high number of evolutionary origins for cooperation. And, indeed, the twenty-three social species that have been documented today appear to represent only a small fraction of the number of social spider species that have existed through evolutionary time, as Agnarrson and her colleagues estimate that most of the spider species that evolved sociality have gone extinct.

This phylogenetic analysis of spider sociality has implications for other evolutionary/behavioral questions. For example, it seems that cooperative nest maintenance, cooperative foraging, and other components of social spider life may be "evolutionary dead ends" in the sense that they are associated with high rates of extinction. Why might this be so? Agnarrson and his colleagues argue that although the short-term benefits of sociality in spiders—increased foraging success and so on—may allow for social spider species to initially prosper, sociality has a long-term cost. Social spiders are very inbred, and display very skewed sex ratios, with females dramatically outnumbering males, sometimes in a 10:1 ratio (Aviles and Maddison, 1991; Aviles et al., 1999, 2000; Johannesen et al., 2002; Lubin and Crozier, 1985; Roeloffs and Riechert, 1988; D. R. Smith

FIGURE 10.27. Communal webs.
A giant web built by a communal spider group. *(Photo credit: AP Photo)*

and Hagen, 1996). In the long run, as inbreeding and skewed sex ratios become more and more pronounced, the probability of extinction may increase.

Interspecific Mutualisms

We have been focusing on cooperation among individuals from the same species. There is, however, much evidence that *interspecific* cooperation (cooperation between individuals from different species) is important in shaping animal social behavior as well (Boucher, 1985; Bronstein, 1994; B'shary and Bronstein, 2004; Connor, 1995, 2010; Kawanabe et al., 1993; Leigh, 2010; J. Thompson, 1982). Such cooperation *between species* is usually referred to as **mutualism**. A classic example of mutualism is that between ants and butterflies. Below we will consider one aspect of this mutualism.

ANTS AND BUTTERFLIES—MUTUALISM WITH COMMUNICATION?

In numerous species of butterflies and ants, a mutualistic relationship has developed in which butterfly pupae and larvae produce a sugary secretion that ants readily consume, and ants protect the larvae from predators such as certain species of wasps and flies (Pierce et al., 2002; Quek et al., 2007). In such mutualistic relationships individuals in both species are better off than they would be otherwise (Eastwood et al., 2006).

Naomi Pierce has been studying the mutualistic relationship between the imperial blue butterfly (*Jalmenus evagoras*) and a species of ants (*Iridomyrmex anceps*; Figure 10.28). The benefits to both parties in this mutualism are large. Pierce and her colleagues have found that when faced by predation in nature,

FIGURE 10.28. Butterflies and ants in a mutualistic relationship. In the mutualism between the butterfly *Jalmenus evagoras* and the ant *Iridomyrmex anceps*, butterfly larvae cannot survive in the absence of ants, and ants receive some of their food from the nectar produced by the butterfly larvae (here an adult *J. evagoras* is shown). *(Photo credit: Naomi Pierce)*

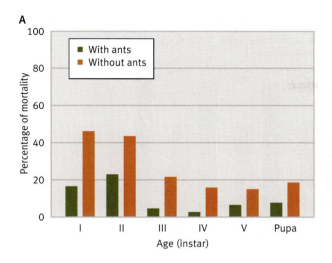

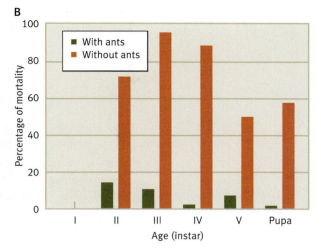

FIGURE 10.29. Butterflies need their ant partners. The probability of survival of *Jalmenus evagoras* larvae and pupae when predators were present was much higher when ants were present than when they were experimentally excluded at two Australian field sites: **(A)** Mt. Nebo and **(B)** Canberra. *(From Pierce et al., 1987, p. 242)*

butterfly larvae are much less likely to survive when ants are experimentally removed from their environment (Figure 10.29). While ants can survive in the absence of the nectar they consume from larvae and pupae, they obtain a significant portion of their nutrients from their butterfly larvae partners (Fiedler and Maschwitz, 1988; Pierce et al., 1987, 2002).

This ant-butterfly mutualism involves costly investment in the other by both parties. Larvae raised in a predator-free laboratory environment pupate later at a much larger size (Pierce et al., 1987), as they are able to modify the amount of nectar they secrete and use the nutrients normally distributed to ants for their own development. Size in both male and female *J. evagoras* is correlated with reproductive success. Pupating early represents a significant cost because smaller body size as a result of earlier pupating leads to reduced reproductive success in the butterflies (Elgar and Pierce, 1988; G. Hill and Pierce, 1989; L. Hughes et al., 2000). Although the costs to ants for protecting butterfly larvae have not been quantified, it is likely an increased risk of detection by their own predators and parasitoids, as well the metabolic costs associated with defense (Pierce et al., 1987).

Travasso and Pierce extended the study of the *J. evagoras/I. anceps* mutualism by examining whether there was interspecific communication in this system. Ants are almost deaf when it comes to airborne sounds, but they are quite sensitive to vibrational cues (Travasso and Pierce, 2000). Given this, Travasso and Pierce hypothesized that vibrational cues may play a role in communication between ants and butterflies. They found that larval sound production (stridulation) was higher when ants were in the vicinity, suggesting that vibrational cues were designed to be used as a way to communicate with ant guards.

In a follow-up experiment, pairs of butterfly pupae were tested together, one of the pair having been muted when the experimenters applied nail polish to its stridulatory organs. Then, using a preference testing device that included two bridges on which the ants could move about, Travasso and Pierce examined whether ants were more attracted to the muted individual in a pair or the individual that was free to produce vibrational communication. They found that ants demonstrated a clear preference for associating with the pupae that could (and did) produce vibrations, providing further evidence that vibrational communication plays a real role in this ant-butterfly mutualistic relationship (Figure 10.30).

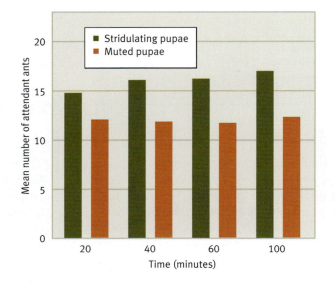

FIGURE 10.30. Stridulating attracts ants.
Stridulating *J. evagoras* pupae attracted more ants
than *J. evagoras* pupae that had been experimentally
muted. *(Based on Travasso and Pierce, 2000)*

MATH BOX 10.1

Evolutionarily Stable Strategies (ESS)

An *evolutionarily stable strategy* is defined as "a strategy such that, if all the members of a
population adopt it, no mutant strategy can invade" (Maynard Smith, 1982). Here *mutant* refers to
a new strategy introduced into a population, and successful invasions center around the relative
fitness of established and mutant strategies. If the established strategy is evolutionarily stable, the
payoff from the established strategy is greater than the payoff from the mutant (new) strategy. To
see this more formally, let's consider two strategies, I and J. We will denote the expected payoff of
strategy I against strategy J as $E(I, J)$, the payoff of J against I as $E(J, I)$, the payoff of I against I as
$E(I, I)$, and the payoff of J against J as $E(J, J)$. Strategy I is an ESS if the following two conditions are
met:

Either

$$E(I, I) > E(J, I) \tag{1}$$

or

$$E(I, I) = E(J, I), \text{ but } E(I, J) > E(J, J) \tag{2}$$

If condition (1) holds true, then I does better against other I's than J does against I. Since we
start with a population full of I's (except for one J mutant), this condition ensures that I is an ESS.

Condition (2) addresses what happens when I and J have the same fitness when interacting with
I. If this is true, then J may now reach higher frequencies by chance, since when J occurs as a mutant
it does just as well as I. I is then an ESS if it does better than J when each is paired up against J, that
is, when $E(I, J) > E(J, J)$.

An ESS Analysis of TFT and the Prisoner's Dilemma Game

Axelrod and Hamilton (1981) examined whether tit for tat (TFT) and the "always defect" (ALLD) strategy were evolutionarily stable—that is, whether they could resist invasion from mutants if they themselves were at a frequency close to 1. They began by proving that if the strategy ALLD or a strategy that alternates D and C (ALTDC) could not invade TFT, then no single pure strategy could invade. They then considered whether ALLD or ALTDC could in fact invade TFT.

Recall the payoffs of the prisoner's dilemma (T, R, P, and S). If we let w equal the probability of interacting with the same player on the next move of a game, then the moves of the game can be represented as a geometric series, and the expected number of interactions with a given opponent is equal to $1/(1 - w)$. So, for example, if $w = 0.9$, the expected number of interactions with a given partner is 10 $[1/(1 - 0.9)]$.

When TFT is close to being at a frequency of 1, virtually all TFT players meet other TFT players, and their payoff is:

$$R + wR + w^2R + w^3R + \ldots, \text{ which sums to } R/(1 - w) \tag{1}$$

An ALLD mutant would have all its interactions with TFT, and its payoff would be:

$$T + wP + w^2P + w^3P + \ldots, \text{ which sums to } T + wP/(1 - w) \tag{2}$$

so TFT can resist invasion from ALLD when:

$$R/(1 - w) > T + wP/(1 - w) \tag{3}$$

Solving the inequality for w, ALLD fails to invade TFT when:

$$w > (T - R)/(T - P) \tag{4}$$

Now, when playing TFT, ALTDC gets a payoff of:

$$T + wS + w^2T + w^3T \ldots, \text{ which sums to } (T + wS)/(1 - w^2) \tag{5}$$

TFT is thus resistant to ALTDC invasion when:

$$R/(1 - w) > (T + wS)/(1 - w^2) \tag{6}$$

Solving the inequality for w, ALTDC fails to invade TFT when:

$$w > (T - R)/(R - S) \tag{7}$$

TFT is resistant to *any* invasion when:

$$w > \text{maximum of these two values: } (T - R)/(T - P) \text{ and } (T - R)/(R - S) \tag{8}$$

Dr. Hudson Kern Reeve

You've worked on cooperation in both wasps and naked mole rats. Isn't that a strange combination? How did you settle on these species?

Superficially, it is indeed a very strange combination—social paper wasps are aerial insects and naked mole rats are subterranean mammals! But there is a deep evolutionary connection between them that has stimulated my interest in both: I think that both manifest common principles of social evolution. The ways that the societies are organized are actually quite similar, with high reproductive skews (near reproductive monopoly by one or a few individuals) occurring among cooperating relatives. In both, the highest-ranking reproductive female behaviorally enforces her reproductive dominance.

I know you spend a good chunk of time engaged in fieldwork each year. Do you find that long hours of behavioral observation and tracking lead you toward new work on cooperation? Could you provide an example?

Yes, I think ultimately that all of the best evolutionary theories are rooted in field observation. For example, Bill Hamilton, the architect of kin selection theory, was a superb naturalist. My own view that social organisms, including social insects, engage in reproductive transactions (as embodied in "transactional skew theory") grew out of repeated observations of resource exchanges and *restraint* of aggression in social wasp colonies. Social wasps are constantly exchanging food and water

and are curiously nonaggressive when each is laying an egg, as if they had the equivalent of a "social contract" over the division of resources, and, ultimately, reproduction. I don't think that I ever would have seriously considered the idea that social wasps are reproductively paying each other to cooperate, had I not watched nearly

a thousand hours of videotapes of field colonies.

Perhaps more theory is devoted to the evolution of cooperation than to any other issue in ethology. Why do you think that is?

I think that it has attracted a great deal of theoretical interest because it represents an evolutionary puzzle that even had Darwin fretting: How could an organism ever achieve a reproductive advantage by enhancing the reproductive output of another organism at the expense of its own? Hamilton provided one important answer to this puzzle through his kin

selection theory. But this solution left unanswered the question of how cooperation could evolve between unrelated organisms—the latter, of course, has been the focus of intense theoretical activity. No doubt, we humans are especially interested in cooperation among nonrelatives, because each of us observes and experiences it regularly (to a varying degree!) every day.

Do you ever encounter the sense that some in the scientific community simply believe animals are not cognitively sophisticated enough to undertake acts of cooperation and altruism?

Yes, unfortunately, this is still a pervasive view. I think that this view is highly questionable. An organism needn't have a huge brain in order to engage in high-order social interactions. (I refer back to the notion that social wasps can have "social contracts" over reproduction!) I think that many behavioral biologists currently underestimate the cognitive complexity of their organisms, and the situation isn't helped by the fact that most theorists find it easiest to model simply behaving organisms!

In fact, there is now a major push to think of social organisms (especially social insects) as self-organizing robots whose rules of social interaction are very simple. I think that the latter view is headed in the wrong direction, because it ignores the evidence that organisms are highly conditional (context-specific) in their social behavior. I think that the evidence will eventually reveal that most social organisms are best viewed as proactive inclusive fitness maximizers; that is, they behave according to their cognitive *projections* of the inclusive fitness consequences of alternative social actions—and are not anything like toasters with a few settings determining simple input-output (stimulus-response) relationships. An inclusive fitness projector will always win evolutionarily over a self-organizing robot, provided that the neural machinery costs are not too great, and it is precisely the latter that I think has been systematically overestimated. Another way to put this is that the difference between a large brain and a small brain is not that the latter results in less sophisticated behavior; rather, the smaller brain still enables very sophisticated behavior, but over a somewhat narrower range of conditions (the ones regularly encountered by its bearers).

How have the fields of mathematical economics and political science contributed to our understanding of the evolution of cooperation?

The influence of economics and political science on evolutionary biology is immense, in large part because the theoretical apparatus of game theory was developed and refined in these two disciplines by people such as von Neumann and Nash and many others. Now, game theory is *the* central theoretical tool in understanding the evolution of social behavior.

Ironically, I think that evolutionary biology, now enriched by game theory, eventually will absorb human economics and political science as

part of itself, because the latter two disciplines are just two subfields of the study of human social behavior, and we evolutionary biologists believe that the only satisfying theory of any organism's social behavior ultimately must be evolutionary! I am sometimes criticized for thinking that social wasps have reproductive transactions on a human economic analogy—surely that is too anthropomorphic! But this criticism gets things backward: If human economic behavior and wasp social behavior are evolved responses to similar selection pressures, then the connection between them is much

> **Game theory is *the* central theoretical tool in understanding the evolution of social behavior.**

deeper than a simple analogy; that is, if they really are manifestations of the same evolutionary principles, then anthropomorphism is the correct stance!

Do you think that the rules governing cooperative behavior are sometimes transmitted culturally in nonhuman primates?

There is growing evidence that this is so, and the important consequence is that, for such primates (and certainly for humans), we need to understand how cultural and biological evolution will interact. Do they proceed independently of each

other, as some believe, or does one somehow entrain the other? This will be a hugely important focus for theoretical and empirical research in the years ahead.

How close are we to having a comprehensive understanding of animal cooperation? What, if anything, remains to be done?

My view is that we are still very far away. There is no shortage of theories, but I suspect the best ones have yet to be developed. What is most limiting are the data that cleanly discriminate among alternative theories. The latter is true in part because theorists have not always been clear about which predictions separate alternative theories and in part because empiricists have not always been good at deriving the right theoretical predictions for their study organisms (or for the contexts in which the latter are studied). This is not to sound pessimistic. On the contrary, it is an extremely exciting time to be a sociobiologist, as we are on the brink of beginning to solve many, many puzzles! In my view, what we cannot afford to lose is the conviction that these puzzles have general, elegant solutions.

Dr. Hudson Kern Reeve is a professor at Cornell University. His work integrates theoretical, empirical, and conceptual approaches to ethology to understand cooperation, kinship, aggression, and the distribution of reproductive opportunities within nonhuman and human groups.

SUMMARY

1. Cooperation typically refers to an *outcome* in which two or more interacting individuals each receives a net benefit from their joint actions, despite the potential costs they may have to pay for undertaking such actions.

2. Cooperation has been studied in taxa ranging from insects and fish to birds and mammals. This work includes studies of cooperation in the context of foraging, predation, antipredator activities, and aggression.

3. In addition to kin-selected cooperation, which we discussed in Chapter 9, three paths to the evolution and maintenance of cooperation in animals have been identified: reciprocity, byproduct mutualism, and group selection. Game theory analyses have been used to model these three paths.

4. Work on cooperation via reciprocity has centered on what is known as the prisoner's dilemma game and a strategy called tit for tat. Recent work in neuroeconomics has shed light on some proximate aspects of human reciprocity by examining which areas of the brain are associated with trust and how individuals respond to cheating by their partners in some economic games.

5. Cooperation via byproduct mutualism occurs when an individual would incur an immediate cost or penalty if it did not act cooperatively, such that the immediate net benefit of cooperating outweighs that of cheating.

6. Group selection models of cooperation have both a within- and between-group component. Within-group selection favors cheating, whereas between-group selection favors cooperation. Every group selection model can be recast as a classic model that averages over different groups, but produces exactly the same result as the corresponding group selection model.

7. Cooperation can also occur in interactions that involve more than two individuals. One example of this type of cooperation is coalition behavior, which is defined as a cooperative action taken by at least two individuals or groups against another individual or group.

8. Phylogenetic analyses can be used to help better understand the distribution of cooperation among related species. The phylogenetic approach does not necessarily conflict with an adaptationist view of cooperative breeding, although it does emphasize the importance of evolutionary history (phylogeny) in our understanding of the distribution of cooperative breeding.

9. In addition to cooperation between members of the same species, there is a great deal of evidence that *interspecific* cooperation—cooperation between members of different species—is also important in shaping animal social behavior.

DISCUSSION QUESTIONS

1. Read G. Wilkinson's (1984) article "Reciprocal Food Sharing in Vampire Bats," in *Nature* (vol. 308, pp. 181–184). Then outline how Wilkinson was able to separate the effects of kinship and reciprocity in his study of vampire bats.

2. Run a small prisoner's dilemma experiment with a few other students. In one group, using coins as payoffs, test pairs of subjects (who cannot communicate with each other in any manner) and tell them beforehand that they will play this game only once. In a second treatment, use pairs of subjects (who cannot communicate with each other), but inform them that they will play this game many, many times together, but do not tell them exactly how many times. In a third and fourth treatment, repeat treatments 1 and 2, but allow the subjects to communicate with each other before the game starts. What sorts of differences and similarities do you predict across treatments? What do the data suggest?

3. Respond to the following statement: Animals aren't capable of human-like thought processes, and therefore they cannot be cooperating.

4. Why do you suppose that work on animal behavior and cooperation draws more attention from other disciplines, such as mathematics, political science, and psychology, than any other area in ethology? What might we learn about human cooperation from studies of animal cooperation? What sorts of things would be difficult to glean about human cooperation by studying animal cooperation?

5. Work from neuroeconomics has shed light on the neurobiology of cooperation in humans. What sorts of evolutionary questions come to mind when you read about these proximate studies in neuroeconomics?

SUGGESTED READING

Axelrod, R., & Hamilton, W. D. (1981). The evolution of cooperation. *Science, 211*, 1390–1396. A classic paper that essentially introduced most ethologists to using game theory to address cooperative behavior.

B'shary, R., & Bergmueller, R. (2008). Distinguishing four fundamental approaches to the evolution of helping. *Journal of Evolutionary Biology, 21*, 405–420. A review of hypotheses about the evolution of helping behavior.

Connor, R. C. (1995). The benefits of mutualism: A conceptual framework. *Biological Reviews of the Cambridge Philosophical Society, 70*, 427–457. A conceptual overview of work on mutualism and byproduct mutualism.

Dugatkin, L. A. (1997a). *Cooperation among animals: An evolutionary perspective.* New York: Oxford University Press. A review of theoretical and empirical work on cooperation.

Kropotkin, P. (1902). *Mutual aid* (3rd ed.). London: William Heinemann. One of the first important books on cooperative behavior in animals.

Packer, C. (1977). Reciprocal altruism in *Papio anubis. Nature, 265*, 441–443. A highly cited study that was one of the first to experimentally study alliances and cooperation.

Zak, P. J. (2011b). The physiology of moral sentiments. *Journal of Economic Behavior and Organization, 77*, 53–65. An introduction to neuroeconomics.

11

Foraging

Searching for and consuming food—**foraging behavior**—is a critical part of every animal's existence. Depending on the circumstances, individuals can survive to reproductive age without being aggressive, they can survive to reproductive age without play, they can survive to reproductive age if they don't cooperate or find a territory. But, they can't survive to reproduce if they never eat. Animals spend a fair share of their available time foraging, whether their food is seeds, nuts, and berries (as in granivores); plants (as in herbivores); living animals (as in carnivores); or dead animals (as in scavengers; Figure 11.1).

Consider the following incredible story of ant foraging. About fifty million years ago, ants began cultivating their own food by entering into a mutually beneficial relationship with certain species of fungi (Caldera et al., 2009; Poulsen and Currie, 2009; Figure 11.2). The ants promote the growth of the fungi (good for the fungi), while also eating the vegetative shoots produced by their fungal partners (good for the ants). Aside from humans, ants are one of the few species on the planet that grow their own food.

Cameron Currie and his colleagues have found that the ant-fungus relationship is incredibly complex (Cafaro and Currie, 2005; Currie, Mueller, et al., 1999; Currie, Scott, et al., 1999). Scientists who study fungus-growing ants have long known of a whitish-gray crust found on and around many ants with fungus food gardens. Research shows that the substance is actually a mass of *Streptomyces* bacteria—a type of bacteria that produces many antibiotics. Currie and his colleagues hypothesized that ants use the bacteria's antibiotics to kill parasites that grow in their fungal gardens (Figure 11.3). That is, they hypothesized that not only have ants evolved a complex relationship with their food source (fungi) but a means to protect their food source from destruction has also evolved in this system.

Four lines of evidence support this claim. First, all twenty species of the fungus-growing ants they examined had *Streptomyces* bacteria associated with them. Second, ants transmit the bacteria across generations, as parents pass the bacteria on to offspring. Third, when male and female reproductive ants are examined (before their mating flights), only females possess the bacteria. This is critical, as only females start new nests that will rely on the bacteria to produce antibiotics, and only females are involved in "cultivating"

A   B

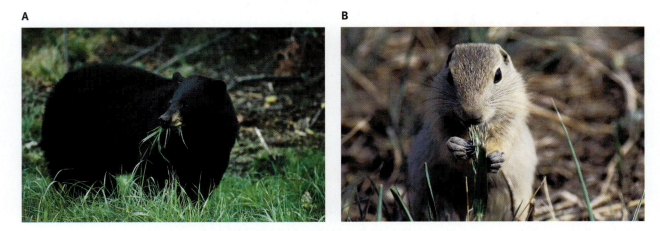

FIGURE 11.1. Foraging. Many animals spend a good deal of their waking hours foraging. Here **(A)** a black bear and **(B)** a Richardson's ground squirrel are foraging. *(Photo credits: Ron Erwin)*

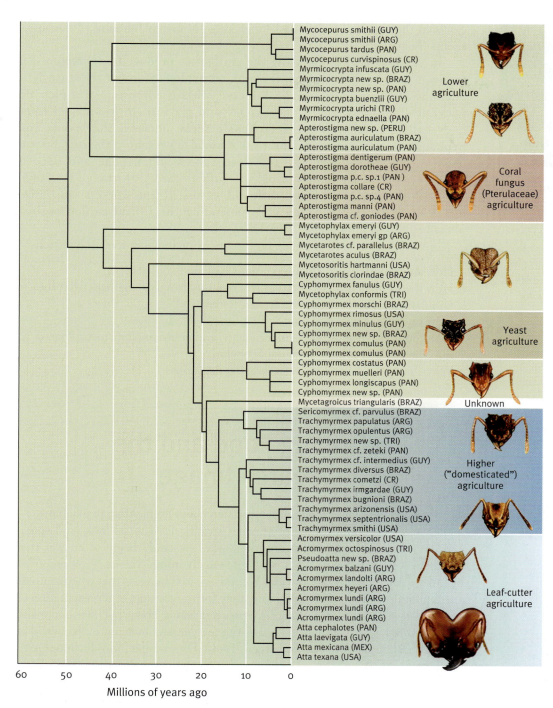

FIGURE 11.2. A phylogeny of fungus-growing ants. The phylogenetic history of five ant "agriculture" systems. *(Adapted from Schultz and Brady, 2008)*

fungus gardens. Fourth, and most important, the bacteria found on fungus-growing ants produce antibiotics that wipe out only *certain* parasitic diseases. When Currie and his colleagues tested the antibiotics produced by the bacteria, they found that they were effective only against *Escovopsis*, a serious parasitic threat to the ants' fungus garden. Other species—those not a danger to fungus-growing ants—were unaffected by *Streptomyces* antibiotics. Recent work has

FIGURE 11.3. Tending the garden.
A worker of the leaf-cutter ant *(Acromyrmex octospinosus)* tending a fungus garden. The thick whitish-gray coating on the worker is the mutualistic bacterium *(Actinomycetous)* that produces the antibiotics that suppress the growth of parasites in the fungus garden. *(Photo credit: Christian Ziegler)*

explored the use of such "natural fungicides" for use by humans as biocontrol agents (Folgarait et al., 2011).

In addition to directly using the antibiotics produced by *Streptomyces* to clean their fungal gardens, the ants meticulously groom these gardens and physically remove fungus infected with *Escovopsis*. Ants pick up parasitic *Escovopsis* spores and hyphae and place them in areas of their body called infrabuccal pockets. Inside these pockets, the spores and hyphae are killed by the antibiotics that are also present in the infrabuccal pockets. The ants then take the dead spores and hyphae and deposit them in a separate pile away from the fungus garden (Little et al., 2003, 2006). Indeed, within nests, different castes of ant workers specialize on different tasks associated with defending the fungal garden against parasites (Abramowski et al., 2011).

In this chapter, we will touch on the following foraging-related questions:

- ► How do animals know what food items look like?
- ► How does foraging theory predict where animals will forage and what they will eat?
- ► How do group social dynamics affect foraging?
- ► How are neurobiology, molecular biology, and endocrinology incorporated into the study of animal foraging behavior?
- ► What role does learning play in foraging decisions?

Finding Food and the Search Image

We begin by addressing a very basic question about animal foraging behavior—namely, how do animals determine what specific food types look (and smell, and feel) like? Although there are many ways we could begin to answer that question, we will focus on what is called **search image** theory. First proposed by Luke Tinbergen in 1960, the idea behind a search image is that when animals encounter a prey type more and more, they form some sort of representation of that target—the prey—and that this representation or image becomes more and more detailed with experience, so that the forager becomes more successful at finding that type of prey (L. Tinbergen, 1960). For example, in Bobwhite Quail (*Colinus virginianus*), the probability of detecting prey increases with each successful prey capture (Gendron, 1986).

There is some debate as to the exact nature of the search image that animals form: Some researchers argue that foragers are keying on one or two salient attributes of the prey (color, movement, pattern), while others argue that the search image formed is closer to some sort of representation of the entire prey item (Langley, 1996; Pietrewicz and Kamil, 1979; Reid and Shettleworth, 1992; Van Leeuwen and Jansen, 2010). In either case, the animals are learning something relevant about their prey, and most animal behaviorists think that the use of the search image likely evolved as a response to the difficulty of finding cryptic prey, and thus is important in assessing what is prey and what is not (Balda et al., 1998; Kamil and Balda, 1990; Zentall, 2005).

Optimal Foraging Theory

In this section, we will examine a class of mathematical models that are collectively known as **optimal foraging theory** (**OFT**; Kamil et al., 1987; Sih and Christensen, 2001; Stephens and Krebs, 1986; Stephens et al., 2007). These models use a form of mathematical analysis known as optimization theory to predict various aspects of animal foraging behavior within a given set of constraints. Optimal foraging theory began in earnest in 1966 with back-to-back papers in the *American Naturalist*, one of which was written by Robert MacArthur and the other by John Emlen (J. M. Emlen, 1966; MacArthur and Pianka, 1966). These papers set the stage for ethological and behavioral ecological models that followed in abundance, especially during the 1970s (see Stephens and Krebs, 1986, for a review of these models).

When animal behaviorists and behavioral ecologists build optimal foraging models, they begin by making a set of assumptions—for example, foragers can't search for more than one item at a time, they only have so many hours in a day to search for food—and then proceed to see which types of food a forager should take, given the constraints. Although there are many optimal foraging models in the animal behavior literature, we will examine four models, which address the following questions:

- What food items should a forager eat?
- How long should a forager stay in a certain food patch?
- How is foraging affected when certain nutrient requirements are in place?
- How does variance in food supply affect a forager's decision about what food types to eat?

WHAT TO EAT

One of the most basic foraging problems faced by an animal is deciding which type of food items should be in its diet and which should be excluded (Charnov and Orians, 1973; Stephens and Krebs, 1986). For example, imagine that a forager could search for, and potentially consume, food types 1, 2, and 3. Should the forager eat all three? Only one? Two of the three? If so, which ones? To tackle this question, animal behaviorists have developed optimality models specifically designed to handle prey choice. We will now examine the underpinnings of this first optimal foraging model and then review a classic test of it.

Let's first consider the simplest possible case—a forager choosing between two different types of food (Caro, 1994a; Figure 11.4). For carnivores, this might be a choice between two prey species; for granivores, the choice could be between two different types of seeds. Indeed, the choice a forager might have to make could even be between two different size classes (small or large) of the same food. In the model, each prey item will have an energy value, encounter rate, and handling time associated with it. For example, let's say that during foraging bouts, one prey type is encountered every three minutes (encounter rate); once it is encountered, such a prey item takes two minutes to kill and ingest (handling time) and provides the forager with 300 calories (energy value). Next, the profitability of a prey item is defined as its energy/handling time. The greater the energy/handling ratio, the greater the profitability of a prey type.

FIGURE 11.4. Foraging decision. Here a female cheetah (the forager) has killed a hare (the prey). In making the decision whether to take hares rather than some other prey into the diet, models of animal foraging behavior assume that foragers will compare the energy value (the amount of calories provided to the forager by eating this prey), encounter rate (how often the prey is encountered by the forager), and handling time (time for the forager to kill and ingest the prey) for each putative prey. *(Based on Caro, 1994a)*

Assume that the prey type with the highest profitability—let's call it prey type 1—is always taken by a forager. This means that the optimal diet problem boils down to two simple questions—should prey type 2 also be taken, and if it should, under what conditions? We begin by examining the assumptions that are made in the basic optimal prey choice model and then proceed to its main conclusion (Charnov and Orians, 1973; Krebs and McCleery, 1984). The model assumes:

▸ Energy intake from prey can be measured in some standard currency (for example, calories).
▸ Foragers can't simultaneously handle one prey item and search for another.
▸ Prey are recognized instantly and accurately.
▸ Prey are encountered sequentially.
▸ Natural selection favors foragers that maximize their rate of energy intake.

With these assumptions in place, a bit of mathematical analysis (Math Box 11.1 at the end of the chapter) produces a fascinating, and somewhat surprising prediction. Recall that we are asking whether a forager should add prey type 2 to its diet. Yet what the model predicts is that the encounter rate with prey type 2—that is, how often a predator encounters this less profitable item—does *not* affect whether it should be added to the diet. Rather, the model predicts that there is a critical encounter rate with the most profitable item (prey type 1)—if the encounter rate the predator experiences for prey 1 is above this critical value, only prey type 1 is taken and, if it isn't, then both prey type 1 and prey type 2 are taken. The decision of whether to add prey type 2 to the diet is not dependent on the forager's encounter rate with prey type 2; rather, it is related to its encounter rate with prey type 1. This basic prediction of optimal diet choice has been tested many times.

John Krebs and his colleagues experimentally tested the predictions of this optimal prey choice model in the great tit, *Parus major* (Krebs, 1978; Krebs et al.,

FIGURE 11.5. Great tit foraging. One classic early experiment using optimal foraging theory had mealworms of different sizes presented on a conveyor belt to great tits.

1978). In a laboratory experiment, great tits were placed in front of a moving conveyor belt (Figure 11.5). Krebs and his team used two different-sized pieces of mealworm as the two different prey item types, and had precise control of the rate at which these two prey types were encountered by the birds, the exact energy provided by both types of prey, and the precise handling time associated with each size of mealworm. With these parameters measured, it was possible to use the model to predict when the birds should take only the most profitable prey types, and when they should take both types of prey.

Krebs and his team found that the optimal prey choice model did a very good job of predicting how the tits would forage, in that it was the encounter rate of the most profitable prey, not the least profitable prey, that determined whether tits took the least profitable items (Figure 11.6). Similar results have

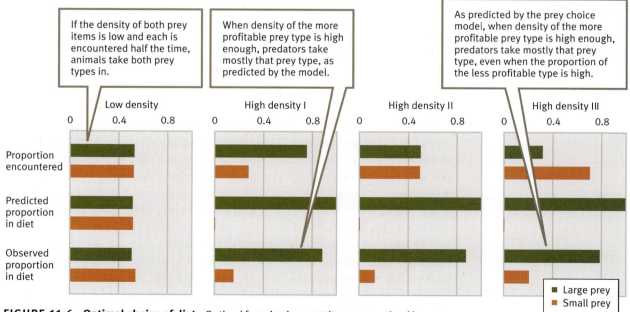

FIGURE 11.6. Optimal choice of diet. Optimal foraging in great tits was examined by Krebs and his colleagues in four density conditions. With a knowledge of exact encounter rates, handling times, and energy values, the researchers were able to predict the birds' optimal diet of larger, more profitable and smaller, less profitable prey. *(Based on Krebs, 1978, p. 31)*

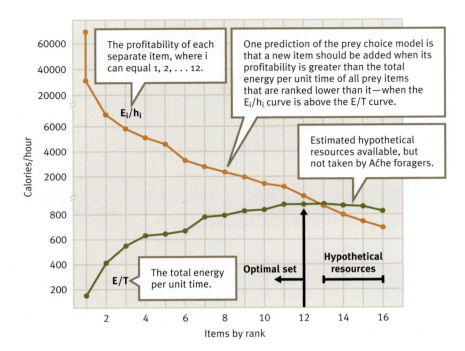

FIGURE 11.7. Optimal prey choice in Aĉhe foragers. Aĉhe foragers select prey items as predicted by the prey choice model. For items ranked 1 through 12, the profitability of a new item is greater than the total energy per unit time for all lower-ranked items. The prey choice model predicts that Aĉhe foragers should take prey items 1 through 12, but not others. The data support this prediction. (*From Hawkes et al., 1982*)

been found when testing the optimal prey choice model in foraging experiments with bluegill sunfish (Werner and Hall, 1974). And this same mathematical model could be used for any type of forager deciding between two prey types—for example, the case of a carnivore predator choosing between two herbivore prey species. This model has also been applied to foraging behavior in humans. For example, Kim Hill has studied foraging in the Aĉhe people—an indigenous tribe of Indians in Paraguay. When Hill examined the prey choices made by Aĉhe foragers, he found that their choices generally matched those of the prey choice model (when the model was expanded to more than two prey items; Hawkes et al., 1982; K. Hill, 2002; K. Hill and Hurtado, 1996; K. Hill et al., 1987; Figure 11.7).

WHERE TO EAT

Another critical decision a forager must make is how long to stay in a patch of food. For example, how long should a hummingbird spend sucking nectar from one flower, given that there are other flowers available, or how long should a bee spend extracting pollen from one flower before moving on to the next one? To address such questions, Eric Charnov developed an optimality model that generated a result known as the **marginal value theorem** (Charnov, 1976; G. Parker and Stuart, 1976).

To see how the marginal value theorem works, imagine a forager feeding in an area that contains different patches of a single type of food. A patch of food for a chimp could be a tree full of fruit; for a bee it might be a flower (Figure 11.8). More generally, a patch is defined as a clump of food that can be depleted by a forager. Once a forager begins feeding in a patch, the *rate* at which it takes in food slows down, as the more the forager eats, the less food remains in the patch. Other less depleted patches will then have relatively large amounts of food availabe, but in order to get to these

FIGURE 11.8. Patch choice. For a bee, different clusters of flowers in a field of flowering plants might represent different patches.

patches, the forager must pay some cost—energy, lost time foraging while traveling, increased rate of predation during travel, and so forth—associated with traveling between patches. The question then becomes how long a forager should stay in a patch that it is depleting before moving on to another patch.

Assuming that we know the rate of food intake, Charnov's model makes clear, testable predictions regarding patch residence time.

▶ A forager should stay in a patch until the marginal rate of food intake—that is, the rate of food intake associated with the next food item in its patch—is equal to that of the average rate of food intake across all patches available. In other words, a forager should stay in a patch T time units, where T is that point in time when its marginal rate of food intake in that patch is equal to the average amount of food it could get in other patches, given that it has to pay a cost to get to such other patches.

▶ The greater the time between patches, the longer a forager should stay in a patch. Increased travel time leads to an increase in the costs associated with such travel, and such costs need to be compensated—remaining in a patch longer is one means by which such compensation can be achieved.

▶ For patches that are already of generally poor quality, the forager should stay longer than if it were foraging in an environment full of more profitable patches. In order to make up for the travel costs associated with a move from a patch, a forager has to stay longer in a poor patch than in a good patch to obtain a fixed amount of energy (Figure 11.9).

Richard Cowie tested one of the predictions of the marginal value theorem using great tits as subjects. Inside a large aviary, Cowie built a series of artificial trees, each of which contained numerous branches

(Cowie, 1977). Attached to some of the branches were sawdust-filled potting baskets. Under the sawdust were mealworms, which were prey for the birds (Figure 11.10A). Cowie was able to calculate the rate of gain in different types of sawdust-filled patches. In addition, he could manipulate the travel time between patches, because each pot had a lid, and the lids could be made to be easily dislodged—creating a short travel time between leaving one patch and starting to forage at another—or the lids could be constructed so as to be very difficult to open, simulating a long travel time. Based on all this, Cowie calculated the optimal time to stay in a patch as a function of travel time between patches. The amount of time birds spent in a patch matched the optimal time predicted by the marginal value theorem (Figure 11.10B).

Any time a resource depletes as a function of use and costs associated with traveling between patches of that resource are present, one can use the marginal value theorem to solve for optimal patch time. This approach

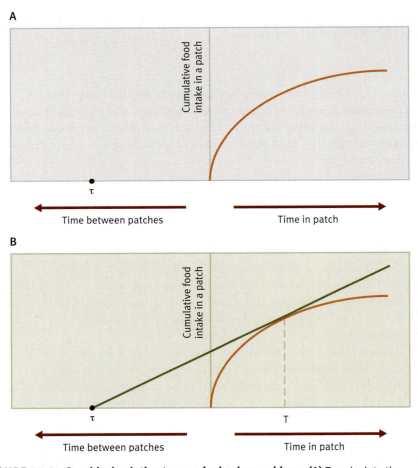

FIGURE 11.9. Graphical solution to marginal value problem. (A) To calculate the optimal time for a forager to remain in a patch, we begin by drawing a curve that represents the cumulative food gain in an average patch in the environment. Then, going west on the *x*-axis we find the average travel time between patches (τ). **(B)** We then draw a straight line from τ that is tangent to the food gain curve. From the point of tangency, we drop a perpendicular (dashed) line to the *x*-axis, which gives us an optimal time (T) for the forager to stay in the patch.

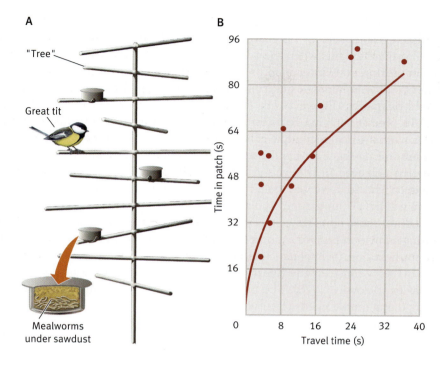

A

"Tree"

Great tit

Mealworms
under sawdust

B

y-axis: Time in patch (s) — 96, 80, 64, 48, 32, 16, 0

x-axis: Travel time (s) — 0, 8, 16, 24, 32, 40

FIGURE 11.10. Optimal time in patch and travel time. In a test of the marginal value theorem, Cowie (1977) constructed **(A)** an artificial tree that allowed him to control both patch quality and travel time. **(B)** The red curve is the predicted optimal time in a patch plotted against the travel time, which was calculated based on the marginal value theorem, while the data points are the observed times the birds stayed in the patch plotted as a function of travel time between patches. (*Based on Krebs, 1978, p. 44*)

has, for example, been used to calculate how long a male should search for and mate with a female (where the female is now the "patch"; G. Parker and Stuart, 1976).

SPECIFIC NUTRIENT CONSTRAINTS

In addition to maximizing their rate of energy intake, animals need to spend time foraging for specific nutrients that are essential for survival. For example, many animals need to take in very small quantities of trace elements like zinc in order to remain healthy, and this requirement may compromise their general foraging rules—rules designed to maximize energy intake per unit time. If looking for a trace element takes time away from other activities—searching for other foods, antipredator behavior, mating, and so forth—fulfilling minimum requirements can be very costly. These constraints appear to be particularly severe in herbivores, that often must take in large amounts of low-energy food to survive (Crawley, 1983; Roguet et al., 1998).

Let's examine this in more detail by focusing on Gary Belovsky's study of nutrient constraints in the moose (*Alces alces*)—a species in which the nutrient constraint is sodium intake (Belovsky, 1978). Sodium is a particularly good candidate for a nutrient constraint study because vertebrates require large amounts of sodium (this chemical is lost in urine), sodium is scarce, and besides water, sodium is the only nutrient for which a "specific hunger" has been documented in animals (Stephens and Krebs, 1986).

Moose have a particular problem in habitats where sodium is restricted to aquatic plants, because such plants are usually inaccessible in the winter when lakes are frozen. This, coupled with the fact that terrestrial plants provide much more energy per unit time than aquatic plants, presents a difficult problem for a foraging moose—namely, how much time to spend eating aquatic plants when the lakes are not frozen. Foraging for such aquatic plants during nonwinter months provides much-needed sodium (which can be stored by the moose), but it takes time away from foraging for energy-rich terrestrial plants (Figure 11.11).

To predict foraging behavior in the above situation, animal behaviorists use a mathematical technique called linear programming. Linear programming models are designed to handle optimality problems in which some optimum must be achieved in the face of a particular constraint. In our case, natural selection should favor any behavioral strategy that provides moose with the most energy per unit time, subject to the constraint that a certain amount of sodium must be ingested. In practice, we need to predict how much time during nonwinter periods a moose should devote to foraging for energy-poor, but relatively sodium-rich, aquatic plants.

Belovsky constructed a linear programming model to predict moose foraging behavior (Belovsky, 1978). The model included the minimum amount of food a day that a moose needs to survive, how quickly a moose digests its food, the energy value of aquatic and terrestrial plants, and the sodium constraint. It turns out that a moose needs to take in about 2.57 grams of sodium a day in the summer to get enough sodium to make it through the winter. Based on all this, the linear programming model that Belovsky built predicted that moose should spend about 18 percent of their summer foraging time on aquatic plants. The actual percentage of time spent foraging on aquatic plants, when such plants are accessible, closely matches the model's prediction. Because of the success of the sodium constraint model, Belovsky has extended his use of linear programming techniques to other herbivore foraging problems.

FIGURE 11.11. Specific foraging constraints. Moose need salt, and they acquire it primarily from energy-poor plants. This takes time away from foraging on energy-rich terrestrial plants. Linear programming models can help predict the moose's foraging behavior.

RISK-SENSITIVE FORAGING

One aspect of foraging that we have ignored so far is the *variability* associated with the rate of obtaining items of food. In the language of statistics, this variance is referred to as *risk*—the term was first used in economics, where more variance implied a greater chance of loss (or gain). The word *risk,* when used in foraging models, is sometimes confused with risk in the sense of danger—but one type of risk need not have anything to do with the other, and it is important to keep them separate in your thinking. When the word *risk* is used in this section, it is used in the economic sense—increased variance in prey availability increases both the odds that an animal will find very little prey and the odds that it will find a bonanza of food.

Consider the following scenario: An animal can choose to forage for six hours in one of two patches. Both patches have the same exact type of food item, and the animal has been trained so that it knows what to expect in each patch. In patch 1, each forager will always receive eight items. In patch 2, there is a 50 percent chance a forager will receive sixteen food items and a 50 percent chance it will get nothing. The mean number of food items that the foragers can expect in both patches is identical (eight), but the variance (risk) in food intake per hour is greater in patch 2. Should our forager take the differences in the variability into account when deciding between patches? This is the sort of question that **risk-sensitive optimal foraging models** address (Caraco, 1980; McNamara and Houston, 1992; Real and Caraco, 1986; Smallwood, 1996; Stephens et al., 2007).

One key component to understanding risk-sensitive foraging is the hunger state of an animal. To see this, consider foragers that can be in one of three different hunger states. Forager 1 has a hunger state in which it values every new food item equally. Whether it is eating its first or fiftieth item, each item provides it with the same reward value. Forager 2 is fairly satiated, and although every additional item it takes in has some value, each additional item is worth less and less. Think about the value to you of a slice of cake after you have just had two ice cream sundaes. It might be worth something, but not what it would have been worth before consuming the sundaes. Forager 3 is starving, and every additional item it eats is worth more and more (to a limit), perhaps providing the difference between surviving or starving to death (Figure 11.12).

Risk-sensitive foraging models, though they can be mathematically complex, make a very straightforward, but powerful, prediction. Returning to our two patches, which have the same mean number of food items but different variances, risk-sensitive models predict that the hunger state of a forager will determine whether it prefers the patch with more or less variance. Our fairly satiated predators are predicted to be *risk averse*—that is, they should choose to forage in patch 1, as that patch has less risk associated with it. On the flip side of the coin, our starving foragers should opt to be *risk prone*, and select to forage in patch 2—the patch with the greater variance. Forager 3 should be indifferent with respect to variance and should not show preference for one patch type or another.

To see why variance is favored in starving birds but avoided by relatively satiated birds, we need to cast the problem in terms of survival probabilities. For the forager that is not all that hungry, each additional piece of food isn't worth that much more, so it should opt for a consistent food source because

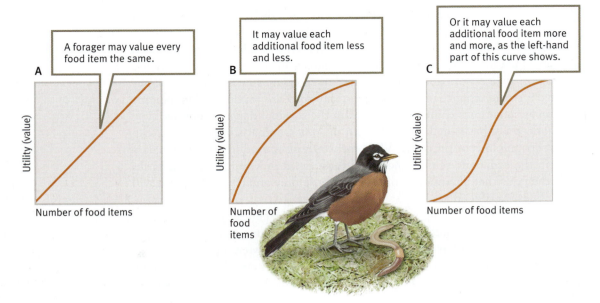

FIGURE 11.12. Utility of food. The bird in **(A)** is predicted to be risk insensitive, the bird in **(B)** is predicted to be risk averse, and the bird in **(C)** is predicted to be risk prone.

sixteen pieces of food isn't worth all that much more than eight food items when the animal is not especially hungry. For our starving bird, eight pieces of food may not be enough to provide it with enough energy to survive the night, but sixteen food items may. In that case, it is worth taking the chance of getting no food at all (one possible outcome in our risky patch) in exchange for the chance of getting sixteen items. The reasoning here is that because eight food items, even if they appear with certainty, would not be enough to pull the starving forager through to the next foraging period, it should opt for the patch with more variance, where there is at least a chance it will receive enough food items to avoid starvation.

As with all the mathematical models in this book, you should not assume that animals make the mental calculations that we just went through, but rather that natural selection favors any "rule of thumb" behavior that allows the animals to solve the problem at hand. In this case, the favored rule of thumb might be "When starving, use patches of food that have high variances."

One of the first, and best, studies of risk-sensitive foraging was that of Thomas Caraco and his team in which they examined foraging behavior in yellow-eyed juncos (Caraco et al., 1980). Caraco and his colleagues presented juncos with two trays containing birdseed, and once a bird made the choice to go to one tray, the other was immediately removed (Figure 11.13). One tray had a "fixed" amount of food—for example, a fixed tray might always have five seeds—while the other tray had a "variable" amount of food—for example, no seeds half the time, and ten seeds half the time. Keep in mind that the variable tray had a mean number of seeds (five) equal to that on the fixed tray. Caraco and his team calculated the value of every new item of food both for hungry birds (that were said to be on a negative energy budget) and for birds that were meeting their daily food requirements (birds on a positive energy budget). Birds on a negative energy budget valued every additional food item much more than those on a positive energy budget (Figure 11.14).

FIGURE 11.13. Optimality models have been tested in juncos. Junco foraging behavior has been used to test numerous optimal foraging models. *(Photo credit: William Leaman/Alamy Images)*

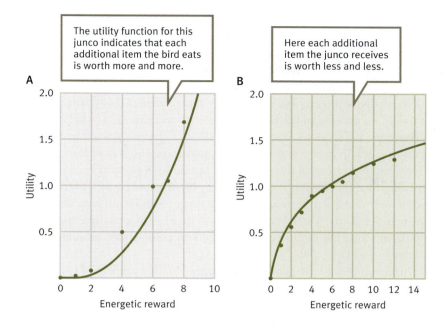

The utility function for this junco indicates that each additional item the bird eats is worth more and more.

Here each additional item the junco receives is worth less and less.

FIGURE 11.14. Utility functions and risk sensitivity. (A) Juncos with this shape of utility function were risk-prone foragers. **(B)** Juncos with this shape of utility function were risk-averse foragers. *(From Caraco et al., 1980)*

When choosing between trays, risk-sensitive theory predicts that birds on a positive energy budget should choose the fixed trays, while very hungry birds should choose the variable trays. The juncos in Caraco's behaved in a fashion very similar to that predicted from theory. Yellow-eyed junco foraging is risk-sensitive, shifting between risk-averse foraging and risk-prone foraging according to energy budgets. Following this study, similar results on risk-sensitive foraging were uncovered in other birds, as well as in mammals and in invertebrates such as honeybees and crayfish. This broad taxonomic distribution of risk-sensitive foraging suggests that variance plays an important role in animal foraging behavior across many species.

Foraging and Group Life

For individuals that spend all or part of their lives in groups, many aspects of group life can affect both foraging behavior and foraging success. With that in mind, we will tackle the following questions in this section:

▸ What is the role of group size on animal foraging?
▸ What role does cooperation within groups play in foraging, and how can this be distinguished from the effect of group size?
▸ How does work on the "public information" available in some groups shed light on animal foraging?

GROUP SIZE

In many species that live in groups, increasing foraging group size increases the amount of food each forager receives (Krause and Ruxton, 2002). This increase may occur because more foragers flush out more prey, or because cooperative

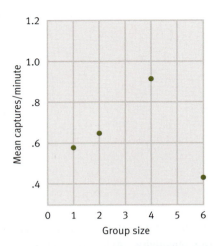

FIGURE 11.15. Group size and foraging success. In bluegill sunfish, the mean rate of prey captured increases with group size until group size reaches about four individuals. *(Based on Mittlebach, 1984, p. 999)*

hunting creates a division of labor between different group members and increases the success rate of the cooperator. Below, we examine an example of each of these possibilities.

FORAGING IN BLUEGILLS. Bluegill sunfish primarily eat small aquatic insects that live in dense vegetation. Such prey items need to be flushed from the vegetation, and success in flushing out prey may increase as a function of group size (Bertram, 1978; Mock, 1980; Morse, 1970). That is, bluegill prey are quite difficult to catch in dense vegetation, but increasing the size of foraging groups may force more prey out of the vegetation. While such prey may not be taken by the specific bluegill chasing them, they may be ingested by other bluegill group members and cause an increase in the average number of prey obtained per group member. Gary Mittlebach examined this hypothesis by experimentally manipulating foraging group size of bluegills in a controlled laboratory setting (Mittlebach, 1984).

Mittlebach placed 300 small aquatic prey (called amphipods) into a large aquarium containing juvenile bluegill sunfish. He examined the success of bluegills that were foraging alone, in pairs, and in groups (ranging from three to six bluegills). Mittlebach measured the number of prey captured per bluegill, and he found a positive relationship between foraging group size and individual foraging success up to a group size of four fish (Figure 11.15). The increased feeding rate per individual was due to two factors. First, more prey were flushed when group size rose. Second, prey clumped together, so when one group member found amphipods, others swam over to this area and then often found food themselves (Figure 11.16).

While increased group size in bluegills does benefit each individual forager in its intake of food (its per capita foraging success), there is no evidence to suggest that bluegill foragers hunt in any coordinated fashion—the flushing effect is merely a byproduct of having more individuals searching for food. The manner in which bluegills increase foraging efficiency in larger groups (that is, by flushing) is only one means of increased foraging efficiency through group living. For example, in a number of species, increased group size reduces the amount of time that any given individual needs to devote to antipredator

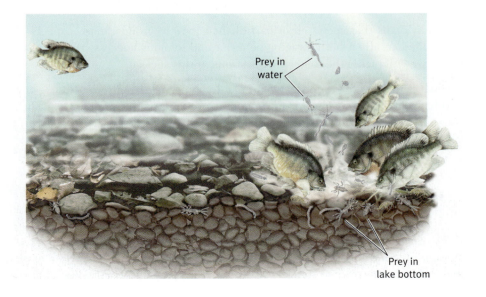

FIGURE 11.16. Bluegill group foraging. When bluegills forage in groups, they flush out more prey and attract other fish to the foraging site.

activities—often, but not always, increasing per capita foraging success (Caraco, 1979; Ens and Goss-Custard, 1984; K. Sullivan, 1984).

The general relationship between group size and foraging success uncovered by Mittlebach has been found over and over in animal studies. For example, Scott Creel ran an analysis on foraging success and group size in seven species that hunt in groups (Creel, 2001). Overall, Creel found a strong positive relationship between per capita foraging success and group size.

DISENTANGLING THE EFFECT OF GROUP SIZE AND COOPERATION ON FORAGING SUCCESS. Individuals may cooperate with one another when hunting in groups. For example, when wild dogs hunt down a prey item, it is a coordinated effort (Creel, 2001; see Chapter 2). Different members of the hunting pack play different roles in the hunt—flushing the prey, making the initial attack, disemboweling the prey, and so forth. In such cases, it is useful to separate the effects of cooperation from that of group size per se. To see how animal behaviorists disentangle such effects, let us examine hunting behavior in chimps.

Cooperative hunting or the lack of it has been examined in chimp populations in (1) the Gombe Preserve in Tanzania (Boesch, 1994b; Busse, 1978; Lawick-Goodall, 1968; Telecki, 1973), (2) the Mahale Mountains in Tanzania (Nishida et al., 1983; S. Uehara et al., 1992), and (3) the Tai National Park in the Ivory Coast (Boesch, 2002; Boesch and Boesch, 1989; Gomes and Boesch, 2009, 2011; Normand and Boesch, 2009). The most comprehensive studies of cooperative hunting among chimps are those of Christophe and Hedwige Boesch, who compared hunting patterns across the Gombe and Tai chimp populations. Major differences in hunting strategies emerged between these populations.

In Tai chimps, hunting success was positively correlated with group size in what is called a nonadditive fashion. That is, adding more hunters to a group did not simply increase the amount of food by a fixed amount for each new hunter added. Rather, with each new hunter, all group members received more additional food than they did when the last new hunter was added to the group (up to a limit). In addition to these group-size effects, Christophe Boesch found evidence of cooperation in Tai chimp hunting behavior (Boesch, 1994a; Gomes and Boesch, 2009, 2011; Figure 11.17). Very complex, but subtle, social rules exist that regulate access to fresh kills and assure hunters—that is, those that actually cooperate during the hunt—greater success than those that fail to join a hunt.

FIGURE 11.17. Cooperative chimps. While we often think of chimps as primarily vegetarian, they readily add meat, when it is available, to their diet. In the Tai Forest (Ivory Coast), chimps cooperate in both capturing and consuming prey. Once a prey is caught, subtle rules for food distribution are invoked. *(Photo credit: Christopher Boesch)*

The situation was quite different with Gombe chimps—no correlation between group size and hunting success was found in chimps from this population. That is, no group-size effect was uncovered in the context of foraging. In addition, unlike in the Tai population, behavioral rules limiting a nonhunter's access to prey were absent, and such individuals received as much food as those that hunted cooperatively (Boesch, 1994a; Goodall, 1986). Part of the difference in hunting behavior in Tai and Gombe chimps may be due to the fact that the success rate for Gombe solo hunters was quite high compared with the individual success rate for chimps in the Tai population, relaxing selection pressure for cooperative hunting in the Gombe population.

GROUPS, PUBLIC INFORMATION, AND FORAGING

Another potential foraging-related benefit associated with group life is the acquisition of public information (Beauchamp et al., 2012; Danchin et al., 2004; Valone, 1989; Valone and Templeton, 2002). The idea behind public information models is that a forager needs to be assessing a whole host of environmental variables—food availability, predators, and so on. One way to update such information is to use the actions of others as a cue to changes in environmental conditions. This public information differs from the sort of information acquired during social learning. In social learning, individuals learn something specific (a new behavior, the preference of others). In public information models, individuals use the actions of others as a means of assessing the condition of the environment, and as such, public information allows group members to reduce environmental uncertainty (Valone and Giraldeau, 1993). While public information models are general and can apply to numerous environmental parameters, here we will focus on how these models have been tested in the context of foraging behavior.

Solitary foragers can reduce environmental variability by acquiring knowledge about patches of food before arriving at a patch and/or by keeping track of how long they have been in a patch and how much they have eaten (Danchin et al., 2004). Foragers in a group, however, can acquire information in these same ways, but they also have access to public information. In particular, a forager in a group can use the foraging success of others as an estimate of patch quality. This public information should allow social foragers access to quicker and better estimates of the productivity of a patch of food than solitary foragers have.

As a case in point, public information models predict that social foragers in poor patches should leave such patches earlier than solitary individuals would, because social foragers can use the failed foraging attempts of their groupmates as additional information about when they themselves should leave a patch of food. Jennifer Templeton and Luc-Alain Giraldeau tested this prediction by examining the use of public information by the starling (*Sturnus vulgaris*). Starlings in this experiment fed at an artificial feeder that had thirty cups that were either empty or contained a few seeds (Templeton and Giraldeau, 1996; Figure 11.18). A given bird (B1) fed from such a feeder either alone or paired with a second bird (B2). Before being paired with B1 partners, B2 birds had been given the chance either to sample a few cups in this feeder or to sample all such cups. Two results support the predictions of public information models. First, when tested on completely empty feeding patches, B1 birds left

FIGURE 11.18. Public information. Social foragers such as starlings have been used to test public information models of foraging behavior. Starlings in this public information experiment were tested using an array of food placed into cups. *(Photo credit: Jennifer Templeton)*

such patches earlier when paired with any B2 bird than when foraging alone. Second, B1 birds left patches earliest of all when paired with B2 birds that had complete information about the patches (as compared with those B2 birds with only partial information).

Natural Selection, Phylogeny, and Seed Caching

Some birds and mammals can remember where they have stored thousands of different food items (Raby and Clayton, 2010; Gibson and Kamil, 2009; Smulders et al., 2010). Watching a squirrel find the nuts it has stored for winter all over a garden often inspires a sense of awe in the observer. And it isn't only squirrels; for example, many bird species in the families *Corvidae* and *Paridae* possess such abilities—some species of birds cache (store) tens of thousands of items per year, and others an astonishing hundreds of thousands of items each year (Balda and Kamil, 1992; Pravosudov and Clayton, 2002). How is that possible? How can an animal find scores of food items that are scattered across its environment, and often hidden over the course of months?

HIPPOCAMPAL SIZE AND CACHING ABILITY

Sue Healey and John Krebs examined hippocampal volume and food-storing abilities in seven species of corvid birds (Healey and Krebs, 1992). They chose

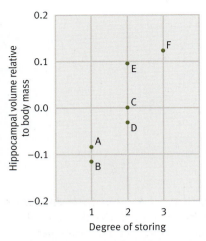

FIGURE 11.19. Foraging and brain size. The volume of the hippocampal region relative to body mass was positively correlated with the extent of food storing in six species of birds: (A) alpine chough, (B) jackdaw, (C) rook and crow combined, (D) red-billed blue magpie, (E) magpie, and (F) European jay. *(Based on Healey and Krebs, 1992)*

to study the hippocampal region because in birds this area of the brain is known to be associated with food retrieval (Krebs et al., 1989; Sherry, 2006; Sherry and Vaccarino, 1989). Moreover, corvids are an ideal group in which to examine the relationship between hippocampal volume and food storage in more detail, because so much variation in food-storing behavior exists within this group (Balda and Kamil, 1989; de Kort and Clayton, 2006; Emery, 2006; D. Goodwin, 1986). Some corvids store no food, while others rely on the food they have stored over the course of nine months.

Healey and Krebs studied two species—jackdaws (*Corvus monedula*) and Alpine choughs (*Pyrrhocorax graculus*)—that rarely if ever stored food; four species—rooks (*Corvus frugilegus*), European crows (*Corvis corone*), European magpies (*Pica pica*), and Asian red-billed blue magpies (*Cissa erythrorhyncha*)—in which food storing plays some role; and one species—the European jay (*Garrulus glandarius*)—in which not only does food storing play an important role but the location of 6,000 to 11,000 seeds must be remembered for nine months. When food-storing behavior was examined in relation to hippocampal volume in these seven species, a strong positive relationship was uncovered (Figure 11.19). Though correlational studies like this cannot tease apart cause and effect, this much is clear—the more food-storing behavior seen in a species, the greater the hippocampal volume.

In addition to the comparative approach adopted by Healey and Krebs, animal behaviorists have also examined the relationship between hippocampal size and caching ability within a single species. For example, researchers have hypothesized that, within a given species, individuals from populations where food resources are relatively scarce would be better at caching and recovering food and would possess a larger hippocampus than individuals from food-rich environments (Pravosudov and Grubb, 1997; Pravosudov et al., 2001). That is, natural selection should favor better caching and retrieval abilities when individuals live in harsh foraging environments, where caching and retrieval is at a premium, and selection should favor larger hippocampal areas and more hippocampal neurons in such individuals. Vladimir Pravosudov and Nicky Clayton tested this idea using black-capped chickadees (*Poecile atricapilla*) from two populations—Colorado (food rich) and Alaska (food poor; Pravosudov and Clayton, 2002).

Pravosudov and Clayton captured fifteen chickadees from a site in Anchorage, Alaska, and twelve birds from a population near Windsor, Colorado, and transported them back to their laboratory. Forty-five days after being brought into the lab, the birds were tested on their ability to retrieve seeds they had cached. When provided with seeds that could be cached, the birds from Alaska (food-scarce population) cached a greater percentage of seeds than the birds from Colorado (food-rich population). Equally important, the Alaskan birds found a greater proportion of their cached seeds than did the Colorado birds, and their searches were more efficient in that they made fewer errors (Figure 11.20).

In terms of hippocampal size, even though the birds from Alaska weighed less and were smaller than the Colorado birds, the hippocampal volumes of the Alaskan birds were greater and their hippocampuses contained more neurons than those of the Colorado birds. What's more, when Pravosudov and Clayton compared birds from Colorado and Alaska that were *not* given the chance to cache seeds, the same hippocampal differences were found, suggesting that the caching experience per se did not increase hippocampus size.

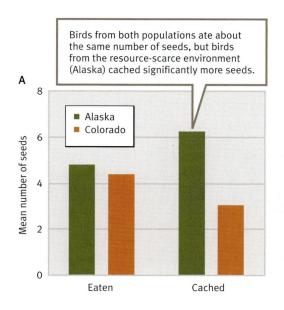

A

Birds from both populations ate about the same number of seeds, but birds from the resource-scarce environment (Alaska) cached significantly more seeds.

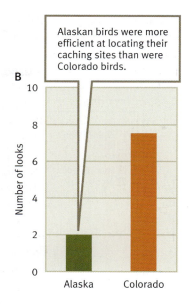

B

Alaskan birds were more efficient at locating their caching sites than were Colorado birds.

FIGURE 11.20. Population differences in food storing. **(A)** Mean number of sunflower seeds eaten and cached. **(B)** Mean number of sites inspected. *(Based on Pravosudov and Clayton, 2002, p. 519)*

PHYLOGENY AND CACHING ABILITY

Another evolutionary approach to seed caching is to study this phenomenon from a phylogenetic perspective (de Kort and Clayton, 2006). In an attempt to understand the evolutionary history of caching behavior in corvids, Selvino de Kort and Nicky Clayton built a phylogeny of forty-six species from this family of birds. Then, each species was categorized on its tendency to cache seeds. Based on a combination of published laboratory and field studies, each of the forty-six species was placed into one of three categories. Non-cachers were defined as species that virtually never cache food. Moderate cachers were those that cached food throughout the course of a year and cached many different types of food, but were never entirely dependent on cached food sources for survival. Finally, specialized cachers were defined as species in which individuals cached a large number of items, the cached items were typically one food type, and caching was seasonal. In addition, specialized cachers often recovered their cached items after long periods of time had passed. Information about caching ability was then mapped onto the phylogeny of the group (Figure 11.21).

A phylogenetic analysis indicates that the ancestral state of caching in corvids is "moderate caching." In other words, the phylogenetic analysis suggests that ancestor species that gave rise to all corvids was a moderate cacher. This result was somewhat surprising. Many animal behaviorists initially argued that the ancestral state in corvids was non-caching. De Kort and Clayton's findings that the ancestral state was moderate caching suggest that some corvid species evolved into specialized cachers, while others lost the caching trait altogether. Indeed, de Kort and Clayton found that evolution toward highly specialized cachers occurred independently at least twice, and perhaps as many as five times. Conversely, over evolutionary time, at least two species completely lost the ability to cache seeds, perhaps because the benefits of caching did not make up for the metabolic costs of maintaining a relatively large hippocampus (Attwell and Laughlin, 2001; Laughlin et al., 1998).

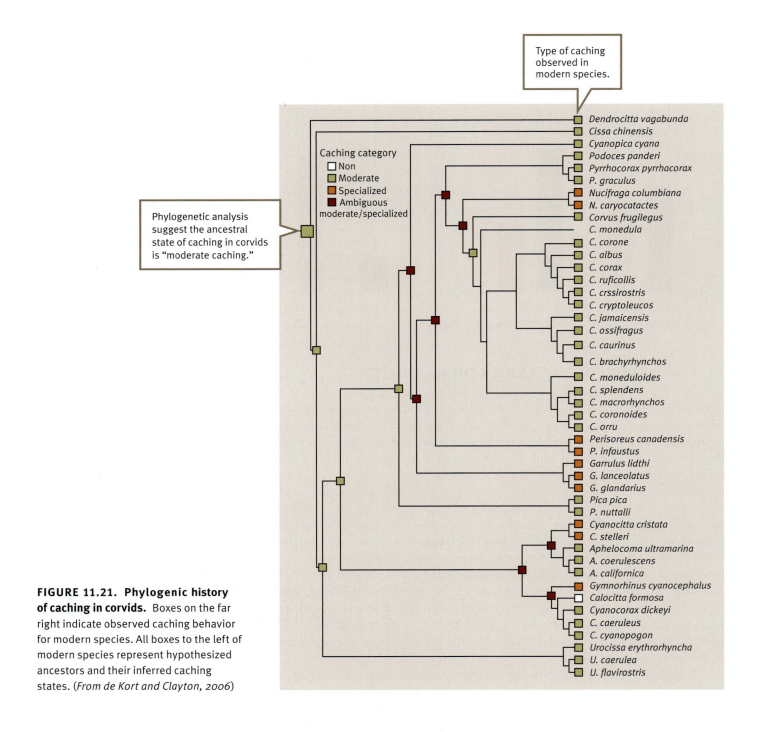

FIGURE 11.21. Phylogenic history of caching in corvids. Boxes on the far right indicate observed caching behavior for modern species. All boxes to the left of modern species represent hypothesized ancestors and their inferred caching states. (*From de Kort and Clayton, 2006*)

Within the figure:

Type of caching observed in modern species.

Phylogenetic analysis suggest the ancestral state of caching in corvids is "moderate caching."

Caching category
- Non
- Moderate
- Specialized
- Ambiguous moderate/specialized

Dendrocitta vagabunda
Cissa chinensis
Cyanopica cyana
Podoces panderi
Pyrrhocorax pyrrhacorax
P. graculus
Nucifraga columbiana
N. caryocatactes
Corvus frugilegus
C. monedula
C. corone
C. albus
C. corax
C. ruficollis
C. crssirostris
C. cryptoleucos
C. jamaicensis
C. ossifragus
C. caurinus
C. brachyrhynchos
C. moneduloides
C. splendens
C. macrorhynchos
C. coronoides
C. orru
Perisoreus canadensis
P. infaustus
Garrulus lidthi
G. lanceolatus
G. glandarius
Pica pica
P. nuttalli
Cyanocitta cristata
C. stelleri
Aphelocoma ultramarina
A. coerulescens
A. californica
Gymnorhinus cyanocephalus
Calocitta formosa
Cyanocorax dickeyi
C. caeruleus
C. cyanopogon
Urocissa erythrorhyncha
U. caerulea
U. flavirostris

Learning and Foraging

While the studies described so far often involve animals learning something about their foraging environment, they were not designed as studies on foraging and learning per se. To do justice to the huge number of studies done on learning and animal foraging would require a series of books. These sorts of studies have been a mainstay in psychology over the last fifty years, and thousands of studies have examined learning in the context of foraging (Balda et al., 1998; Kamil and Yoerg, 1982; Kamil et al., 1987; Pietrewicz and Richards, 1985; Shettleworth, 1998; Stephens and Krebs, 1986; Yoerg, 1991; see the Conservation Connection box). Here, we shall look at a tiny, but important, subsection of the work on foraging and learning by focusing on:

▶ Foraging, learning, and brain size in birds.
▶ Learning and planning for the future.

FORAGING, LEARNING, AND BRAIN SIZE IN BIRDS

Animal behaviorists have hypothesized a neurobiological link between forebrain size and learning abilities in animals (R. Byrne, 1993; Clutton-Brock and Harvey, 1980; Dunbar, 1992; T. Johnstone, 1982; Jolicoeur et al., 1984; Mace et al., 1981; Wyles et al., 1983). The premise underlying this link is as follows: Because the forebrain appears to be associated with behavioral plasticity—including learning—animals with larger forebrains should often be better learners. While this approach is susceptible to numerous potential biases, Louis Lefebvre and his colleagues used its basic principles to test for a relationship between "foraging innovations" and forebrain size in North American and British Isle bird groups (Lefebvre, 2011; Lefebvre et al., 1997a, 1997b; Overington, Griffin, et al., 2011).

Lefebvre and his colleagues defined a *foraging innovation* as "either the ingestion of a new food type or the use of a new foraging technique." For our purposes here, we will focus on the latter types of innovations (which include some examples of social learning). The researchers used descriptions of avian foraging mentioned in nine ornithology journals to gather data on 322 foraging innovations—126 in British Isle birds and 192 in North American birds (Table 11.1). Innovations

TABLE 11.1. Examples of foraging innovations in birds. Lefebvre and his colleagues gathered data on 322 foraging innovations, including those in this list. *(Based on Lefebvre et al., 1997b, pp. 552–553)*

SPECIES	INNOVATION
Cardinal (Florida subspecies)	Nipping off nectar-filled capsule on flower and eating it
Herring gull	Catching small rabbits and killing them by dropping them on rocks or drowning them
Ferruginous hawk	Attracted to gunshot, preys on human-killed prairie dog
Magpie	Digging up potatoes
Storm petrel	Feeding on decaying whale fat
Great skua	Scavenging on roadkill
House sparrow	Using automatic sensor to open bus station door; systematic searching and entering of car radiator grilles for insects
Galápagos mockingbird	Pecking food from sea lion's mouth
Common crow	Using cars as nutcrackers for palm nuts
Osprey	Opening conch shells by dropping them on concrete-filled drum
Turnstone	Raiding gastric cavities of sea anemones
Red-winged blackbirds and Brewer's blackbirds	Following tractor and eating frogs, voles, and insects flushed by it
Sparrowhawk	Drowning blackbird prey
Carrion crow	Landing on floating sheep corpse and feeding from it
Downy woodpecker	Using swaying caused by wind to catch meat hung from a branch

Behavioral Traditions, Foraging, and Conservation in Killer Whales

When social learning affects an animal's choice of food items, it can make the already difficult job of designing a protection program, conservation reserve, or reintroduction program all the more difficult. Conservation biologists tend to think that individuals in the species they are trying to protect will have similar diet preferences, especially if populations are geographically close to one another. But this assumption is inaccurate when diet choice is affected by social learning and foraging "traditions" that differ across populations emerge (Whitehead, 2010). The food choices made by only a few individuals in a group—often older, more experienced individuals—can ripple through a population, creating variation in foraging preferences across groups (recall the case of Imo the Japanese macaque monkey and her potato washing, discussed in Chapter 6).

As an example, consider the killer whale (*Orcinus orca*), a species found on the U.S. Endangered Species List and considered endangered or "depleted" in many places around the world. A survey of the foraging behavior of killer whale populations from around the world showed that they feed on at least 120 species of fishes, as well as cephalopods, sea turtles, sea birds, pinnipeds, and cetaceans (J. Ford and Ellis, 2006). But these data are deceiving, because at the level of the individual population, there is often specialization on one or just a few species of prey, and this prey choice is, in part,

FIGURE 11.22. Foraging in killer whales. The variance in foraging behavior seen across killer whale populations may be in part due to differences in socially learned food preferences. *(Photo credit: Don Johnston/Alamy)*

determined by social learning (J. Ford et al., 1998; Guinet and Bouvier, 1995; Saulitis et al., 2000).

One population of killer whales near British Columbia, Canada, shows a pronounced foraging tradition for specializing on chinook salmon (*Oncorhynchus tshawytscha*; Figure 11.22). This socially learned preference for chinook prey has been maintained for numerous generations, even though other species of salmon—some of which are readily taken as prey in other populations of killer whales—that could sustain killer whales are present.

Any attempt at protecting the British Columbian killer whale population needs to consider their foraging tradition for chinook salmon. Even though there are many *potentially* edible salmon species available for these whales to consume, behavioral traditions limit what the animals will eat. An understanding of the role that social learning plays in the foraging behavior of these whales suggests that researchers pay special attention to ensuring that a sufficient number of chinook salmon are available for individuals in this population.

included behaviors ranging from herring gulls "catching small rabbits and killing them by dropping them on rocks or drowning them" to common crows "using cars as nutcrackers for palm nuts" (Grobecker and Pietsch, 1978; H. Young, 1987). Lefebvre's group then calculated how these innovations were distributed across different orders of birds, at the same time taking into account how common or rare a particular bird order was in Britain or North America.

With the data about foraging innovation in hand, Lefebvre's team obtained information about relative forebrain size in the bird species of interest (Holden and Sharrock, 1988; Portmann, 1947; S. Scott, 1987). In both North American and British Isle birds, relative forebrain size correlated with foraging innovation. Bird orders that contained individuals that possessed larger forebrains were more likely to have high incidences of foraging innovation (Figure 11.23).

Although many of the cases of foraging innovation were based on a single observation, the fact that a larger-scale analysis spanning 322 cases found a striking relationship between forebrain size and one measure of learning (foraging innovation) suggests that this is an important but relatively understudied area in ethology. Lefebvre's group ran a similar analysis on Australasian birds and uncovered very similar results with respect to foraging innovation and forebrain size (Lefebvre et al., 1998).

Daniel Sol, Lefebvre, and their colleagues examined whether there is a fitness advantage associated with being a large-brained bird (Sol, Duncan, et al., 2005; Sol and Lefebvre, 2006; Sol, Lefebvre, et al., 2005; Sol et al., 2006; Sol et al., 2007). One long-held, but largely untested, hypothesis for the fitness advantage associated with large brains (relative to body size) was that large brains would be particularly beneficial when a population was introduced into novel or altered environments, where innovation might be especially important (Allman, 1998; Reader, 2004; Reader and Laland, 2002). Sol and his colleagues put this hypothesis to the test by examining the 646 introductions (involving 196 species) in which humans had placed a population of birds in a novel environment, most often an island or a park outside its native range (Cassey et al., 2004; Cauchard, Overingtton et al., 2011; Sol, Duncan, et al., 2005). The researchers collected data on brain size from 156 of the 196 species, and in the 40 remaining species in which there were no direct measures, they estimated

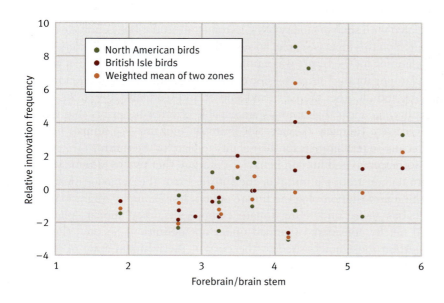

FIGURE 11.23. Brain size and foraging innovation in birds. In North American and British Isle birds, and in the weighted mean of North American and British samples, the relative frequency of foraging innovation is positively correlated with the ratio of forebrain mass to brain stem mass. *(Based on Lefebvre, Whittle, et al., 1997)*

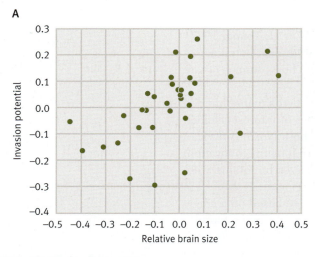

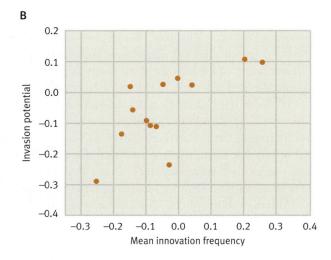

FIGURE 11.24. Brain size, innovation, and survival in birds. (A) The relationship between mean relative brain size and invasion potential (survival in a new environment) across avian families. **(B)** The relationship between mean foraging innovations and invasion potential across avian families. *(From Sol, Duncan, et al., 2005)*

brain size by using the brain size found in the closest phylogenetic relative for which data were available. Because larger-brained birds are generally larger than smaller-brained birds, Sol and his team corrected for this potential bias by using brain size relative to body size as their key parameter.

As discussed in Chapter 6, Sol and his team found a positive relationship between relative brain size and the birds' success in novel environments. What's more, after correcting for potential confounding factors, they found evidence that the success of large-brained species was, at least in part, due to their relatively high use of innovative new foraging techniques, which increased their rate of food intake (Figure 11.24). These findings suggest that there is a fitness correlate with relatively large brains and the innovative abilities associated with them.

PLANNING FOR THE FUTURE

Learning theory rests on the idea that current action is in part a function of prior experience. At the same time, it is also true that if animals could *plan* for the future based on prior experience, as we humans clearly do, there would be fitness benefits associated with such an ability (N. Clayton et al., 2009; Raby and Clayton, 2009). Research on this topic is scarce, however, because for a long time scientists argued that planning for the future was a uniquely human quality—indeed, it was posited that planning for the future was one of the most fundamental differences between humans and all other animals (Suddendorf and Corballis, 1997). Yet over time, it has become clear that animals are capable of other behaviors that we initially believed were uniquely human—for example, tool use—so animal behaviorists have begun to test animals for their ability to plan for the future.

Both animal and human behaviorists argue that two requirements must be met to demonstrate planning for the future. First, the behavior must be novel, not the manifestation of some innate action. For example, migration behavior, which has many underlying innate components, may appear "planned" but would not be considered planned under our definition. Second, the behavior in question must not be tied to the current motivational state of the animal but rather to the anticipated motivational state at some point in the future (Shettleworth, 2007; Suddendorf, 2006). Here, we will examine a study of the ability of western scrub jays (*Aphelocoma californica*) to modify their foraging behavior in an attempt to plan for the future (Figure 11.25).

Western scrub jays are a model system for such experiments, as they display incredible feats of episodic memory—recollecting specific events or episodes—in caching and later uncovering large amounts of food. In addition, these birds not only remember where they have stored items but they remember who was watching them when they cached their food and in such cases will later dig up and rehide the food, presumably to protect their stored food from being stolen (N. Clayton and Dickinson, 1998; Dally et al., 2006; Grodzinski and Clayton, 2010; Grodzinski et al., 2012). With such impressive cognitive abilities, the hypothesis that western scrub jays can plan for the future is a tantalizing one, especially considering that studies in humans suggest that the same neurobiological processes involved in episodic memory are involved in planning for the future (Atance and Meltzoff, 2005; Okuda et al., 2003; Rosenbaum et al., 2005).

Caroline Raby and her team used a simple experimental procedure to examine whether western scrub jays could plan for the future (Raby et al., 2007). On alternate mornings over the course of six days, birds were exposed to one of two compartments—one compartment contained food in the form of ground-up pine nuts (which cannot be cached), and the other compartment contained no food. On the evening before each test, the birds were not fed any food, and they were therefore hungry during their exposure to test compartments. After the six days of exposure to the two compartments, the birds were denied access to any food for two hours before dark, and then they were unexpectedly provided with a bowl of whole pine nuts—which can be cached. They were then allowed to move freely in the area with the whole pine nuts and could access both of the compartments they had been exposed to earlier. But now each of the compartments had a "caching tray" added to it, so the birds could cache food before it got dark, if they so chose.

Raby and her team found that jays cached more nuts in the compartment in which they had consistently received no food in the past, strongly suggesting that they were planning for the fact that, when next exposed to that compartment, there would be no food present unless they cached it there. In an ingenious follow-up experiment, Raby and her colleagues had birds learn that they would be fed peanuts in one compartment, but dog kibble in the other (jays like both of these food types). When an experiment like the one described above was run, jays cached peanuts in the compartment associated with kibble, and kibble in the compartment associated with peanuts, suggesting that they prefer a varied diet, and that they plan for such a diet accordingly (also see Correia et al., 2007). The extent to which other animals can plan for the future remains a relatively unexplored question in ethology.

FIGURE 11.25. Western scrub jays and planning for the future. Researchers have examined whether western scrub jays plan into the future when making foraging decisions. *(Photo credit: David A. Northcott/Corbis)*

SOCIAL LEARNING AND FORAGING

Information about foraging can be culturally transmitted when individuals live in groups. To understand more about such social learning, we will consider social learning and foraging in pigeons (*Columbia livia*). Pigeons are an ideal species in which to examine cultural transmission of feeding behavior (Figure 11.26). Being primarily scavengers feeding on human garbage, pigeons face a same dilemma: Which new food items are safe, and which are dangerous? Louis Lefebvre and his colleagues Luc-Alain Giraldeau and Boris Palemeta have run an intriguing series of experiments examining the strength of cultural transmission to shape diet in the pigeon. This work has focused on three related issues: (1) What type of information

FIGURE 11.26. Scavenging pigeons. Pigeons are scavengers, coming across novel food items all the time. Such a species is ideal for studying foraging and cultural transmission. *(Photo credit: Blickwinkel/Alamy)*

do pigeons transfer about food? (2) How does such information spread or fail to spread through a population of pigeons? and (3) What factors favor the cultural transmission of information over alternative means of acquiring information?

Palemeta and Lefebvre set out to examine cultural transmission in a three-part experiment that used observer and demonstrator animals (Palameta and Lefebvre, 1985). The task that observer pigeons needed to learn was piercing the red half of a half-red/half-black piece of paper covering a box. Under the paper were seeds for the bird.

An observer pigeon was placed in an arena with such a food box—with a half-red/half-black paper cover—and exposed to one of four scenarios. In the first group, birds saw no model on the other side of a clear partition. None of the pigeons in this group learned how to get at the hidden food, suggesting that this was a difficult task to master through individual learning. In a second group, observers saw a model that was eating from a hole in the paper. The hole was made by Palameta and Lefebvre; although observers did see a model eating, they did not see the model solve the hidden food "puzzle"—that is, they did not see the model pierce the red half of the cover. Pigeons in this treatment did learn how to get food from the multicolored box, but the latency to feed from the box was not quite as high. In the final two treatments of the experiment, birds either saw a model pierce the red side of the paper but get no food, or they saw a model both pierce the paper and eat. Birds in the former treatment did not learn to solve the food-finding dilemma. Birds in the latter treatment learned this task and learned fairly quickly (Figure 11.27), suggesting an important role for cultural transmission in the foraging behavior of pigeons.

Pigeon groups contain both producers and scroungers. **Producers** find and procure food, while **scroungers** eat the food that producers have uncovered (Barnard, 1984; Giraldeau and Caraco, 2000; Giraldean and Lefebvre, 1986). And it is the unusual way that producing and scrounging interacts with social learning that makes the pigeon story particularly useful in furthering our understanding of cultural transmission and foraging.

Despite Palameta and Lefebvre's work demonstrating social learning in pigeons, when birds are tested in groups only a few birds seem to learn new feeding behaviors by observing others. Giraldeau and Lefebvre examined whether scrounging behavior somehow inhibited cultural transmission (Giraldeau and Lefebvre, 1987). To do so, they used a different set-up than the one described above (Figure 11.28).

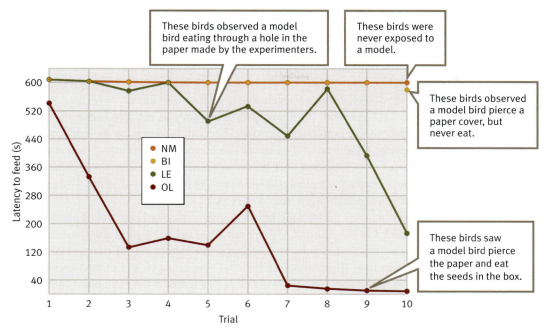

These birds observed a model bird eating through a hole in the paper made by the experimenters.

These birds were never exposed to a model.

These birds observed a model bird pierce a paper cover, but never eat.

These birds saw a model bird pierce the paper and eat the seeds in the box.

FIGURE 11.27. Social learning and foraging in pigeons. Pigeons in this experiment needed to learn to pierce the red half of a paper covering a box of seeds. The graph shows average latency to eating for four groups: NM (no model), BI ("blind" imitation), LE (local enhancement), and OL (observational learning). Pigeons in the NM and BI treatments never learned to feed in the experimental apparatus. The quickest learning occurred in the OL treatment. *(Based on Palameta and Lefebvre, 1985)*

In the new experimental protocol, flocks of pigeons were allowed to feed together. Forty-eight little test tubes were placed in a row and five of these tubes had food. Which five had food was unknown to the birds. To open a tube, an individual had to learn to peck at a stick in a rubber stopper at the top of the tube. When this task was done correctly, it caused the test tube to open and the contents to spread over the floor below. Once the food was out, any bird in the vicinity, not just the one that opened the tube, could eat it.

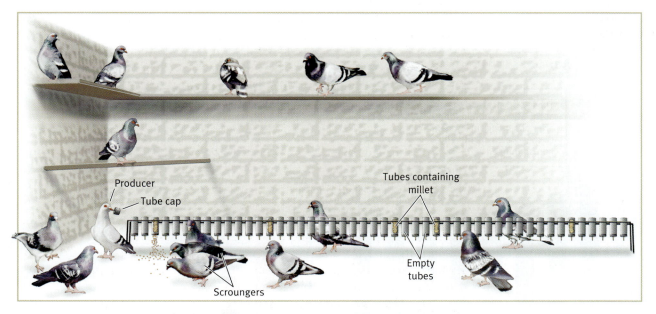

FIGURE 11.28. Producing and scrounging. When a group member finally opens a tube with food in it, the food spills on the floor and is accessible to all. In Giraldeau and Lefebvre's study using such a scenario, out of sixteen pigeons, only two learned to open tubes (these were the producers), while fourteen acted as scroungers. Labels show where millet was hidden, but the birds were not privy to this information.

The results of this experiment were striking. As predicted based on earlier work, only two of the sixteen pigeons in the group learned to open tubes—that is, the flock was composed of two producers and fourteen scroungers. Two additional findings suggested to Giraldeau and Lefebvre that scrounging inhibited an individual from learning how to open tubes via observation. To begin with, scroungers followed producers and seemed more interested in *where* producers were than in *what* producers did to get food; the scroungers' attention was not directed at the actions taken by the producers to open the tubes. Second, by removing the two producers from the group, Giraldeau and Lefebvre discovered that scroungers not only didn't *display* the tube-opening trait, but they also didn't know how to open these tubes. That is, it was not as if scroungers could open the tubes but opted not to; they seem never to have learned to do so from observing the producers.

In order to better understand how scrounging blocked cultural transmission of foraging skills, Giraldeau and Lefebvre ran a second set of controlled experiments. Here, they paired a single observer with a single demonstrator that already knew how to obtain food. If an observer had the chance to view a demonstrator open tubes and obtain food, over time the observer learned how to open tubes. That is, all birds were capable of learning the foraging task. Giraldeau and Lefebvre then set up the experimental cages so that every time the demonstrator opened a tube, the food in that tube slid over to the observer's side of the cage. In these treatments, the observers rarely learned how to open the tubes themselves (Figure 11.29). Their scrounging on the food that was found by others interfered with the transmission of information about how to get the food themselves. Recent game theoretical models have also helped ethologists better understand why producers, which must learn about their environments, are usually found at low frequencies compared with scroungers (Dubois et al., 2010).

Current work on foraging behavior continues in most of the areas we have covered in this chapter—prey choice, patch-leaving rules, risk sensitivity, group effects, planning for the future, and the role of phylogeny in shaping behavior, as well as in the role that both individual and social learning play in making foraging decisions.

FIGURE 11.29. Scrounging prevents social learning. Giraldeau and Lefebvre trained two groups of pigeons. **(A)** One group of birds saw a model bird peck at a stick in a rubber stopper at the end of the tube. This behavior provided food to the model. **(B)** In the second group of birds, when the model pecked at the stick, food was released, but to the observer. *(Based on Giraldeau and Lefebvre, 1987)*

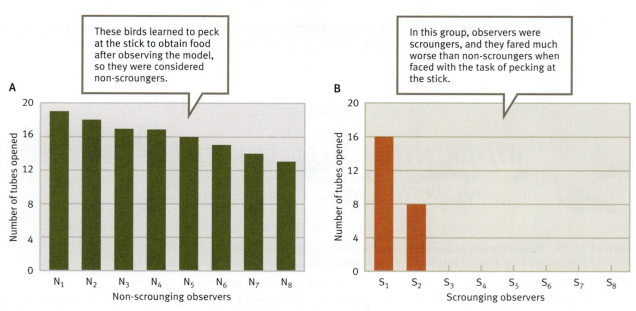

The Optimal Diet Model

Imagine a forager has only two prey types that it consumes (for example, two types of seeds, two species of squirrels, and so on). In our model, a forager must eat one of these prey items (or it will starve), so our question becomes Which item does a forager always take, and under what conditions does a forager take both types of prey? To begin, let us define the following terms:

e_i = energy provided by prey type i
h_i = handling associated with prey type i
λ_i = encounter rate with prey type i
T_s = amount of time devoted to searching for prey
T = total time

We will assume that an animal always takes the prey type that has a higher e_i/h_i value (called the profitability of a prey type), and we will label this prey type as prey 1. The question then becomes whether a forager should take prey 1 alone, or whether a forager should take both prey 1 and prey 2 upon encountering them.

We begin by calculating the total energy (E) associated with prey 1 divided by the total time associated with this prey (T). For prey type 1:

$$\frac{E}{T} = \frac{T_s \lambda_1 e_1}{T_s + T_s \lambda_1 h_1} \tag{1}$$

The numerator here takes the total number of prey type 1 captured ($T_s \lambda_1$) and multiplies it by the energy value (e_1) of each prey, producing the total energy produced by prey type 1. The denominator adds together the total search time (T_s) and the total handling time ($T_s \lambda_1 h_1$), given such a search. This simplifies to:

$$\frac{E}{T} = \frac{\lambda_1 e_1}{1 + \lambda_1 h_1} \tag{2}$$

Now we can ask whether this value is greater than the E/T associated with taking both prey types. To find the E/T of taking both prey type 1 and prey type 2, we calculate the following:

$$\frac{E}{T} = \frac{T_s(\lambda_1 e_1 + \lambda_2 e_2)}{T_s + T_s \lambda_1 h_1 + T_s \lambda_2 h_2} \tag{3}$$

The numerator represents the total energy obtained from prey types 1 and 2, while the denominator adds together the total search time (T_s) and the total handling time for prey types 1 and 2. This simplifies to:

$$\frac{E}{T} = \frac{\lambda_1 e_1 + \lambda_2 e_2}{1 + \lambda_1 h_1 + \lambda_2 h_2} \tag{4}$$

Our question then boils down to when the following inequality is true:

$$\frac{\lambda_1 e_1}{1 + \lambda_1 h_1} > \frac{\lambda_1 e_1 + \lambda_2 e_2}{1 + \lambda_1 h_1 + \lambda_2 h_2} \tag{5}$$

When this inequality holds true, the predator should take only prey type 1. When it does not hold true, the predator should take prey type 1 and prey type 2. Equation 5 can be shown to hold true when:

$$\lambda_1 > \frac{e_2}{e_1 h_2 - e_2 h_1} \tag{6}$$

From equation 6 we can derive two important predictions:

1. Once a critical encounter rate with prey type 1 is reached, it alone should be taken.
2. The decision about whether to take prey type 2 does *not* depend on how common prey type 2 is (that is, on prey type 2's encounter rate). This can be seen in equation 6 by the absence of λ_2 from our inequality.

Dr. John Krebs

How did it come to be that someone whose father won a Nobel Prize for coming up with the Krebs cycle made his initial mark in science by studying animal behavior?

No single factor, but a concatenation of many events led me to a career in animal behavior. I came to zoology as a bird-watcher; as a high school pupil I spent two summers working in the laboratory of the famous ethologist Konrad Lorenz; and as an undergraduate at Oxford, I was lucky enough to be taught by two giants of their generation, David Lack and Niko Tinbergen. Having said all this, my father dissuaded me from my first choice of a university course, namely archaeology, and I ended up as a zoologist, having failed to get in to study medicine! My Ph.D. at Oxford was unusual in that I studied population ecology in Niko Tinbergen's animal behavior group. This meant that right from the start I saw the linkages between ecology and behavior that led to the emergence of "behavioral ecology."

One of the most innovative aspects of your work on foraging in great tits was the use of a conveyer belt to present food to the birds. How did you come up with that idea? What did this allow you to do that otherwise would have been impossible?

Throughout my career I have been extraordinarily fortunate in the exceptional colleagues, students, and co-workers I have had. The "conveyor belt" was not my idea. A graduate student in the animal behavior group

at Oxford, Jon Erichsen, had devised it for psychophysical studies of reaction times in birds. Together we saw its potential for foraging studies. It would enable us to control encounter rates very accurately by having the prey move past the bird, rather than vice versa. I remember vividly posting to (no e-mail in those days) Ric Charnov

the first results from the conveyor belt. These showed that the inclusion or exclusion of the less profitable prey type depended exclusively on the encounter rate with the more profitable prey. The theory actually worked!

Psychologists have been studying the feeding behavior of animals for more than fifty years. Why the lag in ethological studies in this area? Why did they start up in earnest in the 1970s?

Ethologists had in fact been studying feeding behavior for many

years. The important concept of "search image" was well established, and there were many studies of the cues used by animals for detecting food and the development of feeding skills during the life of an individual, as well as of predator avoidance strategies such as crypsis, mimicry, and warning colors. The new approach in the 1970s, known as "optimal foraging theory," resulted from a confluence of ideas from ethology, ecology, evolution, psychology, and economics. As so often happens when new ideas are beginning to crystallize, more than one person independently came to similar conclusions. The seeds of foraging theory were more or less independently sown by J. T. Emlen, R. H. MacArthur and E. R. Pianka, J. D. Goss-Custard, and G. A. Parker. The attraction of foraging theory for me at the time, and I think for many others, was the juxtaposition of a broad theoretical framework, testable predictions from mathematical models, and both field and laboratory data.

How do classical psychologists and ethologists differ in the way they study animal foraging?

Traditionally, psychologists studied the mechanisms of behavior in the laboratory while ethologists and behavioral ecologists studied function or adaptation in the field. In the last fifteen years, these boundaries have become blurred, to the benefit of both disciplines. For example, ethologists have adopted some of the experimental and theoretical tools of operant and classical conditioning as well as of cognitive psychology. Psychologists have increasingly taken on board the potential value of an evolutionary framework for understanding behavior.

Many optimal foraging models involve complex mathematics. Does that mean the animals have to know how to solve mathematical equations to forage efficiently?

Mathematical models of flight are complex, but birds do not need to be mathematicians in order to fly. Nor do they need to do sums in order to forage according to the predictions of optimal foraging models. Mathematics is the universal language used by a scientist to describe and analyze what goes on in nature. There is no implication that nature itself uses mathematics.

A number of foraging models focus on what are called "rules of thumb." Could you explain the basic idea here?

Traditional foraging models were concerned with the end result (or adaptive significance) of foraging—for example, maximizing energy intake, rather than about the mechanisms that control feeding behavior. The notion of "rules of thumb" is that the end result, analyzed in a foraging model, might be achieved by a mechanism that approximates the "right" answer. If, for example, you were trying to find the best textbook of animal behavior, the rule of thumb "read if it is by Dugatkin" might bring you close enough to the optimal solution to be virtually indistinguishable from a more complex search strategy. If students following a different rule of thumb did better in exams, over time, the "Dugatkin rule" would be replaced by

another rule that came closer to the optimal solution. In other words, rules of thumb are themselves subject to evolution by natural selection.

One potential critique of the foraging literature in ethology is that so many of the experiments undertaken are run in the laboratory and not in the field. In your estimation, how serious a problem is this?

If you are studying fundamental properties of animals—for example, decision-making rules—then they can be elucidated in the laboratory as well

> The attraction of foraging theory . . . was the juxtaposition of a broad theoretical framework, testable predictions from mathematical models, and both field and laboratory data.

as in the field. A potential danger is that in the laboratory you may end up finding out more about the properties of your apparatus and experimental set-up than about the animal! This is a criticism that has sometimes been leveled at earlier studies of "schedules of reinforcement" by operant psychologists. Probably the ideal approach is to use a combination of field and laboratory work. Laboratory work certainly has the advantage that it is possible to control the variables more precisely. For laboratory work, intelligent design of the experimental set-up will help to ensure that the essence of the natural situation is captured.

What's "the big question" to be tackled next in the study of animal foraging behavior?

Foraging theory, like any other branch of inquiry, will continue to evolve and change. I do not see a single "big question" on the horizon, but I see several potentially important ways in which the study of foraging behavior could develop. One of these is to apply ideas from foraging behavior to conservation. The pioneering work of J. D. Goss-Custard and W. J. Sutherland has shown how foraging models can be used to predict the effects of habitat loss and disturbance on threatened populations. The key has been to link the foraging of individuals to density-dependent influences on the population as a whole. Another area of recent, and likely continuing, growth consists of the links between behavioral economics, decision theory, and foraging behavior. This is a development of the links between animal psychology and foraging behavior that began about twenty years ago. New insights are flowing in both directions. More generally, I do not think we should regard "foraging behavior" as an ineluctable, distinct field of inquiry. The basic ideas may be reconfigured, remolded, and reinvented so that the categories used to describe behavior today might seem outmoded in ten years' time.

Dr. John Krebs is a professor at Oxford University in England. Dr. Krebs's work on foraging behavior was seminal to the development of optimal foraging theory. Dr. Krebs (along with Dr. Nick Davies; see Chapter 8 interview) is the editor of Behavioural Ecology: An Evolutionary Approach, *4th ed. (Blackwell Science, 1997).*

SUMMARY

1. One way animals determine what sort of food to search for is through the development of a search image. A search image is formed when animals that encounter a prey type more and more form some sort of representation of the prey, and this representation or image becomes more and more detailed with experience, the forager finds itself more successful at finding that type of prey.

2. Optimal foraging theory is a class of models that predicts what and where animals should eat, as well as nutrient constraints and the effect of hunger state on risk-sensitive foraging.

3. Optimal foraging models of prey choice predict that the encounter rate with the less profitable of two prey items does not affect whether that item should be added to the diet or not. Rather, a critical encounter rate with the *most* profitable item is what determines which prey items should be added to a diet.

4. The marginal value theorem predicts how long a forager should stay in a given patch, given that it can leave and travel to feed in less depleted patches. Moving between food patches, however, is assumed to entail some cost, including lost time foraging while traveling and the increased rate of predation while traveling.

5. In addition to maximizing their rate of energy intake, animals are constrained by a need to spend time foraging for specific nutrients that are essential for survival.

6. Theory predicts that hungry foragers will be risk prone—willing to assume greater variance (risk) in food intake than less hungry individuals.

7. Compared with solitary foraging, foraging while in groups often, but not always, increases the foraging success of individuals. Benefits of group life include the ability to learn about various aspects of the foraging environment from observing others.

8. A phylogenetic analysis of caching in corvid birds suggests that a moderate amount of seed caching was the ancestral state of caching behavior in this group. This finding suggests that some corvid species evolved into specialized cachers, whereas others lost the caching trait altogether.

9. Animals often learn how and on what to forage. Learning sometimes manifests itself in foraging innovations; brain size often correlates with the frequency of such foraging innovations. In addition, some evidence now exists that some animals can use what they have learned to plan for the future.

DISCUSSION QUESTIONS

1. Read R. Pulliam's 1973 article "On the Advantages of Flocking," in the *Journal of Theoretical Biology* (vol. 38, pp. 419–422). Based on the article, outline the "many eyes" hypothesis, and discuss how it relates to foraging behavior.

2. Read E. L. Charnov's 1976 article, "Optimal Foraging, the Marginal Value Theorem," in *Theoretical and Population Biology* (vol. 9, pp. 129–136). How might you modify the model to examine other behaviors displayed by animals?

3. Using the model developed on p. 377 as a starting point, construct a foraging model with three prey types. Imagine that you already know that it pays for a forager to eat prey type 1 and prey type 2. What are the conditions under which the forager should add prey type 3?

4. Why do you suppose it took so long for ethologists and psychologists to recognize the larger literature on foraging behavior that exists in each other's field? What do you think were the biggest differences in the way foraging was studied in these two disciplines?

5. Using the graphs in Figure 11.9 (see p. 356) as a starting point, examine what happens to the time an animal should spend in the patch as a function of how profitable that patch is. This will involve changing the shape of the curve that describes food intake as a function of patch residence time and examining what this change does to optimal time in a patch.

6. How does the corvid phylogeny paper discussed in this chapter show that the adaptationist and phylogenetic approaches to studying brain size and caching can complement one another?

SUGGESTED READING

Gibson, B., & Kamil, A. (2009). The synthetic approach to the study of spatial memory: Have we properly addressed Tinbergen's "four questions"? *Behavioural Processes, 80,* 278–287. A thought-provoking piece on the different ways to study memory, with an emphasis on foraging behavior.

Giraldeau, L. A., & Caraco, T. (2000). *Social foraging theory.* Princeton, NJ: Princeton University Press. A comprehensive overview of social foraging theory, with many empirical examples to lead the reader through the math.

Hill, K. (2002). Altruistic cooperation during foraging by the Aćhe, and the evolved human predisposition to cooperate. *Human Nature–an Interdisciplinary Biosocial Perspective, 13,* 105–128. Foraging behavior and its relation to cooperation in the Aćhe Indians of Paraguay.

Raby, C. R., Alexis, D. M., Dickinson, A., & Clayton, N. S. (2007). Planning for the future by western scrub-jays. *Nature, 445,* 919–921. The first evidence that nonprimates can plan for future events in an adaptive manner.

Stephens, D., Brown, J. S., & Ydenberg, R. C (Eds.). (2007). *Foraging: Behavior and Ecology.* Chicago: University of Chicago Press. An edited book on foraging theory and animal behavior.

12

Antipredator Behavior

Robotic, remotely controlled mini-submarines are scouring the ocean floors today, searching for new, undiscovered life forms deep beneath the waves. In 2009, Karen Osborn and her colleagues discovered seven new species of annelid worms—some of which were found swimming as deep as 3,793 meters (more than 12,000 feet) below the ocean surface. Osborn and her colleagues wondered how these species, living in a world with almost no light, defend themselves from the dangers around them. To the researchers' surprise, they found that these newly discovered worms probably startle or distract potential predators with "bombs" of green fluorescent light (Osborn et al., 2009)!

Five of the seven species that Osborn and colleagues discovered produce small bioluminescent sacs (globules)—what the researchers termed *bombs*—that they likely release when encountering predators (Figure 12.1). These bombs light up for a few seconds after the worms release them and may startle a predator long enough to allow the worm to escape the danger. Much remains to be learned about exactly how the bioluminescent bomb system works (Haddock et al., 2010; Widder, 2010), but a phylogenetic analysis based on gene sequences from the newly discovered worm species allowed ethologists to estimate where in the evolutionary history of these worms' bioluminescent sacs evolved (Figure 12.2). What's more, these bioluminescent sacs are homologous with other known anatomical structures (called segmented branchiae) in annelid worms that do not have bioluminescent sacs, so animal behaviorists have an idea what structure the bioluminescent are derived from.

As exciting as it may be, ethologists need not go 12,000 feet below the waves to study antipredator behavior. Richard Coss, Donald Owings, and their

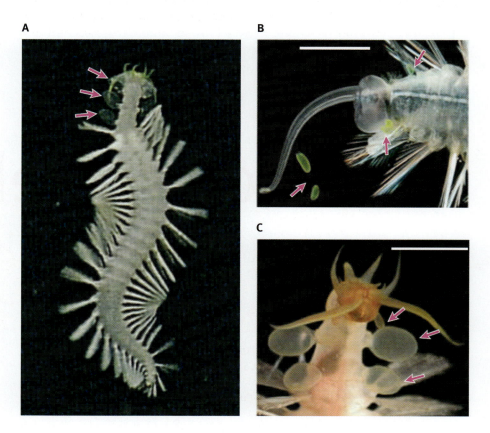

FIGURE 12.1. Luminescent bombs as an antipredator defense. (A) *Swima* sp. 3, with arrows indicating bombs. **(B)** *Swima* sp. 1 with three attached and two unattached bombs. Two bombs are shown at the bottom left and the center arrows. **(C)** *Swima* sp. 3 with three pairs of attached bombs. *(From Osborn et al., 2009)*

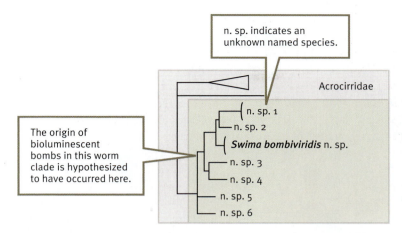

n. sp. indicates an unknown named species.

Acrocirridae

n. sp. 1
n. sp. 2
Swima bombiviridis n. sp.
n. sp. 3
n. sp. 4
n. sp. 5
n. sp. 6

The origin of bioluminescent bombs in this worm clade is hypothesized to have occurred here.

FIGURE 12.2. Phylogeny of bioluminescent bombs in annelid worms. A phylogeny of *Swima* species. *Swima bombiviridus* and unnamed species 1–4 produce "bombs". Unnamed species 5 and 6 do not produce "bombs." *(From Osborn et al., 2009)*

colleagues have been studying predator-prey interactions at a California field site, where gopher snakes (*Pituophis melanoleucus*) and their prey—ground squirrels (*Spermophilus beecheyi*)—have been present for approximately a million years. Over evolutionary time, natural selection has strongly favored ground squirrels that are able to identify their predators and to respond to them with fine-tuned behaviors (Coss, 1991; Coss and Owings, 1985; Owings and Coss, 1977). For example, groups of squirrels will often mob, or converge on, a gopher snake, biting and harassing it until the snake leaves the area (Owings and Coss, 1977). Squirrel antipredator behavior includes throwing dirt, pebbles, and roots at predators, as well as emitting alarm calls that are specifically made when snakes, but not other predators, are present (Owings and Leger, 1980; Figure 12.3). The squirrel's immune system has also evolved in concert with the antipredator behaviors mentioned above. To see how, let us look at some behavioral immunological work undertaken by Naomie Poran and Richard Coss.

FIGURE 12.3. Snakes and squirrels. (A) When they emerge from their burrow, ground squirrels recognize snakes as predators. **(B)** Confrontations with rattlesnakes are common, and **(C)** squirrels sometimes throw dirt and roots at the snake (note the snake's head at the left) to defend itself. Arrows indicate the location of the snake in each photo. *(Photo credits: Richard Coss)*

FIGURE 12.4. Danger at emergence. Ground squirrel pups often face serious predation threat on their first emergence from their burrow.

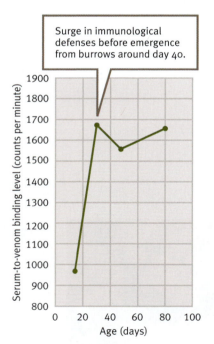

Surge in immunological defenses before emergence from burrows around day 40.

FIGURE 12.5. Immunology and predation. Ground squirrel pups emerge from their burrows at about forty days. There is an increase in serum-to-venom binding levels in the weeks before emergence. *(From Poran and Coss, 1990, p. 237)*

Poran and Coss focused on young California ground squirrels and examined both antipredator behavior and immunological defenses against snakes, using serum-to-venom binding levels as a measure of immunological defenses against snake bites (Biardi et al., 2000, 2005, 2006; Poran and Coss, 1990). To do this, they used radioimmunoassays of serum-to-venom binding levels in squirrels at ages fourteen, twenty, forty-eight, and eighty days. At very young ages, pups have very low serum-to-venom binding levels, but this measure rises to almost adult levels some time between days fourteen and thirty. This occurs even when pups are not exposed to snakes during the experiment. Why is there a spike in serum-to-venom binding levels?

The answer is tied to the natural history of the squirrels. Squirrel pups don't emerge from their natal burrows until they are about forty days old (Linsdale, 1946; Figure 12.4). In their burrows, the pups are relatively safe from snake predation, compared with the threats they face when emerging from their burrows. But just before the pups emerge, when the threat of snake predation will increase dramatically, natural selection has favored a spike in serum-to-venom binding levels (Figure 12.5). Such a spike will help to protect young squirrels from snake bites when they emerge from their burrows.

If an individual makes an error with respect to antipredator tactics, its future reproductive success may well be zero. This obvious, but striking, fact suggests that natural selection should operate very strongly on antipredator behavior and that, in addition, we might expect fine-tuned learning in this behavioral venue as well. Both because interactions between predators and their prey are often spectacular to observe, and because such interactions are so critical for understanding the process of natural selection, ethologists have a long history of studying antipredator behavior (Lind and Cresswell, 2005). Comparative psychologists have also long been interested in predator-prey interactions. But it was not until Robert Bolles developed an explicitly evolutionary approach to this subject that comparative psychologists conceptualized specific antipredator behaviors as specific adaptations (Bolles, 1970).

At the most fundamental level, there are two forms of antipredator behaviors: those that help prey avoid detection by predators, and those that function once a prey encounters a predator. In this chapter, we will explore the two forms of antipredator behavior from both a proximate and an ultimate perspective. Before proceeding, here are two caveats: (1) while these two categories encompass most antipredator behaviors, they are not meant to be exhaustive; and (2) only biotic (living) predators are considered in this chapter. Fire, for example, kills many animals, and there is evidence that reed frogs (*Hyperolius nitidulus*) respond to the sound of oncoming flames and flee (Grafe et al., 2002). We will not consider such instances here.

Avoiding Predators

If putative prey can avoid being detected by their predators, they decrease not only the probability of being captured and eaten but also the costs associated with fleeing, fighting back against a predator, and so on (Brilot et al., 2012; Ruxton et al., 2004; Stanford, 2002). Below, we shall examine three ways that animals can avoid their predators: (1) blending into the environment, (2) being quiet, and (3) choosing safe habitats. Before we begin, it is worth noting that although we are discussing blending into the environment and being quiet as a means to avoid predators, these can also be employed by prey once they are detected by predators.

BLENDING INTO THE ENVIRONMENT

Ethologists have long recognized that one way for animals to avoid predators is to blend into their environment and become **cryptic** (camouflaged), making their detection by predators unlikely. This sort of background matching need not be behavioral per se. For example, in a classic natural history study, Francis Sumner found that populations of the mouse *Peromyscus polionotus* matched the background of the beaches on which they lived (F. B. Sumner, 1929a, 1929b). Here, individual mice were not behaving in any particular way that increased their crypsis. Instead, natural selection had acted so that "beach mouse" variants of *P. polionotus* had color patterns that matched their environment—in particular, "beach mice" displayed light coat pigmentation that better matched the sand on which they lived, and coat pigmentation became darker in populations farther from the beach and closer to the inland areas (Figure 12.6). Indeed, a single amino acid mutation plays a role in this example of crypsis (Hoekstra, 2006).

Animals can increase their ability to remain hidden from their prey by behaviorally changing color to match their environment. Cephalopods—which include octopuses, squids, and cuttlefish—are especially adept at quickly changing color to blend into their background, decreasing their chances of being attacked by a predator (Hanlon and Messenger, 1988; Packard, 1972). While most work on camouflage to stay hidden from predators has been conducted during the day or at sunset, Roger Hanlon and his team have found that a great deal of predation on cephalopods may occur at night, as many predators have sharp night vision.

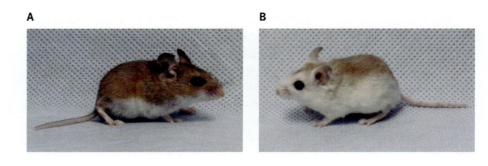

FIGURE 12.6. Coat color variation.
In *Peromyscus polionotus*, mice on inland populations **(A)** tend to have darker coats than mice from beach populations, **(B)** where the sand is light in color. *(Photo credit: (a-b) Hoekstra et al. Science 7 July 2006: Vol. 313. no. 5783, pp. 101–104. Reprinted with permission from AAAS)*

Hanlon and his colleagues used a noninvasive light source—one that relied on a red filter—to observe the nocturnal behavior of the giant Australian cuttlefish (*Sepia apama*), and, in particular, to test whether the cuttlefish used camouflage at night, presumably to hide from predators with good night vision (fur seals, bottlenose dolphins, and some species of fish; Hanlon et al., 2007). The researchers found that cuttlefish were camouflaged in seventy-one of eighty-three (86 percent) nocturnal observations. This is probably an underestimation, because some cuttlefish may have matched the environment so well that the underwater camera employed simply missed them. More important than the raw numbers of how often crypsis was uncovered, Hanlon found that cuttlefish could match their background in one of three ways, and they could change their color and pattern to match their background in a matter of seconds.

The first kind of disguise consisted of a "uniform" camouflage pattern in which cuttlefish adopted a single color for their skin—a color that matched their background (Figure 12.7). Although this form of camouflage was rare in Hanlon's observations, when it occurred it often involved a cuttlefish mimicking rocks. The second kind of crypsis made up almost half of the instances of background matches that Hanlon's team uncovered and involved what are called "mottled camouflage" patterns. Here, the cuttlefish changed their appearance such that their skin had small dark and light splotches all over. The size and color of the splotches often mimicked the cuttlefish's background. This patterning is often seen when the background is composed of small rocks and dark algae.

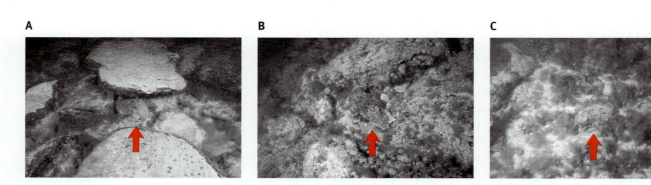

FIGURE 12.7. Cuttlefish camouflaging. (A) Cuttlefish using uniform color to camouflage itself against the rocks; **(B)** cuttlefish using a "mottled" camouflage pattern, with small dark splotches resembling the dark patches on rocks and sand; and **(C)** cuttlefish using a "disruptive" camouflage pattern, with large light and dark areas that enable it to blend in with the background. Arrows indicate where the cuttlefish are located. *(Photo credits: Roger Hanlon)*

The final form of disguise employed by cuttlefish was "disruptive camouflage." This involved the cuttlefish changing color and pattern, and taking on large light and dark stripes. This had the effect of visually breaking up (disrupting) the animal's body, so it did not look like a cuttlefish. On some occasions, though not often, the stripes actually mimicked the pattern of the cuttlefish background.

All of these different forms of camouflage techniques strongly suggest that for large, soft-bodied creatures such as the giant cuttlefish, blending into the environment is an important antipredator behavior (Hanlon and Messenger, 1988).

BEING QUIET

In systems in which predators home in on sounds made by their prey, one of the simplest things that an animal can do to avoid predators is to be quiet (M. Ryan, 1985). With this in mind, Luke Remage-Healey and his colleagues examined the role of sound suppression in the antipredator repertoire of the Gulf toadfish (*Opsanus beta;* Remage-Healey et al., 2006).

Gulf toadfish are preyed on by adult bottlenose dolphins (*Tursiops truncatus*), making up 13 percent of the dolphin's diet (Barros, 1993). Prior work had shown that dolphins orient toward the "boat-whistle" sound produced by male toadfish during breeding season (Gannon et al., 2005); that is, dolphins eavesdrop (see Chapter 15) on the sounds produced by toadfish to better orient toward their prey. Once the dolphin locates a toadfish, it locks onto this prey and tracks it. The question Remage-Healey and his colleagues addressed was whether toadfish listen for sounds associated with bottlenose foraging behavior and then reduce the boat-whistle sounds they produce (Figure 12.8).

Bottlenose dolphins produce a variety of sounds, including high-frequency whistles used in dolphin-to-dolphin social communication (not foraging), as well as two sounds that are particularly associated with foraging—"clicks" and low-frequency "pops" (Janik et al., 2006; Nowacek, 2005; Tyack and Clark, 2000). The low-frequency pops are easiest for toadfish to hear, since this species hears especially well in the low-frequency range, so Remage-Healey and his team focused on these dolphin sounds.

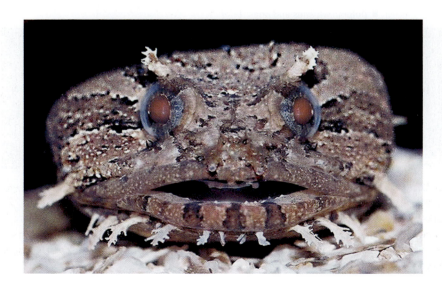

FIGURE 12.8. Gulf toadfish. Gulf toadfish are preyed on by bottlenose dolphins. Dolphins orient toward the "boat-whistle" sound produced by male toadfish during the breeding season. *(Photo credit: Midge Marchaterre)*

The researchers captured toadfish during the breeding season, and they kept individual males in tanks. The males soon began to emit boat-whistle sounds to attract females. At that point, toadfish males were exposed to one of three sounds—the pops associated with dolphin foraging; the high-frequency whistles produced during dolphin social communication; and, as a control, the "snapping" sounds made by snapping shrimp. All sounds were broadcast using underwater speakers, and the activity of toadfish was recorded for the five minutes before sounds were emitted, the five minutes during which the experimental sounds were broadcast, and the five minutes after the sounds were played.

Remage-Healey found no differences in call rate between males before exposure to the experimental sounds (Figure 12.9). Males exposed to pop sounds, however, reduced their call rates by 50 percent. In addition, males exposed to the pop sounds maintained their reduced calling rate for the five minutes following exposure to pops—they eavesdropped on their predators and reduced their activity in a way that made capture by a dolphin less likely. Males in the other treatments showed no changes in boat-whistle call rate when they heard the recorded sounds.

Remage-Healey and his team followed up their behavioral work on call rates and exposure to predators with a hormonal analysis that examined whether dolphin pops produce a stress response in the toadfish. After experimentally exposing the male toadfish to pops or snapping shrimp sounds, the researchers drew blood from the males and measured their cortisol levels. Males exposed to pops not only responded to the pops by reducing their own boat-whistle call, but they also showed higher levels of cortisol than males exposed to the sound of snapping shrimp.

Being quiet in the presence of predators is not unique to toadfish. A similar type of antipredator behavior is displayed by greater wax moths (*Galleria mellonella*) in response to their bat predators (Figure 12.10). Foraging bats emit echolocation

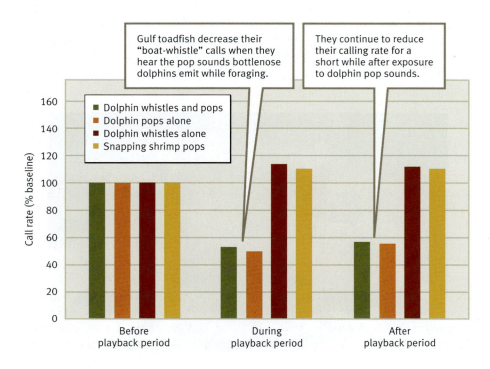

FIGURE 12.9. Gulf toadfish become silent. Gulf toadfish responses to four different calls, before, during, and after hearing these calls. The "before-playback-period" data were standardized to a baseline value of 100 for means of comparison. *(From Remage-Healey et al., 2006)*

pulses, and moths generally reduce their activity when they detect such pulses (Griffin, 1959; G. Jones & Siemers, 2011; G. Jones and Holderied, 2007; G. Jones and Teeling, 2006; G. Jones et al., 2002). Female wax moths, however, face something of a dilemma, in that the temporal and spectral characteristics of bat echolocation pulses are very similar to the sounds produced by courting males.

Male courtship elicits a "wing-fanning" behavior by females—a behavior that would make them a more obvious target for bat predators, that can home in on vibrations associated with wing fanning. Females, then, must be acutely tuned in to the subtle differences between male moth sounds (which should elicit wing fanning) and bat echolocation sounds (which reduce female activity). To examine whether females were able to distinguish between the structurally similar sounds produced by male greater wax moths and bats, Gareth Jones and his colleagues used a playback experiment in which they exposed females to tapes of male great wax moth calls and bat echolocation calls (G. Jones et al., 2002). They found that females were able to distinguish between these types of calls and responded appropriately—that is, fanning their wings when they heard male calls but dramatically decreasing this behavior when they heard bat echolocation calls. In addition, females also distinguished between the pulses produced by bats that were searching for prey versus the pulses emitted by bats that were attacking a moth. The latter produced a greater reduction in wing fanning by female moths.

FIGURE 12.10. Wax moths and predators. Foraging moths typically reduce their activity when bats' echolocation pulses are detected. Nevertheless, bats do capture some moths, as is about to happen here. *(Photo credit: Image courtesy of Wake Forest University)*

CHOOSING SAFE HABITATS

One means by which prey can avoid predators is by selecting habitats that are relatively predator free (Apfelbach et al., 2005; Ferrero et al., 2011; see also the Conservation Connection box). Here we will look at a case study that involves a phylogenetic perspective on habitat selection and predator avoidance in parrots.

PREDATION AND CHOICE OF NESTING SITES IN PARROTS. Some birds build their nests in tree cavities. Other birds nest in cavities of a different kind—for example, some individuals build their nests in old termite mounds, and others, like the white-fronted bee-eaters discussed in Chapter 9, build cavities into the sides of cliffs. From a phylogenetic perspective, some bird taxa contain both species that nest in tree cavities (TC nesters) and species that nest in other sorts of cavities (let's call them OC "other cavity" nesters). This raises a number of questions: Within a clade, which of these behavioral states—nesting in tree cavities or nesting in other sorts of cavities—is the ancestral state? In addition, once we know the ancestral state, what selective force is responsible for the evolution of the other (derived) behavioral state? And especially critical for a discussion of antipredator behavior, did predation drive the choice of nesting area so as to minimize encounters with predators?

Donald Brightsmith addressed these questions in parrot species from two different areas—Australia and the Amazon region. Brightsmith obtained molecular genetic-based phylogenies that contained DNA sequences from six Australian parrot species and dozens of species of Amazon parrots (Brightsmith, 2005a, 2005b; Ribas and Miyaki, 2004; Rowden, 1996; Russello and Amato, 2003). Brightsmith also collected published data on the nesting behavior (TC or OC) of as many of the species in the above-mentioned phylogenies that he could uncover. He then mapped the nesting data onto the phylogenies to examine ancestral and

Co-evolution, Naive Prey, and Introduction Programs

Conservation biologists sometimes use translocation programs, moving individuals from one natural habitat to another, to protect threatened or endangered species (Ewen et al., 2012). To better understand how likely such transplant experiments are to succeed, conservation biologists need to understand that the species they are moving originally evolved in an environment with a particular set of predators, and that **co-evolution** may have been occurring between these predators and prey. Co-evolution occurs when changes to traits in species 1 lead to changes to traits in species 2, which in turn feed back to affect traits in species 1, and so on. Predator-prey co-evolution can lead to an evolutionary **"arms race"** between predator and prey: The prey evolve behaviors that help them protect themselves against the predators, and the predators evolve detection systems that help them find prey. This co-evolutionary dynamic is important for conservation biologists to understand, because when they undertake translocation programs, they may inadvertently introduce the translocated species to predators with which they have had no evolutionary history and to which they are especially susceptible.

To understand how exposure to new predators in a new environment affects prey, Isabel Barrio and her colleagues studied the antipredator behavior of the wild rabbit (*Oryctolagus cuniculus*), which was introduced into Australia by European settlers (Barrio et al., 2010). Rabbits use odor cues to detect many of their predators. In Australia, rabbits are preyed upon by foxes, cats, and ferrets, which have also been introduced there. But rabbits and these other species have a long shared evolutionary history, primarily on the Iberian peninsula (Jaksic and Soriguer, 1981). Wild rabbits are also preyed upon by predators that are native to Australia, such as the spotted-tail quoll (*Dasyurus maculatus*; Glen and Dickman, 2006).

When Barrio and her colleagues exposed the wild rabbits in Australia to the odor of foxes, cats, and ferrets, the rabbits responded by reducing their use of the area with the predator odor. No such response occurred when the rabbits were exposed to the odor of the spotted-tail quoll, with whom they shared no evolutionary history (Figure 12.11). The rabbits' usual first line of defense against predators, odor detection, was ineffective for this new species, leaving the rabbits vulnerable to the quolls.

This work suggests that to maximize success rates, introduction and relocation program managers need to consider whether the species they are trying to protect shares an evolutionary history with the predators in the new environment. The species may not possess evolved antipredator adaptations in either a specific or a general sense. In some cases, translocated species may be under serious risk because they have not evolved antipredator behaviors to a specific predator in their new environment, as in the case of the wild rabbits in Australia. In other cases, the translocated species may not have evolved antipredator behaviors to certain *types* of predators in their new environment (for example, ambush hunters, predators that detect prey by odor, and others). Detailed knowledge of these sorts of issues can help conservation biologists and managers design better programs.

A

B

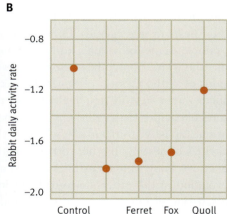

FIGURE 12.11. Rabbits' antipredator behavior and co-evolutionary history. Compared with controls, rabbits decreased their use of areas that had the scent of three predators with which they had an evolutionary history (ferrets, cats, and foxes), but they did not decrease their use of areas that had the scent of the predator that was native to Australia, with which they shared no evolutionary history (the quoll). The *y*-axis data is given in negative numbers as a result of the way researchers transformed the data on activity rate. *(Photo credit: A & J Visage/ Alamy; from Barrio et al., 2010)*

derived nesting behaviors. This analysis showed that the ancestral state was tree cavity nesting, and that nesting in other cavities had evolved independently many times in both Australian and Amazonian parrot species (Figure 12.12).

Returning to the role of avoiding predators and cavity nesting, Brightsmith used the data he had collected to address the following question: What selective forces were responsible for the evolutionary shift from tree cavity nesting to OC nesting? In particular, ecologists had suggested two possible selective forces driving OC nesting: (1) competition for tree cavity nests is typically quite intense, so selection in some species may have favored nesting in other sorts of cavities (Beissinger, 1996; T. E. Martin, 1993; Monkkonen and Orell, 1997; Wiebe, 2011); and (2) predation on eggs and chicks is very high during the nesting period, and shifting from tree cavities to nesting in such places as old termite mounds or the face of cliffs may have decreased predation pressure

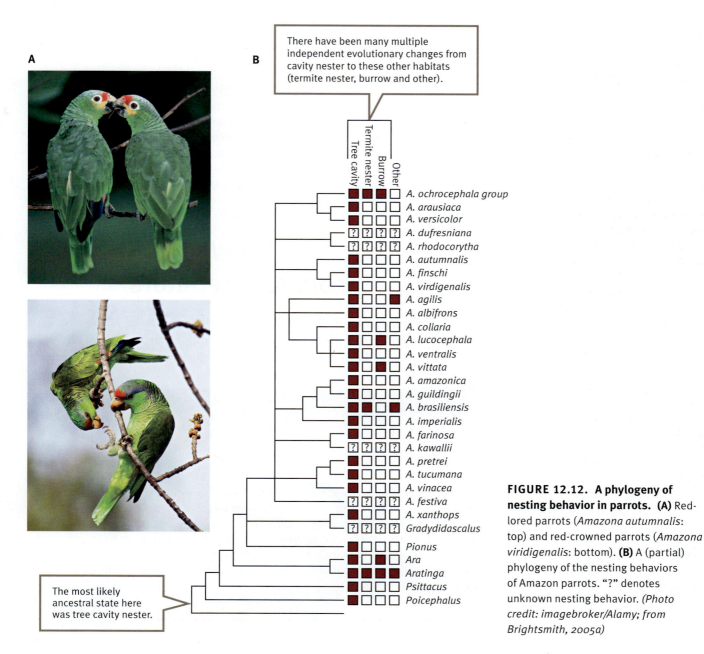

A

B

There have been many multiple independent evolutionary changes from cavity nester to these other habitats (termite nester, burrow and other).

The most likely ancestral state here was tree cavity nester.

FIGURE 12.12. A phylogeny of nesting behavior in parrots. (A) Red-lored parrots (*Amazona autumnalis*: top) and red-crowned parrots (*Amazona viridigenalis*: bottom). **(B)** A (partial) phylogeny of the nesting behaviors of Amazon parrots. "?" denotes unknown nesting behavior. *(Photo credit: imagebroker/Alamy; from Brightsmith, 2005a)*

(Lack, 1968). Predation and competition, of course, are not mutually exclusive, so both forces may have affected the switch in nesting behavior.

To test for the effects of competition and predation on the shift in nesting behavior, Brightsmith reasoned as follows: Prior work suggested that, when birds were released from competition, an increase in clutch size often occurred over evolutionary time. Having had limited opportunities to find nests in the evolutionary past should favor investing heavily in offspring production—that is, producing large clutches when the opportunity arises. No evidence for increased clutch size was found in OC nesters, providing no support for the competition hypothesis.

To examine whether predation was the key force selecting for the shift away from tree cavity nesting, Brightsmith turned to an idea first suggested by David Lack (Lack, 1968). Lack had hypothesized an inverse correlation between nest predation rate and the length of nesting period, and a large-scale study of 101 North American species found such a correlation (T. E. Martin, 1995). As such, Brightsmith hypothesized that if the shift away from tree cavities was due to predation pressure, then OC nesters should have longer nesting periods. The data on both Amazonian and Australian parrot species do, in fact, show that those species that nest in cavities other than tree cavities have longer nesting periods.

With respect to the Amazonian parrots, Brightsmith estimates that almost all of the species that rely on cavities other than trees arose in the late Oligocene–early Miocene geological period, 20 to 30 million years ago (Miyaki et al., 1998). South American mammal communities were undergoing a large change at that time, with rapid increases in the number of nest predators, including both tree rats and primates from Africa (Poirier et al., 1994). These nest predators may have been responsible, in part, for the Amazonian parrots' evolutionary shift away from tree cavity nesting.

What Prey Do When They Encounter Predators

Despite the various means by which prey can avoid predators, encounters with predators are, over the lifetime of an organism, inevitable (Figure 12.13). In this section, we will examine how animals respond when they do encounter

FIGURE 12.13. Encounter with a predator. A skua gull descends from the air in search of penguin eggs or unattended chicks, while these two gentoo penguins attempt to fend it off. *(Photo credit: Seth Resnick/Science Faction)*

predators. Before we go into detail on specific behaviors, it is important to have an understanding of the sort of neuroendocrinological changes that occur when a predator is encountered. Consider, for example, work by David Smith and his colleagues on the proximate effect of predator odor on mice (D. G. Smith et al., 2006). Prior work had shown that the frontal cortex area of the brain regulates the effect of stressors on behavior in rodents and humans, and that this area of the brain may alter neurological and endocrinological responses to stressors such as predators (Amat et al., 2005; Drevets, 2000; Osuch et al., 2000; Spencer et al., 2005). To examine this in more detail, Smith and his team exposed mice to two different stressors. One group of mice was exposed to the odor of a predator—a rat, and mice were exposed to bedding that had been soiled by rats. A second group of mice was exposed to a physical stress—these individuals were immobilized in a device with the wonderful name of a Universal Mouse Restrainer (UMR). A third group of control animals was exposed to neither predator odor nor the UMR.

To measure the neuroendocrinological responses to predator stress (as well as UMR stress), the investigators anesthetized the mice prior to the experiment and implanted a device into their brains that allowed them to draw out a small amount of fluid from the mice's prefrontal cortex. The researchers then took fluid samples from the prefrontal cortex of the experimental mice throughout the course of the experiments. Smith and his group found that both predator odor and the UMR increased the circulation of the neurotransmitters acetylcholine, serotonin, and dopamine within the frontal cortex, but that the increase was greatest in response to predator odor.

While it is always difficult to measure whether a predator induces anxiety in nonhumans, Smith and his team did find that when chlordiazepoxide (Librium®), a drug that reduces anxiety in humans, was administered to mice before exposure to the odor of predators, the increases in acetylcholine, serotonin, and dopamine described above disappeared. This finding suggests that predators may indeed cause anxiety in nonhumans.

Just being exposed to the odor of a predator caused a significant change in the neurological and endocrinological states of the brain in mice—a greater change, in fact, than when they were being physically restrained. Such neuroendocrinological changes will, in turn, produce antipredator behavior such as reduced foraging, burying for safety, and so on (T. Campbell et al., 2003; Koolhaas et al., 1999).

While it is not possible to include all possible behaviors that prey use once they encounter a predator, here we shall examine the following five antipredator actions: (1) fleeing, (2) approaching a predator to obtain information, (3) feigning death, (4) signaling to the predator, and (5) fighting back (see Chapter 9 for a discussion of another example of antipredator behavior—alarm calling).

FLEEING

The most common response of prey that have spotted a predator is to flee for safety (Lima, 1998; Lima and Dill, 1990; Stankowich and Blumstein, 2005; Ydenberg and Dill, 1986; Figure 12.14). Fleeing might involve a squirrel sprinting to its underground den, a bird flying into the trees for safety, a fish heading for cover in a coral reef or, as we shall see in a moment, an embryonic tadpole falling from a branch into the relative safety of the water.

A | B

C | D

FIGURE 12.14. Fleeing. The most common response of prey that have spotted a predator is to flee for safety. *(Photo credits: Johnny Johnson/Getty Images; Tom Brakefield/ Getty Images; Klaus Nigge/NGS Images; Gerry Ellis/Minden Pictures/Getty Images)*

A META-ANALYSIS OF FLIGHT INITIATION BEHAVIOR. Animal behaviorists have measured **flight initiation distance**—that is, how close a predator can approach before prey flee—in a wide variety of species. This is a fairly easy measure for researchers to obtain, both observationally and experimentally, in part because experimentalists can manipulate predator behavior by using trained animals or through the use of "model" predators (for example, a stuffed hawk flown over an area with pigeons).

Ted Stankowich and Dan Blumstein gathered published data from sixty-one studies of flight initiation in mammals, fish, birds, and reptiles. Stankowich and Blumstein analyzed data from a wide variety of taxa, searching for patterns in flight initiation behavior (Stankowich and Blumstein, 2005). They reviewed studies to see how characteristics of the predator, the physical condition of the prey itself, the prey's environment, and the prey's prior experience with predators affected the prey's decision as to when to flee from a predator.

Stankowich and Blumstein found that animals that were far from a refuge (their territory, for example) initiated fleeing from a predator sooner than animals closer to their refuge. In addition to the distance to safety, the researchers found that animals involved in foraging, mating, or fighting were slower to flee from predators than animals that were not currently involved in such behaviors. That is, when animals were distracted by other activities, they were less alert for predators. These sorts of trade-offs between antipredator behavior and foraging, mating, and so on are well documented in the literature (Lima and Dill, 1990).

Stankowich and Blumstein found that the predator's size and speed and the directness of its approach affected the prey's perception of risk. They also examined how morphological traits of the prey affected its behavioral decision of when to flee from a predator. They found that the presence of armor (spines, shells, and so on) reduced a prey's perception of risk—at a given distance, armored animals were not as likely to flee from a predator as prey without armor. They also found that the prey's ability to camouflage itself (cryptic coloration) affected its flight decision. Other factors affecting its decision included the quality of the habitat and the physical condition of the prey (how hungry it was, its size and age, whether it was pregnant, whether it was defending young offspring), but more research is needed to investigate how important these factors are.

Stankowich and Blumstein found that experience and learning played a strong role in when prey initiated flight—prey typically flee at a greater distance as a function of experience with predators. For example, Larry Dill found that zebrafish that had experienced a predator fled at a greater distance from the predator than fish that had no experience with the predator (Dill, 1974). In addition, published work suggests that animals flee humans at a greater distance during hunting season compared with during other times of the year (Croes et al., 2007; Fa and Brown, 2009).

Across many diverse taxa, then, the behavioral decision of when to flee from a predator is affected by a suite of factors, but attributes of the predator (predatory effects), experience with the predator in question, the distance to safety, and the presence or absence of armor (Figure 12.15) are especially important.

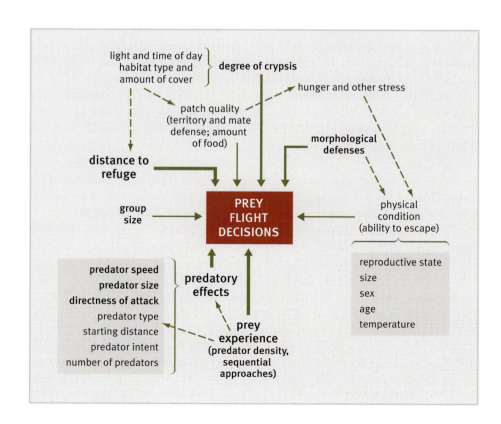

FIGURE 12.15. Flight initiation distance. A summary of the different factors that may influence when an animal decides to flee from a predator. Bolder and larger type and thicker arrows indicate more important factors—that is, there is more evidence for these factors influencing prey flight decisions than for the factors in lighter and smaller type. Dashed lines indicate factors that may have indirect effects on other factors. *(From Stankowich and Blumstein, 2005, p. 2630)*

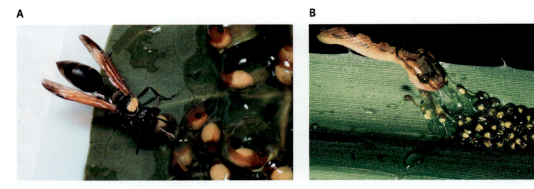

A　　　　　　　　　　**B**

FIGURE 12.16. Predators that feed on treefrogs. Red-eyed treefrogs have numerous predators that specialize in feeding on their eggs. **(A)** Here a wasp forages on treefrog eggs. **(B)** Snakes are another dangerous predator on red-eyed treefrog eggs. *(Photo credits: Karen Warkentin)*

TREEFROG EMBRYOS AND SNAKES. Though we think of the antipredator options available to embryos as limited, some research suggests that even embryos can flee from predation. As a case study of this phenomenon, let us examine Karen Warkentin's work on red-eyed treefrogs (*Agalychnis callidryas*) of Panama and their predators to see how natural selection can produce fleeing behavior on the part of embryos—a behavior that, in fact, lessens their risk of encountering predators (Gomez-Mestre et al., 2010; Touchon et al., 2011; Warkentin, 2000).

Red-eyed treefrogs attach their eggs to the various types of vegetation that hang over water. As soon as eggs hatch, the tadpoles that emerge drop down into their aquatic habitat. Both the terrestrial habitat of the egg/embryo and the aquatic habitat of the tadpole have a set of dangerous, but different, predators that feed on treefrogs. If terrestrial predation from snakes and wasps is weak, eggs hatch late in the season (Warkentin, 1995, 1999; Figure 12.16). This serves two functions: First, it maximizes the time that the eggs/embryos are in a low-predator terrestrial habitat, and second, such late hatching allows the embryos to grow to a size that lowers the levels of fish predation once the eggs finally hatch and the tadpoles fall into the water.

Using the results of mathematical models, Warkentin predicted that treefrog eggs would hatch sooner if predation in the terrestrial environment increased. That is, natural selection should favor embryos that avoid terrestrial predators when such predators are at high frequencies (compared with the aquatic predators that feed on treefrogs; Warkentin, 2000). Both snakes and wasps are terrestrial predators on treefrog eggs—wasps feed on one egg at time, while snakes can eat entire clutches. When predation from snakes and wasps is high, it often pays to mature early and drop into the water, away from heavy terrestrial predation.

To test this idea, Warkentin marked 123 egg clutches of treefrog eggs. She found that less than half the eggs she kept track of survived to hatch, whereas the majority fell prey to wasp attacks. She then examined whether the eggs that survived responded by hatching and having the emerging tadpoles "flee" by dropping to the water. Warkentin found that eggs in clutches that were not disturbed by predators often hatched at about six days. When comparing eggs from these undisturbed clutches to those clutches that had already suffered some predation by wasps, Warkentin found that hatching rates were dramatically different. Eggs hatched at a much quicker rate when their clutch had been the victim of some wasp predation, with most eggs from attacked clutches hatching at four or five days (as opposed to six; Figure 12.17). Indeed, some embryos

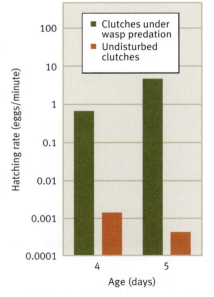

FIGURE 12.17. Wasp predation and development time. Red-eyed treefrogs respond to wasp predation by hatching early. Green bars represent hatching time of clutches that suffered some wasp predation, and orange bars indicate undisturbed clutches. Embryos are capable of hatching sometime during day 4. *(Based on Warkentin, 2000)*

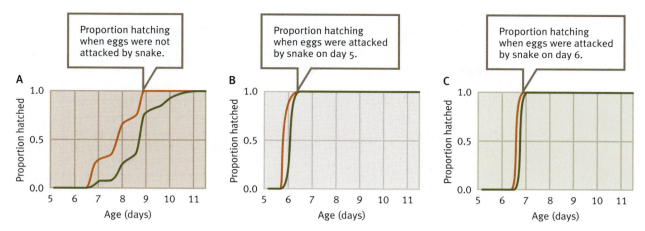

FIGURE 12.18. Snake predation and development time. Red-eyed treefrogs respond to snake predation by hatching earlier than normal. Notice the earlier hatching when clutches were attacked by snakes in (B) and (C). Green and orange lines indicate replicate experiments. *(Based on Warkentin, 1995)*

in clutches that were attacked ruptured their eggs, hatched, and dropped off branches immediately after such attacks. Similar results were found when the effect of snake predation on hatching rates was examined (Figure 12.18).

Warkentin's work on predator avoidance raises an interesting question about proximate cues. What specific cues do the treefrog embryos use to determine when to shift from terrestrial to aquatic habitats? If the developing embryos survive an attack by snakes, survivors drop off branches into the water, but what specific cues are they homing in on? Warkentin hypothesized that the embryos may be using the vibrational cues associated with snake attacks as the proximate cue for when to switch from terrestrial habitats to aquatic ones.

To test her hypothesis, Warkentin released a snake into a cage and, using a device called an accelerometer, she recorded the vibrations associated with this predator. Because Warkentin hypothesized that the vibrations associated with snake attacks—and not just vibrations in general—were the key to a habitat shift away from predators, she also recorded a different vibrational cue. In this case, she recorded the vibrations associated with rainfall. She then played back vibrational recordings of two kinds of snake attacks and of a rainstorm to clutches of developing treefrog eggs (Figure 12.19). As predicted, the cues associated with snakes resulted in treefrogs that hatched earlier than treefrogs that had been played the recording with rainstorm cues (M. S. Caldwell et al., 2009, 2010; Warkentin, 2005; Warkentin et al., 2006, 2007).

APPROACHING PREDATORS

While prey often flee from their predators, animals sometimes approach predators when they initially encounter them. This approach behavior may allow prey to gather important information about putative predators, and reduces their chances of mortality. Approach behavior is often undertaken by healthy adults, but it is not limited to this group. This type of behavior has been extensively documented in vertebrates, particularly in fish, birds, and mammals, and it has been referred to in the literature as approach behavior, boldness, investigative behavior, or predator inspection behavior (Curio et al., 1978; Dugatkin and Godin, 1992; Pitcher et al., 1986). Here, we will examine two examples: approach behavior in gazelles and predator inspection behavior in minnows.

Among vertebrates, the dynamics of approach toward a potential predator are similar. Prey typically approach a potential predator from a distance. The approach

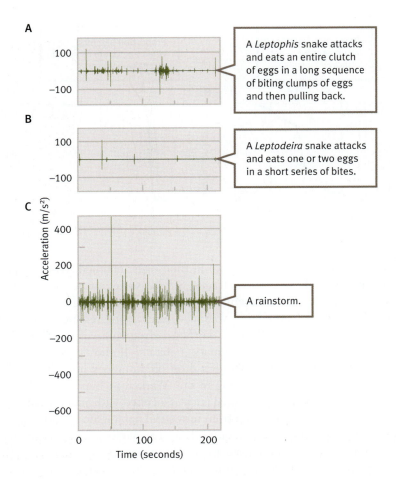

A

A Leptophis snake attacks and eats an entire clutch of eggs in a long sequence of biting clumps of eggs and then pulling back.

B

A Leptodeira snake attacks and eats one or two eggs in a short series of bites.

C

A rainstorm.

FIGURE 12.19. Sounds from red-eyed treefrog nests. Sounds recorded as waveform graphs at the nest of a red-eyed treefrog. *(From Warkentin, 2005, p. 62)*

is characterized by a series of moves toward the predator interrupted by stationary pauses and sometimes alternating with moves away from the predator (Curio et al., 1983; Dugatkin, 1997a; Milinski, 1987; Pitcher et al., 1986). An approach may culminate in a number outcomes, ranging from the prey simply retreating, to the prey rejoining a social group of conspecifics nearby, to an escalation in which prey actually attack the predator (see more on escalation below).

For this sort of antipredator approach behavior to evolve and to be maintained in a population, the average fitness benefits accrued must equal or exceed the associated average fitness costs. To investigate these costs and benefits, we turn to work on Thomson's gazelles.

THE COSTS AND BENEFITS OF THOMSON'S GAZELLES APPROACHING A PREDATOR. Using natural populations of Thomson's gazelles as her experimental subjects, Clare Fitzgibbon undertook one of the most thorough analyses of the costs and benefits of approach behavior (Fitzgibbon, 1994; Figure 12.20). In the Serengeti National Park (in Tanzania), gazelles live in groups that can vary from fairly small (< 10 individuals) to fairly large (> 500), and they interact with four main predators: lions, cheetahs, spotted hyenas, and wild dogs. Although we will focus on the predators that gazelles do approach (lions and cheetahs), it is interesting to note that gazelles rarely approach wild dogs and hyenas. The primary reason for this difference appears to be that cheetahs and lions rely on surprise and short, fast chases, whereas wild dogs and hyenas rely on stamina. Approaching the former group may provide net benefits, whereas approaching the latter likely does not.

With respect to the benefits of approaching a predator, Fitzgibbon examined three possibilities:

1. Approach behavior might decrease the current risk of predation.
2. Approach behavior might allow gazelles to gather information about a potential threat.
3. Approach behavior might serve to warn other group members of the potential danger associated with predators.

In the course of her two-year field study, Fitzgibbon found some evidence for benefit 1. In particular, she found that cheetahs responded to gazelle inspection behavior, which is most common and most pronounced in large gazelle groups (Figure 12.21A), by moving farther between rest periods and between hunting periods (Figure 12.21B). This in turn could cause cheetahs to leave an area sooner than normal as a result of gazelle approach behavior, leading to decreased rates of mortality among potential prey.

FIGURE 12.20. Gazelle antipredator behavior. (A) Gazelles are constantly vigilant for potential predators. (B) Many species, including the cheetah (pictured here chasing a gazelle) hunt gazelles. *(Photo credits: Gerald and Buff Corsi/ Visuals Unlimited; Joe McDonald/Visuals Unlimited)*

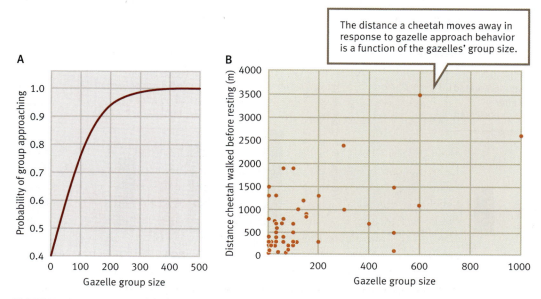

FIGURE 12.21. Approach behavior in gazelles. (A) The probability of approach behavior occurring in gazelles is a function of group size, as indicated by the curve. (B) Cheetahs respond to gazelle approach behavior. *(Based on Fitzgibbon, 1994)*

The cost of gazelle approach behavior is paid primarily in terms of lost time/energy. Through detailed behavioral observations, researchers determined that gazelles actually spend approximately 4 percent of their waking hours involved in approach behavior. This 4 percent could otherwise be devoted to other activities—for example, foraging, mating, resting—and thus it represents a real "opportunity cost" to the animals.

INTERPOPULATION DIFFERENCES IN MINNOW APPROACH BEHAVIOR. Anne Magurran and her colleagues have compared antipredator behavior in two populations of minnows (*Phoxinus phoxinus*). Minnows from the Dorset area of southern England and the Gwynedd area of northern Wales were chosen for study, as the Dorset population of minnows is under strong predation pressure from pike predators, whereas pike are absent from the Gwynedd population of minnows and have apparently never colonized this area (Magurran and Pitcher, 1987).

Both populations of minnows increased their group size when faced with predators in the laboratory, but the Dorset population tended to maintain larger groups. With respect to approaching predators, Dorset minnows engaged in predator inspection more often than did Gwynedd minnows, but they were also much more likely to stop inspecting if a conspecific was eaten by a pike (Figure 12.22). For our purposes here, we will focus on whether these discrepancies between the inspection behavior of the Dorset population, which co-evolved in the presence of pike, and the Gwynedd population, which did not, are a result of natural selection operating differently across these groups.

To address this question, adult minnows from the Gwynedd and Dorset populations were captured in the wild, and their offspring were raised in the laboratory (Magurran, 1990). Four treatments were undertaken. In treatments 1 and 2, at two months of age, offspring from Gwynedd and Dorset minnows were exposed to a "model" pike predator (constructed from wood, but made

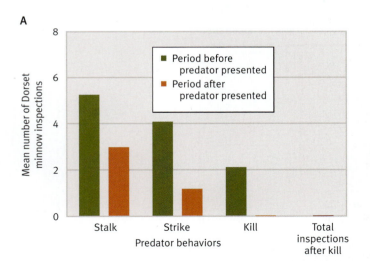
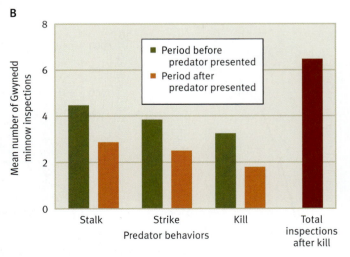

FIGURE 12.22. Inspection across populations. (A) Dorset minnows respond to pike behaviors (stalk, strike, kill another group member) by decreasing their predator inspection behavior, whereas **(B)** no statistically significant decrease in inspections occurs in the Gwynedd minnow population. *(Based on Magurran and Pitcher, 1987)*

to look like a predator) five times during development, and their antipredator behaviors were recorded. In treatments 3 and 4, rearing conditions were identical, except that no minnows in these treatments were exposed to the model predator.

When researchers compared the antipredator behavior of two-month-old fish exposed to predators, they found that young Dorset minnows differed from young Gwynedd minnows in the same manner uncovered in Magurran's earlier work using wild fish. Dorset minnows inspected more often and tended to be found in larger groups, suggesting that natural selection has produced significant differences in antipredator behavior across these two populations (Figure 12.23). The story, however, is more complex.

After raising the minnows in the laboratory for a total of two years, Magurran examined the antipredator tactics of minnows in these four groups. The differences found in two-month-old Dorset and Gwynedd populations were also found in two-year-old fish, but experience also played a key role. Although all Dorset fish—experienced and inexperienced—inspected more and in larger groups than all Gwynedd minnows, there were also interesting between-population differences in the role that experience played in shaping antipredator behavior. In both populations, experienced adults inspected more often than inexperienced adults, but this difference was most pronounced in Dorset fish.

Natural selection seems to have operated in one, and perhaps two, ways on antipredator behavior in minnows. First, as a result of differences in predation pressure, Dorset and Gwynedd fish have genetically based differences in the antipredator tactics they employ. Second, it may be that selection has favored stronger responses based on experience in the Dorset population of minnows from high-predation areas.

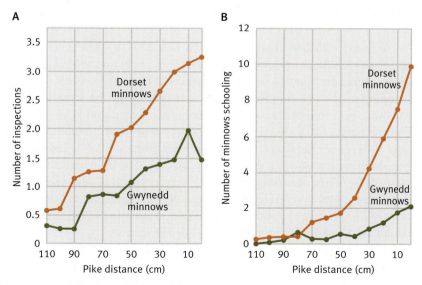

FIGURE 12.23. Reaction to predators. (A) Population differences in inspection behavior were found when comparing two-month-old minnows that had never before been exposed to a predator, from high-predation (Dorset) and low-predation (Gwynedd) sites. **(B)** Population differences in schooling behavior were also found in two-month-old inexperienced fish originating from high-predation (Dorset) and low-predation (Gwynedd) sites. *(Based on Magurran, 1990)*

FEIGNING DEATH

Faking, or feigning, death is an antipredator behavior seen across a spectrum of species. Death feigning occurs in insects when, in response to a predator, an insect falls down and then remains frozen—that is, motionless (sometimes referred to as tonic immobility; Miyatake et al., 2004; Ruxton et al., 2004).

Tatsunori Ohno and Takahisa Miyatake have done extensive work on feigning death in the adzuki bean beetle (*Callosobruchus chinensis*; Ohno and Miyatake, 2007). In this species, when a beetle is on a branch and a predator approaches, beetles can either fly away or feign death, but they cannot do both at the same time. Ohno and Miyatake hypothesized that a negative genetic correlation existed between the intensity of death feigning and the ability to fly—those beetles that feigned death for a long period of time would be poor flyers, and that those beetles that feigned death for shorter time periods would be especially good flyers.

Ohno and Miyatake established two assays—one for feigning death and one for flying ability. For the former, they simply exposed beetles to danger and measured how long the beetles remained frozen, feigning death. For the assay on flying ability, they dropped a beetle through a hole into a cube-shaped apparatus that had a grid on its bottom. The researchers measured whether the beetle dropped straight down (poor flyer), or whether it flew away as it was falling and how far it flew (on a scale of 1–6). Once these assays were established, Ohno and Miyatake ran a classic artificial selection experiment with two treatments. In one treatment—the death-feigning (DF) treatment—a random sample of 100 flies was selected, and the researchers measured the duration of death feigning for each fly. In the second treatment, the researchers selected flies based on their flying abilities.

Two genetic "lines" were established within the DF treatment. In the "long-duration" line, the fourteen individuals (seven males and seven females) that displayed the longest duration of death feigning were selected and allowed to breed with one another. In the "short-duration line," the fourteen individuals (seven males and seven females) with the shortest duration of death feigning were selected and allowed to breed with one another. This was repeated for eight generations—in each generation of the short-duration line, the seven males and seven females with the shortest times for feigning death were selected. The same process of selection and breeding of the seven males and seven females with the longest times for feigning death occurred over eight generations of the long-duration line.

Eight generations of strong artificial selection on death feigning produced dramatic differences between the selected lines (Figure 12.24). Individuals in the long-duration line showed death-feigning times that were about forty times as long as those in the short-duration line. More critically, Ohno and Miyatake found a negative genetic correlation between death feigning and flying abilities. Individuals in the long-duration line were very poor flyers, and, conversely, individuals in the short-duration line were adept flyers.

Ohno and Miyatake also ran a second artificial selection experiment. This time, they created two genetic lines—one in which the best flyers (BF) in every generation bred with one another, and a second in which the worst flyers (WF) bred with one another. The same negative genetic correlation found in the DF treatment was uncovered here. That is, after eight generations, the beetles in the BF line were not only superb flyers but they displayed very short death-feigning times. Conversely, the beetles in the WF line were very poor flyers, but they displayed death feigning for long periods of time.

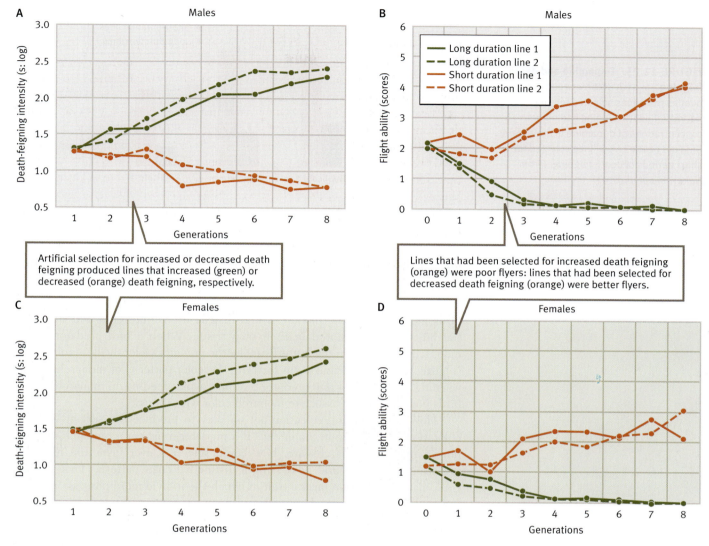

A Males

B Males

Long duration line 1
Long duration line 2
Short duration line 1
Short duration line 2

Artificial selection for increased or decreased death feigning produced lines that increased (green) or decreased (orange) death feigning, respectively.

Lines that had been selected for increased death feigning (orange) were poor flyers: lines that had been selected for decreased death feigning (orange) were better flyers.

C Females

D Females

Ohno and Miyatake's artificial selection experiments are an excellent example of how animal behaviorists can employ classic experimental protocols from genetics to better understand the trade-offs associated with different forms of antipredator behavior, as well as the genetics underlying this trade-off. Recent work by this group has also explored one of the proximate underpinnings of death feigning in beetles. They found that beetles in populations selected for long bouts of death feigning had higher brain concentrations of dopamine than beetles from populations selected for short bouts of death feigning (Nakayama et al., 2012; Figure 12.25).

FIGURE 12.24. Selection for death feigning. Data from two replications (replicate 1 and replicate 2) of the long-duration and short-duration selection experiments on death feigning in the adzuki bean beetle. Points on solid lines represent data from the first replicate; points on dashed lines represent data from the second replicate. There is a negative correlation with flying ability when selection is on death feigning. *(From Ohno and Miyatake, 2007, p. 558)*

SIGNALING TO PREDATORS

Prey sometimes transmit information to a predator, warning the predator about the dangers of contact, or that it has been sighted and will not succeed in capturing a prey, which may deter predator attack. These signals are often visual but can also be auditory, as in the case of tiger moths that can produce warning sounds that cause bats to avoid them because the bats have learned to associate the sounds with a noxious taste (Barber and Conner, 2007). Here we

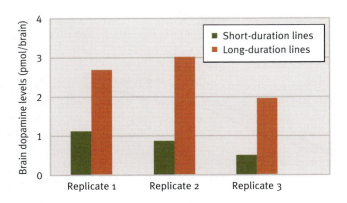

FIGURE 12.25. Dopamine and death feigning. In each of three replicates, beetles from lines selected for long periods of death feigning had higher concentrations of dopamine in their brains than did beetles from lines selected for short periods of death feigning. *(From Nakayama et al., 2012)*

will consider two case studies of signaling: (1) warning coloration in monarch butterflies and (2) tail flagging in ungulates.

WARNING COLORATION IN MONARCH BUTTERFLIES. When they are caterpillars, monarch butterflies (*Danaus plexippus*) ingest milkweed plants, which contain chemicals called cardiac glycosides. These chemicals, which are toxic to birds, do not harm the monarchs and instead are sequestered and stored by the butterflies and stored in their own tissue. When a bird predator eats a monarch, the toxins in the butterfly make the predator violently ill—temporarily (Figure 12.26). From that point forward, the color patterns of monarchs act as warning coloration for *that* predator that now avoids feeding on monarchs. Such birds have learned to associate monarch color with illness.

How could natural selection act on prey to produce the sort of warning coloration that we see in monarchs? If a predator must eat a monarch to learn how dangerous monarchs are, how could selection ever favor the monarch ingesting milkweeds and possessing a distinctive color pattern? That is, how could natural selection favor a trait in which the individual in possession of the trait must die for the predator to learn about the danger? There are a number of ways that this could occur—for example, R. A. Fisher suggested that if prey live in groups, such warning coloration could preferentially aid genetic relatives and so be favored by natural selection (R. A. Fisher, 1930). But the most likely explanation for the evolution of this sort of coloration is that the predator does not always kill the monarch before it senses the toxin—touching the monarch

A **B**

FIGURE 12.26. Monarch warning colors. (A) A bluejay holding the wing of a monarch butterfly that it is about to eat, and **(B)** a bluejay vomiting after eating a monarch butterfly. *(Photo credits: Lincoln Brower)*

may be enough to alert the predator to the presence of the toxin. The presence of the toxin and warning coloration may save the life even of a prey that is the victim of a predator's first encounter with a monarch.

TAIL FLAGGING AS A SIGNAL. Some animals send signals that may serve to warn a predator that it has been spotted. When the predator is an ambush hunter that relies on surprise, such a signal often causes it to move on and leave the area, and clearly benefits the prey. Even when predators aren't strictly ambush hunters, prey may still benefit by signaling predators if signals reduce the probability of capture. Consider tail flagging in ungulates, where individuals "flag" their tails after a predator has been sighted (Figure 12.27). Such flagging occurs as part of a sequence of antipredator behaviors, and often it involves an individual lifting its tail and "flashing" a conspicuous white rump patch. Flagging often, but not always, occurs when a predator is at a relatively safe distance from its potential prey (Hirth and McCullough, 1977).

In principle, tail flagging could serve many different functions. At various points in time, animal behaviorists have hypothesized that this behavior might

FIGURE 12.27. Tail flagging. In white-tailed deer, individuals often display a white patch while running from danger. At least a half dozen hypotheses have been put forth to explain the function of the patch. *(Photo credit: Gary Carter/Visuals Unlimited)*

▶ warn conspecifics (both kin and nonkin) about the presence of a predator (Estes and Goddard, 1967);

▶ "close ranks" and tighten group cohesion, which might make predation less likely (Kitchen, 1972; McCullough, 1969; P. S. Smith, 1991) and ensure group-related foraging and antipredator benefits in the future (R. J. F. Smith, 1986; Stankowitch, 2008; Trivers, 1971);

▶ signal to the predator that it has been sighted and should therefore abandon any attack. This has been dubbed the "pursuit-deterrence" hypothesis (Caro, 1995a; Caro et al., 1995; Woodland et al., 1980);

▶ entice the predator to attack from a greater distance, which may result in a failed attempt at capture (Smythe, 1977);

▶ cause other group members to engage in various antipredator activities, thereby confusing the predator, and making the flagger himself less likely to be the victim of an attack (Charnov and Krebs, 1975); or

▶ serve as a sign for appeasing dominant group members, and only secondarily play a role in antipredator behavior (R. D. Guthrie, 1971).

Some evidence supports the hypothesis that tail flagging causes an increase in group cohesion (Bildstein, 1983; Hirth and McCullough, 1977; P. S. Smith, 1991), but here we will focus on the question whether tail flagging is a signal to the predator. This "pursuit-deterrence" hypothesis about signals is supported by Tim Caro's work on white-tailed deer, as well as the work of other researchers (Bildstein, 1983; Caro, 1994b; Caro et al., 2004; Woodland et al., 1980). Caro and his colleagues found no evidence for a time cost to flagging—that is, flagging did not slow down prey—and no evidence that tail flagging was aimed at conspecifics in the context of cohesion. Rather, they showed that white-tailed deer that run fast flag their tails and are using this signal to communicate to the predator that an attack is unlikely to succeed because the fleeing deer will escape a pursuing predator. This sort of pursuit-deterrence signal is not limited to white-tailed deer or to tail flagging. For example, in a phylogenetic study involving 200 species and seventeen antipredator behaviors, Caro and his colleagues found that, in ungulates, snorting also serves as a signal to predators (Caro et al., 2004). This signal deters attack, presumably because snorting indicates the health and vigor of the signaler.

Although prey are typically physically smaller and weaker than their predators, they do sometimes possess adaptations that allow them to fight back when their life is in danger. Here we will examine two such adaptations: chemical defense in beetles and mobbing behavior in birds.

CHEMICAL DEFENSE IN BEETLES. Thomas Eisner and his colleagues have demonstrated that bombardier beetles use chemical weapons to defend themselves against predators, including swarming ants, orb-weaving spiders, and toads (Dean, 1980a, 1980b; Eisner and Dean, 1976). In the bombardier beetle, *Stenaptinus insignis*, individuals blast potential predators with a highly noxious spray (Eisner and Aneshansley, 1982, 1999; Eisner et al., 2000, 2006). A beetle can discharge its acidic spray twenty times before depleting its supply of chemicals.

Production of *S. insignis*'s noxious spray is an act of engineering brilliance. These beetles have two glands, each of which has two separate compartments. The larger compartment in a gland is referred to as the reservoir, and it contains hydroquanines and hydrogen peroxide. The smaller compartment, called the reaction chamber, holds a variety of catalases and peroxidases. When bombardier beetles are threatened by a predator, they allow the contents of the two compartments to mix, causing a chemical reaction that produces a spray composed of acidic, very noxious *p*-benzoquinones. The heat produced by this chemical reaction causes an audible pop, and the spray shoots out at a temperature of 100° C. Somehow, and it is not yet known how, the beetles themselves are not injured by their own noxious sprays, but predators are.

S. insignis does more than just produce an acidic spray that is released when a predator attacks. Using high-speed photography, Eisner and Daniel Aneshansley have shown that the beetles are quite good at selectively aiming this spray at predators. When they are attacked from the front, they fire the spray forward; when attacked from the rear, they fire the spray backward (Eisner and Aneshansley, 1999; Figure 12.28).

How did such a successful, but complicated, antipredator mechanism evolve? Researchers are still working on that question, but they have found a clue in the spray mechanism of *Metrius contractus*, which is the oldest of all extant species of bombardier beetles (Eisner et al., 2000). Like other bombardier species, it has a reservoir that contains hydroquanines and hydrogen peroxide and a reaction chamber that holds a variety of catalases and peroxidases, suggesting that possession of these traits is an ancestral characteristic. Unlike other species, however, which often discharge their defensive spray as a jet, *M. contractus*'s spray is discharged as a fine mist. Again using high-speed photography, the researchers examined emission of this spray in more detail.

Eisner and his colleagues found that *M. contractus* emits its spray in a unique manner. When it is attacked from the rear, it produces a froth secretion, which builds up on the body of the beetle and wards off predators. When a beetle is attacked from the front, instead of spraying its attacker, it forces the chemical secretion it produces forward, along tracks on its forewings. Although it is not clear why, these mechanisms of discharging chemical weapons appear to lower the temperature of the disseminating chemicals from 100° C to approximately 55° C. This work on *M. contractus* hints that spraying an extremely hot chemical

FIGURE 12.28. Bombardier defenses. When the bombardier beetle, *Stenaptinus insignis*, is threatened, it releases chemicals that ward off predators. This bombardier beetle is being attacked from the front, so it is directing its chemical spray forward. *(Photo credit: Thomas Eisner)*

secretion is a derived trait, whereas frothing and using the forewing tracks to disseminate a somewhat reduced heat spray represent the ancestral version of chemical defense in bombardiers.

Eisner and his team's work shows nicely how animal behaviorists (in this case, in conjunction with collaborators in chemistry) can (1) study the function of a trait, (2) break down the complex anatomical and chemical underpinning of that same trait, and (3) use comparative studies to understand the evolution of behavior.

SOCIAL LEARNING AND MOBBING IN BLACKBIRDS. Chemical weapons are only one way prey can counterattack their predators. Blackbirds (*Turdus merula*), like many other bird species, undertake a form of attack called predator **mobbing** (Altmann, 1956; Sordahl, 1990). Once a flock of blackbirds spots a predator, they join together, fly toward the danger, and aggressively attempt to chase it away. Such group attacks often work well enough to force predators to leave the blackbirds' area.

Eberhard Curio and his colleagues examined whether young, predator-naive blackbirds learn what constitutes a predator by watching which species is mobbed and classifying such a species as predators (Curio et al., 1978). Is mobbing a form of cultural transmission that is useful in the context of antipredator activities? To determine whether it is, Curio and his colleagues ran a fascinating laboratory experiment on mobbing.

In each trial of their experiment, Curio's team began with a "model" and a "naive" bird, each in its own aviary. The experimental apparatus was designed so that each of the blackbirds could see another bird—a noisy friarbird (*Philemon corniculatus*). The friarbird was a species that neither the model nor the naive blackbird had seen before, and it looked nothing like any known predator of blackbirds. The friarbird was presented in such a way that the naive blackbird saw the friarbird alone, but the model saw both a friarbird and, adjacent to it, a little owl, *Athena noctua*—a real predator of blackbirds. From the viewpoint of the naive subject, the little owl was out of sight, so that when the model mobbed the little owl, the naive individual saw it mobbing a friarbird.

Curio's team found that, once naive blackbirds had seen a model apparently mobbing a friarbird, the naive blackbirds themselves were much more likely to mob this odd new creature than if they had not been exposed to the model. The information about what constitutes a danger was transmitted culturally. The researchers then took the experiment one step further and asked whether the (now not so naive) blackbird subject would act as a model for a new naive blackbird. And if that worked, how many times could they get a former naive blackbird to successfully act as a model? That is, how long is the "cultural transmission chain" in blackbirds? The longer such chains are, the more powerful cultural transmission may be in spreading antipredator behaviors through a population. Though their sample was small, Curio and his colleagues found that, in their experiment, the blackbird cultural transmission chain was at least six birds long. After the initial model (bird 1) perceived friarbirds as being dangerous, a new model (bird 2) then saw bird 1 respond to friarbirds. If bird 2 then responded as if friarbirds were dangerous, a new model (bird 3) then observed bird 2 respond to the friarbird as a predator. This procedure went on until six different birds acted as if the friarbird were a predator, and then the chain of information transfer was broken.

Predation and Foraging Trade-offs

Ethologists who study antipredator tactics often study the trade-offs animals make when vigilant for predators. When animals spend time engaged in antipredator activity, they could potentially be doing something else—foraging, mating, resting, playing, and so forth (Lima and Dill, 1990; Figure 12.29). Or, rather than totally curtailing alternative behavior, antipredator behavior could create pressure to perform other behaviors in a different manner—for example, to forage in the vicinity of a refuge, to mate at times when predation is minimal, and so on. In either case, trade-offs between some other behavior and antipredator tactics are often common.

Foraging is one behavior that is especially affected by predation. Predation pressure affects virtually every aspect of foraging—including when a forager begins feeding (Clarke, 1983; Lima, 1988a, 1988b), when it resumes feeding after an interruption (De Laet, 1985; Hegner, 1985), where it feeds (Dill, 1983; Ekman and Askenmo, 1984; Lima, 1985; Schneider, 1984), what it eats (Dill and Fraser, 1984; Hay and Fuller, 1981; Lima and Valone, 1986), and how it handles its prey (Krebs, 1980; Valone and Lima, 1987). As an example, let us consider Steven Lima and Thomas Valone's work on predation and foraging in the gray squirrel (*Sciurus carolinensis*; Lima and Valone, 1986; Figure 12.30).

FIGURE 12.29. Foraging-predation trade-off. When animals are being vigilant for predators, it is often at the cost of other activities. Any bird in the group here can't forage for insects at the same instant that it scans the sky for hawks.

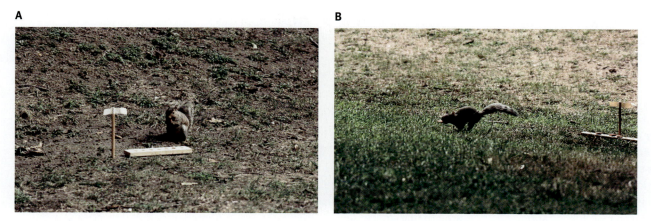

FIGURE 12.30. Foraging/predation trade-off in squirrels. (A) When squirrels forage in open fields, they need to balance the risk of predation (a function of distance to cover) against the benefits associated with various food items. Here a squirrel is foraging at an experimental feeding station. **(B)** A squirrel heads for cover with a food item (part of a cookie) in its mouth. *(Photo credits: Steven Lima)*

Early work by Lima had demonstrated that squirrels alter their foraging choices as a result of predation pressure from redtailed hawks (*Buteo jamaicensis*; Lima, 1985). Squirrels that could either eat their food items where they found them or carry the food to cover were more likely to carry items to an area of safe cover, particularly as the distance to safe cover decreased. The closer the refuge from predation, the more likely squirrels were to use such a shelter when foraging; when it was a quick run to reach safety, squirrels generally chose to do so. Squirrels were also much more likely to carry larger (rather than smaller) items to safe areas before continuing to forage.

Lima and Valone followed up the above study with one in which they presented squirrels with two types of food—a large chunk of cookie (associated with long handling times) or a small chunk of cookie (associated with short handling times; see Chapter 11). Cookie chunks, rather than nuts, were used to avoid the confounding variable of food storage, as nuts are often buried, but cookies are always eaten. A combination of large and small items was placed either close (8 m) to an area of cover or farther from safety (16 m).

In order to make sense of Lima and Valone's results, we need two critical pieces of information. First, the profitability of small food items was greater than the profitability of large food items (see Chapter 11). In the absence of predation, we would expect that squirrels would always take any small food item that they encountered. Second, the total handling time associated with larger food items was great enough that optimal foraging models predicted that larger items should be brought to cover, where it is safe, before being eaten, particularly when the distance to cover was not great. This was not the case for smaller items. Lima and Valone hypothesized that, if faced with predation, squirrels might sometimes pass up the smaller, more profitable food items and continue to search for larger morsels to bring back to cover. This, in fact, is what the squirrels did. Smaller items were rejected in favor of larger items that needed to be brought back to cover—predation pressure redefined which food items should be taken and which should be passed up.

INTERVIEW WITH

Dr. Anne Magurran

Why study antipredator behavior in animals? Are there unique aspects about this sort of behavior?

Predation risk is an uncompromising form of natural selection. An animal that is captured by a predator will be unable to reproduce, so any activity or response that increases an individual's chances of survival is favored. The study of antipredator behavior can therefore tell us a lot about the process of evolution. But it is not simply a matter of prey evolving skills that allow them to evade predators more effectively. Predators will also experience selection to improve their capture success rate. This is sometimes called an "arms race" since better defenses are countered by improved hunting tactics, which in turn will select for improved antipredator behavior. I particularly like Darwin's description of this co-evolutionary game: "Wonderful and admirable as most instincts are, yet they cannot be considered as absolutely perfect: there is a constant struggle going on throughout nature between the instinct of the one to escape its enemy and of the other to secure its prey."

Interpopulational comparisons have proved particularly useful in terms of understanding the evolution of antipredator behavior. Why is that?

I believe there are two main reasons why interpopulational comparisons have been so productive. First, when we compare populations we are, to a large extent, comparing like with like. An investigation of the antipredator behavior of, say, blackbirds and robins could

be confounded by all the other biological and ecological differences between the species. An intraspecies investigation, on the other hand, looks at groups of organisms that are broadly similar. It is a particularly powerful means of understanding the evolution of antipredator behavior when the degree of predation risk is the single most important thing that

differentiates populations. Moreover, if there are multiple populations experiencing each level of predation risk, then we can, by virtue of this replication, be even more confident that the variation in behavior that we observe is related to the activities of predators as opposed to some other variable (such as the history of the population or the temperature of the environment).

Trinidadian guppies, *Poecilia reticulata*, are a classic example of an interpopulational comparison. A number of rivers draining Trinidad's Northern Range have barrier waterfalls that have prevented predators, though not guppies, from moving upstream. This means

that the behavioral characteristics of guppy populations separated by very small distances, and exposed to similar ecological conditions, but differing in the types and numbers of predators that they coexist with, can be attributed to predation. It is also possible to perform simple manipulations in the wild, in which predators are added to or removed from populations, to show that a shift in predation risk does indeed result in heritable change—evolution—of behavior.

The second reason why interpopulational differences are so illuminating is that changes in predation risk—and hence evolution—can be tracked over relatively short time scales. In guppies, for instance, a change in escape response is detectable within about fifteen years of a shift in predation risk. I suspect that behavioral evolution occurs even more quickly than this in the guppy system—no one has looked carefully yet.

What is the most dramatic example of antipredator behavior that you know of?

There are many impressive antipredator behaviors and it is difficult to select a single example, but I think that the evasion tactics of schooling fish are particularly dramatic. Hundreds, or sometimes many thousands, of individuals engage in choreographed maneuvers, such as the flash expansion [in which tightly packed schools of fish "explode," and the fish swim off in all directions], that confuse and outwit their predators. Although it was originally thought that these seemingly cooperative tactics were for the "good of the species," we now know that they are underpinned by individual selection, and that the complex formations that emerge are based on simple decision rules. Unfortunately, this behavior offers little protection against modern fishing vessels since these are equipped with technology to locate and capture large shoals of fish.

Initially a great deal of work on predation focused on foraging/predation trade-offs. How would you summarize the findings from these studies?

One way of looking at this is in relation to what Richard Dawkins and John Krebs called the life-dinner principle. In other words, selection on an individual to remain alive is greater than the selection to find the next meal. So, all other things being equal, animals should invest more in predator avoidance than foraging. However, choices are rarely that simple. An individual that spent all its time in a refuge might avoid predators but would sooner or later die of starvation. This leads to a trade-off between the conflicting demands of foraging and predator avoidance. One resolution of this conflict is for an animal to select the habitat that allows it to minimize the risk of mortality (μ) while maximizing growth rate (g), typically by minimizing the ratio μ/g. Empirical studies—for example, experiments on creek chub, *Semotilus atromaculatus*, by Jim Gilliam and Doug Fraser—support these predictions. Of course, there is considerable variation, among individuals and species, in the trade-offs adopted. Sexually immature animals may place a higher priority on feeding so that they grow faster and reach the size class at which they are able to breed, whereas older ones may be more cautious to help ensure that they remain alive and continue to reproduce.

Your own work on predation pressure suggests that studying this phenomenon may have implications for understanding the process by which new species arise. Can you tell us a little about that?

Our work on guppies uncovered an interesting paradox. Population differences in antipredator behavior (as well as in life history and color pattern) arise over relatively short time periods (typically 10 to 100 generations). Reproductive isolation, in contrast, is slow to emerge and seems to take in excess of a million years to become established. When we looked at its evolution in detail, we found that some postmating isolation has arisen among populations that have been separated for long periods of time. It occurs in the form of gametic isolation (the reduced ability of foreign sperm to compete with native sperm) and as a reduction in hybrid viability. However, there is little evidence for pre-mating isolation, even among long-separated populations. In part this is because of female choice. Although females prefer certain male color patterns, particularly orange spots, they do not seem to discriminate against males from genetically divergent populations. Males show less discrimination of mating partner than females and use sneaky matings as well as consensual courtship to obtain copulations. Sneaky matings occur at a higher rate in populations that co-occur with predators, and are often directed at females preoccupied by predation avoidance.

Where do you see the study of predation and animal behavior heading over the next years?

To date, behavioral studies of predation, along with much of behavioral ecology, have focused on carefully designed, often laboratory-based, experiments. This work has been influential in understanding how animals interact with one another, whether they are predators and prey or individuals of the same species. However, it is now becoming clear that temporal variability in the natural world, such as changes in the presence and the abundance of species in ecological communities across diurnal, seasonal, and annual timescales, can profoundly influence behavior. For example, the predation risk experienced by guppies in their native streams in Trinidad can vary dramatically over short periods of time, but we know relatively little about how the fish respond to this variation in risk. So while it may be unwise to make predictions about future directions in a field like animal behavior, I believe that increasing attention will be paid to natural variability and how it shapes behavior. By learning how animals respond to natural change, we may also be better placed to ameliorate the rapid anthropogenic changes that many species now face.

Studies of this type will continue to use the observational and experimental approaches developed by Tinbergen and other pioneers in the field. However, this is an exciting era, as it is now possible to probe the genetic basis of behavior in real time, by tracking the expression patterns of genes using recently developed genomic tools. These techniques give researchers a much deeper understanding of how animals respond to the nuances of daily life, as well as revealing how the genome interacts with the environment in an increasingly changing world.

Dr. Anne Magurran is a professor at the University of Saint Andrews (Scotland). Dr. Magurran's work, using both guppies and minnows, has helped ethologists better understand the interaction of ecology, evolution, and behavior. She is the author of Measuring Biological Diversity *(Oxford University Press, 2004) and* Evolutionary Ecology: The Trinidadian Guppy *(Oxford University Press, 2005).*

SUMMARY

1. There are two basic forms of antipredator behaviors: those that help prey avoid detection by predators, and those that function once a prey encounters a predator. These two categories encompass most antipredator behaviors, but they are neither exhaustive nor mutually exclusive.

2. Studies using tools from genetics, neurobiology, endocrinology, and chemistry have helped animal behaviorists better understand proximate factors associated with antipredator behavior.

3. Predation pressure not only has an effect on how an animal behaves but also can have a dramatic effect on such life-history variables as development time—for example, when eggs will hatch.

4. If prey can avoid their predators in the first place, they not only can decrease the probability of being captured and eaten but also can reduce the costs associated with fleeing, fighting back against a predator, and so on. Three ways that animals avoid their predators are (a) blending into the environment, (b) being quiet, and (c) choosing a safe habitat.

5. Five types of behaviors that prey use once a predator has been encountered are (a) fleeing, (b) approaching a predator to obtain information, (c) feigning death, (d) signaling to the predator, and (e) fighting back.

6. Comparing a single species in populations with and without a particular predator is a powerful means for studying how selection operates on antipredator behaviors.

7. When there is predation pressure, animals are likely to alter their behavior, especially their foraging behavior, to avoid capture by predators.

DISCUSSION QUESTIONS

1. Pick your animal of choice, and sketch what a normal time budget (how much time it spends feeding, sleeping, mating, and so on) might look like for this animal. Now, besides the direct time spent looking for predators, examine how predation might directly or indirectly affect all the behaviors on your time budget.

2. Some prey, particularly birds, mob their predators and harass them until the predators leave. List some of the costs and benefits associated with such mobbing, and construct a hypothesis for what sorts of environments might favor mobbing.

3. Abrahams and Pratt (2000) used a thyroid hormone to manipulate growth rates of the fathead minnow (*Pimephales promelas*) in the study described in their paper, "Hormonal Manipulations of Growth Rate and Its Influence on Predator Avoidance: Foraging Trade-offs" in Volume 78 (pp. 121–127) of the *Canadian Journal of Zoology*. Thyroid treatment stunted growth rates, and the researchers found that such stunted individuals were less likely to risk exposure to predators to gain access to food. Why might that be? Is it possible to construct an argument that would predict the exact opposite of what was found? Also, what does this study tell you about the relationship between proximate and ultimate factors shaping antipredator behavior?

4. In addition to the three ways we discussed that animals avoid predators, can you think of any other predator avoidance behaviors? How do they work? Why do they work?

5. A number of studies have found that laboratory-raised animals can learn what constitutes danger by watching other animals respond to potential predators. How might such cultural transmission be employed by those interested in wildlife reintroduction programs?

SUGGESTED READING

Caro, T. M., Graham, C. M., Stoner, C. J., & Vargas, J. K. (2004). Adaptive significance of antipredator behaviour in artiodactyls. *Animal Behaviour, 67,* 205–228. A systematic study of the antipredator behavior in one group of large African mammals.

Croes, B. M., Laurance, W. F., Lahm, S. A., Tchignoumba, L., Alonso, A., Lee, M. E., Campbell, P., & Buij, R. (2007). The influence of hunting on antipredator behavior in central African monkeys and duikers. *Biotropica, 39,* 257–263. A review of the effect of human hunting behavior on the antipredator behavior of other primate species.

Lima, S. L. (1998). Stress and decision making under the risk of predation: Recent developments from behavioral, reproductive, and ecological perspectives. *Advances in the Study of Behavior, 27,* 215–290. A very nice review of decision making in the context of antipredator activities, with a particular emphasis on the effects of stress.

Lind, J., & Cresswell, W. (2005). Determining the fitness consequences of antipredation behavior. *Behavioral Ecology, 16,* 945–956. A review of antipredator behavior from an ultimate perspective, with a focus on measuring complex antipredator behavior in relation to estimates of reproductive success.

Ruxton, G., Sherrat, T., & Speed, M. (2004). *Avoiding attack: The evolutionary ecology of crypsis, warning signals and mimicry.* Oxford: Oxford University Press. A comprehensive review of the evolution of antipredator behaviors that allow prey to avoid attack by predators.

Stankowich, T., & Blumstein, D. T. (2005). Fear in animals: A meta-analysis and review of risk assessment. *Proceedings of the Royal Society, B–Biological Sciences, 272,* 2627–2634. A concise overview of factors that affect antipredator behavior.

13

Communication

ervet monkeys living in the Amboseli National Park in southern Kenya face danger from many predators. Leopards hide in the bushes and pounce on passing vervets. Crowned eagles swoop from the sky and pick off vervets, and snakes are also a threat (Cheney and Seyfarth, 1990). Faced with such dangers, it is not all that surprising that vervets communicate information about predators to one another. But they do so in a remarkable fashion. Vervets don't just give an "alarm call" when a predator is sighted; they emit specific alarm calls for specific types of danger and different calls elicit different responses by groupmates (Figure 13.1). Moreover, this sort of complex alarm calling has also been found in other primates such as chimpanzees and tamarins (Crockford and Boesch, 2003; Crockford et al., 2012; Kirchhof and Hammerschmidt, 2006). In blue monkeys (*Cercopithecus mitis*), males not only give specific alarm calls for specific predators but they can gauge the distance of predators from the alarm calls given by others in their group (Papworth et al., 2008; Zuberbuhler, 2009).

When an eagle is spotted, vervets emit a "cough" call. When other vervets hear the cough call, they look into the air or hide in the bushes, where they tend to be safe from avian threat. If a leopard is spotted, a "barking" alarm call is given, and vervets respond by heading up trees, where their agility makes them relatively safe from leopard attacks. That is, rather than respond to the leopard alarm call by looking up or hiding in the bushes—both of which would likely get the animal killed—they respond by heading for safety in the trees. And finally, when a python or cobra is sighted, vervets emit a "chutter" call. Since snakes often hunt vervets by hiding in the tall grass, a chutter call gets other vervets to stand and scan the grass around them for snakes.

Many other examples of animal communication have been documented in the ethological literature. This chapter will address both "how" and "why" questions about animal communication systems.

Ethologists analyze **communication**—defined as the transfer of information from a signaler to a receiver—in order to better understand animal behavior.

A   **B**   **C**

FIGURE 13.1. Vervet alarm calls. Vervets give different alarm calls depending on what type of predator has been sighted. **(A)** Vervets stand up after hearing a "chutter" alarm call indicating that a snake (approaching from the bottom left of the photo) has been spotted. When a leopard **(B)** is detected, vervets give a "barking" alarm call and **(C)** climb trees for safety. *(Photo credits: Richard Wrangham/Anthrophoto; Ian Jones/Alamy)*

Another reason that some people study animal communication is to understand how humans can communicate with animals. For example, people may study animal communication in order to better interact with their pets or domesticated livestock. Here, however, we are not interested in animal-human communication per se, except in terms of how it helps us better understand the evolution of animal communication.

There are a number of reasons not to focus on animal-human communication per se. First, this sort of communication probably only evolved with respect to humans and a small number of species—dogs, cats, horses, and other species that humans have had as pets or livestock for thousands of years. Communication between these species and humans has evolved because only these species have been subject to intense artificial selection. Second, animal-human communications are notoriously difficult to study, primarily because it is very hard to know exactly what is being communicated from the animal side. An example of this sort of problem can be seen in the famous case of a horse named "Clever Hans" (Pfungst, 1911; Sebeok and Rosenthal, 1981).

Around the turn of the twentieth century, a schoolmaster named William von Osten gained fame because of the apparently prodigious ability of his horse, Hans. Hans was seemingly able to do mathematical puzzles, identify music, and answer questions regarding European history. He did this, of course, not by actually speaking, but by using his hoof to count out the right answer to a math problem, or shaking his head "yes" or "no" to a verbal query.

When the Prussian Academy of Science put Hans to the test in a controlled environment, they found something interesting. Hans could only give the right answer when someone in the room knew the right answer. If two people each gave Hans part of a question, but each was ignorant of what the other told Hans, the horse *never* got the correct answer. The reason was that Hans was indeed clever, but not in the way people thought.

Hans's cleverness, it turns out, was that he could pick up on very subtle body cues and facial cues that investigators in the room (unconsciously) emitted while they provided correct and incorrect answers for Hans to select from (Figure 13.2).

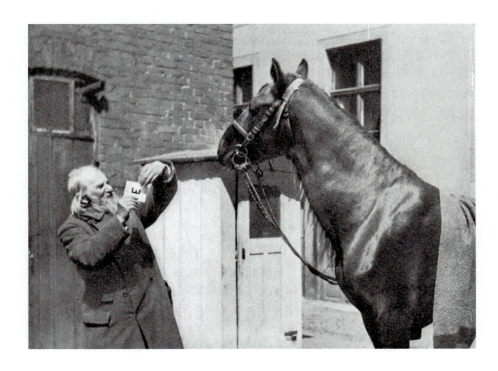

FIGURE 13.2. Clever Hans. Clever Hans, the horse, was thought capable of incredible mental feats. In fact, Clever Hans was picking up very subtle cues from the individual who asked him a question (and consequently who knew the answer to that question). *(Photo credit: © akg-images/The Image Works)*

So, for example, if the problem was "What is 10 + 10?" (asked in German) Hans was able to pick up on subtle cues people emitted when he was approaching 20 hoofbeats on the floor. The lesson here is that human-animal communication can be subtle and extremely difficult to pin down. Of course, the fact that something is difficult to study is no reason not to study it, but this, in conjunction with the first caveat about the evolutionary history of animal-human communication, will keep our focus on evolved animal communication systems.

We will begin our discussion of animal behavior and communication by addressing a fundamental question about communication—Is it honest? That is, are individuals using communication to accurately convey information, or do they use communication to manipulate others? More specifically, we will examine a conceptual framework that will help us predict when communication is expected to be honest and when it is not. Subsequently, at the heart of the chapter, we will analyze communication systems in terms of the problems they solve. Communication is inherently social—that is, it involves more than a single individual. Animals communicate to convey information that is necessary to solve some problem, and we will organize the chapter around this functional, problem-solving approach to communication, examining communication as it relates to problems associated with foraging, mating, and predation.

Communication and Honesty

If communication is the transfer of information from a signaler to a receiver, then an evolutionary approach to this topic immediately raises the question of honesty. Regardless of the problem a communication network is in place to solve, we can ask whether natural selection should favor signalers that are honest in the information they convey, or whether deception is favored. More specifically, we can ask when honesty is favored and when deception is favored. (Dawkins and Krebs, 1978; Grafen 1990a; R. A. Johnstone, 1997; Mock et al., 2011). Consider the following discussion of communication from Richard Dawkins and John Krebs:

> When an animal seeks to manipulate an inanimate object, it has only one recourse—physical power. A dung beetle can move a ball of dung only by forcibly pushing it. But when the object it seeks to manipulate is itself another live animal there is an alternative way. It can exploit the senses and muscles of the animal it is trying to control . . . which are themselves designed to preserve the genes of that other animal. A male cricket does not physically roll a female along the ground and into his burrow. He sits and sings, and the female comes to him under her own power. (Dawkins and Krebs, 1978, p. 282)

Dawkins and Krebs argue that communication is not so much the exchange of information between a signaler and a receiver but rather an attempt by the signaler to manipulate the recipient. They recognize that sometimes what is in the best interest of the signaler is also in the best interest of the recipient. But the signaler's and recipient's interests need not be the same, and when they are, it's usually just an incidental byproduct of what happens to be good for the signaler (see below for some exceptions). When what is good for the signaler is not good for the recipient, natural selection will favor signalers that send signals in whatever way best increases the fitness of the signaler, even if that means manipulating recipients. But natural selection also favor recipients with the ability to unscramble what is honest and what isn't, so it can act in ways that maximize its own fitness. Krebs and Dawkins refer to recipients as "mind readers" and describe an "arms race" between signaler and recipient in which the signaler is selected to better manipulate the receiver, which then is selected to better filter out only that information that benefits it, and so on (Krebs and Dawkins, 1984).

This is a different view from what Dawkins and Krebs refer to as "the classic ethological approach" of communication (Marler, 1968; W. J. Smith, 1968, 1977; N. Tinbergen, 1964). Implicit in the classic approach is that both parties usually benefit from the information exchange, and there is little selection pressure for either to be deceitful. In the classic ethological view, the signaler and receiver have common interests, and selection favors the most economical way to share information.

Krebs and Dawkins recognize that the sort of cooperative signaling that lay at the heart of the classic ethological approach may be occurring in some systems, particularly those involving kin or reciprocal exchanges (Bergstrom and Lachmann, 1997, 1998; Lachmann and Bergstrom, 1998; see Chapters 9 and 10). They offer a way to distinguish between those systems in which there is an arms race between manipulators and mind readers, and those in which cooperative signaling dominates.

Krebs and Dawkins argue that when communication is of the manipulator/mind-reader type, the signals employed should be exaggerated, as one might expect from a salesman attempting to convince a prospective buyer that his product is the top of the line. When cooperative signaling, however, is in play, Krebs and Dawkins hypothesize that natural selection should favor less exaggerated signals—what they refer to as "conspirational whispers." Because signaling often involves some costs—for example, energy costs or drawing attention from predators—natural selection should favor minimizing these costs during conspirational whispers, thus reducing the conspicuousness of the communication itself (De Backer and Gurven, 2006; R. A. Johnstone, 1998; Noble, 1999).

While formal mathematical models of the "conspirational whispers" versus conspicuous display hypotheses have shown that this dichotomy oversimplifies communication systems, it is nonetheless a useful heuristic tool. There is, however, another means besides cooperative signaling by which we might expect communication to be honest. Even within the manipulator/mind-reader view of communication, honesty might evolve if the signals being sent are either impossible or, at the very least, difficult to fake. As an example, imagine that females produce more offspring when they mate with larger males. All males, even small ones, would be favored when they produce signals that made a female believe they were large. But selection should favor females paying attention only to those cues that are honest indicators of large size. Females should cue in on honest signals. This appears to be the case in toads: Deep croaks can be produced only by large males because of the physiology of their vocal system. Because male toads can't fake deep croaks and females prefer larger males as mates, female toads can use croaks as an honest signal when choosing among males (Davies and Halliday, 1978; Figure 13.3).

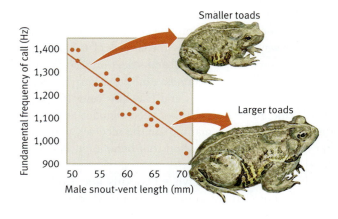

FIGURE 13.3. Toad size and croaks. The relationship between male size (as indicated by his snout-vent length) and the frequency of a male's call. Call frequency may be an honest indicator of size, and hence of fighting ability. *(Based on Davies and Halliday, 1978)*

Amotz Zahavi has suggested that honesty is also possible when traits are not impossible but merely very costly to fake (Grafen, 1990a, 1990b; Zahavi, 1975, 1977, 2003; Zahavi and Zahavi, 1997). Under Zahavi's handicap principle, if a trait is costly to produce, it may be used as an honest signal, because only those individuals that can pay the cost will typically adopt the signal in question. For example, imagine that females are using the length of a male's energy-costly song as a cue for the amount of resources he is able to garner. While a male that is not good at garnering resources could potentially use virtually all of his resources to sing, and thus could give the female a false impression of his resource-garnering skills, most of the time only the males that are genuinely good at gathering resources would be able to afford to sing, as singing is an energy-costly activity. Honest communication may be an outcome even when deception is possible in principle, as long as deception is costly (E. Adams and Mesterton-Gibbons, 1995; R. A. Johnstone, 1995, 1998; Mesterton-Gibbons and Adams, 1998; Zahavi and Zahavi, 1997).

Communication Solves Problems

Communication systems are favored by natural selection when they help solve problems that animals encounter in their natural environments. Here we will examine communication in terms of the problems associated with foraging, predation, and mating. In so doing, we will examine the costs and benefits of communication, the phylogenetic history of communication systems, the role of learning and social learning in shaping communication, and the underlying proximate basis for communication.

Before we look at communication in specific behavioral contexts like foraging and predation, it is worth noting that ethologists have also examined communication at a broader level, not confined to a single behavioral context. For example, Karen McComb and Stuart Semple studied the relationship between vocalization and group size in primates (McComb and Semple, 2005). When these researchers used the published literature to compare vocalization repertoire—the number of types of vocalizations used—and group size, they found a significant positive correlation (Figure 13.4). One possible explanation for this finding is that as group size increases, the benefits of a broad repertoire of

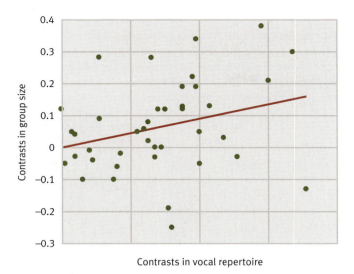

FIGURE 13.4. Vocal repertoire and group size. There is a positive correlation between vocal repertoire and group size in the 42 species of primates examined here. The x-axis and y-axis are measured in "contrasts," which allow a statistical analysis that takes into account the phylogenetic relationship of the species studied (see Chapter 2). (*From McComb and Semple, 2005*)

sounds to communicate with other group members increases (Snowdon, 2009). However, because McComb and Semple's study was correlational, not causal, it is not clear yet whether increased group size favored increased vocalization repertoires or whether increased vocalization repertoire favored the evolution of increased group size (Freeberg et al., 2012).

PROBLEM: HOW TO COORDINATE GROUP FORAGING

When animals forage in social groups, they face coordination problems. For example, when new food sources are found, how can that information be transferred to other group members if such transfer is beneficial to the signaler (Dornhaus et al., 2006; Fernandez-Juricic and Kowalski, 2011; Fernandez-Juricic et al., 2006; Galef and Giraldeau, 2001; J. R. Stevens and Gilby, 2004; Thierry et al., 1995; Torney et al., 2011)?

FOOD CALLS IN BIRDS. Colonial breeding cliff swallows (*Petrochelidon pyrrhonota*) live in nests that serve as "information centers" (C. R. Brown, 1986; Ward and Zahavi, 1973). For some time, researchers thought that individuals living in nests only received "passive" information—that group members simply observed their nestmates and followed them to potential resources. While this sort of system does allow for group foraging, it is not as efficient a solution to coordinating group behavior as an active form of communication about foraging sources—for example, *recruiting* foragers.

The idea that individuals are recruited to food sites was first suggested by Charles Brown and his colleagues (C. Brown et al., 1991). Using both playback experiments (playing tape-recorded bird calls) and provisioning experiments (putting food out to entice birds), Brown and his team found that cliff swallows gave off "squeak" calls, which alerted conspecifics that a new food patch—often a swarm of insects—had been found (Table 13.1). These calls served the specific function of facilitating group foraging, as squeak calls were emitted only in the context of recruiting others to a food site—calls did not appear to be used in any other context.

TABLE 13.1. Squeak calls attract others. The mean number of birds and squeak calls heard two minutes before and two minutes after insects were flushed by foraging birds. For each two-minute period, there were significantly more birds after insects were flushed than before, and significantly more squeak calls were heard after the insects were flushed than before. *(From C. Brown et al., 1991, p. 559)*

PERIOD	MEAN NO. OF BIRDS		MEAN NO. OF CALLS/ BIRD/2 MINUTES		NO. OF TRIALS (*N*)
	BEFORE	AFTER	BEFORE	AFTER	
May 8, 1990	37.8	240.3	0.062	0.267	11
May 9, 1990	9.5	20.5	0.000	0.137	4
May 24, 1990 A.M.	9.7	22.7	0.000	0.100	6
May 24, 1990 P.M.	30.8	88.7	0.053	0.171	9
May 27, 1990	7.9	45.4	0.031	0.133	7
May 29, 1990	8.8	54.8	0.000	0.098	5
May 30, 1990	8.5	38.5	0.033	0.109	2
June 15, 1990	25.9	102.1	0.032	0.134	7

In terms of the costs and benefits of this form of foraging-based communication, those recruited obtain a resource—food. So, it is clearly in the interests of possible recruits to respond to squeak calls. Recruiters also obtain benefits from calling, however, because the increased group size that results from recruiting makes it more likely that some group members will find and track the insect swarm and thus provide further foraging opportunities (C. Brown, 1988; C. Brown et al., 1991). This tracking behavior may be especially critical, as individual swallows must often return to the colony to provision young, and hence they would likely have difficulty relocating an insect swarm without the help of others.

Swallows are not the only birds that emit food calls. When ravens, who are scavengers that can often survive for days if they uncover a large food patch, find a new food source, they often emit a very loud "yell" that attracts other ravens to the caller's newly discovered bounty (Figure 13.5).

Bernd Heinrich and John Marzluff have addressed both proximate and ultimate questions associated with communication and foraging in common ravens (*Corvus corax*; Heinrich, 1988a, 1988b; Heinrich and Marzluff, 1991). On the proximate end, it appears that yelling is, in part, a response to hunger level, as hungry birds call more often than satiated birds (Figure 13.6). In terms of the costs and benefits of calling, it appears that yelling by juvenile ravens attracts other juvenile ravens to a food resource, and this allows them to overpower resident adult ravens. Most of the juvenile ravens are unrelated "vagrants," and the only way an immature vagrant that comes upon a food source that is being defended by an adult territorial male can gain access to this source is to yell (P. G. Parker et al., 1994). Yelling attracts others, that together with the yeller can overpower those originally found at the food source (Heinrich and Marzluff, 1991).

Ravens also communicate about food when they roost together at night, and this form of communication appears to involve some type of learning (Marzluff et al., 1996). While many populations of ravens roost at the same spot for years,

FIGURE 13.5. Raven yells. Under certain conditions, ravens emit a loud "yell" upon uncovering a new food source. Such yells attract other birds. *(Photo credit: Bernd Heinrich)*

Marzluff and his team studied juvenile ravens in the forested mountains of Maine, where individuals form roosts near a newly discovered food source—for example, a large animal carcass. They found that such roosts are very mobile (the ravens move where new prey has been discovered). Marzluff's team hypothesized that these mobile roosts served as information centers that provided roostmates with the chance to share information about prey that they had discovered when away from the roost.

Marzluff and his colleagues ran a series of experiments to test their ideas on communication at evening roosts. Information on naturally occurring mobile roosts suggests that a roost contains both knowledgeable individuals, that know about nearby prey, and naive individuals that do not. Yet, when birds leave the roost in the morning, they all tend to go in the same direction, suggesting that they have communicated with one another, and that one or a few knowledgeable individuals lead the way. In one experiment, Marzluff's team captured ravens and denied them information about the prey environment. When these individuals were released and joined a roost, they tended to follow, rather than lead, other birds out of the roost in the morning.

Marzluff and his team ran a second experiment in which they again captured ravens, but this time they brought the captured birds to a location where there was a new prey item that the researchers themselves had placed into the environment. This information was unavailable to birds that were not captured, as they were not brought to the location of the new food source. When captured birds were released and joined roosts, they tended to be the individuals that led other roostmates out of the roost and to the newly discovered prey item (Figure 13.7). That is, birds that had learned the location of prey acted as leaders. Indeed, Marzluff found that the same individual would act as a leader when it learned the location of new prey, and as a follower when it was denied information that others in the nest knew. While it is not clear exactly how information about who is a knowledgeable forager and who is not is spread at the roost, Marzluff and his team noted that before the ravens departed from the roost in the morning, they emitted a great deal of "honking" sounds. Whether knowledgeable birds were more likely to emit such sounds remains to be tested.

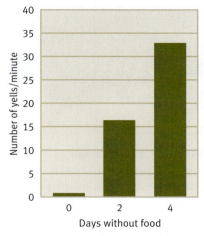

FIGURE 13.6. Yellers are hungry. In ravens, "yelling" is often associated with foraging—in particular, calling others to a food bonanza. Immature ravens yell progressively more as a function of hunger. *(Based on Heinrich and Marzluff, 1991)*

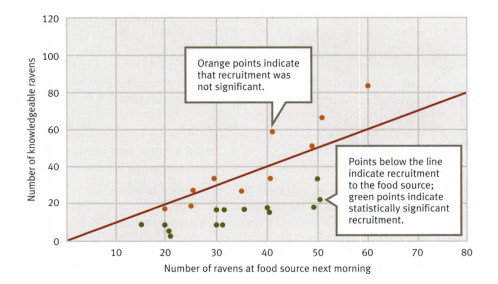

Orange points indicate that recruitment was not significant.

Points below the line indicate recruitment to the food source; green points indicate statistically significant recruitment.

FIGURE 13.7. Raven recruitment. The line denotes when the number of ravens that knew of the prey source equals the number of ravens at the prey source the next day. Points below the line indicate recruitment because they are instances in which more birds arrived at food after roosting than birds that previously knew of the food's location. *(Based on Marzluff et al., 1996, p. 99)*

FIGURE 13.8. Bee foraging.
Honeybee foraging involves a complex communication system, including waggle dances. This dance, along with other informational cues, gives bees in a hive information about the relative position of newly found food sources. *(Photo credit: Leroy Simon/Visuals Unlimited)*

HONEYBEES AND THE WAGGLE DANCE. In honeybees, collecting food for the hive often involves thousands of workers covering large areas of ground, at least from a honeybee's perspective (Visscher and Seeley, 1982; Figure 13.8). Such a system poses enormous logistical problems for honeybees: How do they keep track of the changing distribution of resources (nectar, pollen) through time? How are they able to monitor the needs of others in the colony? How do foragers communicate information about any food they uncover to other members of their hive? The answer to that last question, in part, lies in the now famous **waggle dance** of the honeybee.

The waggle dance of the honeybee was first studied experimentally by Karl von Frisch (von Frisch, 1967). Thomas Seeley describes the waggle dance as

> a unique form of behavior in which a bee, deep inside her colony's nest, performs a miniaturized reenactment of her recent journey to a patch of flowers. Bees following these dances learn the distance, direction and odors of these flowers and can translate this information into a flight to specified flowers. Thus the waggle dance is a truly symbolic message, one which is separated in space and time from both the actions on which it is based and the behaviors it will guide. (Seeley, 1985, pp. 84–85)

The waggle dance has an almost mystical quality to it (Munz, 2005). Fortunately, it is possible to maintain that air of the amazing while studying this behavior in detail, as ethologists have now been doing for more than half a century (Grueter and Farina, 2009; Seeley, 1985, 2012). To see this, imagine that a worker bee has just returned from a cluster of flowers that are approximately 1,500 meters from her nest, and that these flowers are located 40 degrees west of an imaginary straight line running between the worker's nest and the sun (Figure 13.9). How can a worker communicate this information about a new food source to others in the hive?

Upon returning to the nest, the worker bee quickly starts "dancing" up and down a vertical honeycomb within the hive; her sisters stay near her, making as

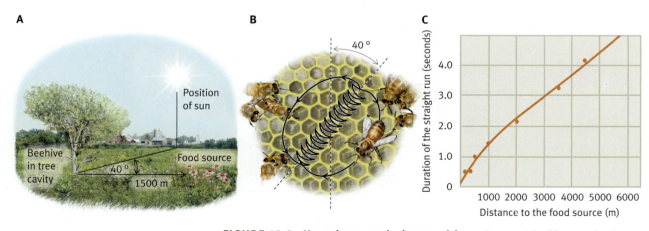

FIGURE 13.9. Honeybee waggle dances. (A) Imagine a patch of flowers that is 1,500 meters from a hive, at an angle 40 degrees west of the sun. **(B)** When a forager returns, the bee dances in a figure-eight pattern. In this case, the angle between a bee's "straight run" (up and down a comb in the hive) and a vertical line is 40 degrees. **(C)** The length of the straight run portion of the dance translates into distance from the hive to the food source. *(Based on Seeley, 1985)*

much physical contact as possible with both her and each other in the process. While dancing vigorously by waggling her abdomen, the worker conveys crucial information to her relatives in the hive. For example, her dance provides topographical information (north, east, south, west, northwest, and so on) for finding the food source from which she has just returned. When compared with a straight up-and-down run along a comb, the angle at which the forager dances provides information about the position of the food source of interest in relation to the hive and to the sun. Further, the longer the bee dances—in a part of the waggle dance known as the "straight line"—the farther away the bounty. Every extra 75 milliseconds of dancing translates into the resource being about an additional 100 meters from the hive. The more precise the information conveyed in the waggle dance, the greater the ability of other bees to find the food source, the more food brought back to the hive, and consequently the greater the inclusive fitness of the forager, because a hive is largely composed of individuals that are closely related to one another.

The honeybee waggle dance has been the subject of intense interest in ethology. For example, Ross Crozier and his colleagues studied the genetics of the honeybee dance by examining the point at which bees shift from other types of dances to the waggle dance (Johnson et al., 2002; Oldroyd and Thompson, 2007; Oxley and Oldroyd, 2010). When resources are close to the hive, honeybee foragers tend to use what is called a round dance. When the resources are at greater distances, bees switch to a sickle dance, and when food is very far from a hive, foragers use the waggle dance (Figure 13.10).

Crozier and his team ran a series of experiments in which they mated individuals from populations of bees that differed in terms of when they shifted from using the round dance to the sickle dance, and finally when they shifted from the sickle dance to the waggle dance. For example, in one population that the researchers studied, bees transitioned from the round to the sickle dance when food was more than 20 meters from the hive, and they shifted from the sickle dance to the waggle dance when food was 60 meters or more from the hive. The results of the genetic crosses undertaken by Crozier and his colleagues suggest that the transition across dance types is under the control of a single gene (Johnson et al., 2002). A single gene, then, can help us understand something about the waggle dance and its importance in foraging and honeybee communication.

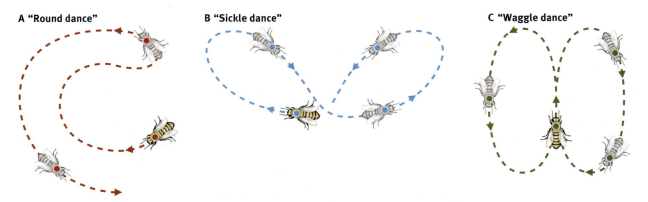

FIGURE 13.10. Different honeybee dances. The three honeybee dances. Each bee at its initial starting point is shown in full color, and the same bee is shown in fainter colors as it moves along the path of its dance. *(Based on Johnson et al., 2002, p. 171)*

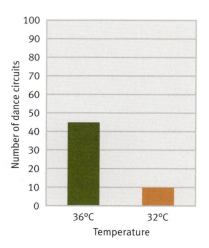

FIGURE 13.11. Number of dance circuits. The number of "figure-eight" circuits in a waggle dance when bees were raised at a temperature of 36°C or 32°C. *(Based on Tautz et al., 2003, p. 7345)*

The honeybee waggle dance has also been examined from a developmental perspective. Jurgen Tautz and his colleagues studied how hive temperature during development affected the waggle dance behavior of bees (Tautz et al., 2003). Tautz's team raised bees in one of three temperatures (all of which were in the normal temperature range within a hive): 32°C, 34.5°C, or 36°C. When bees in these treatments matured, they were individually marked and placed into "foster" hives. When the experimental bees in these hives matured into foragers, they were trained by Tautz and his colleagues to feed at an artificial food source placed 200 meters from the hives. The researchers observed differences in the waggle dance behavior in bees raised at different temperatures.

Bees from the 32°C treatment were less likely to use the waggle dance when they returned to the hive than were bees from either of the other treatments—60 percent of foragers from the 32°C treatment waggle danced versus 90 percent of foragers from the 34.5°C and 36°C treatments (this difference, however, was not statistically significant because of the large amount of variance found within treatments). In addition, bees from the 32°C treatment made significantly fewer "circuits"—trips around the figure eight in the waggle dance—than bees from the 36°C treatment (Figure 13.11).

Tautz and his colleagues found that bees raised in the 36°C treatment fared much better in individual learning tasks than did bees raised in the other treatments. These results suggest that differences in temperature during early development can have important effects for both individual bees and the hive as a whole. When bees were raised in colder temperatures, they were both poor learners and less adept at communicating important foraging-related information to other members of their hive. Because the food that foragers bring into a hive is eventually transferred into energy that keeps the hive temperature high, a dangerous feedback loop is put into play: Lower temperatures lead to bees that are both poor foragers and poor communicators, which in turn leads to less energy for the hive and hence to lower hive temperatures, which then leads to even worse foragers, and so on.

Ethologists have also investigated what chemical cues cause bees to leave a hive to forage in response to the waggle dance. Thom and Dornhaus (2007) found four hydrocarbons—Z-(9)-tricosene, tricosane, Z-(9)-pentacosene, and pentacosane—that were emitted by bees during waggle dances. When synthetic versions of these hydrocarbons were placed in a hive, bees increased their tendency to exit the hive, suggesting a link to foraging. Follow-up work by Thom and his team found that when this compound was added to hives, individually marked bees increased their departures from a hive by almost 50 percent compared with controls, and they more than doubled their visits to a nearby feeder (Gilley et al., 2012; Figure 13.12).

CHEMICAL AND VIBRATIONAL COMMUNICATION IN FORAGING ANTS. Like honeybees, ants often forage in large groups and cover relatively large areas during their foraging bouts. And like honeybees, ants face the problem of how to communicate foraging-related information to those in their nest. Ants solve this communication problem primarily by using chemical markers and sound production through vibrations (stridulation; Billen and Morgan, 1997; Holldobler and Wilson, 1990; Vander Meer et al., 1997; E. O. Wilson, 1971). Here we begin by examining chemical communication in the leaf-cutting ants of the genus *Atta* and then proceed to study vibrational communication in this same species (Holldobler and Roces, 2001).

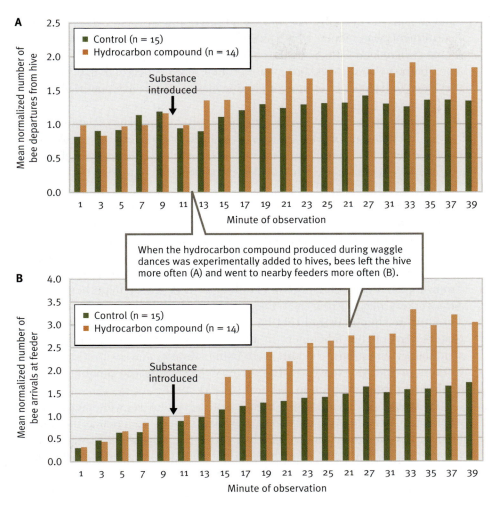

A

Mean normalized number of bee departures from hive

Control (n = 15)
Hydrocarbon compound (n = 14)

Substance introduced

Minute of observation

When the hydrocarbon compound produced during waggle dances was experimentally added to hives, bees left the hive more often (A) and went to nearby feeders more often (B).

B

Mean normalized number of bee arrivals at feeder

Control (n = 15)
Hydrocarbon compound (n = 14)

Substance introduced

Minute of observation

FIGURE 13.12. Hydrocarbon dance compounds. (A) When a hydrocarbon compound produced during waggle dance was added to a hive, more bees exited the hive than in a control treatment. **(B)** More bees were also found at a nearby feeder. (*From Gilley et al., 2012*)

Ants in the genus *Atta* are one of the dominant lifeforms in the neotropics, consuming more vegetation than any other group in that ecosystem. Leaf-cutters subsist entirely on the fungus they grow on leaves and on the sap produced by the plants whose leaves they harvest (Littledyke and Cherrett, 1976). These ants live in nests containing hundreds of thousands of workers (plus a single queen) and have evolved an elaborate, caste-based system for obtaining the leaves on which they subsist. Leaves must first be cut, then carried to the nest, ground up, chewed and treated with enzymes, placed into the "fungus garden," and subsequently cultivated. Bert Holldobler and E. O. Wilson liken the system to an assembly line where different castes handle different tasks (Holldobler and Wilson, 1990; D. S. Wilson, 1980; E. O. Wilson, 1980a, 1980b; E. O. Wilson and Holldobler, 2005a, 2005b).

From a proximate perspective, two chemicals are especially important in long distance foraging communication in *Atta* species: methyl 4-methylpyrrole-2-carboxylate and 3-ethyl-2,5methylpyrazine (Morgan, 1984). These substances are produced in the poison gland of leaf-cutter ants and are used to recruit

FIGURE 13.13. Devastating leaf-cutters. Leaf-cutter ants can ravage foliage in their path. The ants don't attack all the leaves, however, but instead they often strip some leaves to the stalk (for example, those that are most tender or have fewer secondary compounds present), while leaving other leaves untouched. *(Photo credit: Carver Mostardi/Alamy)*

fellow workers to foraging sites that are relatively long distances from their nest. Recruitment pheromones, which fade slowly, are placed along the trails leading to trees where leaves are being harvested. These pheromones are also deposited along branches and twigs, bringing recruited foragers very close to the leaves they need to harvest. Recruitment pheromones are incredibly powerful:

> . . . [T]he discoverers of methyl 4-methylpyrrole-2-carboxylate . . . estimated that one milligram of this substance (roughly the quantity in a single colony), if laid out with maximal efficiency, would be enough to lead a column of ants three times around the earth. (Holldobler and Roces, 2001, p. 94)

When leaf-cutters are in the vicinity of the leaves, they rely on a second mode of communication—stridulation—to determine the precise leaves on which to work. When *Atta* workers are cutting leaves, sections of leaves are not randomly cut throughout a tree (or even a branch) but rather, certain leaves are cut and cut until there is virtually nothing left (Figure 13.13). From a cost/benefit perspective, Holldobler and Roces hypothesized that *Atta* workers were cutting the tenderest leaves (Holldobler and Roces, 2001). They further hypothesized that because *Atta* workers raise and lower their gasters (part of their abdomen) in a manner similar to the way they created stridulatory vibrations, the ants were using vibrational cues to recruit other workers to the best leaves in the vicinity, increasing the amount of high-quality food brought back to the nest (Markl, 1968).

Holldobler and Roces used noninvasive laser Doppler vibrometry to test whether workers in *Atta cephalotes* stridulate while they are cutting leaves. Indeed, workers were stridulating while they were cutting the leaves, and the vibrations were being sent along the length of a leaf in a long series of vibrational "chirps" (Figure 13.14). Next, the researchers offered workers leaves that differed in toughness (tough or tender). In addition, in a follow-up experiment, tough and tender leaves were both dipped in sugar water and offered to ants. The results were clear: While only 40 percent of the ants stridulated when cutting tough leaves (with no sugar water), the number increased to 70 percent when the leaves were tender (with no sugar water) and to almost 100 percent when either type of leaf was dipped in sugar water (Roces et al., 1993). Ants, then, clearly stridulate

Gaster

FIGURE 13.14. Stridulating communication. A schematic of a leaf-cutter ant cutting a leaf and stridulating its gaster up and down. *(Based on Holldobler and Roces, 2001)*

while cutting leaves, and they vary this behavior depending on the value of the leaf in question.

Although these two experiments are powerful, they do not demonstrate that ant workers are recruited by stridulation. To test for the potential recruiting nature of stridulatory communication, Holldobler and Roces hooked one leaf to a vibrator and used a similar ("silent") leaf as a control (Holldobler and Roces, 2001). When given the choice to cut either of the leaves, ant workers clearly preferred the vibrating leaf, demonstrating that these cues were not only produced by workers, but also used to recruit workers (Roces and Holldobler, 1996). Moreover, stridulation was also found to serve another role in communication. In *A. cephalotes*, a caste of very small ants, called minim workers, exists. These minim workers cannot cut leaves, but they are often found hitchhiking rides on leaves on the backs of leaf-cutters. Minims protect these other leaf-cutting workers from attack by parasitic flies, so it is in the interest of leaf-cutting ants to have minims find them (Eibesfeldt and Eible-Eibesfeldt, 1967; Feener and Moss, 1990). Hitchhiking minims apparently use the vibrational cues created by stridulating leaf-cutting nestmates to locate the leaf-cutters (Roces and Holldobler, 1995; Figure 13.15).

FIGURE 13.15. Minim workers hitchhiking on a leaf. The stridulating signals emitted by leaf-cutters are used in numerous contexts. One such venue is between leaf-cutters and "minimum" workers (minims), which use these signals to eventually hitch rides on cut leaves that are carried on leaf-cutters' backs. *(Photo credit: Bert Holldobler)*

PROBLEM: HOW TO FIND AND SECURE A MATE

Animals often solve problems associated with finding a mate by some form of communication (see the Conservation Connection box). In this section, we will examine the role of (1) vocal communication (birdsong) and (2) tactile communication (ripples by insects that live in water), as they relate to problems that animals encounter as a result of both intrasexual and intersexual selection.

BIRDSONG. Some forms of animal communication are more pleasing to humans than others. Go to your local music store and you are bound to find dozens of CDs of birdsong. In addition to its practical, sedative effects on human listeners, birdsong has been studied by naturalists, ethologists, behavioral ecologists, comparative psychologists, and evolutionary biologists for hundreds of years. Indeed, few communication systems in vertebrates have been studied to the extent of birdsong (Beecher, 2008; Berwick et al., 2011; Brumin and Naguib, 2009; Catchpole and Slater, 1995; D. F. Clayton, 2004; Mooney, 2009; Podos and Warren, 2007; Podos et al., 2004; Slater, 2003).

While birdsong has many functions—colony formation, flocking, foraging, and so on (Kroodsma and Byers, 1991)—here we will focus on its role in problems relating to sexual selection. In particular, we will examine how and why birdsong is used to attract and secure mates. We will begin with two studies that address ultimate questions regarding the fitness consequences of birdsong and the phylogeny of birdsong, and then we will examine the proximate underpinnings of birdsong (for an in-depth case study of social learning and birdsong, see the discussion in Chapter 7 of Freeberg's work on cultural transmission in the cowbird).

In most species of songbirds, males don't just learn a single song; they learn *many* different songs. For example, the song sparrow (*Melospiza melodia*) sings approximately ten different songs, the western marsh wren (*Cistothorus palustris*) sings more than a hundred songs, and the brown thrasher (*Toxostoma rufum*) sings an incredible thousand different songs (Beecher and Brenowitz, 2005). Ethologists have hypothesized that females may use the size of a male's song

Anthropogenic Change and Animal Communication

The type of communication system that natural selection favors in a population depends, in part, on the ecology and environment in which that species lives. For example, many forest-dwelling birds that breed in leks (see Chapter 7) will display courtship behavior only when the light breaks through the forest canopy through small gaps at certain times of the day. Altering that environment can disrupt courtship activity.

In the Guianian cock-of-the-rock (*Rupicola rupicola*), exquisite orange-colored males will often display to females when yellow-orange wavelength light from small gaps in the canopy reaches the ground (Endler, 1997; Endler and Thery, 1996). The courtship communication that then takes place is dramatic, as described by Pepper Trail, who studied these birds in Suriname: "The normally silent males burst into ringing choruses of raucous, crowing calls and drop from their resting perches. . . . [E]ach male stands erect and violently beats his wings, flashing the dramatic, usually concealed, black and white primary feathers" (Trail, 1995). Males may then repeat this greeting display (Figure 13.16), and females choose from among the displaying males.

Humans clear-cutting the area of the lekking arena, or even areas in its vicinity, will change the way light enters the lekking arena and might radically disrupt the courtship communication between male and female cock-of-the-rocks (Endler, 1997). Of course, it is difficult to predict exactly how, but such clear-cutting might produce constant light during daytime hours, which could (1) stop males from displaying at all because of increased exposure to predators, (2) induce males to display so often that they become energetically drained, or (3) lead to females no longer being able to assess male quality accurately. Any or all of these effects could decrease population size.

FIGURE 13.16. Communication in cock-of-the-rocks. (A) A male cock-of-the-rock. **(B)** A group of males displaying and singing to attract females. *(Photo credits: Sylvain Cordier/Getty Images; SA Team/Foto Natura/Getty Images)*

repertoire to choose mates if the repertoire size indicates such things as a male's age and/or genetic quality (Hosoi et al., 2005; MacDougal-Shackleton, 1997).

Aki Hosoi, Stephen Rothstein, and Adrian O'Loghlen examined the role of repertoire size in the mating success of male brown-headed cowbirds (*Molothrus ater*). Male cowbirds sing between two and eight different perched songs (songs sung from a perch, generally near the ground, rather than in flight), which they use to attract females during the mating season. Earlier studies indicated a correlation between the size of the perched song repertoire and mating success, but no work had specifically examined whether female mate choice was affected by song repertoire in cowbirds (O'Loghlen, 1995; O'Loghlen and Rothstein, 1993, 1995, 2002). To examine whether male repertoire size had an effect on female mate choice, Hosoi and her colleagues captured female cowbirds from two populations in California—one from Santa Barbara and one from Ventura—and brought them into an experimental aviary (Hosoi et al., 2005). The researchers then subcutaneously implanted the captured females with the hormone estradiol to increase the probability that they would respond to male song during experimental manipulations.

Hosoi and her team exposed each female in the experiment to five different song treatments—a *single* perched song sung three times in succession by a male Santa Barbara cowbird (treatment 1; smaller repertoire) or by a male Ventura cowbird (treatment 2; smaller repertoire), three *different* perched songs sung in quick succession by a male Santa Barbara cowbird (treatment 3; larger repertoire) or by a male Ventura cowbird (treatment 4; larger repertoire), and a control in which females were exposed to the song of males from a different species (song sparrow). The length of time that a female displayed ritualized "copulation-solicitation displays" (CSDs) to different songs was recorded and used as a measure of female choice.

Female cowbirds displayed longer CSDs when they heard cowbird versus song sparrow songs, regardless of how many songs were sung. If we just consider the four treatments involving male cowbird song, females show a marked increase in CSD times when exposed to males with larger song repertoires (Figure 13.17). When females were exposed to males from their own population, they preferred larger song repertoires in the majority of trials. Moreover, females exposed to male cowbirds from a different population also preferred males with larger song repertoires. Larger song repertoire, then, provides benefits to males (Soma and Zsolt-Garamszegi, 2011; but see B. E. Byers and Kroodsma, 2009).

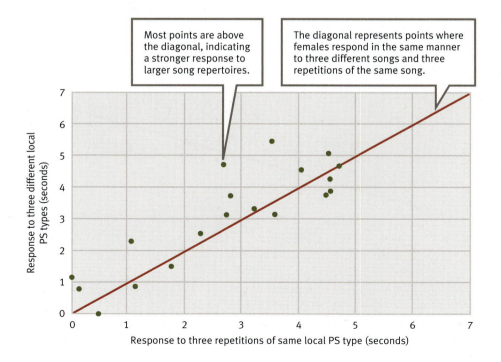

FIGURE 13.17. Same versus different songs. Female cowbirds had longer copulation-solicitation displays (CSDs) when they were exposed to three different songs than to the same song played three times. Each point represents the CSDs of one female. *(From Hosoi et al., 2005, p. 89)*

Phylogenetic studies of birdsong also point to the importance of sexual selection in the evolution of birdsong (Brenowitz, 1997; J. J. Price and Lanyon, 2002a, 2002b, 2004a, 2004b). Many phylogenetic studies of sexual selection look at a single trait—for example, feather length—and map that trait onto a phylogenetic tree. This approach is more difficult to use when song is the trait of interest, as numerous studies show that the many components of song—repertoire size, amplitude, frequency, and so on—can vary across groups. To obtain an accurate assessment, phylogenetic analysis of birdsong and sexual selection needs to incorporate many measures of song—frequency, rate of pauses, maximum note length, and so forth (J. J. Price and Lanyon, 2004a). And, in particular, if researchers are interested in sexual selection and communication, it is especially useful for them to analyze components of song that are costly to produce, and hence hard to fake.

Recognizing the inherent complexity of birdsong, Jordan Price and Scott Lanyon mapped six components of birdsong onto a molecular phylogeny of seventeen species of blackbirds (J. J. Price and Lanyon, 2004a). Price and Lanyon's work focused on caciques and oropendolas—two groups of New World black-birds from Argentina and Mexico. The caciques and oropendolas are well suited for analysis involving sexual selection and song, as tremendous variations in mating systems exist in these groups. Some species are monogamous, while others are polygynous. In addition, some species of caciques and oropendolas are sexually monomorphic for size (males and females are the same size), while others are sexually dimorphic (males are much larger than females). This variance is critical, as much work suggests that sexual selection operates more strongly in polygynous mating systems and in systems that are sexually dimorphic in size (Figure 13.18).

Price and Lanyon's phylogenetic analysis is complicated, but what is most critical is that they uncovered multiple lines of evidence that evolutionary changes in song were most often associated with sexually dimorphic species, where sexual selection is strong. For example, in one oropendola species, *Psarocolius oseryi*, in which males were much larger than females, phylogenetic analysis found changes in three components of song. More generally, in the phylogenetic tree that Price and Larson constructed, in eight of the ten branches

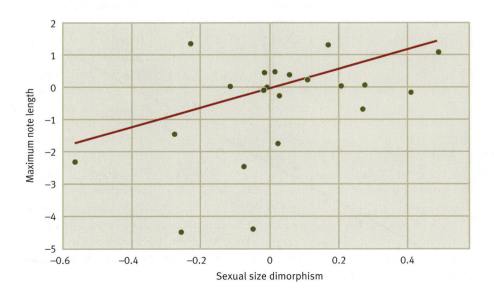

FIGURE 13.18. Sex differences and songs. Across seventeen species of blackbirds, the maximum note length of songs increased as the size difference between males and females increased. The x and y axes have been transformed into independent contrasts and can take negative values. *(From J. J. Price and Lanyon, 2004a, p. 490)*

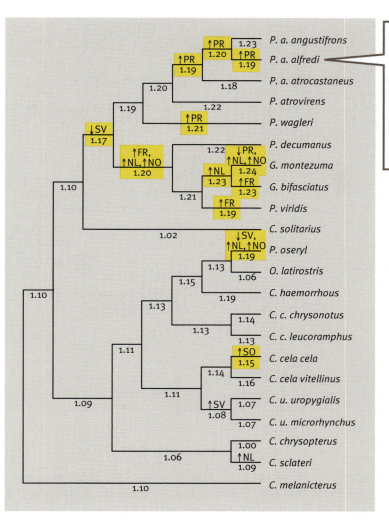

Additions or subtractions to song are indicated by up or down arrows. Numbers below a branch show ratio of male/female size. The larger this number, the greater the presumed effect of sexual selection. Most changes to song were associated with male/female ratios above 1.15—that is, species in which males were at least 15 percent larger than females. These species are highlighted in yellow.

FIGURE 13.19. Phylogeny, sexual selection, and song. A phylogeny of oropendola and cacique birds with changes in song characters mapped on. Above the branches are song characters as they are added or dropped from the song repertoire. SO = song output, SV = song versatility, FR = frequency range, PR = pause rate, NL = maximum note length, and NO = note overlap. Numbers below branches show male/female size ratios. *(From Price and Lanyon, 2004a)*

that showed clear changes in song components, males were at least 15 percent larger than females (Figure 13.19). When sexual selection is powerful, and there is strong competition for access to females, natural selection acts strongly on the many components that make up male song.

In addition to the evolutionary literature on birdsong, a tremendous amount of work has also addressed birdsong from a proximate perspective. Part of this work centers on something of a riddle. As we have seen, birdsong is incredibly diverse in terms of structure, pattern, tempo, frequency, and repertoire size. At the same time, however, the vocal organ used in birds, the syrinx, varies little between different species in this group. How is it possible that morphological (structural) invariance in the syrinx can translate into great diversity in birdsong?

The avian syrinx has two compartments—left and right—and the two sides of a bird's brain can control these compartments independently (Suthers, 1997; Suthers et al., 2004; Suthers and Zollinger, 2004; Table 13.2). This one piece of information sets the stage for moving from *invariance* in the structure of the syrinx to *variance* in the song output. That is, although the syrinx varies little in structure, its *parts* can be manipulated to create new permutations that may give rise to new sounds (Suthers, 1999).

TABLE 13.2. The different ways to sing. The costs and benefits of different song lateralization patterns. *(Based on Suthers, 1999)*

SONG LATERALIZATION PATTERN	SPECIES	ADVANTAGES	DISADVANTAGES
Independent bilateral sound formation	Brown thrasher and gray catbird	Two independent voices increase spectral and phonetic complexity.	Expensive in use of air supply. Best suited for a low syllable repetition rate.
Unilateral dominance	Waterslager canary	Conserves air, favoring shorter minibreaths and longer phrases. Separation of phonatory and inspiratory motor patterns to opposite sides of syrinx. Both of these may facilitate higher syllable repetition rates.	Use of one voice limits frequency range and certain kinds of spectral and temporal complexity.
Alternating lateralization	Brown-headed cowbird	Enhances spectral contrast between notes. Efficient use of air supply. Extended frequency range for overall song.	Two-voice complexity limited to note overlap.
Sequential lateralization	Northern cardinal	Extended frequency range. Conserves air supply.	Lacks spectral complexity of two voices.

Over a hundred years ago, theorists suggested that two different sides of the syrinx contribute differentially to the frequency component of birdsong. But stronger evidence of this phenomenon did not arise until 1968 (Greenewalt, 1968; Stein, 1968). Independence of the two sides of the syrinx allows songbirds to switch off one side at any time. This, in turn, facilitates variation in song production in that it permits some species to

► operate both sides of the syrinx independently throughout their song without one side being dominant, as in brown thrashers (*Toxostoma rufum*) and gray catbirds (*Dumetella carolinensis*; Suthers et al., 1994, 1996; Suthers and Hartley, 1990).
► have one side of the syrinx dominate song generation, as in canaries (*Serinus canaria*; Nottebohm and Nottebohm, 1976).
► alternate which side of the syrinx dominates during a song, as in brown-headed cowbirds (*Molothrus ater*; Allan and Suthers, 1994).
► have one side of the syrinx dominate for certain frequencies, and the other side dominate for the remainder of the frequencies used in a song (sequential lateralization), as in the northern cardinal (*Cardinalis cardinalis*; Suthers, 1997; Suthers and Goller, 1996).

The respiratory muscles that supply air to the syrinx can also be employed to achieve the variance in song output across bird species. Varying the respiratory muscles can have dramatic effects on the structure and timing of birdsong. So, again, the syrinx itself is invariant, but variation around other anatomical structures can produce variation in the output of birdsong.

These sorts of proximate analyses help us to better understand the construction of communication systems that are designed to address problems that arise during sexual selection in songbirds.

RIPPLE COMMUNICATION AND MATE CHOICE IN AQUATIC INSECTS. Vocal signaling, as in birdsong, is only one way for animals to communicate in the context of mating. Indeed, animals communicate information relevant for mate choice in many ways—vocally, visually, chemically, and so on. In 1972, Stim Wilcox discovered a new form of communication—ripple communication—in water striders, insects that live in freshwater lakes, ponds, streams, and small rivers. This form of communication has now been demonstrated in numerous species of water striders, and it has also been found in giant water bugs (Bottger, 1974; Kraus, 1989; R. L. Smith, 1979).

In water striders, ripples are usually produced by an up-and-down movement of the legs, with both right and left legs in synchrony and in constant contact with the water surface (Wilcox, 1995). The water striders produce different patterns of ripples (waves with different amplitudes and frequencies) for different kinds of behaviors, including signals for calling mates, courtship, copulation, postcopulation, sex discrimination, mate guarding, spacing, territoriality, and food defense. Such signals can range from 0.2 seconds during the courtship behavior of the water strider, *Rhagadotarsus anomalus* (Wilcox, 1972), up to 30 seconds in other water strider species (Polhemus, 1990).

During courtship, ripple production appears to have significant benefits associated with it. In *R. anomalus*, males produce ripple signals that can travel more than 60 cm, and females find such ripple signals very attractive. Wilcox undertook a series of playback experiments that demonstrated that females from as far away as 60 cm were attracted to mating ripple signals—ripples that are quite different from those produced by aggressive males (Wilcox, 1972; Figure 13.20). Females would often grasp males and even begin to oviposit (lay eggs) in response to playbacks of calling signals. In addition, Wilcox speculates that ripple calls designed to attract mates also serve as a means for species identification, as *R. anomalus* are often found in the same streams and ponds as other water striders (Polhemus and Karunaratne, 1993).

Ripple signals are also employed by giant water bugs in the context of mating (Bottger, 1974; Kraus, 1989; R. L. Smith, 1979). While performing courtship displays, giant male water bugs pump up and down with their legs, creating ripples, which attract female water bugs. The roots of such pumping behavior may lie in the motion that giant water bugs use to aerate eggs (which are laid on their backs; R. L. Smith, 1979).

FIGURE 13.20. Ripple communication by water striders. (A) The concentric circles of these ripples in a pond are part of the communication used by the water strider, *R. kraepelini*. **(B)** A close-up of the male water strider *(R. kraepelini)* and the ripples he is making to communicate with other water striders. **(C)** An experimental setup to study ripple communication, in which an *A. remigis* female is making a signal via a magnet glued to her leg. In nature, female water striders don't emit such signals. *(Photo credits: Stim Wilcox)*

A

B

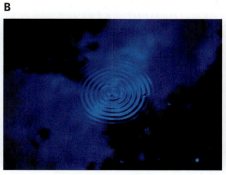

C

FIGURE 13.21. Protecting a mate. While downy woodpeckers don't give alarm calls when they are paired with same-sex partners, they emit such alarm calls when they are paired with a member of the opposite sex. *(Photo credit: Harold R. Stinnette Photo Stock/Alamy)*

PROBLEM: PREDATORS

As the vervet monkey example in the opening to this chapter illustrates, communication can be used to warn conspecifics about predators. Animals that live in social groups often derive benefits from keeping group size above some minimum level, in order, for example, to forage efficiently. In addition, the probability that any given individual will be attacked by a predator often decreases as group size increases (see Chapter 12), yielding another benefit from group living. Predation decreases group size, so animals often must solve a problem: how, when, and why to communicate to other group members that a predator is in the vicinity.

WOODPECKER AND CHICKADEE ALARM CALLS. Kimberly Sullivan has found that in both downy woodpeckers (*Picoides pubescens*) and black-capped chickadees (*Parus atricapillus*), individual alarm calls are emitted to protect *mates*, but not other individuals (Figure 13.21). This form of alarm call communication has been termed "mate investment."

During the winters of 1979–1981, Sullivan observed mixed-species flocks of birds in the Great Swamp National Wildlife Refuge in New Jersey. During her observations, she recorded alarm calls given by downy woodpeckers and black-capped chickadees to three naturally occurring predators: sharp-shinned hawks (*Accipiter straitus*), Cooper's hawks (*Accipiter cooperii*), and American kestrels (*Falco sparverius*). Sullivan also examined alarm calls in response to (stuffed) models of these predators. These models were "flown" over a group of birds on a pulley that ran down a wire between two trees. Sullivan found that downy woodpeckers never gave alarm calls when they were foraging alone (there were 0 alarm calls in 46 instances of solo foraging), when in a flock with no other woodpeckers (0/23), or when with a same-sex downy woodpecker (0/6). The situation was very different, however, when downy woodpeckers were in the presence of a woodpecker of the opposite sex. In seven of nine such instances, they emitted alarm calls (K. Sullivan, 1985). Three of these calls were made by females and four by males, suggesting that both sexes use this form of communication to attract opposite-sex partners.

Sullivan found no evidence that either black-capped chickadees or tufted titmice used alarm calls to attract mates. But other researchers have found evidence that black-capped chickadees do preferentially give alarm calls in the presence of potential mates (Ficken et al., 1980; Witkin and Ficken, 1979). This "mate investment" form of communication may account for alarm calls in other birds as well (Curio and Regelmann, 1985; East, 1981; Hogstad, 1995; Krams et al., 2006; Leopold, 1977; E. S. Morton, 1977).

DEVELOPMENT, LEARNING, AND ALARM CALL COMMUNICATION IN MEERKATS. In species in which parental care is present and lengthy, natural selection will often favor development pathways in which young individuals learn what communicative signals mean from adults (Platzen and Magrath, 2005). This should be especially true when signals are complex, as in the alarm calls of vervets. Indeed, the evidence suggests that young vervets do learn about alarm calls from older individuals (Seyfarth and Cheney, 1986). Meerkats also experience prolonged parental care and learn complex alarms calls for terrestrial and avian predators.

Meerkat pups spend the first three weeks of their lives underground, and then they emerge and join groups of juveniles (six to twelve months old) and adults. Once above ground, pups are subject to intense predation pressure from avian and terrestrial predators. Linda Hollen and Marta Manser examined the development of the response to alarm calls in maturing meerkat pups (Hollen and Manser, 2006). They studied eleven groups of free-ranging meerkats in the Kalahari Desert, and they used a combination of observation of meerkat encounters with predators in natural settings and experimental manipulation through playback experiments in which meerkats were exposed to recorded alarm calls.

Behavior observation indicates that, compared with adults, pups were *initially* more likely to respond to alarm calls in the presence of less dangerous or nondangerous predators, and they were more likely to ignore alarm calls emitted in the presence of dangerous predators. Other evidence also suggests that pups don't react as appropriately to alarm calls as adults do. For example, after hearing an alarm call, pups often moved to shelter when they observed adults only briefly looking up and scanning for aerial predators. And even when pups displayed antipredator behaviors similar to those of adults, they reacted more slowly than did adults (Figure 13.22). As time passed, however, pups began to display more adultlike, adaptive, responses to alarm calls. Follow-up work strongly suggests that pups learn about alarm calls and predators from adults (Hollem et al., 2008).

From a social learning perspective, two things about meerkat alarm calling and development are worth noting. First, female pups (and juveniles) exhibited adultlike responses to alarm calls more quickly than did males, and this difference apparently had a significant impact on fitness, as female pups

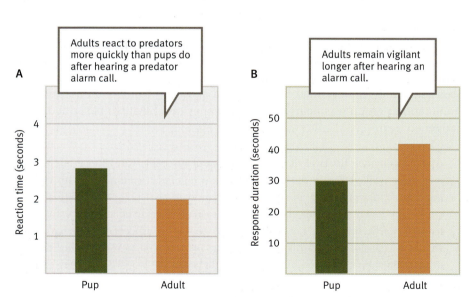
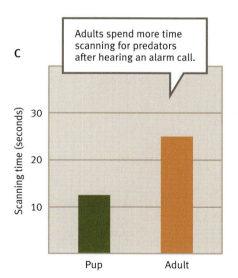

A Adults react to predators more quickly than pups do after hearing a predator alarm call.

B Adults remain vigilant longer after hearing an alarm call.

C Adults spend more time scanning for predators after hearing an alarm call.

FIGURE 13.22. Age differences in reaction to alarm calls. In meerkats, a pup's response to alarm calls was not as strong as the response seen in adults: **(A)** time until reaction after hearing playback of alarm call, **(B)** duration of response to playback of alarm call, and **(C)** length of time spent "scanning" the environment for predators after hearing alarm call. (*Based on Hollen and Manser, 2006, p. 1350*)

suffered less mortality than did males (Russell et al., 2002). Second, throughout development, females remained closer to adults than did males. Together these results suggest that meerkat pups may learn the appropriate response to alarm calls from adult models, and that female pups are more adept at learning than are male pups.

ALARM CALLS AS DECEPTIVE COMMUNICATION. Because alarm calls are a powerful form of communication—failure to listen might lead to death—natural selection should favor paying close attention to such calls. These same selection pressures set up the possibility of using alarm calls in a deceptive manner. As mentioned in Chapter 12, using alarm calls in a deceptive manner is probably the exception rather than the rule. That being said, there is, in fact, some evidence that deceptive use of these calls exists in a number of species.

Cheney and Seyfarth suggest that vervets sometimes use deceptive predator alarm calls during some intergroup encounters—encounters that can lead to serious aggression between group members (Cheney and Seyfarth, 1990). Occasionally, male vervets give an alarm call when encountering a new troop, even though no predator is in the vicinity. It is unlikely that these predator alarm calls are mistakes, as vervets are quite adept when it comes to such calls, and their behavior seems to indicate that they know when a predator is around and when it isn't (see Chapter 12 and the opening section of this chapter). An additional reason that ethologists doubt that such calls are given in error is that the response to alarm calls may provide alarm callers with a benefit: When these calls are emitted, all vervets head for safety (cover), so a potentially dangerous intergroup encounter is avoided and the alarm caller does not have to fight, at least for the moment.

In the 264 intergroup interactions that Cheney and Seyfarth observed, males gave false alarm calls regarding the presence of leopards about 2 percent of the time. Moreover, it was almost always a low-ranking male—an individual near the bottom of a dominance hierarchy that thus stood the most to lose from aggressive intergroup interactions—that gave the call. On the other hand, vervets' deception has its limits. For example, on a number of occasions, one vervet named Kitui emitted an alarm call when an interloper from another group was near his troop. Cheney and Seyfarth described what happened next:

> As if to convince his rival of the full import of his calls, Kitui twice left his own tree, walked across the open plain, and entered a tree adjacent to the interloper's, alarm calling all the while. Kitui acted as if he got only half the story right; he knew that his alarm calls had caused others to believe there was a leopard nearby, but he did not seem to realize that other aspects of his behavior should be consistent with his calls. To leave his tree and to walk toward the other male simply betrayed his own lack of belief in the leopard. (Cheney and Seyfarth, 1990, p. 215)

Deceptive alarm calls are not unique to primates and have been found in birds and mammals (Bro-Jorgensen, 2011; Møller, 1988, 1990; Munn, 1986). In the barn swallow, males may emit false alarm calls when they see their mate engaging in an extrapair copulation that may result in the mate having an offspring by another male (see Chapter 8; Møller, 1990). Anders Møller hypothesized that these alarm calls break up extrapair matings, as a male swallow that is attempting an extrapair mating with the alarm caller's mate should take flight when the call is emitted. To test this idea, when males were away from their nests, Møller chased their mate away from the nest as well. He did so both during the period when extrapair copulations were most likely (during egg laying) and when such pairings were least likely (during egg incubation periods). When females were absent during egg laying, the males that returned to their nests almost always gave false alarm calls, which likely disrupted any extrapair copulation that their mate was undertaking. In the same situation, during periods when extrapair copulations were unlikely, males almost never emitted alarm calls (Figure 13.23). Thus, male swallows that emitted false alarm calls during possible periods of extrapair copulations may be tapping into an evolved response to the real threat that predators pose, co-opting this form of communication and, in the course of such deception, making alarms calls a less reliable signal.

Deceptive alarm calling occurs in a similar context in mammals. Male topi antelopes (*Damaliscus lunatus*), for example, give false alarm calls that serve to keep sexually receptive females within their territories. When a sexually receptive female on a male's territory moves as if to leave, males sometimes emit a "snort" call that otherwise is given only when a predator has been sighted. Females stop when they hear the snort and look in the direction the male is looking when he snorts. This is sometimes enough of a

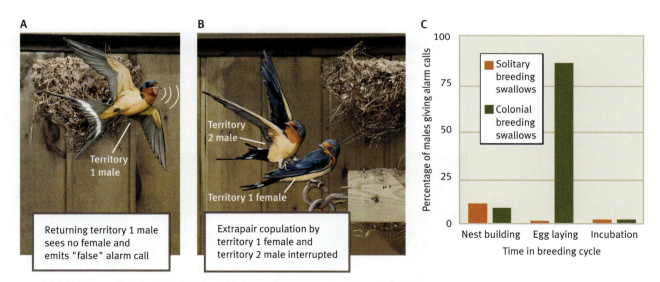

FIGURE 13.23. Dishonest alarm calls in swallows? **(A)** Male barn swallows often give false alarm calls when their fertile mates leave the nest vicinity. **(B)** These false alarm calls sometimes disrupt extrapair copulations (EPCs). **(C)** Møller hypothesized that male swallows would give false alarm calls when they were at the greatest risk of EPCs to disrupt the EPCs. To test this hypothesis, Møller removed a female from the nest at different stages in the breeding cycle for both solitary breeding swallows and colonial breeding swallows. Solitary breeding males almost never emitted alarm calls when their mate was temporarily gone. Colonial breeding males emitted false alarm calls during the period in which EPCs were most likely (during egg laying). *(Based on Møller, 1990)*

A

B

C

FIGURE 13.24. Deceptive alarm calling in topi. (A) A male topi (in background) has given a false alarm snort close to the boundary of his territory and now stares into the distance, as he does when a stalking predator has been detected. A sexually receptive female in the foreground (darker) looks toward the potential danger. **(B)** As the female begins to move away, the male looks toward the female (note the change in the orientation of his gaze and the different position of the ears. **(C)** Soon thereafter, the male mates with the female. *(From Bro-Jorgensen and Pangle, 2010)*

distraction to keep a female on a territory, allowing the male the opportunity to mate with her (Bro-Jorgensen, 2009; Bro-Jorgensen and Pangle, 2010; Figure 13.24).

When signals such as alarm calls become less reliable, natural selection should favor paying less and less attention to them. This raises the question of whether animals that receive inaccurate information from a signaler respond by eventually ignoring the signaler (Beauchamp and Ruxton, 2007; Blumstein et al,. 2004; Hollen and Radford, 2009). James Hare and Brent Atkins experimentally examined this question in Richardson's ground squirrels (*Spermophilus richard-sonii*; Figure 13.25). Hare and Atkins placed juvenile ground squirrels into one of two treatments. In one treatment, juveniles heard a recorded alarm call and then saw a predator (a stuffed badger).

FIGURE 13.25. Alarm calls in Richardson's squirrels. Over time, when predator alarm calls are unreliable, juvenile squirrels begin to ignore such false alarm cues. *(Photo credit: John Cancalosi/naturepl.com)*

They then heard the same alarm call, which was paired with the presentation of the predator nine more times (for a total of ten presentations). In a second treatment, ground squirrels again heard an alarm call, but subsequently they did not see a predator. In this treatment, the same alarm call was repeated ten times but was not followed by presentation of a predator. Individuals in the second group were receiving unreliable—that is, false—alarm calls (Hare and Atkins, 2001).

Hare and Atkins examined how long the squirrels remained vigilant (vigilance duration) and the extent to which juvenile squirrels in the two treatment conditions turned in the direction from which an alarm call emanated (postural change). No differences across the two groups were found in their response to the first alarm call, which was to be expected, as the squirrels had no knowledge beforehand as to which alarm calls would be reliable and which would not. Even after hearing five alarm calls, no differences in vigilance duration were found across the two treatment groups, although there was a difference in postural change: Squirrels that had heard unreliable calls were less likely to look in the direction of the call than were squirrels that had heard reliable calls. After hearing the alarm call ten times, differences in vigilance duration between the two groups of squirrels emerged. Squirrels in the group in which the alarm was paired with a reliable caller (that accurately alerted them to the presence of a predator) responded to alarm calls in a typical manner— by remaining vigilant and looking in the direction of the call. Squirrels in the other treatment—in which the information they had been receiving was inaccurate—essentially ignored the alarm calls, and they were unlikely to look in the direction from which the calls emanated (Figure 13.26). Given enough information, then, Richardson's ground squirrels can distinguish between real and false alarm calls (Pollard and Blumstein, 2012).

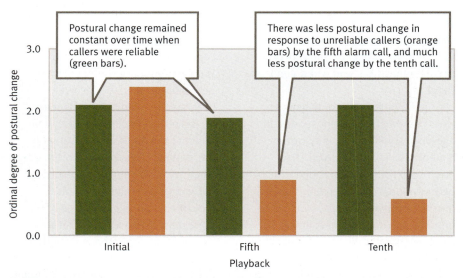

FIGURE 13.26. Responses to reliable and unreliable alarm calls. Postural change— elevation of the head in the direction of the perceived threat—differed depending on whether the alarm caller was a reliable source. Reliable alarm calls are shown in green; unreliable alarm calls are shown in orange. *(From Hare and Atkins, 2001, p. 110)*

Dr. Rufus Johnstone

What do ethologists mean when they speak of communication? How did you come to study this phenomenon?

Different people have defined communication in different ways. I would say, though, that it entails the use of specialized structures or behaviors (signals) by one individual to modify the behavior of others. The key feature of this definition is the emphasis on specialized signals, by which I mean traits that have been favored by natural selection specifically because of their influence on the behavior of receivers. There are plenty of cues that can provide receivers with useful information but have not been selected for that reason. Consider, for instance, a field mouse that rustles the grass as it moves and draws the attention of a predator such as an owl. The rustling noise provides the predator with a cue to the prey's location, but selection has not favored noisy mice for this reason—quite the reverse! So I would not see this as a true instance of communication, nor the rustling as a true signal. By contrast, when a peacock displays its train or a nightingale sings, their behavior has been favored by selection because of the responses it elicits from others.

Animal signals are so striking and so diverse, one can't help being fascinated by them. Just think of a bowerbird's elaborately decorated bower, or the complex song of a humpback whale, or take a look at the gape of a parrot finch chick— what could possibly have driven the evolution of such an extreme display? But more than any individual example,

what most appealed to me about the study of communication was the idea that there might be unifying principles of signal evolution underlying this great diversity of different displays.

What is meant by *honest communication*? When do animal behaviorists expect to find this type of communication, and when don't they?

Again, different people use the word *honesty* in different senses. In everyday speech, it usually carries "intentional" connotations. That is, an honest individual is one who bears no intent to deceive but rather means to convey reliable, truthful information. In this sense, however, the concept of honesty is inapplicable to many animals (though not perhaps to all), who lack any awareness of the responses their signals may evoke (or the meanings they may convey). Nevertheless, biologists often speak of honest and dishonest signals—one

might, for instance, describe the aposematic coloration (bright warning colors) of a palatable hoverfly that mimics a wasp as dishonest. This is not to imply that the hoverfly is engaged in intentional deceit. Rather, a dishonest signal in this sense is one that elicits a response that is beneficial to the signaler, but detrimental to the receiver. In this case, the hoverfly obviously benefits when its bright colors deter a predator, but the predator loses out because it has forgone a potentially palatable meal. Conversely, an honest signal is one that elicits responses that are beneficial for the receiver as well as for the signaler. The aposematic coloration of the wasp model, for instance, unlike that of the hoverfly, is honest—the wasp gains if its black and yellow stripes deter attack, but so too do the predators that avoid it.

When do we expect to find honest communication? To begin with, if there is no evolutionary conflict of interest between signaler and receiver, then there should be no selection for deceitful signals, because what benefits the receiver also benefits the signaler. This is typically the case when we consider communication between close relatives. A nice example is the waggle dance of the honeybee. When a worker returns to the hive with food, the waggle dance she performs conveys information to her fellow workers about the direction and distance of the food source. There is no evolutionary incentive for deceit here, because the worker benefits by recruiting others to the food; all share the same evolutionary interest in the productivity of the colony.

What about communication between individuals with conflicting interests? Even here, we might expect that honest communication is the norm. The reason is that in the long run, selection will favor receivers that ignore a consistently dishonest signal, and that attend only to signals which provide reliable information at least some of the time. Moreover, signals which are costly to produce or maintain are likely to provide such information, because it only pays signalers to employ them if they are of high quality or in great need—this is Zahavi's handicap principle. Of course, there are plenty of examples of dishonesty, but deceitful signalers must typically exploit some underlying system of reliable communication—and dishonesty cannot be too frequent or too costly to receivers if the system is to persist.

How does an ethologist build a mathematical model of something as complex as communication?

Constructing a model always involves a lot of simplification. This is not necessarily a bad thing—in fact, a model that is too detailed and complex is often rather unhelpful because it is likely to be applicable only to one specific system, and it may be very difficult to understand the results it yields. The best models are those that strike a good balance between generality and specificity. The difficulty lies in identifying which factors to include, and which you can omit. In modeling communication, a distinction is often made between efficacy and strategy. The details of signal design have an important influence on the ease with which receivers can detect and identify a signal (its efficacy), but strategic issues of honesty and reliability

depend principally on the costs of signal production, the benefits of eliciting certain responses, and the fitness consequences of those responses for the receiver. So many strategic models ignore the details of signal form and focus solely on costs and benefits. This approach has generated many insights, although in the long run, I suspect it will be necessary to integrate both efficacy and strategy in a more comprehensive analysis.

Do you think communication is more important in some behavioral venues (for example, foraging, aggression, cooperation) than others? If so, why?

I think that communication plays an important role in almost every behavioral interaction. In any social context, individuals are likely to benefit from information about those with whom they interact—their nature, identity, motivation, and physiological state may all be relevant. So we would expect animals to attend to behavioral and morphological cues that provide such information. But whenever receivers do attend to a cue, selection will favor those individuals who tend to elicit the most beneficial responses—leading to the elaboration of cues into true signals. In fact, many behaviors that are not obviously displays may be modified in this way. When a parent bird, for instance, brings food to the nest, it is engaging in parental care. But this behavior may also indicate to its mate that the offspring are hungry, or that the focal parent is a good provider, encouraging the mate to work harder or deterring it from seeking an alternative partner. Indeed, it may be that frequent feeding is partly driven by the need to impress the mate, rather than simply to supply the young. So cooperation over offspring care might entail communication, even if there are no

obvious display behaviors involved. In general, I think it is very difficult to fully understand the evolution of social foraging, aggression, cooperation, or any other type of interaction without thinking about the information that an individual's actions convey to others in this way—and in that sense, communication is always important.

Do you think of human communication via language as fundamentally different from all other forms of communication?

Human language is certainly exceptional. I don't think all of its striking features are individually unique to our species. Other animals, for instance, employ apparently arbitrary signals to convey information about the external world—the usual example is the alarm calling of vervet monkeys, who make distinct vocalizations to alert others to the presence of leopards, eagles, and snakes. Similarly, other species can produce varied signals by combining distinct display elements in different ways—songbirds provide a good example. But there is no animal signaling system even remotely comparable to human language in complexity, flexibility, and scope of reference. What's more, from a strategic perspective talk is cheap; honesty among humans is not maintained by the costs of faking an exaggerated display, but by social sanctions. This is very different from the costly signals with which ethologists are often concerned.

Dr. Rufus Johnstone is a professor at the University of Cambridge (England). Dr. Johnstone's game theory models of animal behavior have shed light on many questions surrounding the evolution of communication.

SUMMARY

1. Communication is defined as the transfer of information from a signaler to a receiver.

2. A question that consistently rises to the surface when studying communication is whether it is "honest." One framework to study this question begins by assuming that the exchange of information between signaler and receiver is an attempt by the signaler to *manipulate* the recipient. Recipients are then selected to be good "mind readers."

3. Zahavi's handicap principle hypothesizes that honest signaling can evolve when traits are very costly to fake.

4. Communication systems are designed to solve problems that animals encounter in their natural environments.

5. When animals forage in social groups, they face coordination problems. Communication systems that solve the problem of recruiting others to food, including "squeak calls" by swallows, the waggle dance by honeybees, and chemical markers and stridulation by leaf-cutter ants, have been favored by natural selection.

6. Animals use communication to attract mates. For example, male birds use birdsong to attract females during the mating season, and male water striders use ripple communication to signal their location to female water striders.

7. Animals use communication to warn others in their family or group about predators. But deceptive communication, such as fake alarm calls, has also been observed in primates and birds, suggesting that animals may adopt communication tactics designed to trick others in a manner beneficial to the deceiver.

8. Work on the costs and benefits of communication, the phylogenetic history of communication systems, the role of learning and social learning in shaping communication, and the underlying proximate basis for communication is providing ethologists with a much improved picture of signaler-receiver dynamics.

DISCUSSION QUESTIONS

1. Bacteria often release chemicals that affect other bacteria in their vicinity. Would you consider this communication? If so, why? If not, would you consider the chemical trails that ants use to direct one another to a food source communication? How does that differ from the bacteria case, if at all?

2. One problem examining whether a communication system more resembles exaggerations or "conspirational whispers" is that it is difficult to know how to define those terms for any given animal social system. How might you overcome this problem? Consider using a comparative study involving many species.

3. Imagine you are studying a group of amphibian species that vary in their habitats, some living in dense, murky water, and others living in very clear ponds. What kind of communication problems exist in each environment? What sorts of differences in communication systems would you expect to see across such species?

4. Suppose you are studying a heretofore unexamined species of primates. During your observations, you note that individuals often throw heavy rocks against trees, causing a large "booming" sound. You speculate that individuals are communicating to one another using this technique. How might you go about testing this hypothesis?
5. How has research on birdsong provided insight into both proximate and ultimate questions regarding communication?

SUGGESTED READING

Beecher, M. D. (2008). Function and mechanisms of song learning in song sparrows. *Advances in the Study of Behavior, 38,* 167–225. A review that touches on many aspects of birdsong.

Bradbury, J. W., & Vehrencamp, S. (1998). *Animal communication.* New York: Oxford University Press. A well-done overview of animal communication systems.

Cheney, D. L., & Seyfarth, R. M. (1990). *How monkeys see the world.* Chicago: University of Chicago Press. A wonderful look into how vervet monkeys perceive the world. Contains what is probably the best discussion of work on animal deception.

Dawkins, R., & Krebs, J. R. (1978). Animal signals: Information for manipulation? In J. R. Krebs & N. B. Davies, *Behavioural ecology* (pp. 282–315). Oxford: Blackwell. Also Krebs, J. R., & Dawkins, R. (1984). Animal signals: Mind reading and manipulation? In J. R. Krebs & N. B. Davies, *Behavioural ecology* (pp. 380–401). Oxford: Blackwell. A pair of book chapters that nicely outline the question of manipulation and mind reading.

McGregor, P. (Ed.). (2005). *Animal communication networks.* Cambridge: Cambridge University Press. An edited volume that focuses on the transfer of information in social groups (networks).

Pollard, K. A., & Blumstein, D. T. (2012). Evolving communicative complexity: Insights from rodents and beyond. *Philosophical Transactions of the Royal Society B–Biological Sciences, 367,* 1869–1878. A review of complex communication systems in rodents.

Seeley, T. D. (2012). Progress in understanding how the waggle dance improves the foraging efficiency of honeybee colonies. In D. Eisenhardt, G. Galizia, & M. Giurfa (Eds.). *Honeybee Neurobiology and Behavior: A Tribute to Randolf Menzel* (pp. 77–88). Dordrecht, Netherlands: Springer. An overview of the honeybee waggle dance and its role in communication and foraging.

14

Habitat Selection, Territoriality, and Migration

Virtually every aspect of an animal's behavior is affected by where it lives. This is true for animals that take up residence in a single place, as well as for those that move from place to place. From an ethological perspective, habitat choice is interesting not only because the habitat of an animal affects its behavior—what sorts of antipredator strategies will be successful, what types of foraging behaviors will yield the most food, what mating opportunities will be available—but also because the behavior of an animal affects its choice of habitat. For mobile animals, individuals can decide where to live, and in that sense they have some control over the ecological and evolutionary forces operating on them (Odling-Smee et al., 2003).

To better understand the relationship between animal behavior and habitat choice, let's walk through a hypothetical scenario. Imagine a male bird that lives near a large marsh (Figure 14.1). Where should he spend his time during the day? To answer that, our bird must balance many factors. If it is the mating season, one question would be Where are the females? Females may prefer the safety of the reeds, so perhaps our male should go there. The reed area, however, may be home to organisms that parasitize this bird species, so there are both costs and benefits to staying in the reeds. Even if the reeds do not have parasites, courting a mate is energetically expensive, and it may be that prey are not found in the reeds, but over the marsh water, so there might be reason to spend time over the marsh first. However, predators may also prefer the area over the marsh, so, as with the reed area, there are costs as well as benefits to spending time over the marsh. The decision-making process is complicated, as it now includes mates,

FIGURE 14.1. Habitat choice. Imagine a red-winged blackbird deciding where to form a territory. All sorts of factors—mates, temperature, predators (such as the hawk in the upper left of the drawing)—play a role in the decision making. Another factor in habitat choice is the availability of prey, such as the presence of dragonflies (not drawn to scale) shown flying above the marsh.

food, and predators. Other factors may also play a role in determining where our bird goes. Temperature will vary both across the day and across the marsh, and this may affect where the bird spends time. The point of all this being that the choices faced by animals are complex and multidimensional.

Let's imagine that in deciding where to spend his time, our bird ends up flying between the reeds and the marsh water but does not spend that much time in any one area and does not stop others from using any of the areas it frequents. In such a case, we might speak of those areas being his **home range**. In other species, it might be that individuals are **nomads**, constantly wandering, rarely returning to the same place with any regularity.

Suppose that our bird flies between the reeds and the marsh area but that he regularly returns to a particular area in the central part of the reeds and actively tries to keep intruders outside of this area. Now we can speak of this bird as having a **territory**—an area occupied and defended by the bird. It might be that while our bird frequents the central marsh area as his home range, he only keeps intruders out of half of that area. In that case, his home range would be twice as big as his territory.

Some species stay in the same territory and home range for long periods of time. Naturally, what constitutes a long period of time will vary dramatically across species, but as an extreme example, some long-lived birds, such as Florida scrub jays, can maintain the same territory for years upon years and pass that territory down to their offspring (Woolfenden and Fitzpatrick, 1978, 1984).

Animal behaviorists have made much progress in understanding habitat choice and territoriality. In some cases, abiotic (nonliving) factors dominate habitat choice; in other cases, biotic (living) factors do. The abiotic factors that affect habitat choice include heat, availability of water, wind, refuge from danger, availability of specific nutrients, and so on. The biotic factors affecting habitat selection may include the location of mates, food, predators, and parasites. We will examine some of these in more depth in the remainder of the chapter.

Ethologists have gone beyond simply creating a laundry list of variables that affect habitat choice. Far more interesting are the studies that delve into how such variables interact in determining habitat selection and territoriality. Researchers are also explicitly incorporating learning into habitat selection and finding fascinating new trends that may pave the way for future work in this area.

Even animals with defined home ranges and territories, however, can make dramatic shifts in the habitats in which they live; that is, they can periodically move long distances—from one region to another—through **migration**. The most dramatic examples of this are the large-scale migrations that are common in insects, reptiles, mammals, and birds. Let's imagine that we continue studying our bird, which wanders about his home range and defends his territory. Six months after watching our subject, we note that he leaves the marsh we have been studying and spends the next half year 2,000 miles south of where we first saw him. At the end of these six months, our bird returns back to the central reed area of the marsh and starts back where he began. In this case, we have home ranges and territories mixed in with large-scale migration. We might, in fact, get any number of permutations here. It may be that 2,000 miles south, our subject has another home range or territory or neither.

Animal behavior researchers have moved well beyond simply observing and recording migration. This is not meant to denigrate the labor-intensive work still underway documenting monumental migrations like those of the monarch butterflies but only to suggest there is a lot more migration work these days that focuses on hypothesis testing per se. For example, we now have a much better handle on the physiology and resultant behavioral changes associated with switching from a "nonmigratory" to a "migratory" mode, as well as the costs and benefits of migration. What seemed impossible to study experimentally a generation ago is now commonplace in the study of migration.

Having introduced habitat selection, territoriality, and migration, this chapter will examine the following topics in more depth:

▶ Models of habitat choice and territoriality
▶ Territoriality and learning
▶ Family dynamics and territoriality
▶ Migration and navigation
▶ Migratory "restlessness"
▶ Parasites and migration
▶ Mapping migration on to a phylogeny

Habitat Choice

Before examining territoriality per se, we will first explore a more general phenomena, that of habitat choice. While territoriality implies the defense of a set area, **habitat choice** is a question of how animals distribute themselves in space and time with respect to some resource in their environment (Bateson, 1990; J. S. Brown and Rosenzweig, 1985; Kacelnik et al., 1992; Kennedy and Gray, 1993; D. W. Morris, 1994; Rosenzweig, 1981, 1985, 1990, 1991). The resource might be food items, mates, refuge from predators, and so on. Ecologists have examined how animals distribute themselves into different habitats. What makes the work in ethology different from earlier work in ecology is that ethologists specifically focus on the costs and benefits of habitat choice and the role of behavioral decision making (how they choose where to live).

THE IDEAL FREE DISTRIBUTION MODEL AND HABITAT CHOICE

Work in natural history has long suggested that animals distribute themselves in relation to the distribution of resources. If two habitats differ in terms of the amount of food available, and one habitat has more food than the other, more animals tend to be found where there is more food. In a sense, this finding is intuitive, but animal behaviorists want to understand why this was such a common finding. What's more, they are interested in what behavioral rules animals use to distribute themselves between habitats (Herrnstein, 1961, 1970). The **ideal free distribution (IFD)** model was developed to address these sorts of issues (J. L. Brown, 1969; Fretwell, 1972; Fretwell and Lucas, 1970; Orians, 1969; G. A. Parker and Stuart, 1976).

The IFD model can be used to (1) predict how individuals should first settle into habitats (Figure 14.2) and (2) predict the equilibrium frequency of individuals in different patches. To see how the IFD model works with respect to the latter, consider the case where individuals choose between only two habitats, which we will call H1 and H2, and these habitats have resources R1 and R2, respectively. For example, we might be looking at two patches that serve as

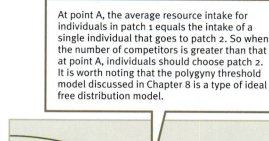

At point A, the average resource intake for individuals in patch 1 equals the intake of a single individual that goes to patch 2. So when the number of competitors is greater than that at point A, individuals should choose patch 2. It is worth noting that the polygyny threshold model discussed in Chapter 8 is a type of ideal free distribution model.

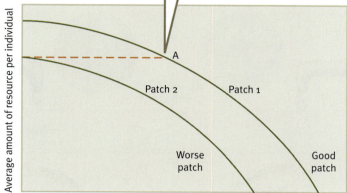

Average amount of resource per individual

Patch 2   Patch 1

Worse patch   Good patch

Number of *other* individuals in patch

FIGURE 14.2. The ideal free distribution. Imagine two different habitat types, or patches: a good patch (1) and a mediocre patch (2). The first individuals that make a decision where to go should choose patch 1. Once the average resource intake at patch 1 drops to the point labeled A, individuals start to fill patch 2. (*Based on Fretwell, 1972*)

possible habitats: One habitat (H1) provides five food units/minute (R1), and the other habitat (H2) provides animals with three food units/minute (R2). Imagine that we are studying a population that has N individuals, that individuals can move freely from habitat 1 to habitat 2, and that moving between food patches has no costs associated with it. How many individuals should end up in habitat H1 and how many in habitat H2?

The IFD model predicts that the equilibrium distribution of individuals into patches should be that distribution at which, if any individual moved from the patch it was in, it would suffer a reduced payoff. That is, at the IFD equilibrium, any individual that moved from H1 to H2, or vice versa, would obtain fewer resources as a result of its move. Mathematically, it can be shown that this translates into individuals settling in habitats in proportion to the resources available in that patch. The equilibrium proportion of individuals in H1 and H2 should be reached when $R_1/N_1 = R_2/N_2$—when the per capita intake rate of individuals in both patches is equal. This solution—$R_1/N_1 = R_2/N_2$—is also known as the resource matching rule, as the distribution of individuals at equilibrium matches the distribution of resources across patches (Herrnstein, 1970; Fagen, 1987; Houston, 2008).

THE IFD MODEL AND FORAGING SUCCESS. The IFD model has been applied to cases in which the resource driving habitat choice ranges from availability of mates to safe refuge from predators (Höglund et al., 1998; G. A. Parker and Stuart, 1976). Here we will focus on two studies where the critical resource is food, and in particular we will address (1) whether animals distribute themselves as the matching resource rule predicts, and (2) at such a distribution, whether all individuals receive approximately the same amount of food, as predicted by the IFD model.

The first direct test of IFD models of foraging was conducted by Manfred Milinski (Milinski, 1979). In an elegantly simple experiment, Milinski had six stickleback fish in a tank. The tank had two feeders that provided food—water fleas (*Daphnia magna*)—at opposite ends of a tank. The foraging behavior of six sticklebacks was then observed in two treatments. In one treatment, water fleas were released from the two feeders in a 5:1 ratio (five times as much food at one feeder than the other); in the second treatment, the ratio of water fleas released from the feeders was 2:1. Milinski found that, after some initial sampling, the fish in both treatments distributed themselves under feeders in a ratio similar to the resource matching rule (Figure 14.3). For example, in

A

Water fleas (*Daphnia*)

5 sticklebacks   1 stickleback

FIGURE 14.3. Ideal free fish.

(A) When sticklebacks are presented with two foraging patches that produce food at a 5:1 ratio (5 water fleas being dropped into the tank from the left feeder versus 1 water flea being dropped into the tank from the right feeder), they distribute themselves in accordance with predictions from the ideal free distribution model. **(B)** Using this setup with sticklebacks foraging from two feeders on opposite sides of an aquarium, Milinski tested the ideal free distribution model for a 5:1 profitability ratio and then for a 2:1 ratio. The arrow on the left indicates the start of feeding; the arrow on the right in panel **(C)** indicates the point at which the profitabilities of the more and less profitable patches were reversed. *(Based on Milinski, 1979)*

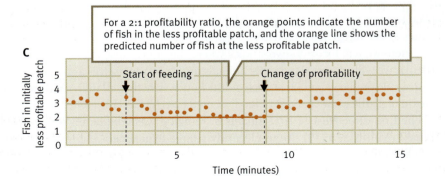

B

For a 5:1 profitability ratio, the green points indicate the number of fish in the less profitable patch, and the green line shows the predicted number of fish at the less profitable patch.

Fish in initially less profitable patch

Start of feeding

Time (minutes)

C

For a 2:1 profitability ratio, the orange points indicate the number of fish in the less profitable patch, and the orange line shows the predicted number of fish at the less profitable patch.

Fish in initially less profitable patch

Start of feeding   Change of profitability

Time (minutes)

the treatment in which five times more food was available at one feeder, five of the six fish were found at that feeder. While Milinski's study was the first experimental work demonstrating the resource matching rule, it was not designed to examine the feeding success of each individual and hence could not determine whether individual foraging success was approximately equal across the two feeding patches.

Using individually recognizable mallards (*Anas platyrhynchos*) living in a pond at Cambridge University, Harper ran an IFD experiment similar to that of Milinski (Harper, 1982). Harper had two observers who were stationed 20 meters apart throw pre-cut, pre-weighed pieces of food (bread) into the pond. The observers acted both as data takers *and* as the food stations themselves. Harper varied the profitability of a patch by varying either the number of pieces of food added to that patch or the weight of each piece (Figure 14.4).

When equal amounts of food were thrown into patches by both observers, ducks quickly distributed themselves in a 1:1 ratio (Figure 14.5). In addition, as predicted by the IFD model, when one patch had twice as much bread as the other, the ducks distributed themselves in a 2:1 ratio.

During trials of his experiment, Harper noticed that some ducks seemed to be very aggressive and tended to receive a disproportionate amount of food within the patches. What this meant was that, although individuals distributed themselves across food patches in a manner similar to the resource matching rule, all individuals were *not* receiving the same amount of food across patches, and as such, the ducks' behavior might be better represented by a modified IFD model that takes into account such aggression. This model is known as the ideal despotic distribution (G. A. Parker and Sutherland, 1986; Sutherland, 1983; Sutherland and Parker, 1985).

AVOIDANCE OF DISEASE-FILLED HABITATS

Another factor that might affect an animal's choice of habitat is the likelihood of exposure to disease in that habitat. Given that virulent nature of many diseases, we might expect that natural selection would act strongly on any behavioral trait that helps minimize an animal's exposure to disease. The two most likely

FIGURE 14.4. Ducks feeding at a pond. (A) One of the first controlled experiments on the ideal free distribution model involved ducks feeding at a pond. **(B)** To test the ideal free distribution model, food was thrown into a pond from two locations, and the distribution of ducks to each feeding station was recorded. *(Photo credit: Marc Epstein/Visuals Unlimited)*

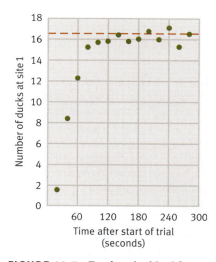

FIGURE 14.5. Testing the ideal free distribution in ducks. The two foraging patches created when two observers threw bread into a pond from different locations had equal profitability. The dashed line represents the predicted number of ducks at site 1. *(Based on Harper, 1982)*

ways this might occur are through (1) the avoidance of habitats that contain pathogens and (2) the avoidance of individuals that are already ill. We will touch on the former here.

One means by which individuals may reduce the risk of infection from parasitic diseases is by producing offspring in habitats that have low parasite levels (Kiesecker and Skelly, 2000). Amphibians are particularly good for testing whether such disease avoidance behaviors are in play, as they host many parasitic pathogens (Blaustein and Bancroft, 2007; Kiesecker et al., 2001, 2004) and are able to distinguish between oviposition sites based on a wide variety of characteristics (Duellman and Trueb, 1994).

Joseph Kiesecker and David Skelly examined whether gray treefrogs (*Hyla versicolor*) base their decisions about where to oviposit (lay eggs) on levels of parasite infection by trematode parasites that are known to affect both larval performance and mortality (Kiesecker and Blaustein, 1997, 1999; Kiesecker and Skelly, 2000). The ponds used by *H. versicolor* are also home to the snail *Pseudosuccinea columella*, which is an intermediate host for a trematode parasite. The presence of this snail may serve as a cue to the frogs that an oviposition area is risky in terms of trematode infections (Figure 14.6).

Kiesecker and Skelly addressed two related questions in their study of ethology and disease avoidance in frogs: (1) Do ovipositing frogs distinguish between sites with and without *P. columella* snails, and (2) Do frogs respond to the density of these snails (Kiesecker and Skelly, 2000)? To address these questions, they set up twenty-five artificial ponds and five treatments: no snails (control), five infected snails, five uninfected snails, ten infected snails, and ten uninfected snails (Figure 14.7). The researchers then assessed oviposition behavior of naturally occurring gray treefrogs. Although controls (ponds with no snails) made up only 20 percent of the ponds, 66.1 percent of all eggs deposited by *H. versicolor* were laid in such ponds, demonstrating clearly that gray treefrogs were favoring ponds with no snails. In addition, the frogs also responded to the density treatments in a manner suggesting that they can distinguish between infected and uninfected snails. Ponds containing uninfected snails received 33.5 percent of the eggs laid, and ponds containing infected snails received only 0.4 percent of the eggs laid.

A

B

FIGURE 14.6. Oviposition and parasite infection level. (A) The gray treefrog (*Hyla versicolor*) bases choice of oviposition site on parasite infection level. **(B)** A trematode parasite uses the snail *Pseudosuccinea columella* as an intermediate host. Frogs attempt to oviposit (lay eggs) in sites with low snail densities, and hence low trematode levels. *(Photo credits: Joseph Kiesecker)*

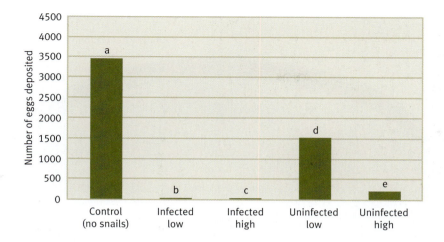

FIGURE 14.7. Parasites and oviposition sites. *Hyla versicolor* laid more eggs at control sites (with no snails) (a) than at sites with experimentally low (b) or high (c) levels of parasites. Frogs also preferred sites with low densities of snails (d) rather than high densities of snails (e), even if the snails were uninfected. *(Based on Kiesecker and Skelly, 2000, p. 2941)*

STRESS HORMONES AND SPATIAL MEMORY IN RATS

When animals make choices about which habitat to occupy, and where within that habitat they will spend time, they often must remember some attributes of the habitats available to them. With respect to decisions about habitat choice, ethologists have studied spatial memory in numerous ways, including by examining the hormones that affect it (de Quervain, 2006; de Quervain et al., 1998, 2000; Joels et al., 2006; Oitzl et al., 2010; Sapolsky, 2003; Schwabe et al., 2010). In many animals, high levels of glucocorticoid hormones—often called stress hormones—interfere with spatial memory skills (McEwen and Sapolsky, 1995; Sapolsky, 1999; Sapolsky et al., 2000).

Dominique de Quervain and his colleagues used a water maze to examine how one glucocorticoid hormone—corticosterone—affects spatial memory and habitat choice in rats (de Quervain et al., 1998). In this experimental setup, rats are put into a large tank of dark and murky water. Some distance from where the rat enters the water, there is a single small platform that is slightly submerged but close enough to the surface to serve as a resting place for the rat: The platform is a place of refuge within the otherwise inhospitable habitat of the water tank. When first put into the tank, rats swim around randomly, but eventually they come across the hidden platform and use it as a refuge. As they are put through more and more trials in the water maze, rats become better and better at finding the platform.

On the basis of prior work on rodents, de Quervain and his colleagues hypothesized that a stress-induced increase in corticosterone would impair a rat's spatial memory regarding the position of the platform in their habitat. To test this hypothesis, they gave rats eight trials in the water maze, and then they divided the rats into four groups (Figure 14.8). In group 1, rats received a shock thirty minutes before their ninth trial in the water maze. In group 2, the shock was administered two minutes before the ninth trial, and in group 3 it was administered four hours before the ninth trial. A fourth group—the control group—received no shock before their ninth trial. De Quervain and his team found that rats in groups 2 and 3 spent about as much time near the platform as did rats in the control group. That is, shocks that were administered two minutes or four hours before a trial did not impede spatial memory for the location of the platform. Rats in group 1—those that experienced the shock

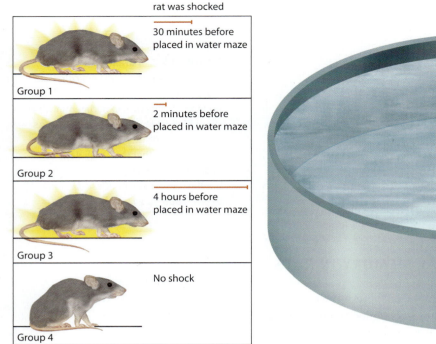

Time at which
rat was shocked

30 minutes before
placed in water maze

Group 1

2 minutes before
placed in water maze

Group 2

4 hours before
placed in water maze

Group 3

No shock

Group 4

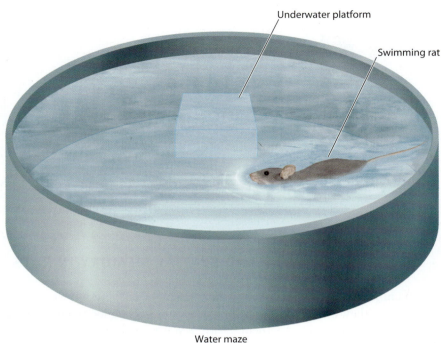

Underwater platform

Swimming rat

Water maze

FIGURE 14.8. Water maze apparatus and learning trials. De Quervain and his colleagues examined stress and learning in rats by using a water maze in murky water that had a platform that the rats could not see but could learn to find. In group 1, rats received a shock thirty minutes before their ninth trial. The shock was administered two minutes before the ninth trial in group 2 rats, and four hours before the ninth trial in group 3 rats. Rats in group 4, the control group, received no shock before their ninth trial.

thirty minutes before the trial—did show impaired memory for the location of the platform (Figure 14.9).

De Quervain and colleagues then measured corticosterone levels in rats in all four groups. As predicted, they found that, compared with the control group, only rats in group 1—the group with decreased ability to find the area of the platform in their tank—had higher levels of corticosterone, suggesting an important memory-inhibiting role for this hormone. These behavioral results show that unlike the effects of other hormones on memory, corticosterone's effects on spatial memory are often displayed about thirty minutes after stress is induced, and then these effects dissipate over the next few hours.

De Quervain and colleagues examined the role of corticosterone on spatial memory and habitat choice in two follow-up experiments. In one of these, they injected rats with a drug called metyrapone, that inhibits the production of corticosterone by blocking certain chemical reactions in the adrenal glands, where corticosterone is produced. When rats were injected with metyrapone, individuals that were shocked before their trial in the water tank did not show impaired memory compared with control animals. When corticosterone production was blocked, rats retained the ability to find the platform, again suggesting that corticosterone was associated with spatial memory tasks in rats.

In their second follow-up experiment, instead of shocking rats thirty minutes before their last trial in the water maze, de Quervain and his colleagues

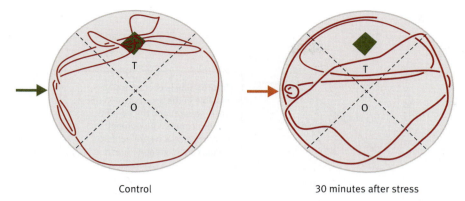

Control 30 minutes after stress

FIGURE 14.9. Shock and spatial memory in rats. The swimming path of a control rat on its ninth trial was consistently near the target (T), indicating that its memory for the location of the platform was not impaired, whereas the swimming path of a rat that was shocked thirty minutes before the ninth trial was random, indicating that its memory for the location of the platform was impaired by the shock. (*From de Quervain et al., 1998*)

injected rats with corticosterone thirty minutes before a trial. If corticosterone changes in response to stress were behind the impairment of spatial memory, then rats that have their corticosterone levels experimentally increased should also show impaired spatial memory in the water maze. Results suggest that this was the case, as rats injected with corticosterone (but not shocked) thirty minutes before being placed in the water maze tank also showed impaired memory when they tried to find the platform.

Taken together, the work by de Quervain and his team is a good example of how the behavioral endocrinology of spatial memory with respect to habitat choice can be studied in a controlled environment. The combination of behavioral experiments with measurements of corticosterone and experimental manipulations of corticosterone, as well as manipulation of chemicals that block the production of corticosterone, allows us a much deeper understanding of the role that endocrinology plays in spatial memory and habitat choice.

Territoriality

Territoriality is typically defined as the occupation and defense of a particular area. Territories can provide their owners with exclusive access to food, mates, and safe haven from predators, and are typically vigorously defended from intruders. Such defense can be costly, in terms of both time and energy. Models of territoriality consider both the costs and benefits of owning a territory. When the benefits are greater than the costs, territory defense is economically feasible and will be favored by natural selection (Figure 14.10).

Most of our discussion of territoriality will focus on territories that contain either a single individual or a family, but territories are sometimes defended by groups of individuals. This sort of group territoriality can often result in dramatic between-group interactions, in which groups attempt to enlarge their own territory by taking over another group's territory. For example, consider the

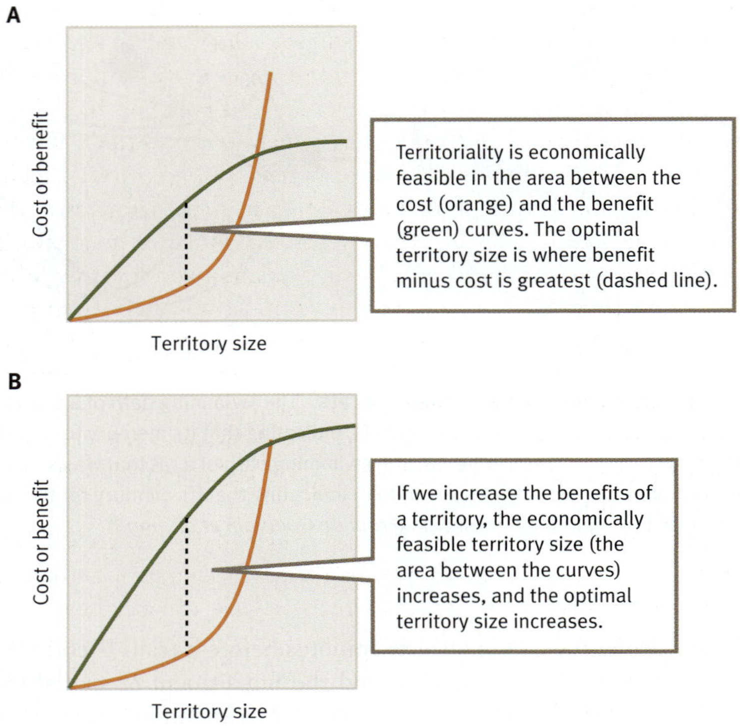

A

FIGURE 14.10. An economic model of territoriality. (A) The zone between the curves indicates where territoriality is economically feasible, and the dashed line shows the optimal territory size. **(B)** When the benefits of territoriality increase, so do the range of economically feasible territory sizes and the optimal territory size.

Territoriality is economically feasible in the area between the cost (orange) and the benefit (green) curves. The optimal territory size is where benefit minus cost is greatest (dashed line).

B

If we increase the benefits of a territory, the economically feasible territory size (the area between the curves) increases, and the optimal territory size increases.

between-group raiding that is occasionally seen in natural populations of chimps. Between-group interactions often appear to be "war-like," and in fact resemble the raiding behavior that is common among many tribes of humans (Boehm, 1992). During raids, all-male chimpanzee patrol groups travel into areas that border their territorial boundaries (Bygott, 1979; Goodall, 1986; Nishida, 1979). In contrast to excursions for food, in which foraging chimps often emit vocalizations, chimps in patrols move in a wary fashion and remain silent (Goodall, 1986). These raids often involve the killing of a small number of members of the raided group and the capture of females. Occasionally, raiding parties from two groups will meet one another. In such instances, rather than engaging in extremely violent interactions, both groups engage in hostile vocalizations and then withdraw (Goodall, 1986). Although all-out warfare does not emerge when two raiding parties meet, raiding can, in the long run, amount to the slow extinction of one group. For example, raiding behavior in the Mahale Mountains of Tanzania led in part to a larger group eradicating a smaller group of chimps (Nishida et al., 1985; see also Goodall, 1986, for evidence of this at Gombe).

TERRITORIALITY AND LEARNING

Most models of territoriality assume that animals are capable of assessing various characteristics about a potential territory—how much food is in the area, how safe the area may be from predators—but these models do not explicitly consider how learning affects the establishment and maintenance of a territory. A few such learning-based models, however, have been constructed, and below we examine two models that examine how learning affects decisions regarding both the acquisition and subsequent maintenance of territories.

TERRITORIALITY AND LEARNING DURING SETTLEMENT. Territoriality in juvenile *Anolis aeneus* lizards has been the subject of an in-depth analysis by Judy Stamps (Stamps, 2001). These lizards form territories early in life, and Stamps has examined, among many other things, where and how juveniles decided to stake out a territory (Figure 14.11). Because much of the work on territoriality has centered on the distribution of food resources, Stamps's initial inclination was to test for the importance of food availability in structuring territoriality in *A. aeneus*. Yet despite numerous experiments manipulating food availability, she did not uncover a clear-cut effect of food availability on territory formation. Rather, in subsequent experiments, Stamps found that safety from predators and suitable temperature appeared to be the most important attributes of a desirable territory. But *how* do juvenile lizards determine which territories are suitable with respect to temperature and predation pressure? More specifically, given that these factors affect what areas lizards choose as territories, do lizards *learn* what areas are best from their interactions with other lizards (Stamps and Krishnan, 1999, 2001)?

From earlier work, Stamps had noticed that juvenile *A. aeneus* watched what other lizards were doing, and she hypothesized that they might be determining territory quality as a result of their interactions with conspecifics. If another individual has already determined a territory is safe and has temperatures that will not cause overheating, then it probably is safe and probably will not cause overheating. This sort of decision-making process, in which individuals use the choices of others to determine the quality of a potential territory, has been called "conspecific cueing."

Stamps tested the conspecific cueing hypothesis by examining whether a territory that had been occupied by an owner sometime in the past but was currently vacant would be viewed as attractive to other lizards (Stamps, 1987a). In this experiment, a juvenile was allowed to observe two very similar territories, one that was currently occupied and one that was vacant. When given the choice between these two areas, with the territory owner now removed, juveniles showed a strong preference for the previously inhabited area (Figure 14.12). Furthermore, juveniles that had not observed the territories during the initial part of the experiment but were given a choice between these territories displayed no preference for the previously occupied territory, suggesting a strong visual component to conspecific cueing in *A. aeneus* (Stamps, 1987b).

FIGURE 14.11. Territorial lizards. **(A)** Juvenile *Anolis aeneus* are very territorial, and their territory formation has been studied in the context of habitat choice and learning. **(B)** *Anolis aeneus* stake out territories in areas such as those depicted here. (*Photo credits: Judy Stamps*)

A

B

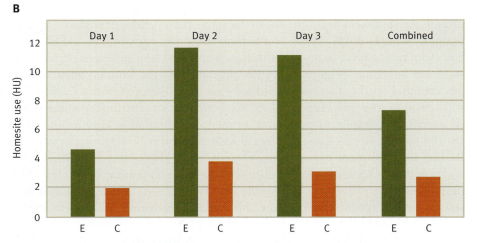

A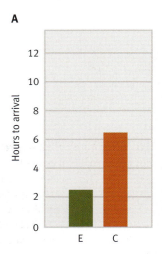

FIGURE 14.12. Conspecific cueing and territorial lizards. (A) Focal juveniles not only spent more time on experimental (E) versus control (C) homesites but also arrived at experimental homesites more quickly. **(B)** Juvenile lizards were drawn to experimentally manipulated homesites (E) over control (C) homesites. Control territories remained empty prior to the focal's choice, while experimental territories had formerly contained a territorial juvenile. *(Based on Stamps, 1987a)*

TERRITORY OWNERS AND SATELLITES

Territory ownership often requires constant vigilance against both predators and conspecific intruders that might attempt to make that territory their own. This relationship between vigilance and territoriality can be complex. Here we will examine this relationship by addressing whether the costs and benefits of territory defense are ever such that it is in a territory holder's interest to allow other individuals to share its territory, at least temporarily.

OWNERS, SATELLITES, AND TERRITORY DEFENSE IN PIED WAGTAILS. From the perspective of a territory holder, intruders onto a territory need not always be a liability. To see why, let's examine the economics of territoriality in pied wagtail birds (*Motacilla alba*; Figure 14.13). During the winter season, pied wagtails defend riverside territories, where they forage on insects that have washed up onto riverbanks (Davies, 1976; Figure 14.14). These insects are a renewable food source in the sense that, even after insects are consumed, more insects eventually will wash up on shore. If a pied wagtail can time its movement within a territory correctly, it can deplete the insects in one part of its territory, proceed to search other parts of its territory for food, and by the time it returns to the original section again, find more insect prey there for it to consume.

Nick Davies and Alasdair Houston tracked the movement and foraging activity of a population of pied wagtails on a riverbank along the Thames River, near Oxford, England, and found that these birds systematically searched their territories and timed their movements such that they did not return to an area on which they had foraged until sufficient time had elapsed for a new crop of insect prey to wash up onto the banks (Davies and Houston, 1981).

While on their winter territories, pied wagtails must often deal with territorial intruders (Davies and Houston, 1981). Under certain conditions, intruders are tolerated, at which point ethologists refer to them as satellites.

FIGURE 14.13. Territorial wagtails. Economic models of territorial ownership and the acceptance of intruders have been tested in pied wagtails. *(Photo credit: Charles McRae/Visuals Unlimited)*

A

40 minutes

B

20 minutes

20 minutes

Under other conditions, intruders are aggressively chased off a territory. Davies and Houston wanted to know what determines when intruders are allowed onto a territory as satellites and when they are not.

To understand how territory owners make such decisions, we need to recognize that a territory is worth more to a pied wagtail owner than to an intruder, because intruders do not know which areas in a territory have recently been cropped (and so have few prey), but owners do. If an owner permits an intruder to stay, it loses some foraging-related benefits, but it will also gain the satellite's assistance in territory defense against other intruders (Table 14.1). Having a satellite around makes a territory owner's residence a bit safer, but not as profitable in terms of food intake.

Davies and Houston built a cost-benefit model to predict when owners would allow satellites on their territories and when they would not. They found that when food availability was particularly high, territory owners tolerated satellites; otherwise, they evicted such intruders. This is a particularly interesting case of conditional cooperation because it suggests that the costs and benefits of byproduct mutualism (see Chapter 10) can vary depending on territory-holding status. Would-be satellites that have no foraging territories of their own always face a harsh environment with respect to food allocation, and they are willing to pay the costs of territory defense in order to have access to food on territories. When food becomes sparse for owners, however, they do not allow intruders in. So intruders are always willing to undertake cooperative territory defense in exchange for food, but the costs and benefits of byproduct mutualism do not always favor owners that permit this cooperative act (J. L. Brown, 1982).

FIGURE 14.14. Pied wagtails and food search. Pied wagtails systemically search for food on their territories along riverbanks. **(A)** A single bird can complete a circuit of the riverbank in 40 minutes and gets all the food it finds. **(B)** When a territory is shared by two birds, however, the circuit is divided up as well, so each bird primarily gets the food that it finds in its 20 minutes of walking the riverbank. *(Based on Davies and Houston, 1984)*

TABLE 14.1. Owners and satellites. Pied wagtails form winter foraging territories along a river. Owners often allow "satellite" individuals to forage on such territories. Some territorial defense is provided by the satellites. *(Based on Davies and Houston, 1981)*

TERRITORY	NUMBER OF DEFENSES BY		PERCENT DEFENSES BY SATELLITE
	OWNER	SATELLITE	
1	13	5	27.8
2	2	1	33.3
4	8	2	20.0
5	6	6	50.0
7	44	22	33.3
8	8	7	46.7
9	13	15	53.6
Total	94	58	38.1

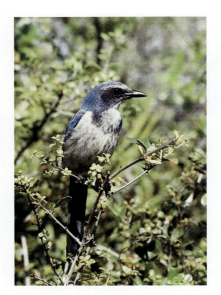

FIGURE 14.15. Scrub jays and territories. In Florida scrub jays, territories are inherited across generations, leading to the establishment of family dynasties. Territory size is increased through a process called budding. *(Photo credit: Charlie Heidecker/ Visuals Unlimited)*

TERRITORIAL "DYNASTIES" IN FLORIDA SCRUB JAYS

Florida scrub jays have an elaborate system of helpers-at-the-nest (see Chapters 3 and 9), resulting in a territorial inheritance system in which territory ownership is passed down across generations, leading to "family dynasties" (J. W. Fitzpatrick and Woolfenden, 1984, 1988; J. S. Quinn et al., 1999; Woolfenden and FitzPatrick, 1990; Figure 14.15). Each Florida scrub jay territory is first occupied by a monogamous pair of birds. In addition, territories contain up to six helpers-at-the-nest, and all individuals, including helpers, defend the territory. When territories are formed, they may be relatively small. But as family size increases, so does territory size (up to a point). This increase typically occurs through a process known as territorial budding (Woolfenden and Fitzpatrick, 1978). At about two and a half years of age, a male scrub jay, with the help of his family members, will expand the boundaries of family territory and actively defend this new area (the "budded" territory) as his own. While there are some aggressive interactions between the male that has acquired this budded territory and the family members in adjacent territories, these are rare compared with the aggressive acts between such a male and any nonfamily members that border his territory.

The budding procedure serves to increase a family's total territory size, provides a male with his own smaller territory, and essentially creates a form of inheritance. This inheritance system, in which larger families outcompete smaller families, may lead to very large territorial family dynasties.

CONFLICT WITHIN FAMILY TERRITORIES

Because they are genetically related, family members that share a territory usually do not differ in terms of what constitutes the best behavioral decisions affecting all group members. But this is not always the case. Under certain conditions, family members on a territory may differ on decisions about when a territory member should leave a territory and breed on its own. For example, it may be in an individual's best interest to leave its natal territory and breed at time 1, while it may be in the best interest of other family members on the territory for that individual to remain at home, not to breed, and to help family members at time 1. Conversely, it may be in an individual's interest not to breed but in the interests of other family members for that individual to breed on its own territory. There is a burgeoning literature on **optimal skew theory**, which studies the distribution of breeding within a group and whether there will be cooperation or conflict over reproductive activities (see Chapter 9). For an example of conflicting interests in terms of breeding, we turn to Steve Emlen's work on parent-offspring conflict over breeding opportunities in white-fronted bee-eaters (*Merops bullockoides*; S. T. Emlen and Wrege, 1992).

Like the young in many communally breeding bird species, young male white-fronted bee-eaters often remain on their natal territory and aid their genetic relatives, usually their parents, in raising young (typically their siblings). When breeding opportunities for young males are rare, no conflict exists between young males and their parents—it is in the best interest of all parties for the young male to remain at home and help his parents. The situation gets more complex when breeding opportunities away from the natal nest are more readily available to such a young male. Under these conditions, it may be in the best interest of a young male to breed on his own territory, but in the best interest of this male's parents for him to remain at home and help raise his siblings. This leads to conflict between offspring and parent.

To understand this conflict of interest, consider the following: Pairs of white-feeted bee-eaters without helpers raise an average of 0.51 offspring per clutch. Every helper a pair has at its service adds approximately 0.47 offspring. Keeping in mind that young males that breed for the first time will rarely have a helper of their own, let's start with the perspective of a young male's parents. If their son attempts to breed, he will, on average, produce 0.51 offspring. But if instead he helps them, he will add, on average, 0.47 offspring to their next clutch. Since the young male's parents are twice as related to their own offspring ($r = 0.5$) as they are to grand-offspring ($r = 0.25$)—that is, the offspring that their son would have produced had he not helped but instead gone off to find a territory and a mate—the parents will have an incentive to keep their son around to help. Therefore, ethologists predict that parents will try to block any attempt on the young male's part to leave.

The genetic accounting is different from the young male's perspective. Because he is equally related ($r = 0.5$) to his own offspring and to his siblings (individuals he would assist if he were to help his parents), there is no incentive for such a male to resist his parents' attempt to suppress his breeding. And, indeed, Emlen and Wrege found that in bee-eaters, parental suppression of offspring breeding is met with little resistance on the part of the son. But when a breeding pair tries to suppress the reproductive efforts of its more distant kin, or even nonkin, their actions are met with much stiffer resistance than when they attempt to suppress the reproduction of their own offspring.

Migration

Consider this eloquent summary of bird migration:

> Twice each year, billions of birds, entire species, swarm across the globe, traveling thousands of miles as they follow the sun to populate regions that are habitable for only part of the year. The spatial scope of these migrations exceeds all other biological phenomena. So fantastic are they that ancient civilizations devised a host of myths to explain the periodic appearance and disappearance of such vast numbers of animals. Those apocryphal stories were concocted in part because what we know to be true seemed then so completely beyond the pale. It seemed more likely that swallows buried themselves in the mud at the bottom of ponds than that they flew all the way from Europe to Africa and back twice each year. But the truth turned out to be more amazing than the myth. (Able, 1999, p. vii)

And migration is not limited to birds. From wildebeests swarming across the African plains to monarch butterflies heading by the tens of millions to Mexico each year, migration is seen across many different groups of animals and is certainly one of the most spectacular of all animal behaviors. In some species, annual migration is obligatory, occurring like clockwork; in other species, it occurs only when conditions become poor in what is called irruptive migration; in still other cases, only a portion of a population will migrate, while the rest stays put (Able, 1999). No matter where a species falls on this migration continuum, questions abound—how do they know where to go, when to go, how to go, how to prepare (Able, 1999; Alerstam, 1990; Baker, 1978; Gauthreaux, 1980; Heape, 1931; Ramenofsky and Wingfield, 2007; Salewsk and Bruderer, 2007; R. Wiltschko and Wiltschko, 2003; Figure 14.16)?

A

B

FIGURE 14.16. Animal migration. In some species of birds and mammals, massive yearly migrations take place. Here we see migration in **(A)** geese and **(B)** gnu. *(Photo credits: William J. Weber/ Visuals Unlimited; Joe McDonald/Visuals Unlimited)*

MIGRATION AND NAVIGATION

To migrate long distances, animals must navigate through the environment. That is, they must use cues to assess where they are in relation to where they are heading. As we are about to see, these cues can come in many forms—the position of the sun, the position of the stars, various landmarks on the ground, the odor of a stream, and so on—but one way or another, migrating animals must determine if they are heading where they want to go, and if not, how they can get back on track (Bruderer and Salewski, 2008; Dingle, 2008; J. L. Gould and Gould, 2012; Zink, 2011; see the Conservation Connection box).

MIGRATION AND THE SUN COMPASS. Both because the sun is such a large, prominent object in the sky and because it provides information about direction, ethologists have long speculated that migrating animals use the sun to help them navigate. Here we shall examine how monarch butterflies use the sun to guide them on their journeys.

Few natural spectacles are as magnificent as seeing tens of millions of monarch butterflies (*Danaus plexippus*) migrating each year from North America to the mountain ranges of central Mexico. During their annual migration, the air is so thick with monarch butterflies that branches of trees have been known to collapse from the weight of too many butterflies (Figure 14.17). Monarch

FIGURE 14.17. Monarch migration. Migration of monarch butterflies can involve tens of millions of individuals. Here we see a tree covered with monarchs. *(Photo credit: Fritz Polking/Visuals Unlimited)*

Migration Patterns, "Stopovers," and Conservation Biology

Information on animal migration patterns, including the location of "stopovers," where migrating individuals feed and rest, is essential for both ethologists and conservation biologists interested in understanding migratory behavior and minimizing human obstacles to such migration (Morales et al., 2010; Sawyer and Kauffman, 2011; Shamoun-Baranes et al., 2012). Obtaining such information is difficult, in part because tracking migrating individuals over long distances can be hard, but also because it is a challenge to use data on a small sample of migrating animals to make accurate estimates of migratory patterns for the *populations* from which the sample animals come.

In 2007, Jon Horne and his colleagues developed a mathematical model that takes global position data (GPS data)

and, using sophisticated mathematical algorithms, estimates where important "stopovers"—those likely associated with foraging, resting, and mating—occur. The model then conglomerates such data from individuals to estimate large-scale population migration patterns (Horne et al., 2007). Hall Sawyer and his colleagues used the model to estimate the migration patterns, including stopovers, of mule deer (*Odocoileus hemionus*) across a pristine area of Wyoming. One of the reasons they chose this population is that the migration area of these deer was being considered as a location for 2,000 gas wells and over 1,600 kilometers of gas pipes. Sawyer and colleagues wanted to make suggestions about what the oil companies could do to minimize the damage that these wells and pipes might have on the mule deer migration (Sawyer et al., 2009).

Sawyer and his team captured forty-seven mule deer, placed a GPS radio collar on each one, and then released the animals. They then tracked the migratory behaviors of these deer and used the BBMM to map out a population-level migration route, which they superimposed onto a map with the proposed sites for the gas wells and pipelines (Figure 14.18). They presented their data on migration paths and stopover sites to the oil industry. This information led these companies to modify the use patterns of gas wells and pipelines (wells and pipelines were used less in important areas during migratory season). The oil companies also made improvements to key habitats along migration routes, especially at highly frequented stopover sites (Sawyer et al., 2009).

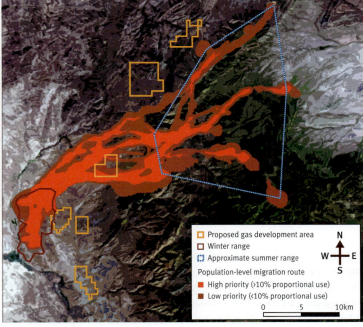

FIGURE 14.18. Migration patterns and oil company gas wells and pipe lines. (A) Mule deer. **(B)** The migratory behavior of mule deer and the proposed placement of gas wells and pipelines are shown on a GPS map. (*Photo credit: Art Wolfe/Photo Researchers; from Kauffman et al., pp. 2016–2025*)

butterflies traverse up to 6,000 miles on their migratory trip, and they almost always navigate successfully without getting lost, even on their first migration. Recent work has begun to study monarch brain structure in relation to migration and data from the monarch butterfly genome project to understand the molecular genetics of monarch migration, but this work is just in its infancy (Heinze and Reppert, 2012; Reppert et al., 2010; Zhan et al., 2011).

As they travel south, monarch butterflies use the position of the sun in relation to the time of day to help them navigate (Mouritsen and Frost, 2002; Perez et al., 1997). To examine the role of solar navigation in this species, Sandra Perez and her colleagues ran a "clock-shift" experiment (Perez et al., 1997). The basic idea in such studies is to experimentally manipulate the amount of daylight and darkness animals are exposed to such that they experience a light-dark cycle that is different from the normal cycle at that time of year. For example, more hours of daylight are typically experienced in the summer than in the winter, but clock-shift experiments allow an experimenter to take an animal during the winter and expose it (in the laboratory) to the light-dark cycle it typically experiences in the summer. This manipulation of its biological clock tricks the animal into acting as if it were summer.

At the start of the experiment, Perez and her team raised one group of monarch butterflies in a laboratory and slowly shifted the butterflies' body clocks back six hours—when the real time, for example, was noon, the clock-shifted monarchs acted as if it were six hours later. The researchers also kept a second control group of butterflies in the laboratory, but these butterflies' body clocks were not shifted. During the period of autumn migration south, Perez and her team released individual monarch butterflies that they had raised in the laboratory. They then watched these individual butterflies and noted their position by using handheld compasses. The control butterflies headed south, just as wild populations of monarch butterflies typically do in the autumn migration (Figure 14.19). The clock-shifted butterflies, however, flew almost due west, which is just where Perez and her team expected them to fly if they were using a sun compass to help them navigate during their migration. That is, the six-hour clock shift caused the butterflies to misinterpret the information that the sun was conveying about direction and to fly west rather than south.

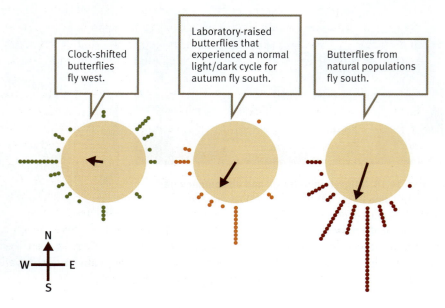

FIGURE 14.19. Clock-shifted monarch butterflies and navigation. The diagrams show the mean body position of monarch butterflies during the first five minutes of autumn migration. *(From Perez et al., 1997, p. 29)*

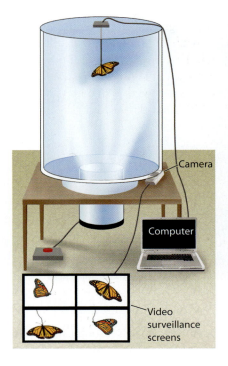

FIGURE 14.20. Insect flight simulator.
A flight simulator for monarchs was constructed in the laboratory, and a butterfly was tethered to this device. The pipe at the bottom of the simulator directed a constant flow of air up toward the butterfly so that it could fly; a video camera was connected to the bottom of the simulator (so that the researchers could see when the butterfly was flying, gliding, or stopping; the four images show the images from different time periods); an encoder (recording the butterfly's direction) was attached to the butterfly from the top of the flight simulator and was connected to a computer that kept a timed record of all the butterfly's movements. With this setup, researchers could track the direction the butterfly was orienting toward and whether it was actively flying or gliding. *(From Mouritsen and Frost, 2002, p. 10163)*

Henrik Mouritsen and Barrie Frost examined in even greater detail monarch butterflies' use of a sun compass to navigate. They built a miniature flight simulator (Figure 14.20) that allowed them to follow the orientation of a tethered monarch butterfly for much longer than the five-minute time frame in the Perez experiment (Mouritsen and Frost, 2002). Using this flight simulation device and the clock-shift protocol discussed above, Mouritsen and Frost confirmed the use of the sun compass during monarch butterfly migration.

INDIGO BUNTINGS AND NAVIGATING BY THE STARS. The majority of passerine birds migrate almost exclusively at night (Able, 1999). What cues might such birds use when cues from the sun are absent? One of the earliest and best-known studies of how migrating individuals use the stars to guide them was undertaken by Stephen Emlen. Building on prior ethological work by Kramer and Sauer, Emlen undertook an ingenious set of experiments involving indigo buntings (*Passerina cyanea*), a nocturnal migratory species that travels 2,000 miles each winter from the northeastern United States to the Bahamas, Mexico, and Panama (S. T. Emlen, 1967, 1970, 1975; Figure 14.21).

Emlen began by creating funnel-shaped test cages for buntings, at the bottom of which he placed an ink pad. The cages were constructed such that each time a bunting tried to fly out, the location of its footprint was marked by ink, so its orientation pattern was easily recorded (Figure 14.22). When these cages were placed under a starlit sky, the buntings oriented their attempts in the direction they would normally migrate—toward the south in September and October, and toward the north in April and May. In addition, these patterns all but disappeared on cloudy nights when stars were not visible, suggesting a role for some sort of star-based navigation system.

Emlen repeated his experiments inside a planetarium, where he could control what the birds saw in a simulated nighttime sky. Results were very similar to those obtained in the field, but Emlen was now able to artificially shift the position of the North Star. In response to this manipulation, buntings

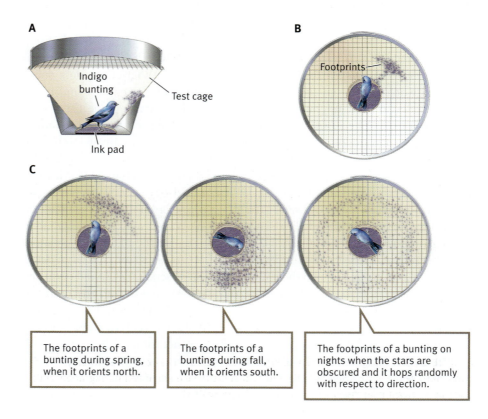

FIGURE 14.21. Bunting migration.
(A) Indigo buntings can use the stars as a navigation tool. **(B)** Planetarium like the one in which indigo buntings in funnel-shaped cages were shown the stars. *(Photo credits: Maslowski/Visuals Unlimited; Cornell University)*

shifted to the "new south" or "new north" (depending on the season). Emlen was able to further demonstrate that buntings appear to use the geometric patterns of stars in northern skies as one guide on their long journey.

Numerous constellations fall within 35 degrees of the North Star, and buntings appear to be able to cue in on more than one of these constellations. Removal of one star from the normal pattern does *not* throw the birds off track, as they apparently fall back to relying on the stellar information that remains.

A

Indigo bunting

Test cage

Ink pad

B

Footprints

C

The footprints of a bunting during spring, when it orients north.

The footprints of a bunting during fall, when it orients south.

The footprints of a bunting on nights when the stars are obscured and it hops randomly with respect to direction.

FIGURE 14.22. Migration and orientation. **(A)** A cross-sectional view and **(B)** a top view of the circular test cage used by Emlen in his orientation/migration work in indigo buntings. The funnel portion of the cage was made of white blotting paper, with an ink pad at the bottom. The entire apparatus was placed in an outdoor cage. The migration tendencies of the buntings were recorded as the bird tried to hop out of the funnel. **(C)** Each time the bunting hopped in one direction, it left black footprints on the blotting paper. *(Based on S.T. Emlen, 1975)*

THE EARTH'S MAGNETIC FIELD AND ANIMAL MIGRATION. The hypothesis that the earth's magnetic field might affect animal navigation and migration has been in the behavioral literature for more than a century, but the first experimental work on this idea was not undertaken until the late 1960s and early 1970s, when it was found that magnets placed on the backs of pigeons disoriented birds navigating across a fifteen- to thirty-mile route (Keeton, 1971; W. Wiltschko, 1968; R. Wiltschko and Wiltschko, 1995, 2003).

Evidence that the magnetic field of the earth is important in migration has been found in a wide diversity of animals, including birds, amphibians, reptiles, and insects (Freake et al., 2006; Maeda et al., 2008; Wajnberg et al., 2010; R. Wiltschko and Wiltschko, 2006). To see how magnetic fields can affect navigation and migration, let's examine the behavior of the bobolink (or rice bird; *Dolichonyx oryzivorus*), which has one of the longest round-trip annual migrations of any animal—12,400 miles. These birds spend the summer months in the northern United States and Canada and then migrate to South America (primarily Brazil, Paraguay, and Argentina) before they return to the Northern Hemisphere (Figure 14.23). Do they use the earth's magnetic field to help guide them on this incredible journey?

To answer that question, Robert Beason and Joan Nichols examined the "headings"—the direction to which the birds oriented—at the time of their fall migration to South America (Beason and Nichols, 1984). They first brought birds into a planetarium and projected the star patterns that were appropriate for the autumn sky in the Northern Hemisphere. In other words, the birds were given correct visual cues with respect to migration, and in response they oriented south (toward South America). Beason and Nichols could also use equipment in the planetarium to manipulate the magnetic polarity that the birds experienced. When the visual cues and the magnetic polarity provided the same information—the northern sky was associated with magnetic north—the birds again oriented in the correct southern direction.

The most critical treatment in the Beason and Nichols experiment was the one in which the visual cues were correct, but the magnetic polarity was

FIGURE 14.23. Magnetic fields and bobolinks. Research has examined whether bobolinks use the earth's magnetic field to help guide them on their annual migration, which involves a 12,400-mile round-trip. *(Photo credit: Jim Fenton)*

reversed—that is, when the appropriate star pattern for autumn in the north was displayed, but the magnetic field was reversed so that the visual cues suggested south was in one direction and the magnetic cues indicated south was in the opposite direction. In this treatment, the birds headed toward the magnetic south, indicating that magnetic cues were critical in the annual round-trip migration.

How are the bobolinks able to sense changes in magnetic polarity? To address this question, when their experiment was complete, Beason and Nichols autopsied the bobolinks, focusing on the head area of the birds. They found high levels of an iron-rich, magnetically sensitive substance, most likely magnetite, in the bobolinks. In particular, this magnetically sensitive substance was consistently found around the olfactory nerves and the bristles that project into the nasal cavity, as well as in the tissue between the nasal cavity and the eyes. Although it is not clear precisely how the birds use magnetite to make navigational decisions, the presence of magnetite strengthens the argument that magnetic orientation is important in bobolink migratory behavior (Freake et al., 2006; Lohmann and Johnsen, 2000).

THE HERITABILITY OF MIGRATORY RESTLESSNESS

Given the broad and varied consequences that migration has on an individual's reproductive success, evolutionary biologists and ethologists have examined whether some aspects of migratory behavior are heritable (Knudsen et al., 2011; Liedvogel et al., 2011; Pulido, 2007). For example, using an array of different techniques, Peter Berthold compared the onset of migratory activity between laboratory and wild birds in eighteen species (Berthold, 1990). He found a strong correlation between onset of migratory activity in both groups of birds, suggesting that the timing of departure for migration may be under genetic control. Such studies cannot directly measure heritability, but Francisco Pulido and his colleagues (including Berthold) did so by designing a selection experiment on autumn migration behavior, choosing the German blackcap bird (*Sylvia atricapilla*) as the species they would study (Pulido et al., 2001).

Pulido and his team brought blackcaps in from the field and measured their migratory restlessness ("zugunruhe") during the autumn migratory season. During migratory restlessness, caged birds in the laboratory show intense movement and activity (including wing-flapping behavior) in their cages; normally, the birds would be migrating during the evenings in the autumn, and this is replaced in the laboratory by extreme activity. Pulido and his team defined the onset of such restlessness as the first night on which a bird was active during at least five thirty-minute periods.

In 1994, the researchers collected forty blackcaps from the field and brought them back to the lab. The ten birds with the latest dates for migratory restlessness were selected and allowed to mate. They produced a total of twenty-six offspring, and from that group four breeding pairs with late-onset migratory restlessness were again allowed to breed, producing fourteen second-generation birds in a line selected for late migratory activity. In only two generations, migratory restlessness was delayed for an average of 7.65 days (Figure 14.24), and Pulido and his co-workers calculated a heritability of 0.72 for the onset of this behavior. In conjunction with the other findings, this suggests that variance in migratory patterns may indeed be due (additive) to genetic variance in migratory behavior rather than to some sort of phenotypically plastic response on the part of birds.

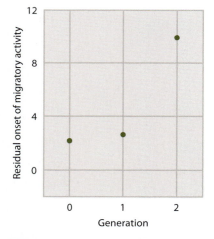

FIGURE 14.24. The heritability of migratory activity. In each generation, German blackcap birds with the latest onset of migratory activity were chosen to breed. In just two generations, the onset of migratory activity was set back more than a week. The *y*-axis here measures a statistical term known as *residuals*, rather than the actual change in absolute days until migration. *(Based on Pulido et al., 2001)*

MIGRATION, TEMPERATURE, AND BASAL METABOLIC RATE

Long migrations require huge amounts of energy on the part of migrants. And, not surprisingly, many animals increase foraging, leading to greater body fat levels, just prior to migration. In birds, there is also a difference in metabolic rates in migrating versus nonmigrating species. Walter Jetz and his team looked at already published data on 135 species of migratory and nonmigratory birds. They found that after correcting for differences in body size, birds' basal metabolic rate (BMR)—defined as the minimum maintenance energy requirement of an endotherm—was significantly higher in migrating species (Jetz et al., 2008; Figure 14.25). Higher metabolic rates *might* help animals maintain the increased metabolic costs associated with large-scale migrations, but Jetz and his colleagues hypothesized that there were other possible explanations for this relationship between increased BMR and migration.

Species that live in colder environments tend to have high BMRs. And species that migrate tend to be found in colder environments than species that are sedentary. So the relation between BMR and migration could just be a by-product of the fact that migratory species tend to live in colder places (Kvist and Lindstrom, 2001; Lindstrom and Klaassen, 2003). To examine this possibility, Jetz and his colleagues used statistical analyses that could evaluate whether migratory behavior or habitat temperature better explained patterns of BMR. Their analysis found that BMR was more tightly correlated with environmental temperature than the tendency to migrate (Jetz et al., 2008). Natural selection favors both migration and higher BMR in bird species in cold environments, thus leading to a spurious correlation between these two variables.

MIGRATION AND DEFENSE AGAINST PARASITES

The energy expended during long migration can reduce immune responsiveness, making animals more susceptible to disease (Buehler et al., 2010; Hoffman-Goetz and Pedersen, 1994; Leffler, 1993). In addition, long-distance migrants also face new parasites and diseases upon arrival at their migratory end point. While nonmigratory birds must combat parasites in one environment, migratory birds face that challenge in two very different environments, and they are thus combating a larger and more diverse array of potential parasites. Anders Møller and Johannes Erritzøe hypothesized that migratory birds should therefore invest more heavily in immune function compared with related resident relatives (Møller and Erritzøe, 1998).

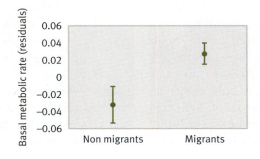

FIGURE 14.25. Basal metabolic rate and migration. Individuals in migrant species have higher basal metabolic rates than those in nonmigratory species. The *y*-axis takes into account differences in body between migratory and nonmigratory species using "residuals" from statistical analysis. *(From Jetz et al., 2008)*

Møller and Erritzøe tested their hypothesis by comparing the size of immune defense organs in pairs of bird species. One member of each pair was from a migratory species, and the other pair member was from a closely related species that was nonmigratory. The immune defense organs that were compared were the spleen and a lymphoid organ called the bursa of Fabricus. The researchers assumed that a larger size in either of these defense organs would provide better immunological resistance to parasites (John, 1994; Toivanen and Toivanen, 1987). Møller and Erritzøe found that, in nine of ten pairwise comparisons, the bursa of Fabricus was larger in birds from the migratory species, while in nine of thirteen pairwise comparisons, the spleen was larger in birds from the migratory species.

Are there other variables besides the tendency to migrate or not that might explain differences in the size of immune defense organs when comparing the pairs of bird species that Møller and Erritzøe tested? The authors themselves note that differences in sexual selection pressure, mating systems, and the pattern of nest use and reuse are also known to affect investment in immune defenses (Møller and Erritzøe, 1996). As such, they reanalyzed their data, using only pairs of species that were known not to differ on any variable (besides migratory tendencies) that might affect the size of immune defense organs. While this comparison lowered their sample size to six pairwise comparisons for both the bursa of Fabricus and the spleen, their initial findings remained unchanged, suggesting stronger selection on the immune of migratory species.

PHYLOGENY AND MIGRATORY BEHAVIOR

Migratory behavior has been studied in a phylogenetic context (Outlaw and Voelker, 2006; Outlaw et al., 2003; Winger et al., 2012; Zink, 2002). Diana Outlaw and Gary Voelker, for example, have examined migratory behavior in the avian family Motacillidae, which includes such species as the pied wagtail (*Motacilla alba*), the golden pipit (*Tmetothylacus tenellus*), and the yellow-throated longclaw (*Macronyx croceus*; Figure 14.26). In particular, one of the hypotheses they set out to examine was the "evolutionary precursor" model of migration. This hypothesis posits that migration will be associated with species that live in open or edge habitats (so-called nonbuffered areas) rather than species that live in

FIGURE 14.26. Migratory behavior in Motacillidae birds. Migratory behavior has been studied in a phylogenetic context in the avian family Motacillidae, which includes such species as **(A)** the pied wagtail (*Motacilla alba*), **(B)** the golden pipit (*Tmetothylacus tenellus*), and **(C)** the yellow-throated longclaw (*Macronyx croceus*). (Photo credits: David Chapman/ Alamy; Tina Manley/Animals, Animals/ Alamy; Arco Images GmbH/Alamy)

A

B

C

forests (buffered areas; Chesser and Levey, 1998; Levey and Stiles, 1992). The underlying logic here is that open and edge habitats exhibit much greater seasonal variation in food resources than do forest habitats, and that this variation might select for migration in the birds that occupy such open and edge habitats.

To test the evolutionary precursor model in a phylogenetic context, Outlaw and Voelker categorized the migratory behavior of forty-nine species in the Motacillidae family as either "migratory" or "sedentary," and their habitat as either open/edge or forest. Using an already published molecular genetic phylogeny of Motacillidae, they found that there was no association between migration and habitat in terms of open/edge versus forest—that is, species that were associated with open/edge habitats were no more likely to migrate than were species that lived in the forest.

Outlaw and Voelker's phylogenetic analysis suggests that the ancestral state of migration in Motacillidae was "sedentary" and that the ancestral habitat was likely open/edge (Figure 14.27). Migratory behavior then evolved independently in many species in Motacillidae, but these species were just as likely to live in

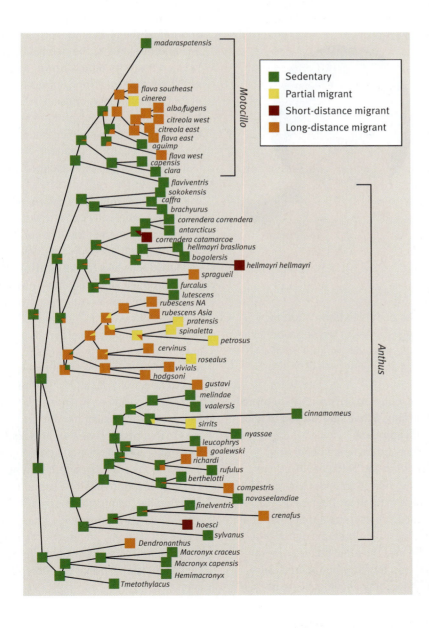

FIGURE 14.27. Phylogeny of migration. A reconstructed phylogeny of migration in Motacillidae birds. Shading at the terminal tips (where there are no branches coming off that taxon) indicates the state of the extant taxon. On other nodes, the portion of color in the square indicate the probabilities of ancestral states. The most primitive state is "sedentary." A similar phylogenetic reconstruction of habitat (not shown here) found open/edge habitats to be the most primitive state on the tree. (*From Outlaw and Voelker, 2006*)

INTERVIEW WITH

Dr. Judy Stamps

Of all the systems you could have chosen to study territoriality, how did you end up working with small lizards?

I originally began studying social behavior in lizards because of a comment made by my professor in an undergraduate animal behavior class. He suggested that lizards might be a good choice for studying stereotyped behavior patterns (displays) because many lizards use stereotyped behavior—head-bobbing patterns—to communicate with each other. So I began by studying the displays of male *Anolis* lizards in captivity. After several years of studying lizards in the laboratory, I became curious about their behavior under natural conditions and made a preliminary field trip to their native habitat, in the West Indies, to observe their behavior there. This first trip not only convinced me that these lizards were doing some pretty interesting things in nature but also that the place they were doing them (the West Indian island of Grenada) had obvious attractions of its own. So my first fieldwork focused on the social behavior of adult lizards in the West Indies.

While observing adults in the field, I noticed that tiny hatchlings exhibited many of the same behavior patterns as adults, including head-bob displays, chasing, fighting, and the defense of territories. In addition, since territory size typically scales with body size, hatchlings that weighed only 2 grams had commensurately small territories, on the order of only 0.5 square meter in size. As a result, it was possible to

study an entire "neighborhood" of territories in a small area in the field and to manipulate structural features of habitats under natural conditions to determine the features that free-living juveniles preferred in their territories. After several months of playing around with juvenile lizards, it was obvious that they were much more amenable to both laboratory and field studies of territorial behavior than adults, and from that point on, I focused on the juveniles.

You've modeled territoriality in terms of learning. What spurred you to take on this approach?

Many hours spent observing juveniles in the field convinced me that I needed to consider the role of learning in territorial behavior. For instance, it was quite apparent that a juvenile who entered a neighborhood for the first time had no idea where any of the territory boundaries were, and that it only became aware of their

locations as a result of being chased from one territory to the next by the residents. Similarly, two juveniles of similar size meeting for the first time engaged in an extended, strenuous fight, involving lengthy exchanges of a wide array of head-bob patterns, and culminating in physical combat, in which the two parties grabbed one another by the jaws, and then banged one another against the perch. It seemed that this "knock down, drag out" fight made a strong impression on both combatants, because when they encountered each other again the next morning, both seemed reluctant to venture near that opponent, and instead they began to exchange head-bob displays from adjacent perches, across an intervening "no-man's-land" that eventually became the border between their two territories. Other observations indicated that if an individual was summarily attacked the first time it ventured into a novel area, it almost never returned to that area. In contrast, if an individual received a comparable attack in an area it had used for an extended period of time, its first response was to flee, but typically it returned to the area within an hour or so, and persisted in using its familiar area in the face of repeated attacks by its opponent. These and other observations suggested that salient experiences, both positive (a snooze on a familiar, comfortable perch with a full stomach) and negative (being attacked by an opponent), had important effects on the subsequent social and spatial behavior of these animals.

Inspection of the psychology literature revealed that scientists studying the effects of positive and aversive stimuli on space use had already described very general behavioral phenomena that could account for the behavior of these lizards. For instance, rodents who receive aversive stimuli (for example, electric shock) as soon as they first venture into an unfamiliar area are unlikely to return to that area, whereas comparable individuals who receive the same stimulus in an area in which they have previously been rewarded are much more likely to return to that area. Translated into territorial terms, these studies suggest that territorial animals may be behaving like little psychologists, delivering punishment to one another in an attempt to dissuade members of their own species from using a particular area.

Does being bigger usually mean you are guaranteed a good territory? If not, what else plays a role in territory acquisition?

In some situations, large body size can be helpful in acquiring a good territory, because if two individuals are competing for an area that is novel to both of them, and if one is larger than the other, the larger one is capable of delivering more punishment to its opponent during aggressive interactions than vice versa. However, size is not everything, because individuals who are already in possession of an area typically retain possession of that area, even when competing with larger opponents, a phenomenon called the "prior residency advantage." The prior residency advantage may be attributable to the fact that a resident knows more about an area

than does a newcomer, as a result of which the resident is likely to fight more vigorously for that area, and persistently return to that area even after losing one or more fights to a larger newcomer.

Regardless of the reason, the prior residency advantage means that animals can acquire space by being the first to settle in an area, rather than by being bigger than their opponents. This means that in many species, the ability to find newly vacant territories, and the ability to settle in them before anyone else finds them, plays an important role in territory acquisition. Thus, the prior residency advantage may explain why some birds establish territories weeks

> **A resident knows more about an area than does a newcomer, as a result of which the resident is likely to fight more vigorously for that area, and persistently return to that area even after losing one or more fights to a larger newcomer.**

or months in advance of the time that the territory will be needed for reproduction: The early bird gets the territory.

Are there any general patterns in territoriality that emerge when you look across taxa?

There are a number of general patterns that occur in territorial species from a wide range of taxa. I have already mentioned the prior

residency advantage, which occurs in virtually all territorial animals. Another generalization is that individuals in territorial species exhibit site tenacity, meaning that a given individual tends to remain in the same area for an extended period of time. Site tenacity is observed in all territorial species, but the reverse is not the case (for example, all species with site tenacity do not defend areas in which they live). The basic elements of territorial behavior (stay in an area and use aggressive behavior to discourage other individuals from remaining in or returning to that area) are exhibited by a very large array of species, ranging from sea anemones to primates, but taxa with a long evolutionary history of territoriality typically add certain refinements to this basic pattern. For instance, territory owners in birds, mammals, frogs, lizards, and insects may produce songs, olfactory signals, conspicuous visual displays, or other conspicuous signals using other sensory modalities. These broadcast signals indicate to other inhabitants of the area that the territory is occupied, as well as providing additional information about the sex, condition, and (sometimes) identity of the territory owner. Territory owners in species with a long evolutionary history of territoriality are also likely to be able to recognize different categories of conspecifics, and tailor their aggressive behavior accordingly.

Dr. Judy Stamps is a professor at the University of California at Davis. Her long-term work on lizards and territoriality has produced fundamentally new ideas on both territoriality and the role of learning in territory formation.

the forest as in open/edge habitats. Outlaw and Voelker's analysis did, however, find one ecological variable that was associated with migration. Species that lived at higher altitudes were much more likely to migrate than species that lived at lower altitudes. Overall, their analysis found numerous independent evolutionary transitions from sedentary to migratory and that these transitions were most likely in species that lived at high latitudes.

SUMMARY

1. Both biotic and abiotic factors affect habitat choice in animals. Abiotic factors include but are not limited to heat, availability of water, wind, refuges from danger, and the availability of specific nutrients, whereas biotic factors include the location of potential mates, food, predators, and parasites.

2. Territories provide exclusive access to food, mates, and safe haven from predators, and are defended from intruders. Defense can be costly, in terms of both time and energy.

3. The ideal free distribution model examines habitat choice as a function of resource distribution. This model predicts that the equilibrium distribution of individuals into patches (or habitats) should be that distribution at which, if any individual moved to a patch (or habitat) it was not in, it would suffer a reduced payoff.

4. Animals use the presence of cues about parasitization when judging the suitability of a habitat.

5. The proximate cues underlying memory and habitat choice and migration are being investigated by ethologists.

6. Animals can learn about territory quality through a process known as conspecific cueing, wherein they judge the suitability of a territory by whether it previously was occupied by an owner sometime in the past.

7. Ethologists studying migration focus not only on the obvious costs (for example, the energy needed for travel) and benefits (breeding in a warmer climate) of migration but also on more subtle factors such as the immunological costs of being a migrant, the various cues that migrants use to locate their breeding and nonbreeding grounds, and the interaction between genetic and environmental factors in shaping migratory behavior.

8. Phylogenetic studies of migration have tested the "evolutionary precursor" model of migration, which posits that migration will be associated with species that live in open or edge habitats rather than species that live in forests.

DISCUSSION QUESTIONS

1. Make a general list of the costs and benefits of territoriality. Using that list, determine what sort of environments would generally favor the formation of long-term territories.

2. The ideal free distribution model predicts that animals will distribute themselves among patches in proportion to resources. What sort of cognitive abilities, if any, does this assume on the part of animals? Would bacteria potentially distribute themselves in accordance with the predictions of IFD models?

3. Suppose that young individuals watch older conspecifics choose their territories and subsequently use such information in their habitat-choice decisions about valuable resources. Outline one scenario by which such observational learning could increase competition for prime habitat sites, and one in which it would decrease competition for such sites.

4. Consider Møller and Erritzøe's work on immune defense organs and migration behavior. Can you make any predictions regarding how a migrating species might fare against local parasites (in both its habitats) as compared to resident species? What is the logic underlying your hypothesis?

SUGGESTED READING

Brown, J. L. (1969). Territorial behavior and population regulation in birds. *Wilson Bulletin, 81*, 293–329. A classic paper that examines the effects of territoriality on population dynamics.

Gould, J. L., & Gould, C. C. (2012). *Nature's compass: The mystery of animal migration*. Princeton, NJ: Princeton University Press. An overview of migration, written in a reader-friendly manner.

Milinski, M., & Parker, G. (1991). Competition for resources. In J. R. Krebs & N. B. Davies (Eds.), *Behavioural ecology: An evolutionary approach* (pp. 137–168). Oxford: Blackwell. A nice overview of all variations of the ideal free distribution model.

Reppert, S., Gegear, R., & Merlin, C. (2010). Navigational mechanisms of migrating monarch butterflies. *Trends in Neurosciences, 33*, 399–406. A current review of the proximate underpinnings of monarch butterfly migration.

Stamps, J. (2001). Learning from lizards. In L. A. Dugatkin (Ed.), M*odel systems in behavioral ecology* (pp. 149–168). Princeton, NJ: Princeton University Press. A review of learning and territory establishment.

15

Aggression

One of the most vivid of all the popular depictions of animal behavior is two rams facing one another, putting their heads down, charging, and butting horns. During the mating season, this sort of fighting behavior is common, and winners are more likely to mate with females. But conspecific aggression—that is, aggression between members of the same species—is not restricted to this sort of dramatic display, nor is it always associated with immediate access to mating opportunities. Animals fight over obtaining food, securing space (territories, home ranges), and providing safety for their family members.

When valuable resources such as space, food, and mates are at stake, animals can be aggressive. The array of weapons that have evolved for use in fights is nothing short of astonishing. From extinct Trilobite arthropods exquisitely preserved in the fossil record to modern arthropods, fish, mammals, and reptiles, ethologists and paleontologists have found an incredible array of physical body parts that are used as weapons (Figure 15.1). Most often these weapons are found on males, and they tend to be associated with the systems in which males can defend some resource that is spatially restricted. Weapons are usually honest indicators of male size and fighting ability (D. J. Emlen, 2008, chapter 7).

Aggression, also called agonistic behavior, occurs when animals either send threating signals (e.g., an animal flashes its canine teeth) and/or engage in some sort of physical combat. Most aggressive behavior studied by ethologists is conspecific aggression, though aggression between members of different species has also been investigated. Ethologists generally do not consider predator-prey interactions as aggression, though there is some debate on this matter. In this chapter, we focus on conspecific aggression. Even with respect to conspecific aggression, the ethological literature is immense, and we can only touch on the tip of the iceberg, focusing on major theoretical and empirical questions.

Aggression has been studied in relatively solitary species in which individuals fight when they occasionally interact at a territory boundary, as well as in species that live in social groups year-round. When aggression occurs in group-living species, and individuals interact with each other many times, we can measure **dominance hierarchies**—rank orderings of the individuals based on the results of pairwise aggressive interactions—in such groups. Where an individual places in a dominance hierarchy can be studied from both proximate and ultimate perspectives. Individuals at the top of hierarchies often have access to more food, more mating opportunities, and safer territories than individuals at the lower end of hierarchies. At the proximate level, ethologists might study how androgen and glucocorticoid levels differ between individuals that hold different ranks in a hierarchy.

Ethologists and naturalists have long been fascinated with animal aggression, and the literature in this area is huge, going back to at least the time of Darwin (Archer, 1988; Drummond, 2006; Huntingford and Turner, 1987; Mock and Parker, 1997). Thomas Henry Huxley, one of the leading intellectual figures of the nineteenth century and Darwin's most vociferous defender, argued that interactions in the animal world resulted in a bloodbath. Huxley thought that aggression, often extreme

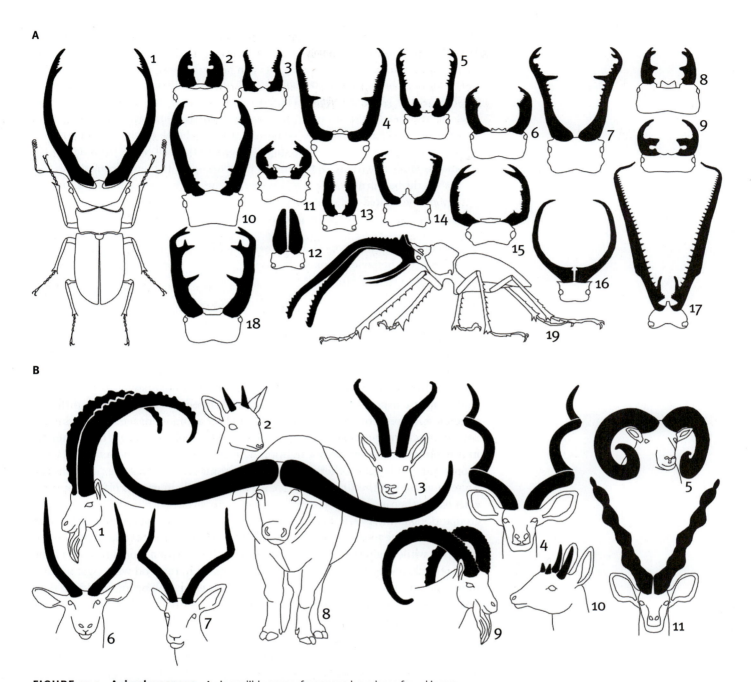

FIGURE 15.1. Animal weapons. An incredible array of weapons have been found in many groups, including **(A)** modern stag beetles: 1. *Cyclommatus elaphus*, 2. *Odontolabis latipennis*, 3. *Prosopocoilus serricornis*, 4. *Hexarthrius mandibularis*, 5. *P. bison*, 6. *Dorcus titanus*, 7. *P. giraffa*, 8. *D. alcides*, 9. *Aegus punctipennis*, 10. *C. giraffa*, 11. *Mesotopus tarandus*, 12. *Colophon primosi*, 13. *P. sericeus*, 14. *Weinreichius perroti*, 15. *Rhaetulus speciosus*, 16. *Sphaenognathus feisthameli*, 17. *Chiasognathus grantii*, 18. *O. femoralis*, 19. *Chiasognathus grantii* (side view); and **(B)** modern Bovids: from left to right. 1. Spanish ibex (*Capra pyrenaica*), 2. dik-dik (*Madoqua kirkii*), 3. Grant's gazelle (*Gazella granti*), 4. kudu (*Tragelaphus strepsiceros*), 5. bighorn sheep (*Ovis canadensis*), 6. waterbuck (*Kobus ellipsiprymnus*), 7. impala (*Aepyceros melampus*), 8. long-horned African buffalo (*Pelorovis antiquus**), 9. Asiatic ibex (*Capra ibex*), 10. chowsingha (*Tetracerus quadricornis*), 11. markhor (*Capra falconeri*). *Asterisks denote extinct species. (From Emlen, 2008)*

aggression, was the norm for animals, and he made his claim in no uncertain terms:

> From the point of view of the moralist, the animal world is on about the same level as the gladiator's show. The creatures are fairly well treated, and set to fight; whereby the strongest, the swiftest and the cunningest live to fight another day. The spectator has no need to turn his thumb down, as no quarter is given . . . the weakest and the stupidest went to the wall, while the toughest and the shrewdest, those who were best fitted to cope with their circumstances, but not the best in any other way, survived. Life was a continuous free fight, and beyond the limited and temporary relations of the family, the Hobbesian war of each against all was the normal state of existence. (T. H. Huxley, 1888, p. 163)

In contrast, Peter Kropotkin, whose book *Mutual Aid* is regarded as a classic early work on behavior, looked around and saw a world that seemed almost antithetical to that of Huxley's:

> . . . [I]n all these scenes of animal life which crossed before my eyes, I saw mutual aid and mutual support carried on to an extent which made me suspect in it a feature of the greatest importance for the maintenance of life, the preservation of each species and its further evolution. (Kropotkin, 1902, p. 18)

To some extent and under certain circumstances, toned-down versions of both views capture aspects of animal behavior at different times and under different conditions. At present, rather than asking whether animals engage in aggression or cooperation, ethologists are interested in the costs and benefits that favor animals in a population fighting or not fighting with one another, or cooperating or not cooperating with each other. It is also important to note that cooperation and aggression are not the flip sides of a coin—individuals in groups often cooperate with each other in order to compete, often aggressively, with individuals in other groups (Dugatkin, 1997a; Gadagkar, 1997). For example, aggression is part and parcel of even the most cooperative of units—the social insect group. In wasps, for example, while cooperation is common within hives, when "foreign" individuals from other hives try to enter a nest, wasp guards often respond with vigorous and sometimes deadly defensive behaviors (Gamboa et al., 1986; Figure 15.2). And this is not unique to the social insects. In green woodhoopoe birds (*Phoeniculus*

FIGURE 15.2. Intruder aggression. When a wasp (left) approaches a nest, guards at the nest determine whether it's a hive mate or an intruder. Intruders are aggressively repelled.

purpureus), individuals increase the amount of grooming they dispense to *others within their group* when they enter areas where conflicts with another group of woodhoopoes is likely. The amount of self-grooming, which is an indicator of stress, did not increase in these situations (Radford, 2003, 2008, 2011). Such grooming may solidify bonds within groups, leading to an increased probability of victory should between-group hostility increase.

This chapter will focus on (1) some proximate mechanisms of aggression; (2) models of aggression, including the hawk-dove game, the sequential assessment model, and the war of attrition model; and (3) the implications of experience, particularly social experience, on aggressive interactions and hierarchy formation, including the effects of winning and losing in one aggressive interaction on subsequent interactions, and the effects of observing or being observed on aggressive bouts.

Fight or Flight?

In the "fight or flight" response (Chapter 3), a surge in adrenaline and norepinephrine produces a quick increase in blood sugar, which, along with oxygen, is delivered to strategic areas such as the brain, skeletal muscles, and heart. Systems such as the digestive and reproductive systems are often temporarily shut down at this point. This response most often occurs when prey must decide to fight off a predator or flee. The same sort of response—albeit not as dramatic—occurs when animals encounter aggressive individuals from their own populations. An individual can choose to fight or flee from a potential antagonistic opponent.

We can think about the decision whether or not to fight against conspecifics at a number of different levels. From an ultimate perspective, this decision revolves around the costs and benefits of aggression. When the benefits of victory, on average, outweigh the costs of fighting, natural selection will favor aggressive behaviors, and ethologists predict that animals will fight; otherwise, they should not (see the Conservation Connection box). From a proximate perspective, we can ask questions about immediate causation and aggression. And, as in the case of responding to predators, when animals opt to fight or flee from others in their own population, ethologists often focus on the endocrinological underpinnings of such behavior and have found similar hormonal responses across different species. Dominant individuals typically display increased androgen levels—for example, increased testosterone—and are more likely to fight than flee. Likewise, winners of fights tend to have their circulating levels of androgens increase— that is, the relationship between androgen level and aggression works both ways. Subordinates, that are more likely than dominants to flee than fight, usually lose aggressive interactions in which they are involved. Subordinates tend to have higher circulating levels of glucocorticoid stress hormones such as cortisol or corticosterone than dominants when going into fights, and even higher levels when a fight is over. But it is not always the case that subordinates and losers are the only individuals that display high levels of circulating stress hormones. For example, dominant individuals defending territories often must expend a large amount of energy fending off competitors leading to high glucocorticoid levels. Similarly, high-ranking individuals in a hierarchy are often challenged by many subordinates in their group, and this too can lead to increases in glucocorticoid levels.

Breeding Programs Can Lead to More Aggressive Animals

Reintroduction plans for endangered species sometimes include captive breeding programs in which animals are bred for many generations in a controlled environment. These breeding programs are especially common in fish, because it is relatively easy to maintain breeding populations of fish bound for reintroduction for many generations. The goal of many of these breeding programs is to both minimize inbreeding and to increase population size to a large enough number that endangered animals can be introduced back in their natural environments. But breeding programs themselves introduce new selection pressures on animals by radically changing the environment from one that animals would experience in the wild. The effects of these new selection pressures on traits like aggressive behavior are being investigated by both ethologists and conservation biologists.

Jennifer Kelley and her colleagues designed an experiment to examine whether breeding programs change the level of aggression in animals compared with that found in natural populations—and if so, how (Kelley et al., 2006). The researchers compared the aggressive behaviors seen in a natural population of the endangered butterfly splitfin fish (*Ameca splendens*) from El Rincon, Mexico, with the aggressive behaviors observed in a captive-bred population that is housed in the London Zoo (Figure 15.3).

Kelley and colleagues first observed the aggressive behavior of males (that can be individually recognized by their yellow and black color bands) to determine whether the general repertoire of aggressive behaviors was similar in the natural population and the captive-bred population they were studying. In general, these behaviors were similar, though some of the captive populations

FIGURE 15.3. Captive breeding and increased aggression. Butterfly splitfin fish have bred in captivity, and this breeding program has inadvertently produced more aggressive individuals. *(Photo credit: blickwink/Alamy)*

displayed territoriality that was not seen in natural populations. Kelley's team next compared the aggressive behavior of males from both populations in a more controlled environment. They exposed males from both the natural and captive-bred population to a structured tank environment containing gravel, aquatic plants, and small shelters or to an unstructured (bare) tank environment. In addition, the number of males in each tank was manipulated to create "high-density" and "low-density" treatments in both the structured and unstructured populations.

Overall, captive-bred males showed much higher levels of aggression than males from the natural population. This increased aggression in captive-bred males was greatest when fish were in high-density populations in the structured environment. Although the experimental protocol used could not disentangle genetic changes underlying increased aggression from the effect of experiences (being raised in either a natural or captive-bred population), it is likely that

both play a role in the differences found (Ruzzante, 1994).

Kelley's work has implications for reintroduction programs in the butterfly splitfin fish and for reintroduction programs in general. In the case of the butterfly splitfin fish, because natural environments will be closer to the structured environment in Kelley and her colleagues' experiment, a reintroduction of captive-bred males into a natural environment could lead to high levels of aggression in the reintroduced population. The fact that butterfly splitfin males are aggressive toward one another does not necessarily mean that they would also be more aggressive toward fish in other species. But if they are, this increased aggression could restructure interspecific interactions.

In general, reintroduction programs based on captive population breeding need to take into account the way that captive breeding programs may inadvertently select for more aggressive animals, and they must consider the implications of introducing such individuals back into natural populations.

If fighting is costly, then once it is clear that an animal is losing a fight, it will often be beneficial for it to signal subordination and reduce future costs (Enquist and Leimar, 1990; Geist, 1974b; Hurd, 1997). It may inhibit its aggressive behaviors, refraining from charging, biting, or rapidly approaching the dominant animal. In addition, it may communicate its subordinate status via color change. Color change may be a particularly good communication vehicle in aggressive contests, because color change can quickly indicate an individual's relative rank in a hierarchy and whether it will engage in aggressive behaviors (S. Rohwer, 1982; T. J. Roper, 1986).

Because fish have tight hormonal and neuronal control over the expansion of pigment cells, they are particularly adept at quick color change over short time periods and have been the subject of numerous experiments on color change and aggression (Baerends et al., 1986; D. M. Guthrie and Muntz, 1993; Rhodes and Schlupp, 2012). For example, researchers have studied Atlantic salmon (*Salmo salar*), a species in which aggression is most often associated with territorial defense. During fights, males go through a series of behaviors, including circling, charging, and biting. Losers, who often have increased levels of cortisol, swim close to the surface to avoid future aggressive interactions with individuals that have defeated them. In terms of signaling relative status in a dominance hierarchy in Atlantic salmon, dominant males develop dark vertical eye bands and subordinate individuals develop darker body color (Keenleyside and Yamamoto, 1962; Figure 15.4). In other species, such as the swordtail (*Xiphoporus helleri*), males display a red lateral stripe when dominant, but a black stripe when subordinate (Rhodes and Schulpp, 2012).

Two caveats are in order before we move to the next section. First, many hormones, not just androgens and glucocorticoids, affect decisions about whether or not to fight. Second, while androgens and glucocorticoids tend to play the same role in aggression and submission across many species, not all chemical messengers associated with aggression and submission have similar effects across species. To see this point more clearly, let us briefly examine the role of the neurotransmitter serotonin in aggressive behavior.

Much research has found that serotonin plays an important role in animal aggression. That role, however, appears to differ across animal systems. In mammals, low serotonin levels are often linked with high levels of aggression but lower social status (Coccaro, 1992; Raleigh et al., 1991; Sheard, 1983). The

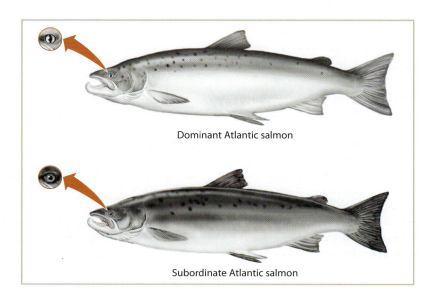

Dominant Atlantic salmon

Subordinate Atlantic salmon

FIGURE 15.4. Color as a signal. In Atlantic salmon, subordinate individuals often assume a much darker body color. Dominants' body color remains light, but they develop dark vertical eye bands.

situation can be complex, however, as the effect of serotonin (and its chemical precursors) may depend on whether animals were raised in social or asocial environments and on what specific type of aggressive behavior is being studied (Balaban et al., 1996; Raleigh and McGuire, 1991).

In fish, as opposed to mammals, enhanced serotonergic function is seen in more subordinate individuals, leading to reduced fighting behavior (Clotfelter et al., 2007; Winberg et al., 1992, 1997). For example, Svante Winberg and his colleagues studied brain serotonergic activity and hierarchy formation in Arctic charr (*Salvelinus alpinus*). In phase 1 of their work they constructed four groups, each consisting of four fish. They then determined the dominance hierarchy of the four fish in each group. After a stable hierarchy had formed in each group, groups were disbanded. In phase 2, Winberg and his team placed all four top-ranked males in one group, all four second-ranked males in a second group, all four third-ranked males in a third group, and all four bottom-ranked males in a fourth group and allowed new hierarchies to form. Correlating serotonin and serotonin-related chemicals with social status, they found that fish that were subordinate in phase 2 showed higher serotonergic activity. The subordinate fish curtailed their aggressive activity, presumably to avoid aggressive interactions with dominant fish in their new group. The researchers suggested that the stress experienced by the subordinate fish in relation to the dominant fish increased the subordinates' serotonergic activity, which in turn inhibited the neural circuitry associated with aggression.

A different picture emerges when we look at the relationship between serotonin, aggression, and social status in crustaceans (D. H. Edwards and Spitzer, 2006; Huber et al., 1997, 2011; Kravitz, 1988). In crustaceans, increased serotonergic function leads to enhanced aggression and high social status. When lobsters are paired up in fights, they generally escalate their aggressive behaviors through a series of ritualized combats (Huber and Kravitz, 1995). Once an individual loses a fight, however, it avoids aggressive interactions for days. But losers can be made more aggressive if they are given injections of serotonin (Huber et al., 1997; Figure 15.5). Furthermore, if fluoxetine (Prozac), an inhibitor of serotonin, is injected into the lobster at the same time as serotonin, this effect disappears, suggesting an important role for serotonin in lobster aggression.

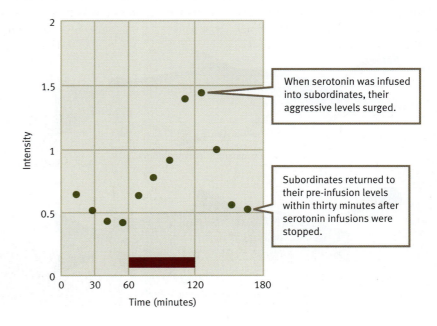

FIGURE 15.5. Serotonin and aggression. A small lobster was made subordinate by being matched against an individual that was 30 percent larger than it was, and then serotonin was continuously infused into subordinates (red bar). The intensity of the aggression over time is shown here. *(Based on Huber et al., 1997)*

When serotonin was infused into subordinates, their aggressive levels surged.

Subordinates returned to their pre-infusion levels within thirty minutes after serotonin infusions were stopped.

While serotonin appears to play an important role in aggression across this wide spectrum—crustaceans, fish, and mammals—the neuroendocrinological effect of serotonin differs dramatically across these groups.

Game Theory Models of Aggression

As discussed in Chapter 10, game theory models are built to examine behavioral evolution when the fitness of an individual depends on both its own behavior and on the behavior of others, as in the case of fighting behavior. If, for example, an individual is willing to fight for a contested resource, its fitness is going to depend on whether its opponent opts to fight or to flee, so we can model fighting using game theory.

Evolutionary game theorists have built a suite of models that examine the evolution of fighting behavior (Riechert, 1998). Here we will focus on the three best-developed game theory models of aggression—the hawk-dove game, the war of attrition model, and the sequential assessment model. All three of these game theory models share certain attributes. For example, a cost to fighting is assumed in hawk-dove, war of attrition, and sequential assessment models. Such a cost can take many different forms, ranging from opportunity costs—the cost associated with not doing something else—to the cost of physical injury, up to and including mortality costs (Enquist and Leimar, 1990).

All game theory models must have some sort of variable that represents the value of the resource being contested. In some cases, the value of a resource will be fairly easy to estimate. When two individuals, for example, are contesting an item of food, the value of the resource will be straightforward to calculate. In other cases, such as a male's access to reproductively active females, resource value can be much more difficult to calculate. The net value of a resource affects not only an animal's decision to fight but also how long and/or how hard it is willing to fight (Figure 15.6).

Two individuals contesting a resource may not assign the same value to that resource. For example, imagine that two animals—one of which is starving, and the other of which is hungry, but not starving—are contesting a one-pound prey item that has just been discovered. One pound may be *valued* differently by

FIGURE 15.6. Deciding to fight.
(A) One of the many resources animals will fight over is food, as shown here by these vultures that are fighting over a carcass. **(B)** Males also fight over females. Here, male elephant seals are fighting over access to reproductively active females. *(Photo credits: Tom Vezo; Jeff Mondragon/ mondragonphoto.com)*

A

B

FIGURE 15.7. Value estimation. When two animals contest a resource, the hungrier animal may fight harder or longer to obtain it. If one of these cats is hungrier than the other, it may be willing to risk more to obtain the remains of the fish.

putative fighters. To a starving animal, it might make the difference between life and death, while to a less hungry animal the value of that pound of meat might be much lower. In that case, the starving animal might fight harder for the food than its opponent does (Figure 15.7). As another example, consider the value of territory to a potential intruder and to a territory holder. Above and beyond the fact that the territory holder might be on a territory because it is a good fighter to begin with, it may well be that a territory holder will value a contested area (that is, its territory) more because it has already invested time and energy in learning where the resources in such a territory are located (Kokko et al., 2006). These sorts of asymmetries in value have been documented many times in the ethological literature (Barnard and Brown, 1984; Elias et al., 2010; Ewald, 1985; Mohamad et al., 2010; Tibbetts, 2008; Verrell, 1986).

THE HAWK-DOVE GAME

The earliest and best-known game theory model of the evolution of aggression is the Maynard Smith and Price's **hawk-dove game** (Maynard Smith, 1982; Maynard Smith and Price, 1973). The simplest form of the hawk-dove game is meant to be a heuristic model. It is meant to spur researchers to think about aggression in a new way. We will first walk through the hawk-dove game, and then we will consider variants of the game that have been tested by ethologists.

Imagine that individuals can adopt one of two behavioral strategies when contesting a resource: (1) hawk, in which a player escalates and continues to escalate until either it is injured or its opponent cedes the resource; or (2) dove, in which a player displays as if it will escalate, but retreats and cedes the resource if its opponent escalates (this strategy was originally labeled "mouse").

If we let V = the value of the contested resource and C = the cost of fighting, we can fill in the potential payoff matrix for the hawk-dove game (Table 15.1). This matrix shows that if a hawk interacts with a dove, the hawk receives the entire value of the resource (V), while the dove receives nothing. To understand the two remaining payoffs—the hawk-hawk and the dove-dove payoffs, we need to be aware of two assumptions underlying the payoff matrix. First, we are assuming that if two doves encounter each other, on average, they receive half the value of the resource (for example, they split a food item in half). Second, when two hawks interact and fight, we are assuming that only the loser pays the cost of fighting (for example, the cost of being injured). This means that hawks have a 50 percent chance of obtaining the resource (V/2) and a 50 percent chance of being injured and not receiving the resource (C/2), so their total expected payoff from an interaction with another hawk is equal to (V − C)/2.

The evolutionarily stable strategy (ESS; see Chapter 10) to this game depends on whether the value of the resource or the cost of fighting is greater. If V > C, then hawk is an ESS, as the hawk-hawk payoff, (V − C)/2, is positive and thus greater than the payoff dove obtains when it meets hawk (0). Dove, however, is not an ESS, as the dove-dove payoff (V/2) is less than the payoff hawk obtains when it meets dove (V). Therefore, when V > C, hawk is the only ESS. This makes sense, since V > C implies that the cost of fighting, paid only by hawks, is low compared with the prize that awaits the winner of any contest (V), and as a result, hawks do well. The situation is a bit more complicated when the cost of fighting is greater than the value of the resource. Now, neither hawk nor dove is an ESS. But it can be shown that some mixture of hawks and doves is an evolutionarily stable strategy in this case (Math Box 15.1).

There are more complicated varieties of the hawk-dove game than the one examined above (Mesterton-Gibbons and Adams, 1998; Riechert, 1998). One interesting version of this game adds in two new strategies called bourgeois and antibourgeois. Using bourgeois strategy, an individual plays hawk if it is a territory holder, but, plays dove if it does not own a territory. The antibourgeois strategy is just the opposite—play dove if you are a territory holder, and hawk if you are not. Below we will look at an example of the bourgeois strategy in butterflies and the antibourgeois strategy in spiders.

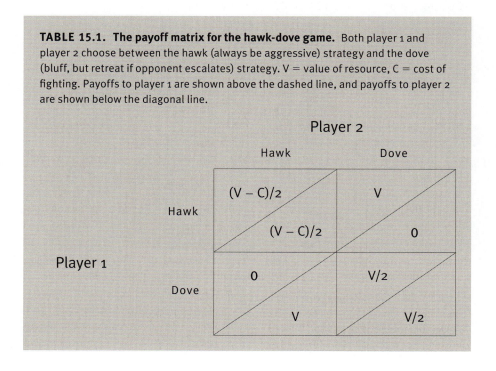

TABLE 15.1. The payoff matrix for the hawk-dove game. Both player 1 and player 2 choose between the hawk (always be aggressive) strategy and the dove (bluff, but retreat if opponent escalates) strategy. V = value of resource, C = cost of fighting. Payoffs to player 1 are shown above the dashed line, and payoffs to player 2 are shown below the diagonal line.

BOURGEOIS BUTTERFLIES. In the speckled wood butterfly (*Pararge aegeria*), territories are not set in space. That is, rather than having a territory with a set place in three dimensions, a male has a territory that is an open patch delineated by well-lit areas that emerge when the sun breaks through the clouds. When a male comes upon an empty well-lit patch, he immediately occupies it and by doing so secures a mating advantage compared with males not in sunlit territories.

When a male speckled wood butterfly comes upon a territory that has another male in it, a contest involving exaggerated spiral flights upward that sometimes include physical combat ensues and is settled as follows: resident wins, intruder retreats. In fact, there is very little aggression at all when males come upon occupied territories, perhaps because a prolonged fight over a short-lived resource, such as a sun patch, is not worth the costs. Rather, once a male is aware that a territory is occupied, he simply leaves (Figure 15.8). What makes the "resident wins" rule so dramatic is that an individual need only be the resident of a sun patch for a few seconds to secure victory over an intruder. In a study by Nick Davies, a male (let's call him M1) was experimentally made a territory owner (Davies, 1978). M1 then always defeated M2, an intruder male. Yet if M1 was removed from his territory and M2 then occupied it, even if M2 was resident for only a short time, M1 would now defer to M2 when he was reintroduced into his original sun patch. The only escalated contests occurred when two males both acted as if they were the rightful owner of a sun patch. This happens naturally when two butterflies come to a sun patch at about the same time but do not notice the other individual for some period (this can

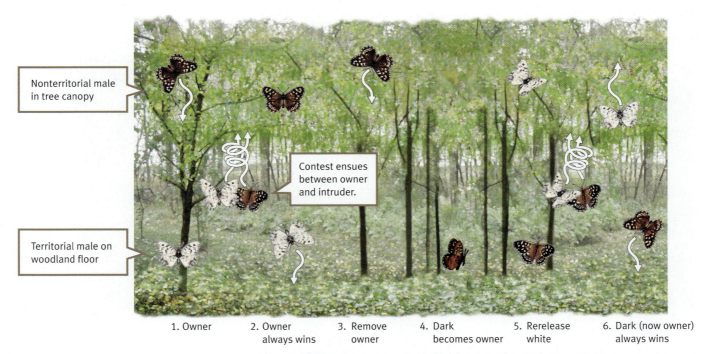

Nonterritorial male
in tree canopy

Contest ensues
between owner
and intruder.

Territorial male on
woodland floor

1. Owner 2. Owner always wins 3. Remove owner 4. Dark becomes owner 5. Rerelease white 6. Dark (now owner) always wins

FIGURE 15.8. Conventional rules. The setup for Davies's experiments on territory ownership and contest rules in the speckled wood butterfly *(Pararge aegeria)*. *(Based on Dawkins and Krebs, 1978, p. 299)*

also be simulated experimentally). In such cases, real fights can and do occur between the speckled wood butterfly males (see Kemp and Wiklund, 2004, for an alternative view).

ANTIBOURGEOIS MEXICAN SPIDERS. The only well-documented instance of the use of the antibourgeois strategy is during territory establishment and home range defense in *Oecobius civitas*, a small, territorial Mexican spider (Burgess, 1976). These spiders establish their territories under rocks, and when an intruder approaches a territory, the territory holder flees rather than fights. The former territory holder then searches for a new territory, and if he comes upon such a territory, the territory owner presently *there* vacates. This produces a potentially long, bizarre series of events in which a territory owner leaves, then challenges a spider in possession of another territory, causing that territory owner to leave, and so on.

Given that in most species, territory holders win fights with an intruder, how can ethologists explain such a seemingly strange cascade of events in *O. civitas*? What factors might be responsible for the antibourgeois strategy in these Mexican spiders (Hodge and Uetz, 1995; Komers, 1989)? The answer can be understood in terms of the costs and benefits of fighting for territories, the number of available territories, and the cost of moving between territories. Imagine that the cost of fighting is relatively high, numerous territories are available, and movement between territories is not particularly dangerous. Under such circumstances, fleeing a territory when an intruder challenges—that is, playing the antibourgeois strategy—may in fact be the most profitable option available to individuals (Mesterton-Gibbons, 1992; Mesterton-Gibbons and Adams, 1998). These ecological and behavioral conditions (high cost of fights, many vacant territories) are what *O. civitas* experiences in nature (Figure 15.9).

FIGURE 15.9. Reverse bourgeois strategy. In *Oecobius civitas*, territory holders relinquish their territories when challenged by an intruder. This behavior can have cascading effects throughout a cluster of *O. civitas* territories. *(Based on Mesterton-Gibbons and Adams, 1998)*

THE WAR OF ATTRITION MODEL

Some agonistic encounters are settled by animals displaying aggressively to one another but not actually fighting, and the victor is the individual that displays the longest. To better understand when natural selection might favor such rules, theoriticians developed the **war of attrition model** (D. T. Bishop et al., 1978; Chatterjee et al., 2012; Hammerstein and Parker, 1982; Maynard Smith, 1974; Maynard Smith and Parker, 1976; G. A. Parker, 1974a; T. Uehara et al., 2007).

The war of attrition model of fighting behavior has three underlying assumptions: (1) individuals can choose to display aggressively for any duration of time; (2) display behavior is costly—the longer the display, the more energy expended; and (3) there are no clear cues such as size, territory possession, and so forth that contestants can use to settle a contest (Riechert, 1998). Let V = the value of the resource being contested, and define x as the length of a contest. The evolutionary stable strategy to the game is a distribution of contest lengths. More technically, the probability that a contest lasts x units of time is equal to $(2/V)e^{-2x/v}$ (G. A. Parker and Thompson, 1980). That is, rather than predicting a set time for aggressive display, the war of attrition model predicts that animals will choose randomly from a specific exponential distribution of contest lengths defined by the probability function $(2/V)e^{-2x/v}$. The model predicts that all contest lengths from this ESS function—any choice of display time from this function—lead to equal fitness gains to individuals.

This predicted distribution of contest lengths matches certain display durations in nature (Broom and Ruxton, 2003; Crowley et al., 1988; S. A. Field et al., 1998; G. A. Parker and Thompson, 1980; Stoewe et al., 2006). Recall the dungflies discussed in Chapter 8. Females arrive at fresh dung patches to lay eggs, and males aggregate at such patches for access to females. The question, in terms of the war of attrition model, is how long should a male stay at a given patch? If he stays too long, he will encounter fewer and fewer females with time. If he leaves too quickly, he will pay the cost of moving and may miss the opportunity to mate with females at the patch he left.

Using detailed measurements of the costs and benefits of dungfly mating, Parker and Thompson found that males' "stay times"—the time they remained at a patch—were exponentially distributed, as predicted from a war of attrition model (G. A. Parker and Thompson, 1980). When Parker calculated the mean time to find and move to a new patch from the dungfly fieldwork, he found that it was approximately four minutes (Curtsinger, 1986; G. A. Parker and Maynard Smith, 1987). When the travel time between patches is about four minutes, in accordance with the predictions of the war of attrition model, the stay times observed in the dungflies translate into approximately equal fitness for all males (Figure 15.10).

THE SEQUENTIAL ASSESSMENT MODEL

A third game theory model of aggression, developed by Magnus Enquist and Olof Leimer, is called the **sequential assessment model** (Enquist and Leimer, 1983, 1987, 1990; Enquist et al., 1990; Leimar and Enquist, 1984). What distinguishes the sequential assessment game from other models of aggression is that it is designed to analyze fights in which individuals continually assess one

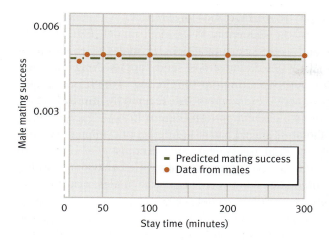

FIGURE 15.10. War of attrition over females. Male dungflies appear to engage in wars of attrition when determining how long to stay on a dung patch where females may alight. Assuming it takes four minutes to move from patch to patch, male mating success appears equal for a wide range of "stay times." *(From Maynard Smith, 1982, p. 31)*

another in a series of "bouts" (Arnott & Elwood, 2009; Hurd & Enquist, 2005). Let us examine how this model works, what predictions it makes, and what experimental evidence is available to test its basic predictions.

In the sequential assessment model, individuals assess their opponent's fighting abilities. Assessing an opponent's fighting ability, Enquist and his colleagues argue, is analogous to a process of statistical sampling (Enquist et al., 1990). A single sample—for example, a single assessment of fighting ability—introduces significant random error. The more sampling (assessment) one does, the lower the error rate, and hence the more confident one can be in whatever is being estimated—in our case, the opponent's fighting ability. After some period of sampling, one individual will eventually determine that its odds of winning a fight are so low that it should end any further aggressive interaction with its opponent.

The sequential assessment game examines contests in which the level of aggression varies from relatively mild to very dangerous. At the evolutionarily stable solution to the sequential assessment game, individuals should begin with the least dangerous type of aggressive behavior (let's call it behavior A) and "sample" each other with respect to behavior A for some period of time. Soon, however, all the information about an opponent with respect to A will be exhausted. At that point, the next most dangerous behavior (B) should be the most common behavior among protagonists. Again, at some point, information about an opponent and behavior B will reach saturation; that is, the animal will now have gathered as much useful information as it can gather about behavior B. Depending on the behavioral repertoire of the animals in question, more and more dangerous behaviors will be added to the sequence. Because gauging your probability of winning a fight is most difficult when your fighting ability is similar to your opponent's fighting ability, the sequential assessment model predicts that the more evenly matched opponents should engage in the more dangerous behaviors.

Studies testing various predictions of the sequential assessment game have produced mixed results. Most studies find partial support of the model's basic predictions (Brick, 1999; DiMarco and Hanlon, 1997; Hack, 1997; Jennions and Backwell, 1996; Jensen and Yngvesson, 1998; Koops and Grant, 1993; McMann, 1993; Molina-Borja et al., 1998; I. C. Smith et al., 1994). Other studies more definitively support the majority of the basic predictions of the sequential assessment game (Leimar et al., 1991). One such study examined contest behavior in the fish *Nannacara anomala*.

FIGURE 15.11. Testing models of aggression in fish. *Nannacara anomala* has proved to be an ideal species for testing various predictions derived from the sequential assessment game. *(Photo credit: Magnus Enquist)*

SEQUENTIAL ASSESSMENT IN *NANNACARA ANOMALA.* *N. anomala* is an ideal species in which to test the sequential assessment model (Figure 15.11). Males form hierarchies, and aggressive interactions in males of this species range from "changing color" (least dangerous) and "approaching" through "tail beating," biting and mouth wrestling, up to "circling," which is the most dangerous of the aggressive activities, wherein fish repeatedly attempt to bite each other while they swim in a circular pattern (Figure 15.12).

Enquist and his collaborators staged pairwise interactions among male fish to test predictions of the sequential assessment game (Enquist et al., 1990). All trials were videotaped to allow a detailed analysis of subtle behavioral changes through time. Overall, interactions between aggressive males matched the predictions of the sequential assessment model very well.

One of the most elementary predictions of the sequential assessment model is that the more evenly matched the opponents are, the longer the fights and the more phases a fight should go through. In *N. anomala,* as in many fish, individuals are able to assess weight asymmetries (Enquist et al., 1987), and weight differences appear to have a very large effect on contest outcome—heavier fish are more likely to emerge victorious from a contest. In accordance with the predictions of the sequential assessment model, fights do indeed take longer when fish are more closely matched for weight than they do when large asymmetries exist.

The sequential assessment model also predicts that when numerous behaviors are used in aggressive contests, they should be used in approximately the same order across all fights. While some fights are predicted to last longer and contain more elements than others, the order in which new aggressive behaviors appear in a fight should be similar in all contests, and shorter contests should simply have fewer types of behavior. Again, the behavior of *N. anomala* matches the model's prediction quite nicely. Males typically begin aggressive interactions with some sort of visual assessment, progress to tail beating, then to biting and mouth wrestling, and occasionally to circling. How many acts in this sequence are played out depends on differences in an opponent's weight, but as predicted by the sequential assessment model, the order in which these acts are displayed tends to be the same, regardless of weight differences.

Here two individuals are engaged in a "mouth wrestle."

FIGURE 15.12. Sequential assessment. Much work on the sequential assessment model has been done studying escalated fighting behavior in *Nannacara anomala*.

Winner, Loser, Bystander, and Audience Effects

Prior experience with an opponent can provide important information that can be used during future potentially aggressive interactions. If animal 1 has recently defeated animal 2, then animal 1 may decide to be more aggressive during the next interaction between these two individuals, and animal 2 may decide to be less aggressive. Indeed, experience with other individuals can be indirect, but still important. Suppose that animal 3 has recently seen animal 4 lose a fight to animal 5. If animals 3 and 4 interact at some point in the future, animal 3 may be more aggressive after seeing animal 4 lose the contest with animal 5. That is, observation, as well as direct interaction, provides information that may affect aggressive interactions. Ethologists have also found that *being* observed can affect the behavior displayed during, and the outcome of, aggressive inter-actions. Here we will examine the effect of direct and indirect experience by focusing on what are referred to as winner effects, loser effects, bystander effects, and audience effects.

WINNER AND LOSER EFFECTS

In sports, individual athletes, as well as teams, go on "winning streaks" and "losing streaks"—that is, winning itself sometimes leads to more winning, and losing to more losing. A similar phenomena exists in nonhumans, wherein winning an aggressive interaction increases the probability of future wins (so-called **winner effects**) and losing an aggressive interaction increases the probability of losing future fights (so-called **loser effects;** Hsu et al., 2005; Landau, 1951a, 1951b; Rutte et al., 2006).

Winner and loser effects are defined as an increased probability of winning at time T, based on victories at times $T - 1$, $T - 2$, and so on, and an increased probability of losing at time T, based on losing at times $T - 1$, $T - 2$, and so on, respectively. Although loser effescts are more common than winner effects, both have been documented many times (Chase et al., 1994; Lindquist and Chase, 2009).

At the proximate level, loser effects are often correlated with increased circulating levels of glucocorticoid stress hormones. Increased glucocorticoids are sometimes also seen in winners, but winners usually, though not always, return to baseline hormone levels much faster than losers. Losers tend to have suppressed levels of testosterone, and winners tend to have increased levels, but the precise effects of testosterone on winning are less understood (Hsu et al., 2005). To get a better flavor for the behavioral dynamics of winner and loser effects, as well as some of the proximate underpinnings of these effects, let us examine some case studies in birds, fish, and snakes.

WINNER AND LOSER EFFECTS IN BLUE-FOOTED BOOBIES. Recall from Chapter 9 that aggressive interactions sometimes occur between siblings at a nest (Mock and Parker, 1997). In a long-term study of nestmate aggression, Hugh Drummond and his colleagues have been investigating such aggressive behaviors during nestmate hierarchy formation in blue-footed boobies (*Sula nebouxii*; Benavides and Drummond, 2007; Drummond, 2006; Drummond and Canales, 1998; Drummond and Osorno, 1992; Gonzalez-Voyer et al.,

2007; Figure 15.13). Drummond's earlier work on the role of experience in shaping hierarchy formation found that if older chicks were aggressive and dominant early on, and younger chicks were submissive and subordinate, the older chicks would later be able to defeat the younger chicks even if the younger chicks were considerably larger than the older chicks (Drummond and Osorno, 1992). While this work hints at the possibility of winner and loser effects in blue-footed boobies, it is not possible to definitively say whether loser effects (in younger chicks), winner effects (in older chicks), neither, or both played a role in establishing hierarchies in this species.

To experimentally examine winner and loser effects, Drummond and Canales followed up the study mentioned above by pitting each dominant and subordinate individual against *inexperienced individuals* (Benavides and Drummond, 2007; Drummond and Canales, 1998). In their experimental protocol, only one of the animals in a trial had prior experience, and it was possible to examine winner and loser effects in isolation from one another. Specifically, Drummond and Canales predicted that when initially subordinate animals were pitted against inexperienced individuals (neutrals), the subordinate bird would have a low probability of winning aggressive bouts. To be conservative, they pitted subordinates against inexperienced partners that were slightly smaller (thus *increasing* a subordinate's chance of victory). The researchers also had initially dominant individuals interact with inexperienced partners (neutrals) and predicted that the dominant individual would have a high probability of winning future aggressive interactions. Again, to be conservative, they pitted dominant individuals against slightly larger inexperienced individuals (*decreasing* the dominant bird's probability of victory).

Drummond and Canales uncovered strong winner and loser effects early on (at about four hours) in their experiment. At this chick stage, subordinates were much less aggressive than were their inexperienced pairmates, and dominants were much more aggressive than their pairmates (Figure 15.14). The winner effect waned with the passage of time, but the loser effect remained intact throughout the ten days of observation (Drummond and Canales, 1998).

FIGURE 15.13. Blue-footed boobies. For blue-footed booby chicks involved in aggression, both winner and loser effects are in play. Winner effects, however, are much shorter lived. *(Photo credit: Hugh Drummond)*

WINNER AND LOSER EFFECTS IN *RIVULUS MARMORATUS*. Yuying Hsu and her colleagues have examined winner and loser effects in the hermaphroditic fish *Rivulus marmoratus* (Hsu and Wolf, 1999, 2001; Hsu et al., 2009; Huang et al., 2011; Lan and Hsu, 2011). They examined not only the effect of wins and losses on the next aggressive interaction in which an individual was involved but also the effect of wins and losses that had occurred two moves back in time (penultimate wins and losses).

Hsu and her team subjected fish to a number of combinations of wins (W), losses (L), or "neutral" (N, no win, no loss) events. The wins and losses an experimental fish received before it was tested were controlled by the researchers.

By comparing fish in the WW versus LW and LL versus WL treatments, Hsu and Wolf were able to document the first experimental evidence that the penultimate (next-to-last) aggressive interaction a fish experiences also affects its current probability of winning or losing (Hsu and Wolf, 1999). To see this, let us focus on the penultimate interactions and their implications for winner effects. Here, fish experiencing WW are significantly more likely to win a fight than fish experiencing LW. If the penultimate interaction had no effect on current aggressive interactions, one would expect no such difference across the WW versus LW treatments. Comparing numerous other treatments, Hsu and Wolf were able to demonstrate that, while penultimate interactions were important, their impact on winning or losing a current interaction was not as powerful as the outcome of the interaction immediately preceding the current interaction. In addition, unlike many other species, *R. marmoratus* showed no asymmetry in winner and loser effects—that is, the loser effect was not stronger or weaker than the winner effect.

WINNER AND LOSER EFFECTS IN COPPERHEAD SNAKES. Gordon Schuett has examined both the behavioral dynamics and endocrinological underpinnings of winner and loser effects in copperhead snakes (*Agkistrodon contortrix*; Schuett, 1997; Schuett and Gillingham, 1989). Because male-male aggression has a significant impact on mating success, Schuett examined whether winner and loser effects had an impact on aggression and mating success.

The experimental protocol involved a trial arena that housed a female in the center and one male at each end. Schuett used males that had had no aggressive interactions for six to twelve months prior to his study, and he pitted two males that differed in size by approximately 10 percent. In all thirty-two trials, the larger male emerged as dominant and gained reproductive access to the female.

Ten winners and ten losers from the "size" contests described above were chosen, and each was matched against a same-sized male copperhead that had no prior experience. Schuett found that prior winners were not more likely to win again, nor were they more likely to win than their opponents that had no experience to obtain access to a female—that is, no winner effects were observed in copperhead males. A loser effect, however, was found. Losers were more likely to lose again and to cede access to reproductively active females to other males. Schuett then examined how individuals that had now lost twice (in the initial "size contests" and then in loser effect treatments) fared against opponents that had no experience and that were about 10 percent smaller. Would the loser effect outweigh the positive size advantage that the losers possessed, or vice versa? Results pointed to the strength of the loser effect in copperheads—two-time losers lost all contests with smaller opponents (Figure 15.15).

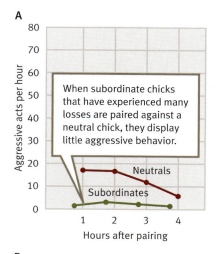

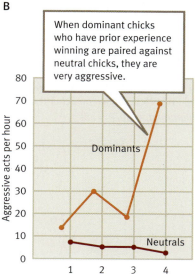

FIGURE 15.14. Winning and losing booby chicks. (A) Subordinate males with a losing experience. **(B)** Dominant males with a winning experience. *(Based on Drummond and Canales, 1998, p. 1672)*

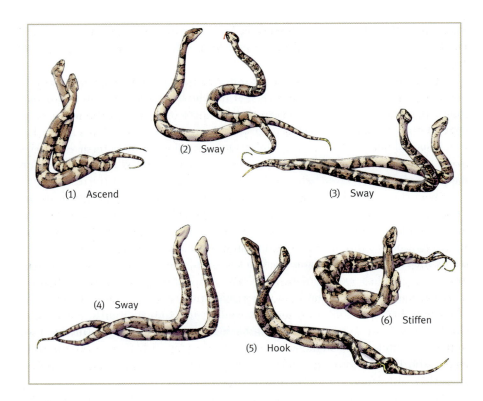

FIGURE 15.15. Winner and loser snakes. In copperhead snakes, losses can have a significant effect on future contest outcome. Snake fights in *Agkistrodon contortrix* include such aggressive behaviors as (1) ascend, (2) ventrad-to-ventrad sway, (3) laterad-to-laterad sway, (4) ventrad-to-dorsad sway, (5) hook, and (6) stiffen. *(Based on Schuett and Gillingham, 1989, p. 248)*

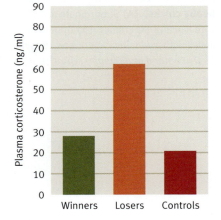

FIGURE 15.16. Hormones, winning, and losing. In copperhead snakes, losers show increased levels of plasma corticosterone compared with controls. No such change was found in winners. *(Based on Schuett et al., 1996)*

In addition to the behavioral work, Schuett and his colleagues also studied endocrinology and aggression in copperheads by examining hormonal correlates of the loser effect (Schuett et al., 1996; Schuett and Grober, 2000). Researchers allowed pairs of males to fight when a female was present, and the snakes were kept together until one male was judged dominant to the other. At that point, the individuals were separated and a blood sample was collected for hormonal analysis. In addition to this treatment, two controls were run: In the first, a lone male in his home cage was used, and in the second, a single male and a female were placed in the arena. The researchers measured plasma corticosterone in winners, losers, and both classes of control males, and they found that plasma corticosterone was significantly greater in losers than in winners or controls (Figure 15.16).

Increased levels of corticosterone produced dramatic effects in terms of both fighting and courting behavior in male copperheads. Not only do males that lose fights, and consequently have raised corticosterone levels, act subordinate and rarely, if ever, challenge other males but they almost never court any females that are in the vicinity of where they fought.

Experimental work in a number of other species suggests that the stress-related hormones, such as glucocorticoids, often inhibit learning and/or memory (de Kloet et al., 1999, 2002; de Quervain, 2006; de Quervain et al., 1998, 2000; Lemaire et al., 2000; Sapolsky et al., 2000). Whether winners and losers differ in their ability to learn as a function of hormonal changes that occur as a result of fights is a fascinating but unexplored question.

MATHEMATICAL MODELS OF WINNER AND LOSER EFFECTS. In addition to empirical work on winner and loser effects, a number of mathematical models examine the implications of winner and loser effects on the formation of animal hierarchies (Beacham, 2003; Bonabeau et al., 1996, 1999; Dugatkin, 1997b; Hock and Huber, 2009; Landau, 1951a, 1951b; Lindquist and Chaser, 2009; Mesterton-Gibbons, 1999).

Recall from the beginning of the chapter that a dominance hierarchy is defined as a rank ordering of individuals in a group based on their aggressive interactions with each other. For example, if individual A wins the majority of fights with B, C, and D, then A is the top-ranked (or α) individual in a hierarchy. If B beats C and D, C beats D, and D beats no one, then a linear hierarchy exists (A > B > C > D), and that hierarchy is said to be transitive, in that if A defeats B and B defeats C, then A defeats C as well.

Part of the interest in hierarchy formation arises from the difficulty in understanding why an animal would ever "accept" a role other than that of highest-ranked individual in a group (Allee, 1951; Dugatkin, 1995; Reeve and Ratnieks, 1993; Vehrencamp, 1983). Initial work on the impact of winner and loser effects on hierarchy formation was undertaken by H. G. Landau over sixty years ago (Landau, 1951a, 1951b). Landau was troubled to find that when he modeled hierarchy formation mathematically, the linear, transitive hierarchies often found in nature did not seem to emerge from his mathematical work— work in which he only considered inherent differences in fighting abilities across individuals. So Landau added winner and loser effects to his models to examine whether linear, transitive hierarchies then emerged. And such hierarchies did emerge from Landau's more complex models.

Despite the importance and impact of Landau's papers on research in the area of dominance hierarchies, a number of critical questions surrounding winner and loser effects and how they interact remained unanswered more than fifty years after the original papers were published. In particular, Landau did not examine winner and loser effects independently but instead only considered the effect on hierarchy formation when both were present. Furthermore, Landau did not take into consideration the fact that animals assess each other's fighting ability, technically referred to as **resource holding power** (RHP). Is it possible that when RHP is assessed, winner effects can have different implications for hierarchy formation than loser effects?

To address these questions, a simple computer simulation was developed that independently examined winner and loser effects when individuals assessed each other's fighting abilities and could choose to fight or flee (Dugatkin, 1997b). The most important result of these models is that the type of hierarchy predicted depends critically on whether winner effects exist alone, loser effects exist alone, or some combination of these forces are acting in a system.

Winner effects alone produce linear hierarchies in which the rank of individuals all the way from top rank (α) to bottom rank can be unambiguously assigned. Loser effects alone produce hierarchies in which a clear α individual exists, but the relationship among other group members is difficult to ascertain. The difference between the hierarchies generated by the winner effect and by the loser effect appears to be due to the fact that winner effects create a situation in which pairs of individuals primarily interact by fighting, which makes assigning position in the hierarchy fairly straightforward. Conversely, loser effects quickly produce individuals that are not going to be aggressive because of their low estimate of their own fighting ability after a few losses. With loser effects, most interactions will end in neither individual opting to fight, making it difficult to assign relative ranks to most individuals in a hierarchy.

BYSTANDER EFFECTS

When the observer of an aggressive interaction between two other individuals changes its assessment of the fighting abilities of those it has observed, the **bystander effect**—sometimes called the "eavesdropper effect"—is in operation. Through observing, bystanders learn beforehand something about the opponents they may face in the future (Coultier et al., 1996; Johnsson and Akerman, 1998; Oliveira et al., 1998). Verbal models and computer simulations have found that bystander effects can have important consequences on the dynamics of hierarchy formation (Chase, 1974, 1982, 1985; Dugatkin, 2001b; Earley, 2010). Experimental work has found bystander effects in birds, mammals, and fish (McGregor, 2005).

Bystander effects have been examined in the green swordtail fish (*Xiphophorus helleri*), a species in which males establish linear dominance hierarchies where rank-order fights and/or attack-retreat sequences are common (Beaugrand and Zayan, 1985; Franck et al., 1998). In these experiments, eavesdroppers first observe aggressive interactions, providing them with an opportunity to extract information from watching those that are fighting. The eavesdroppers are then tested in later aggressive interactions with those they had previously observed.

To experimentally examine the bystander effect, Earley and Dugatkin had eavesdroppers on one side of an experimental tank and a pair of swordtails that were involved in aggressive interactions on the other side. In one treatment, the eavesdropper could observe the pair of fish through a one-way mirror (mirror treatment), in which he could see the fish that were fighting but they could not see him and so would not be affected by the eavesdropper's presence. In the other treatment, the (potential) eavesdropper could not see the pair because of an opaque partition in the tank (opaque treatment; Earley and Dugatkin, 2002, 2005). After this observation period, the eavesdropper was pitted against either the winner of the observed fight or the loser of the observed fight. The opaque partition treatment served as a control for winner and loser effects. Since the eavesdropper could not observe the interaction between pairs of individuals in the opaque partition treatment, the dynamics of the subsequent interaction between the eavesdropper and the winner, or between the eavesdropper and the loser, were the result of winner and loser effects, respectively. In contrast, both bystander effects and winner and loser effects could affect contests between the eavesdropper and winners and losers in the one-way mirror treatment. This means that when we compare the dynamics of contests involving eavesdroppers in the opaque treatment (where there are only winner and loser effects) and the one-way mirror treatment (where there are winner, loser, and bystander effects), we are able to experimentally determine the contribution of bystander effects.

Eavesdroppers that observed a contest in the one-way mirror treatment were much more likely to try to avoid the winner of that contest than were fish in the opaque treatment—eavesdropping per se affected future interactions with winners (Figure 15.17). In addition, eavesdroppers avoided observed winners regardless of how badly they had defeated their prior opponent. The results were quite different when eavesdroppers interacted with losers. In general, eavesdroppers responded in a similar way to all losers, regardless of whether they had witnessed the losers' defeat. But if we break down the data on losers by how badly they lost fights, there is one piece of evidence that suggests that eavesdroppers deal differently with losers because they have

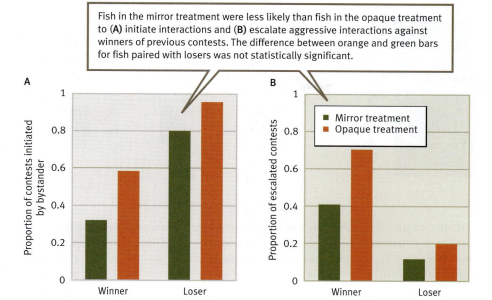

Fish in the mirror treatment were less likely than fish in the opaque treatment to (A) initiate interactions and (B) escalate aggressive interactions against winners of previous contests. The difference between orange and green bars for fish paired with losers was not statistically significant.

FIGURE 15.17. Eavesdroppers, winners, and losers. In one condition, fish could observe aggressive interactions between a pair of other fish (one-way mirror treatment; green bars), but in the other condition such interactions were blocked from view by an opaque partition (opaque treatment; orange bars). *(Based on Earley and Dugatkin, 2005, p. 90)*

observed them lose. Results indicate that in the one-way mirror treatment, eavesdroppers were less likely to initiate aggressive behavior and win against (1) losers that had persisted in their fights or (2) losers that had escalated their aggressive actions. This was not the case for the opaque treatment, suggesting the presence of subtle bystander effects when the eavesdroppers are interacting with losers.

Ethologists are also beginning to understand the underlying proximate mechanisms associated with bystander effects. In the cichlid fish *Oreochromis mossambicus*, for example, when eavesdropping males watch a fight between a pair of other males, their androgen levels rise (Figure 15.18). In particular, Rui Oliveira and his team measured the levels of testosterone in the urine of eavesdroppers before they saw a fight and after they observed a fight, and they found a significant increase in testosterone levels (Oliveira et al., 1998, 2001, 2002, 2005). This increase in testosterone may have better prepared the eavesdropper for future aggressive interactions by indirectly affecting attention, learning, and memory in ways that might prove beneficial to observers. For example, the increase in testosterone may be beneficial for the eavesdropper if the winner of the observed fight subsequently attacks others, including the eavesdropper, in the immediate vicinity.

AUDIENCE EFFECTS

Not only do bystanders change their estimation of the fighting ability of those they observe, but individuals involved in aggressive interactions change their behavior if they are watched. This latter phenomenon is referred to as the **audience**

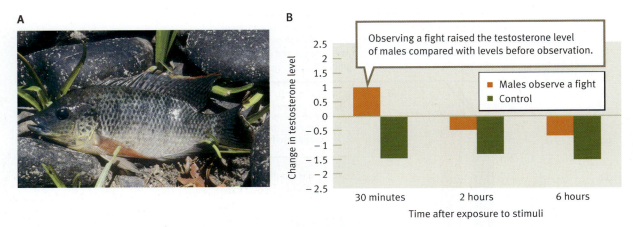

FIGURE 15.18. Eavesdropping and testosterone. (A) *Oreochromis mossambicus.*
(B) The level of testosterone increases after eavesdropping on a fight. Orange bars show the difference in testosterone in experimental males before versus after observing a fight between two other males. A spike in testosterone occurs and lasts at least thirty minutes. The decrease in testosterone in control males is likely due to the normal daily decrease in testosterone from late morning (when experiment was undertaken) to evening (six hours later). *(Photo credit: Johnny Jensen/Image Quest Marine; graph from Oliveira et al., 2001)*

effect, and it has been studied in a number of species (Doutrelant et al., 2001; Dziewerzynski et al., 2012; Evans and Marler, 1994; McGregor and Peake, 2000; Overduin-De Vries et al., 2012; Plath et al., 2009). A particularly interesting example of audience effects has been found in the context of aggression and "recruitment screams" in chimpanzees.

Wild chimpanzees emit screams during pairwise aggressive interactions. The screams of those winning a fight are very different from those losing a fight, and it is possible for researchers to categorize screams as "aggressor screams" and "victim screams" (Slocombe and Zuberbuhler, 2005). Victim screams may, in part, function to entice support from observers—for example, a scream by a victim may entice an observer to intercede and break up the fight. Kate Slocombe and Klaus Zuberbuhler examined this possibility in a population of wild chimpanzees in the Budongo Forest of Uganda (Slocombe and Zuberbuhler, 2007).

Slocombe and Zuberbuhler began their work by categorizing pairwise aggressive interactions between chimps as either mildly or severely aggressive. From their observations, they were able to gather data on eighty-four screaming bouts, and they found evidence suggesting a cognitively complex form of audience effects (Snowdon, 2009). When they compared the screams of victims in mildly aggressive interactions, no differences were found when an audience was present or absent. When interactions involved severe aggression, however, victim screams were sometimes much longer and more intense when there was an audience present (compared with when there was no audience). But this audience effect was only seen when at least one of the audience members held a rank in the hierarchy that was equal to or above the rank of the aggressor (Figure 15.19). This screaming strategy was successful, as victims that emitted longer and more intense screams were able to entice support from high-ranking observers that often intervened and broke up fights.

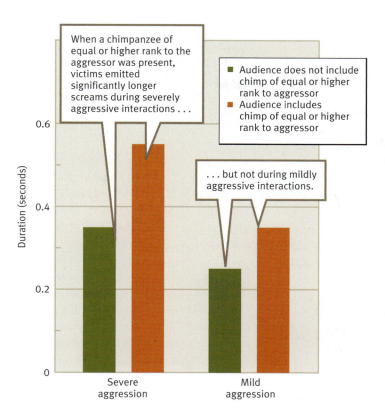

FIGURE 15.19. Audience effects in chimpanzees. Chimpanzees that were victims in severe aggressive interactions emitted distinctive screams. Significantly longer screams were emitted when fights were watched by an audience that included a chimp of equal or higher rank to the aggressor (orange bars) than when the audience did not contain such an individual (green bars). No significant difference was found when victims were involved in mildly aggressive interactions. *(Based on Slocombe and Zuberbuhler, 2007)*

MATH BOX 15.1

The Hawk-Dove Game

Let's look at the case of the hawk-dove game, when $C > V$, in a bit more detail.

If $C > V$, then hawk is not an ESS, as hawk's payoff against other hawks, $(V-C)/2$, is now negative, while dove's payoff against hawk is 0. Using the same calculations as we did for the case of $V > C$, we can see that dove is not an ESS either.

To see if some combination of hawks and doves is an ESS, let p = the frequency of hawks, and $1-p$ be the frequency of doves. Hawk's payoff is then:

$$p(V-C)/2 + (1-p) V, \tag{1}$$

where the first term is hawk's payoff against other hawks and the second is hawk's payoff against dove.

Dove's payoff can be calculated as:

$$p(0) + (1-p)V/2, \tag{2}$$

where the first term is dove's payoff against hawk and the second term is dove's payoff against other doves.

When (1) = (2), the fitness of hawks is equal to the fitness of doves, and we then have our equilibrial frequency of hawks and doves. A little algebra shows that (1) = (2) when $p = V/C$.

INTERVIEW WITH

Dr. Karen Hollis

As a psychologist, what drew you to the study of animal aggression?

I first became interested in aggression as a graduate student working with *Betta splendens*, a territorial freshwater fish native to Southeast Asia. Although my project addressed a question of underlying learning processes, not aggression per se, it afforded me a serendipitous observation of aggressive behavior, one that provided the raw material for years of subsequent research. My project required me to transport each *Betta* male in its home tank to another room where it would have the opportunity to display to a rival. After a few days of this procedure, I noticed that when I approached the shelf on which the males were located and reached for a particular male's tank, it became quite agitated. After a few more days, each male that I selected began to display—at me. And at eye level that display looked quite ferocious! I recognized that some features of my appearance obviously had become a learned cue for the subsequent interaction with a territorial rival. At that time, associative learning in *B. splendens* was pretty well established (even if the experimenter-as-signal was a little unconventional). However, what was even more interesting to me was the realization that this kind of learning could be a wonderfully adaptive mechanism, providing a territory owner with a definitive competitive edge: Confronting a potential usurper with a full-blown aggressive display could be a very effective aggressive strategy for a territory owner. A few

years later, as a postdoctoral student at Oxford, I had the opportunity to conduct those experiments with blue gouramis, *Trichogaster trichopterus*, and, as the data showed pretty convincingly, the best defense is, indeed, a good offense.

What do animals usually fight over? Do they display different types of aggression depending on what the fight is about?

No matter what the situation—from nest-mate aggression and sibling rivalry to parent-offspring conflict and territorial behavior—aggression is all about resources. Moreover, as behavioral ecologists have demonstrated quite convincingly across a variety of taxa, particular forms of aggressive behavior emerge under particular selection pressures. For example, aggressive behavior in the context of resource-defense polygyny is a mating strategy in which individuals of one sex, typically

males, defend a resource that is in short supply, critical for females, and defendable, both in terms of the reliability of its location and its quality. That resource—and, thus, what males fight over—could be anything that meets these criteria. Many species of freshwater fish, including my own study species, blue gouramis, defend a location where aspects of the habitat, such as the temperature and pH of the water, are favorable for the development of eggs and fry. Territorial lizards defend spots safe from predators. Some birds defend shady spots, necessary for the development of eggs. And, of course, many animals, from territorial male bumblebees and green frogs to sunbirds and burrowing owls, defend food sites.

Aggression also appears in another mating strategy, called harem defense, in which a few males defend a group of females directly, rather than a needed resource. Harem defense polygyny is characteristic of grizzly bears, red deer, and elephant seals. Because the benefits of winning are tied more directly to reproductive success whenever males are defending access to reproductively active females than when they are defending only the resources to which females are drawn, and because only a few males, namely those powerful enough to mount this kind of defense, are able to reproduce, the aggressive behavior differs enormously. Males are much more aggressive, and injuries are potentially more lethal, wherever harems are concerned. So, yes, aggressive behavior differs dramatically, depending on what the fight is about, as predicted by evolutionary theory.

How flexible are the aggressive strategies employed by animals? What role does learning play in aggressive behavior?

Many of the learning phenomena studied by animal learning psychologists map nicely onto naturally occurring aggressive behavior and, thus, demonstrate the flexibility of—and modulating influence on—aggression. For example, the "dear enemy effect"—where two adjoining territory holders form a temporary alliance to fight intruders that are threats to both of them—obviously depends upon habituation. Habituation is the progressive decrease in responsiveness resulting from repeated presentation of a stimulus. Although some instances of attenuated responsiveness are merely the result of sensory adaptation or muscular fatigue, true habituation is a phenomenon of neural memory: The response to repeated sensory stimulation is chocked off somewhere in the brain—or nerve net—of the animal. Aggression between territorial neighbors, which is initially both intense and frequent, gradually wanes to a point of sporadic and relatively mild aggressive interactions. Yet, while neighbors spare one another from their territorial aggression, producing dear enemies, newcomers face its full measure. Laboratory experiments on habituated aggression have reproduced successfully the basic components of the dear enemy effect. That is, the habituated response is stimulus-specific: Neighbors, and only neighbors, no longer elicit aggressive behavior. Moreover, habituation itself is neither permanent nor ephemeral, lasting just long enough to handle the time intervals observed in natural settings.

Other examples abound. Many species of territorial fish are capable of recognizing their competitors, even when they do not belong to the same species; fish of other species whose ecological requirements overlap with the territory holder are driven away while noncompetitive species are permitted to remain within the territory. In short, individuals not only learn which species to attack, but also make fine discriminations between different species of the same genera.

Finally, in the same way that animals can use signals to determine when or where an aggressor might appear and, thus, focus their aggression in cost-effective ways, inhibitory learning—the ability to learn that a cue predicts the absence of a particular event for some period of time—provides a mechanism to avoid useless energy expenditure at times when, or places where, rivals are less likely to occur. For example, territory owners often exhibit "spontaneous" aggressive displays, favoring particular locations—locations, as the research shows, where they have encountered rivals in the past—and ignoring places where rivals have not appeared and, thus, are unlikely to pose a threat.

What do you predict will be the next major breakthrough in the study of animal aggression?

To me, science is like a very large construction project, with different teams of professionals arriving on site at particular points to contribute their special expertise, then making way for the next group, and so on. Only, in the case of science, the project never really ends. Not only does the edifice get renovated over and over again, but also those teams of experts return with new and improved technology. In the study of aggression, behavioral ecologists have laid an extensive

and solid foundation of intra- and interspecific behavioral similarities and differences, all of which reflect particular selection pressures. It's now time for geneticists to step on site with their array of exciting new tools.

In particular, I predict that genomic analyses will allow us to explore the vast expanse of questions that lie between genes and the expression of aggressive behavior. Concerning intraspecific differences, what does an ESS look like from a genetic perspective? What are the genetic and developmental differences that separate individuals that become harem defenders from those that are forced to adopt surreptitious strategies to obtain copulations? Single nucleotide polymorphisms (SNPs), locations on the genome where individuals differ by just one chemical compound, already have been implicated in the aggressive threat behavior of monkeys, making this approach a very promising one for addressing individual differences in aggression. At the interspecific level, to what extent do the genomes differ between closely related species that express different types of male-male competition? Conversely, to what extent might the aggressive behavior of convergent species possess similar genetic mechanisms? In short, I predict not only that geneticists and molecular biologists will add a completely new wing to the edifice that is our science of animal behavior, but also that the connectors between these and other groups of behavioral scientists will get easier and easier to traverse.

Dr. Karen Hollis is a professor at Mount Holyoke College. Her work on animal learning and Pavlovian conditioning represents a classic example of how researchers can integrate biological and psychological approaches to animal behavior.

SUMMARY

1. Individuals from many species tend to fight over resources (space, mates, food, and so on).

2. Ethologists and evolutionary game theorists have built a series of models that examine the evolution of fighting behavior. The three best-developed game theory models of aggression are the hawk-dove game, the war of attrition model, and the sequential assessment model. In all three of these game theory models, a cost to fighting is assumed.

3. There are many varieties of the hawk-dove game. One version adds two new strategies: bourgeois and antibourgeois. The bourgeois strategy instructs an individual to play hawk if it is a territory holder and dove if does not own a territory. The antibourgeois strategy codes for the opposite—play dove if you are a territory holder and hawk if you are not.

4. The war of attrition model was created to examine aggression when the choice available to individuals is more continuous—for example, "fight for x seconds, then stop." Rather than establishing a set time for fighting, the war of attrition model predicts an ESS distribution of contest lengths.

5. In the sequential assessment model, individuals constantly assess and update their assessment of their opponents' fighting skills. A single sample (assessment) introduces significant error; the more sampling (assessment) an individual does, the lower the error rate and hence the more confident that individual can be in whatever is being estimated.

6. Winner and loser effects are usually defined as an increased probability of winning at time T, based on victories at times $T - 1$, $T - 2$, and so on, and an increased probability of losing at time T, based on losing at times $T - 1$, $T - 2$, and so on.

7. Bystander effects occur when the observer of an aggressive interaction between two other individuals changes its assessment of the fighting abilities of those it has observed. Not only do bystanders change their estimation of the fighting ability of those they observe, but individuals involved in aggressive interactions change their behavior if they are watched. This latter phenomenon is referred to as the audience effect.

DISCUSSION QUESTIONS

1. In 1990, Enquist and Leimer published a paper, "The Evolution of Fatal Fighting" in *Animal Behaviour* (vol. 39, pp. 1–9). Based on reading their paper, as well as your own thoughts on the costs and benefits of extreme aggression, when do you think fighting to the death might be likely to favored by natural selection?

2. The classic hawk-dove game we examined in this chapter assumes that losers pay a cost (C) that is not paid by winners. In the matrix below, we are assuming that both hawks in a fight pay a cost.

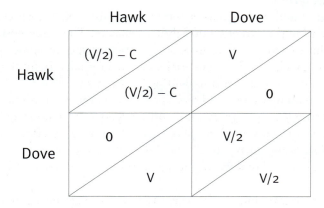

For the case of both V > C and V < C, calculate the ESS for this new game.

3. A number of studies have suggested that loser effects are both more common and more dramatic than winner effects. Construct a hypothesis as to why this might be. How could you test your hypothesis?

4. If stress-related hormones such as cortisol often inhibit learning and/or memory, how might that compound the difficulties subordinate fish face in trying to raise their rank in hierarchies?

SUGGESTED READING

Arnott, G., & Elwood, R. W. (2009). Assessment of fighting ability in animal contests. *Animal Behaviour, 77,* 991–1004. A review of assessment and the role it plays in animal aggression.

Edwards, D. H., & Spitzer, N. (2006). Social dominance and serotonin receptor genes in crayfish. *Current Topics in Developmental Biology, 74,* 177–199. A detailed analysis of serotonin and aggression in crayfish.

Emlen, D. J. (2008). The evolution of animal weapons. *Annual Review of Ecology Evolution and Systematics, 39,* 387–413. A wonderful review of the wide variety of animal weapons and their function.

Hsu, Y. Y., Lee, I. H., & Lu, C. K. (2009). Prior contest information: Mechanisms underlying winner and loser effects. *Behavioral Ecology and Sociobiology, 63,* 1247–1257. A review of the proximate underpinnings of winner and loser effects.

Maynard Smith, J., & Price, G. (1973). The logic of animal conflict. *Nature, 246,* 15–18. The classic paper introducing ethologists to mathematical models (game theory) of aggression.

Riechert, S. (1998). Game theory and animal contests. In L. A. Dugatkin & H. K. Reeve (Eds.), *Game theory and animal behavior* (pp. 64–93). New York: Oxford University Press. A review chapter on game theory and animal behavior.

16

Play

Give a kitten a ball of yarn or throw a new squeeze toy on the floor for your puppy and watch what happens. Kittens and puppies appear to play with these objects. Take two puppies and put them in the same room, and again it looks like they are playing—this time with each other. Pet owners will swear to you that their animals play—indeed, pet play stories are legendary and numerous. But ethologists want more than anecdotes and stories. They want to study play behavior the same way they study every other behavior—using theory and experiment, and from both proximate and ultimate perspectives. Progress is being made on these fronts, but much remains to be done, and the study of play has it own unique challenges associated with it.

Play can take many forms. Evidence from ethology suggests that, while not all species engage in play, it is common, particularly among young individuals. Young animals gently wrestle with each other; toss, kick, and push objects they find in their environment; chase one another for no apparent reason; jump; climb up and down trees over and over; and practice hunting prey that aren't present (Figure 16.1).

While play is probably more common in large-brained vertebrates (such as primates) than in other species, it is not limited to this group nor is play limited to vertebrates, as octopuses also play (Kuba et al., 2006; Mather, 2008a, 2008b; Table 16.1). At the Washington Zoo, an animal named Pigface played with whatever new objects (brown balls, orange balls, hoops, and so on) zookeepers added to his otherwise bland environment (Burghardt, 1998). Pigface approached the new objects, followed them, pushed them around, and so on—he played with them.

Two things make Pigface's story particularly interesting. First, Pigface isn't a dog or a chimp; he's a turtle. Second, play had a profound effect on Pigface's health. Before objects were introduced into his environment, Pigface was in the habit of clawing his own limbs and neck, causing infection and fungal growth. Once the play objects were introduced, this self-destructive behavior decreased dramatically. That is, once the bland environment Pigface once lived in was livened up with potential play items, he chose to play rather than self-mutilate.

Play remains a contentious subject in the ethological literature (Burghardt, 2005; K. L. Graham and Burghardt, 2010). Why? One possible explanation is that pet owners' and zookeepers' stories are usually anecdotal at best, and huge

A

B

FIGURE 16.1. Play. Play behavior in **(A)** sea lions and **(B)** polar bears. *(Photo credits: Marty Snyderman/Visuals Unlimited; Jeff Vanuga/Corbis)*

TABLE 16.1. Distribution of play. Play has been found across many, but not all, major vertebrate groups. (*Based on K. L. Graham and Burghardt, 2010*)

VERTEBRATE GROUP	EVIDENCE OF PLAY
Hagfishes	Play unknown in any
Lampreys	Play unknown in any
Sharks, rays, and skates	Play present in some
Ray-finned fishes	Play present in some
Coelocanth	Play unknown in any
Lungfishes	Play unknown in any
Caecilians	Play unknown in any
Frogs and toads	Play present in some
Salamanders and newts	Play unknown in any
Turtles and tortoises	Play present in some
Birds	Play present in many
Crocodiles and relatives	Play present in some
Lizards, snakes, and relatives	Play present in some
Egg-laying mammals	Play present in some
Kangaroos and relatives	Play present in many
Placental mammals	Play present in many

exaggerations at worst. Though play has been documented in many species, it simply has not been examined *experimentally* until very recently.

A second possible reason that play behavior was (and is) relatively understudied is that animal behaviorists tend to study behaviors that appear to have function. When it comes to play, function is sometimes very hard to determine, and hence there is a tendency to shy away from work in this area. A third, related reason for the relative lack of controlled studies of play is tied to theory. Robert Fagan and others have argued that a lack of concrete *theory* for the function of play can explain why play research lagged behind other areas of interest in ethology (Fagen, 1981). But recently, a theoretical underpinning for play behavior has begun to emerge, and we will examine this more throughout the chapter (Bekoff and Byers, 1998; Burghardt, 2005; Graham & Burghardt, 2010; Spinka et. al., 2001).

The situation with respect to the study of play has improved somewhat since 1975, when E. O. Wilson wrote that "no behavioral concept has proved more ill-defined, elusive, controversial and even unfashionable than play" (E. O. Wilson, 1975). There is now a sizable literature on many aspects of animal play, which seeks to answer various proximate and ultimate questions dealing with play behavior (Bekoff and Byers, 1998; Bruner et al., 1976; Burghardt, 2005; Fagen, 1981; Power, 2000; Symons, 1978). In this chapter, we will examine:

- ► how play is defined,
- ► the different types of play behavior (object, locomotor, and social play) and their purported functions,
- ► the endocrinological and neurobiological bases of play, and
- ► the phylogeny of play.

Defining Play

Since most people know what *foraging* means, when we discuss how animals obtain food, we usually don't run into much in the way of definitional problems. The same holds true for subjects like "parental care" and "antipredator activities" as well as a whole host of other behaviors. Definitional issues, however, do become problematic when it comes to play. Indeed, some of the reluctance of ethologists to study play centers on the notion that play is an amorphous, fuzzy behavior that defies definition.

That said, in the animal behavioral literature there are numerous definitions of **play**. The most widely cited of these definitions is that of Marc Bekoff and John Byers, who propose the following:

> Play is all motor activity performed postnatally that appears to be purposeless, in which motor patterns from other contexts may often be used in modified forms and altered temporal sequencing. If the activity is directed toward another living being it is called social play. (Bekoff and Byers, 1981, pp. 300–301)

This definition centers on the structure of play rather than its function. One problem with it is that behaviors such as repetitive pacing—pacing back and forth for long periods of time in what appears to be a purposeless manner—meets the above definition, but most researchers would argue that this is not really play behavior (Bekoff and Allen, 1998). In addition, not only is it extremely difficult to determine when a behavior is "purposeless," but behaviors that are apparently purposeless may be so for three very different reasons (Heinrich and Smokler, 1998): (1) Observers may simply fail to decipher what the immediate benefit of the play behavior is, (2) the purpose and potential benefit may not be accrued until long after play has occurred (Fagen and Fagen, 2004), and (3) the benefits may be multiple and confounding. The situation is actually more complex, because Bekoff and Byers's definition does not claim that play is purposeless, only that it appears to be so (Heinrich and Smokler, 1998).

So where does that leave us with respect to a definition of play? It seems that those who study play have adopted a definition similar to the United States Supreme Court's definition of pornography—we can't say exactly what it is, but we know it when we see it. This is not quite as bad a state of affairs as it may seem. For example, many ethologists argue that since experimental work on play lags behind controlled work in other areas of animal behavior, it is best to take a "wait and see" approach, wherein observers study play in many species. The hope is that such work will eventually uncover certain commonalities in play behavior, and that such commonalities will then be used to construct a definition (see Conservation Connection box).

Types and Functions of Play

Ethologists generally delineate three different types of play—object, locomotor, and social play. We will examine each of these, looking at the function of these three types of play.

Play Behavior as a Measure of Environmental Stress

The Conservation Connection box in Chapter 7 discussed ways that conservation biologists can use the extent of fluctuating asymmetry present in a population as a cue of environmental stress. Are there other such cues—perhaps more behavioral cues—that can be used to gauge the way populations studied by conservation biologists are under environmental stress of one sort or another? Recently, ethologists have suggested that play behavior may be such a cue.

To understand why, let's first look at a study on play behavior in juvenile squirrel monkeys (*Saimiri sciureus*). Anita Stone observed the foraging behavior, time spent traveling, play behavior, and other social behaviors over a twelve-month period in Eastern Brazilian Amazonia. What she found was that during the dry season, when fruit availability was low, young individuals spent more time foraging and much less time playing (Figure 16.2). This decrease in play was likely due both to the increased time spent foraging and to the high energy expenditure necessary for play—energy that was not available during the dry season (Stone, 2008).

Stone's study suggests that a decrease in play in squirrel monkeys was linked to low food availability and low energy budgets. Other evidence suggests that decreased levels of play may indicate environmental stress in general, not just with respect to decreased food availability. For example, a number of studies have found that when animals are stressed—by, for example, poor habitat quality, increased predation, and other so-called fitness threats—one of the first behaviors that drops from their repertoire is play behavior (Fraser and Duncan, 1998; Lawrence, 1987; Sarti Oliveira et al., 2010; Spinka et al., 2001), though the situation is often not so straightforward and requires future study (Held and Spinka, 2011).

Endocrinological and neurobiological studies are also being employed to understand the relationship between play and stress. Although comfort and pleasure are extraordinarily difficult to measure in nonhumans, some work suggests that play and playlike behaviors may be associated with increased pleasure (Berridge, 2003; Dalgleish, 2004; M. S. Dawkins, 1990, 2008; Mendl, 2010; Paul, 2005; Vanderschuren, 2010). This type of work is still in its infancy, but such studies could, for example, examine whether areas of the brain associated with pleasure in the other contexts, such as feeding and mating, are also active during play behavior.

Changes in play may be a kind of behavioral bellwether that conservation biologists can use to measure whether environmental stressors are affecting individuals.

A

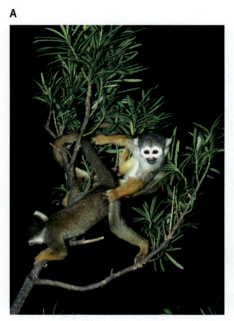

B

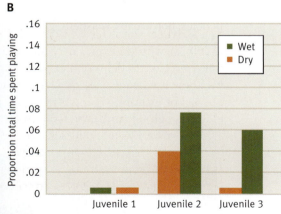

FIGURE 16.2. Seasonality and play behavior. (A) A juvenile squirrel monkey. **(B)** The proportion of time spent in play decreased for most juvenile squirrel monkeys during the dry season in Brazil. Two of the three individuals shown on the graph reduced play behavior during the dry season, when fruit was scarce. *(Photo credit: Francois Gohier/Photo Researchers; graph from Stone, 2008)*

FIGURE 16.3. Leaf play in chimpanzees. Chimpanzees living in the Mahale Mountains of Tanzania play in what is called "leaf-pile pulling." As they go down the slope of a mountain, individuals sometimes walk backward, pulling handfuls of leaves and then stopping and either walking or somersaulting through the pile of leaves. The photo shows a chimp as he gathers a leaf pile. *(Photo credit: Toshi Nishida)*

OBJECT PLAY

Object play involves the use of inanimate objects such as sticks, rocks, leaves, feathers, fruit, and human-provided objects, and the pushing, throwing, tearing, or manipulating of such objects (S. Hall, 1998). A wonderful example of object play has been documented in a chimpanzee population living in the Mahale Mountains of Tanzania. Researchers examining videotapes of the chimps in this population found a behavior they labeled "leaf-pile pulling" (Figure 16.3). As groups of chimps move down the slope of a mountain, an individual will sometimes stop and walk backward, pulling handfuls of leaves along with him as he proceeds. Then, the chimp stops and either walks or somersaults through the pile of the leaves that he created (Nishida and Inaba 2009; Nishida and Wallauer, 2003; Nishida et al., 2009).

Object play has been documented in animals in a wide array of taxa, and is particularly well studied in captive populations—such as zoo animals—where "toys" are given to animals to provide them with new items in an otherwise relatively constant environment (Fraser and Duncan, 1998; Sarti Oliveira et al., 2010). Object play has been distinguished from object exploration, and play often follows exploration (Hutt, 1966; Wood-Gush and Vestergaard, 1991; Figure 16.4). Animals undertaking object exploration discover what an object is, while during object play the animal acts as if it is trying to determine what it can do with this object (Hutt, 1970; Power, 2000).

OBJECT PLAY IN JUVENILE RAVENS. Young animals often have more "free time" than adults to engage in object play (Burghardt, 1988). From a functional perspective, object play in young animals is often associated with some aspect of practice where an animal learns something that will benefit it either in the short or long term—for example, young predators may use object play to practice hunting (B. Beck, 1980; Fagen, 1981; P. Martin and Caro, 1985; P. K. Smith, 1982).

In his delightful book *Mind of the Raven*, Bernd Heinrich describes how play occupies more of a raven's time than one might predict from a general survey of the avian behavioral literature (Heinrich, 1999; Figure 16.5). For example, young ravens play with virtually every new kind of object they encounter—leaves, twigs, pebbles, bottle caps, seashells, glass fragments, and inedible berries (Heinrich and

FIGURE 16.4. Play or exploration? Here a cheetah cub comes upon a novel object: bones. Exploring the bones appears designed to address the "What is it?" question, whereas play appears designed to tackle the "What can I do with this object?" question.

FIGURE 16.5. Play in ravens. Various "hanging games" Heinrich observed in ravens. *(Based on Heinrich, 1999, p. 289)*

Bugnyar, 2007; Heinrich and Smokler, 1998). Heinrich, who has studied these birds for thousands of hours, writes of young birds' seemingly obsessive drive to contact and manipulate any objects they encounter (Heinrich and Smokler, 1998). Ravens play with and display food items to others in their group, and some ethologists have suggested that such behaviors are a rare example of referential gestures in nonprimate species (Bugnyar et al., 2007; Pika and Bugnyar, 2011).

Object play in young ravens sets the stage for what individuals fear, or don't fear, when they mature (Heinrich, 1988a; Heinrich et al., 1996). Adult ravens still manipulate objects after they mature. But compared with how they react to items they played with when they were younger, they treat items they have never encountered before—including potential food sources—with great trepidation. To demonstrate the potential fitness benefit of object play, Heinrich ran an experiment involving both observation and manipulation.

Heinrich examined object play in four young juvenile ravens raised by experimenters in a controlled, but relatively natural, forest environment in Maine (Heinrich, 1995). Young ravens were observed in thirty-minute sessions for more than thirty days. During the first ten observation periods, Heinrich noted all the naturally occurring objects the birds encountered. Nine hundred and eighty naturally occurring items that fell into ninety-five different categories were encountered by the young birds during their first ten trials, and such encounters often involved some combination of exploration and play.

After the first ten trials, Heinrich then added forty-four "novel" items— objects the ravens had never seen before—to the birds' environment and observed the manner in which juveniles interacted with these new objects. He found that exploration and play were directed at novel items. Novelty, per se, rather than other characteristics such as shininess, palatability, or conspicuousness, explained which items they chose. Yet while they did not choose items based on their palatability, they quickly treated inedible items as background material—they did not handle these items much after their initial encounters with them—and edible novel items as their preferred foods. This suggests that juvenile play and exploration in ravens has the important benefit of enabling the juvenile ravens to identify new food sources. Ravens are scavengers (as well as predators), and in a world where many objects may be food, play and curiosity in ravens seem to be selected as a means to decipher what is edible and what isn't (Heinrich, 1999; Heinrich and Bugnyar, 2007).

OBJECT PLAY IN YOUNG CHEETAHS. Tim Caro studied play in cheetah cubs in the Serengeti National Park in Tanzania for three years (Caro, 1995b; Figure 16.6). He observed cubs for more than 2,600 hours, documenting, among other things, many instances of object, social, and locomotor play (Table 16.2). In particular, Caro was

FIGURE 16.6. Cheetah play. Young cheetahs engage in social play often, including play fights, even occasionally while they sit in trees. *(Photo credit: Tim Caro)*

interested in the role of each type of play in shaping how cheetahs would learn to capture prey and what the costs associated with each type of play might be.

Given that cheetahs spend a considerable amount of time playing as cubs (Figure 16.7), Caro wanted to see whether he could measure the costs associated with juvenile play. Estimating the cost of any behavior, particularly a behavior as complex and difficult to measure as play, is difficult, but evidence from a number of studies suggests that under certain conditions play may be very costly (W. Arnold and Trillmich, 1985; J. Berger, 1980; Douglas-Hamilton and Douglas-Hamilton, 1975; R. Harcourt, 1991; Lawick-Goodall, 1968). As an extreme case, while play

TABLE 16.2. Types of cheetah play. Young cheetahs benefit from a wide variety of play activities in nature. *(From Caro, 1995b, p. 335)*

TYPE OF PLAY	BEHAVIOR PATTERN	RECIPIENT
Locomotor play	Bounding gait	No recipient
	Rushing around	
Contact social play	Patting	Any family member
	Biting	
	Kicking	
	Grasping	
Object play	Patting	Object
	Biting	
	Kicking	
	Carrying	
Noncontact social play	Stalking	Any family member
	Crouching	
	Chasing	
	Fleeing	
	Rearing up	

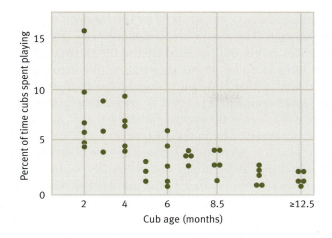

FIGURE 16.7. Play in cheetahs. In cheetahs, play progressively disappears with age. *(From Caro, 1995b)*

accounted for just over 6 percent of young fur seals' time, twenty-two of the twenty-six pups that were killed over the course of long-term observations were engaged in some sort of play at the time they were killed (R. Harcourt, 1991).

To estimate the potential costs of play in young cheetahs, Caro measured both the distance that a cub rushed around or chased during play, and the distance cubs moved from their mothers during object, motor, and social play. The former was an attempt to measure energetic costs of play for the young cheetah cubs, while the latter was an estimate of predation or general injury costs associated with play—the farther the cub was from its mother, the greater was its risk of being harmed either during play or by a predator such as a lion or a spotted hyena.

Caro's data suggest that cheetah play—including object play such as patting, biting, kicking, and carrying—is a very low-cost activity. While it turned out to be nearly impossible to measure the direct energy costs of play, cubs were never seriously injured during play. In addition, while young cheetah cubs that were involved in play were slightly farther from their mothers than cubs not playing, in all cases the cubs' mothers were so close that their cubs were not under any serious predation threat. Costs of play in juvenile cheetahs may still exist, but if such costs do exist, they are in all likelihood minor.

In terms of benefits, increased play led cubs to display increased rates of patting, grasping, and biting live prey that their mother had just released. Such predatory-like behavior on the part of the cubs might make them more successful hunters when they mature.

LOCOMOTOR PLAY

In this section, we will consider **locomotor play**, which is technically referred to as locomotor-rotational play (Power, 2000). Robert Fagan describes this type of play as follows:

> The single most frequent and phylogenetically widespread locomotor act of play must surely be a leap upward. . . . Hops, springs, bounces and bucks are variations on the basic vertical leap. . . . Animals may somersault, roll, flip forward or backward, spin, whirl, pirouette, make handstands, chase their tails, rear and kick up their heels. . . . Often a vertical leap is decorated with body-twists, rear-kicks or head shakes. These acrobatics can be spectacular. (Fagen, 1981, pp. 287–291)

Two hypotheses have been put forth for the function of locomotor play in animals and humans. The first is that locomotor play provides exercise and training for specific motor skills needed later in life (J. Byers, 1984, 1998). The second is that locomotor play provides animals with a better understanding of "the lay of the land"—that is, where things are in relation to one another—and this may provide immediate benefits (Power, 2000; Stamps, 1995; Symons, 1978). Here we will focus on the exercise-related benefits, and we return to "the lay of the land" hypothesis later in the chapter.

Locomotor play has been studied primarily in rodents, primates, and un-gulates, and it includes leaps, jumps, twists, shakes, whirls, and somersaults (Figure 16.8). In an attempt to quantify the possible benefits of locomotor play, John Byers and Curt Walker reviewed a list of nineteen potential anatomical and physiological benefits associated with exercise and physical training (J. Byers and Walker, 1995; Table 16.3). They went through their list of putative benefits and began by searching for benefits that were available to individuals as juveniles, but not as adults, but that were long lasting in their effects. Only two of the nineteen possible benefits met their criteria.

One hypothesized benefit associated with locomotor play is an increase in the creation of synapses in the cerebellum. The cerebellum plays a critical role in limb coordination, movement, postural changes, eye-limb coordination, and many other aspects of movement in mammals (J. Byers and Walker, 1995; Pellegrini and Smith, 1998; Thatch et al., 1992). During development, more cerebellar synapses are created than are used in later life, and some of these synapses may be pruned as a function of experience (M. Brown et al., 1991; Greenough and Juraska, 1979; Jacobson, 1991; Purves and Lichtman, 1980; Pysh and Weiss, 1979).

A

FIGURE 16.8. Pronghorn play.
Pronghorns undertake various forms of locomotor play, including **(A)** "high-speed running," **(B)** "fast turns," and **(C)** "stots" (jumping with all four legs simultaneously off the ground, like the pronghorn in the bottom left corner of the photo). *(Photo credits: John Byers)*

B

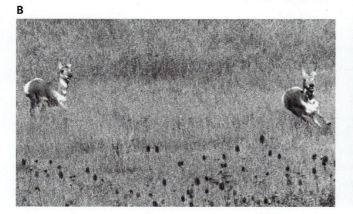

C

TABLE 16.3. Physiological effects of elevated motor activity. In examining the potential benefits of locomotor play, Byers and Walker listed nineteen benefits that might be associated with elevated motor activity. *(Based on J. Byers and Walker, 1995)*

SPECIFIC EFFECT	PRESUMED BENEFIT	EFFECT AVAILABLE TO JUVENILES?	EFFECT PERMANENT?	EFFECT AVAILABLE TO ADULTS?
Increase in maximum oxygen reuptake	Greater endurance	Yes	No	Yes
Decrease in heart rate during exercise	Greater endurance	Yes	No	Yes
Decrease in blood lactate level during exercise	Greater endurance	Yes	No	Yes
Increased heart weight : body weight ratio	Greater endurance	Yes	No	Yes
Increased myoglobin	Greater endurance	Yes	No	Yes
Greater numbers and size of skeletal muscle mitochondria	Greater endurance	Yes	No	Yes
Increased muscle glycogen and triglyceride stores	Greater endurance	Yes	No	Yes
Greater capacity to oxidize fat	Greater endurance	Yes	No	Yes
Greater slow-twitch fiber area	Greater endurance	Yes	No	Yes
Greater total blood volume	Greater endurance	Yes	No	Yes
Greater muscle capillary density	Greater endurance	Yes	No	Yes
Greater maximal ventilation rate	Greater endurance	Yes	No	Yes
Increased maximal muscle blood flow	Greater endurance	Yes	No	Yes
Bone remodeling	Increased strength	Yes	No	Yes
Fast-twitch fiber hypertrophy	Increased strength	Yes	No	Yes
Increased recruitment of motor units	Increased strength	Yes	No	Yes
Modification of cortical areas involved in movement	Increased motor skill/ energetic economy of movement	Yes	No	Yes
Modification of muscle fiber type differentiation	Increased motor skill/ energetic economy of movement	Yes	Probably	Unlikely
Modification of cerebellar synapse distribution	Increased motor skill/ energetic economy of movement	Yes	Yes	Diminished

Researchers have asked how play behavior maps onto cerebellar synapse formation and elimination. With respect to motor play in juvenile mice, the fit is quite good (Figure 16.9A). Mice start playing at about fifteen days of age and peak in their locomotor play activities at nineteen to twenty-five days, and this corresponds with a peak in cerebellar synapse formation. The same general pattern is found when examining locomotor and social play in rats (Figure 16.9B) and social play in cats (Figure 16.9C). A second major developmental change—the differentiation of muscle fibers that will be important for use in foraging and antipredator behaviors—also maps nicely onto the development of play (Close, 1972; Edgerton, 1978; Roy et al., 1988).

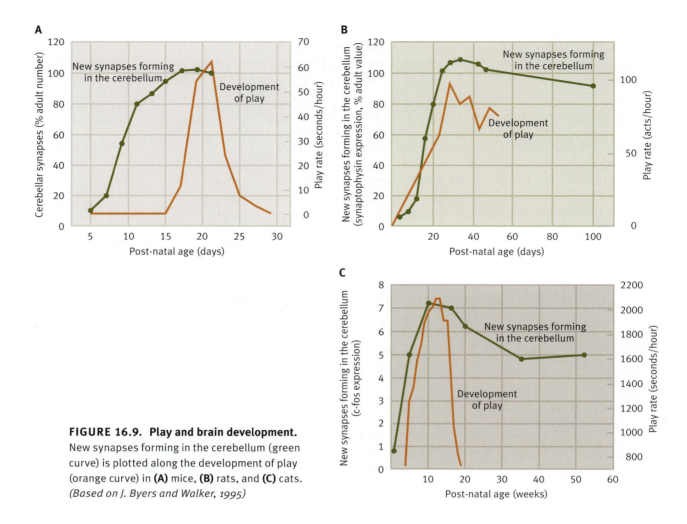

FIGURE 16.9. Play and brain development. New synapses forming in the cerebellum (green curve) is plotted along the development of play (orange curve) in **(A)** mice, **(B)** rats, and **(C)** cats. *(Based on J. Byers and Walker, 1995)*

It is important to realize that this sort of work is correlational. As Byers and Walker note in their study, we do not know if an increase in locomotor play causes an increase in cerebellar synapse formation or if an increase in cerebellar synapse formation leads to an increase in locomotor play. Future work is needed to infer causation.

SOCIAL PLAY

Social play—playing with others—is the most well-studied type of play. A number of functions of social play have been proposed (K. L. Graham and Burghardt, 2010; K. V. Thompson, 1996). (1) Social play may lead to the forging of long-lasting social bonds (Carpenter, 1934). For example, among immature chimps in the Arnhem Zoo (in the Netherlands) male-male social play is especially common (Mendozagranados and Sommer, 1995). A possible benefit of male-male social play sessions is to provide young males with coalition partners that may be important in their adult life (de Waal, 1992; see Chapter 10). (2) Like other forms of play, social play may promote and fine-tune physical skills, such as those relating to fighting, hunting, and mating. (3) Social play may aid in the development of cognitive skills (Bekoff, 2000, 2004, 2007). Let us examine the last two of these benefits of social play in a bit more detail.

SOCIAL PLAY IN BIGHORN SHEEP. In bighorn sheep (*Ovis canadensis*), males compete aggressively for mating opportunities with females (Geist, 1966, 1974a), and male reproductive variance is much higher than that of females—only a few males win many aggressive contests and hence mate, while almost all females have mating opportunities. Joel Berger hypothesized that if males and females differ in the degree to which aggression shapes their adult interactions, then such differences would manifest themselves much earlier, perhaps as early as lamb social play interactions (J. Berger, 1980; Hass and Jenni, 1993).

Berger studied juvenile social play in two populations of bighorn sheep, one in the sloping grassy hills of British Columbia and one in the Colorado Desert. In line with his prediction, Berger found that males engaged in much more contact play than female lambs. This contact play may help males prepare for the aggressive interactions that will play a large factor in their struggle for reproductive success in later life though this idea has not yet been tested against alternative hypotheses.

Three other interesting results emerged from the bighorn sheep study. First, social "contact" play (butting, pushing, and so on) was often preceded by "rotational" movements that appeared to serve as a signal that contact was part of a play sequence (see below for more on play signals). Second, social play behavior in bighorn sheep does not begin at a certain age per se; rather it starts after young individuals have associated with each other for a certain amount of time. Young that stayed by their mothers' side began contact play at later ages than did lambs that associated with other lambs from an early age. Finally, Berger also found differences in the frequency of play across his two populations. In both populations, lambs were involved in play at very early ages, but play tapered off much more quickly in the desert population than in the grassy slope population. One possible explanation for this difference is that young sheep in the desert were more energetically stressed. The data suggest otherwise, however, as lambs in the desert actually receive more milk from their mothers than lambs in the grassy hill population (J. Berger, 1979). Instead, it appears that desert lambs often bump into cactus while playing, and the consequent pain they suffer acts as a negative reinforcement for social play (J. Berger, 1980).

SOCIAL PLAY AND COGNITION. One cognitively related benefit of social play revolves around the idea of "self-assessment." Here, animals use social play as a means to monitor their developmental progress as compared with others. For example, in infant sable antelope (*Hippotragus niger*), individuals prefer same-age play partners. Kaci Thompson has hypothesized that it is primarily a function of young individuals attempting to choose play partners that provide them with a reasonable comparison from which to gauge their own development (K. V. Thompson, 1996; Figure 16.10).

With respect to cognition and social play, one question that ethologists have addressed is, How do animals, especially young animals, know that they are engaged in play (Bekoff, 2000, 2004, 2007)? And, How do they communicate this information to each other? Since many of the behavior patterns seen during play are also common in other contexts—hunting, mating, dangerous aggressive interactions—How do animals know they are playing and not involved in the real activity? Marc Bekoff has proposed three possible solutions to this important, but often overlooked, question (Bekoff, 2000).

One way that animals may distinguish play from related activities is that the order and frequency of behavioral components of play is often quite different

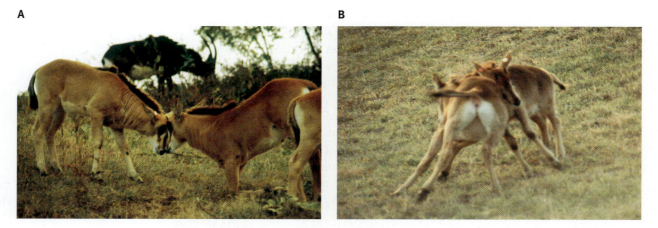

FIGURE 16.10. Antelope play. Young sable antelope like this pair often engage in play, particularly with same-age partners. *(Photo credits: Kaci Thompson)*

from that of the "real" activity (N. Hill and Bekoff, 1977). When play behavior is compared with the adult functional behavior that it resembles, behavioral patterns during play are often exaggerated and misplaced. Young animals may be able to distinguish these exaggerations and misorderings of behavioral patterns by, for example, observing adults that are not involved in play.

A second, somewhat related, means by which animals may be able to distinguish play from other activities is by the placement of **play markers** (Bekoff, 1977, 1995; J. Berger, 1980; Pellis and Pellis, 2011; Petru et al., 2009). These are also known as "play signals" and can serve to initiate play, to indicate the desire to continue playing, and to warn adults that the young are playing and not in danger of injury. In canids, for example, biting and shaking are usually performed during dangerous activities such as fighting and predation. Yet, biting and shaking are also play behaviors of young canids. Bekoff found that play markers such as a "bow" would precede biting and rapid side-to-side shaking of the head to indicate that they were not dangerous behaviors (Bekoff, 1995; Figure 16.11). The bow may communicate that this action should be viewed in a new context—that of play.

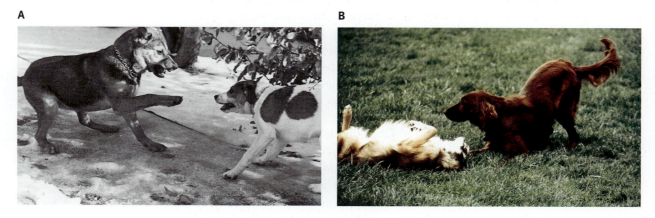

FIGURE 16.11. Play markers. (A) Play signals and canine aggression. The dog on the right growls, while the dog on the left "paws." Pawing is a play signal that can turn a potentially dangerous encounter into a playful one. **(B)** Play bows are also often used as signals that the bowing individual wants to play. The dog in this photo is play bowing to another individual (not visible in the picture). *(Photo credits: Marc Bekoff)*

FIGURE 16.12. Play face in gorillas.
Preceding bouts of aggressive play, juvenile gorillas use a facial gesture called a "play face," which appears to signal that "what is about to occur is play." *(Photo credit: William H. Calvin)*

Another play marker might be a particular kind of vocalization—for example, chirping in a rat, whistling in a mongoose, panting in a wolf or a chimpanzee—before or during a play interaction. Or there might be a distinctive smell that indicates that the animals were engaged in play.

Play markers have been found in primates such as the lowland gorilla (Palagi et al., 2007; Waller and Cherry, 2012). Juvenile lowland gorillas play with each other often, and play ranges from what Elisabetta Palagi and her colleagues call "gentle play" to "rough play." Palagi's team discovered that when juvenile gorillas—particularly males—were involved with rough play, the play was often preceded by a facial gesture they call the "play face" (Figure 16.12). This facial gesture, which is not seen in other contexts, includes slightly lowered eyebrows and an open mouth. In addition to using this facial gesture during rough play, juvenile gorillas also displayed it when a play session was in a place that made escape (leaving) difficult—another context in which it may be important to signal to others that "what is about to occur is play."

Another way in which young animals may be able to distinguish play from related behaviors is by **role reversal**, or **self-handicapping**, on the part of any older playmates they may have (Bekoff and Byers, 1998; Bekoff and Pierce, 2010; K. L. Graham and Burghardt, 2010). In role reversal and self-handicapping, older individuals either allow subordinate younger animals to act as if they are dominant during play, or the older animals perform some act (for example, an aggressive act) at a level clearly below that of which they are capable. Either of these provides younger playmates with the opportunity to recognize that they are involved in a play encounter.

The obvious follow-up question to "How do animals know they are engaged in play?" is "Even when engaged in play, why play fairly?" One possibility is that "cheating" at play—for example, hurting a partner during social play by inflicting damaging bites or swipes of a paw—probably does net an individual many real benefits. In addition, cheating may lead others to ostracize an individual at social playtime. If the benefits of social play are great enough, such ostracism could have a significant negative impact on a cheater.

PLAY FIGHTING AND COGNITIVE TRAINING IN SQUIRREL MONKEYS. Squirrel monkeys (*Saimiri sciureus*) engage in play fighting in both the lab and the field

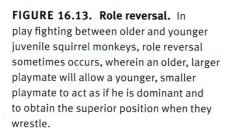

FIGURE 16.13. Role reversal. In play fighting between older and younger juvenile squirrel monkeys, role reversal sometimes occurs, wherein an older, larger playmate will allow a younger, smaller playmate to act as if he is dominant and to obtain the superior position when they wrestle.

(Baldwin, 1969, 1971; Biben, 1986; Dumond, 1968). Play begins at age five weeks old, when infants start interacting with each other while still on their mothers' backs. Young squirrel monkeys prefer playing with same-sex partners, and one fascinating aspect of play fighting in this species is the important position that role reversal plays, particularly in male-male play bouts (Biben, 1986; Figure 16.13).

As in young human children, young male squirrel monkeys prefer to play with others lower in dominance rank (Biben, 1986; Boulton, 1991; Humphreys and Smith, 1987). Nonetheless, while their preference may be to play with those that are subordinate to them, squirrel monkeys often do accept a dominant play partner. But why would subordinate individuals play with someone that is dominant to them (Altmann, 1962)? The answer appears to be that individuals that are clearly dominant *outside* the context of play often allow subordinates to take on the dominant role during play—that is, they engage in role reversal, wherein the subordinate does not defer to the dominant individual as he would outside the play situation, thus providing normally subordinate individuals with an incentive for playing. Somewhat surprisingly, role reversal breaks down in between-sex play. Males, which are typically dominant to females, do not engage in role reversal when playing with females.

Why are dominant males so quick to engage in role reversal when play fighting with same-sex partners? Part of the answer is that role reversal during play does not appear to influence the dominant/subordinate relationship *outside* of play, so the cost of role reversal is probably minimal. Yet, why bother with role reversal in the first place? The answer may be that without role reversal, few play partners would be available because subordinates would not be willing to play with dominants, so any benefits associated with play would be hard to obtain.

What might such benefits be for young male squirrel monkeys? One benefit might be that play fighting trains males for true aggressive behavior later in life. But no evidence for such benefits later in life has been found in squirrel monkeys—it is not as if those who play more win more fights later in life or those who win play fights win real fights. Instead, Maxeen Biben offers three possible benefits to play fighting in squirrel monkeys (Biben, 1998):

1. Behavioral flexibility: Because play involves little in the way of costs to squirrel monkeys, it may be a means for individuals to learn how to be amenable to changing behaviors, and trying new options that they might not otherwise try.

2. Gauging the intentions of others: Real fighting in adult squirrel monkeys can be very dangerous. Play fighting might provide males with training in gauging the intentions of others in adult life.

3. Experience in both the subordinate and dominant roles: Males that end up as dominant in adult life must "work their way up" a dominance hierarchy. As they do, they will lose as well as win many encounters. Play fighting may teach males how to act both as a subordinate and as a dominant—roles they will encounter throughout life (this may be particularly relevant to the role reversal aspect of male play fighting in squirrel monkeys).

Which, if any, of these possible benefits drives play in the squirrel monkey system remains to be tested.

A GENERAL THEORY FOR THE FUNCTION OF PLAY

Marek Spinka and his colleagues have hypothesized that the main function of play is to allow animals to develop the physical and psychological skills to handle unexpected events in which they experience a loss of control. Specifically, they propose that "play functions to increase the versatility of movements used to recover from sudden shocks such as loss of balance and falling over, and to enhance the ability of animals to cope emotionally with unexpected stressful situations" (Spinka et al., 2001, p. 141). So, for example, the loss of control and balance associated with being chased by predators or losing an aggressive interaction may be dealt with more effectively if play allows animals to prepare for such events.

Spinka and his collaborators list twenty-four predictions that emerge from their hypothesis. Here we will touch on a few of these predictions and the evidence available to judge them. At the most general level, Spinka and his colleagues predict that the amount of play experienced will affect an animal's ability to handle unexpected events. While this prediction has not been directly tested, some correlational work in both humans and nonhumans supports it. In rats, for example, individuals deprived of social play often react more negatively to unexpected stimuli than those not deprived of play (Potegal and Einon, 1989). In humans, measures of rough-and-tumble play are sometimes correlated with scores on social problem-solving tests (Pellegrini, 1995; Saunders et al., 1999).

A second prediction that emerges from Spinka and his colleagues' hypothesis is that self-handicapping, where dominant animals allow subordinates to defeat them during play fights, should be ubiquitous in species that play. Self-handicapping is thought to be an excellent means for preparing for the unexpected, as individuals put themselves in a position very different from that in which they normally find themselves. Evidence from many species supports the presence of self-handicapping in species that exhibit play.

During play, the brain must deal with sensory inputs that are different from the sensory inputs from other behaviors and with problems that must be solved by what Spinka and his group call "kinematic improvisation and emotional flexibility." As such, they predict that play should have measurable effects on an animal's somatosensory, motor, and emotion centers. In support of this prediction, rats that have been deprived of social play have long-term changes in opioid receptors, and they have permanently altered levels of dopamine and other neurotransmitters, all of which are important components of the stress response seen in these animals (Paul et al., 2005; Spinka et al., 2001; van den Berg et al., 1999).

One of the more obvious predictions from Spinka's general theory of play is that locomotor play should be most common in species that live in the most variable environments. If locomotor play allows one to experience loss of locomotor control, this effect might be most beneficial in environments that change most rapidly. More generally speaking, individuals that engage in play should be more prepared for the unexpected, which is more likely to occur where there is environmental change. Unfortunately, there is currently very little evidence available to test this particular prediction.

Endocrinological and Neurobiological Bases of Play

With one exception (the work on synapses and locomotor play), so far we have focused our analysis on ultimate questions about play behavior. In this section, we will take a more proximate perspective on play, by focusing on the endocrinological and neurobiological bases of play in rats and squirrels.

PLAY FIGHTING IN YOUNG MALE RODENTS

Male rats fight with one another from early on in development. While fights among adult males can be dangerous and can have significant effects on male reproductive success, aggression among younger rodents, including rats, often takes the form of much less dangerous play fighting (Figure 16.14). This type of social play may prepare rats for the "real" fights they will engage in as adults—and the stress that will arise from such fights. Here we will examine some work on the endocrinology and neurobiology of play fighting behavior in young male rodents.

TESTOSTERONE AND PLAY FIGHTING. Much of the work on the endocrinology of play has focused on testosterone because this hormone has long been linked to male aggression (R. Nelson, 2005). As discussed in earlier chapters, when examining the relationship between testosterone and aggression, it can often be difficult to assign cause and effect. For example, suppose you hypothesize that young male rats with high levels of testosterone are more likely to win play fights than males with lower levels of testosterone. If you took testosterone measures after a fight and found that males with higher levels of testosterone won more aggressive encounters, you would not be able to distinguish cause and effect—winning may be a result of high levels of testosterone, or high testosterone levels may be a result of winning. Even if you had been able to measure testosterone levels in males before they fought, and victors had high levels of prefight testosterone, determining cause and effect would still be difficult, as some other factor might have caused an increase in both testosterone and the probability of victory—that is, testosterone level per se may not have caused an increased probability of winning play fights.

To see how experimental work can distinguish between cause and effect with respect to hormones and play fighting, let us examine the work of Serge Pellis and his colleagues on play fighting and testosterone in rats (Pellis, 2002; Pellis,

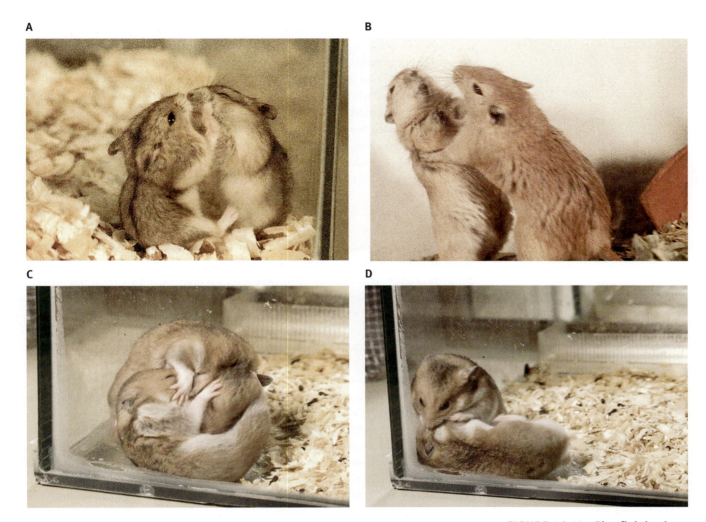

FIGURE 16.14. Play fighting in rodents. (A) Djungarian hamsters (*Phodopus campbelli*); **(B)** fat sand jirds (*Psammomys obesus*); **(C** and **D)** Syrian golden hamsters (*Mesocricetus auratus*). *(Photo credit: © Serge Pellis)*

Pellis, and Kolb, 1992). Play fighting in rodents typically involves attack and defense of the nape area around the neck, and males engage in this sort of activity significantly more often than do females. Early work suggested that sex differences in play fighting were related to testosterone levels, as males that were castrated at birth reduced their play fighting to levels typically seen in juvenile females, and young females whose levels of testosterone were experimentally increased were involved in more play fights (Beatty et al., 1981; Olioff and Stewart, 1978).

To experimentally examine the effect of testosterone levels on play fighting, Pellis and his team injected neonatal male rats with either testosterone propionate (TP) or an oil substance as a control (Pellis, Pellis, and Kolb, 1992). They then compared play fighting of TP-treated and control rats between days thirty and thirty-six. Pellis and his team found that the rate of initiating playful attacks was significantly greater for TP-treated rats.

Evelyn Field, Pellis, and their colleagues next examined whether it was the presence of testosterone or the transformation of testosterone into other substances that affected play fighting in rats. To do so, they worked with a unique strain of rats, called the *tfm* strain (E. F. Field et al., 2006). Male *tfm* rats have testes that secrete normal levels of testosterone, but they lack the gene associated with the production of testosterone receptors, and their appearance is feminized (Purvis et al., 1977;

Yarbrough et al., 1990). Because they lack this gene, the typical process in which some testosterone is transformed into estrogen by the aromatase enzyme in the liver does not occur in male *tfm* rats (McCarthy, 1994; Olesen et al., 2005; K. L. Olsen and Whalen, 1982). Pellis and his team reasoned that work with *tfm* males would allow them to determine whether it was the presence of testosterone or the transformation of testosterone into estrogen that was most important in the development of male play fighting. Their results suggest that both the production of testosterone and its transformation (called aromatization) to estrogen are important in the development of play fighting in males, though some components of play fighting are more tightly tied to the former, and other components are linked more closely to the latter (E. F. Field et al., 2006).

NEUROBIOLOGICAL APPROACHES TO THE STUDY OF PLAY. Ethologists have also examined the neurobiology and neurochemistry of play (Siviy and Panksepp, 2011). Broadly speaking, neuroethologists use one of two techniques to study play behavior. In the first, neurotransmitters are targeted to examine their role in play (inhibition, stimulation, and so on), and in the second, the actual neural pathways involved with a particular form of play are targeted, either by making surgical lesions or by some pharmacological means that "lights up" a neural pathway. Both of these techniques have proved useful in studying the biochemistry of play fighting in rats (Auger and Olesen, 2009; Siviy and Panksepp, 2011).

With respect to examining neurotransmitters and their role in social play, researchers often systemically administer a compound that either blocks or enhances a particular neurotransmitter (Auger and Olesen, 2009). If they do so with enough neurotransmitters, a broad picture of the neurochemistry of play emerges. For example, three neurotransmitter systems—those involved with dopamine, norepinephrine, and serotonin—seem to be involved in rat play fighting. Dopamine inhibitors typically reduce play (Beatty et al., 1984; Holloway and Thor, 1985; Niesink and Van Ree, 1989; Siviy et al., 2011). Rather than looking at dopamine in terms of increasing or decreasing play activities, however, a number of researchers have argued that dopamine's most important function may be to invigorate or "prime" an animal to prepare for play (Blackburn et al., 1992; Salamone, 1994; Siviy, 1998).

Rats can be trained to anticipate play, so it is possible to directly examine whether anticipation of play is linked to changes in dopamine levels (Humphreys and Einon, 1981; Normansell and Panksepp, 1990). Stephen Siviy constructed an experimental apparatus that consisted of two chambers connected by a tube, and he counted the number of times that rats crossed the tube each day (Siviy, 1998; Figure 16.15). Two groups of rats were tested. Those in the "play group"

FIGURE 16.15. Experimental device for studying play. This experimental apparatus served as a "play city" for Siviy's work on rats, play, and neurotransmitters. *(Photo credit: Steven Siviy)*

were allowed to play with another rat in the experimental apparatus for five minutes each day. Those in the control group had five minutes in the apparatus, but no other individual was present, and hence no social play occurred. Rats in the play treatment crossed back and forth in the tube *before* their partner was placed in the experimental apparatus much more than did rats in the control condition at the same point in time.

One interpretation of Siviy's experiment is that rats in the play treatment anticipated the opportunity for play and searched it out, increasing their number of crossings. To further examine this possibility, half the rats were given a dopamine inhibitor drug. These rats reduced their tunnel crossings significantly, but their play behavior, once a partner was present, remained level, providing support for the notion that dopamine acts to increase the anticipation of play (Figure 16.16). Dopamine may also be involved in the increase in the "chirping" sound that is often heard during rat play but not at other times (Panksepp, 2005; Panksepp and Burgdorf, 2003). By neurochemically stimulating dopamine receptors in the rat brain, researchers have increased the rate of chirping, and behavioral geneticists have even bred strains of rats that chirp excessively during play (Burgdorf et al., 2005).

A second question addressed in the neurobiological literature on play fighting in rats is What neural pathway(s) underlie play? One technique used in searching for neural pathways is lesioning various areas of the brain and testing for the effect of such a lesion on play. While this technique has had some limited success (Panksepp et al., 1984; Pellis, Pellis, and Whishaw, 1992; Pellis et al., 1993; Siviy, 1998; Siviy and Panksepp, 1985, 1987), more recent work suggests that modern biochemical tools may provide an even greater window into the neurobiology of play, enabling us to identify areas of the brain that are involved as well as the genes that are active and the changes that are occurring in the neurons. For example, early work had shown that lesions to the parafascicular area (PFA) of a rat's brain reduces play fighting (Siviy and Panksepp, 1985, 1987). Siviy followed up this work using biochemical and immunohistological tools to better understand whether a neural pathway underlying play could be identified in the PFA (Siviy, 1998). He hypothesized that if the PFA was critical to play behavior, then cells in this portion of the rat brain should be very active during play. This activity can be measured by quantifying the amount of a protein product (associated with the *c-fos* gene) that is found in the PFA.

Siviy found that rats that had just been involved in play had much higher neural activity in the PFA than did control rats. Somewhat surprisingly, however, this increased neural activity was not limited to the PFA, but rather it was found in other areas of the brain (in the cortex and hypothalamus) as well.

DEVELOPMENTAL BASIS OF SEXUAL PLAY IN YOUNG BELDING'S GROUND SQUIRRELS

In natural populations of Belding's ground squirrels (*Spermophilus beldingi*), males initiate much more sexual play than do females, but no differences in play behavior are evident across the sexes with respect to play fighting (Figure 16.17). Sexual play in this species involves young individuals assuming the positions and behaviors associated with copulation.

In order to better understand the developmental basis for this difference in sexual play, Scott Nunes and his colleagues treated one group of newborn

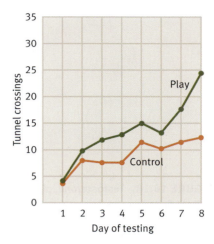

FIGURE 16.16. Anticipating play. Mean number of tunnel crossings in rats. Rats in the "play" treatment were given a five-minute opportunity to play with a same-age partner in a "play city" right before the test. Control animals had no such opportunity. *(Based on Siviy, 1998, p. 229)*

FIGURE 16.17. Squirrel play. Belding's ground squirrels exhibit social play. Social play often resembles adult copulation. *(Photo credit: Scott Nunes)*

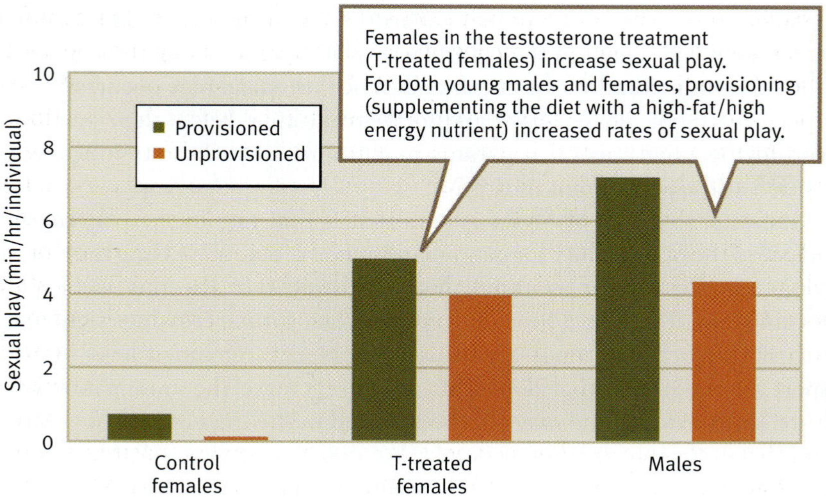

FIGURE 16.18. Testosterone, play, and food. In juvenile Belding's ground squirrels, males typically display much higher levels of sexual play than do females. *(Based on Nunes et al., 1999)*

Figure annotations:
Females in the testosterone treatment (T-treated females) increase sexual play. For both young males and females, provisioning (supplementing the diet with a high-fat/high-energy nutrient) increased rates of sexual play.

Y-axis: Sexual play (min/hr/individual)

Legend: Provisioned, Unprovisioned

X-axis: Control females, T-treated females, Males

females with testosterone (in an oil capsule) and a second control group with a capsule that contained only water (Nunes et al., 1999). In both treatments, total play behavior was highest near weaning and gradually decreased during post-weaning time. Females in the testosterone treatment displayed a significantly increased frequency of sexual play behavior. In fact, testosterone-treated females displayed sexual play behavior at almost the level displayed by same-age males.

In addition to examining developmental underpinnings to play via hormonal manipulation, Nunes and his team manipulated the food available to young squirrels to test whether what they had eaten affected their play behavior. In a "provisioned" treatment, pregnant females and their newborns had their diets supplemented with a high-fat/high-energy nutrient (peanut butter), while controls received no such supplement. The provisioning treatment had a significant effect on juvenile play behavior. In the provisioned groups, rates of play increased, while no change in play was uncovered in the nonprovisioned groups (Figure 16.18).

A Phylogenetic Approach to Play

Social play has been investigated in detail in muroid rodents. Within this family of rodents—a family whose phylogeny is fairly well established—a wide range in the complexity of play has been observed (Jansa and Weksler, 2004; Pellis and Iwaniuk, 1999). Serge Pellis and Andrew Iwaniuk examined this range of complexity in a phylogenetic context and tested whether species with relatively complex play share a common ancestor, and whether species with simpler play repertoires share a different common ancestor that displayed simpler forms of play.

To begin their analysis, Pellis and Iwaniuk developed a composite score for play "complexity" for each of the thirteen muroid rodent species they examined. These composite scores were based on seven different measurements of play seen in rodents—where higher scores indicate greater play complexity. Values for this complexity measure varied from 0 in *Pseudomys shortridgei* to 0.94 in *Rattus norvegicus*. The researchers tested whether similarities in complexity scores were due to common ancestry in the sense that complexity scores mapped well onto the phylogenetic relationship among the species they studied (Maddison and Maddison, 1992; Pellis and Iwaniuk, 1999). None of the aspects of play fighting that were examined could be explained by phylogenetic relationships, in that closely related species of rodents were no more likely to share similar play complexity scores than were species that were much more distantly related.

Pellis and Iwaniuk's phylogenetic analysis suggests that the ancestral state of play in muroids was moderately complex (Figure 16.19). From this state of moderate complexity, species independently evolved either more complex or less complex play-fighting repertoires. This hypothesis remains to be tested, but if it is correct, it will reshape the way play evolution is conceptualized. If supported, ethologists must explain not only what selection forces acted to make play more complex but also how natural selection can favor less complex play as well. Pellis and Iwaniuk's hypothesis forces us to test whether both complex and simple play are adaptive, depending on the costs and benefits to individuals in the species being studied.

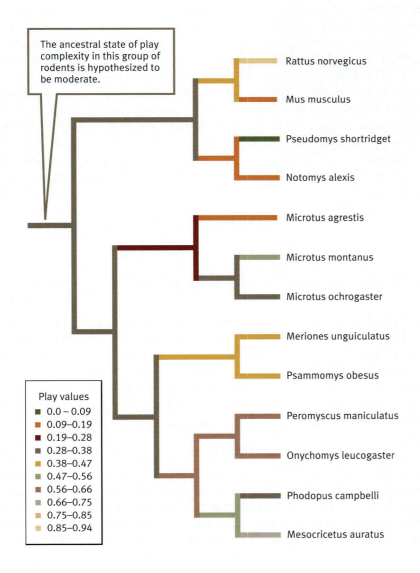

FIGURE 16.19. The phylogeny of play in rodents. The pattern of play complexity was mapped onto a phylogeny of muroid rodents. An analysis of the distribution of play complexity suggests that a moderate level of play (0.28–.38 on a scale of 1) was the ancestral state in this group. *(From Pellis and Iwaniuk, 1999)*

INTERVIEW WITH

Dr. Marc Bekoff

What do researchers have to document before they feel comfortable calling some behavior "play"?

This is a very good question, but I think that it's difficult to know that a behavioral interaction is play until it actually begins. In canids, for example, one would look for a play signal, such as the "bow," to know that at least one of two dogs, coyotes, or wolves wants to play and is saying "I want to play with you, not fight, mate, or eat you," and then see what the recipient of the bow or other signal does. Then, as the interaction ensues one would look for a variety of actions used in sequence and perhaps a bow or other play signal being used to punctuate the interaction to make sure that play remains "the name of the game." When bows, for example, are used during an encounter the message is "I still want to play with you" or "I'm sorry I bit you too hard, let's keep playing." Cooperation, apology, and forgiveness are part of the interaction and our data on dogs, coyotes, and wolves shows this to be so. Of course some play interactions begin spontaneously, especially between animals who know one another, but even then it's most usual to see some sort of play signal. So, it's the presence of a play signal at the beginning and then seeing them used from time to time during an encounter that helps one reliably say this is play.

How do you avoid anthropomorphism —attributing human emotions to nonhumans—when studying play?

Of course it's impossible not to use words that describe human emotions to describe, interpret, and explain the behavior of nonhuman animals so this isn't unique to play. Using solid evolutionary theory about evolutionary continuity as argued by Charles Darwin, it's safe to say that

"If we have something, 'they' (other animals) have it too." The differences among animals are differences in degree not kind. With respect to play, it's clear by watching and very carefully describing what the animals are doing, that they feel safe and secure and are comfortable letting another animal bite or mount them. They trust their play partner that play is the name of the game. It's also clear that animals enjoy playing—it's a voluntary activity and animals can be unrelenting in trying to get another animal to play. Play is also contagious and it's common to see play spread among a group of animals after some begin playing. All mammals, for

example, share the same structures in the limbic system that are important for processing emotions so in fact, the "problem" of anthropomorphism actually disappears *because we're not inserting something human into other animals that they don't already have.* I've written about what I call "biocentric anthropomorphism," which simply means that we need to take into account the nature of the animals about which we're talking and be very careful about the words we use.

Is it possible to experimentally examine whether animals enjoy play?

I think so, but perhaps right now we don't have the ability to do so in a noninvasive way that doesn't change the behavior of the animals involved. One way could involve somehow taking blood from the animals who are "on the run," and other than a pin prick it wouldn't involve any pain. And, when animals are playing I expect they wouldn't feel a pin prick. Another would be to do some brain imaging as the animals are playing to see what areas of the brain light up. However, I think that we already know that there is something very enjoyable about play because it's voluntary, animals seek it out, and they wouldn't seek out something that's not enjoyable, and also because across species we see that clear and unambiguous signals have evolved that allow play to continue without escalating into a fight or attempt to mate. Arguments from analogy appealing to evolutionary continuity go a long way here but I'm optimistic that in the future we will be able to do some really neat noninvasive neural imaging that will clearly show that animals truly enjoy playing.

What is the most unexpected thing you've learned from your own work on play behavior?

When I first began studying play decades ago I was told it was a waste of time—that animals didn't play, that the category of play didn't exist, or that it was so complex it was impossible to study. A small number of people told me that play was a wastebasket category of behavior and when someone didn't really understand what animals were doing they called it "play." I persisted and am thrilled that I did because not only did I learn that variations in play style are related to the life history patterns of various species, in my case members of the dog family, but also that we could learn a lot about the evolution of social communication. I suppose that the most unexpected thing that I've learned is how play is very carefully negotiated to make sure that it remains fair, and that among coyotes, at least, playing fair appears to be related to individual fitness. These data emerged after years of study and thousands of hours of observation but it seems that coyotes who don't play fair don't form strong social bonds with other group members and they often leave their group of their own accord, rather than being forced out. Other coyotes ignore them and they leave, and individuals who leave their group suffer as much as four times higher mortality than those who remain with their group. Of course, that playing fair has a fitness component isn't all that surprising if one understands how natural selection operates.

> It's clear from our own and others' research that play is related to moral development and is the way in which individuals acquire the social skills that are needed to interact fairly and cooperatively, and that this ability is important for maintaining the integrity and stability of social groups such as wolf packs.

What do you see as the next frontier in play research?

It's clear that we need the nitty-gritty details of what animals do when they play. There's no substitute for watching animals play by studying videos of what they do when they're on the run—how they ask another individual to play and how they maintain the play mood. From our own work on play it's clear that play is important in the development of knowing what's right and what's wrong, moral sentiments if you will, what I call "wild justice." I see the next frontier as learning much more about how play is related to the ways in which individuals learn species-specific rules of social engagement. It's clear from our own and others' research that play is related to moral development and is the way in which individuals acquire the social skills that are needed to interact fairly and cooperatively, and that this ability is important for maintaining the integrity and stability of social groups such as wolf packs. What's really exciting is how what we're learning about the development of play and what's right and what's wrong in nonhuman animals is very related to what we're learning about the development of play and moral development in young humans.

Dr. Marc Bekoff *is a professor at the University of Colorado. He is well known for his pioneering work on animal play behavior and animal rights.*

SUMMARY

1. The most widely accepted definition of play is "all motor activity performed postnatally that appears to be purposeless, in which motor patterns from other contexts may often be used in modified forms and altered temporal sequencing. If the activity is directed toward another living being, it is called social play." This definition is not without its problems, including determining when a behavior is purposeless.

2. Ethologists studying play behavior generally delineate three types of play: object, locomotor, and social play.

3. From a functional perspective, object play in young animals tends to be associated with some aspect of "practice" that will benefit the animal either in the short or long term and is tied to learning.

4. One hypothesis for the function of locomotor play in animals and humans is that it both provides general exercise and trains specific motor skills needed later in life.

5. With respect to social play, benefits may include forging long-lasting social bonds; sharpening physical skills, such as those relating to fighting, hunting, and mating; and aiding in the development of cognitive skills.

6. Spinka and his colleagues have hypothesized that play allows animals to develop the physical and psychological skills to handle unexpected events in which they experience a loss of control.

7. Neuroethologists often use two techniques to study play behavior. In the first, certain neurotransmitters are targeted and either inhibited or stimulated to examine their role in play. In the second, the neural pathways involved with a particular form of play are targeted and studied by making surgical lesions or using some pharmacological means to "light up" a neural pathway to determine what brain areas are affected, what gene products are involved, and what changes are occurring in neurons.

8. Play has been studied extensively across many species of rodents, and phylogenetic analysis of play in this group suggests that a moderate level of play is the ancestral state of rodent play in this group.

DISCUSSION QUESTIONS

1. Based on what you have learned in this chapter, try to construct a definition of play. After you have done so, answer the following questions: Does your definition cover all cases of play? Does it cover behaviors that you don't consider to be play?

2. It is very difficult to study whether play is enjoyable to nonhumans. Can you construct an argument that it is, at least in principle, possible to know the answer to this question? If so, what would your argument be? If not, can you give other reasons why you believe we can't know whether animals enjoy play?

3. Recall from the start of the chapter Gordon Burghardt's work with play in turtles. How could this sort of study help in the design of animal habitats in zoos?

4. Think about play in young children. Does reading a book for pleasure count as play? Does watching a movie or television show or playing a video game count as play? If they are considered play, how might these activities fit into Spinka's hypotheses about play?

5. Some researchers have suggested that play facilitates "creativity." After constructing your own definition of creativity, how would you test this hypothesis? Can you construct tests that measure both the behavioral and neurobiological/endocrinological correlates of play as they relate to creativity?

SUGGESTED READING

Bekoff, M., & Byers, J. (Eds.). (1998). *Animal play: Evolutionary, comparative and ecological perspectives*. Cambridge: Cambridge University Press. An edited volume that covers many aspects of the ecology and evolution of play.

Burghardt, G. (2005). *The genesis of animal play: Testing the limits*. Cambridge, MA: MIT Press. A thought-provoking book on ethology and play.

Graham, K. L., & Burghardt, G. M. (2010). Current perspectives on the biological study of play: Signs of progress. *Quarterly Review of Biology, 85*, 393–418. An overview and synthesis of work on animal play.

Held, S. D. E., & Spinka, M. (2011). Animal play and animal welfare. *Animal Behaviour, 81*, 891–899. An examination of ways the study of animal play can inform work on animal welfare.

Spinka, M., Newberry, R., & Bekoff, M. (2001). Mammalian play: Training for the unexpected. *Quarterly Review of Biology, 76*, 141–168. A proposal that training for unexpected events is central to understanding both the proximate and ultimate underpinnings of play behavior.

17

Animal Personalities

Watch almost any group of animals long enough, and you will start noticing individual differences among group members. For example, some primatologists will tell you that chimpanzees have personalities in the same way that you and I do. Once you get to know a chimp's personality, you can even predict, in a general fashion, how it will act when placed in a new behavioral scenario. For example, take a chimp with an "anxious" personality and place it in a novel environment, and it will act differently than a chimp with a "bold" personality. The same general argument can be made with respect to many animals, most of whom are not nearly as closely related to humans as chimps are. For example, in the sticklebacks that are discussed numerous times throughout this book, we can find what might arguably be called "bold" and "inhibited" personalities. If we test a series of sticklebacks over and over again in the presence of danger, we end up with very distinct behavioral types. Some fish are willing to take risks and inspect the source of this danger, and others aren't.

Although there are certainly instances in which a single behavior is the optimal solution to a problem faced by an animal, oftentimes multiple behavioral solutions to problems co-exist. Animals can be cooperative or uncooperative, aggressive or passive, and so on—that is, a combination of different behaviors can often exist concurrently in a population. Here we will extend this concept even further and consider the possibility that consistent behavioral differences among individuals can amount to "personalities," which are sometimes also referred to as **coping styles**, or alternatively as "behavioral syndromes" or "temperaments" in the ethological literature (Bell, 2007; Dall et al., 2004; Dingemanse and Reale, 2005; Reale et al., 2007; Sigh et al., 2004a, 2004b).

As one example of how ethologists study personality, let's look at a study on orangutans that Alexander Weiss and colleagues conducted. For many years, psychologists have used questionnaires to measure subjective well-being in humans. Subjective well-being scales are one way to measure personality traits. Observers are asked to rate a person on a scale (from 1 to 7) with respect to such behaviors as positive versus negative mood, pleasure derived from social interactions, and ability of the individual to achieve his or her goals. These scales have been modified to measure subjective well-being in nonhuman primates (King and Landau, 2003; King et al., 2005; Uher, 2008; Uher and Asendorph, 2008; A. Weiss et al., 2012). For example, Alexander Weiss and his team have measured subjective well-being in 172 orangutans housed in zoos (M. J. Adams et al., 2012; A. Weiss et al., 2006, 2011).

Data from seven years of observation found that some orangutans scored high on subjective well-being and that such orangutans showed low rates of neurotic behavior and high levels of extraversion and were generally agreeable in interactions with other orangutans and their zookeepers. Other orangutans showed the opposite set of traits. In other words, Weiss and his colleagues found two very different personality types in their population of orangutans. When Weiss and his team looked at mortality data on these animals, they found something extraordinary: Orangutans that were rated as scoring high on subjective well-being measures—animals that were happier—lived significantly longer than those that scored lower on subjective well-being. The difference between one standard deviation above and one standard deviation below average scores on subjective well-being translated into an orangutan living an average of 11.34 years longer (A. Weiss et al., 2011; Figure 17.1). Personality differences translate into differences in life expectancy.

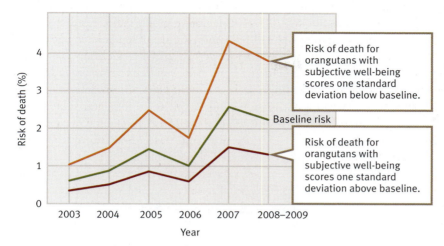

FIGURE 17.1. Subjective well-being and mortality. Orangutans that score high on subjective well-being measures live longer than baseline value; those that score low on subjective well-being measures live shorter than baseline values. The difference between one standard deviation above and one standard deviation below average scores on subjective well-being translated into living an average of 11.34 years longer. *(From A. Weiss et al., 2011)*

We will conceptualize **personality differences** as consistent long-term behavioral differences among individuals (A. M. Bell, 2007; Caro and Bateson, 1986; Carere and Maestripieri, 2013; A. B. Clark and Ehlinger, 1987; Dingemanse and Reale, 2005; Pervin and John, 1997; Stamps, 2007; M. M. Webster and Ward, 2011; A. M. Bell and Aubin-Horth, 2010). Personality traits may be heritable, as well as be affected by individual learning and social learning especially when learning occurs early in development and has long-term effects.

We can cast our discussion of animal personalities in a theoretical mold by viewing individual differences in the language of evolutionary game theory (A. M. Bell, 2007; Dall et al., 2004; Dingemanse and Reale, 2005; Reale et al., 2007; Sih et al., 2004a,b.) Recall that game theory models often produce a solution that contains more than one behavioral strategy—for example, hawks and doves, or cooperators and defectors (Chapters 10 and 15). If individuals adopt strategies for long periods of time, such strategies can be thought of as personality variables. To see how this might work, let's return to the producers and scroungers discussed in Chapter 11. When it comes to foraging in groups, animals often adopt one of two very different strategies. Producers search for food, so they accrue the costs associated with uncovering new food patches. Scroungers, on the other hand, watch producers, and learn where new food patches are by parasitizing the work of producers.

Chris Barnard and Richard Sibly constructed a game theory model of the producer-scrounger scenario that we can use as a baseline model of personality types (producers and scroungers; Barnard and Sibly, 1981; Figure 17.2). The solution to this game is some combination of both producers and scroungers, and the equilibrium frequency of each strategy depends on the exact costs and benefits associated with producing and scrounging. In any game theory model, solutions that contain more than one behavioral strategy

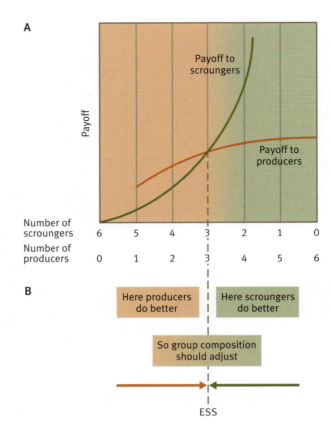

A

Payoff

Payoff to scroungers

Payoff to producers

Number of scroungers	6	5	4	3	2	1	0
Number of producers	0	1	2	3	4	5	6

B

Here producers do better

Here scroungers do better

So group composition should adjust

ESS

FIGURE 17.2. Producers and scroungers. (A) Hypothetical payoff to producers and scroungers as a function of group composition, and **(B)** the ESS, given the payoffs in Panel A. *(From Barnard and Sibly, 1981)*

can be thought of in at least two ways. Imagine that the equilibrium frequency of producers is 50 percent, and the equilibrium frequency of scroungers is 50 percent. This might translate into each individual in a group using the producer strategy 50 percent of the time and the scrounger strategy the remaining 50 percent of the time. Alternatively, we might see 50 percent of the individuals in a group consistently adopting the producer strategy and 50 percent of the individuals adopting the scrounger strategy (Figure 17.3). When this is the case—and it often is—the producer-scrounger game can be thought of in terms of predicting personality traits (Giraldeau and Caraco, 2000).

The producer-scrounger game helps us cast the evolution of personality types in a tractable format. It is important to realize that this model is not meant only for the case of food producers and scroungers per se, but for any situation in which there are personality types with the same kinds of costs and benefits associated with producers and scroungers. For example, in many group-living species, some individuals take on the role of "leaders," directing the orientation and movement of a group, and others are "followers." Recent work in groups of barnacle geese (*Branta leucopsis*) has found consistent leaders, who tended to be bold risk takers, and followers, who tended to be less bold than leaders (Kurvers et al., 2009, 2011; Figure 17.4). A leader is likely to have first access to new resources but may pay various costs—predation, parasitism—that followers do not. In that sense, the payoffs to leaders and followers are similar to those for producers and scroungers.

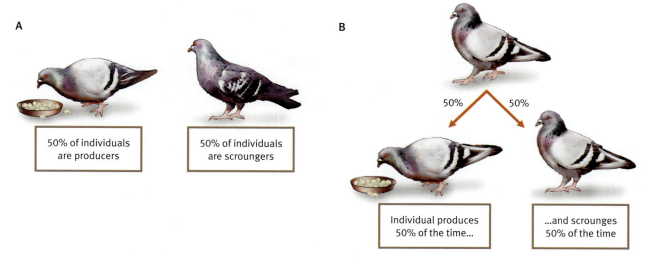

FIGURE 17.3. Behavioral strategies. Suppose a model predicts an equilibrium of 50 percent producers and 50 percent scroungers. This can occur by either **(A)** having 50 percent of the individuals play producer and 50 percent of the individuals play scrounger, or **(B)** having individuals act as producers 50 percent of the time and as scroungers 50 percent of the time.

Producers and scroungers and leaders and followers are only two ways in which personality traits manifest themselves in animals and humans. In this chapter, we will examine

- bold and inhibited personality types,
- a series of case studies of personality across a wide array of species and ecological conditions,
- differences in coping styles,
- practical implications of animal personalities.

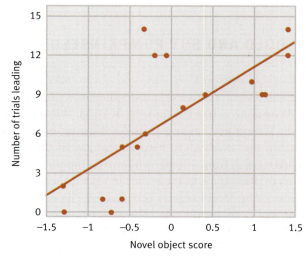

FIGURE 17.4. Leadership and novel objects. In barnacle geese, birds that were leaders—directing the orientation and movement of their groups—also were more willing to explore novel objects than were "follower" geese. *(From Kurvers et al., 2009)*

Boldness and Shyness

Psychologists have long proposed that where a person falls on the continuum from very shy to very bold behavior is one of the most stable personality variables yet studied. If you are shy when you are young, chances are very good that you will be shy when you get older (Kagan, 1994; Kagan et al., 1987, 1988, 1989). While definitions of shyness and boldness vary considerably, **boldness** usually refers to the tendency to take risks in both familiar and unfamiliar situations, while **shyness** refers to the reluctance to take such risks, or even a reluctance to engage in unfamiliar activity at all. In the language of psychology, shyness is similar to behavioral inhibition, while boldness is similar to "sensation seeking" (M. Zuckerman, 1979, 1994).

Psychologists often speak of boldness and inhibition as overarching personality traits—individuals who are bold in one context are likely to be bold in other contexts, and likewise for shy individuals. Ethologists argue, however, that there is no reason that natural selection should necessarily favor boldness or shyness as overarching personality traits. Rather, natural selection might favor being bold in some contexts and shy in others, depending on the costs and benefits associated with the behavior (K. Coleman and Wilson, 1998; Rodel et al., 2006; A. D. M. Wilson and McLaughlin, 2007; A. D. M. Wilson and Stevens, 2005; D. Wilson et al., 1993; D. S. Wilson, 1994). This is the perspective that we will be taking here. Adopting such a view might lead us to think that shyness and boldness are not truly personality traits, as they are not generalizable across different contexts. There is, however, no reason that the context-specific term "bold forager," for example, should not be considered a personality trait, as long as individuals are *consistent* about being bold when foraging.

While a great deal of work has been undertaken to examine both inhibition and boldness from psychological, psychiatric, and physiological perspectives, until recently little in the way of controlled experimental work examined the *costs and benefits* of inhibition and boldness. From an ethological perspective, this omission is striking, as a thorough understanding of the evolution of inhibition and boldness can only be attained through knowledge of the costs and benefits associated with these traits. As such, we will examine two species in which the costs and benefits of boldness and inhibition have been studied in depth.

BOLD AND SHY PUMPKINSEEDS

David Sloan Wilson, Christine Coleman, and their colleagues have studied shyness and boldness in pumpkinseed sunfish (*Lepomis gibbosus*; Figure 17.5). Similar work has been done in bluegill sunfish (A. D. M. Wilson and J. G. Godin, 2009, 2010). D. S. Wilson and his team used a trapping method that allowed them to segregate shy and bold fish in natural populations (D. S. Wilson et al., 1993). They trapped sunfish from a pond using two techniques (Figure 17.6). The first technique was to place traps in the water. In effect, these underwater traps were meant to mimic the "novel object" test that psychologists use to classify humans along the shy–bold continuum. These traps were designed so that a pumpkinseed fish would have to actively swim into them to be collected, but once a fish was inside, the construction of the trap was such that it was very difficult for the fish to get out. Presumably, this technique would primarily capture bold sunfish, who would be the ones willing to enter the trap in the first place.

The second technique for capturing fish involved dragging a large net called a seine through the pond. Seining occurred immediately after all underwater traps

FIGURE 17.5. Bold and shy fish. Pumpkinseed sunfish have been studied extensively in order to understand the evolution of boldness and shyness. *(Photo credit: Dwight R. Kuhn)*

Trap

Trap

Trap

FIGURE 17.6. Experimental setup to study bold and shy fish. To examine boldness and shyness in pumpkinseed sunfish, David Sloan Wilson and his colleagues used two experimental techniques. In one, a large seine was dragged through a pond; in the other, underwater traps were used to capture fish.

were collected, and it was undertaken in such a manner as to capture as many sunfish in the vicinity of the traps as possible. Seining, then, should capture a combination of both inhibited and bold fish, while the trapping method should have captured, on average, bolder fish than the seining method. In addition, the researchers ran a control in which trapped fish were placed in a seine and run through the pond for thirty seconds (approximately how long it takes to run the seine through a pond). This was done to test whether the seining process itself was more traumatic than the trapping process, and hence might account for any differences found between seined and trapped fish. The results indicated that it was not.

Wilson and his colleagues next examined the diets of trapped and seined fish, their parasite loads, and their growth rates. In addition, they tagged and released fish back into the pond, and then made detailed observations of these marked fish. In one experiment in which both trapped and seined fish were tagged and released back into their pond, behavioral observations indicated that trapped fish were less likely than seined fish to flee from human observers, as might be expected from bolder fish. Behavioral observations of pumpkinseeds in their natural habitat suggest that the researchers indeed collected different proportions of bold and inhibited fish using the two techniques.

In ponds, trapped fish were more likely to forage away from other fish, and their diet contained three times as many copepods—small crustaceans that are the usual food of pumpkinseeds—as did the diet of seined fish (Ehlinger and Wilson, 1988; D. S. Wilson et al., 1996). In addition, trapped and seined fish differed in terms of the parasites they carried, suggesting different habitat use based on personality types. The diet and parasite data in conjunction with the behavioral observations suggest that trapped and seined fish, though caught in the same pond, behave quite differently in nature. But behavioral observations suggest that trapped and seined fish were not segregating into groups—that is, forming groups of bold fish and groups of shy fish—when swimming in natural aggregations in ponds.

Once seined and trapped fish were brought into the laboratory, Wilson and his colleagues ran a barrage of behavioral and physiological tests on them over the course of approximately three months (D. S. Wilson et al., 1993). Wilson's team found no

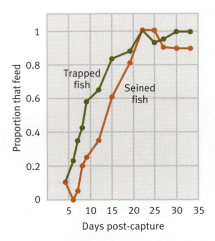

FIGURE 17.7. Feeding and boldness.
Two techniques were used by Wilson and his colleagues to capture pumpkinseed sunfish: trapping, which caught the boldest fish, and seining, which caught a mixture of bold and shy fish. Once brought into the lab, trapped fish acclimated to feeding more quickly than seined fish (this effect disappears with time). *(Based on D. Wilson et al., 1993)*

differences between trapped and seined fish with respect to age or sex (close to a 1:1 ratio was found in both groups). They did find that trapped fish acclimated to feeding in the lab more quickly than did seined fish, as might be expected if in fact trapped fish were bolder (Figure 17.7). Twenty-five days after fish were captured, Wilson and his colleagues ran a series of behavioral and physiological post-acclimation tests of the trapped and seined fish. These tests included a "response to handling" test, a novel object test, aggressive contests between seined and trapped fish, and physiological tests of stress. In no case did the trapped and seined fish differ from each other in any of these experiments. What's more, when fish were housed in individual aquaria in the laboratory, the differences between trapped and seined fish disappeared after thirty days of social and ecological isolation (but in bluegill sunfish these differences persist; Wilson and Godin, 2009). While there are numerous possible explanations for the disappearance of differences across samples over time in the laboratory, D.S. Wilson and his colleagues hypothesize that ecologically and socially relevant cues may solidify the differences between bold and inhibited pumpkinseed fish and that, in the laboratory, where such cues may be absent, differences in boldness and inhibition may either disappear or become too small to notice.

Wilson suggests that a bold/inhibited continuum exists in pumpkinseeds in nature, and that where individuals fall along this continuum affects factors such as diet, safety, and parasite load. Exactly how all the variables measured by Wilson's team interact to produce the frequency of shy and bold individuals seen in nature is complex and difficult to pin down.

GUPPIES, BOLDNESS, AND PREDATOR INSPECTION

Predator inspection behavior in fish (Chapters 2, 10, and 12) is a good behavioral scenario for measuring the costs and benefits of inhibition and boldness (C. Brown et al., 2005; Dugatkin and Alfieri, 2003; Fraser et al., 2001). Individual guppies differ in their tendency to inspect a predator, and both inspectors and non-inspectors are consistent in their behavior when predators are present (Dugatkin and Alfieri, 1991b; Dugatkin and Wilson, 2000; Figure 17.8).

FIGURE 17.8. Variation in risk taking.
Guppies are ranked by their risk-taking scores during predator inspection. Significant differences in boldness exist between individuals, but fish were fairly consistent in their risk-taking tendencies over time. *(Based on Dugatkin and Alfieri, 1991b)*

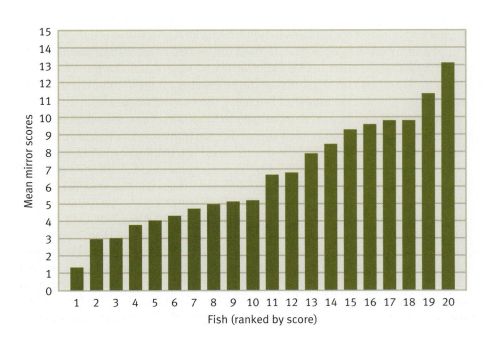

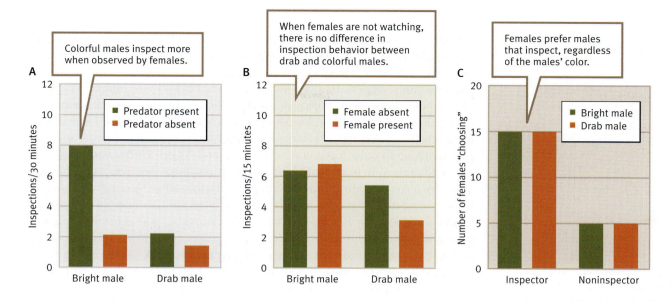

A

Colorful males inspect more when observed by females.

Inspections/30 minutes

- Predator present
- Predator absent

Bright male   Drab male

B

When females are not watching, there is no difference in inspection behavior between drab and colorful males.

Inspections/15 minutes

- Female absent
- Female present

Bright male   Drab male

C

Females prefer males that inspect, regardless of the males' color.

Number of females "choosing"

- Bright male
- Drab male

Inspector   Noninspector

D

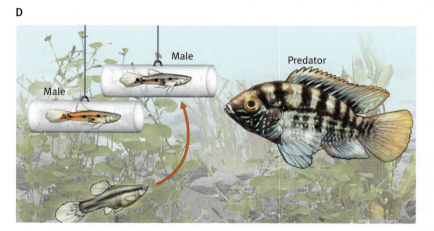

Male

Male

Predator

FIGURE 17.9. Color, boldness, and attractiveness.
(A) While females watched, the inspection behavior of males was examined in the presence and absence of predators. **(B)** Inspection behavior of males was examined in the presence and absence of females. **(C)** The effects of female presence and predator absence on male inspection behavior were experimentally uncoupled. *(Based on Godin and Dugatkin, 1996)* **(D)** To experimentally decouple boldness and color patterns in guppies, a motorized pulley system was built. Using this system, either colorful or drab males could be made bold by placing them in a small tube that was moved back and forth. The orange arrow indicates that the female preferred the bolder male as a mate. *(Based on Dugatkin and Godin, 1998)*

Because of this consistency, we will speak of guppies that approach predators often and closely as bold, and those that do not as inhibited.

Recall from the discussion of cooperation and predator inspection in Chapter 10 that approaching a predator is dangerous, especially for those who inspect the predator most often and most closely (refer back to Figure 10.6). One would hypothesize that there must be some compensating benefit(s) obtained by bold individuals, or natural selection would have culled this trait from the population. Jean-Guy Godin and I uncovered one such benefit in a study of inspection and mate choice (Godin and Dugatkin, 1996). In each trial of this experiment, a pair of males was placed in a large tank. On one side of this tank was another aquarium that served as housing for a pike cichlid predator. In some trials, the predator was present; in other trials, its tank was empty. In addition, in the large tank containing the male guppies, a clear partition was placed on the side of the aquarium farthest from the predator tank. In some trials, females were placed behind this see-through partition, and in some trials, this section remained empty. We found that more colorful male guppies were more likely to inspect predators (Figure 17.9A), but that differences in inspection behavior between colorful and drab males were manifest only when females were observing—no differences in inspection were found when males could not see females observing them (Figure 17.9B). Females also preferred colorful, bold males as potential mates. When we experimentally uncoupled color from boldness, it was a male's boldness, and not his color, that made him attractive to females (Figures 17.9C and 17.9D).

This study suggests that one benefit to being bold (versus inhibited) is that it makes males attractive as potential mates. Are there any other benefits associated with being a bold fish? To explore this question, Michael Alfieri and I tested male guppies both in terms of their boldness during inspection and their ability to associate a cue with a particular food source (Dugatkin and Alfieri, 2003). Two treatments were undertaken. In treatment I, males were evaluated on boldness first, and then they were given the learning task. In treatment II, the order of testing was reversed. Results suggest that bolder male guppies learn more quickly than do behaviorally inhibited individuals. But for this advantage to appear, bold behaviors must be expressed before the learning opportunities. It appears that predator inspection itself either "primes" bold fish to learn better or has a negative effect on learning in behaviorally inhibited inspectors. Whether behaviorally inhibited inspectors have higher levels of stress-related hormones after exposure to a predator than do bold individuals, and whether this difference in hormone levels translates into lowered rates of learning, remains an interesting area for future exploration.

Some Case Studies

Compared with most of the subjects we have examined in this book, animal personality is an understudied area of ethology, and there is no standard methodology used across studies. In addition, though it may be emerging in the growing literature on behavioral syndromes, there is not yet a unified theory for how personality evolves (Carere and Maestripieri, 2013; Sih, 2011; Sih and Bell, 2008; Sih, Bell, and Johnson 2004a; Sih, Bell, Johnson, and Ziema, 2004; but see Uher, 2011). That said, the number of studies on this subject is growing each year, and in this section, we will examine a few case studies to better understand the dynamics of animal personality.

HYENA PERSONALITIES

Spotted hyenas (*Crocuta crocuta*) are native to African savannas and live in stable clans that often engage in cooperative hunting (Figure 17.10). Females hold the alpha (top-ranked) status in groups, dominance is inherited maternally, and the overall dynamics of social life in hyenas are complex (Engh et al., 2000; Frank, 1986a, 1986b; Gosling, 1998; Holekamp and Dloniak, 2010; H. E. Watts et al., 2010; Zabel et al., 1992). This complexity set the stage for Sam Gosling's study of personality in the spotted hyena (Gosling, 1998).

Gosling studied thirty-four hyenas to determine whether personality could be distilled down to a few variables in this group-living species—that is, he wanted to examine whether hyenas that displayed behavior x were more or less likely to display behaviors y, z, and so forth. As far as animal behavior studies go, Gosling's work on hyenas began in a somewhat unconventional manner. At the start of this work, three experts on hyena behavior were provided with a list of more than forty traits that had been employed in other studies on animal personality, and they were asked which traits they believed would apply to hyenas. Then

FIGURE 17.10. Hyena personalities.
Gosling studied forty-four personality traits in spotted hyenas. A detailed analysis revealed that hyena personalities are most easily understood in terms of assertiveness, excitability, human-directed agreeableness, sociability, and curiosity. *(Photo credit: E. R. Degginger/ Color-Pic, Inc.)*

these experts were provided with a list of forty traits used in human personality work, and they were asked which applied to hyenas. Based on the replies of the three experts, Gosling amassed a master list of sixty personality traits. Finally, two other hyena experts were asked to go through this master list and to remove any redundancies (for example, "fearful" and "apprehensive" covered the same trait), and forty-four traits remained after this final culling process.

Four observers then used the traits to score the behavior of thirty-four hyenas. Next, Gosling ran a statistical analysis to determine whether certain traits grouped together, and whether such groupings made any sense in terms of hyena personality attributes. Results of this analysis suggest that hyena personalities are most easily understood in terms of five aggregate traits—assertiveness (which incorporated fifteen of the original forty-four traits), excitability (twelve of the original traits), human-directed agreeableness (seven of the original traits), sociability (four of the original traits), and curiosity (six of the original traits).

Gosling examined whether any of these aggregate personality traits correlated with an individual's sex, age, or rank in the hierarchy. Neither age nor dominance rank correlated with these aggregate traits. Females were more assertive than males, however, as might be expected in a social system with matriarchal dominance hierarchies.

To get a broader picture of what the hyena work meant for personality research in animals, Gosling compared the five aggregate traits that he uncovered in hyenas with major traits found in two studies of rhesus monkeys and in one study of gorillas (Bolig et al., 1992; Gold and Maple, 1994; Stevenson-Hinde and Zunz, 1978; Stevenson-Hinde et al., 1980). Except for "human-directed agreeableness," which was not measured in the primate studies, hyenas and nonhuman primates shared major personality traits, suggesting some cross-species commonalities.

OCTOPUS AND SQUID PERSONALITIES

The vast majority of work on animal personality work has been done on vertebrates. This may reflect an inherent bias in the way we think about the complex nature of personality, but whatever the reason, the result is that a large portion of the animal world—the invertebrates—has largely been ignored. A comparative approach to personality in vertebrates and invertebrates may lead us to new insights on the evolution of personality. For example, Jennifer Mather and her colleagues have proposed that because cephalopods, such as octopuses and squids, display an array of complex behaviors, they are ideal subjects for a sortie into personality in invertebrates (Mather, 1995, 2008; Mather and Anderson, 1993, 1999).

Mather and Anderson tested the response of forty one-year-old *Octopus rubescens* to three treatments that were labeled alert, threat, and feed (Mather and Anderson, 1993). In the "alert" treatment, an experimenter opened the lid to an octopus tank and brought her head down to where the octopus could see it. The "threat" treatment involved using a brush to touch, and presumably frighten, an octopus, and the "food" treatment recorded an octopus's response when a food item (a shore crab) was put into its tank. The nineteen octopus behaviors displayed across these three treatments were subjected to a statistical analysis, and octopuses differed on three aspects of personality—active versus inactive, anxious versus calm, and bold versus inhibited (Table 17.1). Individual octopuses were very consistent with respect to these three personality traits, and these traits accounted for 45 percent of all the behavioral variance uncovered in the experimental treatments (Mather, 1991; Mather and Anderson, 1993; Figure 17.11).

It is interesting to note that the three components underlying octopus personality are quite similar to those documented in human infant development, as well as in rhesus monkey personality, providing further support for some cross-species similarities in terms of personality variables (Buss and Plomin, 1986; Stevenson-Hinde et al., 1980).

Personality traits have also been explored in another cephalopod, the dumpling squid (*Euprymna tasmanica*). David Sinn and his colleagues examined the behavior of these squid in two contexts—threat situations and feeding situations (Sinn and Moltschaniwskyj, 2005; Sinn et al., 2006, 2008, 2010). Each squid was tested four times in each context, and Sinn and his colleagues then examined

TABLE 17.1. Octopus personality traits. Three "factors" (or dimensions) incorporate the personality variables used in Mather and Anderson's work on personality in *Octopus rubescens*. (Based on Mather and Anderson, 1993)

DIMENSION	PREDICTOR BEHAVIOR
Factor 1: Activity (active/inactive)	In den[a]
	At rest[a]
	Grasp[b]
Factor 2: Reactivity (anxious/calm)	Squirt[b]
	Shrink[b]
	Swim[b]
	Crawl[b]
Factor 3: Avoidance (bold/inhibited)	In den[b]
	In den[c]
	Color change[a]
	Ink[b]
	Alert[c]

[a]Alerting test.   [b]Threat test.   [c]Feeding test.

FIGURE 17.11. Octopus personalities. Studies on personality in the red octopus *(Octopus rubescens)* represent some of the only work on invertebrates done in this field. *(Photo credit: Keith Clements/www. emeralddiving.com)*

whether any consistent personality traits could be detected. As in the red octopus, shy versus bold and active versus inactive emerged as two personality types in the dumpling squid. Further work showed that these traits were heritable (Sinn et al., 2006). Personality traits in squid were again context-specific—a squid that was consistently bold in the threat treatment was not any more likely to be bold in the feeding treatment than was a squid that was shy in the threat treatment.

RUFF SATELLITES

Let us return to the alternative male mating strategies of ruff birds (*Philomachus pugnax*), which were discussed in Chapter 2 (Figure 17.12). Here we will reexamine this phenomenon, but from the perspective personality. Recall

FIGURE 17.12. Two types of ruffs. Two male ruffs of different morphs co-occupying a single display court on a lek. The dark male (right) is an independent territory holder, while the white male (left) is a nonterritorial satellite male. As seen here, the independent male often tries to dominate the satellite by standing with his bill, which is a weapon, over the satellite's head. *(Photo credit: Oene Moedt, courtesy of David Lank)*

that "independent" males form leks and actively guard an area within the lek from other independent males. "Satellite" males, in contrast, temporarily share an independent's lek spot (Jaatinen et al., 2010; Lank et al., 1999) with that independent male. Independents and satellites differ not only in mating strategy, but in coloration and body mass, such that satellites are generally smaller and have lighter plumage (Bachman and Widemo, 1999; Hogan-Warburg, 1966). Alliances between independent and satellite males may be favored by natural selection, as females appear to prefer lek areas that contain both morphs (Hugie and Lank, 1997; Widemo, 1998).

Behavioral and morphological differences between independent and satellite ruffs are stable throughout an individual's life—indeed, as discussed in Chapter 2, such differences may be primarily due to genetic variance at a single locus—so we will consider independent and satellite as two different types of ruff personality.

The personality differences discussed above are seen only in male ruffs, but prior work has shown that male mating strategy was *not* inherited via sex chromosomes. This might seem counterintuitive: Since only males display this polymorphism (satellite versus independent), we might assume that the trait would be linked to sex chromosomes, but this need not be the case if females carry the gene for the trait of interest but simply don't express the phenotype associated with that gene. Genetic analysis has suggested that the gene for satellite versus independent is indeed carried, but not expressed, by female ruffs (Lank et al., 1995).

David Lank and his colleagues examined whether there was any way to produce similar phenotypic differences in females, who normally express neither the independent nor satellite personality traits (Lank et al., 1999). Bird systems are particularly good models for these sorts of experiments, as prior work has demonstrated that testosterone implants in female birds can induce phenotypes that are usually only expressed in males (J. T. Emlen and Lorenz, 1942; Witschi, 1961). In addition, in ruffs, males that are castrated fail to molt into breeding plumage, further suggesting a critical role for testosterone in the expression of traits normally only expressed in males (van Oordt and Junge, 1936).

Lank and his colleagues examined females with known pedigrees (lines of descent indicating male and female relatives), that had received testosterone implants, to determine whether they could induce the independent and satellite personality types in females. The use of females with known pedigrees was particularly helpful, as without such knowledge, researchers could only have ascertained whether male phenotypes could be expressed in females, but not whether a particular female expressed the phenotype that one would expect based on the *genotypes* of males in her family. In other words, with a pedigree analysis, Lank and his colleagues could test not just whether females expressed independent or satellite morphology and behavior, but also whether females expressed the morphology and behavior that might be expected based on the independent or satellite genotypes of males in their family. The researchers hypothesized that females with male relatives that were independents would be more likely to express the independent strategy after being injected with testosterone than females that had male relatives that were satellites.

Two days after testosterone implants, female ruffs (known as reeves) were displaying typical male behavior, and a week later they were actually forming leks. After three to five weeks, these reeves had grown neck ruffs and male-like display feathers, clearly demonstrating the effects of testosterone on both female behavior and morphology (Figure 17.13).

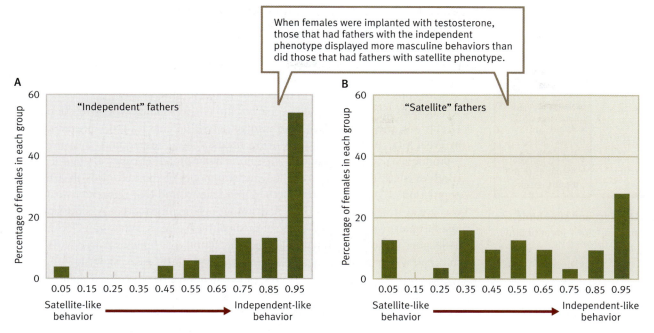

When females were implanted with testosterone, those that had fathers with the independent phenotype displayed more masculine behaviors than did those that had fathers with satellite phenotype.

A

y-axis: Percentage of females in each group (0, 20, 40, 60)

"Independent" fathers

x-axis: 0.05 0.15 0.25 0.35 0.45 0.55 0.65 0.75 0.85 0.95

Satellite-like behavior → Independent-like behavior

B

y-axis: Percentage of females in each group (0, 20, 40, 60)

"Satellite" fathers

x-axis: 0.05 0.15 0.25 0.35 0.45 0.55 0.65 0.75 0.85 0.95

Satellite-like behavior → Independent-like behavior

FIGURE 17.13. Hormones, heritability, and ruff personalities. In ruffs, two male phenotypes, independent and satellite, are well established. No such divergence in females is typically found. When females have testosterone experimentally implanted, however, females whose fathers displayed the "independent" genotype were more likely to behave like independent males. *(Based on Lank et al., 1999, p. 2327)*

Results from the pedigree analysis of the reeves were remarkably similar to those found in Lank's genetic study of male strategies, in which he uncovered two alleles, S and s, associated with the strategy that a male used. S is dominant to s, so SS and Ss males were satellites, whereas only ss males were independents. This means that females with independent fathers should express the independent phenotype, whereas females with satellite fathers should be a mix of satellites and independents. This is, in fact, what Lank and his colleagues found. Females that had independent fathers nearly always displayed the independent personality type. The distribution of personality types from reeves that had satellite fathers was much flatter: some females displaying satellite behavior, and some displaying much more independent traits (compare the shapes of the distributions in Figures 17.13A and 17.13B). These results strongly suggest that, at least in some cases, the inheritance of differences in personality can be understood using a simple genetic model of dominant and recessive genes.

NATURAL SELECTION AND PERSONALITY IN GREAT TITS

In a long-term study of exploratory behavior and reaction to novel objects in great tits (*Parus major*), Pieter Drent, M. E. Verbeek, and their colleagues have uncovered two very different personality types in these birds (Verbeek et al., 1994; Figure 17.14). "Fast" birds quickly approach novel objects and explore new environments rapidly, spending relatively short periods in any particular area. These birds are also aggressive—though they have surprisingly low testosterone

FIGURE 17.14. Personality types in great tit birds. Personality type has been studied in the great tit (*Parus major*). *(Photo credit: Roy Glenn/www.Ardea.com)*

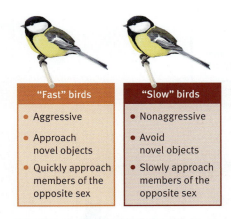

"Fast" birds	"Slow" birds
• Aggressive	• Nonaggressive
• Approach novel objects	• Avoid novel objects
• Quickly approach members of the opposite sex	• Slowly approach members of the opposite sex

FIGURE 17.15. Fast and slow birds. An overview "fast" and "slow" great tit birds.

levels (Chapter 3)—and once they develop a food-searching pattern, they are unlikely to change that pattern (Dingemanse et al., 2012; van Oers et al., 2011; Verbeek et al., 1996, 1999). On the other end of the behavioral spectrum, "slow" birds are reluctant to approach novel objects, vary their foraging routine often, are not physically aggressive (though they have high testosterone levels), and are slow to approach members of the opposite sex (Figure 17.15).

Besides documenting behavioral variation in personality type, Christian Both, Drent, and their colleagues have also gathered data associated with personality in great tits. In terms of measuring fitness, these researchers measured the reproductive success of slow and fast birds in a number of types of environments (Both et al., 2005). Using detailed measurements of marked birds in a natural population of great tits in the Netherlands, they found that reproductive success was greatest when pairs of birds had similar personalities—that is, slow-slow and fast-fast pairings led to production of the healthiest chicks. When van Oers and his colleagues looked at extrapair matings, they found that females paired with males that had a personality type similar to their own (fast-fast or slow-slow) had the highest rates of extrapair matings. One possible benefit of such extrapair partner choice is to increase behavioral variation among offspring (van Oers et al., 2011). Other work by Drent, Kees van Oers, Arie van Noordwijk, and Neils Dingemanse has shown that, in addition to having fitness consequences, personality in the great tit is also a heritable trait (Dingemanse et al., 2002; Drent et al., 2002; van Oers, Drent, de Jong, et al., 2004; van Oers, Drent, de Goece, et al., 2004). Indeed, a recent review of temperament and personality in nonhumans found evidence for heritability of these traits in twenty species (Reale et al., 2007, 2010).

Drent's team has also examined personality in the context of learning. For example, slow birds, while taking longer to explore an environment, also spend more time learning about each aspect of a new environment. To better understand whether other aspects of tit behavior correlate with fast-slow personality types, Chiara Marchetti and Drent examined whether birds could learn about changes in their foraging environment from watching a "tutor" bird (Marchetti and Drent, 2000). The theoretical backdrop for the protocol they devised was that slow birds might be similar to the producers in the producer-scrounger game, while fast birds might map onto the scrounger strategy.

After giving both slow and fast birds the opportunity to see a demonstrator bird forage, they found that although slow birds change feeders often when alone, do not use the information provided by tutors that help them discover new food sources (Marchetti and Drent, 2000). Conversely, fast birds, though reluctant to change feeders once they have established a routine of their own, are quick to change their foraging habits when paired with tutors that provide them with information about new food sources. Slow tits behave like producers, whereas fast birds act more like scroungers.

CHIMPANZEE PERSONALITIES AND CULTURAL TRANSMISSION

In the context of human personality studies, it would not strike anyone as surprising to hear that the society an individual lives in affects his or her personality. In certain limited conditions, this appears to be true for other primates as well. By far the best-documented case of cultural transmission of information affecting animal personality involves seven long-term studies of chimpanzees in Africa, studies that ranged from seven to thirty years in length (Whiten, 2005; Whiten and Boesch, 2001; Whiten et al., 1999, 2005; Lycett et al., 2007). Researchers in these projects constructed a list of sixty-five behaviors that qualified as "cultural variants" that were almost certainly spread by imitation. These variants ranged from using leaves as sponges to picking out bone marrow to displaying at the start of a rainstorm (Figure 17.16). Somewhat remarkably, of these sixty-five behaviors, thirty-nine were present at some sites, but completely absent at others. In addition, this list includes only culturally transmitted behaviors that were, in principle, possible at all sites. For example, since "algae fishing" is not possible when algae are absent, as it is at some chimpanzee sites, algae fishing is not included among these thirty-nine variants.

If you were to visit these seven sites in Africa and watch the chimpanzees for years and years, you would see relatively stable innovative behavior patterns being spread within each group, but across groups you would observe very different suites of behaviors being used. Chimps in different populations use different behaviors to reconcile after fighting, they groom one another in different ways, and they search for food in unique, but different manners. Because we are talking about a long list of behaviors, and because individual behavior is fairly

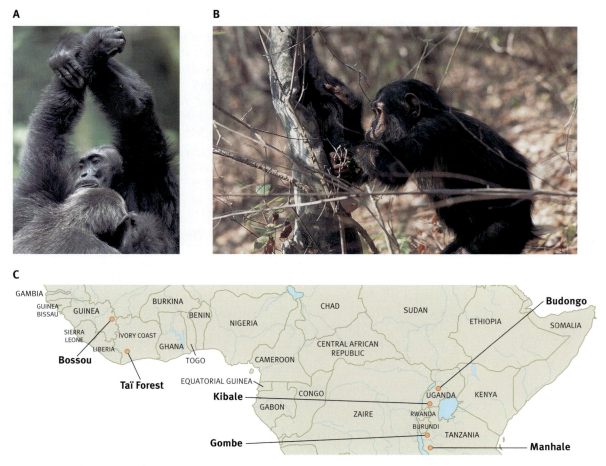

FIGURE 17.16. Chimpanzee culture. Cultural transmission of behaviors affecting animal personality has been examined in seven long-term studies of chimpanzees in Africa (study sites are indicated by orange circles). Variants of similar behaviors are often shown in different populations of chimpanzees. These behaviors include chimpanzees that **(A)** groom while touching their hands above their heads, and **(B)** use leaves as sponges to get water out of tree trunks or out of shallow puddles. **(C)** The sites in Africa where the chimpanzees were studied. *(Photo credits: World Foto/Alamy; Danita Delimont/Alamy)*

consistent within sites, it may be that the personalities of chimps across these seven African sites are very different, and such differences are the result of the cultural transmission of behavior.

It is frustratingly difficult to say more about chimpanzee personality per se than what has just been said. Given the large database available on chimps in the wild, why can't we say more about personality variables that appear to be under cultural control in wild populations? While there are some anecdotes on specific personality traits in specific individuals, controlled experimental studies of personality per se simply have not been undertaken on a large scale for chimps. There are many reasons for this, including the ethical considerations of personality work in chimps, the tendency to want to minimize experimental work in natural populations, and the long generation time of chimpanzees.

Coping Styles

One aspect of an animal's personality is how it handles stress in the environment, because how an animal copes with everyday, as well as more extreme, stressors can have a profound impact on its health, and hence its reproductive success (Bartolomucci, 2007; Cavigelli, 2005; Koolhaas et al., 2007; Overli et al., 2007). In their review of the literature on personality, stress, and coping, Jaap Koolhaas and his colleagues found two coping styles that they labeled **proactive** (sometimes called the active response) and **reactive** (also known as the conservation-withdrawal response; Cannon, 1915; Coppens et al., 2010; Engel and Schmale, 1972; Henry and Stephens, 1977; Koolhaas et al., 1999, 2007, 2010). The proactive personality type is characterized by territorial control and aggression, while the reactive style is characterized by immobility and low levels of aggression (Koolhaas et al., 1999; Table 17.2).

Proactive animals are also more likely to remove negative stimuli from their environment, whereas reactive animals are more likely to hide from any new negative stimulus. Frans Sluyter, for example, studied different strains of rats that had been selected for proactive or reactive coping styles (Sluyter

TABLE 17.2. Proactive and reactive coping styles. A summary of the behavioral differences between proactive and reactive male rats and mice. *(Based on Koolhaas et al., 1999)*

BEHAVIORAL CHARACTERISTICS	PROACTIVE	REACTIVE
Attack latency	Low	High
Defensive burying	High	Low
Nest-building	High	Low
Routine formation	High	Low
Cue dependency	Low	High
Conditioned immobility	Low	High
Flexibility	Low	High

FIGURE 17.17. Proactive and reactive rats. (A) Proactive mice and rats tend to be territorial and aggressive, whereas **(B)** reactive mice and rats tend to be timid and become immobile or hide when threatened. *(Photo credits: Tom McHugh/ Photo Researchers, Inc.; Joe McDonald/ Corbis)*

et al., 1995). First, male rats were exposed to an intruder in an experiment in which another male was placed in an individual's home cage. Proactive rats were very aggressive toward intruders, while reactive rats most often attempted to hide from such intruders (Figure 17.17). The rats were then tested in a "defensive burying" experiment. In the defensive burying experiment, a small electric prod was placed in a male's cage, and if the male investigated the prod and touched it, he received a mild shock. Once shocked, a rat had two ways to avoid future shock: He could either bury the prod under his bedding or significantly curtail his movements. The researchers found that individuals from the proactive strain of mice were much more likely to bury the prod than were mice from the reactive strain, suggesting that the two coping styles map nicely onto personality types.

The underlying proximate mechanisms associated with proactive and reactive strains have also been examined (Sluyter et al., 1995). Proactive and reactive mice were exposed to two different stressors—a five-minute forced swim or long-term stress associated with frequent handling. Alexa Veenema, Ronald de Kloet, Koolhaas, and their colleagues took a series of neuroendocrinological measurements of mice in the proactive and reactive strains, both before and after exposure to stress (Veenema et al., 2004, 2007).

Before exposure to stress conditions, corticosterone levels were fairly stable in reactive mice, but the levels fluctuated in proactive mice. Once they were stressed, reactive mice showed a dramatic increase in corticosterone levels, compared with proactive mice. The "freeze response" and reduced aggression and territoriality in reactive mice might be linked to such increases in corticosterone. Compared with proactive mice, reactive mice also showed reduced cell growth in the hippocampal region of the brain, suggesting that the proximate differences between coping styles may also affect learning and memory (Veenema et al., 2004, 2007).

Proactive and reactive personalities have been found not only in laboratory and natural populations of animals, but also in domesticated animals such as pigs and cows (Hopster, 1998; Prelle et al., 2004; Spoolder et al., 1996; D. Weiss et al., 2004). While data on the costs and benefits of these two coping styles are scant, there is some evidence that proactive and reactive animals differ in their susceptibility to diseases such as hypertension, atherosclerosis, gastric ulcers, and immunosuppressive capabilities (Ely, 1981; Henry et al., 1993; Hessing et al., 1994; Sgoifo et al., 2005; J. Weiss, 1972).

Applications of Animal Personality Research

The study of animal personality has potential practical implications for conservation biology, human-animal interactions, farming, and many other areas (see the Conservation Connection box). Here we will examine the practical implications of animal personality work for aiding people with disabilities.

GUIDE DOG PERSONALITIES

Humans have had a close mutualistic relationship with dogs since we began domesticating grey wolves between fifteen and thirty thousand years ago. In addition to humans' immense practical experience with dog personality, researchers have now completed sequencing the entire genome of the domestic dog, and many studies have been undertaken on the behavioral genetics of dogs (Lindblad-Toh et al., 2005; Saetre et al., 2006). Here we will focus on dog personality as it applies to guide dogs for the blind (Arata et al., 2010; D. L. Duffy and Serpell, 2012; A. C. Jones and Gosling, 2005; Ley and Bennett, 2007; Serpell and Hsu, 2001; Tomkins et al., 2011).

To become a guide dog for the blind, dogs must pass a very rigorous test. Data suggest that most dogs can, in fact, learn the difficult tasks needed to complete their training. Fear often interferes with a dog's performance, however, and is the single most common reason dogs fail to qualify to aid the blind (Baillie, 1972; Goddard and Beilharz, 1982, 1983; J. P. Scott and Bielfelt, 1976). This is not only an interesting finding in itself, but it suggests that a personality trait—fear or fearlessness—underlies which individuals become guide dogs and which do not (Boissy, 1995).

To understand the genetics of fear more completely and to suggest a good breeding program for guide dogs, M. E. Goddard and R. G. Beilharz crossbred four breeds of dogs—Labrador retrievers, Australian kelpies, boxers, and German shepherds (Goddard and Beilharz, 1985). Puppies from all crosses were allowed to remain with their mothers until they were six weeks old, at which point they were weaned and moved to the Guide Dog Association's kennel. At twelve weeks of age, the animals were given to "puppy walkers," who began training them to become guide dogs.

Fear and fearlessness were measured using thirty-eight behaviors that were subsequently transformed into twelve personality components via statistical analysis. The first three of these components map onto "general fearfulness," "fearfulness of objects," and "inhibited response to fear" (Goddard and Beilharz, 1984). Individual dog scores were consistent on these three components when the dogs were tested at six and twelve months of age.

Using Personality to Reduce Human-Animal Conflicts

As a result of species recovery and conservation plans, a number of large carnivores have been successfully reintroduced into the wild or have had their natural population numbers increase dramatically over the last few decades. The downside of this conservation success has been the rekindling of an old rivalry between ranchers and large carnivores that feed on their animals (Blanco et al., 1992; Cozza et al., 1996; Kaczensky, 1996; Oli et al., 1994; Quigley and Cragshaw, 1992). A good example can be found in the problems that arose when wolf populations began to increase in many parts of Europe and began attacking local livestock. A similar sort of problem arose when it was discovered that a small proportion of seal populations that have been the subject of conservation efforts were consuming economically valuable fish like salmon (I. M. Graham et al., 2011; Figure 17.18). The reduction of such conflicts may be facilitated by an understanding of the personality of the carnivores that are causing the problem.

Because there is widespread opposition to large-scale killing of carnivores, one way that the problem of carnivores feeding on livestock has been tackled is to focus on "problem individuals"—carnivore individuals that repeatedly attack and kill ranchers' livestock. The data on specific aspects of the hunting behavior of large carnivores preying on domestic prey are difficult to obtain, but studies of hunting behavior in wolves, cougars, leopards, seals, lions, tigers, bears, and many other species suggest that certain individuals are more likely to prey on domesticated animals (Dickman, 2010; Karlsson and Johansson, 2010; Linnell, 2011; Treves, 2009; Treves and Karanth, 2003).

Attacking and consuming ranchers' livestock is a risky endeavor for a carnivore. The predator must circumvent any fencing or other defensive measures put into place and then risk being killed by humans defending their livestock. Carnivores that consistently attempt

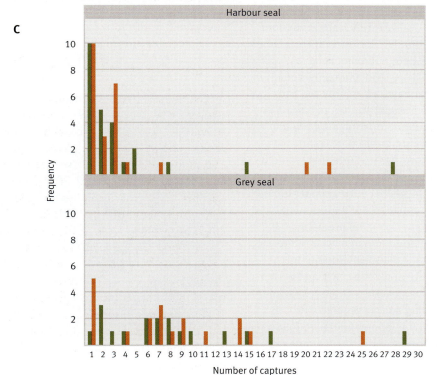

FIGURE 17.18. Seal personality and conservation. (A) A harbor seal (*Phoca vitulina*) and **(B)** a grey seal (*Halichoerus grypus*). In both harbor and grey seals, a few bold individuals are found in rivers and eat economically valuable fish like salmon. These animals (on the right side of each graph) were captured and recaptured many times. Understanding boldness in seals could help reduce conflict between the seals and the people who rely on the salmon that bold seals consume. (*Photo credits: Kenneth M. Highfill/Photo Researchers, Photoshot Holdings Ltd/Alamy*)

to attack such livestock—the so-called problem individuals—display many of the personality traits associated with boldness.

Exactly how to use the information on personality differences and problem individuals is still a matter of debate. One possibility would be to use the growing ethological understanding of bold predators to design new traps specifically for these

sorts of individuals. For example, bold predators might use specific hunting strategies that differ from the strategies of others in the population—they may use different paths to reach prey, hunt at different times, or be more or less attracted to certain stimuli. Traps could then be constructed that were designed with these hunting strategies of bold predators in mind.

Dr. Sam Gosling

Why study animal personality?

I think there are three main reasons to study animal personality: (1) Using animal models to learn about humans: Animal studies have long been used to inform discoveries in human psychology (in the domains of learning, problem solving, language, brain function, sensation, perception, etc.). Now that we have established that animals have personalities and animal personality can be measured reliably, we can also use animal studies to learn a lot about personality more broadly. Animal studies are particularly well poised to offer insights that would be difficult to gain using human research alone; this is because animal studies afford high levels of experimental control, permit measurement and manipulation of a wide range of biological and environmental parameters, can be combined with emerging genetic techniques (such as cloning and transgenic methods), and provide opportunities to follow animals longitudinally in a time frame that is considerably accelerated compared to that for humans. As in all animal research, many of these studies may raise ethical considerations that must be handled responsibly.

(2) Applied contexts: Some humans are better cut out to be librarians and others are better suited to work in sales or be lawyers, and some individual animals are better suited than others to perform the tasks assigned to them (finding explosives, guiding the visually impaired, etc.). Assessments of animal personality can help match individual animals to the tasks for which they are well suited. In addition, knowing about an animal's personality can help promote the welfare of captive animals (for example, by shaping housing conditions to match the personalities of captive animals), can match humans to appropriate

animal companions (for example, in animal shelters), and be used to select animals suitable for relocation in conservation work.

(3) Understanding the forces that drive and maintain individual differences: For too long, individual differences among animals were treated as meaningless variation that could be controlled statistically by averaging findings across numerous individuals. However, in the last decade or so, researchers and theorists have come to realize that variation among individuals is something that needs to be understood, not swept under the rug. Behavioral ecologists trained in evolutionary theory and ethological methods are particularly well positioned to understand the distal and proximal mechanisms that drive individual differences. So research on animal personality is vital to understanding the constraints that help determine how animals behave and, more generally, how variation among animals is maintained.

Why is it important to study animal personality from a comparative perspective?

A comparative perspective on personality can help identify common solutions that different species have taken to solve common problems. Without taking a comparative—and phylogenetically informed—perspective, it is difficult to determine the origin of personality differences and the forces that shaped them.

Darwin wrote *The Expression of Emotions in Man and Animals* more than 125 years ago. Why did it take so long for the experimental study of animal personality to emerge?

In my classes I often ask the students whether they think cows have personality. Those students who grew up in cities generally think it is obvious that cows don't have personality, and those people who grew up on farms generally think it is obvious that cows do have personality. When I first started studying animal personality, I noticed that many animal researchers would often hold both of these positions but at different times. When chatting to one another casually on their tea breaks, the researchers would happily talk about how one animal was friendly or anxious or calm or curious, but when the break was over they would put on their white coats and avoid talking about the animals using such anthropomorphic terms. So there has long been sense among scientists that personality did not offer a legitimate way to characterize animals. As animal behaviorists strove to establish themselves as scientists, they tended to distance themselves from anything that had an aura of anthropomorphism or subjectivity, with personality and emotion being prime examples. At the same time, human personality researchers came from a tradition that, in addition to behavioral traits, incorporated constructs such as identity, values, and motives, which were not such an obvious fit in the animal domain. So both the animal researchers and the personality researchers had their own reasons for not being open to the idea of

personality in animals. But I think the research that we did summarizing and synthesizing all the isolated studies of animal personality, along with the work done by Andy Sih and Alison Bell sketching out some of the evolutionary and ecological

Animal studies have long been used to inform discoveries in human psychology (in the domains of learning, problem solving, language, brain function, sensation, perception, etc.). Now that we have established that animals have personalities and animal personality can be measured reliably, we can also use animal studies to learn a lot about personality more broadly.

implications, gave the topic enough legitimacy for people to start thinking about it seriously. Once they

did that, the benefits of studying personality quickly became clear.

What is the most unexpected thing you've learned from your own work on animal personality?

It sounds odd to say now, but when I started this research I really didn't know if it would be possible to measure personality reliably or that personality assessments would predict behavior. So the biggest surprise was discovering that measures of animal personality typically meet or exceed the standards met by measures of human personality.

What do you see as the next frontier in research on animal personality?

I think the next frontier will be discovering the biological mechanisms underlying personality traits and extending these findings to the human domain.

Dr. Sam Gosling is a Professor in the Department of Psychology at The University of Texas. Among his many research interests are ultimate and proximate questions centering on animal personality. Dr. Gosling is the author of Snoop: What Your Stuff Says About You *(Basic Books, 2008)*

The observations made by the puppy walkers and the individuals who tested the dogs on the guide dog exam suggested that only general fearfulness was a good predictor of which animals would pass their final guide dog exam. This finding is particularly important because only "general fearfulness" has a strong underlying genetic component, whereas environmental factors explain much of the variance in the other two components of fear in these four species (Leotta et al., 2006; Wilsson and Sundgren, 1997). The genetic crosses also show that Labrador retrievers were most easily trained to be guide dogs, and German shepherds were the most difficult to train. Goddard and Beilharz used the genetics of fear, along with the ethologically based information described above, to suggest a breeding program for guide dogs (Goddard and Beilharz, 1985).

SUMMARY

1. Personality differences can be conceptualized as consistent long-term behavioral differences among individuals.
2. Boldness/shyness is one of the most stable personality variables. But animal behaviorists argue that there is no reason that natural selection should necessarily select for boldness or shyness as a general personality trait. Rather, individuals may be bold in some contexts and shy in others, depending on the costs and benefits associated with the behavioral contexts being examined.
3. One example of cultural transmission of information affecting animal personality involves seven long-term studies of chimpanzees in Africa. Stable innovative patterns of behavior are spread within groups of chimpanzees through imitation. But behaviors and personalities across populations are very different.
4. How an animal copes with everyday, as well as more extreme, stressors can have a profound impact on its health. Two general coping styles emerge from studies across a wide variety of animals. These two styles are referred to as the proactive or "active response" and the reactive or "conservation-withdrawal response." The proactive coping style entails territorial control and aggression, whereas the reactive coping style is characterized by immobility and low levels of aggression.
5. The study of animal personality has potential practical implications for conservation biology, human–animal interactions, and many other areas, including preventing the death of livestock and aiding people with disabilities.

DISCUSSION QUESTIONS

1. Go to the monkey exhibit at your local zoo and pick four individual monkeys to observe for at least four hours. Or go to a working farm and pick four individual cows or horses. Record all the information you can on each individual ("grooms," "eats," "attacks," "retreats," "sleeps," "plays," and so on), and if possible, note the proximity of the animals you are studying to others in the group. From your observations, can you suggest a list of behaviors that

you might focus on during a longer, more controlled, study of personality in the population you are observing?

2. How would you construct an experiment to examine whether boldness and/or behavioral inhibition are heritable traits?

3. Besides the ones mentioned in this chapter, what general costs and benefits might you associate with being bold or inhibited? Pick a particular species you are familiar with and create a list of the potential costs and benefits of boldness and shyness in that species.

4. Pick up a few recent issues of the journal *Animal Behaviour* and scan the titles for anything on "alternative strategies." Once you have found one or two such articles, read them—is there any mention of personality in these papers? If not, how might you reanalyze the data to see whether the alternative strategies could be construed in light of the work on animal personalities? What other sorts of data could you collect to better understand whether the alternative strategies studied represent personality types?

5. Besides preventing the death of livestock and selecting guide dogs, can you think of any other practical applications of personality work in animals? How might you construct some experiments to better understand whether the applications you suggest are feasible?

SUGGESTED READING

Carere, C., & Maestripieri, D. (2013). *Animal personalities: Behavior, physiology, and evolution*. Chicago: University of Chicago. An edited volume on ethological approaches to the study of personality.

Dingemanse, N. J., Bouwman, K. M., van de Pol, M., van Overveld, T., Patrick, S. C., Matthysen, E., & Quinn, J. L. (2012). Variation in personality and behavioural plasticity across four populations of the great tit *Parus major*. *Journal of Animal Ecology, 81,* 116–126. A study of personality in four populations of great tit birds.

Jones, A. C., & Gosling, S. D. (2005). Temperament and personality in dogs (*Canis familiaris*): A review and evaluation of past research. *Applied Animal Behaviour Science, 95,* 1–53. An excellent review article on personality in dogs.

Whiten, A., & Boesch, C. (2001). The culture of chimpanzees. *Scientific American*, January 2001, 60–67. Interesting review suggesting how culture can potentially explain some personality traits seen within and across chimp populations.

Wilson, D. S., Clark, A. B., Coleman, K., & Dearstyne, T. (1994). Shyness and boldness in humans and other animals. *Trends in Ecology and Evolution, 9,* 442–446. A short but thought-provoking article on shyness and boldness.

Glossary

adaptation A trait that results in the highest fitness among a specified set of behaviors in a particular environment. Adaptations are typically the result of the process of natural selection.

allele A gene variant; one of two or more alternative forms of a gene.

alliance A long-term coalition. *See also* coalition.

allogrooming The grooming of another individual, usually by scratching or licking an area of skin, often to remove parasites. Also known as social grooming.

analogous traits (analogies) Traits that are similar as a result of similar selection pressures rather than common descent.

appetitive stimulus Any stimulus that is considered positive, pleasant, or rewarding.

arena mating *See* leks.

artificial selection Selection in which humans are the selective agent and choose certain varieties of an organism over others for breeding. *See also* natural selection.

audience effect When individuals involved in social interactions change their behavior as a function of being watched by others.

aversive stimulus Any stimulus (for example, a shock or noxious odor) that is associated with an unpleasant experience.

axons Nerve cell fibers that transmit electrical information from one nerve cell to another.

Bateman's principle The idea that, since eggs require greater energy to produce than sperm, females should be the choosier sex and this should result in greater variance in the reproductive success of males.

behavioral genetics The study of the genetic variance associated with behavior.

blocking When an association between an unconditioned stimulus (US1) and a response prevents an individual from responding to another stimulus (US2) or causes the individual to respond less strongly to the US2.

boldness The tendency to take risks in both familiar and unfamiliar situations.

byproduct mutualism A type of cooperation in which an individual pays an immediate cost or penalty for not acting cooperatively, such that the immediate net benefit of cooperating outweighs that of not cooperating.

bystander effect When observers of an interaction between other individuals change their estimation of the fighting ability of those they are watching as a function of what they observe.

classical conditioning *See* Pavlovian conditioning.

coalition Cooperative action taken by at least two individuals or groups against another individual or group. *See also* cooperation.

communication The transfer of information from a signaler to a receiver.

conceptual approach An approach that usually entails integrating ideas generated in different disciplines and combining them in a new, cohesive way.

conditioned response (CR) The learned response to a conditioned stimulus.

conditioned stimulus (CS) A stimulus that initially fails to elicit a particular response, but comes to do so when it is associated with a second (unconditioned) stimulus.

convergent evolution The process whereby different populations or species converge on the same phenotypic characteristics as a result of similar natural selection pressures.

cooperation An outcome that, despite possible costs to the individual, provides some benefits to others. "To cooperate" means to behave in a way to make cooperation possible.

coping style A set of behavioral and related stress responses that are consistent over time.

copying Behavior that occurs when an observer repeats the actions of a demonstrator. The copier is often rewarded for whatever behavior it has copied.

cross-fostering experiment An experiment undertaken to measure the relative contributions of genetic and environmental variation on the expression of behavioral traits. This experiment often involves removing young individuals from their parent(s), and having them raised by adults that are not their genetic relatives.

cryptic Hidden through camouflage, and thus blending into the environment. Also a form of hidden mate choice (cryptic mate choice), wherein females select which sperm transferred during copulation are used in actual fertilization.

cultural transmission The transfer of information from individual to individual through teaching or social learning. *See also* social learning; teaching.

dendrites Nerve cell fibers that receive electrical information from other cells.

DNA fingerprint A molecular genetic technique often used to examine the genetic relatedness among individuals.

dominance hierarchy The relationship between individuals in a group as a result of aggressive behaviors and the response to aggressive behaviors.

dominant allele An allele that is expressed in heterozygote individuals. Allele A is dominant to allele a if the Aa genotype is identical to the AA genotype. *See also* allele.

dyadic interactions Interactions involving only two individuals.

ecology The study of the interaction of organisms with their environment.

empirical approach An approach that entails gathering data in one form or another and drawing inferences from that data. Empirical work in ethology can take many forms, but most often it is either observational or experimental.

endocrine system A communication network of ductless glands that secrete chemical messengers called hormones.

ethology The scientific study of animal behavior.

eusociality An extreme form of sociality in which there is cooperative brood care, division of labor, and overlapping generations.

evolutionarily stable strategy (ESS) A strategy that, if used by almost all individuals in a population, will not decrease in frequency when new, mutant strategies arise.

excitatory conditioning When a conditioned stimulus leads to an action (for example, searching for food or hiding). Sometimes (incorrectly) used as a synonym for Pavlovian conditioning.

extinction curves Graphs that depict how long an animal will remember some paired association once the pairing itself has stopped.

extrapair copulations Copulations that occur outside the context of a pair bond.

female defense polygyny A mating system in which males aggressively guard females that are found in spatial clusters, obtaining sole reproductive access to such "harems" of females.

fitness Lifetime reproductive success—usually measured in relative terms.

flight initiation distance The distance at which prey begin to flee from a predator.

foraging Feeding and all behaviors associated with feeding.

genetic recombination A recombination of genes during cell division in sexually reproducing organisms that involves sections of one chromosome "crossing over" and swapping positions with sections of the homologous chromosome.

genetic variation Variation caused by genetic differences.

genotype The genetic makeup of an individual.

goal-directed learning *See* instrumental conditioning.

good genes model A model of sexual selection in which females choose to mate with males that possess traits that are indicators of good health and vigor. Females that choose the males with such genes receive "indirect" benefits, in that their offspring receive some of the good genes that led their mother to choose a male as a mate in the first place.

group selection A hierarchical model in which natural selection operates at two levels: within-group selection and between-group selection.

habitat choice How animals distribute themselves in space and time with respect to some resource (for example, food, mates, refuge from predators) in their environment.

habituation Becoming less sensitive to stimuli over time.

hawk-dove game A game theory model of aggression. Hawk is a strategy in which the individual escalates and continues to escalate until either it is injured or its opponent concedes, and dove is a strategy in which the individual bluffs, initially displaying as if it will escalate, but retreating and ceding the resource if its opponent escalates.

heritability The proportion of variance in a trait that is due to genetic variance.

home range A delineated, undefended area in which an animal spends most of its time.

homologous traits (or homologies) Traits that are shared by two or more species as a result of common descent.

homoplasy A trait that is present in two or more species but that is not due to common descent but rather results from natural selection acting independently on each species. *See also* analogous traits.

honest indicators principle The idea that traits that are costly to produce are more difficult to fake and hence truly indicate the genetic quality of an individual.

horizontal cultural transmission Cultural transmission in which information is passed across individuals of the same age or peer group. *See also* cultural transmission.

ideal free distribution (IFD) A mathematical model developed to predict how animals will distribute themselves among habitats with varying levels of resource availability.

imitation The acquisition of a topographically novel response through observation of a demonstrator making that response.

inclusive fitness A measure of fitness that takes into account not only the effect an allele has on its bearer, but also the effect it has on its genetic relatives.

individual learning A relatively permanent change in behavior as a result of experience. Individual learning differs from social learning in that it does not involve learning from others.

inhibitory conditioning When a conditioned stimulus suppresses or inhibits behavior.

instrumental conditioning Learning that occurs when a response made by an animal is reinforced by reward or punishment. An animal must undertake some action or response in order for the conditioning process to produce learning.

intersexual selection A form of sexual selection in which individuals of one sex choose which individuals of the other sex to take as mates. *See also* sexual selection.

intrasexual selection A form of sexual selection whereby members of one sex compete with each other for access to the other sex. *See also* sexual selection.

kamikaze sperm hypothesis The hypothesis that natural selection favors sperm that are designed to kill (or otherwise incapacitate) other males' sperm rather than to fertilize eggs.

kin recognition matching models Models in which an individual uses an internal template to gauge its genetic relatedness to others.

kin selection *See* inclusive fitness.

law of effect If a response in the presence of a stimulus is followed by a satisfying event, the association between the stimulus and the response will be strengthened.

leks Areas in which groups of males set up and defend small, temporary territorial patches and in which they display to females, specifically to attract females to mate with them. The result is called **lekking** and **arena mating**.

local enhancement A phenomenon in which an individual is drawn to a particular area because it observed another individual in that location.

locomotor play Play behaviors in which animals leap, jump, twist, shake, whirl, somersault, roll, or chase their tails. Also known as locomotor rotational play.

loser effect The increased probability of losing a fight at time T, based on losing at time T-1, T-2, and so on.

marginal value theorem A mathematical model developed to predict how long a forager will spend in a patch of food before moving on to new patches.

mate-choice copying The act of copying the mate choice of others.

migration The movement of organisms over long distances (often in a seasonal manner).

mobbing A type of antipredator behavior in which groups of prey join together, approach a predator, and aggressively attempt to chase it away.

monogamous mating system A mating system in which a male and female mate only with one another during a given breeding season.

mushroom bodies A cluster of small neurons located at the front of the brain of some invertebrates and associated with spatial navigation.

mutation Any change in genetic structure.

mutualism Any interaction that benefits all parties involved. Interspecies cooperation is often referred to as a mutualism.

natural selection The process at the heart of Darwin's theory of evolution. Natural selection occurs when variants of a trait that best suit an organism to its environment, and that are heritable, increase in frequency over evolutionary time. This process requires variation, fitness differences, and heritability.

neuroeconomics A collaborative research effort between economists, neurobiologists, and evolutionary biologists that uses brain-imaging technology to understand questions regarding behavior.

neurohormones Hormones that are secreted directly into the bloodstream by nerve cells.

nomads Individuals that lack a home range or a territory and who rarely frequent the same area over time. Occasionally, reference will be made to a nomadic species.

nuptial gifts Prey presented by members of one sex to members of the other sex during courtship.

object play Play that centers on the use of inanimate objects such as sticks, rocks, leaves, feathers, fruit, and human-provided objects, and the pushing, throwing, tearing, or manipulating of such objects.

oblique cultural transmission Cultural transmission in which information is passed across generations, but not from parent to offspring.

operant learning *See* instrumental conditioning.

operant response A learned action that an animal makes to change its environment.

optimal foraging theory (OFT) A family of mathematical models developed to predict animal foraging behavior.

optimal skew theory A family of models that predicts the distribution of breeding within a group, as well as the degree of cooperation or conflict over reproductive activities.

overshadowing A situation in which the learned response to an unconditioned stimulus (US1) is stronger when it is presented alone versus when it is paired with a second unconditioned stimulus (US2).

parental investment The amount of energy parents invest in raising their offspring.

parent-offspring conflict A zone of conflict between how much a given offspring wants in terms of parental resources and how much a parent is willing to give.

parent-offspring regression A technique for measuring heritability that involves measuring a trait in parents and offspring.

parsimony analysis A technique for choosing among alternative phylogenetic trees by selecting the tree that requires the fewest character changes.

Pavlovian conditioning The experimental pairing of a conditioned and unconditioned stimulus. *See also* conditioned stimulus; unconditioned stimulus.

personality differences Consistent long-term phenotypic behavioral differences among individuals.

phenotype The observable characteristics of an organism.

phenotypic plasticity The ability of an organism to produce different phenotypes depending on environmental conditions.

phylogenetic tree The depiction of phylogeny by using branching, treelike diagrams.

phylogeny Evolutionary history via common descent.

play Typically defined as motor activity performed during development that appears to be purposeless.

play markers Behavioral indicators that denote that some action that is about to be undertaken should be considered playful, not dangerous. Also known as play signals, they may serve to initiate play or to indicate the desire to keep playing.

polyadic interactions Interactions involving more than two individuals.

polyandry A mating system wherein females mate with more than one male per breeding season.

polygamy A mating system in which either males or females have multiple mates during a given breeding season.

polygenic Caused by the action of more than one gene.

polygynandry A mating system wherein several males form pair bonds with several females simultaneously.

polygyny A mating system wherein males mate with more than one female per breeding season.

polygyny threshold model A model developed to predict the conditions under which a mating system will move from monogamy to polygyny (or vice versa).

predator inspection An antipredator behavior in which one to a few individuals break away from a group and slowly approach a potential predator to obtain various sorts of information.

preexisting bias *See* sensory exploitation.

prisoner's dilemma A game theory payoff matrix that is used to study the evolution of cooperation.

proactive coping style A personality type characterized by territorial behavior and various forms of aggression; animals with this personality are more likely to remove negative stimuli from their environment. This style is sometimes referred to as the active response.

producers Individuals that find and procure some resource.

promiscuity A type of mating system in which both polyandry and polygyny are occurring. In one form of promiscuity, both males and females mate with many partners and no pair bonds are formed. In the second type of promiscuous breeding system, called polygynandry, several males form pair bonds with several females simultaneously. *See also* polyandry; polygynandry; polygyny.

proximate analysis Analysis based on asking questions that focus on immediate causation.

reactive coping style A personality type characterized by immobility and low levels of aggression; animals with this personality are more likely to become immobile or to hide when exposed to a new negative stimulus. Also known as the conservation-withdrawal response.

recessive allele An allele in heterozygote individuals that is not expressed when combined with a dominant allele. Allele a is recessive to allele A if the Aa genotype is phenotypically identical to the AA genotype.

reciprocal altruism The exchange of altruistic acts.

resource holding power A measure of an animal's fighting ability.

risk-sensitive optimal foraging models A family of mathematical models that examines how variance in food supplies affects foraging behavior.

role reversal Behavior wherein older individuals allow subordinate, younger animals to take on the dominant role during play, or in which older individuals perform some act that is at a level clearly below that of which the older individual is capable. Also known as self-handicapping.

runaway sexual selection A model of sexual selection in which the genes for mate choice in the female and the genes for preferred traits in males become genetically linked.

scroungers Individuals that obtain a portion of their diet by parasitizing the resources that others (producers) have uncovered. *See also* producers.

search image A representation of prey that predators form over time. This representation becomes more detailed with experience.

self-handicapping *See* role reversal.

sensitization Becoming more sensitive to stimuli over time.

sensory bias *See* sensory exploitation.

sensory exploitation A theory of sexual selection that hypothesizes that females may initially prefer male traits that elicit the greatest amount of stimulation from their sensory systems. Also known as sensory exploitation, sensory drive, or preexisting bias.

sequential assessment model A model of fighting in which an individual continually assesses its opponent's fighting ability during different stages of an aggressive interaction, and decides whether to continue in such an interaction based on the assessment it has made.

sexual selection A form of natural selection that, according to Darwin (1871), involves "a struggle between the individuals of one sex, generally the males, for the possession of the other sex."

sexy-son hypothesis A hypothesis that females select among males based on genetic traits in the males that will lead to the production of sons that are attractive to the opposite sex.

shyness The reluctance to take risks or to engage in unfamiliar activity.

sibling rivalry Aggressive interactions among siblings.

social facilitation When the presence of a model, regardless of what it does, facilitates learning on the part of an observer.

social grooming *See* allogrooming.

social learning Learning via the observation of others. Observation is used in a broad sense and can include obtaining information through olfaction and other senses besides sight.

social play Playing with others.

sperm competition A form of sexual selection that occurs directly between sperm after insemination.

teaching A behavior that occurs when one individual serves as an instructor and at least one other individual acts as a student that learns from the instructor. Teachers have a much more active and complicated role than just being a model that others mimic (as in social learning).

territory A delineated, defended area.

theoretical approach An approach that entails the generation of a predictive model—usually a mathematical model.

tit for tat (TFT) A behavioral strategy that instructs a player to initially cooperate with a new partner, and subsequently to do whatever that partner does.

truncation selection experiment An experimental procedure that measures heritability by allowing only those with extreme forms of a trait to breed and then tracking changes in that trait across generations.

ultimate analysis Analysis based on asking questions that relate to the evolution of a trait.

unconditioned stimulus (US) A stimulus that elicits a vigorous response in the absence of training.

vertical cultural transmission Cultural transmission in which information is passed directly from parent(s) to offspring. *See also* cultural transmission.

waggle dance A dance performed by forager bees. The waggle dance provides information on the spatial location of food located at some distance from the hive.

war of attrition model A model of fighting in which animals display aggressively to one another without actually fighting, and the winner of the encounter is the individual that displays the longest.

winner effect The increased probability of winning a fight at time T, based on victories at times T-1, T-2, and so on.

worker policing Behavior that involves workers in a social insect species destroying the eggs laid by other workers.

xenophobia The fear of strangers or those from outside one's group.

References

Abbot, P., Abe, J., Alcock, J., Alizon, S., Alpedrinha, J. A. C., Andersson, M., Andre, J.-B., Balloux, F., Balshine, S., Barton, N., Beukeboom, L. W., Biernaskie, J. M., Bilde, T., Borgia, G., Breed, M., Brown, S., Bshary, R., Buckling, A., Burley, N. T., Burton-Chellew, M. N., Cant, M. A., Chapuisat, M., Charnov, E. L., Clutton-Brock, T., Cockburn, A., Cole, B. J., Colegrave, N., Cosmides, L., Couzin, I. D., Coyne, J. A., Creel, S., Crespi, B., Curry, R. L., Dall, S. R. X., Day, T., Dickinson, J. L., Dugatkin, L. A., El Mouden, C., Emlen, S. T., Evans, J., Ferriere, R., Field, J., Foitzik, S., Foster, K., Foster, W. A., Fox, C. W., Gadau, J., Gandon, S., Gardner, A., Gardner, M. G., Getty, T., Goodisman, M. A. D., Gouyon, P.-H., Grafen, A., Grosberg, R., Grozinger, C. M., Gwynne, D., Harvey, P. H., Hatchwell, B. J., Heinze, J., Helantera, H., Helms, K. R., Hill, K., Jiricny, N., Johnstone, R. A., Kacelnik, A., Kiers, E. T., Kokko, H., Komdeur, J., Korb, J., Kronauer, D., Kuemmerli, R., Lehmann, L., Linksvayer, T. A., Lion, S., Lyon, B., Marshall, J. A. R., McElreath, R., Michalakis, Y., Michod, R. E., Mock, D., Monnin, T., Montgomerie, R., Moore, A. J., Mueller, U. G., Noe, R., Okasha, S., Pamilo, P., Parker, G. A., Pedersen, J. S., Pen, I., Pfennig, D., Queller, D. C., Rankin, D. J., Reece, S. E., Reeve, H. K., Reuter, M., Roberts, G., Robson, S. K. A., Rousset, F., Roze, D., Rueppell, O., Sachs, J. L., Santorelli, L., Schmid-Hempel, P., Schwarz, M. P., Scott-Phillips, T., Shellmann-Sherman, J., Sherman, P. W., Shuker, D. M., Smith, J., Spagna, J. C., Strassmann, B., Suarez, A. V., Sundstrom, L., Taborsky, M., Taylor, P., Thompson, G., Tooby, J., Tsuji, K., Tsutsui, N. D., Turillazzi, S., Ubeda, F., van Baalen, M., Vargo, E. L., Voelkl, B., Wenseleers, T., West, S. A., West-Eberhard, M. J., Westneat, D. F., Wiernasz, D. C., Wild, G., Wrangham, R., Young, A. J., Zeh, D. W., Zeh, J. A., & Zink, A. (2010). Inclusive fitness theory and eusociality. *Nature, 471,* E1–E4.

Able, K. P. (1999). *Gatherings of angels: Migrating birds and their ecology.* Ithaca, NY: Comstock Books.

Abrahams, M. V., & Pratt, T. C. (2000). Hormonal manipulations of growth rate and its influence on predator avoidance: Foraging trade-offs. *Canadian journal of Zoology, 78,* 121–127.

Abramowski, D., Currie, C. R., & Poulsen, M. (2011). Caste specialization in behavioral defenses against fungus garden parasites in *Acromyrmex octospinosus* leaf-cutting ants. *Insectes Sociaux, 58,* 65–75.

Adams, E., & Mesterton-Gibbons, M. (1995). The cost of threat displays and the stability of deceptive communication. *Journal of Theoretical Biology, 175,* 405–421.

Adams, M. J., King, J. E., & Weiss, A. (2012). The majority of genetic variation in orangutan personality and subjective well-being is nonadditive. *Behavior Genetics, 42,* 675–686.

Adkins-Regan, E. (1998). Hormonal control of mate choice. *American Zoologist, 38,* 166–178.

Adler, M. (2010). Sexual conflict in waterfowl: Why do females resist extrapair copulations? *Behavioral Ecology, 21,* 182–192.

Agnarsson, I., Aviles, L., Coddington, J. A., & Maddison, W. P. (2006). Sociality in Theridiid spiders: Repeated origins of an evolutionary dead end. *Evolution, 60,* 2342–2351.

Ågren, G. (1984). Pair formation in the Mongolian gerbil. *Animal Behaviour, 32,* 528–535.

Ahem, T., & Young, L. (2009). The impact of early life family structure on adult social attachment, alloparental behavior, and the neuropeptide systems regulating affiliative behaviors in the monogamous prairie vole (*Microtus ochrogaster*). *Frontiers in Behavioral Neuroscience, 3,* 1–17.

Alatalo, R. V., Hoglund, J., Lundberg, A., & Sutherland, W. J. (1992). Evolution of black grouse leks: Female preferences benefit males in larger leks. *Behavioral Ecology, 3,* 53–59.

Alcock, J. (1998). *Animal behavior* (6th ed.). Sunderland, MA: Sinauer.

Alcock, J. (2001). *The triumph of sociobiology.* New York: Oxford University Press.

Alcock, J. (2003). A textbook history of animal behaviour. *Animal Behaviour, 65,* 3–10.

Alcock, J., & Sherman, P. (1994). The utility of the proximate-ultimate dichotomy in ethology. *Ethology, 96,* 58–62.

Alerstam, T. (1990). *Bird migration.* Cambridge: Cambridge University Press.

Alexander, R. D. (1979). *Darwinism and human affairs.* Seattle: University of Washington Press.

Alexander, R. D. (1991). Social learning and kin recognition. *Ethology and Sociobiology, 12,* 387–399.

Alexander, R. D., & Tinkle, D. W. (Eds.). (1981). *Natural selection and social behavior.* New York: Chiron Press.

Allan, S., & Suthers, R. (1994). Lateralization and motor stereotypy of song production in the brown-headed cowbird. *Journal of Neurobiology, 25,* 1154–1166.

Allee, W. (1951). *Cooperation among animals with human implications.* New York: Henry Schuman.

Allen, G. (1879). *The color sense: Its origin and development.* London: Trubner.

Allman, H. M. (1998). *Evolving brains*. New York: Scientific American Library.

Altmann, S. (1962). Social behavior of anthropoid primates: Analysis of recent concepts. In E. Bliss (Ed.), *Roots of behavior* (pp. 277–285). New York: Harper and Brothers.

Altmann, S. A. (1956). Avian mobbing behavior and predator recognition. *Condor, 58,* 241–253.

Altmann, S. A. (1979). Altruistic behaviour: The fallacy of kin deployment. *Animal Behaviour, 27,* 958–959.

Amat, J., Baratta, M. V., Paul, E., Bland, S. T., Watkins, L. R., & Maier, S. F. (2005). Medial prefrontal cortex determines how stressor controllability affects behavior and dorsal raphe nucleus. *Nature Neuroscience, 8,* 365–371.

Anderson, K. G., Kaplan, H., & Lancaster, J. B. (2007). Confidence of paternity, divorce, and investment in children by Albuquerque men. *Evolution and Human Behavior, 28,* 1–10.

Andersson, M. (1994). *Sexual selection*. Princeton, NJ: Princeton University Press.

Andersson, M., & Simmons, L. W. (2006). Sexual selection and mate choice. *Trends in Ecology & Evolution, 21,* 296–302.

Apfelbach, R., Blanchard, C. D., Blanchard, R. J., Hayes, R. A., & McGregor, I. S. (2005). The effects of predator odors in mammalian prey species: A review of field and laboratory studies. *Neuroscience and Biobehavioral Reviews, 29,* 1123–1144.

Aragona, B. J., Liu, Y., Curtis, T., Stephan, F. K., & Wang, Z. X. (2003). A critical role for nucleus accumbens dopamine in partner-preference formation in male prairie voles. *Journal of Neuroscience, 23,* 3483–3490.

Aragona, B. J., Liu, Y., Yu, Y. J., Curtis, J. T., Detwiler, J. M., Insel, T. R., & Wang, Z. X. (2006). Nucleus accumbens dopamine differentially mediates the formation and maintenance of monogamous pair bonds. *Nature Neuroscience, 9,* 133–139.

Arata, S., Momozawa, Y., Takeuchi, Y., & Mori, Y. (2010). Important behavioral traits for predicting guide dog qualification. *Journal of Veterinary Medical Science, 72,* 539–545.

Archer, J. (1988). *The behavioural biology of aggression*. Cambridge: Cambridge University Press.

Arnold, C., & Taborsky, B. (2010). Social experience in early ontogeny has lasting effects on social skills in cooperatively breeding cichlids. *Animal Behaviour, 79,* 621–630.

Arnold, W., & Trillmich, F. (1985). Time budget in Galapagos fur seals: The influence of mother's presence and absence on pup activity and play. *Behaviour, 92,* 302–321.

Arnott, G., & Elwood, R. W. (2009). Assessment of fighting ability in animal contests. *Animal Behaviour, 77,* 991–1004.

Arnqvist, G., & Rowe, L. (2005). *Sexual conflict*. Princeton, NJ: Princeton University Press.

Atance, C. M., & Meltzoff, A. N. (2005). My future self: Young children's ability to anticipate and explain future states. *Cognitive Development, 20,* 341–361.

Atkinson, D., & Begon, M. (1987). Reproductive variation and adult size in two co-occurring grasshopper species. *Ecological Entomology, 12,* 119–127.

Attwell, D., & Laughlin, S. B. (2001). An energy budget for signaling in the grey matter of the brain. *Journal of Cerebral Blood Flow and Metabolism, 21,* 1133–1145.

Aubin, T., & Jouventin, P. (1998). Cocktail-party effect in king penguin colonies. *Proceedings of the Royal Society of London, Series B, 265,* 1665–1673.

Aubin, T., & Jouventin, P. (2002). How to vocally identify kin in a crowd: The penguin model. *Advances in the Study of Behavior, 31,* 243–277.

Aubin, T., Jouventin, P., & Hildebrand, C. (2000). Penguins use the two-voice system to recognize each other. *Proceedings of the Royal Society of London, Series B, 267,* 1081–1087.

Auger, A. P., & Olesen, K. M. (2009). Brain sex differences and the organisation of juvenile social play behaviour. *Journal of Neuroendocrinology, 21,* 519–525.

Aviles, L., & Maddison, W. (1991). When is the sex ratio biased in social spiders? Chromosome studies of embryos and male meiosis in *Anelosimus* species (Aranae, Theridiidae). *Journal of Arachnology, 19,* 126–135.

Aviles, L., McCormack, J., Cutter, A., & Bukowski, T. (2000). Precise, highly female-biased sex ratios in a social spider. *Proceedings of the Royal Society of London, Series B, 267,* 1445–1449.

Aviles, L., Varas, C., & Dyreson, E. (1999). Does the African social spider *Stegodyphus dumicola* control the sex of individual offspring? *Behavioral Ecology and Sociobiology, 46,* 237–243.

Axelrod, R. (1984). *The evolution of cooperation*. New York: Basic Books.

Axelrod, R., & Hamilton, W. D. (1981). The evolution of cooperation. *Science, 211,* 1390–1396.

Bachman, G., & Widemo, F. (1999). Relationships between body composition, body size and alternative reproductive tactics in a lekking sandpiper, the ruff (*Philomachus pugnax*). *Functional Ecology, 13,* 411–416.

Badyaev, A. V., & Hill, G. E. (2002). Paternal care as a conditional strategy: Distinct reproductive tactics associated with elaboration of plumage ornamentation in the house finch. *Behavioral Ecology, 13,* 591–597.

Baer, B., Dijkstra, M. B., Mueller, U. G., Nash, D. R., & Boomsma, J. J. (2009). Sperm length evolution in the fungus-growing ants. *Behavioral Ecology, 20,* 38–45.

Baerends, G. P., Wanders, J., & Vodegel, R. (1986). The relationship between marking patterns and motivational state in pre-spawning behavior of the cichlid fish *Chromidotilapia guentheri* (Sauvage). *Netherlands Journal of Zoology, 36,* 88–116.

Baerends-van Roon, J., & Baerends, G. (1979). *The morphogenesis of the behavior of the domestic cat, with special emphasis on the development of prey-catching*. Amsterdam: New Holland.

Baillie, J. (1972). The behavioural requirements necessary for guide dogs for the blind in the United Kingdom. *British Veterinary Journal, 128,* 477.

Baker, R. R. (1978). *The evolutionary ecology of animal migration*. New York: Holmes & Meier.

Baker, R. R., & Bellis, M. A. (1988). Kamikaze sperm in mammals. *Animal Behaviour, 36,* 936–939.

Baker, R. R., & Bellis, M. A. (1993). Sperm competition: Ejaculate adjustment by males and the function of masturbation. *Animal Behaviour, 46,* 861–885.

Balaban, E., Alper, J. S., & Kasamon, Y. L. (1996). Mean genes and the biology of aggression: A critical review of recent animal and human research. *Journal of Neurogenetics, 11,* 1–43.

Balda, R., & Kamil, A. (1989). A comparative study of cache recovery by three corvid species. *Animal Behaviour, 38,* 486–495.

Balda, R., & Kamil, A. (1992). Long-term spatial memory in Clark's nutcracker, *Nucifraga columbiana. Animal Behaviour, 44,* 761–769.

Balda, R., Pepperberg, I., & Kamil, A. (Eds.). (1998). *Animal cognition in nature.* San Diego, CA: Academic Press.

Baldwin, J. (1969). The ontogeny of social behavior of squirrel monkeys (*Saimiri sciureus*) in a seminatural environment. *Folia Primologica, 11,* 35–79.

Baldwin, J. (1971). The social organization of a semifree-ranging troop of squirrel monkeys (*Saimiri sciureus*). *Folia Primologica, 14,* 23–50.

Ball, G. F. (1993). The neural integration of environmental information by seasonally breeding birds. *American Zoologist, 33,* 185–199.

Ball, N., Amlaner, C., Shaffery, J., & Opp, M. (1988). Asynchronous eye-closure and unihemispheric quiet sleep of birds. In W. Koella, F. Obal, H. Schulz, & P. Visser (Eds.), *Sleep '86* (pp. 151–153). New York: Gustav Fischer Verlag.

Bandura, A. (1977). *Social learning theory.* Englewood Cliffs, NJ: Prentice Hall.

Bandura, A. (1986). *Social foundations of thought and action.* Englewood Cliffs, NJ: Prentice Hall.

Bandura, A., Ross, D., & Ross, S. (1961). Transmission of aggression through imitation of aggressive models. *Journal of Abnormal and Social Psychology, 63,* 572–582.

Barash, D. P. (1975). Marmot alarm calling and the question of altruistic behavior. *American Midland Naturalist, 94,* 468–470.

Barber, J. R., & Conner, W. E. (2007). Acoustic mimicry in a predator-prey system. *Proceedings of the National Academy of Sciences, 104,* 9331–9334.

Barnard, C. J. (Ed). (1984). *Producers and scroungers.* London: Croom Helm/Chapman Hall.

Barnard, C. J., & Brown, C. A. (1984). A payoff asymmetry in resident-resident disputes between shrews. *Animal Behaviour, 32,* 302–304.

Barnard, C. J., & Sibly, R. M. (1981). Producers and scroungers: A general model and its application to captive flocks of house sparrows. *Animal Behaviour, 29,* 543–550.

Barraza, J. A., McCullough, M. E., Ahmadi, S., & Zak, P. J. (2011). Oxytocin infusion increases charitable donations regardless of monetary resources. *Hormones and Behavior, 60,* 148–151.

Barrio, I. C., Bueno, C. G., Banks, P. B., & Tortosa, F. S. (2010). Prey naivete in an introduced prey species: The wild rabbit in Australia. *Behavioral Ecology, 21,* 986–991.

Barron, A. B., & Robinson, G. E. (2005). Selective modulation of task performance by octopamine in honey bee (*Apis mellifera*) division of labour. *Journal of Comparative Physiology A, 191,* 659–668.

Barros, N. B. (1993). *Feeding ecology and foraging strategies of bottlenose dolphins of the central east coast of Florida.* Miami: University of Miami.

Bartolomucci, A. (2007). Social stress, immune functions and disease in rodents. *Frontiers in Neuroendocrinology, 28,* 28–49.

Bass, A. (1992). Dimorphic male brains and alternative reproductive tactics in a vocalizing fish. *Trends in Neurosciences, 15,* 139–145.

Bass, A. H. (1996). Shaping brain sexuality. *American Scientist, 84,* 352–363.

Bass, A. H. (1998). Behavioral and evolutionary neurobiology: A pluralistic approach. *American Zoologist, 38,* 97–107.

Bass, A., & Baker, R. (1990). Sexual dimorphisms in the vocal control system of a teleost fish: Morphological and physiological identified cells. *Journal of Neurobiology, 21,* 1155–1168.

Bass, A., & Baker, R. (1991). Adaptive modification of homologous vocal control traits in teleost fishes. *Brain, Behavior and Evolution, 38,* 240–254.

Bass, A., & Marchaterre, M. (1989). Sound-generating (sonic) motor system in a teleost fish (*Porichths notatus*): Sexual polymorphism in the ultrastructure of myofibrils. *Journal of Comparative Neurology, 286,* 141–153.

Bass, A. H., & Grober, M. S. (2009). Reproductive plasticity in fish: Evolutionary lability in the patterning of neuroendocrine and behavioral traits underlying divergent sexual phenotypes. In D. Pfaff, A. Arnold, A. Etgen, S. Fahrbach, R. Moss, & R. Rubin (Eds.), *Hormones, brain, and behavior.* Amsterdam, Netherlands: Academic Press.

Bass, A. H., & McKibben, J. R. (2003). Neural mechanisms and behaviors for acoustic communication in teleost fish. *Progress in Neurobiology, 69,* 1–26.

Bass, A. H., & Zakon, H. H. (2005). Sonic and electric fish: At the crossroads of neuroethology and behavioral neuroendocrinology. *Hormones and Behavior, 48,* 360–372.

Bateman, A. J. (1948). Intra-sexual selection in *Drosophila. Heredity, 2,* 349–368.

Bates, L. A., Lee, P. C., Njiraini, N., Poole, J. H., Sayialel, K., Sayialel, S., Moss, C. J., & Byrne, R. W. (2008). Do elephants show empathy? *Journal of Consciousness Studies, 15,* 204–225.

Bateson, P. (1978). Early experience and sexual preferences. In J. B. Hutchinson (Ed.), *Biological determinants of sexual behavior* (pp. 29–53). New York: Wiley.

Bateson, P. (1990). Choice, preference and selection. In M. Bekoff & D. Jamieson (Eds.), *Interpretation and explanation in the study of animal behaviour* (Vol. 1, Interpretation, intentionality and communication, pp. 149–156). Boulder, CO: Westview Press.

Bauer, G. B. (2005). Research training for releasable animals. *Conservation Biology, 19,* 1779–1789.

Baxter, D. A., & Byrne, J. H. (1990). Differential effects of cAMP and serotonin on membrane current, action potential duration, and excitability in somata of pleural sensory neurons of Aplysia. *Journal of Neurophysiology, 64,* 978–990.

Beacham, J. L. (2003). Models of dominance hierarchy formation: Effects of prior experience and intrinsic traits. *Behaviour, 140,* 1275–1303.

Beason, R. C., & Nichols, J. E. (1984). Magnetic orientation and magnetically sensitive material in a transequatorial migratory bird. *Nature, 309,* 151–153.

Beatty, W., Costello, K., & Berry, S. (1984). Suppression of play fighting by amphetamines: Effects of catecholamine antagonists, agonists and synthesis inhibitors. *Pharmacology Biochemistry and Behavior, 20,* 747–755.

Beatty, W. W., Dodge, A. M., Traylor, K. L., & Meaney, M. J. (1981). Temporal boundary of the sensitive period for hormonal organization of social play in juvenile rats. *Physiology & Behavior, 26,* 241–243.

Beauchamp, G., Alexander, P., & Jovani, R. (2012). Consistent waves of collective vigilance in groups using public information about predation risk. *Behavioral Ecology, 23,* 368–374.

Beauchamp, G., & Ruxton, G. D. (2007). False alarms and the evolution of antipredator vigilance. *Animal Behaviour, 74,* 1199–1206.

Beaugrand, J. P., & Zayan, R. (1985). An experimental model of aggressive dominance in *Xiphophorus helleri*. *Behavioural Processes, 10,* 1–52.

Beck, B. (1980). *Animal tool behavior: The use and manufacture of tools by animals.* New York: Garland STPM Press.

Beck, W. S. Liem, K. F., & Simpson, G. G. (1981). *Life, an introduction to biology.* New York: Harper.

Beecher, M. D. (2008). Function and mechanisms of song learning in song sparrows. *Advances in the Study of Behavior, 38,* 167–225.

Beecher, M. D., Beecher, I., & Hahn, S. (1981). Parent-offspring recognition in bank swallows: II. Development and acoustic basis. *Animal Behaviour, 29,* 95–101.

Beecher, M. D., Beecher, I., & Lumpkin, S. (1981). Parent-offspring recognition in bank swallows: I. Natural history. *Animal Behaviour, 29,* 86–94.

Beecher, M. D., & Brenowitz, E. A. (2005). Functional aspects of song learning in songbirds. *Trends in Ecology and Evolution, 20,* 143–149.

Beecher, M. D., Medvin, B., Stoddard, P., & Loesch, P. (1986). Acoustic adaptations for parent-offspring recognition in swallows. *Experimental Biology, 45,* 179–193.

Beissinger, S. R. (1996). On the limited breeding opportunities hypothesis for avian clutch size. *American Naturalist, 147,* 655–658.

Bekoff, M. (1977). Social communication in canids: Evidence for the evolution of a stereotyped mammalian display. *Science, 1971,* 1097–1099.

Bekoff, M. (1995). Play signals as punctuation: The structure of play in social canids. *Behaviour, 132,* 419–429.

Bekoff, M. (2000). Social play behavior: Cooperation, fairness, trust, and the evolution of morality. *Journal of Consciousness Studies, 8,* 81–90.

Bekoff, M. (2004). Wild justice and fair play: Cooperation, forgiveness, and morality in animals. *Biology and Philosophy, 19,* 489–520.

Bekoff, M. (2007). *The emotional lives of animals: A leading scientist explores animal joy, sorrow, and empathy.* Novato, CA: New World Library.

Bekoff, M., & Allen, C. (1998). Intentional communication and social play: How and why animals negotiate and agree to play. In M. Bekoff & J. Byers (Eds.), *Animal play: Evolutionary, comparative and ecological perspectives* (pp. 97–114). Cambridge: Cambridge University Press.

Bekoff, M., & Byers, J. (1981). A critical reanalysis of the ontogeny and phylogeny of mammalian social and locomotor play: An ethological hornet's nest. In K. Immelman, G. Barlow, L. Petrinivich, & M. Main (Eds.), *Behavioral development* (pp. 296–337). Cambridge: Cambridge University Press.

Bekoff, M., & Byers, J. (Eds.). (1998). *Animal play: Evolutionary, comparative and ecological perspectives.* Cambridge: Cambridge University Press.

Bekoff, M., & Pierce, J. (2010). *Wild justice: The moral lives of animals.* Chicago: University of Chicago Press.

Bell, A. M. (2007). Future directions in behavioural syndromes research. *Proceedings of the Royal Society of London, Series B, 274,* 755–761.

Bell, A. M., & Aubin-Horth, N. (2010). What can whole genome expression data tell us about the ecology and evolution of personality? *Philosophical Transactions of the Royal Society, Series B–Biological Sciences, 365,* 4001–4012.

Bell, G. (1997). *Selection: The mechanism of evolution.* New York: Chapman and Hall.

Bellis, M. A., Baker, R. R., & Gage, M. J. G. (1990). Variation in rat ejaculates consistent with the kamikaze sperm hypothesis. *Journal of Mammalogy, 71,* 479–480.

Belovsky, G. E. (1978). Diet optimization in a generalist herbivore: The moose. *Theoretical Population Biology, 14,* 105–134.

Benavides, T., & Drummond, H. (2007). The role of trained winning in a broodmate dominance hierarchy. *Behaviour, 144,* 1133–1146.

Bennett, A., Cuthill, I. C., Partridge, J. C., & Maier, E. (1996). Ultraviolet vision and mate choice in zebra finches. *Nature, 380,* 433–435.

Ben-Shahar, Y., Dudek, N. L., & Robinson, G. E. (2004). Phenotypic deconstruction reveals involvement of manganese transporter malvolio in honey bee division of labor. *Journal of Experimental Biology, 207,* 3281–3288.

Bercovitch, F. (1988). Coalitions, cooperation and reproductive tactics among adult male baboons. *Animal Behaviour, 36,* 1198–1209.

Bereczkei, T., Gyuris, P., Koves, P., & Bernath, L. (2002). Homogamy, genetic similarity, and imprinting: Parental influence on mate choice preferences. *Personality and Individual Differences, 33,* 677–690.

Bereczkei, T., Gyuris, P., & Weisfeld, G. E. (2004). Sexual imprinting in human mate choice. *Proceedings of the Royal Society of London, Series B, 271,* 1129–1134.

Berger, J. (1979). Weaning conflict in desert and mountain bighorn sheep: An ecological interpretation. *Zeitschrift fur Tierpsychologie, 50,* 187–200.

Berger, J. (1980). The ecology, structure and function of social play in bighorn sheep (*Ovis canedensis*). *Journal of Zoology (London), 192,* 531–542.

Berger, R. J., & Phillips, N. H. (1995). Energy conservation and sleep. *Behavioural Brain Research, 69,* 65–73.

Berglund, A., & Rosenqvist, G. (2003). Sex role reversal in pipefish. *Advances in the Study of Behavior, 32,* 131–167.

Bergman, A., & Feldman, M. (1995). On the evolution of learning: Representation of a stochastic environment. *Theoretical Population Biology, 48,* 251–276.

Bergstrom, C. T., & Dugatkin, L. A. (2012). *Evolution.* New York: W. W. Norton.

Bergstrom, C. T., & Lachmann, M. (1997). Signalling among relatives I: Is costly signalling too costly? *Philosophical Transactions of the Royal Society of London, Series B–Biological Sciences, 352,* 609–617.

Bergstrom, C. T., & Lachmann, M. (1998). Signaling among relatives. III: Talk is cheap. *Proceedings of the National Academy of Sciences, U. S. A., 95,* 5100–5105.

Bernasconi, G., & Strassmann, J. E. (1999). Cooperation among unrelated individuals: The ant foundress case. *Trends in Ecology and Evolution, 14,* 477–482.

Berridge, K. C. (2003). Pleasures of the brain. *Brain and Cognition, 52,* 106–128.

Berthold, P. (1990). Wegzugbeginn und Einsetzen der zugunruhe bei 19 Vogelpopulationen: Eine vergleichende untersuchung. In R. van den Elzen, K. Schuchmann, & K. Schmidt-Koenig, K. (Eds.), *Proceedings of the International Centennial Meeting of the DO-G, Current topics in avian biology* (pp. 217–222). Bonn, Germany: DO-G.

Bertram, B. (1978). Living in groups: Predators and prey. In J. R. Krebs & N. B. Davies (Eds.), *Behavioural ecology: An evolutionary approach* (pp. 64–96). Oxford: Blackwell Science.

Berwick, R. C., Okanoya, K., Beckers, G. J. L., & Bolhuis, J. J. (2011). Songs to syntax: The linguistics of birdsong. *Trends in Cognitive Sciences, 15,* 113–121.

Biardi, J. E., Chien, D. C., & Coss, R. G. (2005). California ground squirrel (*Spermophilus beecheyi*) defenses against rattlesnake venom digestive and hemostatic toxins. *Journal of Chemical Ecology, 31,* 2501–2518.

Biardi, J. E., Chien, D. C., & Coss, R. G. (2006). California ground squirrel (*Spermophilus beecheyi*) defenses against rattlesnake venom digestive and hemostatic toxins. *Journal of Chemical Ecology, 32,* 137–154.

Biardi, J. E., Coss, R. G., & Smith, D. G. (2000). California ground squirrel (*Spermophilus beecheyi*) blood sera inhibits crotalid venom proteolytic activity. *Toxicon, 38,* 713–721.

Biben, M. (1986). Individual- and sex-related strategies of wrestling play in captive squirrel monkeys. *Ethology, 71,* 229–241.

Biben, M. (1998). Squirrel monkey playfighting: Making the case for a cognitive function for play. In M. Bekoff & J. Byers (Eds.), *Animal play: Evolutionary, comparative and ecological perspectives* (pp. 161–182). Cambridge: Cambridge University Press.

Bildstein, K. L. (1983). Why white-tailed deer flag their tails. *American Naturalist, 121,* 709–715.

Billen, J., & Morgan, E. D. (1997). Pheromonal communication in social insects: Sources and secretions. In R. Vander Meer, M. Breed, K. Espelie, & M. Winston (Eds.), *Pheromonal communication in social insects* (pp. 3–33). Boulder, CO: Westview Press.

Birkhead, T., & Møller, A. (1992). *Sperm competition in birds.* London: Academic Press.

Birkhead, T., & Møller, A. (1998). *Sperm competition and sexual selection.* London: Academic Press.

Birkhead, T., & Parker, G. (1997). Sperm competition and mating systems. In J. R. Krebs & N. B. Davies (Eds.), *Behavioural ecology: An evolutionary approach* (pp. 121–145). Boston: Blackwell Science.

Birkhead, T. R., & Pizzari, T. (2002). Postcopulatory sexual selection. *Nature Reviews Genetics, 3,* 262–273.

Bisazza, A., Grapputo, A., & Nigro, L. (1997). Evolution of reproductive strategies and male sexual ornaments in poeciliid fishes as inferred by mitochondrial 16 rRNA gene phylogeny. *Ethology Ecology & Evolution, 9,* 55–67.

Bischof, H. J., & Clayton, N. (1991). Stabilization of sexual preferences by sexual experience in male zebra finches, *Taeniopygia guttata castanotis. Behaviour, 118,* 144–154.

Bischof, H. J., & Rollenhagen, A. (1999). Behavioural and neurophysiological aspects of sexual imprinting in zebra finches. *Behavioural Brain Research, 98,* 267–276.

Bishop, D. T., Cannings, C., & Maynard Smith, J. (1978). The war of attrition with random rewards. *Journal of Theoretical Biology, 74,* 377–388.

Bishop, J. M., Jarvis, J. U. M., Spinks, A. C., Bennett, N. C., & O'Ryan, C. (2004). Molecular insight into patterns of colony composition and paternity in the common mole-rat *Cryptomys hottentotus hottentotus. Molecular Ecology, 13,* 1217–1229.

Bitterman, M. (1975). The comparative analysis of learning. *Science, 188,* 699–709.

Blackburn, J., Pfaus, J., & Phillips, A. (1992). Dopamine functions in appetitive and defensive behaviors. *Progress in Neurobiology, 39,* 247–249.

Blackledge, T. A., Kuntner, M., & Agnarsson, I. (2011). The form and function of spider orb webs: Evolution from silk to ecosystems. *Advances in Insect Physiology, 41,* 175–262.

Blanco, J., Reig, S., & Cuesta, L. (1992). Distribution, status and conservation problems of the wolf (*Canis lupis*) in Spain. *Biological Conservation, 60,* 73–80.

Blaustein, A. (1983). Kin recognition mechanisms: Phenotype matching or recognition alleles. *American Naturalist, 121,* 749–754.

Blaustein, A. R., & Bancroft, B. A. (2007). Amphibian population declines: Evolutionary considerations. *Bioscience, 57,* 437–444.

Bloch, G., Rubinstein, C. D., & Robinson, G. E. (2004). Expression in the honey bee brain is developmentally regulated and not affected by light, flight experience, or colony type. *Insect Biochemistry and Molecular Biology, 34,* 879–891.

Bloch, G., Toma, D. P., & Robinson, G. E. (2001). Behavioral rhythmicity, age, division of labor and period expression in the honey bee brain. *Journal of Biological Rhythms, 16,* 444–456.

Blomqvist, D., Andersson, M., Kupper, C., Cuthill, I. C., Kis, J., Lanctot, R. B., Sandercock, B. K., Szekely, T., Wallander, J., & Kempenaers, B. (2002). Genetic similarity between mates and extra-pair parentage in three species of shorebirds. *Nature, 419,* 613–615.

Blomqvist, D., Andersson, M., Kupper, C., Cuthill, I. C., Kis, J., Lanctot, R. B., Sandercock, B. K., Szekely, T., Wallander, J., & Kempenaers, B. (2003). Why do birds engage in extra-pair copulation? Reply. *Nature, 422,* 833–834.

Blumstein, D. T., Verneyre, L., & Daniel, J. C. (2004). Reliability and the adaptive utility of discrimination among alarm callers. *Proceedings of the Royal Society of London, Series B–Biological Sciences, 271,* 1851–1857.

Boehm, C. (1992). Segmentary warfare and management of conflict: A comparison of East African chimpanzees and patrilineal-patrilocal humans. In A. H. Harcourt & F. B. M. de Waal (Eds.), *Coalitions and alliances in humans and other animals* (pp. 137–173). Oxford: Oxford University Press.

Boesch, C. (1994a). Cooperative hunting in wild chimpanzees. *Animal Behaviour, 48,* 653–667.

Boesch, C. (1994b). Hunting strategies of Gombe and Tai chimpanzees. In R. W. Wrangham, W. C. McGrew, F. B. M. de Waal, & P. G. Heltne (Eds.), *Chimpanzee cultures* (pp. 77–91). Cambridge, MA: Harvard University Press.

Boesch, C. (2002). Cooperative hunting roles among Tai chimpanzees. *Human Nature, 13,* 27–46.

Boesch, C., & Boesch, H. (1989). Hunting behavior of wild chimpanzees in the Tai National Park. *American Journal of Physical Anthropology, 78,* 547–573.

Boissy, A. (1995). Fear and fearfulness in animals. *Quarterly Review of Biology, 70,* 165–191.

Bolig, R., Price, C., O'Neill, P., & Suomi, S. (1992). Subjective assessment of reactivity level and personality traits of rhesus monkeys. *International Journal of Primatology, 13,* 287–306.

Bolles, R. (1970). Species-specific defense reactions and avoidance learning. *Psychological Review, 77,* 32–48.

Bonabeau, E., Theraulaz, G., & Deneubourg, J. L. (1996). Mathematical model of self-organizing hierarchies in animal societies. *Bulletin of Mathematical Biology, 58,* 661–717.

Bonabeau, E., Theraulaz, G., & Deneubourg, J. L. (1999). Dominance orders in animal societies: The self-organization hypothesis revisited. *Bulletin of Mathematical Biology, 61,* 727–757.

Bonanni, R., Valsecchi, P., & Natoli, E. (2010). Pattern of individual participation and cheating in conflicts between groups of free-ranging dogs. *Animal Behaviour, 79,* 957–968.

Bonier, F., Martin, P. R., Moore, I. T., & Wingfield, J. C. (2009). Do baseline glucocorticoids predict fitness? *Trends in Ecology & Evolution, 24,* 634–642.

Bonner, J. T. (1980). *The evolution of culture in animals.* Princeton, NJ: Princeton University Press.

Bonte, D., Clerq, N., Zwertvaegher, I., & Lens, L. (2008). Thermal conditions during juvenile development affect adult dispersal in a spider. *Proceedings of the National Academy of Sciences, U. S. A., 105,* 17000–17005.

Bonte, D., De Blauwe, I., & Maelfait, J. P. (2003). Environmental and genetic background of tiptoe-initiating behaviour in the dwarf-spider *Erigone atra. Animal Behaviour, 66,* 169–174.

Bonte, D., Vanden Borre, J., Lens, L., & Maelfait, J. P. (2006). Geographic variation in wolfspider dispersal behaviour is related to landscape structure. *Animal Behaviour, 72,* 655–662.

Borenstein, E., Feldman, M. W., & Aoki, K. (2008). Evolution of learning in fluctuating environments: When selection favors both social and exploratory individual learning. *Evolution, 62,* 586–602.

Borgerhoff-Mulder, M. (1988). Kipsigis bridewealth payments. In L. Betzig, M. Borgerhoff-Mulder, & P. Turke (Eds.), *Human reproductive behaviour: A Darwinian perspective* (pp. 65–82). Cambridge: Cambridge University Press.

Borgerhoff-Mulder, M. (1990). Kipsigis women's preferences for wealthy men: Evidence for female choice in mammals. *Behavioral Ecology and Sociobiology, 27,* 255–264.

Both, C., Dingemanse, N. J., Drent, P. J., & Tinbergen, J. M. (2005). Pairs of extreme avian personalities have highest reproductive success. *Journal of Animal Ecology, 74,* 667–674.

Bottger, K. (1974). Zur biologie von *Sphaerodema grassei ghesquieri.* Studien an zentralafrikanischen belostomatiden (Heteroptera: Insecta) I. *Archive fur Hydrobiologie, 74,* 100–122.

Boucher, D. (Ed.). (1985). *The biology of mutualism: Ecology and evolution.* Oxford: Oxford University Press.

Boulton, M. J. (1991). Partner preferences in middle school children's playful fighting and chasing. *Ethology and Sociobiology, 12,* 177–193.

Boyd, R., & Richerson, P. J. (1985). *Culture and the evolutionary process.* Chicago: University of Chicago Press.

Boyd, R., & Richerson, P. J. (2004). *Not by genes alone.* Chicago: University of Chicago Press.

Boyd, R., & Richerson, P. J. (2005). *The origin and evolution of cultures.* New York: Oxford University Press.

Boyse, E., Beauchamp, G., Yamazaki, K., & Bard, J. (1991). Genetic components of kin recognition in mammals. In P. G. Hepper (Ed.), *Kin recognition* (pp. 162–219). Cambridge: Cambridge University Press.

Bradbury, J. W., & Vehrencamp, S. (1977). Social organization and foraging in emballonurid bats. III. Mating systems. *Behavioural Ecology and Sociobiology, 2,* 1–17.

Brantley, R., & Bass, A. (1994). Alternative male spawning tactics and acoustic signalling in the plainfin midshipman fish, *Porichtys notatus. Ethology, 96,* 213–232.

Brass, M., & Heyes, C. (2005). Imitation: Is cognitive neuroscience solving the correspondence problem? *Trends in Cognitive Sciences, 9,* 489–495.

Braude, S. (2000). Dispersal and new colony formation in wild naked mole-rats: Evidence against inbreeding as the system of mating. *Behavioral Ecology, 11,* 7–12.

Brennan, P. A., & Kendrick, K. M. (2006). Mammalian social odours: Attraction and individual recognition. *Philosophical Transactions of the Royal Society, Series B–Biological Sciences, 361,* 2061–2078.

Brenowitz, E. A. (1997). Comparative approaches to the avian song system. *Journal of Neurobiology, 33,* 517–531.

Bressan, P., & Zucchi, G. (2009). Human kin recognition is self—rather than family—referential. *Biology Letters, 5,* 336–338.

Brick, O. (1999). A test of the sequential assessment game: The effect of increased cost of sampling. *Behavioral Ecology, 10,* 726–732.

Brightsmith, D. J. (2005a). Competition, predation and nest niche shifts among tropical cavity nesters: Phylogeny and natural history evolution of parrots (Psittaciformes) and trogons (Trogoniformes). *Journal of Avian Biology, 36,* 64–73.

Brightsmith, D. J. (2005b). Competition, predation and nest niche shifts among tropical cavity nesters: Ecological evidence. *Journal of Avian Biology, 36,* 74–83.

Brightsmith, D. J., Stronza, A., & Holle, K. (2008). Ecotourism, conservation biology, and volunteer tourism: A mutually beneficial triumvirate. *Biological Conservation, 141,* 2832–2842.

Brilot, B. O., Bateson, M., Nettle, D., Whittingham, M. J., & Read, J. C. A. (2012). When is general wariness favored in avoiding multiple predator types? *American Naturalist, 179,* E180–E195.

Bro-Jorgensen, J. (2011). Intra- and intersexual conflicts and cooperation in the evolution of mating strategies: Lessons learnt from ungulates. *Evolutionary Biology, 38,* 28–41.

Bro-Jorgensen, J., & Pangle, W. M. (2010). Male topi antelopes alarm snort deceptively to retain females for mating. *American Naturalist, 176,* E33–E39.

Bronstein, J. L. (1994). Our current understanding of mutualism. *Quarterly Review of Biology, 69,* 31–51.

Broom, M., & Ruxton, G. D. (2003). Evolutionarily stable kleptoparasitism: Consequences of different prey types. *Behavioral Ecology, 14,* 23–33.

Brown, C. (1988). Social foraging in cliff swallows: Local enhancement, risk sensitivity and the avoidance of predators. *Animal Behaviour, 36,* 780–792.

Brown, C., & Brown, M. B. (2000). Heritable basis for choice of group size in a colonial bird. *Proceedings of the National Academy of Sciences, U.S.A., 97,* 14825–14830.

Brown, C., Brown, M. B., & Shaffer, M. L. (1991). Food-sharing signals among socially foraging cliff swallows. *Animal Behaviour, 42,* 551–564.

Brown, C., Jones, F., & Braithwaite, V. (2005). In situ examination of boldness-shyness traits in the tropical poeciliid, *Brachyraphis episcopi. Animal Behaviour, 70,* 1003–1009.

Brown, C., & Laland, K. N. (2002). Social learning of a novel avoidance task in the guppy: Conformity and social release. *Animal Behaviour, 64,* 41–47.

Brown, C. R. (1986). Cliff swallow colonies as information centers. *Science, 234,* 83–85.

Brown, C. R., & Brown, M. B. (2004a). Empirical measurement of parasite transmission between groups in a colonial bird. *Ecology, 85,* 1619–1626.

Brown, C. R., & Brown, M. B. (2004b). Group size and ectoparasitism affect daily survival probability in a colonial bird. *Behavioral Ecology and Sociobiology, 56,* 498–511.

Brown, J. L. (1969). The buffer effect and productivity in tit populations. *American Naturalist, 103,* 347–354.

Brown, J. L. (1974). Alternate routes to sociality in jays with a theory for the evolution of altruism and communal breeding. *Amerian Zoologist, 14,* 63–80.

Brown, J. L. (1975). *The evolution of behavior.* New York: Norton.

Brown, J. L. (1980). Fitness in complex avian social systems. In H. Markl (Ed.), *Evolution of social behavior* (pp. 115–128). Berlin: Verlag Chemie.

Brown, J. L. (1982). Optimal group size in territorial animals. *Journal of Theoretical Biology, 95,* 793–810.

Brown, J. L. (1983). Cooperation: A biologist's dilemma. In J. S. Rosenblatt (Ed.), *Advances in the study of behavior* (Vol. 13, pp. 1–37). New York: Academic Press.

Brown, J. L. (1987). *Helping and communal breeding in birds.* Princeton, NJ: Princeton University Press.

Brown, J. L., Brown, E., Brown, S. D., & Dow, D. D. (1982). Helpers: Effects of experimental removal on reproductive success. *Science, 215,* 421–422.

Brown, J. L., & Eklund, A. (1994). Kin recognition and the major histocompatibility complex—an integrative review. *American Naturalist, 143,* 435–461.

Brown, J. S., & Rosenzweig, M. L. (1985). Habitat selection in slowly regenerating environments. *Journal of Theoretical Biology, 123,* 151–171.

Brown, M., Hopkins, W., & Keynes, R. (1991). *Essentials of neural development.* Cambridge: Cambridge University Press.

Bruderer, B., & Salewski, V. (2008). Evolution of bird migration in a biogeographical context. *Journal of Biogeography, 35,* 1951–1959.

Brumm, H., & Naguib, M. (2009). Environmental acoustics and the evolution of bird song. *Advances in the Study of Behavior, 40,* 1–33.

Brumm, H., & Slabbekoorn, H. (2005). Acoustic communication in noise. *Advances in the Study of Behavior, 35,* 151–209.

Bruner, J., Jolly, A., & Sylva, K. (Eds.). (1976). *Play: Its role in development and evolution.* New York: Basic Books.

Brush, A. (1990). Metabolism of carotenoid pigments in birds. *Federation of American Societies for Experimental Biology Journal, 4,* 2969–2977.

Brush, A., & Power, D. (1976). House finch pigmentation, carotenoid pigmentation and the effect of diet. *Auk, 93,* 725–739.

Brush, F. R. (2003). Selection for differences in avoidance learning: The Syracuse strains differ in anxiety, not learning ability. *Behavior Genetics, 33,* 677–696.

Brush, F. R., Gendron, C. M., & Isaacson, M. D. (1999). A selective genetic analysis of the Syracuse high- and low-avoidance (SHA/Bru and SLA/Bru) strains of rats (*Rattus norvegicus*). *Behavioural Brain Research, 106,* 1–11.

B'shary, R., & Bergmueller, R. (2008). Distinguishing four fundamental approaches to the evolution of helping. *Journal of Evolutionary Biology, 21,* 405–420.

B'shary, R., & Bronstein, J. L. (2004). Game structures in mutualistic interactions: What can the evidence tell us about the kind of models we need? *Advances in the Study of Behavior, 34,* 59–101.

B'shary, R., Wickler, W., & Fricke, H. (2002). Fish cognition: A primate's eye view. *Animal Cognition, 5,* 1–13.

Buehler, D. M., Tieleman, B. I., & Piersma, T. (2010). How do migratory species stay healthy over the annual cycle? A conceptual model for immune function and for resistance to disease. *Integrative and Comparative Biology, 50,* 346–357.

Bugnyar, T., Schwab, C., Schloegl, C., Kotrschal, K., & Heinrich, B. (2007). Ravens judge competitors through experience with play caching. *Current Biology, 17,* 1804–1808.

Burgdorf, J., Panksepp, J., Brudzynski, S. M., Kroes, R., & Moskal, J. R. (2005). Breeding for 50–kHz positive affective vocalization in rats. *Behavior Genetics, 35,* 67–72.

Burger, J. M. S., Kolss, M., Pont, J., & Kawecki, T. J. (2008). Learning ability and longevity: A symmetrical evolutionary trade-off in *Drosophila*. *Evolution, 62,* 1294–1304.

Burgess, J. W. (1976). Social spiders. *Scientific American,* March 1976, 100–106.

Burghardt, G. (1988). Precocity, play and the ectotherm-endotherm transition. *Behaviour, 36,* 246–257.

Burghardt, G. (1998). The evolutionary origins of play revisited: Lessons from turtles. In M. Bekoff & J. Byers (Eds.), *Animal play: Evolutionary, comparative and ecological perspectives* (pp. 1–26). Cambridge: Cambridge University Press.

Burghardt, G. (2005). *The genesis of animal play: Testing the limits.* Cambridge, MA: MIT Press.

Burley, N. (1986). Sexual selection for aesthetic traits in species with biparental care. *American Naturalist, 127,* 415–445.

Burley, N., Kranntzberg, G., & Radman, P. (1982). Influence of colour-banding on the conspecific preferences of zebra finches. *Animal Behavior, 30,* 444–445.

Burns, J. G., Foucaud, J., & Mery, F. (2011). Costs of memory: Lessons from "mini" brains. *Proceedings of the Royal Society B–Biological Sciences, 278,* 923–929.

Buss, A., & Plomin, R. (1986). The EAS approach to temperament. In R. Plomin & J. Dunn (Eds.), *The study of temperament: Changes, continuities and challenges* (pp. 67–79). Hillsdale, NJ: Lawrence Erlbaum Associates.

Busse, C. (1978). Do chimps hunt cooperatively? *American Naturalist, 112,* 767–770.

Butcher, G., & Rohwer, S. (1989). The evolution of conspicuous and distinctive coloration for communication in birds. *Current Ornithology, 6,* 51–107.

Byers, B. E., & Kroodsma, D. E. (2009). Female mate choice and songbird song repertoires. *Animal Behaviour, 77,* 13–22.

Byers, J. (1984). Play in ungulates. In P. K. Smith (Ed.), *Play in animals and humans* (pp. 43–65). Oxford: Blackwell Science.

Byers, J. (1998). Biological effects of locomotor play: Getting into shape or something more specific. In M. Bekoff & J. Byers (Eds.), *Animal play: Evolutionary, comparative and ecological perspectives* (pp. 205–220). Cambridge: Cambridge University Press.

Byers, J., & Waits, L. (2006). Good genes sexual selection in nature. *Proceedings of the National Academy of Sciences, U.S.A., 103,* 16343–16345.

Byers, J., & Walker, C. (1995). Refining the motor training hypothesis for the evolution of play. *American Naturalist, 146,* 25–40.

Byers, J. A., Moodie, J. D., & Hall, N. (1994). Pronghorn females choose vigorous mates. *Animal Behaviour, 47,* 33–43.

Bygott, J. D. (1979). Agonistic behavior, dominance and social structure in wild chimpanzees of the Gombe National Park. In D. A. Hamburg & E. R. McCown (Eds.), *The great apes.* Menlo Park, CA: Benjamin/Cummings.

Byrne, P. G., Simmons, L. W., & Roberts, J. D. (2003). Sperm competition and the evolution of gamete morphology in frogs. *Proceedings of the Royal Society of London, Series B, 270,* 2079–2086.

Byrne, R. W. (1993). Do larger brains mean greater intelligence? *Behavioral and Brain Sciences, 16,* 696–697.

Byrne, R. W. (2002). Imitation of novel complex actions: What does the evidence from animals mean? *Advances in the Study of Behavior, 31,* 77–105.

Cafaro, M. J., & Currie, C. R. (2005). Phylogenetic analysis of mutualistic filamentous bacteria associated with fungus-growing ants. *Canadian Journal of Microbiology, 51,* 441–446.

Cahan, S., & Julian, G. E. (1999). Fitness consequences of cooperative colony founding in the desert leaf-cutter ant *Acromyrmex versicolor. Behavioral Ecology, 10,* 585–591.

Caldera, E. J., Poulsen, M., Suen, G., & Currie, C. R. (2009). Insect symbioses: A case study of past, present, and future fungus-growing ant research. *Environmental Entomology, 38,* 78–92.

Caldwell, M. S., McDaniel, J. G., & Warkentin, K. M. (2009). Frequency information in the vibration-cued escape hatching of red-eyed treefrogs. *Journal of Experimental Biology, 212,* 566–575.

Caldwell, M. S., McDaniel, J. G., & Warkentin, K. M. (2010). Is it safe? Red-eyed treefrog embryos assessing predation risk use two features of rain vibrations to avoid false alarms. *Animal Behaviour, 79,* 255–260.

Caldwell, R. (1986). Withholding information on sexual condition as a competitive mechanism. In L. Drickamer (Ed.), *Behavioral ecology and population biology* (pp. 83–88). Toulouse, France: Privat.

Caldwell, R. (1992). Recognition, signalling and reduced aggression between former mates in a stomatopod. *Animal Behaviour, 44,* 11–19.

Cameron, E., Day, T., & Rowe, L. (2003). Sexual conflict and indirect benefits. *Journal of Evolutionary Biology, 16,* 1055–1060.

Campbell, C., Steffen-Campbell, J., & Werren, J. H. (1993). Phylogeny of the *Nasonia* species complex (Hymenoptera: Pteromalidae) inferred from internal transcribed spacer (T2) and 28s rDNA sequences. *Insect Molecular Biology, 2,* 225–237.

Campbell, T., Lin, S., DeVries, C., & Lambert, K. (2003). Coping strategies in male and female rats exposed to multiple stressors. *Physiology and Behavior, 78,* 495–504.

Cannon, W. (1915). *Bodily changes in pain, hunger, fear and rage.* New York: Appleton.

Capaldi, E. A., Robinson, G. E., & Fahrbach, S. E. (1999). Neuroethology of spatial learning: The birds and the bees. *Annual Review of Psychology, 50,* 651–682.

Capellini, L., Barton, R. A., McNamara, P., Preston, B. T., & Nunn, C. L. (2008). Phylogenetic analysis of the ecology and evolution of mammalian sleep. *Evolution, 62,* 1764–1775.

Caraco, T. (1979). Time budgeting and group size: A test of theory. *Ecology, 60,* 618–627.

Caraco, T. (1980). On foraging time allocation in a stochastic environment. *Ecology, 61,* 119–128.

Caraco, T., Martindale, S., & Whitham, M. (1980). An empirical demonstration of risk sensitive foraging preferences. *Animal Behaviour, 28,* 820–830.

Carere, C., & Maestripieri, D. (2013). *Animal personalities: Behavior, physiology, and evolution:* Chicago: University of Chicago.

Carlier, P., & Lefebvre, L. (1996). Differences in individual learning between group-foraging and territorial Zenaida doves. *Behaviour, 133,* 1197–1207.

Caro, T. (1980). Effects of the mother, object play and adult experience on predation in cats. *Behavioural and Neural Biology, 29,* 29–51.

Caro, T. (1994a). *Cheetahs of the Serengeti Plains.* Chicago: University of Chicago Press.

Caro, T. M. (1994b). Ungulate antipredator behavior: Preliminary and comparative data from African bovids. *Behaviour, 128,* 189–228.

Caro, T. M. (1995a). Pursuit-deterrence revisited. *Trends in Ecology & Evolution, 10,* 500–503.

Caro, T. M. (1995b). Short-term costs and correlates of play in cheetahs. *Animal Behaviour, 49,* 333–345.

Caro, T. M., & Bateson, P. (1986). Organization and ontogeny of alternative tactics. *Animal Behaviour, 34,* 1483–1499.

Caro, T. M., Graham, C. M., Stoner, C. J., & Vargas, J. K. (2004). Adaptive significance of antipredator behaviour in artiodactyls. *Animal Behaviour, 67,* 205–228.

Caro, T. M., & Hauser, M. D. (1992). Is there teaching in nonhuman animals? *Quarterly Review of Biology, 67,* 151–174.

Caro, T. M., Lombardo, L., Goldizen, A. W., & Kelly, M. (1995). Tail-flagging and other antipredator signals in white deer: New data and synthesis. *Behavioral Ecology, 6,* 442–450.

Carpenter, C. R. (1934). A field study of the behavior and social relations of howling monkeys. *Comparative Psychology Monographs, 10,* 1–168.

Carpenter, C. R. (1942). Sexual behavior of free-ranging rhesus monkeys, *M. mulatta. Journal of Comparative Psychology, 33,* 113–162.

Carroll, S. (2007). *The making of the fittest.* New York: Norton.

Carter, C. S., Devries, A. C., & Getz, L. L. (1995). Physiological substrates of mammalian monogamy: The prairie vole model. *Neuroscience and Biobehavioral Reviews, 19,* 303–314.

Cassey, P., Blackburn, T. M., Sol, S., Duncan, R. P., & Lockwood, J. L. (2004). Global patterns of introduction effort and establishment success in birds. *Proceedings of the Royal Society of London, Series B, 271,* S405–S408.

Catchpole, C., & Slater, P. (Eds.). (1995). *Bird song: Biological themes and variation.* New York: Cambridge University Press.

Cavalli-Sforza, L. L., & Feldman, M. W. (1981). *Cultural transmission and evolution: A quantitative approach.* Princeton, NJ: Princeton University Press.

Cavigelli, S. A. (2005). Animal personality and health. *Behaviour, 142,* 1223–1244.

Charnov, E. L. (1976). Optimal foraging, the marginal value theorem. *Theoretical Population Biology, 9,* 129–136.

Charnov, E. L., & Krebs, J. R. (1975). The evolution of alarm calls: Altruism or manipulation? *American Naturalist, 109,* 107–112.

Charnov, E. L., & Orians, G. (1973). Optimal foraging: Some theoretical explorations. Seattle: University of Washington. Mimeograph.

Chase, I. (1974). Models of hierarchy formation in animal societies. *Behavioral Science, 19,* 374–382.

Chase, I., Bartolomeo, C., & Dugatkin, L. A. (1994). Aggressive interactions and inter-contest interval: How long do winners keep winning? *Animal Behaviour, 48,* 393–400.

Chase, I. D. (1982). Dynamics of hierarchy formation: The sequential development of dominance relationships. *Behaviour, 80,* 218–240.

Chase, I. D. (1985). The sequential analysis of aggressive acts during hierarchy formation: An application of the "jigsaw puzzle" approach. *Animal Behaviour, 33,* 86–100.

Chatterjee, K., Reiter, J. G., & Nowak, M. A. (2012). Evolutionary dynamics of biological auctions. *Theoretical Population Biology, 81,* 69–80.

Cheney, D. L., & Seyfarth, R. M. (1990). *How monkeys see the world.* Chicago: University of Chicago Press.

Chesser, R. T., & Levey, D. J. (1998). Austral migrants and the evolution of migration in new world birds: Diet, habitat, and migration revisited. *American Naturalist, 152,* 311–319.

Chivers, D. P., Wisenden, B. D., & Smith, R. J. F. (1996). Damselfly larvae learn to recognize predators from chemical cues in the predator's diet. *Animal Behaviour, 52,* 315–320.

Chiyo, P. I., Moss, C. J., & Alberts, S. C. (2012). The influence of life history milestones and association networks on crop-raiding behavior in male African elephants. *Plos One, 7.*

Cho, M. M., DeVries, A. C., Williams, J. R., & Carter, C. S. (1999). The effects of oxytocin and vasopressin on partner preferences in male and female prairie voles (*Microtus ochrogaster*). *Behavioral Neuroscience, 113,* 1071–1079.

Choleris, E., Gustafsson, J. A., Korach, K. S., Muglia, L. J., Pfaff, D. W., & Ogawa, S. (2003). An estrogen-dependent four-gene micronet regulating social recognition: A study with oxytocin and estrogen receptor-alpha and -beta knockout mice. *Proceedings of the National Academy of Sciences, U.S.A., 100,* 6192–6197.

Choleris, E., Kavaliers, M., & Pfaff, D. W. (2004). Functional genomics of social recognition. *Journal of Neuroendocrinology, 16,* 383–389.

Christensen, J. W., Rundgren, M., & Olsson, K. (2006). Training methods for horses: Habituation to a frightening stimulus. *Equine Veterinary Journal, 38,* 439–443.

Church, S., Bennett, A., Cuthill, I., & Partridge, J. (1998). Ultraviolet cues affect the foraging behaviour of blue tits. *Proceedings of the Royal Society of London, Series B, 265,* 1509–1514.

Clark, A. B., & Ehlinger, T. J. (1987). Pattern and adaptation in individual behavioral differences. In P. P. G. Bateson & P. H. Klopfer (Eds.), *Perspectives in ethology* (Vol. 7, pp. 1–45). New York: Plenum Press.

Clark, M., Crews, D., & Galef, B. (1991). Concentrations of sex steroid hormones in pregnant and fetal Mongolian gerbils. *Physiology and Behavior, 49,* 239–243.

Clark, M., & Galef, B. (1999). A testosterone mediated trade-off between parental care and sexual effort in male Mongolian gerbils, *Meriones unguiculatus. Journal of Comparative Psychology, 113,* 388–395.

Clark, M., & Galef, B. (2000). Why some male Mongolian gerbils may help at the nest: Testosterone, asexuality and alloparenting. *Animal Behaviour, 59,* 801–806.

Clark, M., vom Saal, F., & Galef, B. (1992). Fetal intrauterine position correlates with endogenous testosterone levels of adult Mongolian gerbils. *Physiology and Behavior, 51,* 957–960.

Clark, M., Vonk, J., & Galef, B. (1998). Intrauterine position, parenting and nest-sites attachment in male Mongolian gerbils. *Developmental Psychobiology, 32*, 177–181.

Clark, M. M., Johnson, J., & Galef, B. G. (2004). Sexual motivation suppresses paternal behaviour of male gerbils during their mates' postpartum oestrus. *Animal Behaviour, 67*, 49–57.

Clark, M. M., Tucker, L., & Galef, B. G. (1992). Stud males and dud males: Intrauterine position effects on the reproductive success of male gerbils. *Animal Behaviour, 43*, 215–221.

Clarke, J. A. (1983). Moonlight's influence on predator/prey interactions between short-eared owls (*Asio flammeus*) and deer mice (*Peromyscus maniculatus*). *Behavioral Ecology and Sociobiology, 13*, 205–209.

Clayton, D. F. (2004). Songbird genomies: Methods, mechanisms, opportunities, and pitfalls. In H. P. Zeigler & P. Marler (Eds.), *Behavioral neurobiology of birdsong* (Vol. 1016, pp. 45–60). New York: New York Academy of Sciences.

Clayton, N. S., & Dickinson, A. (1998). Episodic-like memory during cache recovery by scrub jays. *Nature, 395*, 272–274.

Clayton, N. S., Russell, J., & Dickinson, A. (2009). Are animals stuck in time or are they chronesthetic creatures? *Topics in Cognitive Science, 1*, 59–71.

Clemens, L. G., Gladue, B. A., & Coniglio, L. P. (1978). Prenatal endogenous androgenic influences on masculine sexual behavior and genital morphology in male and female rats. *Hormones and Behavior, 10*, 40–53.

Clements, K. C., & Stephens, D. W. (1995). Testing models of non-cooperation: Mutualism and the prisoner's dilemma. *Animal Behavior, 50*, 527–535.

Clipperton-Allen, A. E., Lee, A. W., Reyes, A., Devidze, M. N., Phan, A., Pfaff, D. W., & Choleris, E. (2012). Oxytocin, vasopressin and estrogen receptor gene expression in relation to social recognition in female mice. *Physiology & Behavior, 105*, 915–924.

Close, R. (1972). Dynamic properties of fast and slow skeletal muscles of the rat during development. *Journal of Physiology, 173*, 74–95.

Clotfelter, E. D., O'Hare, E. P., McNitt, M. M., Carpenter, R. E., & Summers, C. H. (2007). Serotonin decreases aggression via 5-HT1A receptors in the fighting fish *Betta splendens*. *Pharmacology Biochemistry and Behavior, 87*, 222–231.

Clutton-Brock, T. (2009). Cooperation between non-kin in animal societies. *Nature, 462*, 51–57.

Clutton-Brock, T., & Albon, S. (1979). The roaring of red deer and the evolution of honest advertising. *Behaviour, 49*, 145–169.

Clutton-Brock, T., Albon, S., Gibson, S., & Guinness, F. (1979). The logical stag: Adaptive aspects of fighting in the red deer. *Animal Behaviour, 27*, 211–225.

Clutton-Brock, T., & Harvey, P. (1980). Primates, brain and ecology. *Journal of Zoology, London, 190*, 309–323.

Coccaro, E. (1992). Impulsive aggression and central serotonergic system function in humans: An example of a dimensional brain-behavior relationship. *International Clinical Psychopharmacology, 7*, 3–12.

Colborn, T., Saal, F. S. V., & Soto, A. M. (1993). Developmental effects of endocrine-disrupting chemicals in wildlife and humans. *Environmental Health Perspectives, 101*, 378–384.

Coleman, K., & Wilson, D. S. (1998). Shyness and boldness in pumpkinseed sunfish: Individual differences are context-specific. *Animal Behaviour, 56*, 927–936.

Coleman, R., Gross, M., & Sargent, R. C. (1985). Parental investment decision rules: A test in bluegill sunfish. *Behavioral Ecology and Sociobiology, 18*, 59–66.

Connor, R. (1992). Dolphin alliances and coalitions. In A. H. Harcourt & F. B. M. de Waal (Eds.), *Coalitions and alliances in humans and other animals* (pp. 415–443). Oxford: Oxford University Press.

Connor, R. C. (1995). The benefits of mutualism: A conceptual framework. *Biological Reviews of the Cambridge Philosophical Society, 70*, 427–457.

Connor, R. C. (2010). Cooperation beyond the dyad: On simple models and a complex society. *Philosophical Transactions of the Royal Society, Series B–Biological Sciences, 365*, 2687–2697.

Connor, R. C., Heithaus, M. R., & Barre, L. M. (2001). Complex social structure, alliance stability and mating access in a bottlenose dolphin "super-alliance." *Proceedings of the Royal Society of London, Series B, 268*, 263–267.

Connor, R. C., Smolker, R. A., & Richards, A. F. (1992). Two levels of alliance formation among male bottleneck dolphins. *Proceedings of the National Academy of Sciences, U.S.A., 89*, 987–990.

Connor, R. C., Watson-Capps, J. J., Sherwin, W. B., & Krutzen, M. (2011). A new level of complexity in the male alliance networks of Indian Ocean bottlenose dolphins (*Tursiops* sp.). *Biology Letters, 7*, 623–626.

Cook, M., & Mineka, S. (1989). Observational conditioning of fear to fear relevant versus fear-irrelevant stimuli in rhesus monkeys. *Journal of Abnormal Psychology, 98*, 448–459.

Cook, M., Mineka, S., Wolkenstein, B., & Laitsch, K. (1985). Observational conditioning of snake fear in unrelated rhesus monkeys. *Journal of Abnormal Psychology, 94*, 591–610.

Cooke, F., Finney, G. H., & Rockwell, R. (1976). Assortative mating in lesser snow geese (*Anser carulescens*). *Behavioral Genetics, 6*, 127–139.

Cooke, F., & McNally, C. (1975). Mate selection and color preferences in lesser snow geese. *Behaviour, 53*, 151–170.

Cooke, F., Mirsky, P., & Seiger, M. (1972). Color preferences in lesser snow geese and their possible role in mate selection. *Canadian Journal of Zoology, 50*, 529–536.

Coppens, C. M., de Boer, S. F. & Koolhaas, J. M. (2010). Coping styles and behavioural flexibility: Towards underlying mechanisms. *Philosophical Transactions of the Royal Society B–Biological Sciences, 365*, 4021–4028.

Cordero, C. (1996). On the evolutionary origin of nuptial seminal gifts in insects. *Journal of Insect Behavior, 9*, 969–974.

Cords, M. (1995). Predator vigilance costs of allogrooming in wild blue monkeys. *Behaviour, 132*, 559–569.

Coria, J., & Calfucura, E. (2012). Ecotourism and the development of indigenous communities: The good, the bad, and the ugly. *Ecological Economics, 73*, 47–55.

Cornette, R., Farine, J. P., Abed-Viellard, D., Quennedey, B., & Brossut, R. (2003). Molecular characterization of a male-specific glycosyl hydrolase, Lma-p72, secreted on to the abdominal surface of the Madeira cockroach *Leucophaea maderae* (Blaberidae, Oxyhaloinae). *Biochemical Journal, 372*, 535–541.

Correia, S. P. C., Dickinson, A., & Clayton, N. S. (2007). Western scrub-jays anticipate future needs independently of their current motivational state. *Current Biology, 17,* 856–861.

Coss, R. G. (1991). Context and animal behavior III. The relationship between early development and evolutionary persistence of ground squirrel antisnake behavior. *Ecological Psychology, 3,* 277–315.

Coss, R. G., & Owings, D. H. (1985). Restraints on ground squirrel antipredator behavior: Adjustments over multiple time scales. In T. D. Johnston & A. T. Pietrewicz (Eds.), *Issues in the ecological study of learning* (pp. 167–200). Hillsdale, NJ: Lawrence Erlbaum Associates.

Coultier, S., Beaugrand, J. P., & Lague, P. C. (1996). The role of individual differences and patterns of resolution in the formation of dominance orders in domestic hen triads. *Behavioural Processes, 38,* 227–239.

Cowie, R. (1977). Optimal foraging in great tits. *Nature, 268,* 137–139.

Cox, C., & Le Boeuf, B. J. (1977). Female incitation of male competition: A mechanism in sexual selection. *American Naturalist, 111,* 317–335.

Cozza, K., Fico, R., Batistini, L. & Rogers, E. (1996). The damage-conservation interface illustrated by predation on domestic livestock in Italy. *Biological Conservation, 78,* 329–336.

Crane, A. L., & Mathis, A. (2011). Predator-recognition training: A conservation strategy to increase postrelease survival of hellbenders in head-starting programs. *Zoo Biology, 30,* 611–622.

Crawley, M. (1983). *Herbivory.* Oxford: Blackwell Science.

Creel, S. (2001). Cooperative hunting and sociality in African wild dogs, *Lycaon pictus.* In L. Dugatkin (Ed.), *Model systems in behavioral ecology.* Princeton, NJ: Princeton University Press.

Crockford, C., & Boesch, C. (2003). Context-specific calls in wild chimpanzees, Pan troglodytes versus analysis of barks. *Animal Behaviour, 66,* 115–125.

Crockford, C., Wittig, R. M., Mundry, R., & Zuberbuhler, K. (2012). Wild chimpanzees inform ignorant group members of danger. *Current Biology, 22,* 142–146.

Croes, B. M., Laurance, W. F., Lahm, S. A., Tchignoumba, L., Alonso, A., Lee, M. E., Campbell, P., & Buij, R. (2007). The influence of hunting on antipredator behavior in central African monkeys and duikers. *Biotropica, 39,* 257–263.

Croft, D. P., James, R., Thomas, P. O. R., Hathaway, C., Mawdsley, D., Laland, K. N., & Krause, J. (2006). Social structure and co-operative interactions in a wild population of guppies (*Poecilia reticulata*). *Behavioral Ecology and Sociobiology, 59,* 644–650.

Crowl, T., & Covich, A. (1990). Predator-induced life-history traits in a freshwater snail. *Science, 247,* 949–951.

Crowley, P. H., Gillett, S., & Lawton, J. H. (1988). Contests between larval damselflies: Empirical steps toward a better ESS model. *Animal Behaviour, 36,* 1496–1510.

Crozier, R. (1987). Genetic aspects of kin recognition: Concepts, models and synthesis. In D. C. Fletcher & C. D. Michener (Eds.), *Kin recognition in animals* (pp. 55–74). Chichester, UK: Wiley.

Crozier, R. H., & Pamilo, P. (1996). *Evolution of social insect colonies: Sex allocation and kin selection.* Oxford: Oxford University Press.

Curio, E., Ernest, U., & Vieth, W. (1978). Cultural transmission of enemy recognition: One function of mobbing. *Science, 202,* 899–901.

Curio, E., Klump, G., & Regelmann, K. (1983). An antipredator response in the great tit (*Parus major*): Is it tuned to predator risk? *Oecologia, 60,* 83–88.

Curio, E., & Regelmann, K. (1985). The behavioral dynamics of great tits (*Parus major*) approaching a predator. *Zeitschrift fur Tierpsychologie, 69,* 3–18.

Currie, C. R., Mueller, U. G., & Malloch, D. (1999). The agricultural pathology of ant fungus gardens. *Proceedings of the National Academy of Sciences, U.S.A., 96,* 7998–8002.

Currie, C. R., Scott, J. A., Summerbell, R. C., & Malloch, D. (1999). Fungus-growing ants use antibiotic-producing bacteria to control garden parasites. *Nature, 398,* 701–704.

Curtsinger, J. W. (1986). Stay times in *Scatophaga* and the theory of evolutionarily stable strategies. *American Naturalist, 128,* 130–136.

Dalgleish, T. (2004). The emotional brain. *Nature Reviews Neuroscience, 5,* 583–589.

Dall, S. R. X., Houston, A. I., & McNamara, J. M. (2004). The behavioural ecology of personality: Consistent individual differences from an adaptive perspective. *Ecology Letters, 7,* 734–739.

Dally, J. M., Emery, N. J. & Clayton, N. S. (2006). Food-caching western scrub-jays keep track of who was watching when. *Science, 312,* 1662–1665.

Daly, M., & Wilson, M. (1988). *Homicide.* New York: Aldine de Gruyter.

Danchin, E., Giraldeau, L. A., Valone, T. J., & Wagner, R. H. (2004). Public information: From nosy neighbors to cultural evolution. *Science, 305,* 487–491.

Darwin, C. (1845). *The voyage of the Beagle.* London: J. Murray.

Darwin, C. (1859). *On the origin of species.* London: J. Murray.

Darwin, C. (1871). *The descent of man and selection in relation to sex.* London: J. Murray.

Davies, N., & Halliday, T. (1978). Deep croaks and fighting assessment in toads, *Bufo bufo. Nature, 391,* 56–58.

Davies, N. B. (1976). Food, flocking and territorial behavior in the pied wagtail *Motacilla alba* in winter. *Journal of Animal Ecology, 45,* 235–254.

Davies, N. B. (1978). Territorial defence in the speckled wood butterfly (*Parage aegeria*): The resident always wins. *Animal Behaviour, 26,* 138–147.

Davies, N. B. (1986). Reproductive success of dunnocks in a variable mating system. II. Factors influencing provisioning rate, nestling weight and fledgling success. *Journal of Animal Ecology, 55,* 123–138.

Davies, N. B. (1991). Mating systems. In J. R. Krebs & N. B. Davies (Eds.), *Behavioural ecology: An evolutionary approach* (3rd ed., pp. 263–299). Oxford: Blackwell Science.

Davies, N. B. (1992). *Dunnock behaviour and social evolution.* Oxford: Oxford University Press.

Davies, N. B., & Houston, A. I. (1981). Owners and satellites: The economics of territory defence in the pied wagtail *Motacilla alba*. *Journal of Animal Ecology, 50*, 157–180.

Davies, N. B., & Lundberg, A. (1984). Food distribution and a variable mating system in the dunnock. *Journal of Animal Ecology, 53*, 895–912.

Davis, D. (1942). The phylogeny of social nesting in the Crotophaginae. *Quarterly Review of Biology, 17*, 115–134.

Davis, D. E. (1966). *Integral animal behavior.* New York: MacMillan.

Davis, J., & Daly, M. (1997). Evolutionary theory and the human family. *Quarterly Review of Biology, 72*, 407–435.

Dawkins, M. S. (1990). From an animal's point of view: Motivation, fitness, and animal-welfare. *Behavioral and Brain Sciences, 13*, 1–9.

Dawkins, M. S. (2008). The science of animal suffering. *Ethology, 114*, 937–945.

Dawkins, R. (1976). *The selfish gene.* Oxford: Oxford University Press.

Dawkins, R. (1979). Twelve misunderstandings of kin selection. *Zeitschrift fur Tierpsychologie, 51*, 184–200.

Dawkins, R. (2006). *The selfish gene* (30th anniv. ed.). Oxford: Oxford University Press.

Dawkins, R., & Krebs, J. (1979). Arms races between and within species. *Proceedings of the Royal Society of London, Series B, 205*, 489–511.

Dawkins, R., & Krebs, J. R. (1978). Animal signals: Information for manipulation? In J. R. Krebs & N. B. Davies (Eds.), *Behavioural ecology: An evolutionary approach* (pp. 282–309). Oxford: Blackwell Science.

Dawson, A. (2002). Photoperiodic control of the annual cycle in birds and comparison with mammals. *Ardea, 90*, 355–367.

Dayrat, B. (2005). Ancestor-descendant relationships and the reconstruction of the tree of life. *Paleobiology, 31*, 347–353.

Dean, J. (1980a). Effects of thermal and chemical components of bombardier beetle chemical defense: Glossopharyngeal response in two species of toad (*Bufo americanus, Bufo marinus*). *Journal of Comparative Physiology, 135*, 51–59.

Dean, J. (1980b). Encounters between bombardier beetles and 2 species of toads (*Bufo americanus, Bufo marinus*): Speed of prey capture does not determine success. *Journal of Comparative Physiology, 135*, 41–50.

De Backer, C. J. S., & Gurven, M. (2006). Whispering down the lane: The economics of vicarious information transfer. *Adaptive Behavior, 14*, 249–264.

Decety, J., Chaminade, T., Grezes, J., & Meltzoff, A. N. (2002). A PET exploration of the neural mechanisms involved in reciprocal imitation. *Neuroimage, 15*, 265–272.

Dejong, G. (1994). The fitness of fitness concepts and the description of natural selection. *Quarterly Review of Biology, 69*, 3–29.

de Kloet, E. R., Grootendorst, J., Karssen, A. A., & Oitzl, M. S. (2002). Gene x environment interaction and cognitive performance: Animal studies on the role of corticosterone. *Neurobiology of Learning and Memory, 78*, 570–577.

de Kloet, E. R., Oitzl, M. S., & Joels, M. (1999). Stress and cognition: Are corticosteroids good or bad guys? *Trends in Neuroscience, 22*, 422–426.

de Kloet, E. R., Vreugdenhil, E., Oitzl, M. S., & Joels, M. (1998). Brain corticosteroid receptor balance in health and disease. *Endocrine Reviews, 19*, 269–301.

de Kort, S. R., & Clayton, N. S. (2006). An evolutionary perspective on caching by corvids. *Proceedings of the Royal Society of London, Series B, 273*, 417–423.

De Laet, J. (1985). Dominance and antipredator behavior of great tits *Parus major*: A field study. *Ibis, 127*.

Delph, L., & Havens, K. (1998). Pollen competition in flowering plants. In T. Birkhead & A. Møller (Eds.), *Sperm competition and sexual selection* (pp. 149–173). London: Academic Press.

Denison, R. F., Kiers, E. T., & West, S. A. (2003). Darwinian agriculture: When can humans find solutions beyond the reach of natural selection? *Quarterly Review of Biology, 78*, 145–168.

de Quervain, D. J. F. (2006). Glucocorticoid-induced inhibition of memory retrieval: Implications for posttraumatic stress disorder. In R. Yehuda, *Psychobiology of posttraumatic stress disorder: A decade of progress* (Vol. 1071, pp. 216–220). New York: Annals of the New York Academy of Sciences.

de Quervain, D. J. F., Fischbacher, U., Treyer, V., Schelthammer, M., Schnyder, U., Buck, A., & Fehr, E. (2004). The neural basis of altruistic punishment. *Science, 305*, 1254–1258.

de Quervain, D. J. F., Roozendaal, B., & McGaugh, J. L. (1998). Stress and glucocorticoids impair retrieval of long-term spatial memory. *Nature, 394*, 787–790.

de Quervain, D. J. F., Roozendaal, B., Nitsch, R. M., McGaugh, J. L., & Hock, C. (2000). Acute cortisone administration impairs retrieval of long-term declarative memory in humans. *Nature Neuroscience, 3*, 313–314.

DeSalle, R., & Schierwater, B. (Eds.). (1998). *Molecular approaches to ecology and evolution.* Basel, Switzerland: Birkhauser Verlag.

de Waal, F. (1989). *Peacemaking among primates.* Cambridge, MA: Harvard University Press.

de Waal, F. (1997). The chimpanzee's service economy: Food for grooming. *Evolution and Human Behavior, 18*, 375–386.

de Waal, F. B. M. (1992). Coalitions as part of reciprocal relations in the Arnhem chimpanzee colony. In A. Harcourt & F. B. M. de Waal (Eds.), *Coalitions and alliances in humans and other animals* (pp. 233–257). Oxford: Oxford University Press.

Dewey, M. J., & Dawson, W. D. (2001). Deer mice: "The drosophila of North American mammalogy." *Genesis, 29*, 105–109.

Dewsbury, D. (1992). On problems studied in ethology, comparative psychology and animal behavior. *Ethology, 92*, 89–107.

Dewsbury, D. (1994). On the utility of the proximate-ultimate distinction in the study of animal behavior. *Ethology, 96*, 63–68.

Dickman, A. J. (2010). Complexities of conflict: The importance of considering social factors for effectively resolving human-wildlife conflict. *Animal Conservation, 13*, 458–466.

Dill, L. (1974). The escape response of the zebra Danio (*Brachydanio rerio*) II. The effect of experience. *Animal Behaviour, 22*, 723–730.

Dill, L. (1983). Adaptive flexibility in the foraging behavior of fishes. *Canadian Journal of Fisheries and Aquatic Sciences, 40,* 398–408.

Dill, L., & Fraser, A. (1984). Risk of predation and the feeding behaviour of juvenile Coho salmon (*Oncorynchus kisutch*). *Behavioral Ecology and Sociobiology, 16,* 65–72.

DiMarco, F. P., & Hanlon, R. T. (1997). Agonistic behavior in the squid *Loligo plei* (Loliginidae, Teuthoidea): Fighting tactics and the effects of size and resource value. *Ethology, 103,* 89–108.

Dingemanse, N. J., Both, C., Drent, P. J., Van Oers, K., & Van Noordwijk, A. J. (2002). Repeatability and heritability of exploratory behaviour in great tits from the wild. *Animal Behaviour, 64,* 929–938.

Dingemanse, N. J., Bouwman, K. M., van de Pol, M., van Overveld, T., Patrick, S. C., Matthysen, E., & Quinn, J. L. (2012). Variation in personality and behavioural plasticity across four populations of the great tit *Parus major. Journal of Animal Ecology, 81,* 116–126.

Dingemanse, N. J., & Reale, D. (2005). Natural selection and animal personality. *Behaviour, 142,* 1159–1184.

Dingle, H. (2008). Bird migration in the southern hemisphere: A review comparing continents. *Emu, 108,* 341–359.

Dingle, H., & Caldwell, R. (1972). Reproductive and maternal behavior of the mantis shrimp *Gonodactylus bredini* Manning (Crustacea: Stomatopoda). *Biological Bulletin, 142,* 417–426.

Domjan, M. (1992). Adult learning and mate choice: Possibilities and experimental evidence. *American Zoologist, 32,* 48–61.

Domjan, M. (1998). *The principles of learning and behavior* (4th ed.). Pacific Grove, CA: Brooks/Cole.

Domjan, M. (2005). Pavlovian conditioning: A functional perspective. *Annual Review of Psychology, 56,* 179–206.

Domjan, M. (2006). *The principles of learning and behavior: Active learning edition* (5th ed.). Belmont, CA: Thomson/Wadsworth.

Domjan, M., Blesbois, E., & Williams, J. (1998). The adaptive significance of sexual conditioning: Pavlovian control of sperm release. *Psychological Science, 9,* 411–415.

Domjan, M., Cusato, B., & Villarreal, R. (2000). Extensions, elaborations, and explanations of the role of evolution and learning in the control of social behavior. *Behavioral and Brain Sciences, 23,* 269–282.

Domjan, M., & Hollis, K. L. (1988). Reproductive behavior: A potential model system for adaptive specializations in learning. In R. C. Bolles & M. D. Beecher (Eds.), *Evolution and learning* (pp. 213–237). Hillsdale, NJ: Lawrence Erlbaum Associates.

Domjan, M., Lyons, R., North, N., & Bruell, J. (1986). Sexual Pavlovian conditioned approach behavior in male Japanese quail (*Coturnix coturnix japonica*). *Journal of Comparative Psychology, 100,* 413–421.

Donaldson, Z. R., & Young, L. J. (2008). Oxytocin, vasopressin, and the neurogenetics of sociality. *Science, 322,* 900–904.

Dornhaus, A., Collins, E. J., Dechaume-Moncharmont, F. X., Houston, A. I., Franks, N. R., & McNamara, J. M. (2006). Paying for information: Partial loads in central place foragers. *Behavioral Ecology and Sociobiology, 61,* 151–161.

Douglas-Hamilton, I., Bhalla, S., Wittemyer, G., & Vollrath, F. (2006). Behavioural reactions of elephants towards a dying and deceased matriarch. *Applied Animal Behaviour Science, 100,* 87–102.

Douglas-Hamilton, I., & Douglas-Hamilton, O. (1975). *Among the elephants.* Glasgow, Scotland: Collins.

Doutrelant, C., McGregor, P., & Oliveira, R. (2001). The effect of an audience on intra-sexual communication in male Siamese fighting fish, *Betta splendens. Behavioral Ecology, 12,* 283–286.

Drent, P. J., van Oers, K., & van Noordwijk, A. J. (2003). Realized heritability of personalities in the great tit (*Parus major*). *Proceedings of the Royal Society of London, Series B, 270,* 45–51.

Drevets, W. C. (2000). Functional anatomical abnormalities in limbic and prefrontal cortical structures in major depression. In H. B. M. Uylings, C. G. van Eden, J. P. C. de Bruin, M. G. P. Feenstra, & C. M. A. Pennartz (Eds.), *Cognition, emotion and autonomic responses: The integrative role of the prefrontal cortex and limbic structures* (pp. 413–431). Amsterdam: Elsevier.

Drickamer, L. C. (1996). Intra-uterine position and anogenital distance in house mice: Consequences under field conditions. *Animal Behaviour, 51,* 925–934.

Drummond, H. (2006). Dominance in vertebrate broods and litters. *Quarterly Review of Biology, 81,* 3–32.

Drummond, H., & Canales, C. (1998). Dominance between booby nestlings involves winner and loser effects. *Animal Behaviour, 55,* 1669–1676.

Drummond, H., & Osorno, J. L. (1992). Training siblings to be submissive losers: Dominance between booby nestlings. *Animal Behaviour, 44,* 881–893.

Dubois, F., Morand-Ferron, J., & Giraldeau, L. A. (2010). Learning in a game context: Strategy choice by some keeps learning from evolving in others. *Proceedings of the Royal Society, Series B–Biological Sciences, 277,* 3609–3616.

Duckworth, R. A., Mendonca, M. T., & Hill, G. E. (2001). A condition dependent link between testosterone and disease resistance in the house finch. *Proceedings of the Royal Society of London, Series B, 268,* 2467–2472.

Duckworth, R. A., Mendonca, M. T., & Hill, G. E. (2004). Condition-dependent sexual traits and social dominance in the house finch. *Behavioral Ecology, 15,* 779–784.

Duellman, W. E., & Trueb, L. (1994). *Biology of amphibians.* New York: Johns Hopkins University Press.

Duffy, D. L., & Serpell, J. A. (2012). Predictive validity of a method for evaluating temperament in young guide and service dogs. *Applied Animal Behaviour Science, 138,* 99–109.

Duffy, J. E., & Macdonald, K. S. (2010). Kin structure, ecology and the evolution of social organization in shrimp: A comparative analysis. *Proceedings of the Royal Society, Series B–Biological Sciences, 277,* 575–584.

Duffy, J. E., Morrison, C. L., & Rios, R. (2000). Multiple origins of eusociality among sponge-dwelling shrimps (*Synalpheus*). *Evolution, 54,* 503–516.

Dugatkin, L. A. (1991). Dynamics of the tit for tat strategy during predator inspection in guppies. *Behavioral Ecology and Sociobiology, 29,* 127–132.

Dugatkin, L. A. (1992a). Sexual selection and imitation: Females copy the mate choice of others. *American Naturalist, 139,* 1384–1389.

Dugatkin, L. A. (1992b). Tendency to inspect predators predicts mortality risk in the guppy, *Poecilia reticulata. Behavioral Ecology, 3,* 124–128.

Dugatkin, L. A. (1995). Formalizing Allee's idea on dominance hierarchies: An intra-demic selection model. *American Naturalist, 146,* 154–160.

Dugatkin, L. A. (1996a). Copying and mate choice. In C. Heyes & G. Galef (Eds.), *Social learning in animals: The roots of culture* (pp. 85–105). New York: Academic Press.

Dugatkin, L. A. (1996b). The interface between culturally-based preferences and genetic preferences: Female mate choice in *Poecilia reticulata. Proceedings of the National Academy of Sciences, U.S.A., 93,* 2770–2773.

Dugatkin, L. A. (1997a). *Cooperation among animals: An evolutionary perspective.* New York: Oxford University Press.

Dugatkin, L. A. (1997b). Winner effects, loser effects and the structure of dominance hierarchies. *Behavioral Ecology, 8,* 583–587.

Dugatkin, L. A. (1998a). Breaking up fights between others: A model of intervention behaviour. *Proceedings of the Royal Society of London, Series B, 265,* 443–437.

Dugatkin, L. A. (1998b). A model of coalition formation in animals. *Proceedings of the Royal Society of London, Series B, 265,* 2121–2125.

Dugatkin, L. A. (2000). *The imitation factor: Evolution beyond the gene.* New York: Free Press.

Dugatkin, L. A. (2001a). Bystander effects and the structure of dominance hierarchies. *Behavioral Ecology, 12,* 348–352.

Dugatkin, L. A. (Ed). (2001b). *Model systems in behavioral ecology: Integrating conceptual, theoretical and empirical perspectives.* Princeton, NJ: Princeton University Press.

Dugatkin, L. A. (2006). *The altruism equation: Seven scientists search for the origins of goodness.* Princeton, NJ: Princeton University Press.

Dugatkin, L. A., & Alfieri, M. (1991a). Guppies and the tit for tat strategy: Preference based on past interaction. *Behavioral Ecology and Sociobiology, 28,* 243–246.

Dugatkin, L. A., & Alfieri, M. (1991b). Tit for tat in guppies: The relative nature of cooperation and defection during predator inspection. *Evolutionary Ecology, 5,* 300–309.

Dugatkin, L. A., & Alfieri, M. (1992). Interpopulational differences in the use of the Tit for Tat strategy during predator inspection in the guppy. *Evolutionary Ecology, 6,* 519–526.

Dugatkin, L. A., & Alfieri, M. (2003). Boldness, behavioral inhibition and learning: An evolutionary approach. *Ethology, Ecology & Evolution, 15,* 43–49.

Dugatkin, L. A., & Godin, J.-G. J. (1992). Prey approaching predators: A cost-benefit perspective. *Annales Zoologici Fennici, 29,* 233–252.

Dugatkin, L. A., & Godin, J.-G. J. (1998, April). How females choose their mates. *Scientific American,* 46–51.

Dugatkin, L. A., Mesterton-Gibbons, M., & Houston, A. I. (1992). Beyond the prisoner's dilemma: Towards models to discriminate among mechanisms of cooperation in nature. *Trends in Ecology & Evolution, 7,* 202–205.

Dugatkin, L. A., & Reeve, H. K. (1994). Behavioral ecology and levels of selection: Dissolving the group selection controversy. *Advances in the Study of Behaviour, 23,* 101–133.

Dugatkin, L. A., & Reeve, H. K. (Eds.). (1998). *Game theory and animal behavior.* Oxford: Oxford University Press.

Dugatkin, L. A., & Wilson, D. S. (2000). Assortative interactions and the evolution of cooperation in guppies. *Evolutionary Ecology, 2,* 761–767.

Dukas, R. (Ed). (1998a). *Cognitive ecology.* Chicago: University of Chicago Press.

Dukas, R. (1998b). Constraints on information processing and their effects on behavior. In R. Dukas (Ed.), *Cognitive ecology* (pp. 89–127). Chicago: University of Chicago Press.

Dukas, R. (2006). Learning in the context of sexual behaviour in insects. *Animal Biology, 56,* 125–141.

Dukas, R., & Bernays, E. A. (2000). Learning improves growth rate in grasshoppers. *Proceedings of the National Academy of Sciences, 97,* 2637–2640.

Dumond, F. (1968). The squirrel monkey in a seminatural environment. In L. Rosenblum & R. Cooper (Eds.), *The squirrel monkey* (pp. 87–145). New York: Academic Press.

Dunbar, R. (1992). Neocortex size as a constraint on group size in primates. *Journal of Human Evolution, 20,* 469–493.

Dzieweczynski, T. L., Gill, C. E., & Perazio, C. E. (2012). Opponent familiarity influences the audience effect in male-male interactions in Siamese fighting fish. *Animal Behaviour, 83,* 1219–1224.

Earley, R., & Dugatkin, L. A. (2002). Eavesdropping on visual cues in green swordtails (*Xiphophorus helleri*): A case for networking. *Proceedings of the Royal Society of London, Series B, 269,* 943–952.

Earley, R., & Dugatkin, L. A. (2005). Fighting, mating and networking: Pillars of poeciliid sociality. In P. McGregor (Ed.), *Animal communication networks* (pp. 84–113). Cambridge: Cambridge University Press.

Earley, R. L. (2010). Social eavesdropping and the evolution of conditional cooperation and cheating strategies. *Philosophical Transactions of the Royal Society, 365,* 2675–2686.

East, M. (1981). Alarm calling and parental investment in the robin *Erihacus rubecula. Ibis, 123,* 223–230.

Eastwood, R., Pierce, N. E., Kitching, R. L., & Hughes, J. M. (2006). Do ants enhance diversification in lycaenid butterflies? Phylogeographic evidence from a model myrmecophile, *Jalmenus evagoras. Evolution, 60,* 315–327.

Eberhard, W. (1996). *Female control: Sexual selection by cryptic female choice.* Princeton, NJ: Princeton University Press.

Edgerton, V. (1978). Mammalian muscle fibers and their adaptability. *American Zoologist, 18,* 113–125.

Edward, D., & Chapman, T. (2011). The evolution and significance of male mate choice. *Trends in Ecology & Evolution, 26,* 647–654.

Edwards, D. H., & Spitzer, N. (2006). Social dominance and serotonin receptor genes in crayfish. *Current Topics in Developmental Biology, 74,* 177–199.

Edwards, S. V., & Hedrick, P. W. (1998). Evolution and ecology of MHC molecules: From genomics to sexual selection. *Trends in Ecology and Evolution, 13,* 305–311.

Edwards, S. V., & Naeem, S. (1993). The phylogenetic component of cooperative breeding in perching birds. *American Naturalist, 141,* 754–789.

Eggebeen, D., & Hogan, D. (1990). Giving between generations in American families. *Human Nature, 1,* 211–232.

Ehlinger, T. J., & Wilson, D. S. (1988). Complex foraging polymorphism in bluegill sunfish. *Proceedings of the National Academy of Sciences, U.S.A., 85,* 1878–1882.

Eibesfeldt, I., & Eible-Eibesfeldt, E. (1967). Das parasitenabwehren der minima-arbeiterinnen der blattschneider-ameise (*Atta cephalotes*). *Zeitschrift fur Tierpsychologie, 24,* 278–281.

Eichenbaum, H., Stewart, C., & Morris, R. G. M. (1990). Hippocampal representation in place learning. *Journal of Neuroscience, 10,* 3531–3542.

Eisenberg, J. F. (1966). The social organization of mammals. *Handbook of Zoology, 10,* 1–92.

Eisner, T., & Aneshansley, D. (1999). Spray aiming in the bombardier beetle: Photographic evidence. *Proceedings of the National Academy of Sciences, U.S.A., 96,* 9705–9709.

Eisner, T., Aneshansley, D., del Campo, M. L., Eisner, M., Frank, J. H., & Deyrup, M. (2006). Effect of bombardier beetle spray on a wolf spider: Repellency and leg autotomy. *Chemoecology, 16,* 185–189.

Eisner, T., & Aneshansley, D. J. (1982). Spray aiming in bombardier beetles: Jet deflection by the coanda effect. *Science, 215,* 83–85.

Eisner, T., Aneshansley, D. J., Eisner, M., Attygalle, A. B., Alsop, D. W., & Meinwald, J. (2000). Spray mechanism of the most primitive bombardier beetle (*Metrius contractus*). *Journal of Experimental Biology, 203,* 1265–1275.

Eisner, T., & Dean, J. (1976). Ploy and counterploy in predator-prey interactions: Orb-weaving spiders versus bombardier beetles. *Proceedings of the National Academy of Sciences, U.S.A., 73,* 1365–1367.

Ekman, J., & Askenmo, C. (1984). Social rank and habitat use in willow tit groups. *Animal Behaviour, 32,* 508–514.

Elgar, M. (1986). House sparrows establish foraging flocks by giving chirrup calls if the resource is divisible. *Animal Behaviour, 34,* 169–174.

Elgar, M., & Crespi, B. (Eds.). (1992). *Cannibalism: Ecology and evolution among diverse taxa.* Oxford: Oxford University Press.

Elgar, M., & Pierce, N. (1988). Mating success and fecundity in an ant-tended lycaenid butterfly. In T. Clutton-Brock (Ed.), *Reproductive success* (pp. 59–75). Chicago: University of Chicago Press.

Elias, D. O., Botero, C. A., Andrade, M. C. B., Mason, A. C., & Kasumovic, M. M. (2010). High resource valuation fuels "desperado" fighting tactics in female jumping spiders. *Behavioral Ecology, 21,* 868–875.

Ely, D. (1981). Hypertension, social rank and aortic arteriosclerosis in CBA/J mice. *Physiology and Behavior, 26,* 655–661.

Emery, N. J. (2006). Cognitive ornithology: The evolution of avian intelligence. *Philosophical Transactions of the Royal Society of London, Series B, 361,* 23–43.

Emlen, D. J. (2008). The evolution of animal weapons. *Annual Review of Ecology Evolution and Systematics, 39,* 387–413.

Emlen, J. M. (1966). The role of time and energy in food preference. *American Naturalist, 100,* 611–617.

Emlen, J. T., & Lorenz, F. W. (1942). Pairing responses of free-living valley quail to sex hormone pellet tablets. *Auk, 59.*

Emlen, S. T. (1967). Migratory orientation in the Indigo bunting, *Passerina cyanea. II.* Mechanisms of celestial orientation. *Auk, 84,* 463–469.

Emlen, S. T. (1970). Celestial rotation: Its importance in the development of migratory orientation. *Science, 170,* 1198–1201.

Emlen, S. T. (1975). The stellar-orientation system of a migratory bird. *Scientific American, 233,* 102–111.

Emlen, S. T. (1982a). The evolution of helping: An ecological constraints model. *American Naturalist, 119,* 29–39.

Emlen, S. T. (1982b). The evolution of helping: II. The role of behavioral conflict. *American Naturalist, 119,* 40–53.

Emlen, S. T. (1995a). Can avian biology be useful to the social sciences? *Journal of Avian Biology, 26,* 273–276.

Emlen, S. T. (1995b). An evolutionary theory of the family. *Proceedings of the National Academy of Sciences, U.S.A., 92,* 8092–8099.

Emlen, S. T., & Oring, L. W. (1977). Ecology, sexual selection and evolution of mating systems. *Science, 197,* 215–223.

Emlen, S. T., & Wrege, P. H. (1992). Parent-offspring conflict and the recruitment of helpers among bee-eaters. *Nature, 356,* 331–333.

Emlen, S. T., & Wrege, P. H. (2004a). Division of labour in parental care behaviour of a sex-role-reversed shorebird, the wattled jacana. *Animal Behaviour, 68,* 847–855.

Emlen, S. T., & Wrege, P. H. (2004b). Size dimorphism, intrasexual competition, and sexual selection in wattled jacana (*Jacana jacana*), a sex-role-reversed shorebird in Panama. *Auk, 121,* 391–403.

Emlen, S. T., Wrege, P. H., & Webster, M. (1998). Cuckoldry as a cost of polyandry in the sex-reversed wattled jacana, *Jacana jacana. Proceedings of the Royal Society of London, Series B, 265,* 2359–2364.

Enard, W., Przeworski, M., Fisher, S. E., Lai, C. S. L., Wiebe, V., Kitano, T., Monaco, A. P., & Paabo, S. (2002). Molecular evolution of FOXP2, a gene involved in speech and language. *Nature, 418,* 869–872.

Endler, J. (1986). *Natural selection in the wild.* Princeton, NJ: Princeton University Press.

Endler, J. (1995). Multiple trait co-evolution and environmental gradients in guppies. *Trends in Ecology and Evolution, 10,* 22–29.

Endler, J. (1999). Adaptive genetic variation in the wild: Concluding remarks. In T. Mousseau, B. Sinervo, & J. Endler (Eds.), *Adaptive genetic variation in the wild* (pp. 251–260). New York: Oxford University Press.

Endler, J., & McLellan, T. (1988). The process of evolution: Toward a newer synthesis. *Annual Review of Ecology and Systematics, 19,* 395–421.

Endler, J. A. (1997). Light, behavior, and conservation of forest dwelling organisms. In J. Clemmons & R. Buchholz (Eds.), *Behavioral approaches to conservation in the wild* (pp. 329–355). Cambridge: Cambridge University Press.

Endler, J. A., & Thery, M. (1996). Interacting effects of lek placement, display behavior, ambient light, and color patterns in three neotropical forest-dwelling birds. *American Naturalist, 148,* 421–456.

Engel, G., & Schmale, A. (1972). Conservation-withdrawal: A primary regulatory process for organic homeostasis. In R. Porter & J. Knight (Eds.), *Physiology, emotions and psychosomatic illness* (pp. 57–75). Amsterdam: Elsevier.

Engh, A. L., Esch, K., Smale, L., & Holekamp, K. E. (2000). Mechanisms of maternal rank "inheritance" in the spotted hyaena, *Crocuta crocuta*. *Animal Behaviour, 60,* 323–332.

Enquist, M., & Leimar, O. (1983). Evolution of fighting behavior: Decision rules and assessment of relative strength. *Journal of Theoretical Biology, 102,* 387–410.

Enquist, M., & Leimar, O. (1987). Evolution of fighting behavior: The effect of variation in resource value. *Journal of Theoretical Biology, 127,* 185–207.

Enquist, M., & Leimar, O. (1990). The evolution of fatal fighting. *Animal Behaviour, 39,* 1–9.

Enquist, M., Leimar, O., Ljungberg, T., Mallner, Y., & Segardahl, N. (1990). A test of the sequential assessment game: Fighting in the cichlid fish, *Nannacara anomala. Animal Behaviour, 40,* 1–15.

Enquist, M., Ljungberg, T., & Zandor, A. (1987). Visual assessment of fighting ability in the cichlid fish *Nanacara anomala. Animal Behaviour, 35,* 1262–1263.

Ens, B. J., & Goss-Custard, J. (1984). Interference among oyster catchers (*Haematopus ostraleugus*) feeding on mussels (*Mytilus edulis*) on the Exe Estuary. *Journal of Animal Ecology, 53,* 217–231.

Erber, J., Kloppenburg, P., & Scheidler, A. (1993). Neuromodulation by serotonin and octopamine in the honeybee: Behavior, neuroanatomy and electrophysiology. *Experientia, 49,* 1073–1083.

Essapian, F. S. (1963). Observations on abnormalities of parturition in captive bottle-nosed dolphins. *Journal of Mammalogy, 44,* 405–414.

Estes, R. D., & Goddard, J. (1967). Prey selection and hunting behavior of the African dog. *Journal of Wildlife Management, 31,* 52–70.

Evans, C. S., & Marler, P. (1994). Food calling and audience effects in male chickens, *Gallus gallus*: Their relationships to food availability, courtship and social facilitation. *Animal Behaviour, 47,* 1159–1170.

Even, M., Dahr, M., & vom Saal, F. (1992). Transport of steroids between fetuses via amniotic fluid in relation to the intrauterine position in rats. *Journal of Reproduction and Fertility, 96,* 709–712.

Ewald, P. (1985). Influences of asymmetries in resource quality and age on aggression and dominance in black-chinned hummingbirds. *Animal Behaviour, 33,* 705–719.

Ewald, P. W. (1994). *Evolution of infectious diseases.* New York: Oxford University Press.

Ewald, P. W. (2000). *Plague time.* New York: Free Press.

Ewen, J., Armstrong, D., Parker, K., & Seddon, P. (2012). *Reintroduction biology: Integrating science and management.* New York: Wiley-Blackwell.

Ewer, R. (1969). The "instinct to teach." *Nature, 222,* 698.

Ewing, A. W., & Bennet-Clark, H. C. (1968). The courtship songs of Drosophila. *Behaviour, 31,* 288–301.

Fa, J. E., & Brown, D. (2009). Impacts of hunting on mammals in African tropical moist forests: A review and synthesis. *Mammal Review, 39,* 231–264.

Fagen, R. (1987). A generalized habitat matching rule. *Evolutionary Ecology, 1,* 5–10.

Fagen, R., & Fagen, J. (2004). Juvenile survival and benefits of play behaviour in brown bears, *Ursus arctos. Evolutionary Ecology Research, 6,* 89–102.

Fagen, R. M. (1981). *Animal play behavior.* Oxford: Oxford University Press.

Fahrbach, S. E. (2006). Structure of the mushroom bodies of the insect brain. *Annual Review of Entomology, 51,* 209–232.

Farooqui, T. (2007). Octopamine-mediated neuromodulation of insect senses. *Neurochemical Research, 32,* 1511–1529.

Feener, D. H., & Moss, K. A. (1990). Defense against parasites by hitchhikers in leaf-cuttting ants: A quantitative assessment. *Behavioral Ecology and Sociobiology, 26,* 17–29.

Felsenstein, J. (1985). Phylogenies and the comparative method. *American Naturalist, 125,* 1–15.

Felsenstein, J. (2004). *Inferring phylogenies.* Sunderland, MA: Sinauer.

Ferguson, J. N., Aldag, J. M., Insel, T. R., & Young, L. J. (2001). Oxytocin in the medial amygdala is essential for social recognition in the mouse. *Journal of Neuroscience, 21,* 8278–8285.

Ferguson, J. N., Young, L. J., Hearn, E. F., Matzuk, M. M., Insel, T. R., & Winslow, J. T. (2000). Social amnesia in mice lacking the oxytocin gene. *Nature Genetics, 25,* 284–288.

Fernandez, A., & Morris, M. (2007). Sexual selection and trichromatic color vision in primates: Statistical support for the preexisting bias hypothesis. *American Naturalist, 170,* 10–20.

Fernandez-Juricic, E., Gilak, N., McDonald, J. C., Pithia, P., & Valcarcel, A. (2006). A dynamic method to study the transmission of social foraging information in flocks using robots. *Animal Behaviour, 71,* 901–911.

Fernandez-Juricic, E., & Kowalski, V. (2011). Where does a flock end from an information perspective? A comparative experiment with live and robotic birds. *Behavioral Ecology, 22,* 1304–1311.

Fernandez-Teruel, A., Escorihuela, R. M., Gray, J. A., Aguilar, R., Gil, L., Gimenez-Llort, L., Tobena, A., Bhomra, A., Nicod, A., Mott, R., Driscoll, P., Dawson, G. R., & Flint, J. (2002). A quantitative trait locus influencing anxiety in the laboratory rat. *Genome Research, 12,* 618–626.

Ferrari, M. C. O., Capitania-Kwok, T., & Chivers, D. P. (2006). The role of learning in the acquisition of threat-sensitive responses to predator odours. *Behavioral Ecology and Sociobiology, 60,* 522–527.

Ferrari, M. C. O., & Chivers, D. P. (2006). Learning threat-sensitive predator avoidance: How do fathead minnows incorporate conflicting information? *Animal Behaviour, 71,* 19–26.

Ferrari, M. C. O., Messier, F., & Chivers, D. P. (2006). The nose knows: Minnows determine predator proximity and density through detection of predator odours. *Animal Behaviour, 72,* 927–932.

Ferrari, M. C. O., Trowell, J. J., Brown, G. E., & Chivers, D. P. (2005). The role of learning in the development of threat-sensitive predator avoidance by fathead minnows. *Animal Behaviour, 70,* 777–784.

Ferrero, D. M., Lemon, J. K., Fluegge, D., Pashkovski, S. L., Korzan, W. J., Datta, S. R., Spehr, M., Fendt, M., & Liberles, S. D. (2011). Detection and avoidance of a carnivore odor by prey. *Proceedings of the National Academy of Sciences of the United States of America, 108,* 11235–11240.

Ferris, C. F., Stolberg, T., Kulkarni, P., Murugavel, M., Blanchard, R., Blanchard, D. C., Febo, M., Brevard, M., & Simon, N. G. (2008). Imaging the neural circuitry and chemical control of aggressive motivation. *Bmc Neuroscience, 9.* doi:10.1186/1471-2202-9-111

Ficken, M. S., Witkin, S. R., & Weise, C. M. (1980). Associations among members of a black-capped chickadee flock. *Behavioral Ecology and Sociobiology, 8,* 245–249.

Fiedler, K., & Maschwitz, U. (1988). Functional analysis of the myrmecophilous relationships between ants (Hymenoptea: formicidae) and lycaenids (Lepidotera: Lycaenidae). *Oecologia, 75,* 204–206.

Field, E. F., Whishaw, I. Q., Pellis, S. M., & Watson, N. V. (2006). Play fighting in androgen-insensitive tfm rats: Evidence that androgen receptors are necessary for the development of adult playful attack and defense. *Developmental Psychobiology, 48,* 111–120.

Field, S. A., Calbert, G., & Keller, M. A. (1998). Patch defence in the parasitoid wasp *Trissolcus basalis* (Insecta: Scelionidae): The time structure of pairwise contests, and the "waiting game." *Ethology, 104,* 821–840.

Firman, R. C., & Simmons, L. W. (2010). Sperm midpiece length predicts sperm swimming velocity in house mice. *Biology Letters, 6,* 513–516.

Fisher, J., & Hinde, R. (1949). The opening of milk bottles by birds. *British Birds, 42,* 347–357.

Fisher, R. A. (1915). The evolution of sexual preference. *Eugenics Review, 7,* 184–192.

Fisher, R. A. (1930). *The genetical theory of natural selection.* New York: Dover.

Fisher, R. A. (1958). *The genetical theory of natural selection* (2nd ed.). New York: Dover.

Fitzgibbon, C. D. (1994). The costs and benefits of predator inspection behaviour in Thomson's gazelles. *Behavioral Ecology and Sociobiology, 34,* 139–148.

Fitzpatrick, J. W., & Woolfenden, G. E. (1984). The helpful shall inherit the scrub. *Natural History, 93,* 55–63.

Fitzpatrick, J. W., & Woolfenden, G. E. (1988). Components of lifetime reproductive success in the Florida scrub jay. In T. H. Clutton-Brock (Ed.), *Reproductive success* (pp. 305–320). Chicago: University of Chicago Press.

Fitzpatrick, M. J., Ben-Shahar, Y., Smid, H. M., Vet, L. E. M., Robinson, G. E., & Sokolowski, M. B. (2005). Candidate genes for behavioural ecology. *Trends in Ecology & Evolution, 20,* 96–104.

Fleishman, L., Lowe, E., & Leal, M. (1993). Ultraviolet vision in lizards. *Nature, 365,* 397.

Fletcher, D., & Michener, C. (Eds.). (1987). *Kin recognition in animals.* New York: Wiley.

Flint, J., Corley, R., DeFries, J., Fulker, D., Gray, J., Miller, S., & Collins, A. C. (1995). A simple genetic basis for a complex psychological trait in mice. *Science, 269,* 1432–1435.

Flint, J., & Mackay, T. F. C. (2009). Genetic architecture of quantitative traits in mice, flies, and humans. *Genome Research, 19,* 723–733.

Flint, J., & Mott, R. (2008). Applying mouse complex-trait resources to behavioural genetics. *Nature, 456,* 724–727.

Fogarty, L., Strimling, P., & Laland, K. N. (2011). The evolution of teaching. *Evolution, 65,* 2760–2770.

Folgarait, P., Gorosito, N., Poulsen, M., & Currie, C. R. (2011). Preliminary in vitro insights into the use of natural fungal pathogens of leaf-cutting ants as biocontrol agents. *Current Microbiology, 63,* 250–258.

Foltz, D. (1981). Genetic evidence for long-term monogamy in a small rodent, *Peromyscus polionotus. American Naturalist, 117,* 665–675.

Ford, J. K. B., & Ellis, G. M. (2006). Selective foraging by fish-eating killer whales (*Orcinus orca*) in British Columbia. *Marine Ecology 316,* 185–199.

Ford, J. K. B., Ellis, G. M., Barrett-Lennard, L. G., Morton, A. B., Palm, R. S., & Balcomb, K. C. (1998). Dietary specialization in two sympatric populations of killer whales (*Orcinus orca*) in coastal British Columbia and adjacent waters. *Canadian Journal of Zoology 76,* 1456–1471.

Ford, R., & McLauglin, F. (1983). Variation of male fidelity in monogamous birds. In R. Johnstone (Ed.), *Current ornithology* (pp. 329–356). New York: Plenum.

Forslund, P. (2000). Male-male competition and large size mating advantage in European earwigs, *Forficula auricularia. Animal Behaviour, 59,* 753–762.

Fossey, D. (1983). *Gorillas in the mist.* New York: Houghton Mifflin Harcourt.

Fowler, G. S. (1999). Behavioral and hormonal responses of Magellanic penguins (*Spheniscus magellanicus*) to tourism and nest site visitation. *Biological Conservation, 90,* 143–149.

Fragasy, D. M., & Perry, S. (2003). *The biology of traditions: Models and evidence.* Cambridge: Cambridge University Press.

Franck, D., Klamroth, B., Taebel-Hellwig, A., & Schartl, M. (1998). Home ranges and satellite tactics of male green swordtails (*Xiphophorus helleri*) in nature. *Behavioural Processes, 43,* 115–123.

Frank, L. G. (1986a). Social organization of the spotted hyena (*Crocuta crocuta*). 1. Demography. *Animal Behaviour, 34,* 1500–1509.

Frank, L. G. (1986b). Social organization of the spotted hyena (*Crocuta crocuta*). 2. Dominance and reproduction. *Animal Behaviour, 34,* 1510–1527.

Franks, N. R., & Richardson, T. (2006). Teaching in tandem-running ants. *Nature, 439,* 153.

Fraser, D., & Duncan, I. J. H. (1998). "Pleasures," "pains" and animal welfare: Toward a natural history of affect. *Animal Welfare, 7,* 383–396.

Fraser, D. F., Gilliam, J. F., Daley, M., Le, A., & Skalski, G. (2001). Explaining leptokurtic movement distributions: Intrapopulation variance in boldness and exploration. *American Naturalist, 158,* 124–135.

Freake, M. J., Muheim, R., & Phillips, J. B. (2006). Magnetic maps in animals: A theory comes of age? *Quarterly Review of Biology, 81,* 327–347.

Freeberg, T. M. (1998). The cultural transmission of courtship patterns in cowbirds, *Molothrus ater*. *Animal Behaviour, 56,* 1063–1073.

Freeberg, T. M. (2004). Social transmission of courtship behavior and mating preferences in brown-headed cowbirds, *Molothrus ater. Learning & Behavior, 32,* 122–130.

Freeberg, T. M., Dunbar, R. I. M., & Ord, T. J. (2012). Social complexity as a proximate and ultimate factor in communicative complexity. *Philosophical Transactions of the Royal Society, Series B– Biological Sciences, 367,* 1785–1801.

Freeberg, T. M., King, A. P., & West, M. J. (2001). Cultural transmission of vocal traditions in cowbirds (*Molothrus ater*) influences courtship patterns and mate preferences. *Journal of Comparative Psychology, 115,* 201–211.

Fretwell, S. (1972). *Populations in a seasonal environment.* Princeton, NJ: Princeton University Press.

Fretwell, S., & Lucas, H. (1970). On territorial behavior and other factors influencing habitat distribution in birds. I. Theoretical development. *Acta Biotheoretica, 19,* 16–36.

Frommen, J. G., Luz, C., & Bakker, T. C. M. (2007). Kin discrimination in sticklebacks is mediated by social learning rather than innate recognition. *Ethology, 113,* 276–282.

Fruteau, C., Lemoine, S., Helland, E., Van Damm, E., & Noe, R. (2011). When females trade grooming for grooming: Testing partner control and partner choice models of cooperation in two primate species. *Animal Behaviour, 81,* 1223–1230.

Fry, C. (1977). The evolutionary significance of cooperative breeding in birds. In B. Stonehouse & C. Perrins (Eds.), *Evolutionary ecology.* London: MacMillan.

Gabor, C. S., Phan, A., Clipperton-Allen, A. E., Kavaliers, M., & Choleris, E. (2012). Interplay of oxytocin, vasopressin, and sex hormones in the regulation of social recognition. *Behavioral Neuroscience, 126,* 97–109.

Gadagkar, R. (1997). *Survival strategies: Cooperation and conflict in animal societies.* Cambridge, MA: Harvard University Press.

Gaffrey, G. (1957). Zur fortpflanzungbsbiologie bei hunden. *Zoological Garten, Jena, 23,* 251.

Galef, B. G. (1996). Social enhancement of food preferences in Norway rats: A brief review. In C. M. Heyes & B. G. Galef (Eds.), *Social learning in animals: The roots of culture* (pp. 49–64). London: Academic Press.

Galef, B. G. (2004). Approaches to the study of traditional behaviors of free-living animals. *Learning & Behavior, 32,* 53–61.

Galef, B. G. (2009). Strategies for social learning: Testing predictions from formal theory. *Advances in the Study of Behavior, 39,* 117–151.

Galef, B. G., & Giraldeau, L. A. (2001). Social influences on foraging in vertebrates: Causal mechanisms and adaptive functions. *Animal Behaviour, 61,* 3–15.

Galef, B. G., & Laland, K. N. (2005). Social learning in animals: Empirical studies and theoretical models. *Bioscience, 55,* 489–499.

Galef, B. G., Manzig, L., & Field, R. (1986). Imitation learning in budgerigars: Dawson and Foster revisted. *Behavioral Processes, 13,* 191–202.

Galef, B. G., Whiskin, E. E., & Dewar, G. (2005). A new way to study teaching in animals: Despite demonstrable benefits, rat dams do not teach their young what to eat. *Animal Behaviour, 70,* 91–96.

Galef, B. G., & Wigmore, S. (1983). Transfer of information concerning distant foods: A laboratory investigation of the "information-centre" hypothesis. *Animal Behaviour, 31,* 748–758.

Gallagher, J. (1976). Sexual imprinting: Effects of various regimens of social experience on the mate preference in Japanese quail (*Coturnix coturnix japonica*). *Behaviour, 57,* 91–115.

Gallese, V., Fadiga, L., Fogassi, L., & Rizzolatti, G. (1996). Action recognition in the premotor cortex. *Brain, 119,* 593–609.

Gamboa, G. J., Reeve, H. K., & Pfennig, D. W. (1986). The evolution and ontogeny of nestmate recognition in social wasps. *Annual Review of Entomology, 31,* 431–454.

Gannon, D. P., Barros, N. B., Nowacek, D. P., Read, A. J., Waples, D. M., & Wells, R. S. (2005). Prey detection by bottlenose dolphins, *Tursiops truncatus*: An experimental test of the passive listening hypothesis. *Animal Behaviour, 69,* 709–720.

Garcia, J., Ervin, F., & Koelling, R. (1966). Learning with prolonged delay of reinforcement. *Psychonomic Science, 5,* 121–122.

Garcia, J., & Koelling, R. (1966). Relation of cue to consequence in avoidance learning. *Psychonomic Science, 4,* 123–124.

Garcia, J., McGowan, B., & Green, K. (1972). Biological constraints on conditioning. In A. Black & W. Prokasy (Eds.), *Classical conditioning II: Current theory and research* (pp. 3–27). New York: Appleton.

Garcia-Peña, G. E., Thomas, G. H., Reynolds, J. D., & Szekely, T. (2009). Breeding systems, climate, and the evolution of migration in shorebirds. *Behavioral Ecology, 20,* 1026–1033.

Gardner, A., West, S. A., & Wild, G. (2011). The genetical theory of kin selection. *Journal of Evolutionary Biology, 24,* 1020–1043.

Gaston, A. J. (1973). Ecology and behavior of long-tailed tit. *Ibis, 115,* 330–351.

Gaulin, S. J. C., & Fitzgerald, R. W. (1986). Sex-differences in spatial ability: An evolutionary hypothesis and test. *American Naturalist, 127,* 74–88.

Gaulin, S. J. C., & Fitzgerald, R. W. (1989). Sexual selection for spatial-learning ability. *Animal Behaviour, 37,* 322–331.

Gauthreaux, S. (Ed.). (1980). *Animal migration, orientation, and navigation.* New York: Academic Press.

Geary, D. C. (2000). Evolution and proximate expression of human paternal investment. *Psychological Bulletin, 126,* 55–77.

Geist, V. (1966). The evolution of horn-like organs. *Behaviour, 27,* 175–224.

Geist, V. (1974a). The development of social behavior through play in the Stellar sea lion. *American Zoologist, 14,* 205–220.

Geist, V. (1974b). On fighting strategies in animal combat. *Nature, 250,* 354.

Gelowitz, C., Mathis, A., & Smith, R. J. F. (1993). Chemosensory recognition of northern pike (*Esox lucius*) by brook stickleback (*Culaea inconstans*): Population differences and the influence of predator diet. *Behaviour, 127,* 106–118.

Gendron, R. P. (1986). Searching for cryptic prey: Evidence for optimal search rates and the formation of search images in quail. *Animal behavior 34,* 896–912.

George, C. (1960). *Behavioral interactions in the pickerel and the mosquitofish.* Unpublished Ph.D., Harvard University.

Getz, L. L., & Carter, C. S. (1996). Prairie-vole partnerships. *American Scientist, 84,* 56–62.

Giancola, P. R., & Chermack, S. T. (1998). Construct validity of laboratory aggression paradigms: A response to Tedeschi and Quigley (1996). *Aggression and Violent Behavior, 3,* 237–253.

Gianoli, E., & Valladares, F. (2012). Studying phenotypic plasticity: The advantages of a broad approach. *Biological Journal of the Linnean Society, 105,* 1–7.

Gibson, B., & Kamil, A. (2009). The synthetic approach to the study of spatial memory: Have we properly addressed Tinbergen's "four questions"? *Behavioural Processes, 80,* 278–287.

Gibson, R. M., Bradbury, J. W., & Vehrencamp, S. L. (1991). Mate choice in lekking sage grouse: The roles of vocal display, female site fidelity and copying. *Behavioral Ecology, 2,* 165–180.

Gilley, D. C., Kuzora, J. M., & Thom, C. (2012). Hydrocarbons emitted by waggle-dancing honey bees stimulate colony foraging activity by causing experienced foragers to exploit known food sources. *Apidologie, 43,* 85–94.

Giraldeau, L. A., & Caraco, T. (2000). *Social foraging theory.* Princeton, NJ: Princeton University Press.

Giraldeau, L. A., & Lefebvre, L. (1986). Exchangeable producer and scrounger roles in a captive flock of feral pigeons: A case for the skill pool effect. *Animal Behaviour, 34,* 797–803.

Giraldeau, L. A., & Lefebvre, L. (1987). Scrounging prevents cultural transmission of food-finding in pigeons. *Animal Behaviour, 35,* 387–394.

Gleason, J. M., Nuzhdin, S. V., & Ritchie, M. G. (2002). Quantitative trait loci affecting a courtship signal in *Drosophila melanogaster. Heredity, 89,* 1–6.

Glen, A. S., & Dickman, C. R. (2006). Diet of the spotted-tailed quoll (Dasyurus maculatus) in eastern Australia: Effects of season, sex and size. *Journal of Zoology, 269,* 241–248.

Glimcher, P. (2004). *Decisions, uncertainty, and the brain: The science of neuroeconomics.* Cambridge, MA: MIT Press.

Glimcher, P. W., & Rustichini, A. (2004). Neuroeconomics: The consilience of brain and decision. *Science, 306,* 447–452.

Goddard, M. E., & Beilharz, R. (1982). Genetic and environmental factors affecting the suitability of dogs as guide dogs for the blind. *Theoretical and Applied Genetics, 62,* 97–102.

Goddard, M. E., & Beilharz, R. (1983). Genetics of traits which determine the suitability of dogs as guide dogs for the blind. *Applied Animal Ethology, 9,* 299–315.

Goddard, M. E., & Beilharz, R. (1984). A factor analysis of fearfulness in potential guide dogs. *Applied Animal Behaviour Science, 12,* 253–265.

Goddard, M. E., & Beilharz, R. (1985). A multivariate analysis of the genetics of fearfulness in potential guide dogs. *Behavior Genetics, 15,* 69–89.

Godin, J. G. J., & Dugatkin, L. A. (1996). Female mating preference for bold males in the guppy, *Poecilia reticulata. Proceedings of the National Academy of Sciences, U.S.A., 93,* 10262–10267.

Gold, K. C., & Maple, T. (1994). Personality assessment in the gorilla and its utility as a management tool. *Zoo Biology, 13,* 509–522.

Gomendio, M., Harcourt, A., & Roldan, E. (1998). Sperm competition in mammals. In T. Birkhead & A. Moller (Eds.), *Sperm competition and sexual selection* (pp. 667–756). London: Academic Press.

Gomes, C. M., & Boesch, C. (2009). Wild chimpanzees exchange meat for sex on a long-term basis. *Plos One, 4.*

Gomes, C. M., & Boesch, C. (2011). Reciprocity and trades in wild West African chimpanzees. *Behavioral Ecology and Sociobiology, 65,* 2183–2196.

Gomez-Mestre, I., Saccoccio, V. L., Iijima, T., Collins, E. M., Rosenthal, G. G., & Warkentin, K. M. (2010). The shape of things to come: Linking developmental plasticity to post-metamorphic morphology in anurans. *Journal of Evolutionary Biology, 23,* 1364–1373.

Gonzalez-Voyer, A., Szekely, T., & Drummond, H. (2007). Why do some siblings attack each other? Comparative analysis of aggression in avian broods. *Evolution, 61,* 1946–1955.

Goodall, J. (1986). *The chimpanzees of Gombe: Patterns of behavior.* Cambridge, MA: Belknap Press.

Goodson, J. L., & Bass, A. H. (2001). Social behavior functions and related anatomical characteristics of vasotocin/vasopressin systems in vertebrates. *Brain Research Reviews, 35,* 246–265.

Goodwin, D. (1986). *Crows of the world* (2nd ed.). London: British Museum.

Goodwin, N. B., Balshine-Earn, S., & Reynolds, J. D. (1998). Evolutionary transitions in parental care in cichlid fish. *Proceedings of the Royal Society of London, Series B, 265,* 2265–2272.

Goodwin, T. (1950). Carotenoids and reproduction. *Biological Reviews of the Cambridge Philosophical Society, 25,* 391–413.

Gosling, S. D. (1998). Personality dimensions in spotted hyenas (*Crocuta crocuta*). *Journal of Comparative Psychology, 112,* 107–118.

Gosling, S. D. (2001). From mice to men: What can we learn about personality from animal research? *Psychological Bulletin, 127,* 45–86.

Gould, J. L., & Gould, C. C. (2012). *Nature's compass: The mystery of animal migration.* Princeton, NJ: Princeton University Press.

Gould, S. J., & Lewontin, R. (1979). The spandrels of San Marcos and the Paglossian paradigm: A critique of the adaptationist programme. *Proceedings of the Royal Society of London, Series B, 205,* 581–598.

Gowaty, P. A. (2008). Reproductive compensation. *Journal of Evolutionary Biology, 21,* 1189–1200.

Grafe, T. U., Dobler, S., & Linsenmair, K. E. (2002). Frogs flee from the sound of fire. *Proceedings of the Royal Society of London, Series B, 269,* 999–1003.

Grafen, A. (1984). Natural selection, kin selection and group selection. In J. R. Krebs & N. B. Davies (Eds.), *Behavioural ecology: An evolutionary approach* (2nd ed., pp. 62–84). Oxford: Blackwell Science.

Grafen, A. (1988). On the uses of data on lifetime reproductive success. In T. Clutton-Brock (Ed.), *Reproductive success* (pp. 454–471). Chicago: University of Chicago Press.

Grafen, A. (1990a). Biological signals as handicaps. *Journal of Theoretical Biology, 144,* 517–546.

Grafen, A. (1990b). Sexual selection unhandicapped by the Fisher Process. *Journal of Theoretical Biology, 144,* 473–476.

Grafen, A., & Ridley, M. (Eds.). (2006). Richard Dawkins: How a scientist changed the way we think. New York: Oxford University Press.

Graham, I. M., Harris, R. N., Matejusova, I., & Middlemas, S. J. (2011). Do rogue seals exist? Implications for seal conservation in the UK. *Animal Conservation, 14,* 587–598.

Graham, J., & Desjardins, C. (1980). Classical conditioning: Induction of luteinizing and testosterone secretion in anticipation of sexual activity. *Science, 210,* 1039–1041.

Graham, K. L., & Burghardt, G. M. (2010). Current perspectives on the biological study of play: Signs of progress. *Quarterly Review of Biology, 85,* 393–418.

Grant, B. R., & Grant, P. R. (1996). Cultural inheritance of song and its role in the evolution of Darwin's finches. *Evolution, 50,* 2471–2487.

Grant, P. R., & Grant, B. R. (1997). Hybridization, sexual imprinting, and mate choice. *American Naturalist, 149,* 1–28.

Grant, P. R., & Grant, R. (1994). Phenotypic and genetic effects of hybridization in Darwin's finches. *Evolution, 48,* 297–316.

Greenewalt, C. (1968). *Bird song: Acoustics and physiology.* Washington, DC: Smithsonian Press.

Greenough, W., & Juraska, J. (1979). Experience-induced changes in brain fine structure: Their behavioral implications. In M. Hahn, C. Jensen, & B. Dudek (Eds.), *Development and evolution of brain size: Behavioral implications* (pp. 295–320). New York: Academic Press.

Grezes, J., Armony, J. L., Rowe, J., & Passingham, R. E. (2003). Activations related to "mirror" and "canonical" neurons in the human brain: An fMRI study. *Neuroimage, 18,* 928–937.

Grezes, J., & Decety, J. (2001). Functional anatomy of execution, mental simulation, observation, and verb generation of actions: A meta-analysis. *Human Brain Mapping, 12,* 1–19.

Grier, J. W., & Burk, T. (1992) *Biology of animal behavior* (2nd ed.). St. Louis, MO: Times Mirror/Mosby College.

Griffin, D. (1959). *Listening in the dark.* New Haven, CT: Yale University Press.

Griffith, S. C., & Montgomerie, R. (2003). Why do birds engage in extra-pair copulation? *Nature, 422,* 833.

Griffith, S. C., Owens, I. P. F., & Thuman, K. A. (2002). Extra pair paternity in birds: A review of interspecific variation and adaptive function. *Molecular Ecology, 11,* 2195–2212.

Grobecker, D., & Pietsch, T. (1978). Crows use automobiles as nutcrackers. *Auk, 95,* 760.

Grodzinski, U., & Clayton, N. S. (2010). Problems faced by food-caching corvids and the evolution of cognitive solutions. *Philosophical Transactions of the Royal Society, Series B–Biological Sciences, 365,* 977–987.

Grodzinski, U., Watanabe, A., & Clayton, N. S. (2012). Peep to pilfer: What scrub-jays like to watch when observing others. *Animal Behaviour, 83,* 1253–1260.

Gross, M. (1985). Disruptive selection for alternative life histories in salmon. *Nature, 313,* 47–48.

Gross, M., & Charnov, R. (1980). Alternative male life histories in bluegill sunfish. *Proceedings of the National Academy of Sciences, U.S.A., 77,* 6937–6940.

Gross, M. R. (1982). Sneakers, satellites, and parentals: Polymorphic mating strategies in North American sunfishes. *Zeitschrift fur Tierpsychologie, 60,* 1–26.

Gross, M. R., & Sargent, R. C. (1985). The evolution of male and female parental care in fishes. *American Zoologist, 25,* 807–822.

Grueter, C., & Farina, W. M. (2009). Why do honeybee foragers follow waggle dances? *Trends in Ecology & Evolution, 24,* 584–585.

Gubernick, D. J., & Nordby, J. C. (1993). Mechanisms of sexual fidelity in the monogamous California mouse, *Peromyscus californicus. Behavioral Ecology and Sociobiology, 32,* 211–219.

Guinet, C., & Bouvier, J. (1995). Development of intentional stranding hunting techniques in killer whale (*Orcinus orca*) calves at Crozet Archipelago. *Canadian Journal of Zoology, 73,* 27–33.

Guthrie, D. M., & Muntz, W. (1993). Role of vision in fish behavior. In T. Pitcher (Ed.), *Behaviour of teleost fishes.* New York: Chapman and Hall.

Guthrie, R. D. (1971). A new theory of mammalian rump patch evolution. *Behaviour, 38,* 132–145.

Gutierrez, G., & Domjan, M. (1996). Learning and male-male sexual competition in Japanese quail (*Coturnix japonica*). *Journal of Comparative Psychology, 110,* 170–175.

Gutierrez, G., & Domjan, M. (2011). Conditioning of sexual proceptivity in female quail: Measures of conditioned place preference. *Behavioural Processes, 87,* 268–273.

Hack, M. A. (1997). Assessment strategies in the contests of male crickets, *Acheta domesticus* (L). *Animal Behaviour, 53,* 733–747.

Haddock, S. H. D., Moline, M. A., & Case, J. F. (2010). Bioluminescence in the sea. *Annual Review of Marine Science, 2,* 443–493.

Haesler, S., Rochefort, C., Georgi, B., Licznerski, P., Osten, P., & Scharff, C. (2007). Incomplete and inaccurate vocal imitation after knockdown of FoxP2 in songbird basal ganglia nucleus area X. *Plos Biology, 5,* 2885–2897.

Haesler, S., Wada, K., Nshdejan, A., Morrisey, E. E., Lints, T., Jarvis, E. D., & Scharff, C. (2004). FoxP2 expression in avian vocal learners and non-learners. *Journal of Neuroscience, 24,* 3164–3175.

Hagenguth, H., & Rembold, H. (1978). Identification of juvenile hormone 3 as the only JH homolog in all developmental stages of the honey bee. *Zeitschrift fur Naturforschun C, 33,* 847–850.

Haig, D. (1993). Genetic conflicts in human pregnancy. *Quarterly Review of Biology, 68,* 495–532.

Hailman, J. (1982). Ontogeny: Toward a general theoretical framework for ethology. *Perspectives in Ethology, 5,* 133–189.

Hain, T., & Neff, B. (2006). Promiscuity drives self-referent recognition. *Current Biology, 16,* 1807–1811.

Haldane, J. B. S. (1932). *The Causes of Evolution.* New York: Harper and Brothers.

Haldane, J. B. S. (1941). *New paths in genetics*. London: Allen and Unwin.

Hall, K. R., & Devore, I. (1965). Baboon social behavior. In I. Devore (Ed.), *Primate behavior: Field studies of monkeys and apes* (pp. 53–110). New York: Holt, Rinehart & Winston.

Hall, S. (1998). Object play by adult animals. In M. Bekoff & J. Byers (Eds.), *Animal play: Evolutionary, comparative and ecological perspectives* (pp. 45–60). Cambridge: Cambridge University Press.

Hamilton, W. D. (1963). The evolution of altruistic behavior. *American Naturalist, 97*, 354–356.

Hamilton, W. D. (1964). The genetical evolution of social behaviour. I and II. *Journal of Theoretical Biology, 7*, 1–52.

Hamilton, W. D., & Zuk, M. (1982). Heritable true fitness and bright birds: A role for parasites. *Science, 218*, 384–387.

Hammer, M., & Menzel, R. (1998). Multiple sites of associative odor learning as revealed by local brain microinjections of octopamine in honeybees. *Learning and Memory, 5*, 146–156.

Hammerstein, P., & Parker, G. A. (1982). The asymmetric war of attrition. *Journal of Theoretical Biology, 96*, 647–682.

Hammock, E. A. D., & Young, L. J. (2002). Variation in the vasopressin V1a receptor promoter and expression: Implications for inter- and intraspecific variation in social behaviour. *European Journal of Neuroscience, 16*, 399–402.

Hammock, E. A. D., & Young, L. J. (2005). Microsatellite instability generates diversity in brain and sociobehavioral traits. *Science, 308*, 1630–1634.

Hanlon, R. T., & Messenger, J. B. (1988). Adaptive coloration in young cuttlefish (*Sepia officinalis*): The morphology and development of body patterns and their relation to behavior. *Philosophical Transactions of the Royal Society of London, Series B, 320*, 437–487.

Hanlon, R. T., Naud, M. J., Forsythe, J. W., Hall, K., Watson, A. C., & McKechnie, J. (2007). Adaptable night camouflage by cuttlefish. *American Naturalist, 169*, 543–551.

Harcourt, A. H. (1978). Strategies of emigration and transfer by primates, with particular reference to gorillas. *Zeitschrift Fur Tierpsychologie–Journal of Comparative Ethology, 48*, 401–420.

Harcourt, A. H., & de Waal, F. B. M. (Eds.). (1992). *Coalitions and alliances in humans and other animals*. Oxford: Oxford University Press.

Harcourt, R. (1991). Survivorship costs of play in the South American fur seal. *Animal Behaviour, 42*, 509–511.

Hardin, G. (1968). The tragedy of the commons. *Science, 162*, 1243–1248.

Hardy, J. (1961). Studies of behavior and phylogeny of certain New World jays (*Garrulinae*). *University of Kansas Science Bulletin, 42*, 13–149.

Hare, J. F., & Atkins, B. A. (2001). The squirrel that cried wolf: Reliability detection by juvenile Richardson's ground squirrels (*Spermophilus richardsonii*). *Behavioral Ecology and Sociobiology, 51*, 108–112.

Harlow, H. (1959). Learning set and error factor theory. In S. Koch (Ed.), *Psychology: A study of science* (pp. 429–537). New York: McGraw-Hill.

Harper, D. C. (1982). Competitive foraging in mallards: "Ideal free" ducks. *Animal Behaviour, 30*, 575–584.

Hartl, D., & Clark, A. (2006). *Principles of population genetics* (4th ed.). Sunderland, MA: Sinauer.

Harvell, C. (1998). Genetic variation and polymorphism in the inducible spines of a marine bryozoan. *Evolution, 52*, 80–86.

Harvell, C. D. (1991). Coloniality and inducible polymorphism. *American Naturalist, 138*, 1–14.

Harvell, C. D. (1994). The evolution of polymorphism in colonial invertebrates and social insects. *Quarterly Review of Biology, 69*, 155–185.

Hass, C. C., & Jenni, D. A. (1993). Social play among juvenile bighorn sheep: Structure, development, and relationship to adult behavior. *Ethology, 93*, 105–116.

Hatchwell, B. J., Ross, D. J., Fowlie, M. K., & McGowan, A. (2001). Kin discrimination in cooperatively breeding long-tailed tits. *Proceedings of the Royal Society of London, Series B, 268*, 885–890.

Hatchwell, B. J., Russell, A. F., MacColl, A. D. C., Ross, D. J., Fowlie, M. K., & McGowan, A. (2004). Helpers increase long-term but not short-term productivity in cooperatively breeding long-tailed tits. *Behavioral Ecology, 15*, 1–10.

Hauser, M. (1997). *The evolution of communication*. Cambridge, MA: MIT Press.

Hauser, M. (2000). *Wild minds*. New York: Henry Holt.

Hauser, M., & Konishi, M. (Eds.). (1999). *The design of animal communication*. Cambridge, MA: MIT Press.

Hausfater, G., & Blaffer-Hrdy, S. (Eds.). (1984). *Infanticide*. New York: Aldine.

Hawkes, K., Hill, K., & Oconnell, J. F. (1982). Why hunters gather: Optimal foraging and the Aćhe of Eastern Paraguay. *American Ethnologist, 9*, 379–398.

Hawkins, R. D., Kandel, E. R., & Bailey, C. H. (2006). Molecular mechanisms of memory storage in *Aplysia*. *Biological Bulletin, 210*, 174–191.

Hay, M., & Fuller, P. (1981). Seed escape from hetromyids rodents: The importance of microhabitat and seed preference. *Ecology, 62*, 1395–1399.

Healey, S. (1992). Optimal memory: Toward an evolutionary ecology of animal cognition. *Trends in Ecology and Evolution, 8*, 399–400.

Healey, S., & Krebs, J. R. (1992). Food storing and the hippocampus in corvids: Amount and volume are correlated. *Proceedings of the Royal Society of London, Series B, 248*, 241–245.

Heape, W. (1931). *Emigration, migration and nomadism*. Cambridge: Cambridge University Press.

Hegner, R. (1985). Dominance and antipredator behaviour in blue tits (*Parus caeruleus*). *Animal Behaviour, 33*, 762–768.

Heinrich, B. (1988a). Why do ravens fear their food? *Condor, 90*, 950–952.

Heinrich, B. (1988b). Winter foraging at carcasses by three sympatric corvids, with emphasis on recruitment by the raven, *Corvus corax*. *Behavioral Ecology and Sociobiology, 23*, 141–156.

Heinrich, B. (1995). Neophilia and exploration in juvenile common ravens, *Corvus corax*. *Animal Behaviour, 50*, 695–704.

Heinrich, B. (1999). *Mind of the raven*. New York: Harper Collins.

Heinrich, B., & Bugnyar, T. (2007). Just how smart are ravens? *Scientific American, 296*, 64–71.

Heinrich, B., Marzluff, J., & Adams, W. (1996). Fear and food recognition in naive common ravens. *Auk, 112*, 499–503.

Heinrich, B., & Marzluff, J. M. (1991). Do common ravens yell because they want to attract others? *Behavioral Ecology and Sociobiology, 28*, 13–21.

Heinrich, B., & Smokler, R. (1998). Play in common ravens (*Corvus corax*). In M. Bekoff & J. Byers (Eds.), *Animal play: Evolutionary, comparative and ecological perspectives* (pp. 27–44). Cambridge: Cambridge University Press.

Heinze, S., & Reppert, S. (2012). Anatomical basis of sun compass navigation I: The general layout of the monarch butterfly brain. *Journal of Comparative Neurology, 520*, 1599–1628.

Held, S. D. E., & Spinka, M. (2011). Animal play and animal welfare. *Animal Behaviour, 81*, 891–899.

Hemelrijk, C. K., & Ek, A. (1991). Reciprocity and interchange of grooming and "support" in captive chimpanzees. *Animal Behaviour, 41*, 923–935.

Henrich, J., Boyd, R., Bowles, S., Camerer, C., Fehr, E., Gintis, H., McElreath, R., Alvard, M., Barr, A., Ensminger, J., Henrich, N. S., Hill, K., Gil-White, F., Gurven, M., Marlowe, F. W., Patton, J. Q., & Tracer, D. (2005). "Economic man" in cross-cultural perspective: Behavioral experiments in 15 small-scale societies. *Behavioral and Brain Sciences, 28*, 795–855.

Henry, J., Liu, Y., Nadra, W., Qian, G., Mormede, P., Lemaire, V., Ely, D., & Hendley, E. (1993). Psychosocial stress can induce chronic hypertension in normotensive strains of rats. *Hypertension, 21*, 714–723.

Henry, J., & Stephens, P. (1977). *Stress, health and the social environment: A sociobiological approach to medicine*. Berlin: Springer-Verlag.

Hepper, P., & Cleland, J. (1998). Developmental aspects of kin recognition. *Genetica, 104*, 199–205.

Hepper, P. G. (Ed.). (1991). *Kin recognition*. Cambridge: Cambridge University Press.

Herard, F., Keller, M. A., Lewis, W. J., & Tumlinson, J. H. (1988). Beneficial arthropod behavior mediated by airborne semiochemicals. 3. Influence of age and experience on flight chamber responses of *Microplitis demolitor* Wilkinson (Hymenoptera, Braconidae). *Journal of Chemical Ecology, 14*, 1583–1596.

Herbers, J. M. (2009). Darwin's "one special difficulty": Celebrating Darwin 2009. *Biology Letters, 5*, 214–217.

Herrero, A. I., Sandi, C., & Venero, C. (2006). Individual differences in anxiety trait are related to spatial learning abilities and hippocampal expression of mineralocorticoid receptors. *Neurobiology of Learning and Memory, 86*, 150–159.

Herrnstein, R. J. (1961). Relative and absolute strength of responses as a function of frequency reinforcement. *Journal of the Experimental Analysis of Behavior, 4*, 267–272.

Herrnstein, R. J. (1970). On the law of effect. *Journal of the Experimental Analysis of Behavior, 13*, 243–266.

Hessing, M., Hagelso, A., Schouten, W., Wiepkema, P., & Van Beek, J. (1994). Individual, behavioral and physiological strategies in pigs. *Physiology and Behavior, 55*, 39–46.

Heyes, C. M. (1994). Social learning in animals: Categories and mechanisms. *Biological Reviews of the Cambridge Philosophical Society, 69*, 207–231.

Heyes, C. M., & Galef, B. G. (Eds.). (1996). *Social learning in animals: The roots of culture*. London: Academic Press.

Hill, C. E., Akcay, C., Campbell, S. E., & Beecher, M. D. (2011). Extrapair paternity, song, and genetic quality in song sparrows. *Behavioral Ecology, 22*, 73–81.

Hill, G. (1990). Female house finches prefer colourful males: Sexual selection for a condition-dependent trait. *Animal Behaviour, 40*, 563–572.

Hill, G. (1991). Plumage coloration is a sexually selected indicator of male quality. *Nature, 350*, 337–339.

Hill, G. (1992). The proximate basis of variation in carotenoid pigmentation in male house finches. *Auk, 109*, 1–12.

Hill, G. (1993a). Geographic variation in the carotenoid plumage pigmentation of male house finches (*Carpodacus mexicanus*). *Biological Journal of the Linnean Society, 49*, 63–86.

Hill, G. (1993b). Male mate choice and the evolution of female plumage coloration in the house finch. *Evolution, 47*, 1515–1525.

Hill, G. (1993c). The proximate basis of inter- and intra-population variation in female plumage coloration in the house finch. *Canadian Journal of Zoology, 71*, 619–627.

Hill, G., & Pierce, N. (1989). The effect of adult diet on the biology of butterflies, 1: The common imperial blue, *Jalmenus evagoras*. *Oecologia, 81*, 249–257.

Hill, G. E., & Farmer, K. L. (2005). Carotenoid-based plumage coloration predicts resistance to a novel parasite in the house finch. *Naturwissenschaften, 92*, 30–34.

Hill, G. E., Inouye, C. Y., & Montgomerie, R. (2002). Dietary carotenoids predict plumage coloration in wild house finches. *Proceedings of the Royal Society of London, Series B, 269*, 1119–1124.

Hill, H., & Bekoff, M. (1977). The variability of some motor components of social play and agonistic behavior in infant Eastern coyotes, *Canis latrans*. *Animal Behaviour, 25*, 907–909.

Hill, K. (2002). Altruistic cooperation during foraging by the Aćhe, and the evolved human predisposition to cooperate. *Human Nature—an Interdisciplinary Biosocial Perspective, 13*, 105–128.

Hill, K., & Hurtado, A. M. (1996). *Aćhe life history: The ecology and demography of a foraging people*: Hawthorne, NY: Aldine Transaction.

Hill, K., Kaplan, H., Hawkes, K., & Hurtado, A. M. (1987). Foraging decisions among Aćhe hunter-gatherers: New data and implications for optimal foraging models. *Ethology and Sociobiology, 8*, 1–36.

Hillis, D. M., Bull, J. J., White, M. E., Badgett, M. R., & Molineux, I. J. (1992). Experimental phylogenetics: Generation of a known phylogeny. *Science, 255*, 589–592.

Hinde, R. A. (1973). Nobel recognition for ethology. *Nature, 245*, 346.

Hiraiwa, M. (1975). Pebble collecting behavior in Japanese macaques. *Monkey, 19*, 24–25 (in Japanese).

Hirth, D. H., & McCullough, D. R. (1977). Evolution of alarm signals in ungulates with special reference to white-tailed deer. *American Naturalist, 111*, 31–42.

Hock, K., & Huber, R. (2009). Models of winner and loser effects: A cost-benefit analysis. *Behaviour, 146,* 69–87.

Hodge, M. A., & Uetz, G. (1995). A comparison of agonistic behaviour of colonial web-building spiders from desert and tropical habitats. *Animal Behaviour, 50,* 963–972.

Hoekstra, H. E. (2006). Genetics, development and evolution of adaptive pigmentation in vertebrates. *Heredity, 97,* 222–234.

Hoese, H. D. (1971). Dolphin feeding out of water in a salt marsh. *Journal of Mammalogy, 52,* 222–223.

Hoffman-Goetz, L., & Pedersen, B. (1994). Exercise and the immune system: A model of the stress response? *Immunology Today, 15,* 382–387.

Hoffmann, A. (1999). Laboratory and field heritabilities: Some lessons from Drosophila. In T. Mousseau, B. Sinervo, & J. Endler (Eds.), *Adaptive genetic variation in the wild* (pp. 200–218). New York: Oxford University Press.

Hogan, J. (1994). The concept of cause in the study of behavior. In J. Hogan & J. Bolhuis (Eds.), *Causal mechanisms of behavioural development* (pp. 3–15). Cambridge: Cambridge University Press.

Hogan-Warburg, A. J. (1966). Social behavior of the ruff, *Philomachus pugnax (L)*. *Ardea, 54,* 109–229.

Höglund, J. (2003). Lek-kin in birds: Provoking theory and surprising new results. *Annales Zoologici Fennici, 40,* 249–253.

Höglund, J., & Alatalo, R. (1995). *Leks.* Princeton, NJ: Princeton University Press.

Höglund, J., Alatalo, R., Gibson, R., & Lundberg, A. (1995). Mate-choice copying in black grouse. *Animal Behaviour, 49,* 1627–1633.

Höglund, J., Alatalo, R. V., & Lundberg, A. (1990). Copying the mate choice of others? Observations on female black grouse. *Behaviour, 114,* 221–236.

Höglund, J., Alatalo, R. V., Lundberg, A., Rintamaki, P. T., & Lindell, J. (1999). Microsatellite markers reveal the potential for kin selection on black grouse leks. *Proceedings of the Royal Society of London, Series B, 266,* 813–816.

Höglund, J., Widemo, F., Sutherland, W. J., & Nordenfors, H. (1998). Ruffs, *Philomachus pugnax,* and distribution models: Can leks be regarded as patches? *Oikos, 82,* 370–376.

Hogstad, O. (1995). Alarm calling by willow tits, *Paras montanus,* as mate investment. *Animal Behaviour, 49,* 221–225.

Holden, P., & Sharrock, J. (1988). *The Royal Society for the Protection of Birds book. Book of the British Isles.* London: MacMillan.

Holekamp, K. E., & Dloniak, S. M. (2010). Intraspecific variation in the behavioral ecology of a tropical carnivore, the spotted hyena. *Advances in the Study of Behavior, 42,* 189–229.

Holldobler, B., & Roces, F. (2001). The behavioral ecology of stridulatory communication in leaf-cutting ants. In L. A. Dugatkin (Ed.), *Model systems in behavioral ecology.* Princeton, NJ: Princeton University Press.

Holldobler, B., & Wilson, E. O. (1990). *The ants.* Cambridge, MA: Harvard University Press.

Hollen, L. I., Clutton-Brock, T., & Manser, M. B. (2008). Ontogenetic changes in alarm-call production and usage in meerkats (*Suricata suricatta*): Adaptations or constraints? *Behavioral Ecology and Sociobiology, 62,* 821–829.

Hollen, L. I., & Manser, M. B. (2006). Ontogeny of alarm call responses in meerkats, *Suricata suricatta*: The roles of age, sex and nearby conspecifics. *Animal Behaviour, 72,* 1345–1353.

Hollen, L. I., & Radford, A. N. (2009). The development of alarm call behaviour in mammals and birds. *Animal Behaviour, 78,* 791–800.

Hollis, K. L. (1984). The biological function of Pavlovian conditioning: The best defense is a good offense. *Journal of Experimental Psychology–Animal Behavior Processes, 10,* 413–425.

Hollis, K. L., Cadieux, E. L., & Colbert, M. M. (1989). The biological function of Pavlovian conditioning: A mechanism for mating success in blue gourami (*Trichogaster trichopterus*). *Journal of Comparative Psychology, 103,* 115–121.

Hollis, K. L., Dumas, M., Singh, P., & Fackelman, P. (1995). Pavlovian conditioning of aggressive behavior in blue gourami fish (*Trichogaster trichopterus*): Winners become winners and losers stay losers. *Journal of Comparative Psychology, 109,* 123–133.

Holloway, W., & Thor, D. (1985). Interactive effects of caffeine, 2–cholroadenosine and haloperidol on activity, social investigation and play fighting of juvenile rats. *Pharmacology Biochemistry and Behavior, 22,* 421–426.

Holman, L., & Snook, R. R. (2006). Spermicide, cryptic female choice and the evolution of sperm form and function. *Journal of Evolutionary Biology, 19,* 1660–1670.

Holmes, W., & Sherman, P. W. (1983). Kin recognition in animals. *American Scientist, 71,* 46–55.

Holmes, W. G. (2004). The early history of Hamiltonian-based research on kin recognition. *Annales Zoologici Fennici, 41,* 691–711.

Hoogland, J., & Sherman, P. W. (1976). Advantages and disadvantages of bank swallow coloniality. *Ecological Monographs, 46,* 33–58.

Hoogland, J. L. (1983). Nepotism and alarm calls in the black-tailed prarie dog (*Cynomys ludovicianus*). *Animal Behaviour, 31,* 472–479.

Hoogland, J. L. (1995). *The black-tailed prairie dog.* Chicago: University of Chicago Press.

Hopster, H. (1998). *Coping strategies in dairy cows.* Wageningen, Netherlands: Agricultural University.

Horne, J. S., Garton, E. O., Krone, S. M., & Lewis, J. S. (2007). Analyzing animal movements using Brownian bridges. *Ecology, 88,* 2354–2363.

Hosken, D. J., Stockley, P., Tregenza, T., & Wedell, N. (2009). Monogamy and the battle of the sexes. *Annual Review of Entomology, 54,* 361–378.

Hosoi, S. A., Rothstein, S. I., & O'Loghlen, A. L. (2005). Sexual preferences of female brown-headed cowbirds (*Molothrus ater*) for perched song repertoires. *Auk, 122,* 82–93.

Hostetler, C. M., Harkey, S. L., Krzywosinski, T. B., Aragona, B. J., & Bales, K. L. (2011). Neonatal exposure to the D1 agonist SKF38393 inhibits pair bonding in the adult prairie vole. *Behavioural Pharmacology, 22,* 703–710.

Houde, A. E. (1988). Genetic difference in female choice in two guppy populations. *Animal Behaviour, 36,* 510–516.

Houde, A. E. (1992). Sex-linked heritability of a sexually selected character in a natural population of *Poecilia reticulata*. *Heredity, 69,* 229–235.

Houde, A. E. (1994). Effect of artificial selection on male colour patterns on mating preference of female guppies. *Proceedings of the Royal Society of London, Series B, 256,* 125–130.

Houde, A. E. (1997). *Sex, color and mate choice in guppies.* Princeton, NJ: Princeton University Press.

Houde, A. E., & Endler, J. A. (1990). Correlated evolution of female mating preference and male color pattern in the guppy, *Poecilia reticulata. Science, 248,* 1405–1408.

Houston, A. I. (2008). Matching and ideal free distributions. *Oikos, 117,* 978–983.

Houston, A. I., McNamara, J. M., & Steer, M. D. (2007). Do we expect natural selection to produce rational behaviour? *Philosophical Transactions of the Royal Society, Series B–Biological Sciences, 362,* 1531–1543.

Howe, N., & Harris, L. (1978). Transfer of the sea anemone pheromone, anthropleurine, by the nudibranch *Acolidia papillosa. Journal of Chemical Ecology, 4,* 551–561.

Hrdy, S. (1999). *Mother nature: A history of mothers, infants, and natural selection.* New York: Pantheon Books.

Hsu, Y., Earley, R. L., & Wolf, L. (2005). Modulation of aggressive behavior by fighting experience: Mechanisms and contest outcomes. *Biological Reviews of the Cambridge Philosophical Society, 80,* 1–42.

Hsu, Y. Y., Lee, I. H., & Lu, C. K. (2009). Prior contest information: Mechanisms underlying winner and loser effects. *Behavioral Ecology and Sociobiology, 63,* 1247–1257.

Hsu, Y. Y., & Wolf, L. L. (1999). The winner and loser effect: Integrating multiple experiences. *Animal Behaviour, 57,* 903–910.

Hsu, Y. Y., & Wolf, L. L. (2001). The winner and loser effect: What fighting behaviours are influenced? *Animal Behaviour, 61,* 777–786.

Huang, S. P., Yang, S. Y., & Hsu, Y. Y. (2011). Persistence of winner and loser effects depends on the behaviour measured. *Ethology, 117,* 171–180.

Huber, R., & Kravitz, E. (1995). A quantitative analysis of agonistic behavior in juvenile American lobsters (*Homarus americanus*). *Brain, Behavior and Evolution, 46,* 72–83.

Huber, R., Panksepp, J. B., Nathaniel, T., Alcaro, A., & Panksepp, J. (2011). Drug-sensitive reward in crayfish: An invertebrate model system for the study of seeking, reward, addiction, and withdrawal. *Neuroscience and Biobehavioral Reviews, 35,* 1847–1853.

Huber, R., Smith, K., Delago, A., Isakson, K., & Kravitz, E. (1997). Serotonin and aggressive motivation in crustaceans: Altering the decision to retreat. *Proceedings of the National Academy of Sciences, U.S.A., 94,* 5939–5942.

Huffman, M. (1996). Acquisition of innovative cultural behaviors in nonhuman primates: A case study of stone handling, a socially transmitted behavior in Japanese macaques. In C. M. Heyes & B. G. Galef (Eds.), *Social learning in animals: The roots of culture* (pp. 267–289). London: Academic Press.

Hughes, L., Chang, B., Wagner, D., & Pierce, N. (2000). Effects of mating history on ejaculate size, fecundity and copulation duration in the ant-tended lycaenid butterfly, *Jalmenus evagoras. Behavioral Ecology and Sociobiology, 47,* 119–128.

Hughes, W. O. H., Oldroyd, B. P., Beekman, M., & Ratnieks, F. L. W. (2008). Ancestral monogamy shows kin selection is key to the evolution of eusociality. *Science, 320,* 1213–1216.

Hugie, D. M., & Lank, D. B. (1997). The resident's dilemma: A female choice model for the evolution of alternative mating strategies in lekking male ruffs (*Philomachus pugnax*). *Behavioral Ecology, 8,* 218–225.

Hultgren, K. M., & Duffy, J. E. (2011). Multi-locus phylogeny of sponge-dwelling snapping shrimp (*Caridea: Alpheidae: Synalpheus*) supports morphology-based species concepts. *Journal of Crustacean Biology, 31,* 352–360.

Humphreys, A., & Einon, D. (1981). Play as a reinforcer for maze learning in juvenile rats. *Animal Behaviour, 29,* 259–270.

Humphreys, A., & Smith, P. (1987). Rough and tumble play, friendship and dominance in school children: Evidence for continuity and change with age. *Child Development, 58,* 201–212.

Huntingford, F., & Turner, A. (1987). *Animal conflict.* London: Chapman and Hall.

Huntingford, F. A. (1984). Some ethical issues raised by studies of predation and aggression. *Animal Behaviour, 32,* 210–215.

Huntingford, F. A., & Wright, P. J. (1992). Inherited population differences in avoidance conditioning in three-spined sticklebacks, *Gasterosteus aculeatus. Behaviour, 122,* 264–273.

Hurd, P. (1997). Cooperative signalling between opponents in fish fights. *Animal Behaviour, 54,* 1309–1315.

Hurd, P. L. (2006). Resource holding potential, subjective resource value, and game theoretical models of aggressiveness signalling. *Journal of Theoretical Biology, 241,* 639–648.

Hurd, P. L., & Enquist, M. (2005). A strategic taxonomy of biological communication. *Animal Behaviour, 70,* 1155–1170.

Hutchinson, J. M. C., & Gigerenzer, G. (2005). Simple heuristics and rules of thumb: Where psychologists and behavioural biologists might meet. *Behavioural Processes, 69,* 97–124.

Hutt, C. (1966). Exploration and play in children. *Zoological Society of London (Symposium), 18,* 61–81.

Hutt, C. (1970). Specific and diverse exploration. *Advances in Child Development and Behavior, 5,* 119–180.

Huttenlocher, P. (2002). *Neural plasticity: The effects of environment on the development of the cerebral cortex.* Cambridge, MA: Harvard University Press.

Huxley, J. (1938). Darwin's theory of sexual selection and the data subsumed by it, in light of recent research. *American Naturalist, 72,* 416–433.

Huxley, J. (1942). *Evolution: The modern synthesis.* New York: Harper.

Huxley, T. H. (1888). The struggle for existence: A programme. *Nineteenth Century, 23,* 161–180.

Iacoboni, M., Koski, L. M., Brass, M., Bekkering, H., Woods, R. P., Dubeau, M. C., Mazziotta, J. C., & Rizzolatti, G. (2001). Reafferent copies of imitated actions in the right superior temporal cortex. *Proceedings of the National Academy of Sciences, U.S.A., 98,* 13995–13999.

Immelmann, K. (1972). Sexual and other long-term aspects of imprinting in birds and other species. In D. S. Lehrman, R. A. Hinde, & E. Shaw (Eds.), *Advances in the study of behavior* (pp. 147–174). New York: Academic Press.

Immelmann, K., Prove, R., Lassek, R., & Bischof, H. J. (1991). Influence of adult courtship experience on the development of sexual preferences in zebra finches. *Animal Behaviour, 42*, 83–89.

Immler, S., Pitnick, S., Parker, G. A., Durrant, K. L., Lupold, S., Calhim, S., & Birkhead, T. R. (2011). Resolving variation in the reproductive tradeoff between sperm size and number. *Proceedings of the National Academy of Sciences, U.S.A., 108*, 5325–5330.

Ims, R. A. (1987). Responses in social organization and behaviour manipulation of the food resource in the vole *Clethrionomys rufocanus. Journal of Animal Ecology, 56*, 585–596.

Inouye, C. Y., Hill, G. E., Stradi, R. D., & Montgomerie, R. (2001). Carotenoid pigments in male house finch plumage in relation to age, subspecies, and ornamental coloration. *Auk, 118*, 900–915.

Insel, T. R., Wang, Z. X., & Ferris, C. F. (1994). Patterns of brain vasopressin receptor distribution associated with social-organization in microtine rodents. *Journal of Neuroscience, 14*, 5381–5392.

Irwin, D. E., & Price, T. (1999). Sexual imprinting, learning and speciation. *Heredity, 82*, 347–354.

Izuma, K. (2012). The social neuroscience of reputation. *Neuroscience Research, 72*, 283–288.

Jaatinen, K., Lehikoinen, A., & Lank, D. B. (2010). Female-biased sex ratios and the proportion of cryptic male morphs of migrant juvenile ruffs (*Philomachus pugnax*) in Finland. *Ornis Fennica, 87*, 125–134.

Jablonski, P. G., Lee, S. D., & Jerzak, L. (2006). Innate plasticity of a predatory behavior: Nonlearned context dependence of avian flush-displays. *Behavioral Ecology, 17*, 925–932.

Jacobs, G. H. (1992). Ultraviolet vision in vertebrates. *American Zoologist, 32*, 544–554.

Jacobs, L. F., Gaulin, S. J. C., Sherry, D. F., & Hoffman, G. E. (1990). Evolution of spatial cognition: Sex-specific patterns of spatial behavior predict hippocampal size. *Proceedings of the National Academy of Sciences, U.S.A., 87*, 6349–6352.

Jacobson, M. (1991). *Developmental neurobiology*. New York: Plenum Press.

Jaksic, F. M., & Soriguer, R. C. (1981). Predation upon the European rabbit (*Oryctolagus cuniculus*) in Mediterranean habitats of Chile and Spain: A comparative-analysis. *Journal of Animal Ecology, 50*, 269–281.

Janik, V. M., Sayigh, L. S., & Wells, R. S. (2006). Signature whistle shape conveys identity information to bottlenose dolphins. *Proceedings of the National Academy of Sciences, U.S.A., 103*, 8293–8297.

Jansa, S. A., & Weksler, M. (2004). Phylogeny of muroid rodents: Relationships within and among major lineages as determined by IRBP gene sequences. *Molecular Phylogenetics and Evolution, 31*, 256–276.

Jarvis, J. (1981). Eusociality in a mammal: Cooperative breeding in the naked mole-rat. *Science, 212*, 571–573.

Jeffreys, A., Wilson, V., & Thein, S. (1985). Hypervariable "mini-satellite" regions in human DNA. *Nature, 314*, 67–73.

Jenkins, W. M., Merzenich, M. M., Ochs, M. T., Allard, T., & Guicrobles, E. (1990). Functional reorganization of primary somatosensory cortex in adult owl monkeys after behaviorally controlled tactile stimulation. *Journal of Neurophysiology, 63*, 82–104.

Jenni, D., & Collier, G. (1972). Polyandry in the American jacana (*Jacana spinosa*). *Auk, 89*, 743–765.

Jennions, M., & Backwell, P. (1996). Residency and size affect fight duration and outcome in the fiddler crab *Uca annulipes. Biological Journal of the Linnean Society, 57*, 293–306.

Jensen, P., & Yngvesson, J. (1998). Aggression between unacquainted pigs: Sequential assessment and effects of familiarity and weight. *Applied Animal Behavior Science, 58*, 49–61.

Jetz, W., Freckleton, R. P., & McKechnie, A. E. (2008). Environment, migratory tendency, phylogeny and basal metabolic rate in birds. *Plos One, 3*.

Joels, M., Pu, Z. W., Wiegert, O., Oitzl, M. S., & Krugers, H. J. (2006). Learning under stress: How does it work? *Trends in Cognitive Sciences, 10*, 152–158.

Johannesen, J., Hennig, A., Dommermuth, B., & Schneider, J. M. (2002). Mitochondrial DNA distributions indicate colony propagation by single matri-lineages in the social spider *Stegodyphus dumicola* (Eresidae). *Biological Journal of the Linnean Society, 76*, 591–600.

John, J. (1994). The avian spleen: A neglected organ. *Quarterly Review of Biology, 69*, 327–351.

Johnson, R. N., Oldroyd, B. P., Barron, A. B., & Crozier, R. H. (2002). Genetic control of the honey bee (*Apis mellifera*) dance language: Segregating dance forms in a backcrossed colony. *Journal of Heredity, 93*, 170–173.

Johnsson, J., & Akerman, A. (1998). Watch and learn: Preview of the fighting ability of opponents alters contest behaviour in rainbow trout. *Animal Behaviour, 56*, 771–776.

Johnston, T. (Ed.). (1985). *Issues in the ecological study of learning*. Hillsdale, NJ: Lawrence Erlbaum Associates.

Johnstone, R., & Dugatkin, L. A. (2000). Coalition formation in animals and the nature of winner and loser effects. *Proceedings of the Royal Society of London, Series B, 267*, 17–21.

Johnstone, R. A. (1995). Sexual selection, honest advertisement and the handicap principle: Reviewing the evidence. *Biological Reviews of the Cambridge Philosophical Society, 70*, 1–65.

Johnstone, R. A. (1997). The evolution of animal signals. In J. R. Krebs & N. B. Davies (Eds.), *Behavioural ecology: An evolutionary approach* (4th ed., pp. 155–178). Oxford: Blackwell Science.

Johnstone, R. A. (1998). Game theory and communication. In L. A. Dugatkin & H. K. Reeve (Eds.), *Game theory and animal behavior* (pp. 94–117). New York: Oxford University Press.

Johnstone, R. A. (2008). Kin selection, local competition, and reproductive skew. *Evolution, 62*, 2592–2599.

Johnstone, T. D. (1982). The selective costs and benefits of learning. *Advances in the Study of Behavior, 12*, 65–106.

Jolicoeur, P., Pirlot, P., Baron, G., & Stephan, H. (1984). Brain structure and correlation patterns in Insectivora, Chiroptera and primates. *Systematic Zoology, 33*, 14–29.

Jones, A. C., & Gosling, S. D. (2005). Temperament and personality in dogs (*Canis familiaris*): A review and evaluation of past research. *Applied Animal Behaviour Science, 95*, 1–53.

Jones, D., Gonzalez-Lima, F., Crews, D., Galef, B. G., & Clark, M. M. (1997). Effect of intrauterine position on the metabolic capacity of the hypothalamus of female gerbils. *Physiology & Behavior, 61,* 513–519.

Jones, G., Barabas, A., Elliott, W., & Parsons, S. (2002). Female greater wax moths reduce sexual display behavior in relation to the potential risk of predation by echolocating bats. *Behavioral Ecology, 13,* 375–380.

Jones, G., & Holderied, M. W. (2007). Bat echolocation calls: Adaptation and convergent evolution. *Proceedings of the Royal Society of London, Series B, 274,* 905–912.

Jones, G., & Siemers, B. M. (2011). The communicative potential of bat echolocation pulses. *Journal of Comparative Physiology A–Neuroethology Sensory Neural and Behavioral Physiology, 197,* 447–457.

Jones, G., & Teeling, E. C. (2006). The evolution of echolocation in bats. *Trends in Ecology & Evolution, 21,* 149–156.

Jones, T. M. (2001). A potential cost of monandry in the lekking sandfly, *Lutzomyia longipalpis. Journal of Insect Behavior, 14,* 385–399.

Jones, T. M., & Quinnell, R. J. (2002). Testing predictions for the evolution of lekking in the sandfly, *Lutzomyia longipalpis. Animal Behaviour, 63,* 605–612.

Jones, T. M., Quinnell, R. J., & Balmford, A. (1998). Fisherian flies: Benefits of female choice in a lekking sandfly. *Proceedings of the Royal Society of London, Series B, 265,* 1651–1657.

Jouventin, P., & Aubin, T. (2002). Acoustic systems are adapted to breeding ecologies: Individual recognition in nesting penguins. *Animal Behaviour, 64,* 747–757.

Jouventin, P., Aubin, T., & Lengagne, T. (1999). Finding a parent in a king penguin colony: The acoustic system of individual recognition. *Animal Behaviour, 57,* 1175–1183.

Kacelnik, A., Krebs, J., & Bernstein, C. (1992). The ideal free distributions and predator-prey populations. *Trends in Ecology & Evolution, 7,* 50–55.

Kaczensky, P. (1996). *Livestock-carnivore conflicts in Europe.* Munich: Munich Wildlife Society.

Kagan, J. (1994). *Galen's prophecy: Temperament in human nature.* New York: Basic Books.

Kagan, J., Reznick, J. S., & Gibbons, J. (1989). Inhibited and uninhibited types of children. *Child Development, 60,* 838–845.

Kagan, J., Reznick, J. S., & Snidman, N. (1987). Temperamental variation in response to the unfamiliar. In N. A. Krasnegor, E. M. Blass, M. A. Hofer, & W. P. Smotherman (Eds.), *Perinatal development: A psychobiological perspective* (pp. 397–419). New York: Academic Press.

Kagan, J., Reznick, J. S., & Snidman, N. (1988). Biological bases of childhood shyness. *Science, 240,* 167–171.

Kamil, A., & Balda, R. (1990). Spatial memory in food caching birds. *Psychology of Learning and Motivation, 26,* 1–25.

Kamil, A., Krebs, J., & Pulliam, H. R. (Eds.). (1987). *Foraging behavior.* New York: Plenum.

Kamil, A., & Yoerg, S. (1982). Learning and foraging behavior. *Perspectives in Ethology, 5,* 325–364.

Kamin, L. J. (1968). "Attention-like" processes in classic conditioning. In M. R. Jones (Ed.), *Miami symposium on the prediction of behavior: Aversive stimulation* (pp. 9–31). Miami: Miami University Press.

Kamin, L. J. (1969). Predictability, surprise, attention and conditioning. In B. Campbell & R. Church (Eds.), *Punishment and aversive behavior* (pp. 279–296). New York: Appleton-Century-Crofts.

Kaminski, G., Dridi, S., Graff, C., & Gentaz, E. (2009). Human ability to detect kinship in strangers' faces: Effects of the degree of relatedness. *Proceedings of the Royal Society, Series B–Biological Sciences, 276,* 3193–3200.

Karlsson, J., & Johansson, O. (2010). Predictability of repeated carnivore attacks on livestock favours reactive use of mitigation measures. *Journal of Applied Ecology, 47,* 166–171.

Kavaliers, M., Colwell, D. D., Braun, W. J., & Choleris, E. (2003). Brief exposure to the odour of a parasitized male alters the subsequent mate odour responses of female mice. *Animal Behaviour, 65,* 59–68.

Kavaliers, M., Colwell, D. D., & Choleris, E. (2001). NMDA-mediated social learning of fear-induced conditioned analgesia to biting flies. *Neuroreport, 12,* 663–667.

Kavaliers, M., Colwell, D. D., Choleris, E., & Ossenkopp, K. P. (1999). Learning to cope with biting flies: Rapid NMDA-mediated acquisition of conditioned analgesia. *Behavioral Neuroscience, 113,* 126–135.

Kavaliers, M., Ossenkopp, K. P., Galea, L. A. M., & Kolb, B. (1998). Sex differences in spatial learning and prefrontal and parietal cortical dendritic morphology in the meadow vole, *Microtus pennsylvanicus. Brain Research, 810,* 41–47.

Kawai, M. (1965). Newly acquired precultural behavior of the natural troop of Japanese monkeys on Koshima Islet. *Primates, 6,* 1–30.

Kawamura, S. (1959). The process of sub-culture propagation among Japanese macaques. *Primates,* 43–60.

Kawamura, S., Blow, N., & Yokoyama, S. (1999). Genetic analyses of visual pigments of the pigeon (*Columba livia*). *Genetics, 153,* 1839–1850.

Kawanabe, H., Cohen, J., & Iwasaki, K. (Eds.). (1993). *Mutualism and community organization: Behavioural, theoretical, and food-web approaches.* New York: Oxford University Press.

Kawecki, T. J. (2010). Evolutionary ecology of learning: Insights from fruit flies. *Population Ecology, 52,* 15–25.

Keefe, M. (1992). Chemically mediated avoidance behaviour in wild brook trout, *Salvelinus fontinalis*: The response to familiar and unfamiliar predaceous fishes and the influences of diet. *Canadian Journal of Zoology, 70,* 288–292.

Keenleyside, M. (1955). Some aspects of the schooling behavior of fish. *Behaviour, 8,* 183–248.

Keenleyside, M., & Yamamoto, F. (1962). Territorial behaviour of juvenile Atlantic salmon (*Salmo salar L.*). *Behaviour, 19,* 139–169.

Keeton, W. T. (1971). Magnets interfere with pigeon homing. *Proceedings of the American Philosophical Society, 68,* 102–106.

Kelley, J. L., Magurran, A. E., & Garcia, C. M. (2006). Captive breeding promotes aggression in an endangered Mexican fish. *Biological Conservation, 133,* 169–177.

Kemp, D. J., & Wiklund, C. (2004). Residency effects in animal contests. *Proceedings of the Royal Society of London, Series B, 271,* 1707–1711.

Kendal, R. L., Coolen, I., van Bergen, Y., & Laland, K. N. (2005). Trade-offs in the adaptive use of social and asocial learning. *Advances in the Study of Behavior, 35,* 333–379.

Kennedy, M., & Gray, R. (1993). Can ecological theory predict the distribution of foraging animals? A critical analysis of experiments on the ideal free distribution. *Oikos, 68,* 158–166.

Kentner, A. C., Abizaid, A., & Bielajew, C. (2010). Modeling dad: Animal models of paternal behavior. *Neuroscience and Biobehavioral Reviews, 34,* 438–451.

Keverne, E. B., Martel, F. L., & Nevison, C. M. (1996). Primate brain evolution: Genetic and functional considerations. *Proceedings of the Royal Society of London, Series B, 263,* 689–696.

Keverne, E. B., Martenz, N. D., & Tuite, B. (1989). Beta-endorphin concentration in cerebrospinal fluid of monkeys are influenced by grooming relationships. *Psychoneuroendocrinology, 14,* 155–161.

Kiesecker, J., & Blaustein, A. (1997). Influences of egg-laying behavior on pathogenic infection of amphibian eggs. *Conservation Biology, 11,* 214–220.

Kiesecker, J., & Skelly, D. K. (2000). Choice of oviposition site by gray treefrogs: The role of potential parasitic infection. *Ecology, 81,* 2939–2943.

Kiesecker, J. M., Belden, L. K., Shea, K., & Rubbo, M. J. (2004). Amphibian decline and emerging disease. *American Scientist, 92,* 138–147.

Kiesecker, J. M., & Blaustein, A. R. (1999). Pathogen reverses competition between larval amphibians. *Ecology, 80,* 2442–2448.

Kiesecker, J. M., Blaustein, A. R., & Belden, L. K. (2001). Complex causes of amphibian population declines. *Nature, 410,* 681–684.

Kilner, J. M., Paulignan, Y., & Blakemore, S. J. (2003). An interference effect of observed biological movement on action. *Current Biology, 13,* 522–525.

Kim, J. J., & Jung, M. W. (2006). Neural circuits and mechanisms involved in Pavlovian fear conditioning: A critical review. *Neuroscience and Biobehavioral Reviews, 30,* 188–202.

King, J. E., & Landau, V. I. (2003). Can chimpanzee (*Pan troglodytes*) happiness be estimated by human raters? *Journal of Research in Personality, 37,* 1–15.

King, J. E., Weiss, A., & Farmer, K. H. (2005). A chimpanzee (*Pan troglodytes*) analogue of cross-national generalization of personality structure: Zoological parks and an African sanctuary. *Journal of Personality, 73,* 389–410.

Kingsolver, J. G., Hoekstra, H. E., Hoekstra, J. M., Berrigan, D., Vignieri, S. N., Hill, C. E., Hoang, A., Gibert, P., & Beerli, P. (2001). The strength of phenotypic selection in natural populations. *American Naturalist, 157,* 245–261.

Kirchhof, J., & Hammerschmidt, K. (2006). Functionally referential alarm calls in tamarins (*Saguinus fuscicollis* and *Saguinus mystax*): Evidence from playback experiments. *Ethology, 112,* 346–354.

Kirkpatrick, M. (1982). Sexual selection and the evolution of female choice. *Evolution, 36,* 1–12.

Kirkpatrick, M., & Ryan, M. (1991). The evolution of mating preferences and the paradox of the lek. *Nature, 350,* 33–38.

Kiss, A. (2004). Is community-based ecotourism a good use of biodiversity conservation funds? *Trends in Ecology & Evolution, 19,* 232–237.

Kitchen, W. D. (1972). *The social behavior and ecology of the pronghorn.* Ann Arbor: University of Michigan.

Kleiman, D. (1977). Monogamy in mammals. *Quarterly Review of Biology, 52,* 39–69.

Knudsen, E., Linden, A., Both, C., Jonzen, N., Pulido, F., Saino, N., Sutherland, W. J., Bach, L. A., Coppack, T., Ergon, T., Gienapp, P., Gill, J. A., Gordo, O., Hedenstroom, A., Lehikoinen, E., Marra, P. P., Moller, A. P., Nilsson, A. L. K., Peron, G., Ranta, E., Rubolini, D., Sparks, T. H., Spina, F., Studds, C. E., Saether, S. A., Tryjanowski, P., & Stenseth, N. C. (2011). Challenging claims in the study of migratory birds and climate change. *Biological Reviews, 86,* 928–946.

Kodric-Brown, A., & Brown, J. H. (1984). Truth in advertising: The kinds of traits favored by sexual selection. *American Naturalist, 124,* 309–323.

Koenig, W., Pitelka, F. A., Carmen, W. J., Mumme, R. L., & Stanback, M. T. (1992). The evolution of delayed dispersal in cooperative breeders. *Quarterly Review of Biology, 67,* 111–150.

Koenig, W. D., & Pitelka, F. (1981). Ecological factors and kin selection in the evolution of cooperative breeding in birds. In R. Alexander & D. Tinkle (Eds.), *Natural selection and social behavior* (pp. 261–280). New York: Chiron Press.

Koenig, W. D., Walters, E. L., & Haydock, J. (2011). Variable helper effects, ecological conditions, and the evolution of cooperative breeding in the acorn woodpecker. *American Naturalist, 178,* 145–158.

Kokko, H., Brooks, R., Jennions, M. D., & Morley, J. (2003). The evolution of mate choice and mating biases. *Proceedings of the Royal Society of London, Series B, 270,* 653–664.

Kokko, H., & Lindstrom, J. (1996). Kin selection and the evolution of leks: Whose success do young males maximize? *Proceedings of the Royal Society of London, Series B, 263,* 919–923.

Kokko, H., Lopez-Sepulcre, A., & Morrell, L. J. (2006). From hawks and doves to self-consistent games of territorial behavior. *American Naturalist, 167,* 901–912.

Kokko, H., Sutherland, W. J., Lindstrom, J., Reynolds, J. D., & Mackenzie, A. (1998). Individual mating success, lek stability, and the neglected limitations of statistical power. *Animal Behaviour, 56,* 755–762.

Komdeur, J. (1992). Importance of habitat saturation and territory quality for the evolution of cooperative breeding in the Seychelles warbler. *Nature, 358,* 493–495.

Komers, P. E. (1989). Dominance relationships between juvenile and adult black-billed magpies. *Animal Behaviour, 37,* 256–265.

Konopka, R., & Benzer, S. (1971). Clock mutants of *Drosophila melanogaster. Proceedings of the National Academy of Sciences, 68,* 2112–2116.

Koolhaas, J. M., de Boer, S. F., Buwalda, B., & van Reenen, K. (2007). Individual variation in coping with stress: A multidimensional approach of ultimate and proximate mechanisms. *Brain, Behavior and Evolution, 70,* 218–226.

Koolhaas, J. M., Korte, S. M., De Boer, S. F., Van Der Vegt, B. J., Van Reenen, C. G., Hopster, H., De Jong, I. C., Ruis, M. A. W., & Blokhuis, H. J. (1999). Coping styles in animals: Current status in behavior and stress-physiology. *Neuroscience and Biobehavioral Reviews, 23,* 925–935.

Koops, M. A., & Grant, J. W. A. (1993). Weight symmetry and sequential assessment in convict cichlid contests. *Canadian Journal of Zoology, 71,* 475–479.

Korb, J., Weil, T., Hoffmann, K., Foster, K. R., & Rehli, M. (2009). A gene necessary for reproductive suppression in termites. *Science, 324,* 758–758.

Kosfeld, M., Heinrichs, M., Zak, P. J., Fischbacher, U., & Fehr, E. (2005). Oxytocin increases trust in humans. *Nature, 435,* 673–676.

Koski, L., Wohlschlager, A., Bekkering, H., Woods, R. P., Dubeau, M. C., Mazziotta, J. C., & Iacoboni, M. (2002). Modulation of motor and premotor activity during imitation of target-directed actions. *Cerebral Cortex, 12,* 847–855.

Kovalzan, V., & Mukhametov, L. (1982). Temperature variations in the brain corresponding to unihemispheric slow wave sleep in dolphins. *Journal of Evolution and Biochemical Physiology, 18,* 307–309 (in Russian).

Kraaijeveld, K., Kraaijeveld-Smit, F. J. L., & Maan, M. E. (2011). Sexual selection and speciation: The comparative evidence revisited. *Biological Reviews, 86,* 367–377.

Kraak, S. B. M. (2011). Exploring the "public goods game" model to overcome the tragedy of the commons in fisheries management. *Fish and Fisheries, 12,* 18–33.

Krams, I., Krama, T., & Igaune, K. (2006). Alarm calls of wintering great tits Parus major: Warning of mate, reciprocal altruism or a message to the predator? *Journal of Avian Biology, 37,* 131–136.

Kraus, W. F. (1989). Surface wave communication during courtship in the giant water bug, *Abedus-Indentatus* (Heteroptera, Belostomatidae). *Journal of the Kansas Entomological Society, 62,* 316–328.

Krause, J., & Ruxton, G. (2002). *Living in groups.* Oxford: Oxford University Press.

Kravitz, E. (1988). Hormonal control of behavior: Amines and the biasing of behavioral output in lobsters. *Science, 241,* 1775–1781.

Krebs, J. (1978). Optimal foraging: Decision rules for predators. In J. R. Krebs & N. B. Davies (Eds.), *Behavioural ecology: An evolutionary approach* (pp. 23–63). Oxford: Blackwell Science.

Krebs, J. R. (1980). Optimal foraging, predation risk and territory defence. *Ardea, 68,* 83–90.

Krebs, J. R., & Davies, N. B. (1987). *An introduction to behavioural ecology* (2nd ed.). Oxford: Blackwell Science.

Krebs, J. R., & Davies, N. B. (1993). *An introduction to behavioural ecology* (3rd ed.). Oxford: Blackwell Science.

Krebs, J. R., & Dawkins, R. (1984). Animal signals: Mind-reading and manipulation? In J. R. Krebs & N. B. Davies (Eds.), *Behavioural ecology: An evolutionary approach* (2nd ed., pp. 380–402). Oxford: Blackwell Science.

Krebs, J. R., Kacelnik, A., & Taylor, P. (1978). Test of optimal sampling in foraging great tits. *Nature, 275,* 27–31.

Krebs, J. R., & McCleery. (1984). Optimization in behavioural ecology. In J. R. Krebs & N. B. Davies (Eds.), *Behavioural ecology: An evolutionary approach* (2nd ed., pp. 91–121). Oxford: Blackwell Science.

Krebs, J. R., Sherry, D. F., Healey, S. D., Perry, V. H., & Vaccarino, A. L. (1989). Hippocampal specialization of food-storing birds. *Proceedings of the National Academy of Sciences, 86,* 1388–1392.

Kroeber, A. L., & Kluckhohn, C. (1952). Culture, a critical review of the concepts and definitions. *Papers of the Peabody Museum, 47,* 1–223. Cambridge, MA: Peabody Museum of American Archeology and Ethnology.

Kroodsma, D., & Byers, B. (1991). The function(s) of birdsong. *American Zoologist, 31,* 318–328.

Kropotkin, P. (1902). *Mutual aid* (3rd ed.). London: William Heinemann.

Krutzen, M., Mann, J., Heithaus, M. R., Connor, R. C., Bejder, L., & Sherwin, W. B. (2005). Cultural transmission of tool use in bottlenose dolphins. *Proceedings of the National Academy of Sciences, U.S.A. 102,* 8939–8943.

Krutzen, M., Sherwin, W. B., Connor, R. C., Barre, L. M., Van de Casteele, T., Mann, J., & Brooks, R. (2003). Contrasting relatedness patterns in bottlenose dolphins (*Tursiops sp.*) with different alliance strategies. *Proceedings of the Royal Society of London, Series B, 270,* 497–502.

Kuba, M. J., Byrne, R. A., Meisel, D. V., & Mather, J. A. (2006). When do octopuses play? Effects of repeated testing, object type, age, and food deprivation on object play in *Octopus vulgaris. Journal of Comparative Psychology, 120,* 184–190.

Kuczaj, S. A., Gory, J. D., & Xitco, M. J. (1998). Using programs to solve problems: Imitation versus insight. *Behavioral and Brain Sciences, 21,* 695.

Kuhn, T. (1962). *The structure of scientific revolutions.* Chicago: University of Chicago Press.

Kunz, T. H., Allgaier, A. L., Seyjagat, J., & Caligiuri, R. (1994). Allomaternal care: Helper-assisted birth in the Rodrigues fruit bat, *Pteropus rodricensis* (Chiroptera: Pteropodidae). *Journal of Zoology, London, 232,* 691–700.

Kurtz, J., Kalbe, M., Aeschlimann, P. B., Haberli, M. A., Wegner, K. M., Reusch, T. B. H., & Milinski, M. (2004). Major histocompatibility complex diversity influences parasite resistance and innate immunity in sticklebacks. *Proceedings of the Royal Society of London, Series B, 271,* 197–204.

Kurvers, R., Adamczyk, V., van Wieren, S. E., & Prins, H. H. T. (2011). The effect of boldness on decision-making in barnacle geese is group-size-dependent. *Proceedings of the Royal Society, Series B–Biological Sciences, 278,* 2018–2024.

Kurvers, R., Eijkelenkamp, B., van Oers, K., van Lith, B., van Wieren, S. E., Ydenberg, R. C., & Prins, H. H. T. (2009). Personality differences explain leadership in barnacle geese. *Animal Behaviour, 78,* 447–453.

Kutsukake, N., & Clutton-Brock, T. H. (2006). Social functions of allogrooming in cooperatively breeding meerkats. *Animal Behaviour, 72,* 1059–1068.

Kvist, A., & Lindstrom, A. (2001). Basal metabolic rate in migratory waders: Intra-individual, intraspecific, interspecific and seasonal variation. *Functional Ecology, 15,* 465–473.

Kyriacou, C., Oldroyd, M., Wood, J., Sharp, M., & Hill, M. (1990). Clock mutations alter developmental timing in Drosophila. *Heredity, 64,* 395–401.

Lacey, E. A., & Sherman, P. W. (1991). Social organization of naked mole-rats: Evidence for divisions of labor. In P. W. Sherman, J. Jarvis, & R. D. Alexander (Eds.), *The biology of the naked mole-rat* (pp. 275–336). Princeton, NJ: Princeton University Press.

Lachlan, R. F., & Servedio, M. R. (2004). Song learning accelerates allopatric speciation. *Evolution, 58,* 2049–2063.

Lachmann, M., & Bergstrom, C. T. (1998). Signalling among relatives. II: Beyond the tower of Babel. *Theoretical Population Biology, 54,* 146–160.

Lack, D. (1968). *Ecological adaptations for breeding in birds.* London: Methuen.

Laland, K. N., & Janik, V. M. (2006). The animal cultures debate. *Trends in Ecology & Evolution, 21,* 542–547.

Laland, K. N., & Williams, K. (1998). Social transmission of maladaptive information in the guppy. *Behavioral Ecology, 9,* 493–499.

Lan, Y. T., & Hsu, Y. Y. (2011). Prior contest experience exerts a long-term influence on subsequent winner and loser effects. *Frontiers in Zoology, 8.* Online only doi10.1186/1742-9994-8-28

Landau, H. G. (1951a). On dominance relations and the structure of animal societies: I. Effects of inherent characteristics. *Bulletin of Mathematical Biophysics, 13,* 1–19.

Landau, H. G. (1951b). On dominance relations and the structure of animal societies: II. Some effects of possible social causes. *Bulletin of Mathematical Biophysics, 13,* 245–262.

Lane, J. E., Forrest, M. N. K., & Willis, C. K. R. (2011). Anthropogenic influences on natural animal mating systems. *Animal Behaviour, 81,* 909–917.

Langley, C. M. (1996). Search images: Selective attention to specific visual features of prey. *Journal of Experimental Psychology–Animal Behavior Processes, 22,* 152–163.

Lank, D. B., Coupe, M., & Wynne-Edwards, K. E. (1999). Testosterone-induced male traits in female puffs (*Philomachus pugnax*): Autosomal inheritance and gender differentiation. *Proceedings of the Royal Society of London, Series B, 266,* 2323–2330.

Lank, D. B., Smith, C. M., Hanotte, O., Burke, T., & Cooke, F. (1995). Genetic polymorphism for alternative mating behavior in lekking male ruff *Philomachus pugnax. Nature, 378,* 59–62.

Laughlin, S. B., van Steveninck, R. R. D., & Anderson, J. C. (1998). The metabolic cost of neural information. *Nature Neuroscience, 1,* 36–41.

Lawick-Goodall, J. (1968). The behaviour of free-living chimpanzees in the Gombe Stream Reserve. *Animal Behavior Monographs, 1,* 161–311.

Lawrence, A. (1987). Consumer demand theory and the assessment of animal-welfare. *Animal Behaviour, 35,* 293–295.

Leadbeater, E., Raine, N. E., & Chittka, L. (2006). Social learning: Ants and the meaning of teaching. *Current Biology, 16,* R323–R325.

Leamy, L. J., & Klingenberg, C. P. (2005). The genetics and evolution of fluctuating asymmetry. *Annual Review of Ecology Evolution and Systematics, 36,* 1–21.

Le Boeuf, B. J. (1978). Social behaviour in some marine and terrestrial carnivores. In E. Reese & F. Lighter (Eds.), *Contrasts in behavior* (pp. 251–279). New York: Wiley.

Le Boeuf, B. J., & Reiter, J. (1988). Lifetime reproductive success on Northern elephant seals. In T. Clutton-Brock (Ed.), *Reproductive success.* Chicago: University of Chicago.

Leca, J. B., Gunst, N., & Huffman, M. (2011). Complexity in object manipulation by Japanese macaques (*Macaca fuscata*): A cross-sectional analysis of manual coordination in stone handling patterns. *Journal of Comparative Psychology, 125,* 61–71.

Lee, P. C. (1987). Allomothering among African elephants. *Animal Behaviour, 35,* 278–291.

Lee, P. C., Poole, J. H., Njiraini, N., & Moss, C. J. (2011). Male social dynamics: Independence and beyond. In C. J. Moss, H. J. Croze, & P. C. Lee (Eds.), *The Amboseli elephants: A long-term perspective on a long-lived mammal* (pp. 260–271). Chicago: University of Chicago Press.

Lefebvre, L. (2011). Taxonomic counts of cognition in the wild. *Biology Letters, 7,* 631–633.

Lefebvre, L., Gaxiola, A., Dawson, S., Timmermans, S., Rosza, L., & Kabai, P. (1997a). Feeding innovations and forebrain size in Australasian birds. *Behaviour, 135,* 1077–1097.

Lefebvre, L., Gaxiola, A., Dawson, S., Timmermans, S., Rosza, L., & Kabai, P. (1998). Feeding innovations and forebrain size in Australasian birds. *Behaviour, 135,* 1077–1097.

Lefebvre, L., Reader, S. M., & Sol, D. (2004). Brains, innovations and evolution in birds and primates. *Brain Behavior and Evolution, 63,* 233–246.

Leffler, J. (1993). *An introduction to free radicals.* New York: Wiley.

Lehmann, L., & Keller, L. (2006). The evolution of cooperation and altruism: A general framework and a classification of models. *Journal of Evolutionary Biology, 19,* 1365–1376.

Leigh, E. G. (2010). The evolution of mutualism. *Journal of Evolutionary Biology, 23,* 2507–2528.

Leimar, O., Austad, S., & Enquist, M. (1991). A test of the sequential assessment game: Fighting in the bowl and doily spider *Frontinella pyramitela. Evolution, 45,* 862–874.

Leimar, O., & Enquist, M. (1984). Effects of asymmetries in owner-intruder conflicts. *Journal of Theoretical Biology, 111,* 475–491.

Leisler, B., Winkler, H., & Wink, M. (2002). Evolution of breeding systems in acrocephaline warblers. *Auk, 119,* 379–390.

Lemaire, V., Koehl, M., Le Moal, M., & Abrous, D. N. (2000). Prenatal stress produces learning deficits associated with an inhibition of neurogenesis in the hippocampus. *Proceedings of the National Academy of Sciences, U.S.A., 97,* 11032–11037.

Lengagne, T., Aubin, T., Jouventin, P., & Lauga, J. (2000). Perceptual salience of individually distinctive features in the calls of adult king penguins. *Journal of the Acoustical Society of America, 107,* 508–516.

Lens, L., Van Dongen, S., & Matthysen, E. (2002). Fluctuating asymmetry as an early warning system in the critically endangered Taita thrush. *Conservation Biology, 16,* 479–487.

Leonard, J. E., & Boake, C. R. B. (2006). Site-dependent aggression and mating behaviour in three species of *Nasonia* (Hymenoptera: Pteromalidae). *Animal Behaviour, 71*, 641–647.

Leopold, A. S. (1977). *The California quail*. Los Angeles: University of California Press.

Leotta, R., Voltini, B., Mele, M., Curadi, M. C., Orlandi, M., & Secchiari, P. (2006). Latent variable models on performance tests in guide dogs. 1. Factor analysis. *Italian Journal of Animal Science, 5*, 377–385.

Lesku, J. A., Roth, T.C., Rattenborg, N. C., Amlaner, C. J., & Lima, S. L. (2008). Phylogenetics and the correlates of mammalian sleep: A reappraisal. *Sleep Medicine Reviews, 12*, 229–244.

Lesku, J. A., Vyssotski, A. L., Martinez-Gonzalez, D., Wilzeck, C., & Rattenborg, N. C. (2011). Local sleep homeostasis in the avian brain: Convergence of sleep function in mammals and birds? *Proceedings of the Royal Society, Series B–Biological Sciences, 278*, 2419–2428.

Levey, D. J., & Stiles, F. G. (1992). Evolutionary precursors of long-distance migration: Resource availability and movement patterns in neotropical birds. *American Naturalist, 140*, 447–476.

Levins, R. (1968). *Evolution in changing environments*. Princeton, NJ: Princeton University Press.

Levitan, D. (2000). Sperm velocity and longevity trade off each other and influence fertilization in the sea urchin *Lytechinus variegatus*. *Proceedings of the Royal Society of London, Series B, 267*, 531–534.

Levitis, D. A., Lidicker, W. Z., & Freund, G. (2009). Behavioural biologists do not agree on what constitutes behaviour. *Animal Behaviour, 78*, 103–110.

Levrero, F., Gatti, S., Menard, N., Petit, E., Caillaud, D., & Gautier-Hion, A. (2006). Living in nonbreeding groups: An alternative strategy for maturing gorillas. *American Journal of Primatology, 68*, 275–291.

Ley, J. M., & Bennett, P. C. (2007). Understanding personality by understanding companion dogs. *Anthrozoos, 20*, 113–124.

Lieberman, D., Tooby, J., & Cosmides, L. (2007). The architecture of human kin detection. *Nature, 445*, 727–731.

Liedvogel, M., Akesson, S., & Bensch, S. (2011). The genetics of migration on the move. *Trends in Ecology & Evolution, 26*, 561–569.

Liers, E. (1951). Notes on the river otter (*Luta canadensis*). *Journal of Mammalogy, 32*, 1–9.

Lim, M. M., Wang, Z. X., Olazabal, D. E., Ren, X. H., Terwilliger, E. F., & Young, L. J. (2004). Enhanced partner preference in a promiscuous species by manipulating the expression of a single gene. *Nature, 429*, 754–757.

Lima, S. (1988a). Initiation and termination of daily feeding in dark-eyed juncos: Influences of predation risk and energy reserves. *Oikos, 53*, 3–11.

Lima, S. (1988b). Vigilance during the initiation of daily feeding in dark-eyed juncos. *Oikos, 53*, 12–16.

Lima, S., & Dill, L. (1990). Behavioral decisions made under the risk of predation: A review and prospectus. *Canadian Journal of Zoology, 68*, 619–640.

Lima, S. L. (1985). Foraging-efficiency-predation-risk trade-off in the grey squirrel. *Animal Behaviour, 33*, 155–165.

Lima, S. L. (1998). Stress and decision making under the risk of predation: Recent developments from behavioral, reproductive, and ecological perspectives. *Advances in the Study of Behavior, 27*, 215–290.

Lima, S. L., Rattenborg, N. C., Lesku, J. A., & Amlaner, C. J. (2005). Sleeping under the risk of predation. *Animal Behaviour, 70*, 723–736.

Lima, S. L., & Valone, T. J. (1986). Influence of predation risk on diet selection: A simple example in the grey squirrel. *Animal Behaviour, 34*, 536–544.

Lind, J., & Cresswell, W. (2005). Determining the fitness consequences of antipredation behavior. *Behavioral Ecology, 16*, 945–956.

Lindblad-Toh, K., Wade, C. M., Mikkelsen, T. S., Karlsson, E. K., Jaffe, D. B., Kamal, M., Clamp, M., Chang, J. L., Kulbokas, E. J., Zody, M. C., Mauceli, E., Xie, X. H., Breen, M., Wayne, R. K., Ostrander, E. A., Ponting, C. P., Galibert, F., Smith, D. R., deJong, P. J., Kirkness, E., Alvarez, P., Biagi, T., Brockman, W., Butler, J., Chin, C. W., Cook, A., Cuff, J., Daly, M. J., DeCaprio, D., Gnerre, S., Grabherr, M., Kellis, M., Kleber, M., Bardeleben, C., Goodstadt, L., Heger, A., Hitte, C., Kim, L., Koepfli, K. P., Parker, H. G., Pollinger, J. P., Searle, S. M. J., Sutter, N. B., Thomas, R., Webber, C., & Lander, E. S. (2005). Genome sequence, comparative analysis and haplotype structure of the domestic dog. *Nature, 438*, 803–819.

Lindenfors, P., Tullberg, B. S., & Biuw, M. (2002). Phylogenetic analyses of sexual selection and sexual size dimorphism in pinnipeds. *Behavioral Ecology and Sociobiology, 52*, 188–193.

Lindquist, W. B., & I. D. Chase. (2009). Data-based analysis of winner-loser models of hierarchy formation in animals. *Bulletin of Mathematical Biology, 71*, 556–584.

Lindstrom, A., & Klaassen, M. (2003). High basal metabolic rates of shorebirds while in the arctic: A circumpolar view. *Condor, 105*, 420–427.

Linnell, J. D. C. (2011). Can we separate the sinners from the scapegoats? *Animal Conservation, 14*, 602–603.

Linsdale, J. (1946). *The California ground squirrel*. Berkeley: University of California Press.

Little, A. E. F., Murakami, T., Mueller, U. G., & Currie, C. R. (2003). The infrabuccal pellet piles of fungus-growing ants. *Naturwissenschaften, 90*, 558–562.

Little, A. E. F., Murakami, T., Mueller, U. G., & Currie, C. R. (2006). Defending against parasites: Fungus-growing ants combine specialized behaviours and microbial symbionts to protect their fungus gardens. *Biology Letters, 2*, 12–16.

Littledyke, M., & Cherrett, J. (1976). Direct ingestion of plant sap from leaves cut by the leaf-cutting ants *Atta cepholotes* and *Acromymex octospinus*. *Bulletin of Entomological Research, 66*, 205–217.

Liu, Y., & Wang, Z. X. (2003). Nucleus accumbens oxytocin and dopamine interact to regulate pair bond formation in female prairie voles. *Neuroscience, 121*, 537–544.

Lohmann, K. J., & Johnsen, S. (2000). The neurobiology of magnetoreception in vertebrate animals. *Trends in Neurosciences, 23*, 153–159.

Long, T. A. F. (2005). The influence of mating system on the intensity of parent-offspring conflict in primates. *Journal of Evolutionary Biology, 18*, 509–515.

Lorenz, K. (1935). Der Kumpan in der Umwelt des Vogels. *Journal of Ornithology, 83*, 289–413.

Loup, F., Tribollet, E., Dubois-Dauphin, M., & Dreifuss, J. J. (1991). Localization of high-affinity binding sites for oxytocin and vasopressin in the human brain: An autoradiography study. *Brain Research*, 555, 220–232.

Low, B. (2000). *Why sex matters: A Darwinian look at human behaviour*. Princeton, NJ: Princeton University Press.

Lubin, Y. D., & Crozier, R. H. (1985). Electrophoretic evidence for population differentiation in a social spider, *Achaearanea wau* (Theridiidae). *Insectes Sociaux*, 32, 297–304.

Lucas, N. S., Hume, E. M., & Henderson, H. (1937). On the breeding of the common marmoset (*Hapale jacchus*) in captivity when irradiated with ultra-violet rays. II. A ten-year family history. *Proceedings of the Zoological Society of London, Series A*, 107, 205–211.

Lucas, P. W., Darvell, B. W., Lee, P. K. D., Yuen, T. D. B., & Choong, M. F. (1998). Colour cues for leaf food selection by long-tailed macaques (*Macaca fascicularis*) with a new suggestion for the evolution of trichromatic colour vision. *Folia Primatologica*, 69, 139–152.

Lucas, P. W., Dominy, N. J., Riba-Hernandez, P., Stoner, K. E., Yamashita, N., Loria-Calderon, E., Petersen-Pereira, W., Rojas-Duran, Y., Salas-Pena, R., Solis-Madrigal, S., Osorio, D., & Darvell, B. W. (2003). Evolution and function of routine trichromatic vision in primates. *Evolution*, 57, 2636–2643.

Lundstrom, J. N., Boyle, J. A., Zatorre, R. J., & Jones-Gotman, M. (2009). The neuronal substrates of human olfactory based kin recognition. *Human Brain Mapping*, 30, 2571–2580.

Lurling, M., & Scheffer, M. (2007). Info-disruption: Pollution and the transfer of chemical information between organisms. *Trends in Ecology & Evolution*, 22, 374–379.

Lutz, C. C., Rodriguez-Zas, S. L., Fahrbach, S. E., & Robinson, G. E. (2012). Transcriptional response to foraging experience in the honeybee mushroom bodies. *Developmental Neurobiology*, 72, 153–166.

Lyamin, O. I., & Chetyrbok, I. S. (1992). Unilateral EEG activation during sleep in the cape fur seal, *Arctocephalus pusillus*. *Neuroscience Letters*, 143, 263–266.

Lycett, S. J., Collard, M., & McGrew, W. C. (2007). Phylogenetic analyses of behavior support existence of culture among wild chimpanzees. *Proceedings of the National Academy of Sciences, U.S.A.*, 104, 17588–17592.

MacArthur, R., & Pianka, E. (1966). On optimal use of a patchy environment. *American Naturalist*, 100, 603–609.

MacDougal-Shackleton, S. (1997). Sexual selection and the evolution of song repertoires. In V. Nolam, E. Ketterson, & C. Thompson (Eds.), *Current ornithology* (pp. 81–124). New York: Plenum.

Mace, G., Harvey, P., & Clutton-Brock, T. (1981). Brain size and ecology in small mammals. *Journal of Zoology*, 193, 333–354.

Mackenzie, A., Reynolds, J. D., Brown, V. J., & Sutherland, W. J. (1995). Variation in male mating success on leks. *American Naturalist*, 145, 633–652.

Macnair, M. R., & Parker, G. A. (1978). Models of parent-offspring conflict. 2. Promiscuity. *Animal Behaviour*, 26, 111–122.

Maddison, P., & Maddison, W. (1992). *MacClade*. Sunderland, MA: Sinauer.

Maeda, K., Henbest, K. B., Cintolesi, F., Kuprov, I., Rodgers, C. T., Liddell, P. A., Gust, D., Timmel, C. R., & Hore, P. J. (2008). Chemical compass model of avian magnetoreception. *Nature*, 453, 387–391.

Maestripieri, D. (1993). Vigilance costs of allogrooming in macaque mothers. *American Naturalist*, 141, 744–753.

Maestripieri, D. (1995). Maternal encouragement in nonhuman primates and the question of animal teaching. *Human Nature*, 6, 361–378.

Magurran, A. E. (1990). The inheritance and development of minnow antipredator behaviour. *Animal Behaviour*, 39, 834–842.

Magurran, A. E. (2005). *Evolutionary ecology: The Trinidadian guppy*. Oxford: Oxford University Press.

Magurran, A. E., & Higham, A. (1988). Information transfer across fish shoals under predator threat. *Ethology*, 78, 153–158.

Magurran, A. E., & Pitcher, T. J. (1987). Provenance, shoal size and the sociobiology of predator-evasion in minnow shoals. *Proceedings of the Royal Society of London, Series B*, 229, 439–465.

Magurran, A. E., & Seghers, B. H. (1990). Population differences in predator recognition and attack cone avoidance in the guppy *Poecilia reticulata*. *Animal Behaviour*, 40, 443–452.

Magurran, A. E., & Seghers, B. H. (1991). Variation in schooling and aggression amongst guppy populations (*Poecilia reticulata*) in Trinidad. *Behaviour*, 118, 214–234.

Magurran, A. E., Seghers, B. H., Carvalho, G. R., & Shaw, P. W. (1992). Behavioral consequences of an artificial introduction of guppies (*Poecilia reticulata*) in N. Trinidad: Evidence for the evolution of antipredator behaviour in the wild. *Proceedings of the Royal Society of London, Series B*, 248, 117–122.

Magurran, A. E., Seghers, B. H., Carvalho, G. R., & Shaw, P. W. (1993). Evolution of adaptive variation in antipredator behaviour. *Marine Behaviour and Physiology*, 23, 29–44.

Magurran, A. E., Seghers, B. H., Shaw, P. W., & Carvalho, G. R. (1995). The behavioral diversity and evolution of guppy, *Poecilia reticulata*, populations in Trinidad. *Advances in the Study of Behavior*, 24, 155–202.

Mahometa, M. J., & Domjan, M. (2005). Classical conditioning increases reproductive success in Japanese quail, *Coturnix japonica*. *Animal Behaviour*, 69, 983–989.

Mank, J. E., Promislow, D. E. L., & Avise, J. C. (2005). Phylogenetic perspectives in the evolution of parental care in ray-finned fishes. *Evolution*, 59, 1570–1578.

Mann, J., & Sargeant, B. L. (2003). Like mother, like calf: The ontogeny of foraging strategies in the wild Indian Ocean bottlenose dolphins (*Tursiops sp.*). In D. M. Fragasy & S. Perry (Eds.), *The biology of traditions: Models and evidence* (pp. 236–266). Cambridge: Cambridge University Press.

Manning, C. J., Dewsbury, D. A., Wakeland, E. K., & Potts, W. K. (1995). Communal nesting and communal nursing in house mice, *Mus musculus domesticus*. *Animal Behaviour*, 50, 741–751.

Manning, C. J., Wakeland, E. K., & Potts, W. K. (1992). Communal nesting patterns in mice implicate MHC genes in kin recognition. *Nature*, 360, 581–583.

Marchetti, C., & Drent, P. J. (2000). Individual differences in the use of social information in foraging by captive great tits. *Animal Behaviour*, 60, 131–140.

Margulis, S. W., Nabong, M., Alaks, G., Walsh, A., & Lacy, R. C. (2005). Effects of early experience on subsequent parental behaviour and reproductive success in oldfield mice, *Peromyscus polionotus. Animal Behaviour, 69*, 627–634.

Marinesco, S., Duran, K. L., & Wright, W. G. (2003). Evolution of learning in three aplysiid species: Differences in heterosynaptic plasticity contrast with conservation in serotonergic pathways. *Journal of Physiology–London, 550*, 241–253.

Markl, H. (1968). Die verstandigung durch stridulationssignale bei blattschneiderameisesn. II Erzeugung und eigenschaften der signale. *Zeitschrift fur Vergleichende Physiologie, 60*, 103–150.

Marler, P. (1968). Visual systems. In T. Sebeok (Ed.), *Animal communications* (pp. 103–126). Bloomington: Indiana University Press.

Marler, P., & Griffin, D. R. (1973). The 1973 Nobel prize for physiology or medicine. *Science, 182*, 464–466.

Martel, F. L., Nevison, C. M., Rayment, F. D., Simpson, M. J. A., & Keverne, E. B. (1993). Opioid receptor blockade reduces maternal effect and social grooming in rhesus monkeys. *Psychoneuroendocrinology, 18*, 307–321.

Martin, J. K., & Martin, A. A. (2007). Resource distribution influences mating system in the bobuck (*Trichosurus cunninghami: Marsupialia*). *Oecologia, 154*, 227–236.

Martin, P., & Caro, T. (1985). On the functions of play and its role in behavioral development. *Advances in the Study of Behavior, 15*, 59–103.

Martin, T. E. (1993). Evolutionary determinants of clutch size in cavity nesting birds: Nest predation or limited breeding opportunities? *American Naturalist, 142*, 937–946.

Martin, T. E. (1995). Avian life history in relation to nest sites, nest predation and food. *Ecological Monographs, 65*, 101–127.

Marzluff, J. M., Heinrich, B., & Marzluff, C. (1996). Raven roosts are mobile information centres. *Animal Behaviour, 51*, 89–103.

Masseti, M. (2000). Did the study of ethology begin in Crete 4000 years ago? *Ethology Ecology & Evolution, 12*, 89–96.

Mather, J., & Anderson, R. (1993). Personalities of octopuses (*Octopus rubescens*). *Journal of Comparative Psychology, 107*, 336–340.

Mather, J. A. (1991). Foraging, feeding and prey remains in middens of juvenile *Octopus vulgaris* (Mollusca: Cephalopoda). *Journal of Zoology–London, 224*, 27–39.

Mather, J. A. (1995). Cognition in cephalopods. *Advances in the study of behavior, 24*, 317–353.

Mather, J. A. (2008a). Cephalopod consciousness: Behavioural evidence. *Consciousness and Cognition, 17*, 37–48.

Mather, J. A. (2008b). To boldly go where no mollusc has gone before: Personality, play, thinking, and consciousness in cephalopods. *American Malacological Bulletin, 24*, 51–58.

Mather, J. A., & Anderson, R. C. (1999). Exploration, play, and habituation in octopuses (*Octopus dofleini*). *Journal of Comparative Psychology, 113*, 333–338.

Mathis, A., & Smith, R. J. F. (1993a). Chemical labeling of northern pike (*Esox lucius*) by the alarm pheromone of fathead minnows (*Pimephales promelas*). *Journal of Chemical Ecology, 19*, 1967–1979.

Mathis, A., & Smith, R. J. F. (1993b). Fathead minnows, *Pimephales promelas*, learn to recognize northern pike, *Esox lucius*, as predators on the basis of chemical stimuli from minnows in the pike's diet. *Animal Behaviour, 46*, 645–656.

Maynard Smith, J. (1974). The theory of games and the evolution of animal conflicts. *Journal of Theoretical Biology, 47*, 209–221.

Maynard Smith, J. (1982). *Evolution and the theory of games.* Cambridge: Cambridge University Press.

Maynard Smith, J., & Price, G. (1973). The logic of animal conflict. *Nature, 246*, 15–18.

Maynard Smith, J., & Parker, G. A. (1976). The logic of asymmetric contests. *Animal Behaviour, 24*, 159–175.

Mayr, E. (1961). Cause and effect in biology. *Science, 134*, 1501–1506.

Mayr, E. (1982). *The growth of biological thought.* Cambridge, MA: Harvard University Press.

Mayr, E. (1983). How to carry out the adaptationist program. *American Naturalist, 121*, 324–334.

Mays, H. L., & Hill, G. E. (2004). Choosing mates: Good genes versus genes that are a good fit. *Trends in Ecology & Evolution, 19*, 554–559.

McCarthy, M. M. (1994). Molecular aspects of sexual differentiation of the rodent brain. *Psychoneuroendocrinology, 19*, 415–427.

McComb, K., & Semple, S. (2005). Coevolution of vocal communication and sociality in primates. *Biology Letters, 1*, 381–385.

McCullough, D. R. (1969). The tule elk: Its history, behavior and ecology. *University of California Publications in Zoology, 88*, 1–209.

McEwen, B., & Sapolsky, R. M. (1995). Stress and cognitive function. *Current Opinions in Neurobiology, 5*, 205–210.

McGraw, K. J., Stoehr, A. M., Nolan, P. M., & Hill, G. E. (2001). Plumage redness predicts breeding onset and reproductive success in the house finch: A validation of Darwin's theory. *Journal of Avian Biology, 32*, 90–94.

McGraw, L. A., & Young, L. J. (2010). The prairie vole: An emerging model organism for understanding the social brain. *Trends in Neuroscience, 33*, 103–109.

McGregor, P. (2005). *Animal communication networks.* Cambridge: Cambridge University Press.

McGregor, P. K., & Peake, T. (2000). Communication networks: Social environments for receiving and signalling behaviour. *Acta Ethologica, 2*, 71–81.

McGuire, B., & Novak, M. (1984). A comparison of maternal behavior in the meadow vole (*Microtus-Pennsylvanicus*), prairie vole (*Microtus-Ochrogaster*) and pine vole (*Microtus-Pinetorum*). *Animal Behaviour, 32*, 1132–1141.

McGuire, M., & Troisi, A. (1998). *Darwinian psychiatry.* New York: Oxford University Press.

McKinney, F., Cheng, K., & Bruggers, D. (1984). Sperm competition in apparently monogamous birds. In R. Smith (Ed.), *Sperm competition and the evolution of animal mating systems* (pp. 523–545). New York: Academic Press.

McMann, S. (1993). Contextual signaling and the structure of dyadic encounters in *Anolis carolinensis. Animal Behaviour, 46*, 657–668.

McNab, B. K. (1973). Energetics and distribution of vampires. *Journal of Mammalogy, 54,* 131–144.

McNamara, J. M., & Houston, A. I. (1992). Risk sensitive foraging: A review of the theory. *Bulletin of Mathematical Biophysics, 54,* 355–378.

Mead, L. S., & Arnold, S. J. (2004). Quantitative genetic models of sexual selection. *Trends in Ecology & Evolution, 19,* 264–271.

Melis, A. P., Hare, B., & Tomasello, M. (2006). Chimpanzees recruit the best collaborators. *Science, 311,* 1297–1300.

Mello, C., Nottebohm, F., & Clayton, D. (1995). Repeated exposure to one song leads to a rapid and persistent decline in an immediate early gene's response to that song in zebra finch telencephalon. *Journal of Neuroscience, 15,* 6919–6925.

Mello, C. V., Velho, T. A. F., & Pinaud, R. (2004). A window on song auditory processing and perception. In H. P. Zeigler & P. Marler (Eds.), *Behavioral neurobiology of birdsong* (Vol. 1016, pp. 263–281). New York: Annals of the New York Academy of Sciences.

Mello, C. V., Vicario, D. S., & Clayton, D. F. (1992). Song presentation induces gene expression in the songbird forebrain. *Proceedings of the National Academy of Sciences, U.S.A., 89,* 6818–6822.

Mendl, M., Burman, O. H. P., & Paul, E. S. (2010). An integrative and functional framework for the study of animal emotion and mood. *Proceedings of the Royal Society, Series B–Biological Sciences, 277,* 2895–2904.

Mendozagranados, D., & Sommer, V. (1995). Play in chimpanzees of the Arnhem Zoo: Self-serving compromises. *Primates, 36,* 57–68.

Mercer, A. R., Emptage, N. J., & Carew, T. J. (1991). Pharmacological dissociation of modulatory effects of serotonin in aplysia sensory neurons. *Science, 254,* 1811–1813.

Mery, F., & Burns, J. G. (2010). Behavioral plasticity: An interaction between evolution and experience. *Evolution Ecology, 24,* 571–583.

Mesoudi, A., Whiten, A., & Laland, K. N. (2006). Towards a unified science of cultural evolution. *Behavioral and Brain Sciences, 29,* 329–381.

Mesterton-Gibbons, M. (1992). Ecotypic variation in an asymmetric hawk–dove game: When is bourgeois an ESS? *Evolutionary Ecology, 6,* 198–222.

Mesterton-Gibbons, M. (1999). On the evolution of pure winner and loser effects: A game-theoretic model. *Bulletin of Mathematical Biology, 61,* 1151–1186.

Mesterton-Gibbons, M., & Adams, E. (1998). Animal contests as evolutionary games. *American Scientist, 86,* 334–341.

Mesterton-Gibbons, M., & Dugatkin, L. A. (1992). Cooperation among unrelated individuals: Evolutionary factors. *Quarterly Review of Biology, 67,* 267–281.

Mesterton-Gibbons, M., Gavrilets, S., Gravner, J., & Akcay, E. (2011). Models of coalition or alliance formation. *Journal of Theoretical Biology, 274,* 187–204.

Miklosi, A. (1999). The ethological analysis of imitation. *Biological Reviews of the Cambridge Philosophical Society, 74,* 347–374.

Milinski, M. (1977). Do all members of a swarm suffer the same predation? *Zeitschrift fur Tierpsychologie, 45,* 373–378.

Milinski, M. (1979). An evolutionarily stable feeding strategy in sticklebacks. *Zeitschrift fur Tierpsychologie, 51,* 36–40.

Milinski, M. (1987). Tit for tat and the evolution of cooperation in sticklebacks. *Nature, 325,* 433–435.

Milinski, M. (2006). The major histocompatibility complex, sexual selection, and mate choice. *Annual Review of Ecology, Evolution, and Systematics, 37,* 159–186.

Milinski, M., & Bakker, T. (1990). Female sticklebacks use male coloration in mate choice and hence avoid parasitized sticklebacks. *Nature, 344,* 330–333.

Milinski, M., Griffiths, S., Wegner, K. M., Reusch, T. B. H., Haas-Assenbaum, A., & Boehm, T. (2005). Mate choice decisions of stickleback females predictably modified by MHC peptide ligands. *Proceedings of the National Academy of Sciences, U.S.A.,102,* 4414–4418.

Milinski, M., Kulling, D., & Kettler, R. (1990). Tit for tat: Sticklebacks "trusting" a cooperating partner. *Behavioral Ecology, 1,* 7–12.

Milinski, M., Luthi, J., Eggler, R., & Parker, G. (1997). Cooperation under predation risk: Experiments on costs and benefits. *Proceedings of the Royal Society of London, Series B, 264,* 831–837.

Milinski, M., & Wedekind, C. (2001). Evidence for MHC-correlated perfume preferences in humans. *Behavioral Ecology, 12,* 140–149.

Mills, A. D., Crawford, L. L., Domjan, M., & Faure, J. (1997). The behavior of the Japanese or domestic quail, *Coturnix japonica. Neuroscience and Biobehavioral Reviews, 21,* 261–281.

Mineka, S., & Cook, M. (1986). Immunization against the observational conditioning of snake fear in rhesus monkeys. *Journal of Abnormal Psychology, 95,* 307–318.

Mineka, S., & Cook, M. (1988). Social learning and the acquisition of snake fear in monkeys. In T. Zentall & B. G. Galef (Eds.), *Social learning: Psychological and biological perspectives* (pp. 51–73). Hillsdale, NJ: Lawrence Erlbaum Associates.

Mineka, S., Davidson, M., Cook, M., & Keir, R. (1984). Observational conditioning of snake fear in rhesus monkeys. *Journal of Abnormal Psychology, 93,* 355–372.

Mitchell, W. A., & Valone, T. J. (1990). Commentary: The optimization approach—studying adaptations by their function. *Quarterly Review of Biology, 65,* 43–52.

Mitman, G. (1992). *The state of nature: Ecology, community and American social thought, 1900–1950.* Chicago: University of Chicago Press.

Mittlebach, G. (1984). Group size and feeding rate in bluegills. *Copeia, 4,* 998–1000.

Miyaki, C. Y., Matioli, S. R., Burke, T., & Wajntal, A. (1998). Parrot evolution and paleogeographical events: Mitochondrial DNA evidence. *Molecular Biology and Evolution, 15,* 544–551.

Miyatake, T., Katayama, K., Takeda, Y., Nakashima, A., Sugita, A., & Mizumoto, M. (2004). Is death-feigning adaptive? Heritable variation in fitness difference of death-feigning behaviour. *Proceedings of the Royal Society of London, Series B, 271,* 2293–2296.

Mock, D. (1980). White-dark polymorphism in herons. In D. L. Drawe (Ed.), *Proceedings of the first Welder Wildlife Symposium* (pp. 145–161). Sinton, TX: Welder Wildlife Foundation.

Mock, D. (2004). *More than kin and less than kind: The evolution of family conflict.* Cambridge, MA: Belknap Press.

Mock, D., & Parker, G. (1997). *The evolution of sibling rivalry.* New York: Oxford University Press.

Mock, D. W., Dugas, M. B., & Strickler, S. A. (2011). Honest begging: Expanding from signal of need. *Behavioral Ecology, 22,* 909–917.

Mohamad, R., Monge, J. P., & Goubault, M. (2010). Can subjective resource value affect aggressiveness and contest outcome in parasitoid wasps? *Animal Behaviour, 80,* 629–636.

Molenberghs, P., Cunnington, R., & Matttingley, J. B. (2009). Is the mirror neuron system involved in imitation? A short review and meta-analysis. *Neuroscience and Biobehavioral Reviews, 33,* 975–980.

Molenberghs, P., Cunnington, R., & Matttingley, J. B. (2012). Brain regions with mirror properties: A meta-analysis of 125 human fMRI studies. *Neuroscience and Biobehavioral Reviews, 36,* 341–349.

Molina-Borja, M., Padron-Fumero, M., & Alfonso-Martin, T. (1998). Morphological and behavioural traits affecting the intensity and outcome of male contests in *Gallotia galloti galloti* (family Lacertidae). *Ethology, 104,* 314–322.

Møller, A., & Jennions, M. (2001). How important are direct benefits of sexual selection? *Naturwissenschaften, 88,* 401–415.

Møller, A. P. (1988). False alarm calls as a means of resource usurpation in the great tit *Parus major. Ethology, 79,* 25–30.

Møller, A. P. (1990). Deceptive use of alarm calls by male swallows, *Hirundo rustica:* A new paternity guard. *Behavioral Ecology, 1,* 1–6.

Møller, A. P., & Erritzøe, J. (1996). Parasite virulence and host immune defense: Host immune response is related to nest re-use in birds. *Evolution, 50,* 2066–2072.

Møller, A. P., & Erritzøe, J. (1998). Host immune defence and migration in birds. *Evolutionary Ecology, 12,* 945–953.

Mollon, J. D. (1989). "Tho' she knnel'd in that place where they grew . . . " *Journal of Experimental Biology, 146,* 21–38.

Monkkonen, M., & Orell, M. (1997). Clutch size and cavity excavation in parids (Paridae): The limited breeding opportunities hypothesis tested. *American Naturalist, 149,* 1164–1174.

Montague, R. (2006). *Why choose this book? How we make decisions.* New York: Dutton.

Montgomerie, R., Lyon, B., & Holder, K. (2001). Dirty ptarmigan: Behavioral modification of conspicuous male plumage. *Behavioral Ecology, 12,* 429–438.

Mooney, R. (2009). Neural mechanisms for learned birdsong. *Learning & Memory, 16,* 655–669.

Moore, H. D. M., Martin, M., & Birkhead, T. R. (1999). No evidence for killer sperm or other selective interactions between human spermatozoa in ejaculates of different males in vitro. *Proceedings of the Royal Society of London, Series B, 266,* 2343–2350.

Moore, T. (2012). Review: Parent-offspring conflict and the control of placental function. *Placenta, 33,* S33–S36.

Morales, J. M., Moorcroft, P. R., Matthiopoulos, J., Frair, J. L., Kie, J. G., Powell, R. A., Merrill, E. H., & Haydon, D. T. (2010). Building the bridge between animal movement and population dynamics. *Philosophical Transactions of the Royal Society, 365,* 2289–2301.

Morgan, E. (1984). Chemical words and phrases in the language of pheromones for foraging and recruitment. In T. D. Lewis (Ed.), *Communication* (pp. 169–194). London: Academic Press.

Morris, D. W. (1994). Habitat matching: Alternatives and implications to populations and communities. *Evolutionary Ecology, 8,* 387–406.

Morris, R. G. M., Garrud, P., Rawlins, J. N. P., & Okeefe, J. (1982). Place navigation impaired in rats with hippocampal lesions. *Nature, 297,* 681–683.

Morse, D. H. (1970). Ecological aspects of some mixed species foraging flocks of birds. *Ecological Monographs, 4,* 119–168.

Morton, E. S. (1977). On the occurrence and significance of motivation-structural rules in some bird and animal sounds. *American Naturalist, 111,* 855–869.

Moser, M. B., Trommald, M., & Andersen, P. (1994). An increase in dendritic spine density on hippocampal CA1 pyramidal cells following spatial learning in adult rats suggests the formation of new synapses. *Proceedings of the National Academy of Sciences, U.S.A., 91,* 12673–12675.

Mouritsen, H., & Frost, B., J. (2002). Virtual migration in tethered flying monarch butterflies reveals their orientation mechanisms. *Proceedings of the National Academy of Sciences, U.S.A., 99,* 10162–10166.

Mousseau, T., Sinervo, B., & Endler, J. (Eds.). (1999). *Adaptive genetic variation in the wild.* New York: Oxford University Press.

Mousseau, T. A., & Roff, D. A. (1987). Natural selection and the heritability of fitness components. *Heredity, 59,* 181–197.

Muhlau, M., Hermsdorfer, J., Goldenberg, G., Wohlschlager, A. M., Castrop, F., Stahl, R., Rottinger, M., Erhard, P., Haslinger, B., Ceballos-Baumann, A. O., Conrad, B., & Boecker, H. (2005). Left inferior parietal dominance in gesture imitation: An fMRI study. *Neuropsychologia, 43,* 1086–1098.

Mukhametov, L., Supin, A., & Lyamin, O. (1988). Interhemispheric asymmetrie of the EEG during sleep in marine mamals. In T. Oniani (Ed.), *Neurobiology of sleep-wakefulness cycle* (pp. 147–159). Tbilisi, Georgia: Metsniereba.

Mulder, R., Dunn, P., Cockburn, A., Lazenby-Cohen, K., & Howell, M. (1994). Helpers liberate female fairy-wrens from constraints on extra-pair mate choice. *Proceedings of the Royal Society of London, Series B, 255,* 223–229.

Mungall, E. C. (1978). The Indian blackbuck antelope: A Texas view. In Kleberg, *Studies in natural resources.* College Station, TX.

Munn, C. A. (1986). Birds that "cry wolf." *Nature, 319,* 143–145.

Munz, T. (2005). The bee battles: Karl von Frisch, Adrian Wenner and the honey bee dance language controversy. *Journal of the History of Biology, 38,* 535–570.

Nahallage, C., & Huffman, M. (2007). Age-specific functions of stone handling, a solitary-object play behavior, in Japanese macaques (*Macaca fiscata*). *American Journal of Primatology, 69,* 267–281.

Nakahashi, W. (2010). Evolution of learning capacities and learning levels. *Theoretical Population Biology, 78,* 211–224.

Nakayama, S., Sasaki, K., Matsumura, K., Lewis, Z., & Miyatake, T. (2012). Dopaminergic system as the mechanism underlying personality in a beetle. *Journal of Insect Physiology, 58,* 750–755.

Nash, S., & Domjan, M. (1991). Learning to discriminate the sex of conspecifics in male Japanese quail (*Coturnix coturnix japonica*): Tests of "biological constraints." *Journal of Experimental Psychology: Animal Behavior Processes, 17,* 342–353.

Neff, B. D., Fu, P., & Gross, M. R. (2003). Sperm investment and alternative mating tactics in bluegill sunfish (*Lepomis macrochirus*). *Behavioral Ecology, 14,* 634–641.

Nelissen, M. (1985). Structure of the dominance hierarchy and dominance determining "group factors" in *Melanochromatis auratus. Behaviour, 94,* 85–107.

Nelson, D. A., Khanna, H., & Marler, P. (2001). Learning by instruction or selection: Implications for patterns of geographic variation in bird song. *Behaviour, 138,* 1137–1160.

Nelson, R. (2005). *An introduction to behavioral endocrinology* (3rd ed.). Sunderland, MA: Sinauer.

Nelson, R. (2011). *An introduction to behavioral endocrinology* (4th ed.). New York: Sinauer.

Nesse, R. M., Bergstrom, C. T., Ellison, P. T., Flier, J. S., Gluckman, P., Govindaraju, D. R., Niethammer, D., Omenn, G. S., Perlman, R. L., Schwartz, M. D., Thomas, M. G., Stearns, S. C., & Valle, D. (2010). Making evolutionary biology a basic science for medicine. *Proceedings of the National Academy of Sciences, U.S.A. 107,* 1800–1807.

Nesse, R., & Williams, G. C. (1995). *Why we get sick: The new science of Darwinian medicine.* New York: Vintage Books.

Nesse, R. M. (1988). Life table tests of evolutionary theories of senescence. *Experimental Gerontology, 23,* 445–453.

Neudorf, D. L. H. (2004). Extrapair paternity in birds: Understanding variation among species. *Auk, 121,* 302–307.

Newton-Fisher, N. E., & Lee, P. C. (2011). Grooming reciprocity in wild male chimpanzees. *Animal Behaviour, 81,* 439–446.

Niesink, R., & Van Ree, J. (1989). Involvement of opioid and dopaminergic systems in isolation-induced pinning and social grooming of young rats. *Neuropharmacology, 28,* 411–418.

Ninnes, C. E., Waas, J. R., Ling, N., Nakagawa, S., Banks, J. C., Bell, D. G., Bright, A., Carey, P. W., Chandler, J., Hudson, Q. J., Ingram, J. R., Lyall, K., Morgan, D. K. J., Stevens, M. I., Wallace, J., & Mostl, E. (2011). Environmental influences on Adelie penguin breeding schedules, endocrinology, and chick survival. *General and Comparative Endocrinology, 173,* 139–147.

Nishida, T. (1979). The social structure of chimpanzees of the Mahale Mountains. In D. A. Hamburg & E. R. McCown (Eds.), *The great apes.* Menlo Park, CA: Benjamin Cummings.

Nishida, T., Hiraiwa-Hasegawa, M., Hasegawa, T., & Takahata, Y. (1985). Group extinction and female transfer in wild chimpanzees in the Mahale National Park, Tanzania. *Zeitschrift fur Tierpsychologie, 67,* 284–301.

Nishida, T., & Inaba, A. (2009). Pirouettes: The rotational play of wild chimpanzees. *Primates, 50,* 333–341.

Nishida, T., Matsusaka, T., & McGrew, W. C. (2009). Emergence, propagation or disappearance of novel behavioral patterns in the habituated chimpanzees of Mahale: A review. *Primates, 50,* 23–36.

Nishida, T., Uehara, S., & Nyondo, R. (1983). Predatory behaviour among wild chimpanzees of the Mahale Mountains. *Primates, 20,* 1–20.

Nishida, T., & Wallauer, W. (2003). Leaf-pile pulling: An unusual play pattern in wild chimpanzees. *American Journal of Primatology, 60,* 167–173.

Noble, J. (1999). Cooperation, conflict and the evolution of communication. *Adaptive Behavior, 7,* 349–369.

Noe, R. (1986). Lasting alliances among adult male Savannah baboons. In J. Else & P. Lee (Eds.), *Primate ontogeny, cognition and social behaviour* (pp. 381–392). Cambridge: Cambridge University Press.

Noe, R., & Hammerstein, P. (1995). Biological markets. *Trends in Ecology & Evolution, 10,* 336–339.

Noe, R., Van Hooff, J., & Hammerstein, P. (2002). *Economics in nature: Social dilemma, mate choice and biological markets.* Cambridge: Cambridge University Press.

Nonacs, P., & Hager, R. (2011). The past, present and future of reproductive skew theory and experiments. *Biological Reviews, 86,* 271–298.

Normand, E., & Boesch, C. (2009). Sophisticated Euclidean maps in forest chimpanzees. *Animal Behaviour, 77,* 1195–1201.

Normansell, L., & Panksepp, J. (1990). Effects of morphine and naloxone on play-rewarded social discrimination in juvenile rats. *Developmental Psychobiology, 23,* 75–83.

Nottebohm, F., & Nottebohm, M. E. (1976). Left hypoglossal dominance in control of canary and white-crowned sparrow song. *Journal of Comparative Physiology, 108,* 1368–1370.

Novick, L. R., & Catley, K. M. (2007). Understanding phylogenies in biology: The influence of a Gestalt perceptual principle. *Journal of Experimental Psychology–Applied, 13,* 197–223.

Nowacek, D. P. (2005). Acoustic ecology of foraging bottlenose dolphins (*Tursiops truncatus*), habitat-specific use of three sound types. *Marine Mammal Science, 21,* 587–602.

Nowak, M. A. (2006). Five rules for the evolution of cooperation. *Science, 314,* 1560–1563.

Nowak, M. A., Tarnita, C. E., & Wilson, E. O. (2010). The evolution of eusociality. *Nature, 466,* 1057–1062.

Nunes, S., Muecke, E. M., Anthony, J., & Batterbee, A. S. (1999). Endocrine and energetic mediation of play behavior in free-living Belding's ground squirrels. *Hormones and Behavior, 36,* 153–165.

Obanda, V., Ndeereh, D., Mijele, D., Lekolool, I., Chege, S., Gakuya, F., & Omondi, P. (2008). Injuries of free ranging African elephants (*Loxodonta africana africana*) in various ranges of Kenya. *Pachyderm, 44,* 54–58.

Odeen, A., Hastad, O., & Alstrom, P. (2011). Evolution of ultraviolet vision in the largest avian radiation—the passerines. *Bmc Evolutionary Biology, 11.*

Odeen, A., Pruett-Jones, S., Driskell, A. C., Armenta, J. K., & Hastad, O. (2012). Multiple shifts between violet and ultraviolet vision in a family of passerine birds with associated changes in plumage coloration. *Proceedings of the Royal Society, Series B–Biological Sciences, 279,* 1269–1276.

Odling-Smee, F. J., Laland, K., & Feldman, M. (2003). *Niche construction: The neglected process in evolution.* Princeton, NJ: Princeton University Press.

Ohno, T., & Miyatake, T. (2007). Drop or fly? Negative genetic correlation between death-feigning intensity and flying ability as alternative antipredator strategies. *Proceedings of the Royal Society of London, Series B, 274,* 555–560.

Oitzl, M. S., Champagne, D. L., van der Veen, R., & de Kloet, E. R. (2010). Brain development under stress: Hypotheses of glucocorticoid actions revisited. *Neuroscience and Biobehavioral Reviews, 34,* 853–866.

O'Keefe, J., & Nadel, L. (1978). *The hippocampus as a cognitive map.* Oxford: Claredon.

Okuda, J., Fujii, T., Ohtake, H., Tsukiura, T., Tanji, K., Suzuki, K., Kawashima, R., Fukuda, H., Itoh, M., & Yamadori, A. (2003). Thinking of the future and past: The roles of the frontal pole and the medial temporal lobes. *NeuroImage, 19,* 1369–1380.

Oldroyd, B. P., & Thompson, G. J. (2007). Behavioural genetics of the honey bee *Apis mellifera. Advances in Insect Physiology, 33,* 1–49.

Olesen, K. M., Jessen, H. M., Auger, C. J., & Auger, A. P. (2005). Dopaminergic activation of estrogen receptors in neonatal brain alters progestin receptor expression and juvenile social play behavior. *Endocrinology, 146,* 3705–3712.

Oli, M., Taylor, K., & Rogers, M. (1994). Snow leopard (*Panthera uncia*) predation of livestock: An assessment of local perceptions in the Annapurna conservation area, Nepal. *Biological Conservation, 68,* 63–68.

Olioff, M., & Stewart, J. (1978). Sex differences in play behavior of prepubescent rats. *Physiology and Behavior, 20,* 113–115.

Oliveira, R. F., Carneiro, L. A., & Canario, A. V. M. (2005). No hormonal response in tied fights. *Nature, 437,* 207–208.

Oliveira, R. F., Hirschenhauser, K., Carneiro, L. A., & Canário, A. V. M. (2002). Social modulation of androgen levels in male teleost fish. *Comparative Biochemistry and Physiology B–Biochemistry & Molecular Biology, 132,* 203–215.

Oliveira, R. F., Lopes, M., Carneiro, L. A., & Canário, A. V. M. (2001). Watching fights raises fish hormone levels: Cichlid fish wrestling for dominance induce an androgen surge in male spectators. *Nature, 409,* 475.

Oliveira, R. F., McGregor, P. K., & Latruffe, C. (1998). Know thine enemy: Fighting fish gather information from observing conspecific interactions. *Proceedings of the Royal Society of London, Series B, 265,* 1045–1049.

Oliveras, D., & Novak, M. (1986). A comparison of paternal behavior in the meadow vole *Microtus pennsylvanicus,* the pine vole *Microtus pinetorum* and the prairie vole *Microtus ochrogaster. Animal Behaviour, 34,* 519–526.

O'Loghlen, A. L. (1995). Delayed access to local songs prolongs vocal development in dialect populations of brown-headed cowbirds. *Condor, 97,* 402–414.

O'Loghlen, A. L., & Rothstein, S. I. (1993). An extreme example of delayed vocal development: Song learning in a population of wild brown-headed cowbirds. *Animal Behaviour, 46,* 293–304.

O'Loghlen, A. L., & Rothstein, S. I. (1995). Culturally correct song dialects are correlated with male age and female song preferences in wild populations of brown-headed cowbirds. *Behavioral Ecology and Sociobiology, 36,* 251–259.

O'Loghlen, A. L., & Rothstein, S. I. (2002). Ecological effects on song learning: Delayed development is widespread in wild populations of brown-headed cowbirds. *Animal Behaviour, 63,* 475–486.

Olsen, E. M., Heino, M., Lilly, G. R., Morgan, M. J., Brattey, J., Ernande, B., & Dieckmann, U. (2004). Maturation trends indicative of rapid evolution preceded the collapse of northern cod. *Nature, 428,* 932–935.

Olsen, K. L., & Whalen, R. E. (1982). Estrogen binds to hypothalamic nuclei of androgen-insensitive (Tfm) rats. *Experientia, 38,* 139–140.

Olson, J. A., Olson, J. M., Walsh, R. E., & Wisenden, B. D. (2012). A method to train groups of predator-naive fish to recognize and respond to predators when released into the natural environment. *North American Journal of Fisheries Management, 32,* 77–81.

Oppliger, A., Naciri-Graven, Y., Ribi, G., & Hosken, D. J. (2003). Sperm length influences fertilization success during sperm competition in the snail *Viviparus ater. Molecular Ecology, 12,* 485–492.

Ord, T. J., Martins, E. P., Thakur, S., Mane, K. K., & Borner, K. (2005). Trends in animal behaviour research (1968–2002): Ethoinformatics and the mining of library databases. *Animal Behaviour, 69,* 1399–1413.

Orians, G. (1962). Natural selection and ecological theory. *American Naturalist, 96,* 257–263.

Orians, G. (1969). On the evolution of mating systems in birds and mammals. *American Naturalist, 103,* 589–603.

Ortega, C. (1998). *Cowbirds and other brood parasites.* Tucson: University of Arizona Press.

Osborn, K. J., Haddock, S. H. D., Pleijel, F., Madin, L. P., & Rouse, G. W. (2009). Deep-sea, swimming worms with luminescent "bombs." *Science, 325,* 964.

Osuch, E. A., Ketter, T. A., Kimbrell, T. A., George, M. S., Benson, B. E., Willis, M. W., Herscovitch, P., & Post, R. M. (2000). Regional cerebral metabolism associated with anxiety symptoms in affective disorder patients. *Biological Psychiatry, 48,* 1020–1023.

Outlaw, D. C., & Voelker, G. (2006). Phylogenetic tests of hypotheses for the evolution of avian migration: A case study using the Motacillidae. *Auk, 123,* 455–466.

Outlaw, D. C., Voelker, G., Mila, B., & Girman, D. J. (2003). Evolution of long-distance migration in and historical biogeography of *Catharus* thrushes: A molecular phylogenetic approach. *Auk, 120,* 299–310.

Overduin-De Vries, A. M., Massen, J. J. M., Spruijt, B. M., & Sterck, E. H. M. (2012). Sneaky monkeys: An audience effect of male rhesus macaques (*Macaca mulatta*) on sexual behavior. *American Journal of Primatology, 74,* 217–228.

Overington, S. E., Cauchard, L., Cote, K. A., & Lefebvre, L. (2011). Innovative foraging behaviour in birds: What characterizes an innovator? *Behavioural Processes, 87,* 274–285.

Overington, S. E., Griffin, A. S., Sol, D., & Lefebvre, L. (2011). Are innovative species ecological generalists? A test in North American birds. *Behavioral Ecology, 22,* 1286–1293.

Overli, O., Sorensen, C., Pulman, K. G. T., Pottinger, T. G., Korzan, W., Summers, C. H., & Nilsson, G. E. (2007). Evolutionary background for stress-coping styles: Relationships between physiological, behavioral, and cognitive traits in non-mammalian vertebrates. *Neuroscience and Biobehavioral Reviews, 31,* 396–412.

Owings, D. H., & Coss, R. G. (1977). Snake mobbing by California ground squirrels: Adaptive variation and ontogeny. *Behaviour, 62,* 50–69.

Owings, D. H., & Leger, D. W. (1980). Chatter vocalization of California ground squirrels: Predator- and social-role specificity. *Zeitschrift fur Tierpsychologie, 54,* 163–184.

Oxley, P. R., & Oldroyd, B. P. (2010). The genetic architecture of honeybee breeding. *Advances in Insect Physiology, 39,* 83–118.

Packard, A. (1972). Cephalopods and fish: Limits of convergence. *Biological Reviews of the Cambridge Philosophical Society, 47,* 241–307.

Packer, C. (1977). Reciprocal altruism in *Papio anubis. Nature, 265,* 441–443.

Packer, C., Lewis, S., & Pusey, A. (1992). A comparative analysis of non-offspring nursing. *Animal Behaviour, 43,* 265–281.

Palagi, E., Antonacci, D., & Cordoni, G. (2007). Fine-tuning of social play in juvenile lowland gorillas (*Gorilla gorilla gorilla*). *Developmental Psychobiology, 49,* 433–445.

Palameta, B., & Lefebvre, L. (1985). The social transmission of a food-finding technique in pigeons: What is learned? *Animal Behaviour, 33,* 892–896.

Panksepp, J. (2005). Beyond a joke: From animal laughter to human joy? *Science, 308,* 62–63.

Panksepp, J., & Burgdorf, J. (2003). "Laughing" rats and the evolutionary antecedents of human joy? *Physiology & Behavior, 79,* 533–547.

Panksepp, J., Siviy, S., & Normansell, L. (1984). The psychobiology of play: Theoretical and methodological considerations. *Neuroscience and Biobehavioral Reviews, 8,* 465–492.

Papaj, D., & Lewis, A. (Eds.). (1993). *Insect learning: Ecological and evolutionary perspectives.* New York: Chapman and Hall.

Papaj, D., Lewis, A., & Prokopy, R. (1989). Ecological and evolutionary aspects of learning in phytophagus insects. *Annual Review of Entomology, 34,* 315–340.

Papworth, S., Bose, A. S., Barker, J., Schel, A. M., & Zuberbuhler, K. (2008). Male blue monkeys alarm call in response to danger experienced by others. *Biology Letters, 4,* 472–475.

Parker, G. (1970a). Sperm competition and its evolutionary consequences in insects. *Biological Reviews of the Cambridge Philosophical Society, 45,* 525–567.

Parker, G. (1970b). Sperm competition and its evolutionary effect on copula duration in the fly *Scatophagua stercoraria. Journal of Insect Physiology, 16,* 1301–1328.

Parker, G. A. (1974a). Assessment strategy and the evolution of fighting behaviour. *Journal of Theoretical Biology, 47,* 223–243.

Parker, G. A. (1974b). Courtship persistence and female-guarding as male investment strategies. *Behaviour, 48,* 157–184.

Parker, G. A. (2001). Golden flies, sunlit meadows: A tribute to the yellow dungfly. In L. A. Dugatkin (Ed.), *Model systems in behavioral ecology: Integrating conceptual, theoretical and empirical perspectives* (pp. 3–26). Princeton, NJ: Princeton University Press.

Parker, G. A., & Macnair, M. R. (1979). Models of parent-offspring conflict. 4. Suppression: Evolutionary retaliation by the parent. *Animal Behaviour, 27,* 1210–1235.

Parker, G. A., & Maynard Smith, J. (1987). The distribution of stay times in *Scatophaga*: A reply to Curtsinger. *American Naturalist, 129,* 621–628.

Parker, G. A., & Pizzari, T. (2010). Sperm competition and ejaculate economics. *Biological Reviews, 85,* 897–934.

Parker, G. A., & Stuart, R. A. (1976). Animal behavior as a strategy optimizer: Evolution of resource assessment strategies and optimal emigration thresholds. *American Naturalist, 110,* 1055–1076.

Parker, G. A., & Sutherland, W. J. (1986). Ideal free distribution when individuals differ in competitive ability: Phenotype-limited ideal free models. *Animal Behaviour, 34,* 1222–1242.

Parker, G. A., & Thompson, E. (1980). Dung fly struggles: A test of the war of attrition. *Behavioral Ecology and Sociobiology, 7,* 37–44.

Parker, P. G., Waite, T. A., Heinrich, B., & Marzluff, J. M. (1994). Do common ravens share ephemeral food resources with kin? DNA fingerprinting evidence. *Animal Behaviour, 48,* 1085–1093.

Parmesan, C. (2006). Ecological and evolutionary responses to recent climate change. *Annual Review of Ecology Evolution and Systematics, 37,* 637–669.

Patterson, E. M., & Mann, J. (2011). The ecological conditions that favor tool use and innovation in wild bottlenose dolphins (*Tursiops sp.*). *Plos One, 6.*

Paul, E. S., Harding, E. J., & Mendl, M. (2005). Measuring emotional processes in animals: The utility of a cognitive approach. *Neuroscience and Biobehavioral Reviews, 29,* 469–491.

Pavlov, I. P. (1927). *Conditioned reflexes.* New York: Oxford University Press.

Pedersen, C. A., & Boccia, M. L. (2002). Oxytocin maintains as well as initiates female sexual behavior: Effects of a highly selective oxytocin antagonist. *Hormones and Behavior, 41,* 170–177.

Pellegrini, A. D. (1995). Boys' rough-and-tumble play and social competence: Contemporaneous and longitudinal relations. In A. D. Pellegrini (Ed.), *The future of play theory: A multidisciplinary inquiry into the contributions of Brian Sutton-Smith* (pp. 107–126). Albany: State University of New York Press.

Pellegrini, A. D., & Smith, P. K. (1998). Physical activity play: The nature and function of a neglected aspect of play. *Child Development, 69,* 577–598.

Pellis, S., Casteneda, E., McKenna, M., Tran-Nguten, L., & Whishaw, I. B. (1993). The role of the striatum in organizing sequences of play fighting in neonatally dopamine depleted rats. *Neuroscience Letters, 158,* 13–15.

Pellis, S., & Iwaniuk, A. (1999). The roles of phylogeny and sociality in the evolution of social play in muroid rodents. *Animal Behaviour, 58,* 361–373.

Pellis, S., Pellis, V., & Kolb, B. (1992). Neonatal testosterone augmentation increases juvenile play fighting but does not influence the adult dominance relationships of male rats. *Aggressive Behavior, 18,* 437–442.

Pellis, S., Pellis, V., & Whishaw, I. B. (1992). The role of the cortex in play fighting by rats: Developmental and evolutionary implications. *Brain, Behavior and Evolution, 39,* 270–284.

Pellis, S. M. (2002). Sex differences in play fighting revisited: Traditional and nontraditional mechanisms of sexual differentiation in rats. *Archives of Sexual Behavior, 31,* 17–26.

Pellis, S. M., & Pellis, V. C. (2011). To whom the play signal is directed: A study of headshaking in black-handed spider monkeys (*Ateles geoffroyi*). *Journal of Comparative Psychology, 125*, 1–10.

Penn, D., & Potts, W. (1998). How do major histocompatibility complex genes influence odor and mating preferences? *Advances in Immunology, 69*, 411–436.

Penn, D., & Potts, W. (1999). The evolution of mating preferences and the major histocompatibility complex genes. *American Naturalist, 153*, 145–164.

Pepper, J. W., Braude, S. H., Lacey, E. A., & Sherman, P. W. (1991). Vocalizations of the naked mole-rat. In P. W. Sherman, J. Jarvis, & R. D. Alexander (Eds.), *The biology of the naked mole-rat* (pp. 243–274). Princeton, NJ: Princeton University Press.

Pereira, M. E., Rosat, R., Huang, C. H., Godoy, M. G. C., & Izquierdo, I. (1989). Inhibition by diazepam of the effect of additional training and of extinction on the retention of shuttle avoidance behavior in rats. *Behavioral Neuroscience, 103*, 202–205.

Perez, S. M., Taylor, O. R., & Jander, R. (1997). A sun compass in monarch butterflies. *Nature, 387*, 29.

Pervin, L., & John, O. P. (1997). *Personality: Theory and research.* New York: Wiley.

Petrie, M., Krupa, A., & Burke, T. (1999). Peacocks lek with relatives even in the absence of social and environmental cues. *Nature, 401*, 155–157.

Petru, M., Spinka, M., Charvatova, V., & Lhota, S. (2009). Revisiting play elements and self-handicapping in play: A comparative ethogram of five old world monkey species. *Journal of Comparative Psychology, 123*, 250–263.

Pfennig, D. W. (1995). Absence of joint nesting advantage in desert seed harvester ants: Evidence from a field experiment. *Animal Behaviour, 49*, 567–575.

Pfennig, D. W. (1999). Cannibalistic tadpoles that pose the greatest threat to kin are most likely to discriminate kin. *Proceedings of the Royal Society of London, Series B–Biological Sciences, 266*, 57–61.

Pfennig, D. W., Collins, J. P., & Ziemba, R. E. (1999). A test of alternative hypotheses for kin recognition in cannibalistic tiger salamanders. *Behavioral Ecology, 10*, 436–443.

Pfennig, D. W., Reeve, H. K., & Sherman, P. W. (1993). Kin recognition and cannibalism in spadefoot toads. *Animal Behaviour, 46*, 87–94.

Pfennig, D. W., & Sherman, P. W. (1995, June). Kin recognition. *Scientific American*, 98–103.

Pfennig, D. W., Wund, M. A., Snell-Rood, E. C., Cruickshank, T., Schlichting, C. D., & Moczek, A. P. (2010). Phenotypic plasticity's impacts on diversification and speciation. *Trends in Ecology & Evolution, 25*, 459–467.

Pfungst, O. (1911). *Clever Hans: The horse of Mr. Von Osten.* New York: Holt, Rinehart & Winston.

Phelps, S. M., & Young, L. J. (2003). Extraordinary diversity in vasopressin (V1a) receptor distributions among wild prairie voles (*Microtus ochrogaster*): Patterns of variation and covariation. *Journal of Comparative Neurology, 466*, 564–576.

Philipp, D., & Gross, M. (1994). Genetic evidence for cuckoldry in bluegill *Lepomis macrochirus. Molecular Ecology, 3*, 563–569.

Pierce, N., Kitchling, R., Buckley, R., Talor, M., & Benbow, K. (1987). The costs and benefits of cooperation between the Australian lycaenid butterfly, *Jalmenus evagoras*, and its attendant ants. *Behavioral Ecology and Sociobiology, 21*, 237–248.

Pierce, N. E., Braby, M. F., Heath, A., Lohman, D. J., Mathew, J., Rand, D. B., & Travassos, M. A. (2002). The ecology and evolution of ant association in the *Lycaenidae* (Lepidoptera). *Annual Review of Entomology, 47*, 733–771.

Pietrewicz, A., & Richards, J. (1985). Learning to forage: An ecological perspective. In T. Johnston & A. Pietrewicz (Eds.), *Issues in the ecological study of learning* (pp. 99–119). Hillsdale, NJ: Lawrence Erlbaum Associates.

Pietrewicz, A. T., & Kamil, A. C. (1979). Search image formation in the blue jay (*Cyanocitta cristata*). *Science, 204*, 1332–1333.

Pika, S., & Bugnyar, T. (2011). The use of referential gestures in ravens (*Corvus corax*) in the wild. *Nature Communications, 2.*

Pitcher, T. (1986). Functions of shoaling behaviour. In T. Pitcher (Ed.), *The behavior of teleost fishes* (pp. 294–338). Baltimore: John Hopkins University Press.

Pitcher, T. (1992). Who dares wins: The function and evolution of predator inspection behaviour in shoaling fish. *Netherlands Journal of Zoology, 42*, 371–391.

Pitcher, T. J., Green, D. A., & Magurran, A. E. (1986). Dicing with death: Predator inspection behavior in minnow shoals. *Journal of Fish Biology, 28*, 439–448.

Plath, M., Kromuszczynski, K., & Tiedemann, R. (2009). Audience effect alters male but not female mating preferences. *Behavioral Ecology and Sociobiology, 63*, 381–390.

Platzen, D., & Magrath, R. D. (2005). Adaptive differences in response to two types of parental alarm call in altricial nestlings. *Proceedings of the Royal Society of London, Series B, 272*, 1101–1106.

Pleszczynska, W. K., & Hansell, R. (1980). Polygeny and decision theory: Testing of a model in lark buntings, *Calamospiza melanocorys. American Naturalist, 116*, 821–830.

Plomin, R., DeFries, J. C., McClearn, G. E., & McGuffin, P. (2008). Behavioral genetics (5th ed.). New York: Worth.

Plotnik, J. M., Lair, R., Suphachoksahakun, W., & de Waal, F. B. M. (2011). Elephants know when they need a helping trunk in a cooperative task. *Proceedings of the National Academy of Sciences, U. S. A., 108*, 5116–5121.

Podos, J., Huber, S. K., & Taft, B. (2004). Bird song: The interface of evolution and mechanism. *Annual Review of Ecology, Evolution, and Systematics, 35*, 55–87.

Podos, J., & Warren, P. S. (2007). The evolution of geographic variation in birdsong. *Advances in the Study of Behavior, 37*, 403–458.

Poirier, F. E., Stini, W. A., & Wreden, K. B. (1994). *In search of ourselves: An introduction to physical anthropology.* New York: Prentice Hall.

Polhemus, J. (1990). Surface wave communication in water striders. Field observations of unreported taxa (Heteroptera: Gerridae, Veliidae). *Journal of the New York Entomological Society, 98*, 383–384.

Polhemus, J., & Karunaratne, P. (1993). A review of the genus *Rhagadotarsus*, with descriptions of three new species (Heteroptera: Gerridae). *Raffles Bulletin of Zoology, 41*, 95–112.

Pollard, K. A., & Blumstein, D. T. (2012). Evolving communicative complexity: Insights from rodents and beyond. *Philosophical Transactions of the Royal Society, Series B–Biological Sciences, 367,* 1869–1878.

Pollet, T., & Nettle, D. (2009). Market forces affect patterns of polygyny in Uganda. *Proceedings of the National Academy of Sciences, U. S. A., 106,* 2114–2117.

Pollock, G. B., Cabrales, A., & Rissing, S. W. (2004). On suicidal punishment among *Acromyrmex versicolor* co-foundresses: The disadvantage in personal advantage. *Evolutionary Ecology Research, 6,* 891–917.

Poppleton, F. (1957). The birth of an elephant. *Oryx, 4,* 180–181.

Poran, N. S., & Coss, R. G. (1990). Development of antisnake defenses in California ground squirrels (*Spermophilus beecheyi*): I. Behavioral and immunological relationships. *Behaviour, 112,* 222–245.

Portmann, A. (1947). Etudes sur la cerebralisation des oiseaux: Les indices intra-cerebraux. *Alauda, 15,* 1–15.

Potegal, M., & Einon, D. (1989). Aggressive behaviors in adult rats deprived of playfighting experience as juveniles. *Developmental Psychobiology, 22,* 159–172.

Poulsen, M., & Currie, C. R. (2009). On ants, plants and fungi. *New Phytologist, 182,* 785–788.

Poundstone, W. (1992). *Prisoner's dilemma: Jon Von Neuman, game theory and the puzzle of the bomb.* New York: Doubleday.

Power, T. G. (2000). *Play and exploration in children and animals.* Hillsdale, NJ: Lawrence Erlbaum Associates.

Pravosudov, V. V., & Clayton, N. S. (2002). A test of the adaptive specialization hypothesis: Population differences in caching, memory, and the hippocampus in black-capped chickadees (*Poecile atricapilla*). *Behavioral Neuroscience, 116,* 515–522.

Pravosudov, V. V., & Grubb, T. C. (1997). Energy management of passerine birds during the non-breeding season. *Current Ornithology, 14,* 189–234.

Pravosudov, V. V., Kitaysky, A. S., Wingfield, J. C., & Clayton, N. S. (2001). Long-term unpredictable foraging conditions and physiological stress response in mountain chickadees (*Poecile gambeli*). *General and Comparative Endocrinology, 123,* 324–331.

Prelle, I., Phillips, C. J. C., da Costa, M. J. P., Vandenberghe, N. C., & Broom, D. M. (2004). Are cows that consistently enter the same side of a two-sided milking parlour more fearful of novel situations or more competitive? *Applied Animal Behaviour Science, 87,* 193–203.

Prete, F. R. (1990). The conundrum of the honey-bees: One impediment to the publication of Darwin's theory. *Journal of the History of Biology, 23,* 271–290.

Prete, F. R. (1991). Can females rule the hive: The controversy over honey-bee gender roles in British beekeeping texts of the 16th–18th centuries. *Journal of the History of Biology, 24,* 113–144.

Price, J. J., & Lanyon, S. M. (2002a). Reconstructing the evolution of complex bird song in the oropendolas. *Evolution, 56,* 1514–1529.

Price, J. J., & Lanyon, S. M. (2002b). A robust phylogeny of the oropendolas: Polyphyly revealed by mitochondrial sequence data. *Auk, 119,* 335–348.

Price, J. J., & Lanyon, S. M. (2004a). Patterns of song evolution and sexual selection in the oropendolas and caciques. *Behavioral Ecology, 15,* 485–497.

Price, J. J., & Lanyon, S. M. (2004b). Song and molecular data identify congruent but novel affinities of the green oropendola (*Psarocolius viridis*). *Auk, 121,* 224–229.

Price, T., & Schulter, D. (1991). On the low heritability of life-history traits. *Evolution, 45,* 853–861.

Price, T., Schulter, D., & Heckman, N. E. (1993). Sexual selection when the female benefits directly. *Biological Journal of the Linnean Society, 48,* 187–211.

Price, T. A. R., Hodgson, D. J., Lewis, Z., Hurst, G. D. D., & Wedell, N. (2008). Selfish genetic elements promote polyandry in a fly. *Science, 322,* 1241–1243.

Profet, M. (1993). Menstruation as a defense against pathogens transported by sperm. *Quarterly Review of Biology, 68,* 335–386.

Pruett-Jones, S., & Lewis, M. (1990). Sex ratio and habitat limitation promote delayed dispersal in superb fairy-wrens. *Nature, 348,* 541–542.

Pruett-Jones, S. G. (1992). Independent versus non-independent mate choice: Do females copy each other? *American Naturalist, 140,* 1000–1009.

Pulido, F. (2007). The genetics and evolution of avian migration. *Bioscience, 57,* 165–174.

Pulido, F., Berthold, P., Mohr, G., & Querner, U. (2001). Heritability of the timing of autumn migration in a natural bird population. *Proceedings of the Royal Society of London, Series B, 268,* 953–959.

Pulliam, R. (1973). On the advantages of flocking. *Journal of Theoretical Biology, 38,* 419–422.

Purves, D., & Lichtman, J. (1980). Elimination of synapses in the developing nervous system. *Science, 210,* 153–157.

Purvis, K., Haug, E., Clausen, O. P. F., Naess, O., & Hansson, V. (1977). Endocrine status of testicular feminized male (Tfm) rats. *Molecular and Cellular Endocrinology, 8,* 317–334.

Pysh, J., & Weiss, G. (1979). Exercise during development induces an increase in Purkinje cell dendritic tree size. *Science, 206,* 230–231.

Quader, S. (2005). Mate choice and its implications for conservation and management. *Current Science, 89,* 1220–1229.

Quek, S. P., Davies, S. J., Ashton, P. S., Itino, T., & Pierce, N. E. (2007). The geography of diversification in mutualistic ants: A gene's-eye view into the neogene history of Sundaland rain forests. *Molecular Ecology, 16,* 2045–2062.

Queller, D. C. (1992). Quantitative genetics, inclusive fitness and group selection. *American Naturalist, 139,* 540–558.

Quigley, H., & Cragshaw, P. (1992). A conservation plan for the jaguar (*Panthera onca*) in the Pantanal region of Brazil. *Biological Conservation, 61,* 149–157.

Quinn, J. S., Woolfenden, G. E., Fitzpatrick, J. W., & White, B. N. (1999). Multi-locus DNA fingerprinting supports genetic monogamy in Florida scrub-jays. *Behavioral Ecology and Sociobiology, 45,* 1–10.

Raby, C. R., Alexis, D. M., Dickinson, A., & Clayton, N. S. (2007). Planning for the future by western scrub-jays. *Nature, 445,* 919–921.

Raby, C. R., & Clayton, N. S. (2009). Prospective cognition in animals. *Behavioural Processes, 80,* 314–324.

Radford, A. N. (2003). Territorial vocal rallying in the green wood-hoopoe: Influence of rival group size and composition. *Animal Behaviour, 66*, 1035–1044.

Radford, A. N. (2008). Duration and outcome of intergroup conflict influences intragroup affiliative behaviour. *Proceedings of the Royal Society, Series B–Biological Sciences, 275*, 2787–2791.

Radford, A. N. (2011). Preparing for battle? Potential intergroup conflict promotes current intragroup affiliation. *Biology Letters, 7*, 26–29.

Raleigh, M., & McGuire, M. (1991). Bidirectional relationships between tryptophan and social behavior in vervet monkeys. In R. Schwarcz, S. N. Young, & R. R. Brown (Eds.), *Kynurenine and serotonin pathway: Progress in tryptophan research, advances in experimental medicine and biology* (pp. 289–298). New York: Plenum Press.

Raleigh, M., McGuire, M., Brammer, G., Pollack, D., & Yuwiler, A. (1991). Serotonergic mechanisms promote dominance in adult vervet monkeys. *Brain Research, 559*, 181–190.

Ramenofsky, M., & Wingfield, J. C. (2007). Regulation of migration. *Bioscience, 57*, 135–143.

Rammensee, H., Bachmann, J., & Stefanovic, S. (1997). *MHC ligands and peptide motifs*. Georgetown, TX: Landes Biosciences.

Rands, S. A. (2011). Approximating optimal behavioural strategies down to rules-of-thumb: Energy reserve changes in pairs of social foragers. *Plos One, 6*.

Ratnieks, F. L. (1995). Evidence for queen-produced egg-marking pheromone and its use in worker policing in the honey bee. *Journal of Apicultural Research, 34*, 31–37.

Ratnieks, F. L., & Visscher, P. K. (1988). Reproductive harmony via mutual policing by workers in eusocial hymenoptera. *American Naturalist, 132*, 217–236.

Ratnieks, F. L., & Visscher, P. K. (1989). Worker policing in the honeybee. *Nature, 342*, 796–797.

Ratnieks, F. L. W., Foster, K. R., & Wenseleers, T. (2011). Darwin's special difficulty: The evolution of "neuter insects" and current theory. *Behavioral Ecology and Sociobiology, 65*, 481–492.

Ratnieks, F. L. W., & Helantera, H. (2009). The evolution of extreme altruism and inequality in insect societies. *Philosophical Transactions of the Royal Society, Series B–Biological Sciences, 364*, 3169–3179.

Rattenborg, N., Lima, S., & Amlaner, C. (1999a). Facultative control of avian unihemispheric sleep under the risk of predation. *Behavioral Brain Research, 105*, 163–172.

Rattenborg, N., Lima, S., & Amlaner, C. (1999b). Half-awake to the risk of predation. *Nature, 397*, 397–398.

Rattenborg, N. C., Amlaner, C. J., & Lima, S. L. (2000). Behavioral, neurophysiological and evolutionary perspectives on unihemispheric sleep. *Neuroscience and Biobehavioral Reviews, 24*, 817–842.

Rattenborg, N. C., Martinez-Gonzales, D., Roth, T. C., & Pravosudov, V. V. (2011). Hippocampal memory consolidation during sleep: A comparison of mammals and birds. *Biological Reviews, 86*, 658–691.

Raven, P. H., & Johnson, G. G. (1989). *Biology*. St. Louis, MO: Times Mirror/Mosby College.

Reader, S. M. (2004). Don't call me clever. *New Scientist, 183*, 34–37.

Reader, S. M., Kendal, J. R., & Laland, K. N. (2003). Social learning of foraging sites and escape routes in wild Trinidadian guppies. *Animal Behaviour, 66*, 729–739.

Reader, S. M., & Laland, K. (Eds.). (2003). *Animal innovation*. Oxford: Oxford University Press.

Reader, S. M., & Laland, K. N. (2002). Social intelligence, innovation and enhanced brain size in primates. *Proceedings of the National Academy of Sciences, U.S.A., 99*, 4436–4441.

Real, L., & Caraco, T. (1986). Risk and foraging in stochastic environments. *Annual Review of Ecology and Systematics, 17*, 371–390.

Reale, D., Dingemanse, N. J., Kazem, A. J. N., & Wright, J. (2010). Evolutionary and ecological approaches to the study of personality. *Philosophical Transactions of the Royal Society, Series B–Biological Sciences, 365*, 3937–3946.

Reale, D., Reader, S. M., Sol, D., McDougall, P. T., & Dingemanse, N. J. (2007). Integrating animal temperament within ecology and evolution. *Biological Reviews, 82*, 291–318.

Reby, D., McComb, K., Cargnelutti, B., Darwin, C., Fitch, W. T., & Clutton-Brock, T. (2005). Red deer stags use formants as assessment cues during intrasexual agonistic interactions. *Proceedings of the Royal Society of London, Series B, 272*, 941–947.

Reeve, H. K. (1989). The evolution of conspecific acceptance thresholds. *American Naturalist, 133*, 407–435.

Reeve, H. K. (1992). Queen activation of lazy workers in colonies of the eusocial naked mole-rat. *Nature, 358*, 147–149.

Reeve, H. K., & Ratnieks, F. L. (1993). Queen-queen conflicts in polygynous societies: Mutual tolerance and reproductive skew. In L. Keller (Ed.), *Queen number and sociality in insects* (pp. 45–86). Oxford: Oxford University Press.

Reeve, H. K., & Sherman, P. W. (1991). Intracolonial aggression and nepotism by the breeding female naked mole rat. In P. W. Sherman, J. Jarvis, & R. D. Alexander (Eds.), *The biology of the naked mole-rat* (pp. 337–357). Princeton, NJ: Princeton University Press.

Reeve, H. K., & Sherman, P. W. (1993). Adaptation and the goals of evolutionary research. *Quarterly Review of Biology, 68*, 1–32.

Reeve, H. K., Westneat, D. F., Noon, W. A., Sherman, P. W., & Aquadro, C. F. (1990). DNA "fingerprinting" reveals high levels of inbreeding in colonies of the eusocial naked mole rat. *Proceedings of the National Academy of Sciences, U.S.A., 87*, 2496–2500.

Reeve, H. K., Westneat, D. F., & Queller, D. C. (1992). Estimating average within-group relatedness from DNA fingerprints. *Molecular Ecology, 2*, 223–232.

Reid, P. J., & Shettleworth, S. J. (1992). Detection of cryptic prey: Search image or search rate. *Journal of Experimental Psychology: Animal Behavior Processes, 18*, 273–286.

Remage-Healey, L., & Bass, A. H. (2007). Plasticity in brain sexuality is revealed by the rapid actions of steroid hormones. *Journal of Neuroscience, 27*, 1114–1122.

Remage-Healey, L., Nowacek, D. P., & Bass, A. H. (2007). Dolphin foraging sounds suppress calling and elevate stress hormone levels in a prey species, the Gulf toadfish. *Journal of Experimental Biology, 209*, 4444–4451.

Reppert, S., Gegear, R., & Merlin, C. (2010). Navigational mechanisms of migrating monarch butterflies. *Trends in Neurosciences, 33*, 399–406.

Reusch, T. B. H., Haberli, M. A., Aeschlimann, P. B., & Milinski, M. (2001). Female sticklebacks count alleles in a strategy of sexual selection explaining MHC polymorphism. *Nature, 414,* 300–302.

Reznick, D. (1996). Life history evolution in guppies: A model system for the empirical study of adaptation. *Netherlands Journal of Zoology, 46,* 172–190.

Reznick, D., Brygaa, H., & Endler, J. (1990). Experimentally induced life-history evolution in a natural population. *Nature, 346,* 357–359.

Rhodes, S. B., & Schlupp, I. (2012). Rapid and socially induced change of a badge of status. *Journal of Fish Biology, 80,* 722–727.

Ribas, C. C., & Miyaki, C. Y. (2004). Molecular systematics in Aratinga parakeets: Species limits and historical biogeography in the "solstitialis" group, and the systematic position of Nandayus nenday. *Molecular Phylogenetics and Evolution, 30,* 663–675.

Richards, R. J. (1987). *Darwin and the emergence of evolutionary theories of mind and behavior.* Chicago: University of Chicago Press.

Ridley, M. (1996). *Evolution* (2nd ed.). Oxford: Blackwell Science.

Riechert, S. (1998). Game theory and animal contests. In L. A. Dugatkin & H. K. Reeve (Eds.), *Game theory and animal behavior* (pp. 64–93). New York: Oxford University Press.

Rilling, J., Gutman, D., Zeh, T., Pagnoni, G., Berns, G., & Kilts, C. (2002). A neural basis for social cooperation. *Neuron, 35,* 395–405.

Rilling, J. K., DeMarco, A. C., Hackett, P. D., Thompson, R., Ditzen, B., Patel, R., & Pagnoni, G. (2012). Effects of intranasal oxytocin and vasopressin on cooperative behavior and associated brain activity in men. *Psychoneuroendocrinology, 37,* 447–461.

Rilling, J. K., & Sanfey, A. G. (2011). The neuroscience of social decision-making. *Annual Review of Psychology, 62,* 23–48.

Rissing, S., & Pollock, G. (1986). Social interaction among pleometric queens of *Veromessor pergandei* during colony foundation. *Animal Behaviour, 34,* 226–234.

Rissing, S., & Pollock, G. (1987). Queen aggression, pleometric advantage and brood raiding in the ant *Veromessor pergandei. Animal Behaviour, 35,* 975–982.

Rissing, S., & Pollock, G. (1991). An experimental analysis of pleometric advantage in *Messor pergandei. Insectes Sociaux, 63,* 205–211.

Rissing, S., Pollock, G., Higgins, M., Hagen, R., & Smith, D. (1989). Foraging specialization without relatedness or dominance among co-founding ant queens. *Nature, 338,* 420–422.

Rissing, S. W., Pollock, G. B., & Higgins, M. R. (1996). Fate of ant foundress associations containing "cheaters." *Naturwissenschaften, 83,* 182–185.

Ritchie, M. G., Halsey, E. J., & Gleason, J. M. (1999). Drosophila song as a species-specific mating signal and the behavioural importance of Kyriacou & Hall cycles in *D. melanogaster* song. *Animal Behaviour, 58,* 649–657.

Rizzolatti, G., & Craighero, L. (2004). The mirror-neuron system. *Annual Review of Neuroscience, 27,* 169–192.

Rizzolatti, G., Fadiga, L., Gallese, V., & Fogassi, L. (1996). Premotor cortex and the recognition of motor actions. *Cognitive Brain Research, 3,* 131–141.

Rizzolatti, G., Fogassi, L., & Gallese, V. (2001). Neurophysiological mechanisms underlying the understanding and imitation of action. *Nature Reviews Neuroscience, 2,* 661–670.

Rizzolatti, G., Fogassi, L., & Gallese, V. (2006). Mirrors in the mind. *Scientific American, 295,* 54–61.

Robbins, M. M. (1996). Male-male interactions in heterosexual and all-male wild mountain gorilla groups. *Ethology, 102,* 942–965.

Roberts, M., & Shapiro, M. (2002). NMDA receptor antagonists impair memory for nonspatial, socially transmitted food preference. *Behavioral Neuroscience, 116,* 1059–1069.

Roberts, R. L., Williams, J. R., Wang, A. K., & Carter, C. S. (1998). Cooperative breeding and monogamy in prairie voles: Influence of the sire and geographical variation. *Animal Behaviour, 55,* 1131–1140.

Robinson, G. (1985). Effects of a juvenile hormone analog on honey bee foraging behavior and alarm pheromone. *Journal of Insect Physiology, 31,* 277–282.

Robinson, G. (1987). Modulation of alarm pheromone perception in the honey bee: Evidence for division of labor based on hormonally regulated response thresholds. *Journal of Comparative Physiology. A, 160,* 613–619.

Robinson, G. E. (1992). Regulation of division of labor in insect societies. *Annual Review of Ecology and Systematics, 37,* 637–665.

Robinson, G. E., Grozinger, C. M., & Whitfield, C. W. (2005). Sociogenomics: Social life in molecular terms. *Nature Reviews Genetics, 6,* 257–271.

Robinson, G. E., Page, R. E., Strambi, C., & Strambi, A. (1989). Hormonal and genetic control of behavioral integration in honeybee colonies. *Science, 246,* 109–112.

Robinson, S., & Smotherman, W. (1991). Fetal learning: Implications for the development of kin. In P. G. Hepper (Ed.), *Kin recognition* (pp. 308–334). Cambridge: Cambridge University Press.

Roces, F., & Holldobler, B. (1995). Vibrational communication between hitchhikers and foragers in leaf-cutting ants (*Atta cephalotes*). *Behavioral Ecology and Sociobiology, 37,* 297–302.

Roces, F., & Holldobler, B. (1996). Use of stridulation in foraging leaf-cutting ants: Mechanical support during cutting or short-range recruitment signal? *Behavioral Ecology and Sociobiology, 39,* 293–299.

Roces, F., Tautz, J., & Holldobler, B. (1993). Stridulation in leaf-cutting ants: Short-range recruitment through plant-borne substances. *Naturwissenschaften, 80,* 521–524.

Rodd, F. H., Hughes, K. A., Grether, G. F., & Baril, C. T. (2002). A possible non-sexual origin of mate preference: Are male guppies mimicking fruit? *Proceedings of the Royal Society of London, Series B, 269,* 475–481.

Rodel, H. G., Monclus, R., & von Holst, D. (2006). Behavioral styles in European rabbits: Social interactions and responses to experimental stressors. *Physiology and Behavior, 89,* 180–188.

Roeder, T. (1999). Octopamine in invertebrates. *Progress in Neurobiology, 59,* 533–561.

Roeloffs, R., & Riechert, S. E. (1988). Dispersal and population genetic structure of the cooperative spider, *Agelena consociata* in West African rainforests. *Evolution, 42,* 173–183.

Roguet, C., Dumont, B., & Prache, S. (1998). Selection and use of feeding sites and feeding stations by herbivores: A review. *Annales De Zootechnie, 47,* 225–244.

Rohwer, S. (1982). The evolution of reliable and unreliable badges of fighting ability. *American Zoologist, 22,* 531–546.

Rollenhagen, A., & Bischof, H. J. (1991). Rearing conditions affect neuron morphology in a telecephalic area of the zebra finch. *Neuroreport, 2,* 711–714.

Rollenhagen, A., & Bischof, H. J. (1994). Phase specific morphological changes induced by social experience in two forebrain areas of the zebra finch. *Behavioural Brain Research, 65,* 83–88.

Romanes, G. J. (1884). *Mental evolution in animals.* New York: AMS Press.

Romanes, G. J. (1889). *Mental evolution in man.* New York: Appleton.

Romanes, G. J. (1898). *Animal intelligence* (7th ed.). London: Kegan Paul, Trench, Trubner and Co.

Ronce, R. (2007). How does it feel to be like a rolling stone? Ten questions about dispersal evolution. *Annual Review of Ecology and Systematics, 38,* 231–253.

Roper, K., Kaiser, D., & Zentall, T. (1995). True directed forgetting in pigeons may occur only when alternative working memory is required on forget-cue trials. *Animal Learning and Behavior, 23,* 280–285.

Roper, T. J. (1986). Badges of status in avian societies. *New Scientist, 109,* 38–40.

Rosenbaum, R. S., Kohler, S., Schacter, D. L., Moscovitch, M., Westmacott, R., Black, S. E., Gao, F. Q., & Tulving, E. (2005). The case of KC: Contributions of a memory-impaired person to memory theory. *Neuropsychologia, 43,* 989–1021.

Rosenthal, T., & Zimmerman, B. (1978). *Social learning and cognition.* New York: Academy Press.

Rosenzweig, M. (1981). A theory of habitat selection. *Ecology, 62,* 327–335.

Rosenzweig, M. (1985). Some theoretical aspects of habitat selection. In M. Cody (Ed.), *Habitat selection in birds* (pp. 517–540). New York: Academic Press.

Rosenzweig, M. (1990). Do animals choose habitats? In M. Bekoff & D. Jamieson (Eds.), *Interpretation and explanation in the study of animal behaviour* (pp. 157–179). Boulder, CO: Westview Press.

Rosenzweig, M. (1991). Habitat selection and population interactions. *American Naturalist, 137,* S5–S28.

Roth, T. C., LaDage, L. D., & Pravosudov, V. V. (2010). Learning capabilities enhanced in harsh environments: A common garden approach. *Proceedings of the Royal Society, Series B–Biological Sciences, 277,* 3187–3193.

Rothstein, S., & Pirotti, R. (1987). Distinctions among reciprocal altruism, kin selection and cooperation and a model for the initial evolution of beneficent behavior. *Ethology and Sociobiology, 9,* 189–209.

Rowden, J. (1996). The evolution of display behavior in the parrot genus *Neophema* (Aves: Psittaciformes). Durham, NC: Duke University.

Rowland, W. J., & Sevenster, P. (1985). Sign stimuli in three-spined sticklebacks, (*Gasterosteus aculeatus*): A reexamination of some classic experiments. *Behaviour, 93,* 241–257.

Roy, R., Graham, S., & Peterson, J. (1988). Fiber type composition of the plantaflexors of giraffes (*Giriffa camelopardalis*) at different postnatal stages of development. *Comparative Biochemistry and Physiology A–Physiology, 91,* 347–352.

Rueppell, O. (2009). Characterization of quantitative trait loci for the age of first foraging in honey bee workers. *Behavior Genetics, 39,* 541–553.

Russell, A. F., Clutton-Brock, T. H., Brotherton, P. N. M., Sharpe, L. L., McIlrath, G. M., Dalerum, F. D., Cameron, E. Z., & Barnard, J. A. (2002). Factors affecting pup growth and survival in co-operatively breeding meerkats *Suricata suricatta. Journal of Animal Ecology, 71,* 700–709.

Russello, M., & Amato, G. (2003). A molecular phylogeny of *Amazona*: Implications for Neotropical biogeography, taxonomy and conservation. *Molecular Phylogenetics and Evolution, 30,* 421–437.

Rutte, C., Taborsky, M., & Brinkhof, M. W. G. (2006). What sets the odds of winning and losing? *Trends in Ecology & Evolution, 21,* 16–21.

Ruxton, G., Sherrat, T., & Speed, M. (2004). *Avoiding attack: The evolutionary ecology of crypsis, warning signals and mimicry.* Oxford: Oxford University Press.

Ruzzante, D. E. (1994). Domestication effects on aggressive and schooling behavior in fish. *Aquaculture, 120,* 1–24.

Ryan, B. C., & Vandenbergh, J. G. (2002). Intrauterine position effects. *Neuroscience and Biobehavioral Reviews, 26,* 665–678.

Ryan, K. K., & Altmann, J. (2001). Selection for male choice based primarily on mate compatibility in the oldfield mouse, *Peromyscus polionotus rhoadsi. Behavioral Ecology and Sociobiology, 50,* 436–440.

Ryan, M. (1985). *The tungara frog: A study in sexual selection and communication.* Chicago: University of Chicago Press.

Ryan, M. J. (1990). Sexual selection, sensory systems and sensory exploitation. *Oxford Surveys in Evolutionary Biology, 7,* 157–195.

Ryan, M. J., Fox, J., Wilczynski, W., & Rand, S. (1990). Sexual selection for sensory exploitation in the frog, *Physaleamus pustulosus. Nature, 343,* 66–67.

Ryan, M. J., & Rand, A. S. (1993). Sexual selection and signal evolution: Ghost of biases past. *Philosophical Transactions of the Royal Society of London, Series B, 340,* 187–195.

Ryti, R. T., & Case, T. J. (1984). Spatial arrangement and diet overlap between colonies of desert ants. *Oecologia, 62,* 401–404.

Sachs, J., Mueller, U. G., Wilcox, T., & Bull, J. J. (2004). The evolution of cooperation. *Quarterly Review of Biology, 79,* 135–160.

Saetre, P., Strandberg, E., Sundgren, P. E., Pettersson, U., Jazin, E., & Bergstrom, T. F. (2006). The genetic contribution to canine personality. *Genes, Brain and Behavior, 5,* 240–248.

Salamone, J. (1994). The involvement of nucleus accumbens dopamine in appetitive and aversive motivation. *Behavioural Brain Research, 61,* 117–133.

Salcedo, E., Zheng, L. J., Phistry, M., Bagg, E. E., & Britt, S. G. (2003). Molecular basis for ultraviolet vision in invertebrates. *Journal of Neuroscience, 23,* 10873–10878.

Salewsk, V., & Bruderer, B. (2007). The evolution of bird migration: A synthesis. *Naturwissenschaften, 4,* 268–279.

Sandi, C. (2004). Stress, cognitive impairment and cell adhesion molecules. *Nature Reviews Neuroscience, 5,* 917–930.

Sanfey, A. G., Loewenstein, G., McClure, S. M., & Cohen, J. D. (2006). Neuroeconomics: Cross-currents in research on decision-making. *Trends in Cognitive Sciences, 10,* 108–116.

Sansone, M. (1975). Benzodiazepines and amphetamines: Avoidance behavior in mice. *Archives Internationales de Pharmacodynamie et de Therapie, 218,* 125–132.

Sapolsky, R. M. (1999). Glucocorticoids, stress, and their adverse neurological effects: Relevance to aging. *Experimental Gerontology, 34,* 721–732.

Sapolsky, R. M. (2003). Stress and plasticity in the limbic system. *Neurochemical Research, 28,* 1735–1742.

Sapolsky, R. M., Romero, L. M., & Munck, A. U. (2000). How do glucocorticoids influence stress responses? Integrating permissive, suppressive, stimulatory, and preparative actions. *Endocrine Reviews, 21,* 55–89.

Sargeant, B. L., & Mann, J. (2009). Developmental evidence for foraging traditions in wild bottlenose dolphins. *Animal Behaviour, 78,* 715–721.

Sargeant, B. L., Mann, J., Berggren, P., & Krutzen, M. (2005). Specialization and development of beach hunting, a rare foraging behavior, by wild bottlenose dolphins (*Tursiops sp.*). *Canadian Journal of Zoology, 83,* 1400–1410.

Sarti Oliveira, A. F., Rossi, A. O., Romualdo Silva, L. F., Lau, M. C., & Barreto, R. E. (2010). Play behaviour in nonhuman animals and the animal welfare issue. *Journal of Ethology, 28,* 1–5.

Sasvari, L. (1985). Keypeck conditioning with reinforcement in two different locations in thrush, tit and sparrow species. *Behavioural Processes, 11,* 245–252.

Saulitis, E., Matkin, C., Barrett-Lennard, L., Heise, K., & Ellis, G. (2000). Foraging strategies of sympatric killer whale (*Orcinus orca*) populations in Prince William Sound, Alaska. *Marine Mammal Science, 16,* 94–109.

Saunders, I., Sayer, M., & Goodale, A. (1999). The relationship between playfulness and coping in preschool children: A pilot study. *American Journal of Occupational Therapy, 53,* 221–226.

Sawaguchi, T., & Kudo, H. (1990). Neocortical development and social structure in primates. *Primates,* 283–290.

Sawyer, H., & Kauffman, M. J. (2011). Stopover ecology of a migratory ungulate. *Journal of Animal Ecology, 80,* 1078–1087.

Sawyer, H., Kauffman, M. J., Nielson, R. M., & Horne, J. S. (2009). Identifying and prioritizing ungulate migration routes for landscape-level conservation. *Ecological Applications, 19,* 2016–2025.

Schaller, G. (1967). *The deer and the tiger.* Chicago: University of Chicago Press.

Scheiner, R., Pluckhahn, S., Oney, B., Blenau, W., & Erber, J. (2002). Behavioural pharmacology of octopamine, tyramine and dopamine in honey bees. *Behavioural Brain Research, 136,* 545–553.

Schenkel, R. (1966). Play, exploration and territoriality in the wild lion. *Symposia of the Zoological Society of London, 18,* 11–22.

Schilcher, F. (1976a). The function of pulse song and sine song in the courtship of *Drosophila melanogaster. Animal Behaviour, 24,* 622–625.

Schilcher, F. (1976b). The role of auditory stimuli in the courtship song of *Drosophila melanogaster. Animal Behaviour, 24,* 18–26.

Schino, G. (2007). Grooming and agonistic support: A meta-analysis of primate reciprocal altruism. *Behavioral Ecology, 18,* 115–120.

Schino, G., & Aureli, F. (2007). Grooming reciprocation among female primates: A meta-analysis. *Biology Letters, 4,* 9–11.

Schino, G., & Aureli, F. (2009). Reciprocal altruism in primates: Partner choice, cognition, and emotions. *Advances in the Study of Behavior, 39,* 45–69.

Schino, G., Scuccio, S., Maestrippi, D., & Turillazzi, P. G. (1988). Allogrooming as a tension reduction mechanism: A behavioral approach. *American Journal of Primatology, 16,* 43–50.

Schlomer, G. L., Del Giudice, M., & Ellis, B. J. (2011). Parent-offspring conflict theory: An evolutionary framework for understanding conflict within human families. *Psychological Review, 118,* 496–521.

Schmoll, T. (2011). A review and perspective on context-dependent genetic effects of extra-pair mating in birds. *Journal of Ornithology, 152,* 265–277.

Schneider, K. (1984). Dominance, predation and optimal foraging in white-throated sparrow flocks. *Ecology, 65,* 1820–1827.

Schuett, G. (1997). Body size and agonistic experience affect dominance and mating success in male copperheads. *Animal Behaviour, 54,* 213–224.

Schuett, G. W., & Gillingham, J. C. (1989). Male-male agonistic behavior of the copperhead, *Agkistrodon contortrix. Amphibia-Reptilia, 10,* 243–266.

Schuett, G. W., & Grober, M. S. (2000). Post-fight levels of plasma lactate and corticosterone in male copperheads, *Agkistrodon contortrix* (Serpentes, Viperidae): Differences between winners and losers. *Physiology & Behavior, 71,* 335–341.

Schuett, G. W., Harlow, H. J., Rose, J. D., van Kirk, E. A., & Murdoch, W. J. (1996). Levels of plasma corticosterone and testosterone in male copperheads (*Agkistrodon contortrix*) following staged fights. *Hormones and Behavior, 30,* 60–68.

Schultz, T. R., & Brady, S. G. (2008). Major evolutionary transition in ant agriculture. *Proceedings of the National Academy of Sciences, U. S. A., 105,* 5435–5440.

Schulz, D. J., Elekonich, M. M., & Robinson, G. E. (2003). Biogenic amines in the antennal lobes and the initiation and maintenance of foraging behavior in honey bees. *Journal of Neurobiology, 54,* 406–416.

Schwabe, L., Wolf, O. T., & Oitzl, M. S. (2010). Memory formation under stress: Quantity and quality. *Neuroscience and Biobehavioral Reviews, 34,* 584–591.

Scott, J. P., & Bielfelt, S. W. (1976). Analysis of the puppy training program. In C. J. Pfaffenberger, J. P. Scott, J. L. Fuller, B. E. Ginsburg, & S. W. Bielfelt (Eds.), *Guide dogs for the blind: Their selection, development and training* (pp. 39–75). Amsterdam: Elsevier.

Scott, S. (1987). *Field guide to the birds of North America.* Washington, DC: National Geographic Society.

Searby, A., Jouventin, P., & Aubin, T. (2004). Acoustic recognition in macaroni penguins: An original signature system. *Animal Behaviour, 67,* 615–625.

Sebeok, T., & Rosenthal, R. (Eds.). (1981). *The Clever Hans phenomenon: Communication with horses, whales, apes, and people.* New York: New York Academy of Sciences.

Seehausen, O., van Alphen, J. J. M., & Witte, F. (1997). Cichlid fish diversity threatened by eutrophication that curbs sexual selection. *Science, 277,* 1808–1811.

Seeley, T. (1985). *Honeybee ecology: A study of adaptation in social life.* Princeton, NJ: Princeton University Press.

Seeley, T. (1995). *The wisdom of the hive*. Cambridge, MA: Harvard University Press.

Seeley, T. (1997). Honey bee colonies are group-level adaptive units. *American Naturalist, 150 (Suppl.)*, S22–S41.

Seeley, T. D. (2012). Progress in understanding how the waggle dance improves the foraging efficiency of honey bee colonies. *Honeybee neurobiology and behavior, 77–87*.

Seeley, T. D., & Tarpy, D. R. (2007). Queen promiscuity lowers disease within honeybee colonies. *Proceedings of the Royal Society of London, Series B, 274, 67–72*.

Seger, J. (1989). All for one, one for all, that is our device. *Nature, 338, 374–375*.

Seghers, B. H. (1973). An analysis of geographic variation in the antipredator adaptations of the guppy, *Poecilia reticulata*. Vancouver: University of British Columbia.

Selander, R., Smith, M., Yang, S., Johnson, W., & Gentry, J. (1971). Biochemical polymorphism and systematics in the genus *Peromyscus*: I. Variation of the oldfield mouse (*Peromyscus polionotus*). *Studies in Genetics, 6, 49–90*.

Seligman, M., & Hager, J. (1972). *Biological boundaries of learning*. New York: Appleton.

Selye, H. (1936). A syndrome produced by diverse nocuous agents. *Nature, 138, 32–35*.

Semple, S., & McComb, K. (2000). Perception of female reproductive state from vocal cues in a mammal species. *Proceedings of the Royal Society of London, Series B–Biological Sciences, 267, 707–712*.

Serpell, J. A., & Hsu, Y. Y. (2001). Development and validation of a novel method for evaluating behavior and temperament in guide dogs. *Applied Animal Behaviour Science, 72, 347–364*.

Seyfarth, R. M., & Cheney, D. L. (1986). Vocal development in vervet monkeys. *Animal Behaviour, 34, 1640–1658*.

Seyfarth, R. M., Cheney, D. L., & Marler, P. (1980). Vervet monkey alarm calls: Semantic communication in a free-ranging primate. *Animal Behaviour, 28, 1070–1094*.

Sgoifo, A., Costoli, T., Meerlo, P., Buwalda, B., Pico'Alfonso, M. A., De Boer, S., Musso, E., & Koolhaas, J. (2005). Individual differences in cardiovascular response to social challenge. *Neuroscience and Biobehavioral Reviews, 29, 59–66*.

Shamoun-Baranes, J., van Loon, E. E., Purves, R. S., Speckmann, B., Weiskopf, D., & Camphuysen, C. J. (2012). Analysis and visualization of animal movement. *Biology Letters, 8, 6–9*.

Sharp, S. P., & Hatchwell, B. J. (2005). Individuality in the contact calls of cooperatively breeding long-tailed tits (*Aegithalos caudatus*). *Behaviour, 142, 1559–1575*.

Shaw, P. W., Carvalho, G. R., Seghers, B. H., & Magurran, A. E. (1992). Genetic consequences of an artificial introduction of guppies (*Poecilia reticulata*) in N. Trinidad. *Proceedings of the Royal Society of London, Series B, 248, 111–116*.

Shaw, K. L., & Lesnick, S. C. (2009). Genomic linkage of male song and female acoustic preference QTL underlying a rapid species radiation. *Proceedings of the National Academy of Sciences, U. S. A., 106, 9737–9742*.

Shawkey, M. D., Pillai, S. R., & Hill, G. E. (2009). Do feather-degrading bacteria affect sexually selected plumage color? *Naturwissenschaften, 96, 123–128*.

Sheard, M. (1983). Aggressive behavior: Effects of neural modulation by serotonin. In E. Simmell, M. Hahn, & J. Walters (Eds.), *Aggressive behavior* (pp. 167–181). Hillsdale, NJ: Lawrence Erlbaum Associates.

Sheehan, M. J., & Tibbetts, E. A. (2011). Specialized face learning is associated with individual recognition in paper wasps. *Science, 334, 1272–1275*.

Shen, S., Vehrencamp, S., Johnstone, R., Chen, H. S., Chan, S. F., Liao, W. Y., Lin, K., & Yuan, H. W. (2011). Unfavourable environment limits social conflict in *Yuhina brunneiceps*. *Nature Communications, 3, 885*. doi:10.1038/ncomms1894

Shen-Feng, S., Reeve, H. K., & Vehrencamp, S. (2011). Parental care, cost of reproduction and reproductive skew: A general costly young model. *Journal of Theoretical Biology, 284, 24–31*.

Sherman, P. (1988). The levels of analysis. *Animal Behaviour, 36, 616–619*.

Sherman, P. W. (1977). Nepotism and the evolution of alarm calls. *Science, 197, 1246–1253*.

Sherman, P. W. (1980). The meaning of nepotism. *American Naturalist, 116, 604–606*.

Sherman, P. W. (1981). Kinship, demography, and Belding's ground squirrel nepotism. *Behavioral Ecology and Sociobiology, 8, 251–259*.

Sherman, P. W. (1985). Alarm calls of Belding's ground squirrels to aerial predators: Nepotism or self-preservation? *Behavioral Ecology and Sociobiology, 17, 313–323*.

Sherman, P. W., & Holmes, W. (1985). Kin recognition: Issues and evidence. *Fortschritte der Zoologie, 31, 437–460*.

Sherman, P. W., Jarvis, J., & Alexander, R. (Eds.). (1991). *The biology of the naked mole-rat*. Princeton, NJ: Princeton University Press.

Sherry, D. F. (2006). Neuroecology. *Annual Review of Psychology, 57, 167–197*.

Sherry, D. F., & Vaccarino, A. (1989). Hippocampus and memory for food caches in black-capped chickadees. *Behavioral Neuroscience, 103, 308–318*.

Shettleworth, S. J. (1998). *Cognition, evolution and behavior*. New York: Oxford University Press.

Shettleworth, S. J. (2007). Animal behaviour: Planning for breakfast. *Nature, 445, 825–826*.

Shettleworth, S. J. (2010). *Cognition, evolution, and behavior*. (2nd ed.). New York: Oxford University Press.

Shettleworth, S. J. (2012a). *Fundamentals of comparative cognition*. New York: Oxford University Press.

Shettleworth, S. J. (2012b). Modularity, comparative cognition, and human uniqueness. *Philosophical Transactions of the Royal Society (London), 367, 2794–2802*.

Shi, Y. S., Radlwimmer, F. B., & Yokoyama, S. (2001). Molecular genetics and the evolution of ultraviolet vision in vertebrates. *Proceedings of the National Academy of Sciences, U.S.A., 98, 11731–11736*.

Shi, Y. S., & Yokoyama, S. (2003). Molecular analysis of the evolutionary significance of ultraviolet vision in vertebrates. *Proceedings of the National Academy of Sciences, U.S.A., 100, 8308–8313*.

Shuster, S. M. (2009). Sexual selection and mating systems. *Proceedings of the National Academy of Sciences, U. S. A., 106, 10009–10016*.

Shuster, S. M., & Caldwell, R. (1989). Male defense of the breeding cavity and factors affecting the persistence of breeding pairs in the stomatopod, *Gonodactylus bredini (Crustacea: Hoplocarida)*. *Ethology, 82*, 192–207.

Sigg, D., Thompson, C. M., & Mercer, A. R. (1997). Activity-dependent changes to the brain and behavior of the honey bee, *Apis mellifera* (L.). *Journal of Neuroscience, 17*, 7148–7156.

Sih, A. (2011). Effects of early stress on behavioral syndromes: An integrated adaptive perspective. *Neuroscience and Biobehavioral Reviews, 35*, 1452–1465.

Sih, A., & Bell, A. M. (2008). Insights for behavioral ecology from behavioral syndromes. *Advances in the Study of Behavior, 38*, 227–281.

Sih, A., Bell, A., & Johnson, J. C. (2004). Behavioral syndromes: An ecological and evolutionary overview. *Trends in Ecology & Evolution, 19*, 372–378.

Sih, A., Bell, A., Johnson, J. C., & Ziema, R. (2004). Behavioral syndromes: An integrative approach. *Quarterly Review of Biology, 79*, 241–277.

Sih, A., & Christensen, B. (2001). Optimal diet theory: When does it work, and when it does it fail? *Animal Behaviour, 61*, 379–390.

Silk, J. B. (2007). Social components of fitness in primate groups. *Science, 317*, 1347–1351.

Simmons, L. W. (2001). *Sperm competition and its evolutionary consequences*. Princeton, NJ: Princeton University Press.

Sinn, D. L., Apiolaza, L. A., & Moltschaniwskyj, N. A. (2006). Heritability and fitness-related consequences of squid personality traits. *Journal of Evolutionary Biology, 19*, 1437–1447.

Sinn, D. L., & Moltschaniwskyj, N. A. (2005). Personality traits in dumpling squid (*Euprymna tasmanica*): Context-specific traits and their correlation with biological characteristics. *Journal of Comparative Psychology, 119*, 99–110.

Sinn, D. L., Moltschaniwskyj, N. A., Wapstra, E., & Dall, S. R. X. (2010). Are behavioral syndromes invariant? Spatiotemporal variation in shy/bold behavior in squid. *Behavioral Ecology and Sociobiology, 64*, 693–702.

Sinn, D. L., Gosling, S. D., & Moltschaniwskyj, N. A. (2008). Development of shy/bold behaviour in squid: Context-specific phenotypes associated with developmental plasticity. *Animal Behaviour, 75*, 433–442.

Siviy, S. (1998). Neurobiological substrates of play behavior: Glimpses into the structure and function of mammalian playfulness. In M. Bekoff & J. Byers (Eds.), *Animal play: Evolutionary, comparative and ecological perspectives* (pp. 221–242). Cambridge: Cambridge University Press.

Siviy, S., & Panksepp, J. (1985). Dorsomedial diencephalic involvement in the juvenile play of rats. *Behavioral Neuroscience, 99*, 1103–1113.

Siviy, S., & Panksepp, J. (1987). Juvenile play in the rat: Thalamic and brain stem involvement. *Physiology and Behavior, 41*, 103–114.

Siviy, S. M., Crawford, C. A., Akopian, G., & Walsh, J. P. (2011). Dysfunctional play and dopamine physiology in the Fischer 344 rat. *Behavioural Brain Research, 220*, 294–304.

Siviy, S. M., & Panksepp, J. (2011). In search of the neurobiological substrates for social playfulness in mammalian brains. *Neuroscience and Biobehavioral Reviews, 35*, 1821–1830.

Skinner, B. F. (1938). *The behavior of organisms*. New York: Appleton-Century-Crofts.

Skinner, B. F. (1959). A case history in scientific method. In S. Koch (Ed.), *Psychology: A study of science* (pp. 359–379). New York: McGraw-Hill.

Slabbekoorn, H., & Smith, T. B. (2002). Bird song, ecology and speciation. *Philosophical Transactions of the Royal Society of London, Series B, 357*, 493–503.

Slagsvold, T., Hansen, B. T., Johannessen, L. E., & Lifjeld, J. T. (2002). Mate choice and imprinting in birds studied by cross-fostering in the wild. *Proceedings of the Royal Society of London, Series B–Biological Sciences, 269*, 1449–1455.

Slansky, F., & Scriber, J. (1985). Food consumption and utilization. In G. A. Kerkut & L. I. Gilbert (Eds.), *Comprehensive insect physiology, biochemistry and pharmacology* (pp. 87–163). Oxford: Pergamon Press.

Slater, P. J. B. (2003). Fifty years of bird song research: A case study in animal behaviour. *Animal Behaviour, 65*, 633–639.

Slocombe, K., & Zuberbuhler, K. (2007). Chimpanzees modify recruitment screams as a function of audience composition. *Proceedings of the National Academy of Sciences, U.S.A., 104*, 17228–17233.

Slocombe, K. E., & Zuberbuhler, K. (2005). Agonistic screams in wild chimpanzees (*Pan troglodytes schweinfurthii*) vary as a function of social role. *Journal of Comparative Psychology, 119*, 67–77.

Sluyter, F., Bult, A., Lynch, C., Oortmerssen, G., & Koolhaus, J. (1995). A comparison between house mouse lines selected for attack latency or nest building: Evidence for genetic basis for alternative behavioral strategies. *Behavior Genetics, 25*, 247–252.

Smallwood, P. D. (1996). An introduction to risk sensitivity: The use of Jensen's inequality to clarify evolutionary arguments of adaptation and constraint. *American Zoologist, 36*, 392–401.

Smith, A. P., & Alcock, J. (1980). A comparative study of the mating systems of Australian eumenid wasps (Hymenoptera). *Zeitschrift fur Tierpsychologie, 53*, 41–60.

Smith, C., Barber, I., Wootton, R. J., & Chittka, L. (2004). A receiver bias in the origin of three-spined stickleback mate choice. *Proceedings of the Royal Society of London, Series B, 271*, 949–955.

Smith, C. R., Toth, A. L., Suarez, A. V., & Robinson, G. E. (2008). Genetic and genomic analyses of the division of labour in insect societies. *Nature Reviews Genetics, 9*, 735–748.

Smith, D. G., Davis, R. J., Gehlert, D. R., & Nomikos, G. G. (2006). Exposure to predator odor stress increases efflux of frontal cortex acetylcholine and monoamines in mice: Comparisons with immobilization stress and reversal by chlordiazepoxide. *Brain Research, 1114*, 24–30.

Smith, D. R., & Hagen, R. H. (1996). Population structure and interdemic selection in the cooperative spider *Anelosimus eximius*. *Journal of Evolutionary Biology, 9*, 589–608.

Smith, I. C., Huntingford, F. A., Atkinson, R. J. A., & Taylor, A. C. (1994). Strategic decisions during agonistic behaviour in the velvet swimming crab, *Necora puber*. *Animal Behaviour, 47*, 885–894.

Smith, P. K. (1982). Does play matter? Functional and evolutionary aspects of animal and human play. *Behavioral and Brain Sciences, 5*, 139–184.

Smith, P. S. (1991). Ontogeny and adaptiveness of tail-flagging behavior in white-tailed deer. *American Naturalist, 138,* 190–200.

Smith, R. J. F. (1986). Evolution of alarm signals: Role of benefits of retaining group members or territorial neighbors. *American Naturalist, 128,* 604–610.

Smith, R. L. (1979). Paternity assurance and altered roles in the mating behavior of a giant water bug, *Abedus herberti* (Heteroptera: Belostomatidae). *Animal Behaviour, 27,* 716–725.

Smith, W. J. (1968). Message meaning analysis. In T. Sebeok (Ed.), *Animal communications* (pp. 44–60). Bloomington: Indiana University Press.

Smith, W. J. (1977). *The behavior of communicating: An ethological approach.* Cambridge, MA: Harvard University Press.

Smolker, R., Richards, A., Connor, R., Mann, J., & Berggren, P. (1997). Sponge carrying by dolphins (Delphinidae, *Tursiops sp.*): A foraging specialization involving tool use? *Ethology, 103,* 454–465.

Smuts, B. (1985). *Sex and friendship in baboons.* New York: Aldine.

Smythe, N. (1977). The function of mammalian alarm advertising: Social signals or pursuit invitation. *American Naturalist, 111,* 191–194.

Snowdon, C. T. (2009). Plasticity of communication in nonhuman primates. *Advances in the Study of Behavior, 40,* 239–276.

Sober, E. (1987). What is adaptationism? In J. Dupre (Ed.), *The latest on the best: Essays on evolution and optimality* (pp. 105–118). Cambridge, MA: MIT Press.

Sober, E., & Wilson, D. S. (1998). *Unto others: The evolution and psychology of unselfish behavior.* Cambridge, MA: Harvard University Press.

Sol, D., Duncan, R. P., Blackburn, T., Cassey, P., & Lefebvre, L. (2005). Big brains, enhanced cognition, and response of birds to novel environments. *Proceedings of the National Academy of Sciences, U.S.A., 102,* 5460–5465.

Sol, D., Garcia, N., Iwaniuk, A., Davis, K., Meade, A., Boyle, W. A., & Szekely, T. (2010). Evolutionary divergence in brain size between migratory and resident birds. *Plos One, 5.*

Sol, D., & Lefebvre, L. (2006). Behavioral flexibility and response in birds to changes in the environment. *Journal of Ornithology, 147,* 34.

Sol, D., Lefebvre, L., & Rodriguez-Teijeiro, J. D. (2005). Brain size, innovative propensity and migratory behaviour in temperate Palaearctic birds. *Proceedings of the Royal Society of London, Series B, 272,* 1433–1441.

Sol, D., Szekely, T., Liker, A., & Lefebvre, L. (2006). Big-brained birds survive better in nature. *Journal of Ornithology, 147,* 254.

Sol, D., Szekely, T., Liker, A., & Lefebvre, L. (2007). Big-brained birds survive better in nature. *Proceedings of the Royal Society of London, Series B, 274,* 763–769.

Solberg, L. C., Valdar, W., Gauguier, D., Nunez, G., Taylor, A., Burnett, S., Arboledas-Hita, C., Hernandez-Pliego, P., Davidson, S., Burns, P., Bhattacharya, S., Hough, T., Higgs, D., Klenerman, P., Cookson, W. O., Zhang, Y. M., Deacon, R. M., Rawlins, J. N. P., Mott, R., & Flint, J. (2006). A protocol for high-throughput phenotyping, suitable for quantitative trait analysis in mice. *Mammalian Genome, 17,* 129–146.

Solomon, N. G., & French, J. A. (Eds.). (1996). *Cooperative breeding in mammals.* Cambridge: Cambridge University Press.

Soma, M., & Zsolt-Garamszegi, L. (2011). Rethinking birdsong evolution: Meta-analysis of the relationship between song complexity and reproductive success. *Behavioral Ecology, 22,* 363–371.

Sordahl, T. A. (1990). The risks of avian mobbing and distraction behavior: An anectodal review. *Wilson Bulletin, 102,* 349–352.

Sparks, J. (1967). Allogrooming in primates: A review. In D. Morris (Ed.), *Primate ethology* (pp. 148–175). London: Weidenfeld and Nicolson.

Spencer, S. J., Buller, K. M., & Day, T. A. (2005). Medial prefrontal cortex control of the paraventricular hypothalamic nucleus response to psychological stress: Possible role of the bed nucleus of the stria terminalis. *Journal of Comparative Neurology, 481,* 363–376.

Spieth, H. T., & Ringo, J. M. (1983). Mating behavior and sexual selection in Drosophila. In M. Ashburn, H. Carson, & J. L. Thompson (Eds.), *The genetics and biology of Drosophila* (pp. 224–284). London: Academic Press.

Spinka, M., Newberry, R., & Bekoff, M. (2001). Mammalian play: Training for the unexpected. *Quarterly Review of Biology, 76,* 141–168.

Spinks, A. C., O'Riain, M. J., & Polakow, D. A. (1998). Intercolonial encounters and xenophobia in the common mole rat, *Cryptomys hottentotus hottentotus* (Bathyergidae): The effects of aridity, sex and reproductive status. *Behavioral Ecology, 9,* 354–359.

Spoolder, H., Burbidge, J., Lawrence, A., Simmins, P., & Edwards, S. (1996). Individual behavioral differences in pigs: Intra- and inter-test consistency. *Applied Animal Behavior Science, 49,* 185–198.

Spoon, T. R., Millam, J. R., & Owings, D. H. (2006). The importance of mate behavioural compatibility in parenting and reproductive success by cockatiels, *Nymphicus hollandicus. Animal Behaviour, 71,* 315–326.

Spooner, C. (1964). Observations on the use of the chick in pharmacological investigation of the cental nervous system. Los Angeles: University of California at Los Angeles.

Stacey, P., & Koenig, W. (Eds.). (1990). *Cooperative beeding in birds: Long-term studies of ecology and behavior.* Cambridge: Cambridge University Press.

Stacey, P. B., & Ligon, J. D. (1987). Territory quality and dispersal options in the acorn woodpecker, and a challenge to the habitat-saturation model of cooperative breeding. *American Naturalist, 130,* 654–676.

Stammbach, E. (1988). Group responses to specially skilled individuals in a *Macaca fascicularis* group. *Behaviour, 107,* 687–705.

Stamps, J. (1995). Motor learning and the adaptive value of familiar space. *American Naturalist, 146,* 41–58.

Stamps, J. (2001). Learning from lizards. In L. A. Dugatkin (Ed.), *Model systems in behavioral ecology* (pp. 149–168). Princeton, NJ: Princeton University Press.

Stamps, J. (2003). Behavioural processes affecting development: Tinbergen's fourth question comes of age. *Animal Behaviour, 66,* 1–13.

Stamps, J., & Krishnan, V. V. (1999). A learning-based model of territorial assessment. *Quarterly Review of Biology, 74,* 291–318.

Stamps, J. A. (1987a). Conspecifics as cues to territory quality: A preference of juvenile lizards (*Anolis aeneus*) for previously used territories. *American Naturalist, 129,* 629–642.

Stamps, J. A. (1987b). The effect of familiarity with a neighborhood on territory acquisition. *Behavioral Ecology and Sociobiology, 21,* 273–277.

Stamps, J. A. (2007). Growth-mortality tradeoffs and "personality traits" in animals. *Ecology Letters, 10,* 355–363.

Stamps, J. A., & Krishnan, V. V. (2001). How territorial animals compete for divisible space: A learning-based model with unequal competitors. *American Naturalist, 157,* 154–169.

Stanford, C. B. (2002). Avoiding predators: Expectations and evidence in primate antipredator behavior. *International Journal of Primatology, 23,* 741–757.

Stankowich, T. (2008). Tail-flicking, tail-flagging, and tail position in ungulates with special reference to black-tailed deer. *Ethology, 114,* 875–885.

Stankowich, T., & Blumstein, D. T. (2005). Fear in animals: A meta-analysis and review of risk assessment. *Proceedings of the Royal Society of London, Series B, 272,* 2627–2634.

Stankowich, T., & Sherman, P. W. (2002). Pup shoving by adult naked mole-rats. *Ethology, 108,* 975–992.

Starr, C., & Taggart, R. (1992). *Biology: The unity and diversity of life.* Belmont, CA: Wadsworth.

Stearns, S. C., & Ebert, D. (2001). Evolution in health and disease: Work in progress. *Quarterly Review of Biology, 76,* 417–432.

Stearns, S. C., Nesse, R. M., Govindaraju, D. R., & Ellison, P. T. (2010). Evolutionary perspectives on health and medicine. *Proceedings of the National Academy of Sciences, U.S.A., 107,* 1691–1695.

Stein, R. C. (1968). Modulation in bird sound. *Auk, 94,* 229–243.

Stephens, D. (1991). Change, regularity and value in the evolution of learning. *Behavioral Ecology, 2,* 77–89.

Stephens, D., Brown, J. S., & Ydenberg, R. C. (Eds.). (2007). *Foraging: Behavior and ecology.* Chicago: University of Chicago Press.

Stephens, D., & Krebs, J. (1986). *Foraging theory.* Princeton, NJ: Princeton University Press.

Stephens, D. W. (1993). Learning and behavioral ecology: Incomplete information and environmental predictability. In D. Papaj (Ed.), *Insect learning* (pp. 195–218). New York: Chapman Hall.

Stephens, D. W., Anderson, J. P., & Benson, K. E. (1997). On the spurious occurrence of tit-for-tat in pairs of predator-approaching fish. *Animal Behaviour, 53,* 113–131.

Stevens, A., & Price, J. (1996). *Evolutionary psychiatry: A new beginning.* New York: Routledge.

Stevens, J. R., & Gilby, I. C. (2004). A conceptual framework for nonkin food sharing: Timing and currency of benefits. *Animal Behaviour, 67,* 603–614.

Stevenson-Hinde, J., Stillwell-Barnes, R., & Zunz, M. (1980). Subjective assessment of rhesus monkeys over four successive years. *Primates, 21,* 66–82.

Stevenson-Hinde, J., & Zunz, M. (1978). Subjective assessment of individual rhesus monkeys. *Primates, 19,* 473–482.

Stockley, P., Gage, M., Parker, G., & Møller, A. (1997). Sperm competition in fishes: The evolution of testis size and ejaculate characteristics. *American Naturalist, 149,* 933–954.

Stoewe, M., Bugnyar, T., Heinrich, B., & Kotrschal, K. (2006). Effects of group size on approach to novel objects in ravens (*Corvus corax*). *Ethology, 112,* 1079–1088.

Stone, A. I. (2008). Seasonal effects on play behavior in immature *Saimiri sciureus* in eastern Amazonia. *International Journal of Primatology, 29,* 195–205.

Strassmann, B. (1996). The evolution of endometrial cycles and menstruation. *Quarterly Review of Biology, 71,* 181–220.

Stratakis, C., & Chrousos, G. (1955). Neuroendocrinology and pathophysiology of the stress system. *Annals of the New York Academy of Sciences, 771,* 1–18.

Struhsaker, T. T. (1967). Auditory communication among vervet monkeys (*Cercopithecus aethiops*). In S. A. Altmann (Ed.), *Social communication among primates* (pp. 281–324). Chicago: University of Chicago Press.

Suddendorf, T. (2006). Foresight and evolution of the human mind. *Science, 312,* 1006–1007.

Suddendorf, T., & Corballis, M. C. (1997). Mental time travel and the evolution of the human mind. *Genetic Social and General Psychology Monographs, 123,* 133–167.

Sugita, S., Baxter, D. A., & Byrne, J. H. (1994). Camp-independent effects of 8–(4–Parachlorophenylthio)-Cyclic Amp on spike duration and membrane currents in pleural sensory nmeurons of *Aplysia. Journal of Neurophysiology, 72,* 1250–1259.

Sullivan, J., Jassim, O., Fahrbach, S., & Robinson, G. (2000). Juvenile hormone paces behavioral development in the adult worker honey bee. *Hormones and Behavior, 37,* 1–14.

Sullivan, J. P., Fahrbach, S. E., Harrison, J. F., Capaldi, E. A., Fewell, J. H., & Robinson, G. E. (2003). Juvenile hormone and division of labor in honey bee colonies: Effects of allatectomy on flight behavior and metabolism. *Journal of Experimental Biology, 206,* 2287–2296.

Sullivan, K. (1984). The advantages of social foraging in downy woodpeckers. *Animal Behaviour, 32,* 16–32.

Sullivan, K. (1985). Selective alarm calling by downy woodpeckers in mixed-species flocks. *Auk, 102,* 184–187.

Summers, K., Weigt, L. A., Boag, P., & Bermingham, E. (1999). The evolution of female parental care in poison frogs of the genus Dendrobates: Evidence from mitochondrial DNA sequences. *Herpetologica, 55,* 254–270.

Summers-Smith, J. D. (1963). *The house sparrow.* London: Collins.

Sumner, F. B. (1929a). The analysis of a concrete case of intergradation between two subspecies. *Proceedings of the National Academy of Sciences, U.S.A., 15,* 110–120.

Sumner, F. B. (1929b). The analysis of a concrete case of intergradation between two subspecies: Additional data and interpretation. *Proceedings of the National Academy of Sciences, U.S.A., 15,* 481–493.

Sumner, P., & Mollon, J. D. (2000). Catarrhine photopigments are optimized for detecting targets against a foliage background. *Journal of Experimental Biology, 203,* 1963–1986.

Sumner, P., & Mollon, J. D. (2003). Colors of primate pelage and skin: Objective assessment of conspicuousness. *American Journal of Primatology, 59,* 67–91.

Sumner, S. (2006). Determining the molecular basis of sociality in insects: Progress, prospects and potential in sociogenomics. *Annales Zoologici Fennici, 43,* 423–442.

Sutherland, W. J. (1983). Aggregation and the "ideal free" distribution. *Journal of Animal Ecology, 52,* 821–828.

Sutherland, W. J., & Parker, G. (1985). Distribution of unequal competitors. In R. Sibly & R. Smith (Eds.), *Behavioural ecology* (pp. 255–274). Oxford: Blackwell Science.

Suthers, R. A. (1997). Peripheral control and lateralization of birdsong. *Journal of Neurobiology, 33,* 632–652.

Suthers, R. A. (1999). The motor basis of vocal performance in songbirds. In M. D. Hauser & M. Konishi (Eds.), *The design of animal communication* (pp. 37–62). Cambridge, MA: MIT Press.

Suthers, R. A., & Goller, F. (1996). Respiratory and syringeal dynamics of song production in northern cardinals. In M. Burrows, T. Matheson, P. Newland, & H. Schuppe (Eds.), *Nervous systems and behavior: Proceedings of the 4th International Congress on Neuroethology* (p. 333). Stuttgart: Georg Thieme Verlag.

Suthers, R. A., Goller, F., & Hartley, R. S. (1994). Motor dynamics of song production by mimic thrushes. *Journal of Neurobiology, 25,* 917–936.

Suthers, R. A., Goller, F., & Hartley, R. S. (1996). Motor stereotypy and diversity in songs of mimic thrushes. *Journal of Neurobiology, 30,* 231–245.

Suthers, R. A., & Hartley, R. S. (1990). Effect of unilateral denervation on the acoustic output from each side of the syrinx in the singing mimic thrushes. *Society for Neuroscience (Abst), 16,* 1249.

Suthers, R. A., Vallet, E., Tanvez, A., & Kreutzer, M. (2004). Bilateral song production in domestic canaries. *Journal of Neurobiology, 60,* 381–393.

Suthers, R. A., & Zollinger, S. A. (2004). Producing song: The vocal apparatus. In H. P. Zeigler & P. Marler (Eds.), *Behavioral neurobiology of bird song. Annals of the New York Academy of Sciences, 1016,* 109–129.

Swaddle, J. P. (2003). Fluctuating asymmetry, animal behavior, and evolution. *Advances in the Study of Behavior, 3,*169–205.

Symons, D. (1978). *Play and aggression: A study of rhesus monkeys.* New York: Columbia University Press.

Tada, T., Altun, A., & Yokoyama, S. (2009). Evolutionary replacement of UV vision by violet vision in fish. *Proceedings of the National Academy of Sciences, U.S.A., 106,* 17457–17462.

Tai, Y. F., Scherfler, C., Brooks, D. J., Sawamoto, N., & Castiello, U. (2004). The human premotor cortex is "mirror" only for biological actions. *Current Biology, 14,* 117–120.

Takagi, K., Ono, N., & Wright, W. (2010). Interspecific variation in palatability suggests cospecialization of antipredator defenses in sea hares. *Marine Ecology Progress Series, 416,* 137–144.

Talbot, C. J., Nicod, A., Cherny, S. S., Fulker, D. W., Collins, A. C., & Flint, J. (1999). High-resolution mapping of quantitative trait loci in outbred mice. *Nature Genetics, 21,* 305–308.

Taub, D. (1980). Female choice and mating strategies among wild Barbary macaques. In D. Lindburg (Ed.), *The macaques: Studies in ecology, behavior and evolution.* New York: van Nostrand Reinhold.

Taub, D. (1984). Male caretaking behavior among wild Barbary macaques. In D. Taub (Ed.), *Primate paternalism.* New York: van Nostrand Reinhold.

Tautz, J., Maier, S., Groh, C., Rossler, W., & Brockmann, A. (2003). Behavioral performance in adult honey bees is influenced by the temperature experienced during their pupal development. *Proceedings of the National Academy of Sciences, U.S.A., 100,* 7343–7347.

Taylor, M. (1986). Receipt of support from family among black Americans: Demographic and familial differences. *Journal of Marriage and the Family, 48,* 67–77.

Telecki, G. (1973). *The predatory behavior of chimpanzees.* Lewisburg, PA: Bucknell University Press.

Templeton, J. J., & Giraldeau, L.-A. (1996). Vicarious sampling: The use of personal and public information by starlings foraging in a simple patchy environment. *Behavioral Ecology and Sociobiology, 38,* 105–114.

ten Cate, C., & Vos, D. R. (1999). Sexual imprinting and evolutionary processes in birds: A reassessment. *Advances in the Study of Behavior, 28,* 1–31.

Teramitsu, I., Kudo, L. C., London, S. E., Geschwind, D. H., & White, S. A. (2004). Parallel FoxP1 and FoxP2 expression in songbird and human brain predicts functional interaction. *Journal of Neuroscience, 24,* 3152–3163.

Terry, R. L. (1970). Primate grooming as a tension reduction mechanism. *Journal of Psychology, 76,* 129–136.

Teskey G., Campbell, G., & Pellis S. M. (2010). Preface to special issue on behavioural and neural plasticity. *Behavioral Brain Research, 214,* 1–2.

Thatch, W., Goodkin, H., & Keating, J. (1992). The cerebellum and adaptive coordination of movement. *Annual Review of Neuroscience, 15,* 403–442.

Thierry, B., Theraulaz, G., Gautier, J. Y., & Stiegler, B. (1995). Joint memory. *Behavioural Processes, 35,* 127–140.

Thiessen, D., & Yahr, P. (1977). *The gerbil in behavioral investigations.* Austin: University of Texas Press.

Thom, C., & Dornhaus, A. (2007). Preliminary report on the use of volatile compounds by foraging honey bees in the hive (Hymenoptera: Apidae: *Apis mellifera*). *Entomologia Generalis, 29,* 299–304.

Thompson, J. N. (1982). *Interaction and coevolution.* New York: Wiley.

Thompson, K. V. (1996). Play-partner preferences and the function of social play in infant sable antelope, *Hippotragus niger. Animal Behaviour, 52,* 1143–1152.

Thorndike, E. L. (1898). Animal intelligence: An experimental study of the association processes in animals. *Psychological Review Monographs,* Whole (issue).

Thorndike, E. L. (1911). *Animal intelligence: Experimental studies.* New York: MacMillan.

Thornhill, R. (1976). Sexual selection and nuptial feeding behavior in *Bittacus apicalis. American Naturalist, 110,* 529–548.

Thornhill, R. (1979a). Adaptive female-mimicking in a scorpionfly. *Science, 295,* 412–414.

Thornhill, R. (1979b). Male and female sexual selection and the evolution of mating systems in insects. In M. Blum & N. Blum (Eds.), *Sexual selection and reproductive competition in insects.* New York: Academic Press.

Thornhill, R. (1980a). Mate choice in *Hylobittacus apicalis* and its relation to some models of female choice. *Evolution, 34,* 519–538.

Thornhill, R. (1980b). Sexual selection in the black-tipped hangingfly. *Scientific American, 242,* 162–172.

Thornhill, R., & Alcock, J. (1983). *The evolution of insect mating systems.* Cambridge, MA: Harvard University Press.

Thornton, A., & Clutton-Brock, T. (2011). Social learning and the development of individual and group behaviour in mammal societies. *Philosophical Transactions of the Royal Society, Series B–Biological Sciences, 366,* 978–987.

Thornton, A., & Malapert, A. (2009a). Experimental evidence for social transmission of food acquisition techniques in wild meerkats. *Animal Behaviour, 78,* 255–264.

Thornton, A., & Malapert, A. (2009b). The rise and fall of an arbitrary tradition: An experiment with wild meerkats. *Proceedings of the Royal Society, Series B–Biological Sciences, 276,* 1269–1276.

Thornton, A., & McAuliffe, K. (2006). Teaching in wild meerkats. *Science, 313,* 227–229.

Thornton, A., Samson, J., & Clutton-Brock, T. (2010). Multigenerational persistence of traditions in neighbouring meerkat groups. *Proceedings of the Royal Society, Series B–Biological Sciences, 277,* 3623–3629.

Thorpe, W. H. (1956). *Learning and instinct in animals.* London: Methuen.

Thorpe, W. H. (1963). *Learning and instinct in animals* (2nd ed.) London: Methuen.

Tibbetts, E. A. (2008). Resource value and the context dependence of receiver behaviour. *Proceedings of the Royal Society, Series B–Biological Sciences, 275,* 2201–2206.

Tinbergen, L. (1960). The natural controls of insect in pinewoods. I. Factors influencing the intensity of predation by songbirds. *Archives Neerlandaises de Zoologie, 13,* 265–343.

Tinbergen, N. (1963). On aims and methods of ethology. *Zeitschrift fur Tierpsychologie, 20,* 410–440.

Tinbergen, N. (1964). The evolution of signalling devices. In W. Etkin (Ed.), *Social behavior and organization among vertebrates* (pp. 206–230). Chicago: University of Chicago Press.

Tinghitella, R. M., Wang, J. M., & Zuk, M. (2009). Preexisting behavior renders a mutation adaptive: Flexibility in male phonotaxis behavior and the loss of singing ability in the field cricket *Teleogryllus oceanicus. Behavioral Ecology, 20,* 722–728.

Toft, S. (1995). Two functions of gossamer dispersal in spiders? *Acta Jutlandica, 70,* 257–268.

Toivanen, P., & Toivanen, A. (Eds.). (1987). Avian immunology. Boca Raton, FL: CRC Press.

Tollrian, R., & Harvell, C. (Eds.). (1998). *The ecology and evolution of inducible defenses.* Princeton, NJ: Princeton University Press.

Toma, D., Bloch, G., Moore, D., & Robinson, G. (2000). Changes in *period* mRNA levels in the brain and division of labor in honey bee colonies. *Proceedings of the National Academy of Sciences, U.S.A., 97,* 6914–6919.

Tomaru, M., & Oguma, Y. (1994). Genetic basis and evolution of species specific courtship song in the *Drosophila auraria* complex. *Genetical Research, 63,* 11–17.

Tomkins, L. M., Thomson, P. C., & McGreevy, P. D. (2011). Behavioral and physiological predictors of guide dog success. *Journal of Veterinary Behavior: Clinical Applications and Research, 6,* 178–187.

Torney, C. J., Berdahl, A., & Couzin, I. D. (2011). Signalling and the evolution of cooperative foraging in dynamic environments. *Plos Computational Biology, 7.*

Touchon, J. C., Urbina, J., & Warkentin, K. M. (2011). Habitat-specific constraints on induced hatching in a treefrog with reproductive mode plasticity. *Behavioral Ecology, 22,* 169–175.

Tourmente, M., Gomendio, M., & Roldan, E. R. S. (2011). Sperm competition and the evolution of sperm design in mammals. *Bmc Evolutionary Biology, 11.*

Tourmente, M., Gomendio, M., Roldan, E. R. S., Giojalas, L. C., & Chiaraviglio, M. (2009). Sperm competition and reproductive mode influence sperm dimensions and structure among snakes. *Evolution, 63,* 2513–2524.

Trail, P. (1985). A lek's icon: The courtship display of a Guianan cock-of-the-rock. *American Birds, 39,* 235–240.

Travasso, M., & Pierce, N. (2000). Acoustics, context and function of vibrational signalling in a lycaenid butterfly-ant mutualism. *Animal Behaviour, 60,* 13–26.

Treves, A. (2009). Hunting for large carnivore conservation. *Journal of Applied Ecology, 46,* 1350–1356.

Treves, A., & Karanth, K. U. (2003). Human-carnivore conflict and perspectives on carnivore management worldwide. *Conservation Biology, 17,* 1491–1499.

Trivers, R. (1972). Parental investment and sexual selection. In B. Campbell (Ed.), *Sexual selection and the descent of man* (pp. 136–179). Chicago: Aldine.

Trivers, R. (1985). *Social evolution.* Menlo Park, CA: Benjamin Cummings.

Trivers, R. L. (1971). The evolution of reciprocal altruism. *Quarterly Review of Biology, 46,* 189–226.

Trivers, R. L. (1974). Parent-offspring conflict. *American Zoologist, 14,* 249–265.

Trivers, R. L., & Hare, H. (1976). Haploidploidy and the evolution of the social insect. *Science, 191,* 249–263.

Trut, L., Oskina, I., & Kharlamova, A. (2009). Animal evolution during domestication: The domesticated fox as a model. *Bioessays, 31,* 349–360.

Trut, L. N. (1999). Early canid domestication: The farm-fox experiment. *American Scientist, 87,* 160–169.

Turri, M. G., DeFries, J. C., Henderson, N. D., & Flint, J. (2004). Multivariate analysis of quantitative trait loci influencing variation in anxiety-related behavior in laboratory mice. *Mammalian Genome, 15,* 69–76.

Tyack, P. L., & Clark, C. W. (2000). Communication and acoustic behavior of dolphins and whales. In W. W. Au, A. Popper, & R. Ray (Eds.), *Hearing by whales and dolphins* (pp. 156–224). New York: Springer.

Uehara, S., Nishida, T., Hamai, M., Hasegawa, T., Hayaki, H., Huffman, M., Kawanaka, K., Kobayashi, S., Mitani, J., Takahata, Y., Takasaki, H., & Tsukahara, T. (1992). Characteristics of predation by the chimpanzees in the Mahale National Park, Tanzania. In T. Nishida, W. C. McGrew, P. Marler, M. Pickford, & F. de Waal (Eds.), *Topics in primatology* (pp. 143–158). Tokyo: University of Tokyo Press.

Uehara, T., Iwasa, Y., & Ohtsuki, H. (2007). ESS distribution of display duration in animal contests to assess an opponent before fighting or fleeing. *Evolutionary Ecology Research, 9*, 395–408.

Uher, J. (2008). Comparative personality research: Methodological approaches. *European Journal of Personality, 22*, 427–455.

Uher, J. (2011). Individual behavioral phenotypes: An integrative meta-theoretical framework. Why "behavioral syndromes" are not analogs of "personality." *Developmental Psychobiology, 53*, 521–548.

Uher, J., & Asendorphf, J. B. (2008). Personality assessment in the great apes: Comparing ecologically valid behavior measures, behavior ratings, and adjective ratings. *Journal of Research in Personality, 42*, 821–838.

Valone, T., & Lima, S. (1987). Carrying food items to cover for consumption: The behavior of ten bird species under the risk of predation. *Oecologia, 71*, 286–294.

Valone, T. J. (1989). Group foraging, public information and patch estimation. *Oikos, 56*, 357–363.

Valone, T. J., & Giraldeau, L.-A. (1993). Patch estimation by group foragers: What information is used? *Animal Behaviour, 45*, 721–728.

Valone, T. J., & Templeton, J. J. (2002). Public information for the assessment of quality: A widespread social phenomenon. *Philosophical Transactions of the Royal Society of London, Series B, 357*, 1549–1557.

van Baaren, J., Outreman, Y., & Boivin, G. (2005). Effect of low temperature exposure on oviposition behaviour and patch exploitation strategy in parasitic wasps. *Animal Behaviour, 70*, 153–163.

van den Berg, C., van Ree, J., Spruijt, B., & Kitchen, I. (1999). Effects of juvenile isolation and morphine treatment on social interactions and opioid receptors in adult rats: Behavioural and autoradiographic studies. *European Journal of Neuroscience, 11*, 3023–3032.

van Leeuwen, E., & Jansen, V. A. A. (2010). Evolutionary consequences of a search image. *Theoretical Population Biology, 77*, 49–55.

van Oers, K., Buchanan, K. L., Thomas, T. E., & Drent, P. J. (2011). Correlated response to selection of testosterone levels and immunocompetence in lines selected for avian personality. *Animal Behaviour, 81*, 1055–1061.

van Oers, K., de Jong, G., van Noordwijk, A. J., Kempenaers, B., & Drent, P. J. (2005). Contribution of genetics to the study of animal personalities: A review of case studies. *Behaviour, 42*, 1185–1206.

van Oers, K., Drent, P. J., de Goede, P., & van Noordwijk, A. J. (2004). Realized heritability and repeatability of risk-taking behaviour in relation to avian personalities. *Proceedings of the Royal Society of London, Series B, 271*, 65–73.

van Oers, K., Drent, P. J., de Jong, G., & van Noordwijk, A. J. (2004). Additive and nonadditive genetic variation in avian personality traits. *Heredity, 93*, 496–503.

van Oers, K., Drent, P. J., Dingemanse, N. J., & Kempenares, B. (2008). Personality is associated with extrapair paternity in great tits, Parus major. *Animal Behaviour, 76*, 555–563.

van Oordt, G., & Junge, G. (1936). Die hormonal Wirkung der Gonaden auf Sommer and Prachtklied. III. Der Einfluss der kastration suf mannliche Kampflaufer (*Philomachus pugnax*). *Wilhem Roux' Arch. Entwicklungsmech Org., 134*, 112–121.

Vander Meer, R., Breed, M., Espelie, K., & Winston, M. (Eds.). (1997). *Pheromone communication in social insects.* Boulder, CO: Westview Press.

Vanderschuren, L. J. M. (2010). How the brain makes play fun. *American Journal of Play, 2*, 315–337.

Veenema, A. H. (2012). Toward understanding how early life social experiences alter oxytocin- and vasopressin-regulated social behaviors. *Hormones and behavior, 61*, 310–312.

Veenema, A. H., de Kloet, E. R., de Wilde, M. C., Roelofs, A. J., Kawata, M., Buwalda, B., Neumann, I. D., Koolhaas, J. M., & Lucassen, P. J. (2007). Differential effects of stress on adult hippocampal cell proliferation in low and high aggressive mice. *Journal of Neuroendocrinology, 19*, 489–498.

Veenema, A. H., Koolhaas, J. M., & De Kloet, E. R. (2004). Basal and stress-induced differences in HPA axis, 5–HT responsiveness, and hippocampal cell proliferation in two mouse lines. *Stress: Current neuroendocrine and genetic approaches* (Vol. 1018, pp. 255–265). New York: Annals of the New York Academy of Sciences.

Vehrencamp, S. (1983). A model for the evolution of despotic versus egalitarian societies. *Animal Behaviour, 31*, 667–682.

Verbeek, M., Boone, A., & Drent, P. (1996). Exploration, agonistic behaviour and dominance in juvenile male great tits. *Behaviour, 133*, 945–963.

Verbeek, M., Drent, P., & Wiepkema, P. (1994). Consistent individual differences in early exploration behaviour of male great tits. *Animal Behaviour, 48*, 1113–1121.

Verbeek, M. E. M., De Goede, P., Drent, P. J., & Wiepkema, P. R. (1999). Individual behavioural characteristics and dominance in aviary groups of great tits. *Behaviour, 136*, 23–48.

Verlinden, H., Vleugels, R., Marchal, E., Badisco, L., Pfluger, H. J., Blenau, W., & Broeck, J. V. (2010). The role of octopamine in locusts and other arthropods. *Journal of Insect Physiology, 56*, 854–867.

Verrell, P. (1986). Wrestling in the red spotted newt: Resource value and contest asymmetry determine contest duration and outcome. *Animal Behaviour, 34*, 398–402.

Viitala, J., Korpimaki, E., Palokangas, P., & Koivula, M. (1995). Attraction of kestrels to vole scent marks visible in ultraviolet light. *Nature, 373*, 425–427.

Villarreal, R., & Domjan, M. (1998). Pavlovian conditioning of social-affiliative behavior in the Mongolian gerbil (*Meriones unguiculatus*). *Journal of Comparative Psychology, 112*, 26–35.

Visalberghi, E., & Addessi, E. (2000). Seeing group members eating a familiar food enhances the acceptance of novel foods in capuchin monkeys. *Animal Behaviour, 60*, 69–76.

Visscher, P. K., & Seeley, T. (1982). Foraging strategies of honeybee colonies in a temperate forest. *Ecology, 63*, 1790–1801.

vom Saal, F. (1989). Sexual differentiation in litter-bearing mammals: Influence of sex of adjacent fetuses in utero. *Journal of Animal Science, 67*, 1824–1840.

vom Saal, F. S., & Bronson, F. H. (1978). In utero proximity of female mouse fetuses to males: Effect on reproductive performance during later life. *Biology of Reproduction, 19,* 842–853.

vom Saal, F. S., Grant, W. M., McMullen, C. W., & Laves, K. S. (1983). High fetal estrogen concentrations: Correlation with increased adult sexual activity and decreased aggression in male mice *Science, 220,* 1306–1309.

von Frisch, K. (1967). *The dance language and orientation of bees.* Cambridge, MA: Harvard University Press.

Voultsiadou, E., & Tatolas, A. (2005). The fauna of Greece and adjacent areas in the age of Homer: Evidence from the first written documents of Greek literature. *Journal of Biogeography, 32,* 1875–1882.

Wagener-Hulme, C., Kuehn, J. C., Schulz, D. J., & Robinson, G. E. (1999). Biogenic amines and division of labor in honey bee colonies. *Journal of Comparative Physiology A, 184,* 471–479.

Wajnberg, E., Acosta-Avalos, D., Alves, O. C., de Oliveira, J. F., Srygley, R. B., & Esquivel, D. M. S. (2010). Magnetoreception in eusocial insects: An update. *Journal of the Royal Society Interface, 7,* S207–S225.

Waldman, B. (1987). Mechanisms of kin recognition. *Journal of Theoretical Biology, 128,* 159–185.

Walker, B. G., Boersma, P. D., & Wingfield, J. C. (2005). Field endocrinology and conservation biology. *Integrative and Comparative Biology, 45,* 12–18.

Walker, P. (Ed.). (1989). *Cambridge dictionary of biology.* Cambridge: Cambridge University Press.

Wallace, M. (1997). Conservation and ontogeny of behavior. In L. Gosling & W. Sutherland (Eds.), *Behaviour and Conservation* (pp. 300–314). Cambridge: Cambridge University Press.

Wallace, R., Sanders, G. P., & Ferl, R. J. (1991). *Biology: The science of life* (3rd ed.). New York: Harper Collins.

Waller, B. M., & Cherry, L. (2012). Facilitating play through communication: Significance of teeth exposure in the gorilla play face. *American Journal of Primatology, 74,* 157–164.

Walters, J. R., Doerr, P. D., & Carter, J. H. I. (1992). Delayed dispersal and reproduction as a life-history tactic in cooperative breeders: Fitness calculations from red-cockaded woodpeckers. *American Naturalist, 139,* 623–643.

Wang, Z. X., Ferris, C. F., & Devries, G. J. (1994). Role of septal vasopressin innervation in paternal behavior in prairie voles (*Microtus-Ochrogaster*). *Proceedings of the National Academy of Sciences, U. S. A., 91,* 400–404.

Ward, P., & Zahavi, A. (1973). The importance of certain assemblages of birds as "information centres" for finding food. *Ibis, 115,* 517–534.

Warkentin, K. (1995). Adaptive plasticity in hatching age: A response to predation risk trade-offs. *Proceedings of the National Academy of Sciences, U.S.A., 92,* 3507–3510.

Warkentin, K. (1999). The development of behavioral defenses: A mechanistic analysis of vulnerability in red-eyed tree frog hatchlings. *Behavioral Ecology, 10,* 251–262.

Warkentin, K. (2000). Wasp predation and wasp-induced hatching of red-eyed treefrog eggs. *Animal Behaviour, 60,* 503–510.

Warkentin, K. M. (2005). How do embryos assess risk? Vibrational cues in predator-induced hatching of red-eyed treefrogs. *Animal Behaviour, 70,* 59–71.

Warkentin, K. M., Caldwell, M. S., & McDaniel, J. G. (2006). Temporal pattern cues in vibrational risk assessment by embryos of the red-eyed treefrog, *Agalychnis callidryas. Journal of Experimental Biology, 209,* 1376–1384.

Warkentin, K. M., Caldwell, M. S., Siok, T. D., D'Amato, A. T., & McDaniel, J. G. (2007). Flexible information sampling in vibrational assessment of predation risk by red-eyed treefrog embryos. *Journal of Experimental Biology, 210,* 614–619.

Warren, W. C., Clayton, D. F., Ellegren, H., Arnold, A. P., Hillier, L. W., Kunstner, A., Searle, S., White, S., Vilella, A. J., Fairley, S., Heger, A., Kong, L. S., Ponting, C. P., Jarvis, E. D., Mello, C. V., Minx, P., Lovell, P., Velho, T. A. F., Ferris, M., Balakrishnan, C. N., Sinha, S., Blatti, C., London, S. E., Li, Y., Lin, Y. C., George, J., Sweedler, J., Southey, B., Gunaratne, P., Watson, M., Nam, K., Backstrom, N., Smeds, L., Nabholz, B., Itoh, Y., Whitney, O., Pfenning, A. R., Howard, J., Voelker, M., Skinner, B. M., Griffin, D. K., Ye, L., McLaren, W. M., Flicek, P., Quesada, V., Velasco, G., Lopez-Otin, C., Puente, X. S., Olender, T., Lancet, D., Smit, A. F. A., Hubley, R., Konkel, M. K., Walker, J. A., Batzer, M. A., Gu, W. J., Pollock, D. D., Chen, L., Cheng, Z., Eichler, E. E., Stapley, J., Slate, J., Ekblom, R., Birkhead, T., Burke, T., Burt, D., Scharff, C., Adam, I., Richard, H., Sultan, M., Soldatov, A., Lehrach, H., Edwards, S. V., Yang, S. P., Li, X. C., Graves, T., Fulton, L., Nelson, J., Chinwalla, A., Hou, S. F., Mardis, E. R., & Wilson, R. K. (2010). The genome of a songbird. *Nature, 464,* 757–762.

Watts, D. P. (1990). Mountain gorilla life histories, reproductive competition, and sociosexual behavior and some implications for captive husbandry. *Zoo Biology, 9,* 185–200.

Watts, H. E., Blankenship, L. M., Dawes, S. E., & Holekamp, K. E. (2010). Responses of spotted hyenas to lions reflect individual differences in behavior. *Ethology, 116,* 1199–1209.

Webster, M. M., & Hart, P. J. B. (2006). Subhabitat selection by foraging threespine stickleback (*Gasterosteus aculeatus*): Previous experience and social conformity. *Behavioral Ecology and Sociobiology, 60,* 77–86.

Webster, M. M., & Ward, A. J. W. (2011). Personality and social context. *Biological Reviews, 86,* 759–773.

Webster, M. S., & Westneat, D. F. (1998). The use of molecular markers to study kinship in birds: Techniques and questions. In R. DeSalle & B. Schierwater (Eds.), *Molecular approaches to ecology and evolution* (pp. 7–35). Basel, Switzerland: Birkhauser.

Wedekind, C., & Furi, S. (1997). Body odour preferences in men and women: Do they aim for specific MHC combinations or simply heterozygosity? *Proceedings of the Royal Society of London, Series B, 264,* 1471–1479.

Wedekind, C., Seebeck, T., Bettens, F., & Paepke, A. J. (1995). MHC dependent mate preferences in humans. *Proceedings of the Royal Society of London, Series B, 260,* 245–249.

Wegner, K. M., Kalbe, M., Kurtz, J., Reusch, T. B. H., & Milinski, M. (2003). Parasite selection for immunogenetic optimality. *Science, 301,* 1343.

Wegner, K. M., Reusch, T. B. H., & Kalbe, M. (2003). Multiple parasites are driving major histocompatibility complex polymorphism in the wild. *Journal of Evolutionary Biology, 16,* 224–232.

Weigensberg, I., & Roff, D. (1996). Natural heritabilities: Can they be reliably estimated in the laboratory? *Evolution, 50,* 2149–2157.

Weil, T., Rehli, M., & Korb, J. (2007). Molecular basis for the reproductive division of labour in a lower termite. *Bmc Genomics, 8.*

Weinstock, M. (1997). Does prenatal stress impair coping and regulation of hypothalamic-pituitary-adrenal axis? *Neuroscience and Biobehavioral Reviews, 21,* 1–10.

Weiss, A., Adams, M. J., & King, J. E. (2011). Happy orangutans live longer lives. *Biology Letters, 7,* 872–874.

Weiss, A., Inoue-Murayama, M., King, J. E., Adams, M. J., & Matsuzawa, T. (2012). All too human? Chimpanzee and orangutan personalities are not anthropomorphic projections. *Animal Behaviour, 83,* 1355–1365.

Weiss, A., King, J. E., & Perkins, L. (2006). Personality and subjective well-being in orangutans (*Pongo pygmaeus* and *Pongo abelii*). *Journal of Personality and Social Psychology, 90,* 501–511.

Weiss, D., Helmreich, S., Mostl, E., Dzidic, A., & Bruckmaier, R. M. (2004). Coping capacity of dairy cows during the change from conventional to automatic milking. *Journal of Animal Science, 82,* 563–570.

Weiss, J. (1972). Influence of psychological variables on stress-induced pathology. In R. Porter & J. Knight (Eds.), *Physiology, emotion and psychosomatic illness.* Amsterdam: Elsevier.

Wells, J. C. K. (2007). The thrifty phenotype as an adaptive maternal effect. *Biological Reviews, 82,* 143–172.

Wenegrat, B. (1984). *Sociobiology and mental disorders: A new view.* Menlo Park, CA: Addison-Wesley.

Wenseleers, T., & Ratnieks, F. L. W. (2006). Enforced altruism in insect societies. *Nature, 444,* 50.

Werner, E. E., & Hall, D. J. (1974). Optimal foraging and the size selection of prey by the bluegill sunfish *Lepomis macrochirus. Ecology, 55,* 1042–1055.

West, H., Clancy, A., & Michael, P. (1992). Enhanced responses of nucleus accumbens neurons in male rats to novel odors associated with sexually receptive males. *Brain Research, 585,* 49–55.

West, M. J., King, A. P., & Freeberg, T. M. (1998). Dual signaling during mating in brown-headed cowbirds (*Molothrus ater,* family Emberizidae/Icterinae). *Ethology, 104,* 250–267.

West, M. J., King, A. P., & White, D. J. (2003). The case for developmental ecology. *Animal Behaviour, 66,* 617–622.

West, S. A., El Mouden, C., & Gardner, A. (2011). Sixteen common misconceptions about the evolution of cooperation in humans. *Evolution and Human Behavior, 32,* 231–262.

West-Eberhard, M. J. (1975). The evolution of social behavior by kin selection. *Quarterly Review of Biology, 50,* 1–35.

West-Eberhard, M. J. (1979). Sexual selection, social competition and evolution. *Proceedings of the American Philosophical Society, 123,* 222–234.

West-Eberhard, M. J. (1981). Intragroup selection and the evolution of insect societies. In R. D. Alexander & D. W. Tinkle (Eds.), *Natural selection and social behavior* (pp. 3–17). New York: Chiron Press.

West-Eberhard, M. J. (1989). Phenotypic plasticity and the origins of diversity. *Annual Review of Ecology and Systematics, 20,* 249–278.

West-Eberhard, M. J. (2003). *Developmental plasticity and evolution.* New York: Oxford University Press.

Westneat, D. F. (1987a). Extra-pair copulations in a predominantly monogamous bird: Genetic evidence. *Animal Behaviour, 35,* 877–886.

Westneat, D. F. (1987b). Extra-pair copulations in a predominantly monogamous bird: Observations of behaviour. *Animal Behaviour, 35,* 865–876.

Westneat, D. F., Frederick, P. C., & Wiley, R. H. (1987). The use of genetic markers to estimate the frequency of successful alternative reproductive tactics. *Behavioral Ecology and Sociobiology, 21,* 35–45.

Westneat, D. F., Sherman, P. W., & Morton, M. L. (1990). The ecology and evolution of extra-pair copulations in birds. In D. Power (Ed.), *Current ornithology* (pp. 331–369). New York: Plenum Press.

Westneat, D. F., Walters, A., McCarthy, T., Hatch, M., & Hein, W. (2000). Alternative mechanisms of nonindependent mate choice. *Animal Behaviour, 59,* 467–476.

Wheeler, J., & Rissing, S. (1975). Natural history of *Veromessor pergandei* I. The nest. *Pan-Pacific Entomologist, 51,* 205–216.

White, L. (1994). Coresidence and leaving home: Young adults and their parents. *Annual Review of Sociology, 20,* 81–102.

White, L., & Reidmann, A. (1992). Ties among adult siblings. *Social Forces, 71,* 85–102.

Whitehead, H. (2010). Conserving and managing animals that learn socially and share cultures. *Learning & Behavior, 38,* 329–336.

Whiten, A. (1992). On the nature and evolution of imitation in the animal kingdom: Reappraisal of a century of research. *Advances in the Study of Behavior, 21,* 239–283.

Whiten, A. (2005). The second inheritance system of chimpanzees and humans. *Nature, 437,* 52–55.

Whiten, A. (2011). The scope of culture in chimpanzees, humans, and ancestral apes. *Philosophical Transactions of the Royal Society, Series B–Biological Sciences, 366,* 997–1007.

Whiten, A., & Boesch, C. (2001, January). The culture of chimpanzees. *Scientific American,* 60–67.

Whiten, A., Goodall, J., McGrew, W., Nishida, T., Reynolds, V., Sugiyama, Y., Tutin, C., Wrangham, R., & Boesch, C. (1999). Cultures in chimpanzees. *Nature, 399,* 682–685.

Whiten, A., Horner, V., & de Waal, F. B. M. (2005). Conformity to cultural norms of tool use in chimpanzees. *Nature, 437,* 737–740.

Whitfield, C. W., Ben-Shahar, Y., Brillet, C., Leoncini, I., Crauser, D., Le Conte, Y., Rodriguez-Zas, S., & Robinson, G. E. (2006). Genomic dissection of behavioral maturation in the honey bee. *Proceedings of the National Academy of Sciences, U.S.A., 103,* 16068–16075.

Whitfield, C. W., Cziko, A. M., & Robinson, G. E. (2003). Gene expression profiles in the brain predict behavior in individual honey bees. *Science, 302,* 296–299.

Whitten, P. L. (1987). Infants and adult males. In B. B. Smuts, D. L. Cheney, R. M. Seyfarth, R. Wrangham, & T. T. Struhsaker (Eds.), *Primate societies* (pp. 343–357). Chicago: University of Chicago Press.

Widder, E. A. (2010). Bioluminescence in the ocean: Origins of biological, chemical, and ecological diversity. *Science, 328,* 704–708.

Widemo, F. (1998). Alternative reproductive strategies in the ruff (*Philomachus pugnax*): A mixed ESS. *Animal Behaviour, 56,* 329–336.

Wiebe, K. L. (2011). Nest sites as limiting resources for cavity-nesting birds in mature forest ecosystems: A review of the evidence. *Journal of Field Ornithology, 82,* 239–248.

Wilcox, R. S. (1972). Communication by surface waves: Mating behavior of a water strider (Gerridae). *Journal of Comparative Physiology, 80,* 255–266.

Wilcox, R. S. (1995). Ripple communication in aquatic and semi-aquatic insects. *Ecoscience, 2,* 109–115.

Wilkinson, G. (1984). Reciprocal food sharing in vampire bats. *Nature, 308,* 181–184.

Wilkinson, G. (1985). The social organization of the common vampire bat. I. Patterns and causes of association. *Behavioral Ecology and Sociobiology, 17,* 111–121.

Wilkinson, G. S. (1993). Artificial sexual selection alters allometry in the stalk-eyed fly *Cyrtodiopsis dalmanni* (Diptera: Diopsidae). *Genetical Research, 62,* 213–222.

Wilkinson, G. S., Kahler, H., & Baker, R. H. (1998). Evolution of female mating preferences in stalk-eyed flies. *Behavioral Ecology, 9,* 525–533.

Wilkinson, G. S., & Reillo, P. R. (1994). Female choice response to artificial selection on an exaggerated ale trait in a stalk-eyed fly. *Proceedings of the Royal Society of London, Series B, 255,* 1–6.

Williams, G. (1966). *Adaptation and natural selection.* Princeton, NJ: Princeton University Press.

Willing, E. M., Bentzen, P., van Oosterhout, C., Hoffmann, M., Cable, J., Breden, F., Weigel, D., & Dreyer, C. (2010). Genome-wide single nucleotide polymorphisms reveal population history and adaptive divergence in wild guppies. *Molecular Ecology, 19,* 968–984.

Willis-Owen, S. A. G., & Flint, J. (2006). The genetic basis of emotional behaviour in mice. *European Journal of Human Genetics, 14,* 721–728.

Willmer, P. G., & Stone, G. N. (2004). Behavioral, ecological, and physiological determinants of the activity patterns of bees. *Advances in the Study of Behavior, 34,* 347–466.

Wills, C. (1991). Maintenance of multiallelic polymorphism at the MHC region. *Immunologica Reviews, 124,* 165–220.

Wilson, A. D. M., & Godin, J. G. J. (2009). Boldness and behavioral syndromes in the bluegill sunfish, *Lepomis macrochirus. Behavioral Ecology, 20,* 231–237.

Wilson, A. D. M., & Godin, J. G. J. (2010). Boldness and intermittent locomotion in the bluegill sunfish, *Lepomis macrochirus. Behavioral Ecology, 21,* 57–62.

Wilson, A. D. M., & McLaughlin, R. L. (2007). Behavioural syndromes in brook charr, *Salvelinus fontinalis*: Prey-search in the field corresponds with space use in novel laboratory situations. *Animal Behaviour, 74,* 689–698.

Wilson, A. D. M., & Stevens, E. D. (2005). Consistency in context-specific measures of shyness and boldness in rainbow trout, *Oncorhynchus mykiss. Ethology, 111,* 849–862.

Wilson, D., Coleman, K., Clark, A., & Biederman, L. (1993). Shy-bold continuum in pumpkinseed sunfish (*Lepomis gibbosus*): An ecological study of a psychological trait. *Journal of Comparative Psychology, 107,* 250–260.

Wilson, D. J., & Lefcort, H. (1993). The effects of predator diet on the alarm response of red-legged frog, *Rana aurora,* tadpoles. *Animal Behaviour, 46,* 1017–1019.

Wilson, D. S. (1980). *The natural selection of populations and communities.* Menlo Park, CA: Benjamin Cummings.

Wilson, D. S. (1990). Weak altruism, strong group selection. *Oikos, 59,* 135–140.

Wilson, D. S., Clark, A. B., Coleman, K., & Dearstyne, T. (1994). Shyness and boldness in humans and other animals. *Trends in Ecology & Evolution, 9,* 442–446.

Wilson, D. S., Muzzall, P. M., & Ehlinger, T. J. (1996). Parasites, morphology, and habitat use in a bluegill sunfish (*Lepomis macrochirus*) population. *Copeia,* 348–354.

Wilson, D. S., & Wilson, E. O. (2007). Rethinking the theoretical foundation of sociobiology. *Quarterly Review of Biology, 82,* 327–348.

Wilson, E. O. (1971). *The insect societies.* Cambridge, MA: Harvard University Press.

Wilson, E. O. (1975). *Sociobiology: The new synthesis.* Cambridge, MA: Harvard University Press.

Wilson, E. O. (1980). Caste and division of labor in leaf-cutter ants (Hymenoptera: Fromicidae: Atta). I. The ergonomic optimization of leaf cutting. *Behavioral Ecology and Sociobiology, 7,* 157–165.

Wilson, E. O., & Holldobler, B. (2005a). Eusociality: Origin and consequences. *Proceedings of the National Academy of Sciences, U.S.A., 102,* 13367–13371.

Wilson, E. O., & Holldobler, B. (2005b). The rise of the ants: A phylogenetic and ecological explanation. *Proceedings of the National Academy of Sciences, U.S.A., 102,* 7411–7414.

Wilsson, E. O., & Sundgren, P. E. (1997). The use of a behaviour test for selection of dogs for service and breeding. 2. Heritability for tested parameters and effect of selection based on service dog characteristics. *Applied Animal Behaviour Science, 54,* 235–241.

Wiltschko, R., & Wiltschko, W. (1995). *Magnetic orientation in animals.* Berlin: Springer-Verlag.

Wiltschko, R., & Wiltschko, W. (2003). Avian navigation: From historical to modern concepts. *Animal Behaviour, 65,* 257–272.

Wiltschko, R., & Wiltschko, W. (2006). Magnetoreception. *Bioessays, 28,* 157–168.

Wiltschko, W. (1968). Uber den Einflu-2 statischer Magentfelder auf die Zugorientierung der Rotkehlchen (*Erithacus rubecula*). *Zeitschrift fur Tierpsychologie, 65,* 257–272.

Winberg, S., Nilsson, G., & Olsen, K. H. (1992). Changes in brain serotonergic activity during hierarchy behavior in Artic charr (*Savelinus alpinus* L.) are socially mediated. *Journal of Comparative Physiology, 170,* 93–99.

Winberg, S., Winberg, Y., & Fernald, R. (1997). Effect of social rank on brain monoaminergic activity in a cichlid fish. *Brain, Behavior and Evolution, 49,* 230–236.

Winger, B. M., Lovette, I. J., & Winkler, D. W. (2012). Ancestry and evolution of seasonal migration in the Parulidae. *Proceedings of the Royal Society, 279,* 610–618.

Wingfield, J. C., Lynn, S. E., & Soma, K. K. (2001). Avoiding the "costs" of testosterone: Ecological bases of hormone-behavior interactions. *Brain, Behavior and Evolution, 57,* 239–251.

Withers, G., Fahrbach, S., & Robinson, G. (1993). Selective neuroanatomical plasticity and division of labour in the honeybee. *Nature, 354,* 238–240.

Witkin, S. R., & Ficken, M. S. (1979). Chickadee alarm calls: Does mate investment pay dividends? *Animal Behaviour, 27,* 1275–1276.

Witschi, E. (1961). Sex and secondary sexual characteristics. In A. Marshall (Ed.), *Biology and comparative physiology of birds* (pp. 115–168). New York: Academic Press.

Witte, K., Hirschler, U., & Curio, E. (2000). Sexual imprinting on a novel adornment influences mate preferences in the Javanese mannikin *Lonchura leucogastroides. Ethology, 106,* 349–363.

Wood-Gush, D. G. M., & Vestergaard, K. (1991). The seeking of novelty and its relation to play. *Animal Behaviour, 42,* 599–606.

Woodland, D. J., Jaafar, Z., & Knight, M.-L. (1980). The "pursuit deterrent" function of alarm calls. *American Naturalist, 115,* 748–753.

Woodroffe, R., & Vincent, A. (1994). Mother's little helpers: Patterns of male care in mammals. *Trends in Ecology & Evolution, 9,* 294–297.

Woolfenden, G. E., & Fitzpatrick, J. W. (1978). Inheritance of territory in group-breeding birds. *Bioscience, 28,* 104–108.

Woolfenden, G. E., & Fitzpatrick, J. W. (1984). *The Florida scrub jay: Demography of a cooperative-breeding bird.* Princeton, NJ: Princeton University Press.

Woolfenden, G. E., & Fitzpatrick, J. W. (1990). Florida scrub jays: A synopsis after 18 years. In P. B. Stacey & W. D. Koenig (Eds.), *Cooperative breeding in birds* (pp. 241–266). Cambridge: Cambridge University Press.

Wright, W. G. (1998). Evolution of nonassociative learning: Behavioral analysis of a phylogenetic lesion. *Neurobiology of Learning and Memory, 69,* 326–337.

Wright, W. G. (2000). Neuronal and behavioral plasticity in evolution: Experiments in a model lineage. *Bioscience, 50,* 883–894.

Wright, W. G., Kirschman, D., Rozen, D., & Maynard, B. (1996). Phylogenetic analysis of learning-related neuromodulation in molluscan mechanosensory neurons. *Evolution, 50,* 2248–2263.

Wyatt, G., & Davey, K. (1996). Cellular and molecular action of juvenile hormones: II. Roles of juvenile hormone in adult insects. *Advances in Insect Physiology, 26,* 2–155.

Wyles, J., Kunkel, J., & Wilson, A. (1983). Birds, behaviour and anatomical evolution. *Proceedings of the National Academy of Sciences, 80,* 4394–4397.

Xitco, M. J., & Roitblat, H. L. (1996). Object recognition through eavesdropping: Passive echolocation in bottlenose dolphins. *Animal Learning & Behavior, 24,* 355–365.

Yalcin, B., Willis-Owen, S. A. G., Fullerton, J., Meesaq, A., Deacon, R. M., Rawlins, J. N. P., Copley, R. R., Morris, A. P., Flint, J., & Mott, R. (2004). Genetic dissection of a behavioral quantitative trait locus shows that Rgs2 modulates anxiety in mice. *Nature Genetics, 36,* 1197–1202.

Yamamoto, I. (1987). Male parental care in the raccoon dog (*Nyctereutes procyonoides*) during the early rearing stages. In Y. Eto, J. L. Brown, & J. Kikkawa (Eds.), *Animal societies* (Vol. 4, pp. 189–196). Tokyo: Japan Societies Scientific Press.

Yarbrough, W. G., Quarmby, V. E., Simental, J. A., Joseph, D. R., Sar, M., Lubahn, D. B., Olsen, K. L., French, F. S., & Wilson, E. M. (1990). A single base mutation in the androgen receptor gene causes androgen insensitivity in the testicular feminized rat. *Journal of Biological Chemistry, 265,* 8893–8900.

Ydenberg, R. C., & Dill, L. M. (1986). The economics of fleeing from predators. *Advances in the Study of Behavior, 16,* 229–249.

Yoerg, S. I. (1991). Ecological frames of mind: The role of cognition in behavioral ecology. *Quarterly Review of Biology, 66,* 288–301.

Yokoyama, S. (2002). Molecular evolution of color vision in vertebrates. *Gene, 300,* 69–78.

Yokoyama, S., Radlwimmer, B., & Blow, N. (2000). Ultraviolet pigments in birds evolved from violet pigments by a single amino acid change. *Proceedings of the National Academy of Sciences, U.S.A., 97,* 7366–7371.

Young, H. (1987). Herring gull preying on rabbits. *British Birds, 80,* 363.

Young, L. J., Young, A. Z. M., & Hammock, E. A. D. (2005). Anatomy and neurochemistry of the pair bond. *Journal of Comparative Neurology, 493,* 51–57.

Zabel, C., Glickman, S., Frank, L., Woodmansee, K., & Keppel, G. (1992). Coalition formation in a colony of prepubertal spotted hyenas. In A. H. Harcourt & F. B. M. de Waal (Eds.), *Coalitions and alliances in humans and other animals* (pp. 112–135). Oxford: Oxford University Press.

Zahavi, A. (1975). Mate selection: A selection for a handicap. *Journal of Theoretical Biology, 53,* 205–214.

Zahavi, A. (1977). The cost of honesty (futher remarks on the handicap principle). *Journal of Theoretical Biology, 67,* 603–605.

Zahavi, A. (2003). Indirect selection and individual selection in sociobiology: My personal views on theories of social behaviour. *Animal Behaviour, 65,* 859–863.

Zahavi, A., & Zahavi, A. (1997). *The handicap principle.* New York: Oxford University Press.

Zajonc, R. B. (1965). Social facilitation. *Science, 269*–274.

Zak, P. J. (2004). Neuroeconomics. *Philosophical Transactions of the Royal Society of London, Series B, 359,* 1737–1748.

Zak, P. J. (2011a). Moral markets. *Journal of Economic Behavior & Organization, 77,* 212–233.

Zak, P. J. (2011b). The physiology of moral sentiments. *Journal of Economic Behavior & Organization, 77,* 53–65.

Zak, P. J., Kurzban, R., & Matzner, W. T. (2005). Oxytocin is associated with human trustworthiness. *Hormones and Behavior, 48,* 522–527.

Zamble, E., Hadad, G., Mitchell, J., & Cutmore, T. (1985). Pavlovian conditioning of sexual arousal: First- and second-order effects. *Journal of Experimental Psychology: Animal Behavior Processes, 11,* 598–610.

Zamble, E., Mitchell, J., & Findlay, H. (1986). Pavlovian conditioning of sexual arousal: Parametric and background manipulations. *Journal of Experimental Psychology: Animal Behavior Processes, 12,* 403–411.

Zentall, T. R. (2005). Selective and divided attention in animals. *Behavioural Processes, 69,* 1–15.

Zentall, T. R., & Galef, B. G. (1988). *Social learning: Psychological and biological perspectives.* Hillsdale, NJ: Lawrence Erlbaum Associates.

Zhan, S., Merlin, C., Boore, J., & Reppert, S. (2011). The monarch butterfly genome yields insights into long-distance migration. *Cell, 147,* 1171–1185.

Zhang, S., Amstein, T., Shen, J., Brush, F. R., & Gershenfeld, H. K. (2005). Molecular correlates of emotional learning using genetically selected rat lines. *Genes, Brain and Behavior, 4,* 99–109.

Ziegler, A., Kentenich, H., & Uchanska-Ziegier, B. (2005). Female choice and the MHC. *Trends in Immunology, 26,* 496–502.

Zink, R. M. (2002). Towards a framework for understanding the evolution of avian migration. *Journal of Avian Biology, 33,* 433–436.

Zink, R. M. (2011). The evolution of avian migration. *Biological Journal of the Linnean Society, 104,* 237–250.

Zuberbuhler, K. (2009). Survivor signals: The biology and psychology of animal alarm calling. *Advances in the Study of Behavior, 40,* 277–322.

Zuckerman, M. (1979). *Sensation seeking.* Hillsdale, NJ: Lawrence Erlbaum Associates.

Zuckerman, M. (1994). *Behavioral expressions and biosocial bases of sensation seeking.* Cambridge: Cambridge University Press.

Zuckerman, M. (1996). The psychobiological model for impulsive unsocialized sensation seeking: A comparative approach. *Neuropsycholobiology, 34,* 125–129.

Zuckerman, S. (1932). *The social life of monkeys and apes.* New York: Harcourt Brace.

Zuk, M., & Kolluru, G. R. (1998). Exploitation of sexual signals by predators and parasitoids. *Quarterly Review of Biology, 73,* 415–438.

Zuk, M., Rotenberry, J. T., & Tinghitella, R. M. (2006). Silent night: Adaptive disappearance of a sexual signal in a parasitized population of field crickets. *Biology Letters, 2,* 521–524.

Credits

Index

Note: Page numbers in *italics* refer to illustrations and tables.

canaries (*Serinus canaria*), 436
Canis lupus (wolves), 331
Cannon, W., 556
Capaldi, E. A., 94
Capellini, L., 98
Capitania-Kwok, T., 149
capuchin monkey (*Cebus paella*), *174*, 174–75
Caraco, T., 173, 359, 360, 361, 363, 373, 374, 542
Cardinalis cardinalis (northern cardinal), 436
Carere, C., 541, 548
Carlier, P., 145, 146
Caro, T., 182, 183, 184, 185, 351, 407, 516, 517, 518, 519, 541
carotenoid-based food, plumage coloration and, 72–74
Carpenter, C. R., 311
Carpodacus mexicanus (house finch), 71–75, *72, 73*
Carter, C. S., 122, 241
cartilaginous fish, phylogeny, 55
Case, T. J., 328
Cassey, P., 371
Catchpole, C., 431
Catley, K. M., 55
cats, learning in, 140
Cauchard, L., 371
causality, 22–23
Cavalli-Svorza, L. L., 185
Cavigelli, S. A., 556
Cebus paella (capuchin monkey), *174*, 174–75
cephalopods, 387, 550
Cercopithecus mitis (blue monkeys), 312
Cervus elaphus (red deer), 201, *203*, 225–27, *226*
cetaceans:
    alliances in, 334
    See also bottle-nosed dolphins
Chapman, T., 240
Charnov, E., 351, 352, 354, 355, 407
Charnov, R., 229
Chase, I. D., 497, 501, 502
Chatterjee, K., 494
cheater problem, 301
cheetah (*Actinonyx jubatus*), 184, *184*, 331
    object play in, 517–19, *518, 519*
    predator inspection by gazelles and, 400–402, *401*
chemical communication, 152
chemical defense, 149, 150, 151, 152, 408–9
chemical signaling, queen-worker and, 106
Cheney, D. L., 418, 438, 440
Cherrett, J., 429
Cherry, L., 525
Chesser, R. T., 475
Chetyrbok, I. S., 102
Chicago School of Animal Behavior, 313

chimpanzees:
    aggression and "recruitment screams" in, 504, *505*
    coalition forming and, *331*
    cultural transmission in, 169
    leaf play in, *516, 516*
    personality in, 554–56
    See also primates
chinook salmon (*Oncorhynchus tshawytscha*), 370
Chivers, D., 149, 150, 151
Chiyo, P., 176
Cho, M. M., 86
Choleris, E., 223
Christensen, B., 351
Christensen, J. W., 134
chromosomes, 37
Chrousos, G., 82
Church, S., 114
churr calls, learned, 154
cichlid fish (*Neolamprologus pulcher*):
    early nest development and behavior in, 123
    Oreochromis mossambicus, 503, *504*
Cissa erythrorhyncha (Asian red-billed blue magpies), 366
Cistohorus palustris (western marsh wren), 431
Clark, A., 39
Clark, A. B., 541
Clark, C. W., 389
Clark, M., 83
Clarke, J. A., 410
classical conditioning. *See* Pavlovian conditioning
Clayton, D., 116, 431
Clayton, N. S., 219, 365, 366, 367, 372, 373
Cleland, J., 298
Clemens, L. G., 83
Clements, K., 324, 326
Clethrionomys rufocanus (grey-sided vole), 252
Clever Hans, *419*, 419–20
cliff swallow (*Petrochelidon pyrrhonota*), *41*, 41–42, 173, 423–24
climate change:
    development, dispersal and, 120, *120*
    mating systems and, 253, *253*
Clipperton-Allen, A. E., 223
Close, R., 521
Clotfelter, E. D., 488
Clutton-Brock, T., 182, 184, 225, 226, 227, 312, 313, 369
coaching, 185
coalition, defined, 565
coalition formation, 331–33, 334
    alliances and, *331*, 331–32, 334
    defined, 331
    in primates, 333–34
coatis (*Nasua narica*), 331
Coccaro, E., 487
cock-of-the-rocks (*Rupicola rupicola*), communication in, 432, *432*

co-evolution, naive prey, introduction programs and, 392, *392*
cognition, social play and, 523, 536
Colborn, T., 253
Coleman, K., 544
Coleman, R., 229
Colinus virginianus (Bobwhite Quail), 350
collaboration, 100
Columbia livia (pigeons), 373–76, *375, 376*
common mole rat (*Cryptomys hottentotus*), *10*, 10–11, *11*
common raven (*Corvus corax*), *424*, 424–25, *425*
communal nesting, 301
communication, 418–47
    alarm calls and, 440–43
    animal-human, 419
    anthropogenic change and, 432
    "arms race" in, 421
    birdsong as, 431–37
    by chemical signaling, 428–31
    classical approach to, 421
    deceptive, 420–21, 422, 446
    defined, 418–19, 444, 565
    foraging and, 423–24
    handicap principle and, 422, 445, 466
    honesty in, 420–22, 444, 445, 446
    learning and, 438–40, 446
    mathematical model of, 445
    mating and, 431–37
    modes of, 418
    predation and, 438–43
    by ripples, 437
    by vibrations, 428–31
comparative method, 58
competition, predation and, 394
conceptual approach:
    defined, 565
    to ethology, *17*, 18–19, 26
conditioned response (CR), 136, *136*, 565
conditioned stimulus (CS), 135, 136, 162, 565
Conner, W. E., 405
Connor, R., 324, 334, 338
conopressin, evolutionary history of, 85
conservation biology, 25
    genetic diversity, genetic quality and, 202
    symmetry as indicator of risk and, 44, *44*
conspecific cueing hypothesis, 461, *462*, 478
"conspirational whispers" hypothesis, 421
convergent evolution, 57
    defined, 565
    in wing structure, 57
Cook, M., 187
Cooke, F., 218

cooperation, 308–44, 342, 484
    alliances and, 331
    byproduct mutualism and, 324–26
    coalitions and, 331, 333–34
    defined, 309, 565
    food sharing and, *308*, 308–9
    foraging and, 329–30, 363–64
    game theory and, 313–14
    genetic relatedness and, 50–52
    group selection and, 327–31
    helping in birthing process and,
        310–11
    interspecific mutualism and, 338–39
    paths to, 312–31, *313*
    phylogenetic approach to, 334–38
    reciprocity and, 313–24
    social grooming and, 311–12
    tragedy of the commons,
        overharvesting and, 332, *332*
cooperative breeding, 285–87, 289–91.
        See also helpers–at–the–nest
Cooper's hawk (*Accipiter cooperii*), 438
copepods, 545
coping styles, 540, *556*, 556–58, 565
Coppens, C. M., 556
copperhead snakes (*Agkistrodon
        contortrix*), winner and loser
        effects in, 499–500, *500*
copulation-solicitation displays (CSDs),
        433
copying:
    cultural transmission and, 196
    defined, 565
Corballis, M. C., 372
Cordero, C., 205
Cords, M., 312
Coria, J., 78
Cornette, R., 106
corpora pedunculata, 94
Correia, S. P. C., 373
corticosterone, 78, 158, 159, 457, 458,
        459, 485, 500, 557
corticotropin-releasing hormone (CRH),
        77, 82
cortisol, 485, 487
Corvidae, 365
Corvus:
    C. corax (common raven), *424,*
        424–25, *425*
    C. corone (European crow), 366
    C. frugilegus (rook), 366
    C. mondeula (jackdaw), 366
Coss, R., 286, 384, 385
Coultier, S., 502
courtship behavior, phylogeny and,
        62–63, *63*
courtship song, 203
Covich, A., 150
covids, moderate caching in, 367, 380
cowbirds, 304
    song learning and mate choice in,
        226–27
Cowie, R., 355, 356
cows, coping style in, 557

Cox, C., 227, 228
Cozza, K. C., 559
CR. *See* conditioned response (CR)
Cragshaw, P., 559
Craighero, L., 179
Crane, A. L., 150
Crawley, M., 357
Creel, S., 33
Crenicichla alta (pike cichlid), *46,* 316
Crespi, B., 299
Cresswell, W., 386
CRH. *See* corticotropin-releasing
        hormone (CRH)
crickets, Hawaiian, 8–10, *9*
Crockford, C., 418
crocodilians, phylogeny, 55
Crocuta crocuta (spotted hyena), 331,
        548–49, *549*
Croes, B. M., 397
crop raiding behavior, in elephants, 176,
        176
cross-fostering, 224, *224*
    defined, 565
    experiment, 42
Crowl, T., 150
Crowley, P. H., 494
Crozier, R., 299, 337, 427
crustaceans, aggression in, 488
cryptic, defined, 566
cryptic antipredator behavior, 387
cryptic mate choice, 263, 265
Cryptobrancus alleganeinsis
        (hellbenders), 150, *150*
Cryptomys hottentotus (common mole
        rat), *10,* 10–11, *11*
Cryptotermes secundus, 107
CS. *See* conditioned stimulus (CS)
CSDs. *See* copulation-solicitation
        displays (CSDs)
cuckoldry, male-male competition by,
        229–31
cuckoo birds, 304
cultural transmission, 8, 15–17, *170,*
        194–95, 196
    of birdsong, 189–91, *190*
    brain size and, 192–93, *193*
    in chimpanzees, 169
    copying and, *180,* 180–82, *181,* 196
    defined, 166, 169, 566
    in finches, 186, *186*
    genes and, 189–91, 196
    horizontal, 185, 188–89, 196
    in humans, *170, 171*
    imitation and, 177–80, *178,* 196
    importance of, 170–71
    local enhancement and, *172,* 172–73,
        196
    in macaques, 166–68, *167, 168,* 171
    mate choice and, 221–25
    mobbing and, 409
    modes of, 185–89
    oblique, 185, 187, 196
    personality and, 554–56
    scavenging and, 169

social facilitation and, *173,* 173–75,
        174, 196
    teaching and, 171, *171,* 183–85, *184,*
        196
    traditions and, 182, 196
    vertical, 185, 186–87, 196
culture, 169, 194
Curio, E., 399, 400, 409, 438
Currie, C. R., 348, 349
Curtsinger, J. W., 494
Cyanocitta cristata (blue jay), 324–26,
        326
Cziko, A.-M., 113

Dalgleish, T., 515
Dall, S. R. X., 540, 541
Dally, J. M., 373
Daly, M., 273, 274, 279, 282, 283, 285,
        290, 303
Damaliscus lunatus (topi antelope),
        441–42, *442*
damselfly (*Enallagma spp.*), 150–52, *151*
Danaus plexippus (monarch butterflies),
        406–7, 466, *466, 468,* 468–69,
        469
Danchin, E., 15, 166, 364
Daphnia magna (water fleas), 453–54,
        454
Darwin, C., 8, 24, 30, 31, 32, 38, 52, 53,
        54, 64, 65, 200, 201, 234, 265,
        312, 342, 482, 534, 560, 562
Dasyurus maculatus (spotted-tail quoll),
        392
Davey, K., 87
Davies, N. B., 145, 238, 249, 250, 251, 255,
        263, 266, 267, 421, 462, 463, 492
    interview with, 264–65. *264*
Davis, D., 7
Davis, J., 279, 282, 283, 285, 290
Dawkins, M. S., 328, 420, 421
Dawkins, R., 43, 64, 304
Dawson, A., 79
Dawson, W. D., 39
Dayrat, B., 54
Dean, J., 408
"dear enemy effect," 507
death feigning, as antipredator behavior,
        404–5, *405*
De Backer, C. J. S., 421
Decety, J., 179
defensive response, copying and, *181,*
        181–82
de Goede, P., 36, 554
de Jong, G., 36, 38, 554
de Kloet, E. R., 158, 500, 557
de Kort, S. R., 366, 367
De Laet, J., 410
deletion mutations, 37
Delph, L., 259
dendrites, 89, *89,* 566
dendritic spines, 89, *89*
    learning and, 219
    sex differences in number of, 93, *93*
Denison, R. F., 31

fossil record, determining polarity and, 57

FOXP2 gene, song learning and, 115, *115*

Fragasy, D. M., 183

Frank, L. G., 548

Franks, N. R., 182

Fraser, D. F., 515, 516, 546

Freake, M. J., 471, 472

Freeberg, T., 189, 224, 225, 423

French, J. A., 154

Fretwell, S., 452

friarbird (*Philemon corniculatus*), 409

frogs, 216, *216*
    ethology and disease avoidance in, 456
    See also specific frogs

Frommen, J. G., 300

Frost, B., 469

fruit fly (*Drosophila mercatorum*), 148, 203

Fruteau, C., 312

Fuller, P., 410

functional magnetic resonance imaging (fMRI), 91

Furi, S., 210, 211

fur seal, unihemispheric sleep in, *99*

Gabor, C. S., 223

Gadagkar, R., 154, 484

Gaffrey, G., 311

Galápagos Islands, finch studies on, 189–91

Galef, A., 423

Galef, B. G., 15, 83, 166, 170, 175, 178, 179, 183, 185

Galef, J., 166

Gallagher, J., 220

Galleria mellonella (greater wax moths), 390–91, *391*

Gallese, V., 180

Gamboa, G. J., 484

game theory, 313–14, 343, 344
    models of aggression, 489–96, 508
    personality and, 541–42
    See also prisoner's dilemma game

Gannon, D. P., 389

Garcia, J., 141, 142

García-Peña, G. E., 253

Garrulus glandarius (European jay), 366

Gasterosteus aculeatus (sticklebacks), 47, 146–47, 210, 211, *211, 212,* 453, 454, *454*

Gaulin, S. J. C., 92

Gauthreaux, S., 465

gazelle fawn, mathematical optimality and foraging in, *20*

gazelles, approach behavior in, 400–402, *401*

Geary, D. C., 294

geese, migration of, *466*

Geist, V., 487, 523

Gelowitz, C., 150

genes:
    cultural transmission and, 189–91, 196
    mRNA, honeybee foraging and, 112–14
    parasite resistance and, 208–9
    for polygenic traits, locating, 109–11
    proximate analysis and, 106
    termite workers, queens and, 106, *107*

Genetical Theory of Natural Selection (Fisher), 65

genetic diversity, migration and, 37

genetic recombination, 37, 566

genetic relatedness, calculating, 277, 277–78, *278*

genetics, mate choice and, 202

genetic variation, 36, 566

Genfron, R. P., 350

genotype, defined, 34, 566

Geospiza:
    G. fortis (ground finch), 189, 190
    G. scandens (cactus finch), 189, 190

gerbils, in utero exposure to high testosterone levels, 84

German blackcap bird (*Sylvia atricapilla*), 472

gestational diabetes, 295

Getz, L. L., 122

GHRH. *See* growth hormone-releasing hormone (GHRH)

Gianoli, E., 131

giant water bugs, 437

Gibson, B., 365

Gibson, R., 223

Gigerenzes, G., 301

Gilby, I. C., 423

Gilley, D. C., 428

Gillingham, J. C., 499

Giraldeau, L.-A., 173, 364, 373, 374, 375, 423, 542

Gleason, J. M., 203

Gleitman, H., 160

Glen, A. S., 392

Glimcher, P. W., 321

glucocorticoids, 158, 457, 485, 487, 497

gnu, migration of, *466*

goal-directed learning, 139–40, 566

Goddard, J., 407

Goddard, M. E., 558, 562

Godin, J.-G., 47, 318, 399, 544, 546, 547

Gold, K. C., 549

golden pipit (*Tmetothylacus tenellus*), 474, *474*

Goller, F., 436

Gomendio, M., 261

Gomes, C. M., 363

Gomez-Mestre, I., 398

Gonodactylus bredini (stomatopods), 144, *144*

Gonzalez-Voyer, A., 497

Goodall, J., 364, 460

good genes model, 204, 207–12, 234
    defined, 566
    Hamilton-Zuk hypothesis and, 208, 209
    harem size and, 207
    MHC and, 209–12
    parasite resistance and, 208–9

Goodwin, D., 366

Goodwin, N. B., 60

gopher snakes (*Pituophis melanoleucus*), 385

gorillas, play markers in, 525, *525*

Gosling, S. D., 548, 549, 558
    interview with, *560*, 560–61

Goss-Custard, J. D., 363, 378, 379

Gould, C. C., 466

Gould, J. L., 466

Gould, S. J., 38, 43

gourami fish, 153, *153*

Gowaty, P. A., 240

Grafen, A., 38, 43, 278, 420, 422
    interview with, *64*, 64–65

Graft, T. U., 387

Graham, I. M., 559

Graham, J., 152

Graham, K. L., 512, 513, 522, 525

Grant, B. R., 186

Grant, J. W. A., 495

Grant, P. R., 186, 189, 190, 202

Grant, R., 189, 190, 202

grasshopper (*Schistoceria americana*), 8, 12–15, *13, 14, 15*

Gray, R., 452

gray catbird (*Dumetella carolinensis*), 436, *436*

gray squirrel (*Sciurus carolinensis*), 410–11, *411*

gray treefrog (*Hyla versicolor*), 456, *456*

greater wax moth (*Galleria mellonella*), 390–91, *391*

Great Swamp National Wildlife Refuge, New Jersey, 438

great tit (*Parus major*), 352–53, *353,* 355–56, 553, 553–54

Greenewalt, C., 436

Greenough, W., 520

green swordtail fish (*Xiphophorus helleri*), 487, 502

green woodhoopoe birds (*Phoeniculus purpureus*), 484–85

grey seal (*Halichoerus grypus*), 559

grey-sided vole (*Clethrionomys rufocanus*), 252

Grezes, J., 179

Grier, J. W., 7

Griffin, A. S., 369

Griffin, D., 86, 391

Griffith, S. C., 256

Grobecker, D., 371

Grober, M. S., 95, 500

Grodzinski, U., 73

Gross, M. R., 229, 230

ground finch (*Geospiza fortis*), 189, 190

ground squirrel (*Spermophilus beecheryi*),
   385–86, *386*
group hunting:
   fitness benefits to, *35*
   natural selection for, *34*, 34–35
group selection, 344
   cooperation and, 327–31
   defined, 566
   within-group and between-group,
      328–31
group size:
   foraging and, 361–64, *362*
   vocalization and, 422–23, *423*
growth hormone-releasing hormone
      (GHRH), 82
Grubb, T. C., 366
Grueter, C., 426
guide dogs, personality in, 558, 562
Guinet, C., 370
Gulf toadfish (*Opsanus beta*), 389,
      389–90, *390*
guppy (*Poecilia reticulata*), 412
   antipredator behavior in, *43*, *45*,
      45–49, *46*
   boldness and predator inspection in,
      546–48, *547*
   courtship behavior in, 62–63
   horizontal cultural transmission in,
      188–89
   mate-choice copying in, *180*, 180–81,
      191–92, *192*
   predator inspection and TFT in, *316*,
      316–19, *317*, 319
   "retaliation" in, *319*
   risk-taking in, *546*
   sexual selection in, 232–33
Gurven, M., 421
Guthrie, D. M., 487
Guthrie, R. D., 407
Gutierrez, G., 153

habitat choice, 450, *450*, 452–59
   abiotic factors in, 451, 478
   avoidance of disease-filled habitats,
      455–56, 478
   biotic factors in, 451, 478
   defined, 566
   foraging success and, 453–55
   home range in, 451
   ideal free distribution model and,
      452–55, *453*, *455*, 478
   predation and, 391, *393*, 393–94
   resource matching rule and, 454, 455
   spatial memory and, 457–59
habituation, 116, *117*, 133, 134, *134*, 162,
      566
Hack, M. A., 495
Haddock, S. H. D., 384
Haeckel, E., 54
Haesler, S., 115
Hagen, R. H., 338
Hagenguth, H., 87
Hager, J., 142
Hager, R., 279

Haig, D., 294, 295
Hailman, J., 6
Hain, T., 293
Haldane, J. B. S., 275
Halichoerus grypus (grey seal), 559
Hall, D. J., 354
Hall, K. R., 187
Hall, S., 516
Halliday, T., 421
Hamilton, W. D., 18, 19, 20, 50, 65, 208,
      264, 274, 275, 278, 302, 304, 313,
      314, 315, 341, 342
Hamilton's Rule, 278, 279, 286
Hamilton-Zuk hypothesis, 208, 209
Hammer, M., 88
Hammerschmidt, K., 418
Hammerstein, P., 312, 494
Hammock, E. A. D., 118
handicap principle, 422, 445, 446
Hanlon, B. T., 495
Hanlon, R. T., 387, 388, 389
Hansell, R., 255
haplodiploidy, 287
harbor seal (*Phoca vitulina*), 559
Harcourt, A. H., 281, 331, 333
Harcourt, R., 518, 519
Hardin, G., 332
Hare, H., 287
Hare, J., 442, 443
harem defense, aggression and, 506
Harlow, H., 141
Harper, D. C., 455
Harris, L., 150
Hartl, D., 39
Hartley, R. S., 436
Harvell, C. D., 131
Harvey, P., 369
Haskins, C. P., 47
Hass, C. C., 523
Hatchwell, B. J., 154
Hauser, M., 182, 183, 184, 185
Havens, K., 259
Hawaiian crickets, 214
hawk-dove game model, 485, 489,
      490–93, 505, 508
   antibourgeois strategy in, 491, 493, 508
   bourgeois strategy in, 491, 492, 508
   defined, 566
   payoff matrix for, *491*
   resource value in, 491
Hawkes, K., 354
Hay, M., 410
Healey, S., 365, 366
Heape, W., 465
Hedrick, P. W., 211
Hegner, R., 410
Heinrich, B., 424, 514, 516, 517
Heinze, S., 468
Held, S. D. E., 515
hellbenders (*Cryptobrancus alleganeinsis*),
      150, *150*
helping, in Mongolian gerbils, 83
helping-at-the-nest, 123, 154, 278, 279,
      279, 282, 464

Hemelrijk, C. K., 334
Henrich, J., 171
Henry, J., 556, 558
Hepper, P. G., 154, 298
Herard, F., 119
Herbers, J. M., 287
herding behavior
   artificial selection and, 33
   in cetaceans, 334
heritability, 39, 40, 566
Herrero, A., 158, 159
Herrnstein, R. J., 452
Hessing, M., 558
Heterocephalus glaber (naked mole rat),
      49–52, *50*, *51*, *52*
Heyes, C., 133, 166, 172, 175, 178, 179,
      194–95
   interview with, *194*, 194–95
Higham, A., 317
Hill, C. E., 257
Hill, G., 71, 72, 73, 74, 207, 339
   interview with, *100*, 100–101
Hill, K., 354
Hill, N., 524
Hillis, D. M., 58
Hinde, R., 86, 178, 179
hippocampus, 94, 157, 158, 159
   caching ability and, 365–66
   learning in voles and, 92, 93
   sex difference in size of, 93
Hiraiwa, M., 167
Hirth, D. H., 407
Hock, K., 501
Hodge, M. A., 493
Hoekstra, H. E., 387
Hoese, H. D., 186
Hoffman-Goetz, L., 473
Hoffmann, A., 41
Hogan, D., 285
Hogan, J., 6
Hogan-Warburg, A. J., 108, 552
Höglund, J., 222, 244, 245
Hogstad, O., 438
Holden, P., 371
Holderied, M. W., 391
Holdobler, B., 328
Holekamp, K. E., 548
Hölldobler, B., 25, 428, 429, 430, 431
Hollen, L., 439, 442
Hollis, K., 152, 153, 155, 156, 506,
      506–7
   interview with, *506*, 506–7
Holloway, W., 530
Holman, L., 261, 263
Holmes, W., 298, 301
home range, 451, 566
homicide, in humans, 273–74, *275*
homing pigeons, 30, *31*
homologous traits, defined, 566
homology, 56, 57
homoplasy, 56, 57, 566
honest indicators, of male genetic
      quality, 208
honest indicators principle, defined, 566

honeybee (*Apis mellifera*), *51*, 94–95, 124
    foraging by, 86–89, *87*, 94–95, 112–14
    waggle dance of, *87*, *426*, 426–29, *427*, *428*, *429*, 446
    worker policing in, *289*, 289–90
Hoogland, J., 273, 304
Hopster, H., 557
horizontal cultural transmission, 188–89, 196, 566
hormones:
    aggression and, 485, 487–89
    behavior systems during early development and, *81*, 81–82
    classification of, 76
    day length, behavior and, 79, *79*
    honeybee foraging and, 86–89, *87*
    integration of sensory input/output and, 80–82
    learning and, 158–59
    long-term effects of in utero exposure to, 82–84
    play and, 528
    proximate causation and, 75–77, 79–80
    winner and loser effects and, 497, 500, *500*
Horne, J., 467
Hosken, D. J., 266
Hosoi, A., 432
Hostetler, C. M., 242
Houde, A., *232*, 232–33
Houde, A. E., 43, 47, 63, 191
Houde, A., interview with, *232*, 232–33
house finch (*Carpodacus mexicanus*), 71–75, *72*, *73*
house mice (*Mus musculus domesticus*), 301
house sparrow (*Paser domesticus*), 326, *327*
Houston, A., 301, 359, 462, 463
Howe, N., 150
Hsu, Y. Y., 497, 499, 558
Huber, R., 488, 501
Huffman, M., 167, 168, 169
Hughes, L., 339
Hughes, W., 287
Hugie, D. M., 552
Hultgren, K. M., 336
human-animal conflicts, reducing, 559
human chorionic gonadotropin, 295
human placental lactogen, 295
humans:
    artificial selection and, 31
    cultural transmission in, *170*, *171*
    dispersal and residence patterns in, 282–83
    female mate choice in, 210–11
    homicide in, 273–74, *275*
    horizontal cultural transmission in, 188
    mate choice in, 52–53, *53*
    mirror neurons in, 180
    reciprocity in, 321–23
    sexual imprinting on faces in, 219–20
    sperm number in, 262–63, *263*
    in utero conflicts in, *294*, 294–95
    wealth and kinship in, 284–85

Humphreys, A., 526, 530
hunger, risk-sensitive foraging ad, 359
Huntingford, F. A., 134, 146, 482
Huntingford, R. A., 155
Hurd, P. L., 155, 487, 495
Hurtado, A. M., 354
Hutchinson, J. M. C., 301
Huxley, J., 6, 31, 203
Huxley, T. H., 194, 482, 484
hybridization, increasing, 202
hydrocarbon dance compounds, 428, *428*
hyena (*Crocuta crocuta*), 331
Hyla versicolor (gray treefrog), 456, *456*
Hylobittacus apicalis (scorpionfly), *205*, 205–7
hymenoptera, 286–87
Hyperolius nitidulus (reed frog), 387
hypothalamic-pituitary-adrenal (HPA) axis, 77

Iacoboni, M., 179
ideal despotic distribution, 455
ideal free distribution (IFD) model:
    defined, 566
    habitat choice and, 452–55, *453*, *455*, 478
identical by descent alleles, 276–77
IFD model. *See* ideal free distribution (IFD) model
imitation (observational learning), *171*, 183
    birdsong and, 178–79, 183
    cross-fostering and, 224, *224*
    cultural transmission and, 177–80, *178*, 196
    defined, 566
    dolphins and, 186–87
    guppies and, 188–89
    rhesus monkeys and, 187
    social learning and, 409
Immelmann, K., 219
immunology, predation and, 385–86, *386*
imperial blue butterfly (*Jalmenus evagoras*), 338
Inaba, A., 516
inclusive fitness (kin selection) theory, 19, 274–79, 284, 302, 304, 566
    nonbreeding groups and, 281, *281*
    relatedness and, 276–79, 304
    See also family dynamics
indigo buntings (*Passerina cyanea*), 256, 256–57, *257*, 469–70, *470*
indirect benefit models, 207, *208*
indirect fitness, 18, *19*
individual learning, 8, 12–15, 131–33, 567
inducible defenses, 131, *132*
information-center hypothesis, 15
inhibition, 544
inhibitory conditioning, 136, 567
innovation, brain size and, 193, 196
Inouye, C. Y., 72
input systems, hormones and, *80*, 80–82

insects:
    aggression in, 484, *484*
    convergent evolution in wing structure of, 57
    death feigning in, 404
    dispersal strategies, climate change and, 120, *120*
    fitness consequences of learning in, 12–15
    haplodiploidy in, 287
    polyandry in, 247–49
    relationship between naked mole rats and, *51*
    worker altruism in, 303
    See also specific species
Insect Societies, The (Wilson), 24
Insel, T. R., 118
instrumental (operant) conditioning, 139–41, 162, 567
intersexual relation, defined, 200
intersexual selection, 204, 234
    defined, 567
    See also mate choice; sexual selection
interspecific cooperation, 344
interspecific mutualism, 338–39
intrasexual selection, 203, 204. *See also* mate choice; sexual selection, 234
    defined, 200, 567
intrinsic factors, aggression and, 155
Iridomyrmex anceps, 338–39
irruptive migration, 465
Irwin, D. E., 147
Iwaniuk, A., 532, 533
Izuma, K., 322

Jaatinen, K., 552
Jablonski, P., 132, 133
Jacana jacana (wattled jacana), 242
jackdaw (*Corvus mondeula*), 366
Jacobs, G. H., 114
Jacobs, L., 93
Jacobson, M., 520
Jaksic, F. M., 392
Jalmenus evagoras (imperial blue butterfly), 338, 339, *339*
Janik, V. M., 166, 389
Jansa, S. A., 532
Jansen, V. A. A., 350
Japanese quail, learning and mate choice in, 153, *153*, 220, 221
Jarvis, J., 50
Jenni, D. A., 523
Jennions, M., 205, 495
Jensen, P., 495
Jetz, W., 473
Joels, M., 457
Johannesen, J., 337
Johansson, O., 559
John, J., 474
John, O. P., 541
Johnsen, S., 472
Johnson, G. G., 7
Johnson, J. C., 548
Johnsson, J., 502

population comparisons and, 145–47
predation and, 149–52, *150*
in rats, 142, 156–58, *156*
sensitization and, 133, *133, 134*
single-stimulus experience and,
    133–34
social. *See* social learning
spatial, 158
temperature, egg laying and, 121, *121*
territory and territoriality and,
    460–61, 476–77
in voles, 92–93
within-species studies of, 141–44
Le Boeuf, B. J., 227, 228, 252
Leca, J. B., 168
Lee, P. C., 176, 312
Lefcort, H., 150
Lefebvre, L., 145, 146, 193, 369, 371,
    373, 374, 375
Leffler, J., 473
Leger, D. W., 385
Lehmann, L., 313
Leigh, E. G., 338
Leimar, O., 487, 489, 494
Leisler, B., 246
leks (arena mating), 222, 223, 243–46, 432
    defined, 567
    peacocks on, *245*, 245–46
Lemaire, O., 500
Lengagne, T., 298
Lens, L., 44
Leopold, A. S., 438
Leotta, R., 562
Lepomis gibbosus (pumpkinseed sunfish),
    544, 544–45
Lepomis macrochirus (bluegill sunfish),
    229–31, *230, 231*
Lesku, J. A., 98
Levero, F., 281
Levey, D. J., 475
Levins, R., 131
Levitan, D., 260, 261
Levitis, D., 7
Lewis, A., 119
Lewis, M., 282
Lewontin, R., 38, 43
Ley, J. M., 558
Librium, 395
Lichtman, J., 520
Lieberman, D., 298
Liedvogel, M., 472
Liers, E., 184
Ligon, D., 284
Lim, M. M., 118
Lima, S., 98, 395, 410, 411
Lind, J., 386
Lindblad-Toh, K., 558
Lindenfors, P., 228, 229
Lindquist, W. B., 497, 501
Lindstrom, A., 473
Lindstrom, J., 245
Linnell, J. D. C., 559
Linsdale, J., 386
lions (*Panthera leo*), 331

Littledyke, M., 429
Liu, Y., 241
lizard (*Anolis aeneus*), 461, *461, 462*
lobe-finned fish, phylogeny, 55
lobsters, aggression in, 488, *488*
local enhancement:
    cultural transmission and, *172,*
        172–73, 196
    defined, 567
    social facilitation and, 174, *174*
locomotor play, 519–22, 536
    benefits of, *521*
    cerebral synapse development and,
        520–22, *522*
    defined, 567
Lohmann, K. J., 472
Lonchura leucogastroides (mannikin
    bird), 218–19
Long, T. A. F., 293
long-tailed tit (*Aegithalos caudatus*), 154,
    154
long-tailed weasel (*Mustela frenata*), 272
Lorenz, F. W., 552
Lorenz, K., 21, 86, 218, 378
loser effects, 155, 156, 567
Loup, F., 323
lowland gorilla (*Gorilla gorilla gorilla*),
    281, *281, 525*, 525
Lubin, Y. D., 337
Lucas, H., 452
Lucas, N. S., 311
Lucas, P. W., 215
Lundberg, A., 266, 267
Lundstrom, J. N., 298
Lurling, M., 253
Lutz, C.C., 95
Lutzomyia longipalpi (sandfly), 244
Lyamin, O. I., 102
Lycaon pictus (African hunting dog), *33,*
    33–35, *35*
Lycett, S. J., 555

Macaca:
    M. mulatta (rhesus macaque), 312
    M. sylvanus (Barbary macaque), 249
macaque monkeys:
    potato washing and, 166–67, *167*
    stone play and, 167–68, *168*
MacArthur, R. H., 351, 378
MacDonald, K. S., 336
MacDougal-Shackleton, S., 432
Mace, G., 369
Mackey, T. F. C., 109
Macnair, M. R., 293
Macronyx croceus (yellow-throated
    longclaw), 474, *474*
Maddison, P., 533
Maddison, W., 337, 533
Maeda, K., 471
Maestripieri, D., 183, 312, 541, 548
Magellanic penguin (*Spheniscus
    magellanicus*), 78, *78*
magnetic field, migration and, *471,* 471–72
Magrath, R. D., 438

Magurran, A., 43, 45, 46, 47, 48, 63, 191,
    317, 402
    interview with, *412,* 412–13
Mahometa, M. J., 220
major histocompatibility complex
    (MHC):
    good genes and, 209–12, *211, 212*
    kinship, templates, and, 300, 304
Malapert, A., 182
male-male competition, 201, *203, 204,*
    225–31, 234
    by cuckoldry, 229–31
    by interference, 227–28
    phylogeny and, 228–29
    red deer roars and, 225–27, *226*
male mating fertilization strategies, in
    Poecillinae fish, 63, *63*
mallard (*Anas platyrhynchos*), 98, 98–99,
    99, 455, 455
Malurus cyaneus (superb fairy wren),
    257, 282, *282, 283*
malvolio, manganese transport to
    honeybee brain and, 113
mammals:
    aggression in, 482
    locomotor play in, 519–22
    object play in, 518–19
    phylogeny, 55
    social grooming in, 312
    See also specific species
manganese, pollen foragers *vs.* nectar
    foragers and, *113,* 113–14
Mank, J. E., 60, 61, 62
Mann, J., 186, 187
mannikin bird (*Lonchura
    leucogastroides*), 218–19
Manning, C. J., 301
Manser, M., 439
Maple, T., 549
Marchaterre, M., 97
Marchetti, C., 554
marginal value theorem, 354–55, *356,*
    357, 380, 567
Margulis, S., 123, 125
marker loci, 109
Marler, P., 86, 504
Martel, F. L., 312
Martin, A. A., 253
Martin, J. K., 253
Martin, P., 516
Martin, T. E., 393, 394
Marzluff, J., 424, 425
Maschwitz, U., 339
Masseti, M., 4
matching-to-self hypothesis, 220
mate choice, *203,* 234
    boldness and, *547,* 547–48
    copying and, *180,* 180–81
    in cowbirds, 226–27
    cryptic, 263, 265
    cultural transmission and, 191–92,
        192, 221–25
    direct benefits model of, 204, *205,*
        205–7, *206,* 234

evolutionary models of, 204–18
female, 204
genetics and, 202
good genes model of, 204, 207–12, 234
in humans, 52–53, *53*
learning and, *12, 12,* 152–53, 218–20
plumage coloration and, 74
polygyny threshold model and, 252, 254–55, 255
runaway selection model of, 204, 234
sensory exploitation model of, 204, 234
See also sexual selection
mate-choice copying, 221–24
in cowbirds, 224–25
defined, 221–22, 567
in grouse, 222–23, *223*
in mice, 223–24
mathematical optimality theory, foraging and, *20,* 20–21, 351–361
Mather, J., 512, 549, 550
Mathis, A., 150
mating systems, 267
anthropogenic effects on, 253
battle of the sexes and, 266
extrapair copulations (EPCs) and, 256–58
forms of, 238–67, *239,* 267
mate aggression and, 83
multiple, 263, 266–67
parent-offspring conflict and, 293–94
PTM and, 252–55
sperm competition in. *See* sperm competition
See also specific systems
Maynard Smith, J., 20, 264, 315, 340, 490, 494
Mayr, E., 6, 38, 43, 70, 144
Mays, H. L., 207
McAuliffe, K., 182, 184, 185
McCarthy, M. M., 530
McCleery, R.H., 352
McComb, K., 249, 422
McCullough, D. R., 407
McEwen, B., 457
McGraw, K., 74
McGraw, L. A., 241
McGregor, P., 502, 504
McGuire, B., 86
McGuire, M., 488
McKibben, J. R., 96
McKinney, F., 256
McLaughlin, R. L., 544
McLauglin, F., 256
McLellan, T., 214
McMann, S., 495
McNab, B. K., 319
McNally, C., 218
McNamara, J. M., 359
Mead, L. S., 222
meadow vole (*Microtus pennsylvanicus*), 85–86, *86,* 118
learning in, 92–93
sex differences in water maze trials, *93*

meerkats, 182–85, *183, 184,* 438–40, *439*
Melis, A. P., 308
Mello, C., 116, 117
Melospiza melodia (song sparrow), 431
Meltzoff, A. N., 373
Membranipora membranacea (bryozoan), 131, *132*
memory:
coping style and, 557
natural selection and, 144
neural plasticity and, 182
optimal, 144, *144*
planning for the future and, 373, *373*
See also learning
Mendel, G., 32, 108, 166
Mendel's laws, 107, 108–9
Mendl, M., 515
Mendozagranados, D., 522
menstruation, 250, *251*
Menzel, R., 88
Meriones unguiculatus (Mongolian gerbil), 152, 153
Merops bullockoides (white-fronted bee-eaters), 285–86, *286,* 464–65
Mesocricetus auratus (Syrian golden hamsters), 529
Mesoudi, A., 15, 16
Messenger, J. B., 387, 389
Messier, F., 149
Messor pergandei (desert seed harvester ant), 328, *328, 329,* 331
Mesterton-Gibbons, M., 309, 331, 422, 491, 493
methoprene, allatectomized bees and, 88
Metrius contractus (bombardier beetle), 408
Mexican jay, 19
Mexican spider (*Oecobius civitas*), 493, *493*
MHC. *See* major histocompatibility complex (MHC)
mice, *81,* 81–82, 111, *181,* 181–82, 223–24, *239, 239, 240,* 301, 387, 395, 521, 557, *557*
Michener, C., 154, 298
Microtus ochrogaster (prairie vole), 85–86, *86,* 118, 122, *122,* 241
Microtus pennsylvanicus (meadow vole), 85–86, *86,* 118
Microtus pinetorum (pine vole), 93
migration, 451, 465–75
basal metabolic rate and, 473, *473*
climate change, mating systems and, 253, *253*
as defense against parasites, 473–74
defined, 567
"evolutionary precursor" model of, 474–75, 478
genetic diversity and, 37
heritability in, 472, *472*
irruptive, 465
magnetic orientation in, *417,* 471–72
navigation and, 466, 468–69
phylogeny and, 474–75, *475*

stellar navigation and, 469–72, *470*
"stopovers," conservation biology and, 467, *467*
sun compass and, 466, 468–69
"zugunruhe" (restlessness) in, 472
Miklosi, A., 178
Milinski, M., 209, 211, 212, 317, 318, 319, 400, 453, 454, 455
Mind of the Raven (Heinrich), 516
Mineka, S., 187
minnow (*Phoxinus phoxinus*), 402–3, *403*
effects of predation and, 47
predator inspection and, *318*
Minoan wall paintings, of "white antelopes," 4–5, *5*
Mirounga angustirostris (elephant seal), 227, 227–28, 228
mirror neurons, 179, 180
Mitchell, W. A., 43
Mitman, G., 313
Mittlebach, G., 362
Miyaki, C. Y., 391, 394
Miyatake, T., 404, 405
mobbing behavior:
defined, 567
social learning and, 409
Mock, D., 293, 295, 296, 362, 420, 482, 497
mode of inheritance, natural selection and, 38–39
Mohamad, R., 490
molecular genetics, 65
animal behavior and, 107–18
of learning in rats, 156–58
proximate causes and, 106
song acquisition in birds and, 107, 115–17
ultraviolet vision in birds and, 107, 114–15
within-family interactions in voles and, 118
Molenberghs, P., 179, 180
Molina-Borja, M., 495
Møller, A., 205, 258, 262, 440, 441, 473, 474
Mollon, J. D., 215
Molothrus ater (brown-headed cowbird), 432–33, *433,* 436
Moltschaniwskyj, N. A., 550
monarch butterfly (*Danaus plexippus*), 406–7, *466,* 466, 468–69, *468, 469*
Mongolian gerbil (*Meriones unguiculatus*), 83, 152, 153
Monkkonen, M., 393
monogamous mating system, 238–42, *239,* 264, 265, 266, 267, 287
defined, 204, 238, 567
fitness consequences and, 240–41
genetic relatedness and, 293
in oldfield mouse, 239, *239, 240*
proximate underpinnings of, 241–42, 267
serial, 238
social, 256

Sharrock, J., 371
Shaw, P. W., 48
Shawkey, M. D., 74
Sheard, M., 487
Sheehan, M., 130, 131
Shen, S., 324
Shen-Feng, S., 279, 324
Sherman, P. W., 6, 38, 43, 50, 70, 272, 298, 301, 304
Sherry, D. F., 366
Shettleworth, S., 131, 141, 144, 145, 350, 369, 372
    interview with, *160*, 160–61
Shi, Y. S., 114, 115
shrimp, phylogeny and cooperation in, *336*, *336*
Shuster, S. M., 144
shyness (inhibition):
    defined, 544, 568
    in fish, 544–46, *545*
sibling rivalry, 295–97, *296*, 304, 568
Sibly, R., 541
Siemers, B. M., 391
Sigg, D., 95
signaling to predators, 405–9
    tail flagging and, *407*, 407–8
    warning coloration and, *406*, 406–7
Sih, A., 351, 540, 541, 548, 561
silent mutations, 37
Simmons, L. W., 203, 262
simultaneous polygamy, 242
Sinn, D., 550, 551
Siviy, S., 530, 531
Skelly, D. K., 456
Skinner, B. F., 140, 141
Skinner box, 140, *140*, 141, 324
Slabbekoorn, H., 189, 298
Slagsvold, T., 202
Slansky, F., 15
Slater, P., 431
sleep:
    in dolphins, *102*
    predation and, 98, 99, *99*
Slocombe, K. E., 504
slow-wave sleep, 99
Sluyter, F., 556, 557
Smallwood, P. D., 359
Smith, A. P., 243
Smith, C., 214
Smith, D. G., 395
Smith, D. R., 337
Smith, I. C., 495
Smith, P. K., 520, 526
Smith, P. S., 407
Smith, R. J. F., 150, 407
Smith, R. L., 437
Smith, T. B., 189
Smith, W. J., 421
Smokler, R., 514, 517
Smotherman, W., 299
Smythe, N., 407
snails, 456, 457
snake aversion, in rhesus monkeys, 187, *188*

snakes, 385–86, *386*
    phylogeny, 55
    predation and, 399, *399*
Snook, R. R., 261, 263
Snowdon, C. T., 423, 504
Sober, E., 43, 327
social facilitation:
    cultural transmission and, *173*, 173–75, *174*, 196
    defined, 569
social grooming, cooperation and, *311*, 311–12
social learning, 8, 131, 183, 186
    brain size and, 193
    crop raiding, elephants, and, 176, *176*
    defined, 569
    foraging and, 370, 373–75
    mobbing behavior and, 409
    personality and, 541
    See also cultural transmission
social monogamy, 256
social pair bonding, 256
social play, 522–27, 536
    in bighorn sheep, 523
    cognition and, 523–25, 536
    functions of, 522
    neurotransmitters and, 530
    role reversal in, 525–26, *526*
    self-handicapping in, 525
sociobiology:
    birth of, 24
    defined, 43
Sociobiology: The New Synthesis (Wilson), 24, 25, 303
Sol, D., 193, 371, 372
Solberg, L. C., 111
Solomon, N. G., 154
Soma, M., 433
Sommer, V., 522
song acquisition, in birds, *115*, 115–17, *116*, *117*
song learning, mate choice in cowbirds and, 226–27
song sparrow (*Melospiza melodia*), 431
sonic muscles, vocalization and, 97
Sordahl, T. A., 409
Soriguer, R. C., 392
spadefoot toad tadpoles (*Scaphiopus bombifrons*), *299*, 299–300, *300*, *301*
Sparks, J., 311
spatial memory, stress hormones and, 457–59, *458*, *459*
specialization, 100
speckled wood butterfly (*Pararge aegeria*), *492*, 492–93
Spencer, S. J., 395
sperm competition, 258–63, 293
    cryptic choice in, 263
    defined, 569
    in dungflies, *259*, 259–60
    kamikaze sperm hypothesis and, 261–62
    last male precedence in, 260

number of sperm per ejaculate, 262
in sea urchins, 260–61
sperm morphology and, 262, *262*
sperm velocity and, 261, *261*
Spermophilus beecheryi (ground squirrel), 385–86, *386*
Spermophilus beldingi (Belding's ground squirrels), *531*, 531–32, *532*
Spermophilus richard-sonii (Richardson's ground squirrels), *442*, 442–43
sperm production, Pavlovian conditioning and, *218*
Spheniscus magellanicus (Magellanic penguin), 78, *78*
spiders, 120, *120*, 337, 337–38
Spieth, H. T., 203
Spinka, M., 513, 515, 527
Spinks, A., 10
Spitzer, N., 488
Spoolder, H., 557
Spoon, T. R., 240
Spooner, C., 98
spotted hyena (*Crocuta crocuta*), 548–49, *549*
spotted-tail quoll (*Dasyurus maculatus*), 392
squirrel monkey (*Saimiri sciureus*), 515, *515*, 525–27, *526*
stable fly (*Stomoxys calcitrans*), *181*, 181–82
Stacey, P., 154, 284, 335
stag beetles, 201
    weapons in, *483*
stalk-eyed flies, runaway selection and, *213*, 213–14
Stammbach, E., 312
Stamps, J., 119, 461, 541
    interview with, *476*, 476–77
Stanford, C. B., 387
Stankowich, T., 50, 395, 396, 397, 407
starling (*Sturnus vulgaris*), 364, 365
Starr, C., 7
Stearns, S. C., 295
Stein, R. C., 436
Stenaptinus insignis (bombardier beetle), 408, *408*
Stephens, D., 147, 148, 324, 326, 351, 357, 359, 369
Stephens, P., 556
steroid hormones, 76–77
Stevens, E. D., 544
Stevens, J. R., 423
Stevenson-Hinde, J., 549
Stewart, J., 529
stickleback (*Gasterosteus aculeatus*), 47, 146–47, *210*, 211, *211*, *212*, 453, 454, *454*, 540
Stiles, F. G., 475
Stoewe, M., 494
stomatopods (*Gonodactylus bredini*), 144, *144*
Stomoxys calcitrans (stable fly), *181*, 181–82

Stone, A., 515
Stone, G. N., 94
stone play, in macaques, 167–68, *168*
Strassman, B. I., 250
Strassmann, J. E., 328
Stratakis, C., 82
stress:
  coping styles and, *556*, 556–58
  ecotourism and, 78, *78*
  learning and, 158
  play and, 515
  spatial memory in rats and stress
    hormones, 457–59, *458*, *459*
stridulation, *430*, 430–31
Struhsaker, T. T., 187
Stuart, R. A., 260, 354, 452, 453
Sturnus vulgaris (starling), 364, *365*
subjective well-being, mortality and,
  540, *541*
Suddendorf, T., 372
Sula nebouxii (blue footed boobies),
  497–98, *498*, *499*
Sullivan, J., 87, 88
Sullivan, K., 363, 438
Summers, K., 60
Summers-Smith, J. D., 326
Sumner, F. B., 387
Sumner, P., 113, 215
Sundgren, P. E., 562
superb fairy wren (*Malurus cyaneus*),
  257, 282, *282*, 283
Sutherland, W. I., 379
Sutherland, W. J., 455
Suthers, R. A., 435, 436
Swaddle, J. P., 44
Swima sp., bioluminescent bombs and,
  384, *384*
Sylvia atricapilla (German blackcap
  bird), 472
symmetry, as indicator of risk, 44, *44*
Symons, D., 513, 520
Synalpheus shrimp, 336
synaptic terminals, 89
Syrian golden hamster (*Mesocricetus
  auratus*), 529
syrinx, avian, 435–36

Taborsky, B., 123
Tada, T., 114
Taeniopygia guttata (zebra finch), *114*,
  114–15
Taggart, R., 7
Tai, Y. F., 179
Tai chimps, cooperative hunting and,
  363, 363–64
tail flagging, *407*, 407–8
Tai National Park, Ivory Coast, 363
Taita thrush (*Turdus helleri*), 44, *44*
Talbot, C. J., 111
Tarpy, D. R., 248
Tatolas, A., 5
Taub, D., 249
Tautz, J., 428
Taylor, M., 285

teaching, 183–85, *184*, 186
  cultural transmission and, 171, *171*, 196
  defined, 569
Teeling, E. C., 391
Telecki, G., 363
Teleogryllus oceanicus (field cricket), *9*,
  9–10
temperament, 540
template matching, in tadpoles, *299*,
  299–300
Templeton, J. J., 364
ten Cate, C., 218
tension reduction, through social
  grooming, 311–12
Teramitsu, I., 115
termites:
  queens and developmental pathway
    in, 197
  sociality in, 106
territoriality, defined, 459
territory, defined, 451, 569
territory and territoriality, 156, 459–65
  antibourgeois strategy and, 491, 493
  bourgeois strategy and, 492
  "budding" of, 464, *464*
  defense of, 464, 478
  dynasty-building hypothesis and, 284
  extrapair copulations and, 256
  family conflict and, 464–65
  female choice of territories, *254*,
    254–55
  learning, foraging, group living and,
    145, 145–46
  learning and, 460–61, 476–77
  owners, satellites, pied wagtails and,
    462, 462–63, *463*
  prior residency advantage in, 477
  raiding behavior and, 460
  resource value and, 490
  sneaker and satellite behavior and,
    229–31, *231*
Terry, R. L., 311
testes size, parental investment and, 293,
  294
testosterone, 76
  aggression and, 485, 528
  day length, behavior and, *79*, 79–80
  eavesdropping and, 503, *504*
  intrauterine position and, 80–81, *81*,
    82–83
  male parental care and, 84
  personality and, 552
  play fighting and, 528–30, *529*
  sexual play behavior and, 532
  winner and loser effects and, 497
TFT strategy. *See* tit for tat (TFT)
  strategy
Thatch, W., 520
Thery, M., 432
Thierry, B., 423
Thiessen, D., 152
Thom, C., 428
Thompson, E., 494
Thompson, G. J., 427

Thompson, K. V., 522, 523
Thor, D., 530
Thorndike, E., 140, 141
Thornhill, R., 205, 206, 207, 243, 248, 249
Thornton, A., 182, 184, 185
Thorpe, W., 172
thyroid gland, 76
thyrotropin-releasing hormone (TRH),
  82
Tibbetts, E., 130, 131, 490
tiger salamander (*Ambystoma tigrinum*),
  299
Tinbergen, L., 350
Tinbergen, N., 6, 21, 26, 70, 86, 119,
  161, 421
Tinghitella, R. M., 10
Tinkle, D. W., 43
tit for tat (TFT) strategy, 315–19, *316*,
  319, 341, 569
Tmetothylacus tenellus (golden pipit),
  474, *474*
toad size, croaks and, 421, *422*
Toft, S., 120
Toivanen, A., 474
Toivanen, P., 474
Tollrian, R., 131
Toma, A., 112, 113
Tomaru, M., 203
topi antelope (*Damaliscus lunatus*),
  441–42, *442*
Torney, C. J., 423
Touchon, J. C., 398
Tourmente, M., 261, 262
Toxostoma rufum (brown thrasher), 431, 436
traditions
  cultural transmission and, 196
  rise and fall of, 182
tragedy of the commons, cooperation,
  overharvesting and, 332, *332*
Trail, P., 432
trait-group selection models, 327
traits, fitness consequences of, 38, *39*
Travasso, M., 339
tree format, phylogeny, 55
tree representation, phylogeny, 55
Treves, A., 559
TRH. *See* thyrotropin-releasing hormone
  (TRH)
Trichogaster trichopterus (blue gourami
  fish), 155–56
Trichosurus cunninghami (brushtail
  possum), 253
tricolor vision, fruits, sensory bias and,
  214–16, *216*
Trillmich, F., 518
Trivers, R., 20, 52, 200, 264, 287, 291,
  313, 314, 407
Trueb, L., 227, 456
truncation selection experiment, 39–41,
  569
trust game, 322–23, *323*
Trut, L. N., 32
"tumbler" pigeons, 30, *31*
Turdus helleri (Taita thrush), 44, *44*

Turdus merula (blackbird), 409
Turner, A., 155, 482
Tursiops truncatus (bottle-nosed
    dolphin), 186, *186*, 187, *187*, 331,
    331, 334, 389, *389*
turtles, phylogeny, 55
tutors, song learning and, 224
Tyack, P. L., 389

Uehara, S., 363
Uehara, T., 494
Uetz, G., 493
Uher, J., 540, 548
ultimate analysis, 6, 8, 26, 31, 63, 70, 71,
    72, 74, 100, 569
ultraviolet vision, in birds, 114–15
unconditioned stimulus (US), 135, 136,
    137, 569
ungulates, tail flagging in, 407
unihemispheric sleep, 98, 99, *99*, *102*
unlinked loci, 108
US. *See* unconditioned stimulus (US)

Vaccarino, A., 366
Valladores, F., 131
Valone, T. J., 43, 364, 410, 411
vampire bat (*Desmodus rotundus*), 319,
    320, 321
van Baaren, J., 119, 121
van den Berg, C., 527
Vandenbergh, J. G., 83
Vander Meer, R., 428
Vanderschuren, L. J. M., 515
Van Leeuwen, E., 350
van Noordwijk, A., 554
van Oers, K., 36, 554
van Oordt, G., 552
van Osten, W., 419
Van Ree, J., 530
variance:
    foraging and, 360–61
    in reproductive success, polygyny and,
        242, *242*
variation:
    natural selection and, 36–37, *37*
    parental care and, 122, *122*
    polygenic traits and, 109–11
vasopressin:
    homologs of, 84
    sociality in voles and, 84–86
vasopressin receptors, in prairie and
        meadow voles, 86, *86*
vasotocin, evolutionary history of, 85
Veenema, A., 557
Vehrencamp, S., 251
Verbeek, M. E., 553
Verlinden, H., 88
Verrell, P., 490
vertical cultural transmission, 185,
        186–87, 196, 569
vervet monkeys, 438
    alarm calls in, 418, *418*, 440
    deceptive alarm calls in, 440
    fitness in, *19*

Vestergaard, K., 516
Viitala, J., 114
Villarreal, R., 152, 153
Vincent, A., 241
Visalberghi, E., 174
Visscher, P. K., 94, 289, 426
vocalization:
    group size and, 422–23, *423*
    in plainfin midshipman fish, 95–98,
        96, *97*
Voelker, G., 474, 475
voles, vasopressin and sociality in,
        84–86, 118
vom Saal, F. S., 83
von Frisch, K., 21, 86, 426
Vos, D. R., 218
Voultsiadou, E., 5

Wagener-Hulme, C., 88
waggle dance, 87, *426*, 426–29, *427*, *428*,
        429, 446, 569
Waits, L., 207
Wajnberg, E., 471
Waldman, B., 301
Walker, B., 78
Walker, C., 520
Walker, P., 131
Wallace, M., 202
Wallace, R., 7
Wallauer, W., 516
Waller, B. M., 525
Walters, J. R., 282
Wang, Z. X., 86, 241
warbler mating systems, phylogeny of,
        246–47, *247*
Ward, A. J. W., 541
Ward, P., 15, 423
Warkentin, K., 398, 399
warning coloration, *406*, 406–7
war of attrition model, 485, 489, 494,
        495, 508, 569
Warren, P. S., 431
Warren, W. C., 117
wasps, *288*, *290*, *398*, 398
water flea (*Daphnia magna*), 453–54, *454*
water strider (*Rhagadotarsus anomalus*),
        ripple communication by, 437
wattled jacana (*Jacana jacana*), 242–43,
        243
Watts, D. P., 281
Watts, H. E., 548
Webster, M. M., 541
Webster, M. S., 257
Wedekind, C., 210, 211
Wegner, K. M., 211
Weigensberg, I., 41
Weil, T., 106
Weinstock, M., 158
Weiss, A., 540
Weiss, D., 557
Weiss, G., 520
Weiss, J., 558
Weksler, M., 532
Wells, J. C. K., 295

Wenseleers, T., 290
Werner, E. E., 354
West, H., 152
West, M., 119, 225
West, S., 313
West-Eberhard, M. J., 286
western marsh wren (*Cistohorus
        palustris*), 431
western scrub jay (*Aphelocoma
        californica*), 373, *373*
Westneat, D., 221, 256, 257, 258
Whalen, R. E., 530
Wheeler, J., 328
Whishaw, I. B., 531
White, L., 285
white-fronted bee-eater (*Merops
        bullockoides*), 285–86, *286*,
        464–65
Whiten, A., 169, 178, 555
white-tailed deer, tail flagging in, *407*
Whitfield, C., 113
Widder, E. A., 384
Widemo, F., 108, 552
Wiebe, K. L., 393
Wigmore, S., 15, 170
Wiklund, C., 493
Wilcox, R. S., 437
wild dog (*Lycaon pictus*), 33, 33–35, *35*
wildebeest, *33*
wild rabbit (*Oryctolagus cunuculus*), 392
Wilkinson, G., 213, 319
Williams, G. C., 20, 32, 35, 64, 65, 295
Williams, K., 188, 189
Willis-Owen, S. A. G., 111
Willmer, P. G., 94
Wilson, A. D. M., 544, 546
Wilson, D. J., 150
Wilson, D. S., 319, 327, 328, 429, 544,
        545, 546
Wilson, E. O., 43, 192, 303, 327, 328,
        428, 429, 513, 562
    interview with, *24*, 24–25
Wilson, M., 273, 274, 303
Wilson, W. D., 313
Wiltschko, R., 465, 471
Wiltschko, W., 465, 471
Winberg, S., 488
Winger, B. M., 474
Wingfield, J. C., 79, 465
wing flapping, 132–33
wing structure, convergent evolution in, 57
winner and loser effects, 497–501, 508
    in blue-footed boobies, 497–99
    in copperhead snakes, 499–500
    defined, 497
    mathematical models of, 500–501
winner effects, 155, 156, 569
Withers, G., 94, 95
Witkin, S. R., 438
Witschi, E., 552
Witte, K., 218, 219
Wolf, L. L., 499
wolf (*Canis lupus*), 331
Wood-Gush, D. G. M., 516